ENHANCED SEVENTH EDITION

SOCIOLOGY
A Global Perspective

ENHANCED SEVENTH EDITION

SOCIOLOGY
A Global Perspective

JOAN FERRANTE
NORTHERN KENTUCKY UNIVERSITY

WADSWORTH
CENGAGE Learning™

Australia • Brazil • Japan • Korea • Mexico • Singapore • Spain • United Kingdom • United States

WADSWORTH
CENGAGE Learning™

Sociology: A Global Perspective,
Enhanced Seventh Edition
Joan Ferrante

Senior Publisher: Linda Schreiber

Managing Development Editor: Jeremy Judson

Assistant Editor: Melanie Cregger

Editorial Assistant: Rachael Krapf

Media Editor: Melanie Cregger

Marketing Manager: Andrew Keay

Marketing Assistant: Jillian Meyers

Marketing Communications Manager:
Laura Localio

Content Project Manager: Cheri Palmer

Creative Director: Rob Hugel

Art Director: Caryl Gorska

Print Buyer: Becky Cross

Rights Acquisitions Account Manager, Text:
Roberta Broyer

Rights Acquisitions Account Manager, Image:
Leitha Etheridge-Sims

Production Service: Lachina Publishing Services

Text Designer: Diane Beasley

Photo Researcher: Jill Engebretson and
Catherine Schnurr

Copy Editor: Robert Green and Sue Nodine

Illustrator: Lachina Publishing Services

Cover Designer: RHDG

Cover Image: Elie Bernager, Getty Images

Compositor: Lachina Publishing Services

For product information and technology assistance, contact us at
Cengage Learning Customer & Sales Support, 1-800-354-9706.

For permission to use material from this text or product,
submit all requests online at **www.cengage.com/permissions.**
Further permissions questions can be e-mailed to
permissionrequest@cengage.com.

Library of Congress Control Number: 2009934043

ISBN-13: 978-0-8400-3204-1

ISBN-10: 0-8400-3204-8

Wadsworth
20 Davis Drive
Belmont, CA 94002-3098
USA

Cengage Learning is a leading provider of customized learning solutions with office locations around the globe, including Singapore, the United Kingdom, Australia, Mexico, Brazil, and Japan. Locate your local office at **www.cengage.com/global.**

Cengage Learning products are represented in Canada by Nelson Education, Ltd.

To learn more about Wadsworth, visit **www.cengage.com/wadsworth**

Purchase any of our products at your local college store or at our preferred online store **www.ichapters.com.**

Printed in the United States of America
1 2 3 4 5 6 7 13 12 11 10 09

To my mother, Annalee Taylor Ferrante

and in memory of my father, Phillip S. Ferrante
(March 1, 1926–July 8, 1984)

Brief Contents

Contents

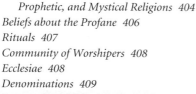

16 Social Change 458
With Emphasis on Greenland

Boxes

Intersection of Biography and Society

Global Comparisons

No Borders, No Boundaries

Working For Change

Preface

Over the seven editions of *Sociology: A Global Perspective,* I have never done (or ever thought about doing) an update to an edition until now. Because this sociology textbook is written from a global perspective, it is imperative that it be updated to reflect the dramatic economic downturn that is affecting not just the United States, but every country in the world. All indicators suggest that what is being called a global recession will continue for some time—perhaps for as long as a decade—as the United States and other consumption-oriented countries attempt to transition from economies based on borrowing (credit) and overconsumption to economies based on living within one's means, saving, and sustainability. Literally every area of life is being affected in some way by this new economic reality.

Three chapters have been substantially revised: The opening chapter (The Sociological Imagination) has been completely revised to reflect this new economic reality. Chapter 8 (Social Stratification) has also been thoroughly revised to show how existing inequalities within and between countries worldwide are expected to become even more pronounced. Finally, the economics portion of Chapter 11 (Economics and Politics) has been revised to describe the U.S. economy in light of this new economic context. Strategic changes have been made to charts and graphs in other chapters to capture the current economic realities. The changes described will give context to issues discussed in other chapters—which by the author's assessment are still relevant as written if contextualized within the new economic reality. Specific changes to each of the three revised chapters are listed below.

Major Changes to Updated Edition

The revision of Chapter 1—The Sociological Imagination—was informed by the ongoing economic crisis in the United States in the context of a larger global economic crisis. This crisis will not be short-lived, if only because 24 million Americans—16 percent of the workforce—are currently unemployed or working fewer hours than desired. If we consider that even in the best of times the U.S. economy creates 2.5 million or so jobs each year, it will take at least a decade to return to a normal labor environment. Among other things, this chapter

- considers the social forces that have created the global economic crisis—which is primarily a complex crisis of overextended credit, indebtedness, and the desire of financial investors and related organizations to increase returns.
- introduces seven classic sociological theorists but now applies each theorist's ideas to the current economic crisis.
- replaces Harriet Marineau as one of the seven classic theorists with Jane Addams, who advocated for sympathetic knowledge, or first-hand knowledge gained by living and working among those being studied, because "knowing one another better reinforces the common connection of people such that the potential for caring and empathetic moral actions increase" (Addams 1912, p. 7).
- refines and expands the definition and discussion of globalization and global interdependence.
- includes an "Intersection of Biography and Society" box in which college students write about the experience of taking out college, car, and home loans between 2001 and 2007—the years leading up to the economic crisis.

Like Chapter 1, the revision of Chapter 8—Social Stratification: With Emphasis on the World's Richest and Poorest—has been informed by the global economic crisis, and it now

- incorporates the experiences of one student of sociology, who upon graduation joins the Peace Corps and is assigned to Mauritania (an African country). This recent college graduate is from one of the world's richest countries—the United States—and he now finds himself working in one of the world's poorest countries. The chapter opens with that student's letter to his sociology professor describing his self-death (in a symbolic sense); that is, his transition from a person who had a strong sense of self-directed destiny to someone who is now oriented toward just surviving. This student's experiences are incorporated throughout the chapter to help convey what life is like in one of the world's poorest countries.

- includes new sections on the distribution of global household wealth, types of social mobility, the symbolic interactionist view of social inequality, modernization theory, and dependency theory.
- updates the progress on the United Nations' plan to significantly reduce global inequality by 2015 and how that plan is being impacted by the global economic crisis.
- incorporates an "Intersection of Biography and Society" box in which college students write about the impact of the economic crisis on their work life and job prospects.
- incorporates examples that are particularly relevant to the global economic crisis, such as the large divide between the haves and the have nots, including a new section on the indebted, which expands on the new social divide: debt-free versus indebted.
- includes a strong overview of the post-industrial society.

The economics half of Chapter 11—Economics and Politics: With Emphasis on Iraq—has been revised so that it now

- opens with a memorable photograph of an American soldier and two small Iraqi boys born after the beginning of the U.S. military presence in Iraq (2003). Whether they know it or not, the soldier and children have something in common—both depend on oil. The economy of the United States depends on oil, 60 percent of which is imported from foreign sources. The economy of Iraq depends on oil exports—more than 85 percent of its revenue comes from oil exports. Iraq's economy is oil and the U.S. economy is driven by oil. There is little doubt that oil shapes the economy and politics of both countries.
- includes a section on the U.S. economy (the largest in the world) that emphasizes the characteristics relevant to the current and ongoing economic crisis and incorporates sections on the U.S. national debt, consumer debt, dependence on oil and other minerals from foreign sources, and the influence of the financial sector on the overall economy.
- offers a strong overview of U.S. dependence on oil, with emphasis on the efficacy of drilling for oil in Alaska as a way of reducing dependence on foreign oil imports.

Finally there are a number of miscellaneous changes, such as including new photos that acknowledge the change in U.S. presidents (replacing photos of President George W. Bush with photos of President Barack Obama), updating the ten largest global corporations (since the major automobile industries have fallen from that list), and updating the map showing Internet connectivity by country.

Organizing Features of *Sociology: A Global Perspective*

More than 20 years ago, I signed a contract to write *Sociology: A Global Perspective*. That moment launched my academic career and research agenda, which since that time has been devoted to finding the most effective and interesting way to introduce sociology to those new to the discipline. In fact, it is fair to say that my area of academic specialization is introduction to sociology. The book's essential purpose has remained unchanged: to introduce students to the discipline of sociology in such a way that the final product is not an encyclopedic overview. The book presents sociological concepts and theories as more than a list of terms and definitions to be memorized. Rather, the book presents them as powerful tools for analyzing events occurring in one's personal life (biography), community, country, and world.

The book retains its unique approach: Twelve of the sixteen chapters pair a sociological topic with a country or territory—more specifically, with an issue centered on the chosen place but having critical relevance for understanding the United States and its position in the world order. The other four chapters pair a sociological topic not with a country, but with a broad theme related to global interdependence, such as McDonaldization, the world's richest and poorest countries, and the global story behind the peopling of the United States.

Each chapter includes a "Why Focus On" opening to explain the pairing. The choice of pairing is in some ways arbitrary, because sociological concepts are robust enough to be used to analyze any social situation. However, some countries and territories do lend themselves more than others to a particular sociological topic. The 100-plus years of conflict between Palestinians and Jews in what is now Israel and the Palestinian territories is particularly relevant to socialization, the focus of Chapter 4. For a conflict to last more than a century, it must be passed on from one generation to the next. The 20 or so concepts covered in that chapter are woven together to explain how this transmission occurs.

Each chapter is divided into eight to ten "core conceptual principles" to organize the material. Imagine being a student new to the discipline, skimming a chapter and seeing 25 to 30 key terms highlighted. Unfortunately, many students feel overwhelmed by such a presentation, and they set out to memorize the terms. Each core conceptual principle functions as an umbrella, pulling a variety of key terms and theories under it. The conceptual principles also drive end-of-chapter summaries titled "Visual Summary of Core Concepts." Here each core concept is paired with an image and a short summary.

The book features four types of boxes, each with a clear pedagogical focus: (1) Intersection of Biography and Society, (2) Global Comparisons, (3) No Borders, No Boundaries, and (4) Working for Change.

- *Intersection of Biography and Society:* The agreed-on objective of Introduction to Sociology courses is to instill in students the sociological imagination—a quality of mind that allows people to make connections between biography and seemingly remote and impersonal social forces and historical events. This box reinforces that objective by giving concrete examples of such intersections.

- *Global Comparisons:* Insight about the United States comes from comparing it with other countries. For example, we know that for every 1,000 babies born in the United States, approximately 6 do not survive the first year of life. Knowing that babies in 41 other countries have a better chance of surviving the first year of life puts this statistic in perspective. Likewise, using the comparative approach offers insight about the countries over which the United States exerts the greatest influence. For example, 1.3 million U.S. servicemen and servicewomen are stationed in various locations worldwide. Knowing that more than 120,000 of those troops are stationed in Iraq and 414 are stationed in Honduras provides a rough measure of the United States' relative influence in each country.
- *No Borders, No Boundaries:* This textbook takes a global perspective, which means that it emphasizes the ongoing and ever-increasing flow of goods, services, money, people, information, and culture across political borders. These boxes offer a sample of the various boundary-crossing activities and events that make up the phenomenon known as globalization.
- *Working for Change:* Sociologists critique existing social arrangements and social structures. The critical approach may leave some students apathetic, thinking that everything in the world is a mess or that nothing can be changed. This box focuses on solutions—specifically, on people who do something to change the system or who persuade others to change their behavior to benefit society.

Ancillary Materials

Sociology: A Global Perspective is accompanied by a wide array of supplements prepared to create the best learning environment inside and outside the classroom for both the instructor and the student.

ebank Test Bank. This enhanced and updated test bank consists of 70–100 multiple choice questions and 20–25 true-or-false questions per chapter, all with answers and page references. The test bank also includes 5 concept application questions, 15–20 short answer questions, and 1–2 essay questions per chapter.

ExamView®. Create, deliver, and customize tests and study guides (both print and online) in minutes with this easy-to-use assessment and tutorial system. ExamView offers both a Quick Test Wizard and an Online Test Wizard that guide you step by step through the process of creating tests, while its "what you see is what you get" interface allows you to see the test you are creating on the screen exactly as it will print or display online.

Instructor's Resource Manual. The instructor's manual offers the instructor suggestions for creating the course syllabus (including in-class student activities), background information on focus countries, and suggestions for enhancing the key sociological concepts in each chapter. Additionally, the manual includes chapter summaries in question-answer outline form (which provide responses to the study questions in the student study guide) and detailed lecture outlines.

Study Guide. This student learning tool includes 15–25 study questions for each chapter to guide reading, 5 concept application scenarios, practice tests containing 20–25 multiple-choice and 5–10 true-or-false questions, suggested film and Internet resources to enhance chapter material, and additional background information on the focus country, territory, or theme for each chapter.

More resources available from Wadsworth. Also available are supplements such as Practice Tests, WebTutor on Blackboard®, Microsoft® PowerPoint®, Audio Study Tools, CengageNOW™, and more. Please contact your local Cengage Learning sales representative for more information.

Acknowledgments

This seventh-edition update builds on the efforts of those who helped me with the previous editions. Five people stand out as particularly influential: Sheryl Fullerton (the editor who signed this book in 1988), Serina Beauparlant (the editor who saw the first and second editions through to completion), Eve Howard (the editor who managed the third, fourth, and fifth editions), and Chris Caldeira (the editor who developed the revision plan for the seventh edition and this update, who is now in the Ph.D. sociology program at the University of California–Davis). Of course, any revision plan depends on thoughtful, constructive, and thorough reviewer critiques. In this regard, I wish to extend my deepest appreciation to those who have reviewed this edition and/or its update:

Monique Balsam, Shawnee State University
Amy Bellone-Hite, Xavier University of Louisiana
John Brenner, Southwest Virginia Community College
Michelle Bemiller, Kansas State University
Lois Sabol, Yakima Valley Community College
Heather Dalmage, Roosevelt University
Brian Moss, Oakland Community College
Kristie Vise, Northern Kentucky University
Elizabeth Watson, Humboldt State University
LaQueta L. Wright, Richland College

When only one name—the author's—graces the cover of this textbook, it is difficult to convey just how many people were involved with its production. Their names appear in the most unassuming manner on the copyright page, belying the significant role they played in shaping the book. Perhaps the least recognized of those named on the copyright pages are production editors. I have been fortunate to work with Cheri Palmer and Bonnie Briggle, who take care of an overwhelming number of details associated with the book, including coordinating the work of the copyeditor, photo researcher, designer, proofreader, author, and others into a textbook ready to go to press. Both handle this pressure in ways that seemed effortless. But then such a style is a sign of true professionals—making something very few people can do seem effortless.

Apart from the support I received from Wadsworth/Cengage Learning on this update and past editions, I also received ongoing support and interest from several faculty members teaching in the Texas Community College system: Kay Coder, Pam Gaiter, Rachell Shane, Valerie Smith, Debbie White, and LaQueta Wright. I would also like to thank Loyd Ganey (Western International University), Nan McBlane (Thompson Rivers University), and Janice Proctor (Ohio University) for their long-standing support and encouragement. I also received valuable support from several colleagues at my university, who currently use or have adopted my textbook at one time or another: Prince Brown Jr., Molly Blenk, David Brose, Kris Hehn, Boni Li, J. Robert Lilly (who was one of my undergraduate professors), Jamie McCauley, and Mel Posey. And then there is Gale Largey (professor emeritus at Mansfield University), who wrote several boxes for this textbook and re-introduced me to the sociology of Lester Ward, an influential yet neglected and, some would say, forgotten sociologist.

As always, I wish to express my appreciation to my best friend—my mother, Annalee Taylor Ferrante—who keeps my files, clips relevant newspaper articles, and provides ideas that help me organize chapters (in this edition, especially Chapter 16, which focuses on Greenland). She also cooks meals for my husband and me several times a week. The care with which my mother prepares food and the exquisite results have no parallel. My mother has resisted the false promises of processed foods as a quick and easy source of sustenance—no small feat in a society that seems not to care about preserving the nutritional quality of food and seems resigned to processed food's destructive effect on the body and mind.

For the past two editions and this update, I have had the privilege of working with Missy Gish, a sociology major who recently completed the Master of Liberal Studies Program at NKU. Missy worked behind the scenes, assisting with photo research, updating tables and charts, checking references, and preparing chapters and manuals for production. On the surface, Missy's job description may seem simple, but I must emphasize that these tasks require an alertness, attention to detail, and ability to handle the stress associated with meeting deadlines that very few people possess.

I would like to thank the following NKU students, graduates, and professors who contributed photos to this edition: Renee McCafferty (Dance), Kim Bo-Kyung and Kevin Kirby (Math and Computer Science), Billy Carter (class of 2007), Ray Elfers (class of 2001), Missy Gish (class of 2005), James F. Hopgood (Anthropology), Boni Li (Sociology), Prince Brown Jr. (Sociology), Robert K. Wallace (Literature), Steve McCafferty (class of 1977), Yvonne DuPont Dressman (class of 1990), Beth Lorenz (class of 2003), Tom Kaelin (class of 2003), Kristie Vise (class of 1998), Noriko Ikarashi (class of 2003), Ted Weiss (Geography), and Mary Ann Weiss (First Year Programs).

I also received photos from Christopher Brown and Rob Williams. Special thanks goes to Lisa Southwick, who took several photographs especially for this update. We plan to collaborate in the future on identifying photographs taken of people in natural (versus posed) settings that effectively capture and illustrate the sociological perspective.

In closing, I acknowledge, as I have done in all editions of this and other books, the tremendous influence of Horatio C. Wood IV, MD, on my philosophy of education. As time passes, my feelings of warmth and gratitude toward him only deepen. Finally, I express my love for the most important person behind this book: my husband, colleague, friend, and greatest supporter, Robert K. Wallace.

1

The Sociological Imagination

The fascination of sociology lies in the fact that its perspective makes us see
in a new light the very world in which we have lived all our lives. . . . It can be
said that the first wisdom of sociology is this—things are not what they seem.

—Peter L. Berger, *Invitation to Sociology* (1963, pp. 21, 23)

Lisa Southwick

▲ Between 2000 and 2008, loan officers handed keys to millions of people who took out loans for
homes and cars that they could not afford and that the lenders knew that they could not afford. In
addition, banks and other financial institutions issued credit cards and college loans under terms that
would keep many borrowers in debt the rest of their lives. These millions of transactions between
lender and borrower were part of a larger financial process that triggered a global economic crisis.

Why Focus On

The Sociological Imagination?

THE **SOCIOLOGICAL IMAGINATION** is a quality of mind that allows people to see how larger social forces, especially their place in history and the ways in which society is organized, shape their life stories or biographies. A **biography** consists of all the day-to-day activities from birth to death that make up a person's life. Those activities can include taking out a loan to pay for college, a car, or a house; charging something on a credit card; or looking for a job. **Social forces** are any human-created ways of doing things that influence, pressure, or force people to behave, interact with others, and think in specified ways. Social forces are considered remote and impersonal because, for the most part, people have no hand in creating them, nor do they know those who did. People can embrace social forces, be swept along or bypassed by them, and most importantly challenge them.

One social force that contributed to the current economic crisis was the 1950s invention of the universal credit card (e.g., VISA, MasterCard, Discover, Capital One)—a bank-issued credit card that can be used to defer payment for products and services. This human-created invention became a "social force" that encouraged unprecedented numbers of people to spend money ahead of their earnings. While credit cards afforded those who could acquire them opportunities to delay paying for things they needed or wanted, it took special effort, discipline, and/or an advantaged position in life to resist using them.

In the 1980s another social force emerged that also encouraged many people to live beyond their means. Dur-

ing this time, banks moved away from a system in which they had made loans and issued credit cards to borrowers living in the surrounding community only after doing careful credit checks documenting real income, job stability, and credit history. Banks shifted to a system in which they knowingly issued loans to those with poor credit histories, gave loans larger than many borrowers could realistically afford to repay, and extended spending limits on credit cards to levels that many consumers found hard to resist.

In this chapter we will use the sociological imagination to think more about the social forces that shaped borrowing, lending, and spending, especially between 2000 and 2008—the years leading up to the economic crisis. Why is it important to develop the sociological imagination—a point of view that allows us to identify the seemingly remote and impersonal social forces that shape our lives? The payoff for those who acquire the sociological imagination is that they can (1) better understand their own biography by locating it in a broader context, (2) recognize the responses available to them by becoming aware of the many individuals who share (and do not share) their situation and response, and (3) position themselves to resist destructive forces and to change society for the better. (Awareness is the first step in the long, difficult process of personal and societal change.)

FACTS TO CONSIDER

- After three decades of using credit or debt to fuel the growth of the U.S. and global economy, the total amount of money owed by American consumers, the government, and businesses was $39 trillion, an amount three times as large as the gross domestic product of the United States (Federal Reserve 2009).

- The current economic crisis, which some call the Great Recession, resulted in more than 24 million Americans—16 percent of the workforce and growing—looking for work, working fewer hours than they would like, or becoming so discouraged that they gave up looking for work. Putting this many people back to work will take years even in the event of a recovery (Uchitelle 2009).

3

What Is Sociology?

■ **CORE CONCEPT 1: Sociologists focus on the social forces that shape human activity.** **Sociology** is the study of human activity as it is affected by social forces emanating from groups, organizations, societies, and even the global community. Human activity is simply the things people do with and to one another. The activities sociologists study are too many to name, but they can include people using credit cards, college students using loans to pay for education, parents serving their children a glass of apple juice, good friends walking in public, the unemployed searching for ways to secure an income, or a child gazing in a mirror and wondering why she appears to be a different race from her father. These activities may involve just one or two people or as many as several billion. The important thing is that the activities studied are affected by social forces (see Intersection of Biography and Society: "Six Social Forces Shaping Human Activity").

The Study of Social Facts

It seems, then, that on some level social forces exist outside the consciousness of individuals. French sociologist Émile Durkheim called such forces social facts. More specifically, Durkheim defined **social facts** as ideas, feelings, and ways of behaving "that possess the remarkable property of existing outside the consciousness of the individual" (Durkheim 1982, p. 51). That is, for the most part, social facts do not originate with the people experiencing them. From the time we are born, the people around us seek to impose upon us ways of thinking, feeling, and acting that we had no hand in creating. The words and gestures people use to express thoughts; the monetary and credit system used to pay debts; the rules governing games such as soccer and basketball; the beliefs and rituals of the religions people follow—all were created before they came on the scene. Thus, social facts have a life that extends beyond individuals.

Not only do social facts exist outside individuals, but they also have coercive power. When people freely and unthinkingly conform to social facts, that power "is not felt or felt hardly at all" (Durkheim 1982, p. 55). Only when people resist do they come to know and experience the power of social facts. Durkheim wrote that he was not forced to speak French or to use the legal currency, but it was impossible for him to do otherwise. "If I tried to escape the necessity, my attempt would fail miserably. . . . Even when in fact I can struggle free from these rules and successfully break them, it is never without being forced to fight against them" (Durkheim 1982, p. 51). In other words, even when people challenge them, social facts make their power known by the difficulty people experience trying to do things and think in different ways. Still, it is impressive that most of us can think of at least one example in which we fought against social facts.

These remote and impersonal social forces extend even to the ways we relate to good friends. In this regard, one Senegalese student in my class expressed dismay because now that he was in the United States, it was no longer normal for him to hold his best friend's hand when they walked to class or elsewhere. He wrote: "Out of habit, I reached for the hands of other male students I came to like and had to deal with their surprised looks and rejection. But I got used to this imposed distance between friends and started thinking and acting like an American. Now when I return home, I know I will be uncomfortable when my best friend tries to take my hand." Coincidentally, an American student in my class who traveled to Ghana on business was taken off guard when a man he was with took his hand. The American student wrote: "In the United States it is typically unacceptable for two men to hold hands. I spent some time in Ghana, Africa, several years ago and one of the first cultural differences I noticed was that men, including the men I was with, hold hands. This cultural difference definitely hit home when one day one of the men I was with took my hand as we walked. In order not to offend him, I followed through with this until an appropriate opportunity allowed me to disengage our hands. Even though I was in a country where this was perfectly acceptable, I still felt extremely uneasy with this tradition."

For Durkheim, social facts also included what he called **currents of opinion**, the state of affairs with regard to some way of being. The intensity of these currents is broadly reflected in rates summarizing various behaviors—for example, marriage, suicide, or birth rates. Durkheim believed the rates at which people around us marry, take

---

sociological imagination A quality of mind that allows people to see how larger social forces, especially their place in history and the ways in which society is organized, shape their life stories or biographies.

biography All the day-to-day activities from birth to death that make up a person's life.

social forces Any human-created ways of doing things that influence, pressure, or force people to behave, interact with others, and think in specified ways.

sociology The study of human activity as it is affected by social forces emanating from groups, organizations, societies, and even the global community.

social facts Ideas, feelings, and ways of behaving "that possess the remarkable property of existing outside the consciousness of the individual."

currents of opinion The state of affairs with regard to some way of being expressed through rates (suicide, marriage, savings).

their own life, or give birth to children both influence and reflect others' thinking and behavior on these matters. Table 1.1 compares household savings rates across countries in 1990 and 2009. Durkheim would argue that such rates offer insights about the overall value a society places on saving or spending. The intensity of that current of opinion shapes the behavior of people who live in the society.

The Sociological Consciousness

■ **CORE CONCEPT 2:** Sociologist Peter L. Berger offers the best description of the sociological consciousness: "The first wisdom of sociology is this—things are not what they seem." In his classic book *Invitation to Sociology*, Berger (1963) equates sociologists with curious observers walking the neighborhood streets of a large city, fascinated with what they cannot see taking place behind the building walls. The buildings themselves offer few clues beyond hinting at the architectural tastes of the people who built the structures and who may no longer live there. According to Berger, the wish to look inside and learn more is analogous to the sociological perspective.

The discipline of sociology offers us theories, concepts, and methods needed to look beyond popular meanings and interpretations of what is going on around us. Berger (1963) points out that sociologists, by the very logic of their discipline, are driven to debunk the social systems they study. One should not mistake this drive as being located in a sociologist's temperament or personal inclination. Apart from his or her field of study, a sociologist may be "disinclined to disturb the comfortable assumptions on which he rests his own social existence" (p. 29). Nevertheless, the sociological perspective compels sociologists to explore levels of reality that dig below the surface. The logic of the discipline presupposes a "measure of suspicion about the way in which human events are officially interpreted by the authorities, be they political, juridical or religious in character" (p. 29).

The Intersection of Biography and Society box on pages 6 and 7 offers a preview of the kinds of social forces that we will address in upcoming chapters. Whatever human activity sociologists study, they are compelled to ask questions about the nature and origin of the social force(s) shaping it. Those questions include:

- What are the social forces shaping the human activity under question?
- Under what circumstances do people resist and challenge social forces?
- What is the reach of a social force—is it confined to a specific group of people, or does it affect human activity on a local, regional, national, or global scale?
- How are social forces initiated? Who benefits from a particular social force and at whose expense?
- What are the anticipated and unanticipated consequences of social forces?

Table 1.1 Percent of Disposable Household Income That Is Saved by the United States and Other Wealthy Economies, 1990 versus 2009

Until just recently, the current household savings rate in the United States was 1.2 percent of disposable household income (defined as income after taxes). France, Germany, and Spain currently has the highest rates of household savings—at least 10 percent of disposable income is saved. However, savings rates had declined over the past 20 years for all countries shown in the chart below. In the United States, savings rates have dropped by 5.8 percent, down from 7 percent of disposable income saved in 1990. In some countries, household savings has declined even more dramatically over the past 20 years. Durkheim argues that these rates offer clues about the level of intensity by which a society conveys messages about the importance of saving or spending. It is important to point out that as a response to the economic crisis, the American rate of saving approached 7.0 percent in May 2009, suggesting that the "currents of opinions" about saving have changed.

Country	1990	2009
Australia	8.2	2.5
Austria	10.3	9.8
Canada	13.0	1.1
Finland	1.9	−2.2*
France	9.4	12.3
Germany	13.7	10.6
Italy	21.7	6.8
Japan	13.9	2.6
Korea	22.5	2.5
Netherlands	18.1	6.4
Norway	2.7	1.4
Sweden	3.9	7.8
Switzerland	9.6	9.5
United States	7.0	1.2

*A negative savings rate suggests that there is no overall savings and money is being withdrawn from existing savings.

Source: OECD (2009)

Troubles and Issues

■ **CORE CONCEPT 3:** Sociologists distinguish between *troubles*, which can be resolved by changing the individual, and *issues*, which can be resolved only by addressing the social forces that created them. C. Wright Mills (1959) defines **troubles** as personal needs, problems, or difficulties that

troubles Personal needs, problems, or difficulties that can be explained as individual shortcomings related to motivation, attitude, ability, character, or judgment.

INTERSECTION OF BIOGRAPHY AND SOCIETY

Six Social Forces Shaping Human Activity

Missy Gish

Lisa Southwick

1.   It seems like most of how we go about our day and live our lives has been "planned" out for us. This is because humans create *institutions*, relatively stable and predictable arrangements among people that have emerged over time to coordinate human activity in ways that meet some human need, such as school systems to pass on accumulated knowledge to new generations. In the United States, 66 percent of college graduates leave with an average loan to repay valued at $22,500; a loan that they will be paying on 20 or more years (Bernard 2009). The debt burden U.S. students assume is not something that students attending college in the 27 countries that make up the European Union face (see Chapter 13).

2.   The cell phone is a *technology* humans invented to free them from landline phones and to allow them to communicate with others while on the move. There is no question that this social force has changed the way people communicate. Because cell phones are typically not shared (most people have their own phone), callers do not have to speak to a party other than the person he or she is calling. There is no need to make "small talk" to a third person who answers. In this sense, cell phones expand individuality and privacy. With the cell phone, a parent–calling to see if a child playing at a friend's house is coming home–talks to the child directly and misses an opportunity to talk to that friend's parent. On the surface, it seems efficient to avoid a third party; on the other hand, the ties among parents are weakened (see Chapter 3).

Lisa Southwick

© L. Lartigue/USAID

3.   Humans have assigned great *symbolic meaning* to diamonds–the stones are a sign of engagement, marriage, love, and wealth. That meaning is a social force that creates an insatiable demand for the stones, such that low-wage diamond miners working in the Democratic Republic of the Congo and elsewhere produce some 800 million gem- and industrial-quality diamonds per year. U.S. consumers–about 4.6 percent of the world's population–buy almost 50 percent of these stones (see Chapter 7). (National Geographic 2002)

U.S. photo by Mass Communication Specialist 2nd Class Kimberly Williams/Released

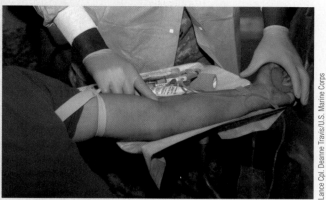

Lance Cpl. Deanne Travis/U.S. Marine Corps

4. If you think about it, we spend much of our time interacting with strangers. Once we determine a stranger's *social status* in relation to our own social status, we know what to say and do. This customer does not need to know the sales clerk to interact with him, nor does the clerk need to know him. Associated with the social statuses of customer and clerk are *roles* specifying how those occupying each status are expected to behave toward the other and what each can expect of the other in return. Statuses and associated roles are social forces in the sense that they guide and "pressure" people to behave and interact in expected ways (see Chapter 5).

5. *Globalization* is a social force that is largely invisible to most of us but affects our lives in countless ways. Globalization encompasses the ever-increasing flow of goods, services, money, people, technology, information, and other cultural items that routinely move across national borders. As one example, it is very likely the plasma people sell to one of the 500 for-profit plasma centers in the United States will be exported to another country. In fact, U.S. plasma donors supply 60 percent of global demand, with Japan being one of the largest trading partners and acquiring 90 percent of its plasma from American donors (National Plasma Center 2009). Some of the largest plasma companies operating in the United States are foreign-owned with headquarters in Japan, West Germany, Austria, and Canada. One may wonder why the United States is the OPEC of blood plasma—one answer is that, of the world's countries, the U.S. has the most liberal guidelines regarding plasma donations—a person can give/sell plasma every two weeks or 24 times a year. In addition, many countries ban the sale of blood plasma (see Chapter 5). (National Plasma Center 2009; Gaul 1989)

Department of Defense photo by SSGT Lanie McNeal, USAF

6. *Race* is not a biological reality but a socially created way of categorizing people. As such, it is a social force of immense significance. Simply consider that in the United States, a parent and his or her biological offspring can be classified as different races. This soldier returning home from Iraq meets his son for the first time. Both father and son will have to come to terms with the fact that each is considered a different race—the son has a black father and the father has a white son. The couple will have to come to terms with questions, often unspoken, as to whether this man is really the father (see Chapter 8).

ISSUES	TROUBLES
Focus Outside the individual	**Focus** Personal needs, problems, and difficulties
Cause Large social forces	**Cause** Individual shortcomings
Example of Causes Financial system with products created to profit from debt; a culture that values consumption over living within means	**Example of Causes** Lack of motivation Bad attitude Flawed character Weak skills
Resolution Identify and counteract larger social forces	**Resolution** Change individual shortcomings
Change Strategies Regulate financial services industry so that it does not reward lenders who are making risky loans	**Change Strategies** Seek therapy Find new friends Take mood-altering medications Look for a job/change jobs Try harder

▲ **Figure 1.1**   Issues and Troubles

can be explained as individual shortcomings related to motivation, attitude, ability, character, or judgment. The resolution of a trouble, if it can indeed be resolved, lies in changing the individual in some way. Mills states that when only one man or woman is unemployed in a city of 100,000, that situation is his or her personal trouble. For its relief, we properly look to that person's character, skills, and immediate opportunities (that is, we think, "She is lazy," "He has a bad attitude," "He didn't try very hard in school," or "She had the opportunity but didn't take it").

By comparison, an **issue** is a matter that can be explained only by factors outside an individual's control and immediate environment. When 24 million men and women are unemployed or underemployed in a nation with a workforce of 156 million, that situation is an issue. Clearly, we cannot hope to solve this kind of employment crisis by focusing on the character flaws of 24 million individuals. According to Mills, an accurate description of the problem and of the possible solutions to it requires us to think beyond individual shortcomings and to consider the underlying social forces that create them (see Figure 1.1).

Mills argues that many people cannot see the intricate connection between their personal situations or "trou-

issue A matter that can be explained only by factors outside an individual's control and immediate environment.

bles" and larger social forces. When prospective home-owners with no money for a down payment, an income too low to manage loan payments, and/or a poor credit history received loans for homes that were not affordable, they probably did not see themselves as one of millions to whom banks were awarding subprime loans that they would eventually bundle and sell to investors who would assume risk of default.

Mills also argues that most people cannot (or do not want to) see how their successes connect to others' so-called failures in life. The "success" loan officers felt upon meeting performance measures related to the number of loans closed was built upon putting many borrowers into financial situations they could not sustain with the incomes they earned (see Intersection of Biography and Society: "The Personal Experience of Securing a Loan"). Mills believes that most people are unaware of how their successes are built upon others' failures because they do not possess a quality of mind that enables them to grasp the interplay between self and world and between biography and the large social forces pressuring them to think and behave as they do.

In *The Sociological Imagination*, Mills (1959) asks, "Is it any wonder that ordinary people feel they cannot cope with the larger worlds with which they are so suddenly confronted?" (pp. 4–5). Is it any wonder that people often feel trapped by the social forces that affect them? As Mills pointed out, we live in a world in which informa-

INTERSECTION OF BIOGRAPHY AND SOCIETY

The Personal Experience of Securing a Loan

BETWEEN 2000 AND early 2008, many banks made loans to people who were high credit risks and issued loans for amounts larger than the borrowers requested. My students shared their experiences with lenient lending practices during this time period and the consequences of such practices.

• I took out a car loan in 2005. At that time I had no credit history, and the dealership asked me if I could find a co-signer; I explained that both my parents were deceased and I could absolutely not find a co-signer. Several hours later, I walked away with my brand new car and a loan for $17,000, with no money down and no co-signer. It was a bad deal for me; my interest rate was horrible and the length of the loan was long. I still look back and ask how the car company could give me a loan.

• We bought the house that we live in now in 2003. I was pregnant with our third child and thought that it was time to move to a bigger house. At the time, we made about $80,000 a year. We had several credit card bills as well and we were worried that we may not qualify for a loan. We had to show two weeks of pay stubs, W2's from the previous year, and any savings or 401(k) statements. That's it!!! We completed the application and, to our surprise, within two days the bank called and offered us more money than we had asked. They were willing to give us $250,000 with no down payment. We needed $175,000, an amount we knew was going to be a stretch for our budget but, being naïve, we took $200,000.

• My family purchased three houses in the last seven years, all of which we bought without a down payment and after a minimal credit check. We sold two of the houses fairly easily. The last house, which we had built, cost us $200,000. Now, due to the economy and a housing slump, brand-new homes on our street are selling for between $160,000 and $180,000. This drop in prices means we have to sell this house for a price below what we paid for it.

• When I was looking to buy a home in January 2007, I talked to a mortgage company and a local bank. The mortgage company rep (who worked on commission) approved me for an amount of money that I knew I couldn't afford. When I told that rep I couldn't afford it he responded that "on paper" I could definitely afford it. I told him that I didn't live on paper and I knew that I couldn't possibly make those monthly payments. He then offered me an interest only loan (where I wouldn't have to pay on the principle) with an adjustable rate to make the monthly payments low enough. I told him thanks and I would get back to him. I went to a local bank (where the loan officers don't work on commission). After talking to them, they approved me for an amount I knew I could afford with a 30-year, fixed rate.

---

tion dominates our attention and overwhelms our capacity to make sense of all we hear, see, and read every day. Consequently, we may be exhausted by the struggle to learn from that information about the forces that shape our daily lives. According to Mills, people need "a quality of mind that will help them to use information" so they can think about "what is going on in the world and what may be happening within themselves" (p. 5). Mills calls this quality of mind the *sociological imagination*. Those who possess a sociological imagination can view their inner life and human career in the context of larger social forces. The payoff for those who demonstrate this quality of mind is that they can understand their own experiences and fate by locating themselves in an historical and social context; they can recognize the responses available to them by becoming aware of the many individuals who share their situations.

The sociological imagination is evident in the work of the earliest and most influential sociologists. In fact, one can make the case that the discipline of sociology emerged as part of an effort to understand how a social force known as the Industrial Revolution changed human activity and interaction in an endless number of ways.

The Industrial Revolution

■ **CORE CONCEPT 4:** Sociology emerged in part as a reaction to the Industrial Revolution, an ongoing and evolving social force that transformed society, human behavior, and interaction

in incalculable ways. The **Industrial Revolution** is the name given to the changes in manufacturing, agriculture, transportation, and mining that transformed virtually every aspect of society. The defining feature of the Industrial Revolution was **mechanization**, the process of replacing human and animal muscle as a source of power with external sources derived from burning wood, coal, oil, and natural gas. The new sources of power eventually replaced hand tools with power tools, sailboats with freighters, and horse-drawn carriages with trains. Mechanization changed how goods were produced and how people worked. It turned workshops into factories, skilled workers into machine operators, and hand-made goods into machine-made goods.

Before mechanization, goods were produced and distributed at a human pace, as illustrated by the effort required to make glass by hand: Skilled workers "endured the tremendous heat to coax beautiful forms from the fire using nothing more than their breath and a few simple tools. They worked hard to polish their skills to uniformity and precision, but even so each creation was as individual as the maker" (Thrall 2007). With industrialization, products previously crafted by a few skilled people were now standardized and assembled by many relatively unskilled workers, each involved in part of the overall production process. Now no one person could say, "I made this; this is a unique product of my labor." The factory owners gained power over the artisans as machines rendered them obsolete. Now people with little or no skill could do the artisan's work—and at a faster pace. With regard to glass making, mechanically pressing hot glass replaced "the time-consuming handcrafting." Because of the faster pace, more glass products could be produced faster and things made of glass became common household items (Thrall 2007).

Industrialization did more than change the nature of work; it affected virtually every aspect of daily life in incalculable ways. A social order that had existed for centuries "vanished," and a new order—familiar in its outline to us today—appeared (Lengermann 1974). A series of developments—the railroad, the steamship, the cotton gin, the spinning jenny, running water, central heating, electricity, the telegraph, and mass-circulation newspapers—transformed how people lived their daily lives and with whom

Library of Congress Prints and Photographs Division Washington, D.C.

Before industrialization, human and/or animal muscle powered hand tools and pulled loads. Imagine the human muscle and the hours of labor required to fell a 16-foot cedar with axes. The gasoline-powered chain saw changed all that in 1926–dramatically reducing the time it took to cut down trees.

Industrial Revolution Changes in manufacturing, agriculture, transportation, and mining that transformed virtually every aspect of society.

mechanization The addition of external sources of power, such as that derived from burning coal and oil, to muscle-powered tools and modes of transportation.

they interacted. Coal-powered trains, for example, turned a month-long trip by stagecoach into a day-long one. These trains permitted people and goods to travel day and night; in rain, snow, or sleet; and across smooth and rough terrain. Railroads increased the human and freight traffic to and from previously remote and unconnected areas. The railroad caused people to believe they had "annihilated" time and space (Gordon 1989). In addition, railroads facilitated an unprecedented degree of economic interdependence, competition, and upheaval. Now people in one area could be priced out of a livelihood if people in another area could provide goods and materials at a lower price (Gordon 1989).

The Industrial Revolution drew people from even the most remote parts of the globe into a process that produced unprecedented quantities of material goods. The historical period known as the Age of Imperialism (1880–1914) involved the most rapid industrial and colonial expansion in history. During this time, rival European powers (such as Britain, France, Germany, Belgium, Portugal, the Netherlands, and Italy) competed to secure colonies and, by extension, the labor and natural resources within those colonies. By 1914, for example, all of Africa had been divided into European colonies. By that year, 84 percent of the world's land area had been affected by colonization, and an estimated 500 million people were living as members of European colonies (Random House Encyclopedia 1990).

The Industrial Revolution changed everything—the way in which goods were produced, the ways in which people negotiated time and space, the relationships between what were once geographically separated peoples, the ways in which people made their livings, the density of human populations (e.g., urbanization), the relative importance and influence of the home in people's lives, access to formal education (the rise of compulsory and mass education), and the emergence of a consumption-oriented economy and culture. The accumulation of wealth became a valued and necessary pursuit. In *The Wealth of Nations*, Adam Smith argued that invisible hand of the free market (capitalism) embodied in private ownership and self-interested competition held the key to a nation's advancement and prosperity. The unprecedented changes caught the attention of the early sociologists who wrote in the 19th and early 20th centuries. In fact, sociology emerged out of their efforts to document and explain the effects of the Industrial Revolution on society.

■ **CORE CONCEPT 5:** Early sociologists were witnesses to the transforming effects of the Industrial Revolution. They offered

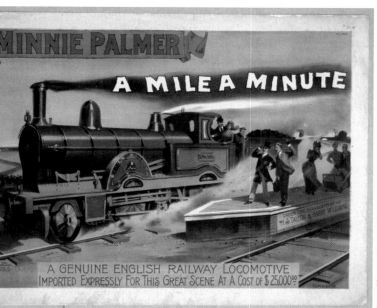

Coal-powered locomotives celebrated in this 1891 poster permitted people to travel day and night; in rain, snow, or sleet; across smooth and rough terrain—turning month-long trips into day-long ones. Railroads increased opportunities for personal mobility and boosted the freight and passenger traffic to and from previously remote areas.

Library of Congress Prints and Photographs Division Washington, D.C.

lasting conceptual frameworks for analyzing the ongoing social upheavals. Sociology emerged as an effort to understand the dramatic and almost immeasurable effects of the Industrial Revolution on human life across the globe. Although the early sociologists wrote in the 19th and early 20th centuries, their observations remain relevant. In fact, many insights into the character of contemporary society can be gained by reading their writings, because those who witness and adjust to a significant event are intensely familiar with its consequences in daily life. Because most of us living today know only an industrialized life, we lack the insights that came from living during the transitional period following the Industrial Revolution and observing its consequences. To grasp the significance of these observations, consider the following anecdote. In a recent interview, a scientist maintained that we are close to understanding the mechanisms governing aging, and that people might soon live to be 150 years old. If aging mechanisms are in fact controlled, the first people to witness the change will have to make the greatest adjustment. In contrast, people born after this discovery will know only a life in which they can expect to live 150 years. If these post-discovery humans are curious, they may wish to understand how living to age 150 shapes their lives. To fully understand this subject, they will have to look to those who recorded life before the change and who made sense of their adjustments to the so-called advancement. So it is with industrialization: to understand how it shaped and continues to shape human life (and how it has shaped sociology), we can look to six of the early sociologists.

Three of the six sociologists covered are nicknamed the "big three." Those three are Karl Marx, Émile Durkheim, and Max Weber. There is universal agreement among sociologists that these three are the giants in the field and that their writings form the heart of the discipline. It is safe to say that all sociologists who have come after Marx, Durkheim, and Weber have been deeply influenced by their ideas even as they expand, refine, and challenge them (Appellrouth and Edles 2007).

We also include three other central figures: Auguste Comte, because he gave sociology its name, and Jane Addams and W.E.B. DuBois. DuBois focused attention on the color line, and Jane Addams championed sympathetic knowledge or knowledge gained from living and working among those being studied. The color line and sympathetic knowledge are certainly core ideas within sociology. Upon discussing the work of each of these six sociologists, we consider what features of the global economic crisis each would emphasize if he or she were alive today.

Auguste Comte (1798–1857)

The French philosopher Auguste Comte, known as the father of positivism, gave sociology its name in 1839.

Positivism holds that valid knowledge about the world can be derived only from *sense experience* or knowing the world through the senses of sight, touch, taste, smell, and hearing and from empirical associations (e.g., evidence of cause and effect must be observable by the senses). Comte advanced the "Law of Three Stages," which maintains that societies develop according to three stages: (1) the theocratic, (2) the metaphysical, and (3) the positive. In the theocratic stage, people explain the events going on in the world as the work of personified deities—those deities may be objects such as the sun or trees, a variety of gods, or a supreme deity. Deities possess supernatural qualities that allow them to exert their will over humans and nature. In the metaphysical stage, people draw upon abstract and broad concepts to define features of reality that cannot be observed through the senses or direct experience. Metaphysics deals with big philosophical questions such as the nature of the human mind, the meaning of life, and good versus evil. In the positive stage (stage 3)—the conceptually superior stage according to Comte—people use scientific explanations grounded in observation and experimental designs to understand the world. Comte placed sociology in this third stage of thinking; he maintained that sociologists were scientists who studied the results of the human intellect (DeGrange 1938). What did he mean by this?

First, sociology is a science and only those sociologists who follow the scientific method can presume to have a voice in describing and guiding human affairs. The scientific method rejects personal opinions and political agendas in favor of disciplined and objective strategies in thinking about and addressing social issues and making effective policies. Second, sociologists study the things humans have created: an idea, an invention, or a way of behaving and the effects those creations have on society. Comte recommended that sociologists study **social statics**, the forces that hold societies together such that they endure over time, and **social dynamics**, the forces that cause societies to change. Comte's preoccupation with order, change, and the things humans have created is not surprising given that he was writing at a time when the Industrial Revolution was transforming society in unprecedented ways.

Comte on the Economic Crisis. If Auguste Comte were alive today, he would emphasize the dramatic and far-reaching changes associated with the economic crisis. At the same time, he would also consider that, in spite of these changes, for the most part, societies around the world are holding together such that they do not collapse into something beyond recognition. Comte might choose to illustrate the forces holding societies together (social statics) in the midst of change (social dynamics) by comparing the U.S. and European response to the global debt crisis. One way some German and other European companies have handled the economic slowdown is by asking their employees to work fewer hours and by allowing their governments to compensate employees for hours not worked out of an existing fund supported by payroll deductions and employer contributions (Kulish 2009).

While, in the United States, some states offer wage compensation, few employers are aware that such programs exist. In addition, the state programs have a number of requirements that discourage participation, such as compensating only a percentage of the lost wages and requiring employers to continue paying for health and other benefits. Finally, most states with the programs do not have the "bureaucratic infrastructure" in place to run such programs (Greenhouse 2009). These different approaches explain why the Europeans generally opposed new government spending programs as a way of stimulating their economies. From the European point of view, the social safety nets they have in place are the spending programs; they don't have to create them on the spot during economic downturns as the United States does. European employees are able to maintain their wages for at least 18 months in the face of reduced hours. In addition, most Europeans do not have to worry about paying for health care upon losing their jobs nor about companies taking away such benefits as a cost-saving measure. While the United States has unemployment benefits, it does not have an ongoing safety net in place, which explains the stimulus spending programs instituted by the federal government. Comte would note that there is something broadly predictable about the way the United States and European countries respond to economic change. It is the "predictable" component that holds the respective societies together even as they undergo dramatic change.

Karl Marx (1818-1883)

The political philosopher Karl Marx was born in Germany but spent much of his professional life in London, working and writing in collaboration with Friedrich Engels. Two of Marx and Engels's most influential treatises are *Das Kapital* and *The Communist Manifesto*. *Das Kapital*, a massive multivolume work published in 1867, 1885, and 1894, is critical of the capitalist system and predicts its defeat by a more humane and more cooperative economic system:

positivism A theory stating that valid knowledge about the world can be derived only from *sense experience* or knowing the world through the senses of sight, touch, taste, smell, and hearing and from empirical associations.

social statics The forces that hold societies together such that they endure over time.

social dynamics The forces that cause societies to change.

socialism. The *Communist Manifesto* is a 23-page pamphlet that was issued in 1848 and has since been translated into more than 30 languages (Marcus 1998). Upon reading it today, more than 150 years later, one is "struck by the eerie way in which its 1848 description of capitalism resembles the restless, anxious and competitive world of today's global economy" (Lewis 1998, p. A17).

The *Manifesto* includes these famous lines: "The workers have nothing to lose but their chains; they have a whole world to gain. Workers of all countries, unite." In an essay marking the 150th anniversary of *The Communist Manifesto*, John Cassidy (1997) wrote that "in many ways, Marx's legacy has been obscured by the failure of Communism, which wasn't his primary interest. In fact, . . . Marx was a student of capitalism, and that is how he should be judged" (p. 248).

Marx sought to analyze and explain **conflict**, the major force that drives social change. The character of conflict is shaped directly and profoundly by the means of production, the resources (land, tools, equipment, factories, transportation, and labor) essential to the production and distribution of goods and services. Marx viewed every historical period as characterized by a system of production that gave rise to specific types of confrontation between an exploiting class and an exploited class. For Marx, class conflict was the vehicle that propelled people from one historical epoch to another.

From Marx's perspective, the Industrial Revolution was accompanied by the rise of two distinct classes, creating a fundamental divide: the **bourgeoisie**, the owners of the means of production, and the **proletariat**, those individuals who must sell their labor to the bourgeoisie. Marx expressed profound moral outrage over the plight of the proletariat, who, at the time of his writings, were unable to afford the products of their labor and suffered from deplorable living conditions. Marx devoted his life to documenting and understanding the causes and consequences of this inequality, which he connected to a fatal flaw in the organization of production (Lengermann 1974).

Karl Marx believed that the pursuit of profit was behind the explosion of technological innovation and the never-before-seen increase in the amount of goods and services produced during the Industrial Revolution. In a capitalist system, profit is the most important measure of success. Marx described class conflict as an antagonism that grows out of the opposing interests held by these two parties. The bourgeoisie's interest lies with making a profit and the proletariat's with increasing wages. To maximize profit, the bourgeoisie work to cut labor costs with labor-saving technologies, employ the lowest-cost workers, and find the cheapest materials to make products.

The capitalist system is a vehicle of change in that it requires technology and products to be revolutionized constantly. Marx believed that capitalism was the first economic system capable of maximizing the immense productive potential of human labor and ingenuity. He also felt, however, that capitalism ignored too many human needs and that too many people could not afford to buy the products of their labor. Marx believed that if this economic system were in the right hands—the hands of the workers or the proletariat—public wealth would be more than abundant and would be distributed according to need. Instead, according to Marx (1887), capitalism survived and flourished by sucking the blood of living labor. The drive is a "boundless thirst—a werewolf-like hunger—that takes no account of the health and the length of life of the worker unless society forces it to do so" (p. 142). That thirst for profit "chases the bourgeoisie over the whole surface of the globe" in search of the lowest-cost labor and resources to make products (Marx 1881, p. 531).

In the *Class Struggles of France 1848–1850*, Marx named another class, the finance aristocracy, who lived in obvious

Karl Marx's writings have had a tremendous influence on the discipline of sociology. At his funeral, Friedrich Engels spoke these words: "On the 14th of March 1883, at a quarter to three in the afternoon, the greatest living thinker ceased to think."

conflict The major force that drives social change.

bourgeoisie The owners of the means of production who exploit the labor of the proletariat.

proletariat Those individuals who must sell their labor to the bourgeoisie.

luxury among masses of starving, low-paid, and unemployed workers (Bologna 2008, Proudhon 1847). The finance aristocracy includes bankers and stockholders seemingly detached from the world of "work." Marx (1856) described this source of income as "created from nothing—without labor and without creating a product or service to sell in exchange for wealth." The finance aristocracy speculates or employs financial advisors to speculate for them. "But while speculation has this power of inventiveness, it is at the same time also a gamble and a search for the 'easy life'; as such it is the art of getting rich without work." According to Marx (1856), the financial aristocracy appropriate to themselves "public funds or private funds without giving anything equivalent in exchange; it is the cancer of production, the plague of society and of states."(Bologna 2008, Proudhon 1847)

Marx on the Economic Crisis. If Karl Marx were alive today, he would emphasize the size and power of the financial sector, which accounts for 21 percent of the gross domestic product (GDP) of the United States. This sector includes corporations such as Bank of America and Citigroup that are considered too big to fail. By contrast, the manufacturing sector accounts for 12 to 13 percent of GDP. Marx would also point out that the growth of the financial sector was fueled by consumer debt—whether that debt is derived from mortgages, college loans, car loans, credit cards, or commercial borrowing. Specifically, Marx would emphasize the fact that the financial elite created and lobbied for minimally regulated financial products that allowed lenders to assume no risk from making bad loans. Those products included securitization and credit default swaps (CDSs). Securitization, as it relates to loans, involves lenders packaging hundreds to hundreds of thousands of loans of varying risk of default together and selling them to investors. In selling the loans, the lender makes an immediate profit and walks away from the loan; the investors assume the risk of borrowers defaulting. However, the investors lowered their risk through credit default swaps, an insurance-like system in which investors take out policies to protect themselves from loan defaults by shifting the risk onto a third party (an insurance company). To complicate matters, those who purchase credit default swaps do not have to own the insured loan; other investors can purchase that "insurance" as a bet that borrowers will default. If that happens, the buyer collects. Hedge fund managers who purchased CDSs as bets that high-risk borrowers would default on their home mortgages represent one example of high-stake investors being able to profit even during the debt crisis. One of the highest-

solidarity The ties that bind people to one another in a society.

To avoid a run on the banks reminiscent of the Great Depression, in 2008 and 2009 the U.S. government bailed out banks and other financial institutions. One of the most publicized bailouts went to American International Group (AIG), which sold credit protection on collateralized debt obligations (CDOs). When credit default rates increased and home values fell, many investors filed claims to collect on failed investments. The U.S. government gave AIG several hundred billion dollars so it could meet its credit default swap obligations to those banks and companies that had taken out policies. Its payouts included $8.1 billion to Goldman Sachs, $5.4 billion to Deutsche Bank, $4.9 billion to Merrill Lynch, and $700 million to Royal Bank of Scotland (Walsh 2009).

compensated hedge fund managers who bet in this way earned $3.7 billion in one year (Anderson 2008).

Marx's description of the finance aristocracy applies to those who seek to increase their wealth without creating a product or service in exchange for money earned. During the years preceding the debt crisis, these high-yield financial products disproportionately rewarded the wealthiest 1 to 2 percent of the U.S. and global populations (Frank 2009).

Émile Durkheim (1858-1918)

To describe the Industrial Revolution and its effects, Frenchman Émile Durkheim focused on the division of labor and solidarity. The division of labor is the way a society divides and assigns day-to-day tasks. Durkheim was interested in how the division of labor affected **solidarity**, the system of social ties that connects people to one another and to the wider society. This system of social ties acts as "cement" binding people to each other and to the society. Durkheim observed that industrialization changed the division of labor from relatively simple to complex and, by extension, changed the nature of solidarity. Durkheim believed that the sociologist's task is to analyze and explain the solidarity. Durkheim's preoccupation with the ties that bind is evident in his writings on education, deviance, the division of labor, and suicide (see Chapters 4, 5, 7, and 12). By way of introduction to Durkheim's emphasis, we turn to his writings on suicide.

In *Suicide*, Durkheim argued that it is futile to study the immediate circumstances that lead people to kill themselves, because an infinite number of such circumstances exist. For example, one person may kill herself in the midst of newly acquired wealth, while another kills herself in the lap of poverty. One may kill himself because he is unhappy in his marriage and feels trapped, while another kills himself because his unhappy marriage has just ended in divorce. In one case, a person kills himself after losing a business; in another case, a lottery winner kills herself because she cannot tolerate her family and friends fighting one another to share in her newfound fortune. Because almost any personal circumstance can serve as a pretext for suicide, Durkheim concluded that there is no situation that could not serve as an occasion for someone's suicide.

Durkheim also reasoned that no central emotional quality was common to all suicides. We can point to cases in which people live on through horrible misfortune while others kill themselves over seemingly minor troubles. Moreover, we can cite examples in which people renounce life at times when it is most comfortable or at times of great achievement. Given these conceptual difficulties, Durkheim offered a definition of suicide that goes beyond its popular meaning (the act of intentionally killing oneself). This definition takes the spotlight off the victim and points it outward toward the ties that bind (or fail to bind) people to others in the society. In short, Durkheim defined **suicide** as the severing of relationships. To make his case, he argued that every group has a greater or lesser propensity for suicide. The suicide rates for various age, sex, and race groups in the United States, for example, show that for some categories of people—the elderly, males, 15- to 19-year-olds—suicide is more prevalent than for other categories. From a sociological point of view, these differences in suicide rates cannot be explained by pointing to each victim's immediate circumstances.

Instead, Durkheim examined the social ties that bind or fail to bind social categories to others. For example, all people who suddenly find themselves in the unemployed category must adjust to life without a job. That adjustment may entail finding a way to live on a reduced budget, trying to stay cheerful while hunting for a job, or feeling uncomfortable around friends who have a job. According to Durkheim, it is inevitable that a certain number of those in the unemployed category will succumb to social pressures and choose to sever the relationships from which such pressures emanate.

Durkheim identified four types of social ties, each of which describes a different kind of relationship to the group: egoistic, altruistic, anomic, and fatalistic. **Egoistic** describes a state in which the ties attaching the individual to others in the society are weak. When individuals are detached from others, they encounter less resistance to suicide. The lives of the chronically ill, for example, are often characterized by excessive individuation if friends, family, and other acquaintances avoid interacting with the ill person out of fear of upsetting the patient or themselves.

Altruistic describes a state in which the ties attaching the individual to the group are such that he or she has no life beyond the group. In these situations, a person's sense of self cannot be separated from the group. When such people commit suicide, it is on behalf of the group they love more than themselves. The classic example is members of a military unit: the first quality of soldiers is a sense of selflessness. Soldiers must be trained to place little value on the self and to sacrifice themselves for the unit and its larger purpose.

Anomic describes a state in which the ties attaching the individual to the group are disrupted due to dramatic changes in social circumstances. Durkheim gave particular emphasis to economic circumstances such as a recession, a depression, or an economic boom. In all cases, a reclassification occurs that suddenly casts individuals into a lower or higher status than before. When people are cast into a lower status, they must reduce their requirements, restrain their needs, and practice self-control. When individuals are cast into a higher status, they must adjust to increased prosperity, which unleashes aspirations and expands desires to an unlimited extent. A thirst to acquire goods and services arises that cannot be satisfied.

Fatalistic describes a state in which the ties attaching the individual to the group involve discipline so oppressive it offers no chance of release. Under such conditions, individuals see their futures as permanently blocked. Durkheim asked, "Do not the suicides of slaves, said to be frequent under certain conditions, belong to this type?" (1951, p. 276).

---

suicide The act of severing relationships.

egoistic A state in which the ties attaching the individual to others in the society are weak.

altruistic A state in which the ties attaching the individual to the group are such that he or she has no life beyond the group and strives to blend in with the group to have a sense of being.

anomic A state in which the ties attaching the individual to the group are disrupted due to dramatic changes in economic circumstances.

fatalistic A state in which the ties attaching the individual to the group involve discipline so oppressive it offers no chance of release.

Lance Cpl. Manuel F. Guerrero/United States Marine Corps

The debt crisis was preceded by a period of easy credit in which people who could not afford a home in a particular price range were able to obtain loans from those banks engaging in the subprime lending. The easy credit unleashed aspirations and expanded desires. The debt crisis cast many people into a lower status, where they had to reduce their consumption, constrain needs, and practice self-control.

Durkheim on the Economic Crisis. If Durkheim were alive today, he would focus on the ties that bind lenders and borrowers to one another (solidarity) with particular emphasis on how that relationship has changed as a result of new technologies such as computers, the Internet, and new financial products (such as securitization and CDSs). Broadly speaking, banking shifted away from a system in which the neighborhood bank dominated loan making to one in which regional, national, international, and global-scale banks with many branches and online financial services dominated. In addition, a shadow banking industry emerged—nonbanks such as investment firms and hedge funds that offered "bank-like" products and services with higher returns than did savings and checking accounts. Now, instead of banking at a community-based bank, consumers deal with financial corporations in which they remain largely anonymous. Durkheim would be most interested in how the personal, face-to-face relationship between borrowers and lenders was replaced by a system in which they remained largely anonymous.

Durkheim's concept of anomie is also useful for thinking about how both easy credit and debt crisis affect people and their relationships to others. The era of easy credit and high returns on investments gave many people access to lifestyles that they could not afford, casting them into a higher status and giving them a false sense of prosperity. Durkheim argues that when individuals are cast into

social action Actions people take in response to others.

a higher status, they must adjust to increased prosperity, which unleashes aspirations and expands desires to an unlimited extent, thus creating a thirst to acquire goods and services that cannot be satisfied. The debt crisis, on the other hand, pushed people into a lower status and drastically lowered their sense of prosperity. Now, many people find themselves in situations where they must reduce their requirements, restrain their needs, and practice self-control.

Max Weber (1864-1920)

The German scholar Max Weber made it his task to analyze and explain how the Industrial Revolution affected **social action**—actions people take in response to others—with emphasis on the forces that motivate people to act. In this regard, Weber suggested that sociologists focus on the broad reasons that people pursue goals, whatever those goals may be. He believed that social action is oriented toward one of four ideal types—ideal, not in the sense of being the most desirable, but as a gauge against which actual behavior can be compared. In the case of social action, an ideal type is a deliberate simplification or caricature of what motivates people to act, in that it exaggerates and emphasizes the distinguishing characteristics that make one type of action distinct from another. In reality, social action is not so clear-cut but involves some mixture of the four types.

1. *Traditional*—a goal is pursued because it was pursued in the past (i.e., "that is the way it has always been").
2. *Affectional*—a goal is pursued in response to an emotion such as revenge, love, or loyalty (a soldier throws himself/herself on a grenade out of love and sense of duty for those in his/her unit).
3. *Value-rational*—a valued goal is pursued with a deep and abiding awareness of the "symbolic meaning" of the actions taken to pursue the goal. "There can be no compromises or cost-accounting, no rational weighing of one end against another" (Weintraub and Soares 2005). Instead, action is guided by codes of conduct that prohibit certain kinds of behavior and permit others. With value-rational action, the manner in which people go about achieving a goal is valued as much as the goal itself—perhaps even more so as, in an effort to stay true to a code of conduct, the goal may not be realized (Weintraub and Soares 2005).
4. *Instrumental-rational*—a valued goal is pursued by the most efficient means, often without considering the appropriateness or consequences of those means. It is result-oriented action. In the context of the Industrial Revolution, the valued goal is profit and the most efficient means are the cost-effective ones taken without regard for their consequences to workers or the environment. In contrast to value-rational action, this type of action does

not require or prohibit any manner by which people go about achieving goals—any way of achieving the desired end is allowed. One might equate this type of action with an addiction in the sense that the person will work to acquire a drug or other desired state at any cost to self or to others. There is an inevitable self-destructive quality to this form of action (Henri 2000). In the short run, the instrumental-rational action (with no constraints on behavior) will defeat the value-rationally motivated actors. However, in the long run the "anything goes" approach will eventually collapse on itself.

Weber maintained that in the presence of industrialization, behavior was less likely to be motivated by tradition or emotion and was more likely to be instrumental-rational. Weber was particularly concerned about instrumental-rational action because he believed that it could lead to disenchantment, a great spiritual void accompanied by a crisis of meaning.

Weber on the Economic Crisis. Weber's concept of instrumental-rational action in which a valued goal is pursued by the most efficient means, often without considering the appropriateness or consequences of those means, best applies to those lenders, borrowers, and investors who focused exclusively on material gain by whatever means necessary. In an effort to achieve the American dream—which includes home ownership and economic success—many Americans took out home loans they could not afford as the fastest way to achieve that success. In addition, Americans opened a record number of credit cards with generous credit limits but high interest rates. Once in credit card debt, banks advised them to consolidate that debt with a lower-interest home equity loan. Of course, many banks gave equity loans that equaled and even exceeded the inflated value of the indebted homes, which were being used as collateral. When housing prices dropped, many borrowers owed more than the value of their home.

The exclusive focus on material gain caused lenders to abandon the proven way of making safe loans—that is, to make loans to creditworthy borrowers with a good credit history and those able to put up collateral/equity as security. In addition, many lenders arranged terms that exceeded lenders' actual capacity to pay. Securitization and credit default swaps took the risk out of making and investing in bad loans. Without risk there is moral hazard, a situation in which lenders believe that they can make risky loans, and even loans they know will result in default, because they will not have to absorb losses incurred from borrowers who default. Abandoning these lending practices also meant that home builders and buyers could secure easy money to build and buy houses, creating a glut of overvalued houses on the market. This, coupled with high rates of default on mortgages, eventually caused home values to fall and a major source of collateral/equity backing up the

W.E.B. DuBois's writings on the "strange meaning of being black" in America were no doubt influenced by his French, African, and Dutch ancestry.

loans to disappear. Now, when borrowers defaulted, the value of the home could not cover the cost of their debts.

W.E.B. DuBois (1868–1963)

Another voice that was initially ignored and then later "discovered" as important to sociology is that of the U.S. educator and writer W.E.B. DuBois. DuBois wrote about the "strange meaning of being black" and about the color line. In *The Souls of Black Folk* (1903)—a book that has been republished in 119 editions (Gates 2003)—DuBois announced his preoccupation with the "strange meaning of being black here in the dawning of the Twentieth Century." The strange meaning of being black in America includes a **double consciousness** that DuBois defined as "this sense of always looking at one's self through the eyes of others, of

double consciousness According to DuBois, "this sense of always looking at one's self through the eyes of others, of measuring one's soul by the tape of a world that looks on in amused contempt and pity." The double consciousness includes a sense of two-ness: "an American, a Negro; two souls, two thoughts, two unreconciled strivings; two warring ideals in one dark body, whose dogged strength alone keeps it from being torn asunder."

measuring one's soul by the tape of a world that looks on in amused contempt and pity." The double consciousness includes a sense of two-ness: "an American, a Negro; two souls, two thoughts, two unreconciled strivings; two warring ideals in one dark body, whose dogged strength alone keeps it from being torn asunder." DuBois's preoccupation with the "strange meaning of being black" was no doubt affected by the facts that his father was a Haitian of French and African descent and his mother was an American of Dutch and African descent (Lewis 1993). Historically in the United States, a person has been considered "black" even when his or her parents are of different or blended "races." To accept this idea, we must act as if whites and blacks do not marry each other or produce offspring together and as if one parent, the "black" one, contributes a disproportionate amount of genetic material—so large that it negates the genetic contribution of the other parent.

In addition to writing about the "strange meaning of being black" and about racial mixing, DuBois also wrote about the **color line**, a barrier supported by customs and laws separating nonwhites from whites, especially with regard to their place in the division of labor. The color line originated with the colonial expansion that accompanied the Industrial Revolution. That expansion involved rival European powers (Britain, France, Germany, Belgium, Portugal, the Netherlands, and Italy) competing to secure colonies, and by extension, the labor and natural resources within those colonies. The colonies' resources and labor fueled European and American industrialization. DuBois (1970) traced the color line's origin to the scramble for Africa's resources, beginning with the slave trade upon which the British empire and American republic were built, costing black Africa "no less than 100,000,000 souls" (p. 246). DuBois maintained that the world was able "to endure this horrible tragedy by deliberately stopping its ears and changing the subject in conversation" (p. 246). He further maintained that an honest review of Africa's history could only bring us to conclude that Western governments and corporations coveted Africa for its natural resources and for the cheap labor needed to extract them.

DuBois on the Economic Crisis. If W.E.B. DuBois were alive today, he would emphasize that while the debt crisis is global in scale, the focus is primarily on the world's richest countries and their efforts to fix their own financial system through various kinds of stimulus packages, or by turning to institutionalized safety nets. On the other hand, there is no global plan in place to help the world's poorest people, whose precarious economies (legacies of colonization) are in jeopardy. The poorest countries are experiencing a catastrophic loss of income because the demand from wealthier countries for their commodities has decreased, bringing about a corresponding decline in the price of those exports. For example, in 2009 Botswana, a country dependent on diamond exports, lost 90 percent of that revenue. Cameroon, a country dependent on timber exports, lost $300 million in annual revenue (Woods 2009; Stearns 2009). In the face of such crisis, the poorest countries do not have the resources to institute a stimulus package. In addition, wealthier countries have reduced the amount of aid they are sending and remittances—money sent home by those living abroad—have declined sharply in the wake of the global economic crisis. Without revenue from exports, foreign aid, and remittances, poor countries have to reduce their already inadequate public services. As a result, significant numbers of teachers, nurses, and police have been laid off. Children—especially girls—stop attending school. Finally, the emergency measures governments in richer countries are employing to protect their economies take a toll on the poorest countries as well. For instance, government subsidies provided by richer countries to favored industries, or policies encouraging banks to lend at home, disadvantage poor countries seeking to sell their products or trying to secure loans on international markets (Woods 2009).

Jane Addams (1860-1935)

In 1889 Jane Addams (with Ellen Gates Starr) co-founded one of the first settlement houses in the United States, the Chicago Hull House. Settlement houses, which originated in London, were community centers that provided services to the poor and other marginalized populations. Settlements were supported by the wealthy donors from the surrounding community and by university faculty and college students who lived with, served, and learned from these populations. Hull House, considered one of the two largest and most influential settlements in the United States, offered educational, cultural, and social services to immigrants and other diverse populations of Chicago. At the time of Hull House's founding, immigrants constituted almost 50 percent of Chicago's population, which was second in size only to New York's immigrant population. Hull House was established at a time when Chicago was industrializing and in the midst of unprecedented growth that started in 1860 when the city's population was 10,000; over the course of the next 50 years, it grew to 2 million. The dramatic increase was accompanied by a variety of social problems, including homelessness, substandard housing, unemployment, and exploitive and unsafe working conditions.

Hull House's facilities included a night school for adults, morning kindergarten classes, clubs for girls and boys, a

color line A barrier supported by customs and laws separating nonwhites from whites, especially with regard to their place in the division of labor.

U.S. Navy photo

The global debt crisis caused consumption of commodities such as timber and diamonds to drop in the United States, Europe, and other wealthy countries. Cargo ships carried less product from poorer to richer countries. This drop in consumption affected countries such as Botswana, for which 66 percent of exports are diamonds, and Cameroon, for which timber is a major export.

public kitchen, an art gallery, a coffeehouse, recreation facilities (including a swimming pool), a music school, a drama group, and a library. The Hull House had strong ties with the University of Chicago School of Sociology, and, through her community-based work, Jane Addams influenced sociological thought and work in an area of the discipline now known as "public sociology." Not only did she establish a variety of programs and services to address the needs of Chicago's urban population, but Addams also worked to give those populations a voice and to change the way society operated with regard to child labor, juvenile justice, industrial safety, working hours, women's and minority rights, and a variety of other areas (Deegan 1986; Hamington 2007).

The Chicago sociologists considered her work "social work" and an appropriate specialization for women. For that reason, female students tended to intern and live at Hull House. Addams, however, did not consider her work social work. She maintained that the settlements were equivalent to an applied university where knowledge about how to change the situation of people could be applied and tested. Addams advocated for **sympathetic knowledge**, first-hand knowledge gained by living and working among those being studied because "knowing one another better reinforces the common connection of people such that the potential for caring and empathetic moral actions increase" (Addams 1912, p. 7). Addams made a point of never addressing a "Chicago audience on the subject of the Settlement and its vicinity without inviting a neighbor to go with me, that I might curb my hasty generalization by the consciousness that I had an auditor who knew the conditions more intimately than I could hope to do" (Addams 1910, p. 80).

As one measure of Addams's influence and popularity in American society, consider that in the results of one newspaper poll asking, "Who among our contemporaries are of the most value to the community?" she was voted second, after Thomas Edison (U.S. Library of Congress 2008). However, when Addams publicly opposed U.S. involvement in World War I (1914–1918), she was branded a traitor and unpatriotic and was expelled from the Daughters of the American Revolution. Many years later, in 1931, Jane Addams was awarded the Nobel Peace Prize for her work to promote peace, which included her opposition to the war and her humanitarian efforts assisting President Herbert Hoover in providing food aid and other relief to those living in enemy nations (Abrams 1997). Jane Addams's influence on the discipline of sociology is apparent, as she helped to found the American Sociological Association (see Working for Change).

Addams on the Economic Crisis. If Jane Addams were alive today, she would employ sympathetic knowledge to understand the debt crisis's effects on poor and marginalized peoples. She would also emphasize the response of community centers and other nonprofit organizations in addressing the increased numbers of newly poor—"the next layer of people—a rapidly expanding roster of childcare workers, nurse's aides, real estate agents and secretaries facing a financial crisis for the first time" (Bosman 2009). Addams would consider how "the face of those who need food is changing." As one food bank staff member explained it: "When I came here, it was mostly people on a very fixed income or people who were very poor. Now we are seeing more people who consider themselves middle class or even slightly above middle class" (O'Donoghue 2009). Feeding America (2009), a network of 200 food banks across the United States, surveyed its members to learn what factors are behind the rise in demand for emergency food assistance. Respondents pointed to rising costs of food, increased unemployment and underemployment, the cost of fuel, unmanageable mortgage and rent costs, and an inadequate amount of food stamps. The survey found that 72 percent of food banks were not able to meet this demand without reducing the amount of food distributed per person. The survey asked respondents to describe the new populations requesting emergency food assistance.

- 99.4% reported seeing more first-time users.
- 74% reported seeing more newly unemployed persons.
- 73% reported seeing increased need among existing clients (more repeat visits).
- 59% reported seeing more employed persons.
- 48% reported seeing more children.

sympathetic knowledge First-hand knowledge gained by living and working among those being studied.

WORKING FOR CHANGE

The American Sociological Association

IN DECEMBER 1905, a group of sociologists, many of whom were members of the American Economic Association, met at Johns Hopkins University to organize the American Sociological Society (renamed the American Sociological Association in 1959). Their goal was to foster the discipline of sociology and to increase its practical influence upon American society. Lester F. Ward, the recognized founder of American sociology, was unanimously elected president. The founding members included:

- Charlotte Perkins Gilman, a highly influential humanist whose writings have been an inspiration for generations of feminists; a woman who dedicated her book *Man Made World* to Lester Ward.
- Jane Addams, a pioneer in social work; a renowned leader of the Woman's International League for Peace and Freedom and Nobel Laureate in 1931.
- Emily Balch, a noted sociology professor and activist; a woman who changed her pacifist beliefs to defend fundamental human rights against Nazism. She was a Nobel Peace Laureate.
- Mary McDowell, a vigorous advocate for the rights of minorities and the poor, a staunch trade union orga-

Vitaliy Vozniyak

nizer in Chicago who became known as the Angel of the Stockyards.

In 1906, the American Sociological Society held its first official meeting at Brown University in Providence, Rhode Island. The meeting was a great success and marked the beginning of a steady growth of American sociology.

In 1909, six of the founding members of the American Sociological Society helped organize the National Association for the Advancement of Colored People (NAACP). Since then, American sociologists have continued and expanded their effort to promote understanding and respect of racial and ethnic minorities.

In 1936, the first issue of the *American Sociological Review* was published. Today, the organization produces 10 journals and a wide range of reports and position statements on important social issues.

Since its inception, the organization has grown from 115 members to nearly 14,000. But more importantly, over the past 100 years, it has had a significant role in the understanding and shaping of American society.

Source: Adapted from *A Century of Progress: Presidential Reflections 1905–2005,* American Sociological Association, Gale Largey, writer/director.

The Importance of a Global Perspective

■ **CORE CONCEPT 6:** A global perspective assumes that social interactions do not stop at political borders and that

global interdependence A situation in which social activity transcends national borders and in which one country's problems—such as unemployment, drug abuse, water shortages, natural disasters, and the search for national security in the face of terrorism—are part of a larger global situation.

globalization The ever-increasing flow of goods, services, money, people, information, and culture across political borders.

the most pressing social problems are part of a larger, global situation. **Global interdependence** is a situation in which human interactions and relationships transcend national borders and in which social problems within any one country—such as unemployment, drug addiction, water shortages, natural disasters, or the search for national security—are shaped by social forces and events taking place outside the country, indeed in various parts of the globe. Global interdependence is part of a dynamic process known as **globalization**—the ever-increasing flow of goods, services, money, people, technology, information, and other cultural items across political borders (Held, McGraw, Goldblatt, and Perraton 1999). This flow has become more dense and quick moving as space- and time-related constraints separating people in various locations seemingly dissolve. As a result of globalization, no longer are people, goods, services, technologies, money, and images fixed to specific geographic locations (see No Borders, No Boundaries).

The classic sociologists were wide ranging and comparative in their outlooks. They did not limit their observations to a single academic discipline, a single period in history, or a single society. They were particularly interested in the transformative powers of history, and they located the issues they studied according to time and place. All lived at a time when Europe was colonizing much of Asia and Africa; when Europeans were migrating to the United States, Canada, South Africa, Australia, New Zealand, and South America; and when enslaved people and/or indentured servants were moving to new areas to fill demands for cheap labor. Sociologist Patricia M. Lengermann (1974) describes the significance of European expansion and movement of peoples on the discipline of sociology:

> Explorers, traders, missionaries, administrators, and anthropologists recorded and reported more or less accurately the details of life in the multitudes of new social groupings which they encountered. . . . Never had man more evidence of the variety of answers which his species could produce in response to the problems of living. This knowledge was built into the foundations of sociology—indeed, one impulse behind the emergence of the field must surely have been Western man's need to interpret this evidence of cultural variation. (1974, p. 37)

This textbook continues this tradition by incorporating a global perspective throughout. It applies sociological concepts and theories to a wide range of critical issues, international relationships, and events affecting the United States that cannot be separated from a larger global context (Held, McGraw, Goldblatt, and Perraton 1999). A global perspective is guided by the following assumptions:

- Globalization is not new, although the scale of global interdependence changed dramatically with the Industrial Revolution, which created a production process that draws unprecedented amounts of labor and raw materials from even the remotest corners of the world to produce unprecedented quantities of material goods and services, which are distributed unevenly.
- Globalization has been further intensified by the Internet and related technologies, which allow people around the world to communicate instantaneously. In addition, it has increased global competition for jobs that involve processing, managing, and analyzing information.
- Globally established social arrangements that we never see deliver to us products and services, including apple juice containing concentrate from Austria, China,

Globalization involved two seemingly opposing trends embodied in these photographs of activity along the Mexico-U.S. border. On one hand, border patrol officers seek to process travelers and cargo moving across the border as quickly as possible and, at the same time, close access to real and imagined threats. Each year there are an estimated 300 million border crossings from Mexico into the United States. In hopes of preventing this massive exchange of people, the United States is constructing 700 miles of strategically placed fences along that border, including reinforced fencing, physical barriers, lighting, cameras, and sensors to stop illegal crossings. (Dinan 2007)

Turkey, and other countries (Lemert 1995; Zaniello 2007).
- The global exchange of goods, services, and influences is uneven, with some countries—most notably, the United States—generally being the more dominant trading partners.
- Multinational and global corporations are key forces in structuring social relationships that transcend national boundaries (Harvey, Rail, and Thibault 1996).
- Efforts to open and erase national boundaries are accompanied by simultaneous efforts to protect and enforce boundaries. Gatekeepers such as airport security and border patrol officers seek to process travelers

No Borders, No Boundaries

Globalization

GLOBALIZATION INVOLVES ECONOMIC, political, and cultural transformations. There are at least four positions on the nature of these transformations (Appelrouth and Edles 2007).

Position 1: Globalization is producing a homogeneous world characterized by (1) a belief that freedom of expression and appreciation of, and respect for, human and cultural differences should be universally valued and (2) a fusion of distinct cultural practices into a new world culture. This respect and fusion is embodied in trends such as world beat, world cuisine, and world cinema. Globalization includes the emergence of the global citizen, who thinks of the world as one community and feels a responsibility to the planet. The size of the 2007 Live Earth concerts, which engaged 2 billion people worldwide, is evidence that the concept of a global citizen, even if not fully developed, is emerging.

Position 2: Globalization is producing a homogeneous world by destroying variety or the local cultures that get in the way of progress or simply cannot compete against large corporations. The engines of cultural destruction—sometimes referred to as McWorld and Coca-colonization—are consumerism and corporate capitalism. How is globalization destroying local cultures? When people eat a Big Mac or drink a Coke, they are consuming more than a burger or a drink; they are also consuming American/Western images and their associated values. Those values relate to importance placed on food (the time to prepare it and eat), the nature of the relationship between the cook and the person eating (personal versus anonymous), and the place of the individual in relationship to the group (i.e., I can eat whatever I want whenever I want *versus* I eat what others are eating at standard times of the day).

Position 3: Globalization actually brings value to and appreciation for local products and ways of doing things.

Consumption of goods and services is not a one-way exchange in which the buying culture simply accepts a foreign product as it is known and used in the exporting culture(s). While the products of corporate capitalism penetrate local markets, they do not eliminate demand for local ingredients and products. Moreover, local tastes are incorporated into corporate offerings. Coca-Cola, for example, offers 450 different brands in 200 countries, many brands that we may not have heard about, such as Inca Ko, a sparkling beverage available in South America; Samurai, an energy drink available in Asia; and Vita, an African juice drink. Just because a Big Mac or a Coke can be found anywhere in the world does not mean that locally, regionally, or nationally inspired products vanish.

Position 4: Globalization and its interconnections intensify cultural differences by actually "sparking religious, ethnic, and cultural conflicts as people fight to preserve their identity and particular way of life" to resist Western influences that have dominated globalization to date, to assert an identity that "clashes" with Western ideals (i.e., individualism, freedom of expression, democracy), or to protect and enforce boundaries even as they are opened and erased (Appelrouth and Edles 2007, p. 568). Gatekeepers such as airport security and border patrol officers seek to process travelers and cargo from around the world as quickly as possible and, at the same time, close access to real and imagined threats. As a case in point, each year there are an estimated 300 million border crossings from Mexico into the United States (one indicator of global interdependence). In hopes of preventing this massive exchange of people, the U.S. is constructing 700 miles of strategically placed fences along that border, including reinforced fencing, physical barriers, lighting, cameras, and sensors to stop illegal crossings (Dinan 2007).

and cargo from around the world as quickly as possible and yet close access to high-risk threats.

- As part of the pursuit of profit, multinational corporations are increasingly gaining and solidifying control over scarce and valued basic life-sustaining resources, such as water, seeds, human organs and tissue, and DNA (Zaniello 2007).

Depending on where you live and who you are, globalization plays out differently. On the one hand, it connects

the economically, politically, and educationally advantaged to one another while pushing to the sidelines those who are not so advantaged. On the other hand, it connects those working at the grassroots level to protect, restore, and nurture the environment and to enhance access for the disadvantaged to the basic resources they need to live a dignified existence (Calhoun 2002; Brecher, Childs, and Cutler 1993).

■ VISUAL SUMMARY OF CORE CONCEPTS

■ CORE CONCEPT 1: Sociologists focus on the social forces that shape human activity.

Sociology is the study of human activity as it is affected by social forces emanating from groups, organizations, societies, and even the global community. The activities sociologists study are too many to name, but they can include people using credit cards, college students using loans to pay for education, individuals donating blood and selling plasma, good friends walking in public, the unemployed searching for ways to secure an income, or a child gazing in a mirror and wondering why she is considered a different race from her father. On some level, social forces exist outside the consciousness of individuals. French sociologist Émile Durkheim believed that social facts are ideas, feelings, and ways of behaving "that possess the remarkable property of existing outside the consciousness of the individual."

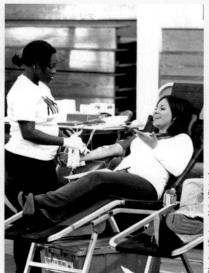

Pfc. Jared Eastman/United States Army

■ CORE CONCEPT 2: Sociologist Peter L. Berger offers the best description of the sociological consciousness: "The first wisdom of sociology is this—things are not what they seem."

Berger equates sociologists with curious observers walking the neighborhood streets of a large city, fascinated with what they cannot see taking place behind the building walls. The wish to look inside and learn more is analogous to the sociological perspective. The discipline of sociology offers us theories and concepts needed to look beyond popular meanings and interpretations of what is going on around us.

Lisa Southwick

■ CORE CONCEPT 3: Sociologists distinguish between *troubles*, which can be resolved by changing the individual, and *issues*, which can be resolved only by addressing the social forces that created them.

Lance Cpl. Manuel F. Guerrero/United States Marine Corps

Troubles are personal needs, problems, or difficulties that can be explained in terms of individual shortcomings in motivation, attitude, ability, character, or judgment. The resolution of a trouble, if it can indeed be resolved, lies in changing the individual in some way. By comparison, an issue is a matter that can be explained only by factors outside an individual's control and immediate environment. Issues can only be resolved by implementing solutions that change or offset the influence of underlying social forces. The sociological imagination is a quality of mind that allows people to make connections between biography and seemingly remote and impersonal social forces and historical events.

■ **CORE CONCEPT 4:** Sociology emerged in part as a reaction to the Industrial Revolution, an ongoing and evolving social force that transformed society, human behavior, and interaction in incalculable ways.

The Industrial Revolution is the name given to the changes in manufacturing, agriculture, transportation, and mining that transformed virtually every aspect of society. The defining feature of the Industrial Revolution was mechanization. The Industrial Revolution changed everything—the way in which goods were produced, the ways in which people negotiated time and space, the relationships between what were once geographically separated peoples, the ways in which people made their livings, the density of human populations (e.g., urbanization), the relative importance and influence of the home in people's lives, access to formal education (the rise of compulsory and mass education), and the emergence of a consumption-oriented economy and culture. The accumulation of wealth became a valued and necessary pursuit.

■ **CORE CONCEPT 5:** Early sociologists were witnesses to the transforming effects of the Industrial Revolution. They offered lasting conceptual frameworks for analyzing the ongoing social upheavals.

Auguste Comte invented the name "sociology" during the most dramatic period of the Industrial Revolution. Karl Marx sought to analyze and explain conflict, which he saw as being shaped by the means of production. Émile Durkheim wrote about solidarity—the ties that bind people to one another—and about how the Industrial Revolution profoundly changed those ties. Max Weber set out to analyze and explain the course and consequences of social actions. Weber maintained that in the presence of industrialization, behavior was less likely to be guided by tradition or emotion and more likely to be instrumental-rational. W.E.B. DuBois wrote about the origins of the color line and about the "strange meaning of being black" in America. Jane Addams advocated for sympathetic knowledge—first-hand knowledge gained by living and working among those being studied—because "knowing one another better reinforces the common connection of people such that the potential for caring and empathetic moral actions increase." Each of the six theorists has left us with concepts and theories that apply to the world in which we live today.

■ **CORE CONCEPT 6:** A global perspective assumes that social interactions do not stop at political borders and that the most pressing social problems are part of a larger, global situation.

This textbook incorporates a global perspective throughout. It applies sociological concepts and theories to a wide range of critical issues, international relationships, and events affecting the United States that cannot be separated from a global context.

Resources on the Internet

 Sociology: A Global Perspective Book Companion Web Site

www.cengage.com/sociology/ferrante

Visit your book companion Web site, where you will find flash cards, practice quizzes, Internet links, and more to help you study.

CENGAGE**NOW**™

Just what you need to know NOW!

Spend time on what you need to master rather than on information you have already learned. Take a pre-test for this chapter, and CengageNOW will generate a personalized study plan based on your results. The study plan will identify the topics you need to review and direct you to online resources to help you master those topics. You can then take a post-test to help you determine the concepts you have mastered and what you will need to work on. Try it out! Go to www.cengage.com/login to sign in with an access code or to purchase access to this product.

Key Terms

altruistic   15
anomic   15
biography   3
bourgeoisie   13
color line   18
conflict   13
currents of opinion   4
double consciousness   17
egoistic   15
fatalistic   15

global interdependence   20
globalization   20
Industrial Revolution   10
issue   8
mechanization   10
positivism   12
proletariat   13
social action   16
social dynamics   12
social facts   4

social forces   3
social statics   12
sociological imagination   3
sociology   4
solidarity   14
suicide   15
sympathetic knowledge   19
troubles   5

2 Theoretical Perspectives and Methods of Social Research

With Emphasis on Mexico

Sociologists view theory and research as interdependent, because (1) theory inspires research; (2) research inspires theory creation; (3) theory is used to interpret research findings, and (4) research findings are used to support, disprove, or modify theory.

U.S. Army photo by Sgt. Jim Greenhill

▲ In 2006 Congress passed the Secure Fence Act mandating the construction of 700 miles of fence—some of it virtual (i.e. radar, sensors, and cameras)—strategically placed along the U.S.-Mexican border. The 700 miles of fence is almost complete. There is a debate over whether more fencing should be erected over the approximately 1,200 miles of unfenced border.

WHY FOCUS ON

Mexico?

THE UNITED STATES and Mexico share a 2,000-mile border (which millions of people cross each week to work, shop, socialize, and vacation). The border includes fences and other barriers to prevent illegal immigrants from crossing into the United States from Mexico. The fences, referred to as the Wall of Shame in Mexico, are known in the United States by such names as Operation Gatekeeper in California (launched 1994), Operation Hold-the-Line in Texas (launched 1993), and Operation Safeguard in Arizona (launched 1994). In 2006 Congress passed the Secure Fence Act, authorizing the construction of at least 700 miles of strategically placed fences, including "two layers of reinforced fencing, the installation of additional physical barriers, roads, lighting, cameras and sensors" between Tecate, California, and Brownsville, Texas. Today about 40 per-

cent of the 2,000-mile border is actually and virtually fenced. Virtually fenced means that radar, sensors, and cameras serve as "fences."

We draw upon the three major sociological theories and methods of social research to assess the causes and consequences of the proposed and existing border fences. The research methods offer guidelines for collecting and analyzing data related to the fences' impact on illegal immigrants and other affected parties. Although we are focusing on border fences in this chapter, we can apply the perspectives and methods of research to any issue.

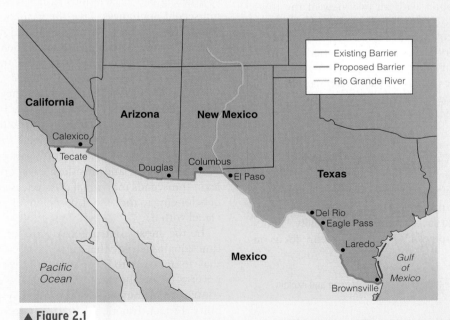

▲ **Figure 2.1**

This map shows the locations of existing and proposed fences. Notice how the Rio Grande forms the border between Texas and Mexico. How will fence construction affect animal and human access to the river?

- About 97 percent of all apprehensions of undocumented immigrants occur on the U.S.-Mexican border (Nunez-Neto 2006).

- About 90 percent of the 10,000-plus U.S. Border Patrol agents are stationed along the border with Mexico.

- An estimated 12 million undocumented immigrants live in the United States.

- In 2006 Congress passed the Secure Fence Act, authorizing the construction of at least 700 miles of strategically placed fences on the U.S.-Mexican border.

Sociological Theories

■ **CORE CONCEPT 1:** Sociological theories offer a set of guiding questions and key concepts that address how societies operate and how people relate to one another. The sociological approach to understanding important social issues and events involves two interdependent and essential parts: theory and research. In the most general sense, a **theory** is a framework that can be used to think about what is going on around us. A **sociological theory** is a set of core assumptions and core concepts that examine how societies operate and how people in them relate to one another and respond to their environment.

We begin this chapter with an overview of the three major theories that dominate the discipline of sociology: functionalist, conflict, and symbolic interactionist. We outline the framework that each theory offers to interpret any social issue or event. Each theory offers a central question to help guide thinking and a vocabulary or set of concepts for answering that question. We turn first to an overview of the functionalist theory.

The Functionalist Theory

■ **CORE CONCEPT 2:** Functionalists focus on how the "parts" of society contribute in expected and unexpected ways to social order and stability and to social disorder and instability. Functionalists, who are inspired by Émile Durkheim, focus on

theory A framework that can be used to comprehend and explain events.

sociological theory A set of principles and definitions that tell how societies operate and how people in them relate to one another and respond to their surroundings.

function The contribution part of a society makes to order and stability within the society.

order and stability in society. Like Durkheim they define society as a system of interrelated, interdependent parts. To illustrate this vision, functionalists use the human body as an analogy for society. The human body is composed of parts such as bones, cartilage, ligaments, muscles, a brain, a spinal cord, nerves, hormones, blood, blood vessels, a heart, a spleen, kidneys, and lungs. All of these body parts work together in impressive harmony. Each functions in a unique way to maintain the entire body, but it cannot be separated from other body parts that it affects and that in turn help it function.

Society, like the human body, is made up of parts, such as schools, automobiles, sports teams, funeral rites, ways of greeting people, religious rituals, laws, languages, and household appliances. Like the various body parts, each of society's parts is interdependent and *functions* to maintain a larger system. Functionalists define a **function** as the contribution a part makes to order and stability within the society.

Consider sports teams—whether they be Little League, grade school, high school, college, city, Olympic, or professional teams. Sports teams function to draw audiences whose members are often extremely different from one another economically, culturally, linguistically, politically, religiously, and in other ways. Loyalty to a sports team transcends individual differences and fosters a sense of belonging to the school, company, city, or country associated with it.

In the most controversial form of this perspective, functionalists argue that all parts of society—even those that do not seem to serve a constructive purpose, such as poverty, crime, illegal immigration, and drug addiction—contribute in some way to the larger system's overall stability. In fact, functionalists maintain that a part would cease to exist if it did not serve some function. Thus they strive to identify how parts—even seemingly problematic ones—contribute to the stability of the larger society. Consider one function of poverty: poor people often "volunteer" for over-the-counter and prescription drug tests. Most new drugs, from AIDS vaccines to allergy medicines,

Sports teams function to transcend individual differences and foster a sense of belonging to the school, company, city, or country associated with it.

Department of Defense photo by Master Sgt. Robert W. Valenca, U.S. Air Force

must eventually be tried on healthy human subjects to determine their potential side effects (for example, rashes, headaches, vomiting, constipation, and drowsiness) and appropriate dosages. The chance to earn money motivates subjects to volunteer for these clinical trials. Because payment is relatively low, however, the tests attract a disproportionate share of low-income, unemployed, or underemployed people as subjects (Morrow 1996).

This function of poverty shows why a part of the society that everyone agrees is problematic and should be eliminated remains intact: it contributes to the stability of the overall system. Therefore, the pharmaceutical and medical systems would be seriously strained if we completely eliminated poverty (see "Functions of Poverty" in Chapter 8). As you might imagine, early functionalists were criticized for defending existing social arrangements. To address some of this criticism, sociologist Robert K. Merton (1967) introduced other concepts to the functionalist perspective that help us think about a part's overall effect on society, not just its contribution to order and stabil-

ity. Those concepts are manifest and latent functions and dysfunctions.

Manifest and Latent Functions and Dysfunctions

Merton distinguished between two types of functions that contribute to order and stability in society: manifest functions and latent functions. **Manifest functions** are a part's anticipated or intended effects on order and stability. **Latent functions** are the unanticipated or unintended effects on order and stability. To illustrate this distinction, consider the manifest and latent functions associated with annual community-wide celebrations, such as fireworks displays on the Fourth of July and concerts in the park. Corporate sponsors often join with city government to mount such events. Three manifest functions readily come to mind: The community celebration functions (1) as a marketing and public relations event for the city and for corporate sponsors, (2) as an occasion to plan activities with family and friends, and (3) as an experience that draws the community together for celebration.

At the same time, several unanticipated, or latent, functions are associated with community celebrations. First, such celebrations put the spotlight on public transportation systems as people take buses or ride trains to avoid traffic jams. Second, such events function to break down barriers across neighborhoods. People who do drive may find that they must park some distance from the event, often in neighborhoods that they would not otherwise visit. Consequently, after they park, people have the opportunity to walk through such neighborhoods and observe life up close instead of at a distance.

Merton also points out that parts of a social system can have **dysfunctions**; that is, they can have disruptive consequences for order and stability or for some segment of society. Like functions, dysfunctions can be either manifest or latent. **Manifest dysfunctions** are a part's anticipated disruptions to order and stability. Anticipated disruptions that seem to go hand in hand with community-wide celebrations include traffic jams, closed streets, piles of garbage, and a shortage of clean public toilets.

manifest functions Intended or anticipated effects that part of a society has on order and stability within the society.

latent functions Unintended or unanticipated effects that part of a society has on order and stability within the society.

dysfunctions Disrupted consequences to society or to some segment in society

manifest dysfunctions A part's anticipated disruptions to order and stability

In contrast, **latent dysfunctions** are unanticipated or unintended disruptions to order and stability. For instance, community-wide celebrations often have some unanticipated negative consequences. Sometimes police departments and other city workers choose to negotiate contracts with the host city just before the celebration, thereby using the event as a bargaining tool to secure a good contract. (Actually, one might argue that this development is a latent function for the police and a latent dysfunction for the city.) In addition, many people celebrate so vigorously that the celebration has the unintended consequence of lowering worker productivity, as people miss class or work the next day to recover.

From this brief analysis of community-wide celebrations, you can readily see that the concepts of manifest and latent functions and dysfunctions provide a more balanced framework than does the concept of function alone. We now use the functionalist theory to analyze the intended and unintended consequences of the border fences.

The Functionalist Perspective on United States–Mexico Border Fences

To see how the functionalist theory can be applied to a specific issue, we will consider how functionalists analyze the U.S.-Mexico border fences. Functionalists ask, Why do fences exist on the U.S.-Mexico border? What are the anticipated and unintended consequences of the border fences for American and Mexican societies? Functionalists use the concepts of manifest and latent functions and dysfunctions to answer these questions. The answers that follow rely on the documented consequences associated with the existing 80 miles of border fences. The purpose of this functionalist analysis is not to generate an exhaustive list of functions and dysfunctions associated with the border fences, but to apply the functionalist perspective.

Manifest Functions

To identify the manifest functions (*anticipated* effects on social order and stability) of the border fences, we need to understand why the United States constructed them in the first place. In the mid-1990s three major border cities constructed 80 miles of fences: San Diego (Operation Gatekeeper), El Paso (Operation Hold-the-Line), and Nogales (Operation Safeguard). The three operations were a response to an increased flow of illegal immigrants resulting from the devaluation of the peso and the subsequent economic crisis in Mexico. The idea for barriers was a mid-1990s response to the real or imagined belief that the United States was being overrun by illegal immigrants. In addition, the overall crime rate along the border was 30 percent higher than the national average. The San Diego area, the site of one of the first fences, accounted for more than 40 percent of the 1.3 million illegal immigrants apprehended along the Southwest border. Newspaper accounts at that time "described large groups of immigrants, serviced by Mexican food and drink vendors in a carnival atmosphere," waiting on the Mexican side of the border for an after-dark surge into San Diego. Border Patrol agents there were overwhelmed by as many as 6,000 immigrants crossing into the city at one time (U.S. Department of Justice 2007). In response, what was then the Immigration and Naturalization Service (INS) shifted its emphasis from apprehension after entry to prevention, erecting barriers and increasing the number of border agents in areas believed to have the highest numbers of illegal entries.

In 2006 President George W. Bush deployed 6,000 National Guard members (in Operation Jump Start) to the U.S.-Mexico border to temporarily assist Border Patrol agents with surveillance, fence construction, and logistics until more agents could be trained (Kruzel 2007). Also in 2006, Congress passed the Secure Fence Act, authorizing construction of more strategically placed fences. This act was a response to reports that some 20 million illegal immigrants were living in the United States and to post-9/11 priorities of achieving operational control "over the entire international land and maritime borders of the United States." The act gave highest priority to the Southwest border, calling for 700 miles of fencing and security improvements between the Pacific Ocean and the Gulf of Mexico (Secure Fence Act of 2006).

Department of Defense photo by Sgt. Jim Greenhill, U.S. Army

Six thousand National Guard members were sent to the Southwest border in 2006 to temporarily assist Border Patrol agents with surveillance, fence construction, and logistics until more agents could be trained.

latent dysfunctions Unintended, unanticipated disruptions to order and stability

Department of Defense photo by Mass Communication Specialist Seaman Orlando Ramos, U.S. Navy

A latent function of the border fences was the creation of the highly trained Border Patrol Search, Trauma, and Rescue Team. This team helps not only illegal immigrants injured from trying to cross inhospitable terrain but anyone in distress. Here team members are assisting a child critically injured in a car accident.

The manifest functions associated with constructing the fences have included the following:

1. A decrease in the number of illegal immigrants apprehended crossing the now-fenced area of the border from Mexico into heavily populated areas on the U.S. side
2. Success in forcing illegal entries away from now-fenced urban areas to less populated areas and through rough terrain and climates (such as steep mountains, deep canyons, thick brush, the extreme cold of winter, and the searing heat of summer) to give Border Patrol agents a strategic advantage
3. An overall drop in the crime rate along the border from 30 percent higher than the national average to 12 percent

Latent Functions

The construction of barriers along the border has had these latent functions (*unanticipated* effects on social order and stability):

1. Cooperation between Mexican and U.S. officials in launching the Border Safety Initiative Program to prevent injuries and fatalities of those crossing the desert and other rough terrain to enter the United States
2. The creation of the Border Patrol Search, Trauma, and Rescue team, which responds to *all* incidents involving people in distress, not just incidents involving illegal immigrants (The team has rescued almost 4,000 people in three years.)

3. A fence that doubles as a volleyball net, allowing U.S. and Mexican volleyball players to face off as part of goodwill festivals and other cross-border celebrations (stuff.co.nz 2007)
4. The emergence of humanitarian groups that provide food, drinking water, and medical supplies to distressed illegal immigrants crossing difficult terrain (thereby preventing immigrant deaths)

Manifest Dysfunctions

The construction of the barriers is associated with several manifest dysfunctions (anticipated disruptions of social order and stability), including increases in the following:

1. Increased apprehensions of illegal immigrants in border counties not protected by fences
2. The crime rate above the national average in thinly populated, unfenced counties
3. Fatalities along the border as illegal immigrants seek to enter the United States through the desert and other inhospitable terrain
4. Illegal immigrants paying organized smugglers, or "coyotes," to guide them through areas where U.S. immigration policies are most strictly enforced

Latent Dysfunctions

Several latent dysfunctions (unanticipated disruptions of social order and stability) have followed the construction of the fences:

1. The emergence of humanitarian groups that save the lives of many illegal immigrants, but in doing so, help people circumvent the law
2. Dramatic disruptions of grazing, hunting, watering, and migration patterns of wildlife ("If it doesn't fly, it's not getting across" [Pomfret 2006].)
3. Some ranchers, farmers, and sport fishers denied access to Rio Grande (The river, which runs for 1,254 miles and serves as a natural border between Texas and Mexico, is used for watering herds, crop irrigation, and fishing [Pomfret 2006].)
4. Longer, and perhaps permanent, stays in the United States by migrant laborers who normally would work seasonally or part of the year here but who do not return home for fear that they will be unable to get back into the United States
5. Redirected flows of illegal immigrants to areas unaccustomed to this movement, fueling the perception that illegal immigration to the United States is out of control (Massey 2006)
6. Disruptions of the efficient exchange of goods, services, and people between border communities that are now separated by a fence

Tomas Castelazo

The border fences have forced illegal immigrants to enter the United States through desert and other inhospitable terrain, resulting in a latent dysfunction of increasing fatalities among illegal immigrants. The monument was erected in memory of people who have died crossing the border since 1994, the year the first fence was built dividing Tijuana from San Diego.

Department of Defense photo by Sgt. 1st Class Gordon Hyde, U.S. Army

The border fence separates Nogales, Arizona, and Sonora, Mexico. How do you think the construction of the border fence affected interactions among people in the two cities?

means of production The land, machinery, buildings, tools, labor, and other resources needed to produce and distribute goods and services.

bourgeoisie The owners of the means of production (such as land, machinery, buildings, and tools), who purchase labor.

proletariat A social class composed of workers who own nothing of the production process and who sell their labor to the bourgeoisie.

facade of legitimacy An explanation that members of dominant groups give to justify their actions.

The Conflict Theory

■ **CORE CONCEPT 3:** The conflict perspective focuses on conflict over scarce and valued resources and the strategies dominant groups use to create and protect social arrangements that give them an advantage over subordinate groups. In contrast to functionalists, who emphasize order and stability, conflict theorists focus on conflict as an inevitable fact of social life and as the most important agent for social change. Conflict can take many forms, including physical confrontations, exploitation, disagreement, tension, hostility, and direct competition. In any society, dominant and subordinate groups compete for scarce and valued resources (access to material wealth, education, health care, well-paying jobs, and so on). Those who gain control of these resources strive to protect their own interests against the competing interests of others.

Conflict theorists ask this basic question: Who benefits from a particular social pattern or arrangement, and at whose expense? In answering this question, they try to identify dominant and subordinate groups as well as practices that the dominant groups have established, consciously or unconsciously, to promote and protect their interests. Exposing these practices helps explain why access to valued and scarce resources remains unequal. Not surprisingly, the privileged or socially advantaged seek to protect their position while the relatively disadvantaged seek to change their position.

Conflict theorists draw their inspiration from Karl Marx, who focused on the **means of production** (the land, machinery, buildings, tools, labor, and other resources needed to produce and distribute goods and services). In Chapter 1 we learned that Marx identified two social classes, with class membership determined by one's relationship to the means of production. The more powerful class is the **bourgeoisie**, or the owners of the means of production and the purchasers of labor. The bourgeoisie, motivated by a desire for profit, search for ways to expand the markets for their products, to make the production process more efficient and less dependent on human labor (by using machines, robots, and automation), and to find the cheapest labor and raw materials. The less powerful class, the **proletariat**, own nothing of the production process except their labor. Conflict exists between the two classes because the bourgeoisie exploit workers by paying the workers only a fraction of the profits they make from the workers' labor.

Exploitation is disguised by a **facade of legitimacy**—an explanation that members of dominant groups give to justify their actions. On close analysis, however, this explanation turns out to be based on "misleading arguments, incomplete analyses, unsupported assertions, and implausible premises" (Carver 1987, pp. 89–90). To illustrate,

These workers, who own only their labor, sit in a clean room, wearing glasses to shield their eyes from bright lights, which are used to complete a labor-intensive procedure (photo-cure epoxy) that holds fiber-optic components in place. One of the best known corporations involved with epoxy is Dow Chemical.

consider that the bourgeoisie justify their exploitation of workers by stating that workers are free to take their labor elsewhere if they are dissatisfied with their working conditions, wages, or benefits. On close analysis, we see that this explanation does not hold. If the capitalist and the worker cannot reach a labor agreement, the capitalist "can afford to wait, and live upon his capital. The workman cannot. He has but wages to live upon, and must therefore take work when, where, and at what terms he can get it. The workman . . . is fearfully handicapped by hunger" (Engels 1886).

Two of the most common facades of legitimacy are (1) blaming the exploited by proposing that character flaws impede their chances of financial success and (2) emphasizing that the less successful really benefit from the system established by the powerful (arguing, for example, that a $2.00-per-hour job—or even a $0.48-per-hour job—is better than no job). Consider the justification a Denver woman gave to *MacNeil/Lehrer Newshour* correspondent Tom Bearden (1993) for hiring an illegal immigrant to care for her children. In particular, pay attention to how the woman justifies her economic relationship with the woman she employs:

> *MR. BEARDEN:* There are some that believe that people who hire undocumented aliens gain an unfair power over them; it gives them influence over them because they're, in a sense, collaborating in something that's against the law. Do you agree with that, or have any thoughts about that?
>
> *DENVER WOMAN:* I guess I would disagree with that. The one thing that you get in undocumented child care or the biggest thing that you probably get, my woman from Mexico was available to me 24 hours a day. I mean, her cost of living in Mexico and quality of life in Mexico compared to what she got in my household were two extremes. When we hired her, she said, "I'll be available all hours of the day, I'll clean the house,

I'll cook." They do everything. And if you hire someone from here in the States, all they're going to do is take care of your children. So not only do you have a differentiation in price, you have a differentiation in services in your household. I have to admit that was, at that point, with a newborn infant, wonderful to have someone who was so available. . . .

> *MR. BEARDEN:* And it's not like indentured servitude?
>
> *DENVER WOMAN:* That crossed my mind, and after she had been here for six months or so, we went to a schedule where she finished at 6 or 7 o'clock at night. And I don't think I ever really took advantage of her. Once a week I'd have her get up with the baby, so I didn't. . . . She was available to me, but I don't feel like I really took advantage of her, other than the fact that I paid her less and she was certainly more available. But she got paid more here than she would have gotten paid if she'd stayed where she was. (p. 8)

Conflict theorists take issue with the logic that this Denver woman uses to justify hiring an undocumented worker at a low salary. When it comes right down to it, the Denver woman is protecting and promoting her interests (having someone available at all hours of the day to cook, clean, and provide child care) at the expense of the illegal worker.

The Conflict Perspective on United States–Mexico Border Fences

In analyzing the construction of border fences along the U.S.-Mexico border, conflict theorists would point out that the fences divide a high-wage economy from a low-wage one. The fences have been constructed to stop, or at least control, the free movement of labor from the low-wage side to the high-wage side. Many illegal immigrants risk life and limb to escape an economy in which they are being paid about $4.50 per day to enter one that pays $60–$80 per day. Upon entry, illegal immigrants assume the status of undocumented workers: dishwashers, farm workers, meat packers, maids, day laborers, roofers, caretakers of children and elderly people, and so on (Judis 2006).

Conflict theorists point out that the legal and illegal migration of labor from Mexico to the United States has been going on steadily since at least 1880. The social forces both pushing and pulling Mexican workers to the United States are deeply institutionalized and multigenerational. In some Mexican communities, 22–75 percent of adult residents have worked or are working in the United States. Not surprisingly, such communities actually specialize in sending workers to the United States. Mexican youth grow up seeing their parents and grandparents leaving to work there and have come to view employment across the border as their only option for supporting themselves and their families. In addition, the would-be illegal immigrant

The construction industry is the largest employer of undocumented workers. About 12 percent (1.4 million) of construction workers are believed to be illegal workers (Pew Hispanic Center, 2006).

learns about job opportunities from existing family and other social networks based in the United States (Kochhar 2005). Many Mexican households have come to rely on remittance income (Pew Hispanic Center 2005; see Working For Change: "Remittance Income from Migrants").

While conflict theorists would acknowledge the push factors from the Mexican side of the border, they would give special attention to the social forces pulling the Mexican worker into the United States. Since 1880, it has been the employers (those who purchase labor) on the U.S. side that have determined the size and destination of migration flows to the United States. In fact, "all available evidence indicates that there will continue to be a substantial number of Mexican workers in the United States labor market, whether or not the general public or United States officials like it. Indeed, there is no evidence that any step taken by the United States government in the last one hundred years to restrict immigration from Mexico has had any appreciable effect on the underlying, structural demand for Mexican labor in the United States economy" (Cornelius 1981, p. 75).

Why is this the case? U.S. employers and consumers depend on foreign labor, especially labor from Mexico and other Central American countries. The Pew Hispanic

Research Center (2006) estimates that between 11.5 and 12 million unauthorized workers are in the United States; 6.3 million are believed to be from Mexico (Kochhar 2005). Employers depend on the increased profits they make from the labor of low-wage workers; consumers depend on low prices of goods and services. This dependence is fueled by a number of factors, including (1) the internationalization of the labor market, which has an overall depressing effect on wages, as employers seek to keep labor-related production costs low; (2) a shortage of U.S.-born workers to fill low-skill, low-status, physically demanding entry-level jobs; and (3) a domestic labor supply that is insufficiently mobile or willing to respond to seasonal, cyclical low-paying job opportunities that are hundreds or thousands of miles away (Cornelius 1981).

The most common justification (facade of legitimacy) for the construction of fences is that such physical barriers prevent illegal workers from entering the country. But a close examination of the facts shows the flaws of this argument. First, in the popular imagination, illegal immi-

For more than 100 years the United States has both invited and turned away workers from the Mexican side of the border. One photo, taken in 1926, shows Border Patrol officers in Laredo, Texas, forming a "wall" with cars and guns to prevent illegal immigrants from crossing. The other photo shows Mexican workers, recruited by the U.S. Farm Security Program in 1943, traveling by train to Arkansas, Colorado, Nebraska, and Minnesota to harvest beets.

WORKING FOR CHANGE

Remittance Income from Migrants

INTERNATIONAL REMITTANCES ARE monies earned by people living or working in one country and sent to someone (usually family and friends) in a home country or other country. An estimated 150 million migrants worldwide send home more than $111 billion to help 500 million people pay for food, medicine, clothing, housing, education, and land. Because senders do not always use official channels (such as banks or Western Union) to send this money, it is likely that the actual amount remitted exceeds the $111 billion estimate by at least another $10 billion (Robinson 2003). As one measure of how widespread this practice is, consider that an estimated 20 million people born in Latin American countries live in a foreign country (*Stalker's Guide to International Migration* 2003). Half of these 20 million people send home an estimated $23 billion per year (Van Doorn 2003). An Inter-American Development Bank poll found that almost one in five adult Mexican residents receives money from relatives working in the United States. One in every four Guatemalan and El Salvadoran adults receives such money (Suro 2003b; Thompson 2003). Other estimates, categorized by sending and receiving countries, are listed below.

These sums represent the "monetary expression of a profound human bond" between migrants (who are mostly low-wage laborers) and the families they left behind (Suro 2003a, p. 2). Taken together, remittances represent an important source of external income for developing countries. For many small island economies, remittances—along with foreign aid and tourism—represent one of the only sources of income. Many critics of foreign aid policies and programs believe that remittance "aid" represents an ideal altruistic self-help model (Kapur 2003, p. 10).

While remittance aid clearly has many positive effects, it would be naive to think that remittances alone could eliminate poverty, drive economic development, and reduce budget deficits (Lowell, De la Garza, and Hogg 2000). The potential positive effects are reduced by the cost of sending remittances. This cost can exceed 15 percent of a remittance when check-cashing fees, money transfer fees, currency conversion fees, and fees on the receiving end are considered. Some critics estimate that reducing fees by just 5 percent could generate another $3.5 billion in remittance aid. Banks and other financial institutions have taken notice of this money flow and are competing with Western Union and MoneyGram to offer transfer services. Such competition may work to reduce transfer fees. Nevertheless, many migrants cannot use banks' services because of the migrants' illegal status and because of high checking account and electronic transfer fees.

Top Five Countries Receiving Annual Remittances (in U.S. $)	
India	$10.0 billion
Mexico	$9.9 billion
France	$9.2 billion
Philippines	$6.3 billion
Germany	$4.1 billion

Top Five Countries Sending Annual Remittances (in U.S. $)	
United States	$28.4 billion
Saudi Arabia	$15.4 billion
Germany	$8.8 billion
Switzerland	$8.1 billion
France	$4.9 billion

grants sneak across the Southwest border, when in fact more likely routes to illegal status are (1) entering legally and overstaying one's visa, (2) entering by using border cards that allow the holder to stay in the United States for 72 hours and then not returning to the home country, and (3) using official ports of entry and evading border guards' detection. These facts suggest that the real purpose of the border fences is political: constructing the fence gives the appearance that government leaders are taking action to stop illegal immigration, when in reality the unauthorized

low-wage laborers continue to enter the country (Singer and Massey 1998). Employers benefit from the existence of 12 million undocumented workers, as the sheer size of this workforce helps suppress wages in general. The Pew Hispanic Research Center estimates that about 60 percent of unauthorized workers earn $300 or less per week and that many face periods of unemployment that last at least a month (Kochhar 2006).

Conflict theorists therefore ask, Who benefits from the fences, at whose expense? For these theorists, the answer is clear: the winners are U.S. employers and consumers. They benefit at the expense of unauthorized workers, who must leave their families and communities to earn wages. These wages, while certainly higher than what they can earn in their home country, still qualify as poverty-level wages in the United States. Besides employers and consumers, another group that benefits from fence construction and increased border surveillance is private contractors, such as Boeing, which was awarded a multibillion-dollar contract to (1) supply small unmanned aerial surveillance vehicles that can be launched from Border Patrol truck beds and (2) equip as many as 1,800 watchtowers with cameras, heat and motion detectors, and other sensors, the first of which will be installed in Tucson, Arizona (Witte 2006).

We turn now to a third major sociological theory: symbolic interactionism.

The Symbolic Interactionist Theory

■ **CORE CONCEPT 4:** Symbolic interactionists focus on social interaction and related concepts of self-awareness/reflexive thinking, symbols, and negotiated order. Symbolic interactionists draw much of their inspiration from American sociologists George Herbert Mead, Charles Horton Cooley, and Herbert Blumer (who coined the term *symbolic interactionism*). We will learn more about these sociologists in Chapters 4 and 7. In contrast to functionalists (who ask how parts of society contribute to order and stability) and to conflict theorists (who ask who benefits, at whose expense, from a particular social arrangement), symbolic interactionists focus on **social interaction** (everyday events in which people communicate, interpret, and respond to each other's words and actions). Symbolic interactionists ask, How do involved parties experience, interpret, influence, and respond to what they and others are doing while interacting? Symbolic interactionists draw upon the following concepts to help them address this question: (1) self-awareness/reflexive thinking, (2) symbols, and (3) negotiated order.

Symbolic interactionists study people in interaction. They would be interested, for example, in how street vendors working the Tijuana side of the border interact with potential customers and how those customers respond.

Self-awareness takes place through *reflexive thinking*, the process of observing and evaluating the self from another's viewpoint. During interaction, people interpret

Symbolic interactionists study people in interaction. They would be interested, for example, in how street vendors working the Tijuana side of the border interact with potential customers and how those customers respond.

the actions, appearances, motives, and words of those with whom they are interacting. At the same time, they imagine how others view their actions, evaluate their appearance, attach meaning to their motives, and interpret their words. In imagining others' reactions, they may decide to make adjustments (apologize, change facial expressions, lash out, and so on). Symbolic interactionists maintain that people interpret others' actions, words, and gestures *first* and then respond based on their interpretations (Blumer 1962). This interpretation-and-response process suggests that interaction between people depends on shared symbols.

The importance of sharing a symbol system becomes most evident when the parties involved speak different languages. In such cases an interpreter can help the two parties communicate. Here an interpreter helps a U.S. army dental technician explain to Iraqi women how to floss teeth.

---

social interaction Everyday events in which two people communicate, interpret, and respond to each other's words and actions.

A **symbol** is any kind of physical phenomenon (such as a word, an object, a color, a sound, a feeling, an odor, a piece of jewelry, a gesture, or a bodily movement) to which people assign a name, meaning, or value (White 1949). This name, meaning, or value is not evident from the physical phenomenon alone, however. If someone we are talking to glances at his watch, we may interpret eye movement toward the watch to mean he is in a hurry and take action to end the conversation. That person may notice that we noticed his glance at the watch and say, "I have 20 minutes before I need to go," signaling that it is OK to continue the conversation.

Without some shared meanings, encounters with others would be very confusing. When we enter into interaction with others, most of us are aware (consciously or unconsciously) that a system of expected behaviors and shared meanings is already in place to guide the interaction. College students know, for example, that when they enter a classroom on the first day of class, they should not walk to the front of the room and give instructions to the class. Likewise, professors know that on the first day of class, students expect them to give an overview of the course. An already established social order is in place, guiding interaction. Usually, however, some room for negotiation exists; that is, the parties involved have the option of negotiating a social order. The *negotiated order* is the sum of existing and newly negotiated expectations, rules, policies, agreements, understandings, pacts, contracts, and other working arrangements (Strauss, 1978). On the first day of class, a professor may negotiate with students by asking them to take a role in structuring the course—say, by voting on topics to be covered. Professors know that they cannot "negotiate" a social order in which students pay money to receive a desired grade.

While many Mexicans see the border as U.S imposed, many Americans view the border as something to protect, as symbolized by the eagle and flag painted on military helicopters used to patrol the border.

The Symbolic Interactionist Perspective on United States–Mexico Fences

As we have learned, symbolic interactionists study people as they engage in social interaction. Thus, with regard to the border fence, symbolic interactionists immerse themselves in the border world by studying the interactions between border control agents and those crossing legally and illegally; the way Border Patrol agents are recruited and trained; the interactions between illegal immigrants and contacts in the United States; the way employers knowingly or unknowingly hire illegal immigrants; the strategies illegal immigrants use to blend into American society upon entry; and the strategies illegal immigrants use to escape detection when passing though official border crossings. As one example of the symbolic interactionist emphasis on the close-up and personal, consider the following description of border agents inspecting cars and their passengers entering one of the world's busiest border crossings in California. The description helps us understand how many unauthorized immigrants manage to blend in with the crowds passing through official ports of entry (see No Borders, No Boundaries: "Interaction That Transcends the U.S. Mexican Border".)

> Primary inspectors had to dispose of entering vehicles at an average of one per 45 seconds. In this time the officer had to enter the license plate number into the Customs computer system, read the results (to see if there was any previous record of smuggling or illegal entry, or indeed any "lookout" rumors

When Americans think of the United States as symbolized by a map, they probably do not think about one-third of the territory as once belonging to Mexico.

symbol Any kind of physical phenomenon to which people assign a name, meaning, or value.

in the system), verbally ask about nationality and, if occupants had brought anything back from Mexico, inspect any immigration documents provided and examine the occupants of the car for their comportment (to see for example if they were nervous or sitting rigidly). Using this set of clues, some clear and others quite vague, the officer either cleared the car into the U.S. without further inspection, or sent the vehicle to the side for an appropriate secondary inspection. If an admitted car was later stopped and found to have some narcotics or illegal immigrants, the computer records would reveal who allowed the vehicle to enter. On the other hand, if the inspector took too long making decisions, it would back up traffic, with

NO BORDERS, NO BOUNDARIES

Interaction That Transcends the U.S.-Mexican Border

THE BORDER BETWEEN Mexico and the United States stretches for 2,000 miles. The border region extends 60 miles south into Mexico and 60 miles north into the United States and includes 12 million people (Brown 2004). "From the perspective of the border, borderlines are not lines of sharp demarcation, but broad scenes of intense interactions in which people from both sides work out everyday accommodations based on face-to-face relationships. Each crosser seeks something that exists on the other side of the border. Each needs to make herself or himself understood on the other side of the border, to get food or gas or a job" (Thelen 1992, p. 437).

This map shows the tremendous two-way traffic across the border as well as attempts by the United States to control that traffic.

Sources: U.S. Department of Homeland Security (2006a, 2006b), Migration Immigration Source (2006), U.S. Department of Transportation (2007)

Imports from Mexico to the United States
$190.3 billion

Exports to Mexico from the United States
$124.1 billion

20 metric tons of cocaine 2.02 metric tons of marijuana intercepted	
$9.9 billion in cash remittance by Mexican workers in the United States to relatives in Mexico	4.1 million commercial trucks cross per year
2,810 *maquiladoras*	1.17 million apprehensions of unauthorized immigrants
11,114 border patrol agents and 6,000+ National Guard members assigned to border	50 million pedestrians, 90 million vehicle crossings each year with 240 million passengers per year

▲ Figure 2.2   Selected Inventory of Border Activity Per Year

consequences including exacerbating air pollution and slowing down the interchange of people and goods on the Tijuana–San Diego corridor, the most important passage in the Mexico-US free trade system. (Heyman 1999, p. 626)

Based on this description it is easy to apply the core concepts that drive the symbolic interactionist analysis of interactions: self-awareness/reflexive thinking, symbols, and negotiated order. In such situations both drivers/passengers and agents are imagining how the other party is viewing their actions, evaluating their appearance, attaching meaning to their motives, and interpreting their words. Both parties are aware of the existing expectations, rules, and policies for crossing the border but do on-the-spot "negotiations" to speed up inspection.

Critique of Three Sociological Theories

To this point, we have seen how each of the three sociological theories guides analysis of the border fences and related topics through its central question and key concepts. Apparently, no single theory can give us a complete picture of a situation. Therefore, we can acquire a more complete picture by applying all three. Of course, each theory has its strengths and weaknesses.

Concerning the border fences, one strength of the functionalist theory is that it gives a balanced overview of the fences' intended and unintended contributions both to social order and stability *and* to social disorder and instability. One weakness of the functionalist theory is that it leaves us wondering about the fences' *overall* effect on the United States and Mexico. That is, do the manifest and latent functions outweigh the manifest and latent dysfunctions?

One strength of the conflict theory is that it forces us to look beyond popular justifications for fence construction and explore questions about whose interests are being protected and promoted and at whose expense. A weakness of the conflict theory is that it presents a simplistic view of the relationship between dominant and subordinate groups: dominant groups are portrayed as all-powerful and capable of imposing their will without resistance from subordinate groups, who are portrayed as exploited victims.

One strength of the symbolic interactionist theory is that it encourages firsthand, extensive knowledge about how the border fences shape interactions between Border Patrol agents and legal and illegal immigrants. If the fences stop unauthorized movement into the United States, how do illegal immigrants manage to evade the Border Patrol? One weakness of the approach is that we cannot be sure if the observations are unique to those being observed or apply to all interactions between Border Patrol agents and legal and illegal immigrants.

We have reviewed the three major theoretical perspectives (see Figure 2.3). Now we turn to the various ways of data gathering and analysis that sociologists (no matter what their perspective) use to formulate and answer meaningful research questions.

	Functionalist Perspective	Conflict Perspective	Symbolic Interactionist Perspective
Focus:	Order and stability	Conflict over scarce and valued resources	Social Interaction
Key Terms:	Function, dysfunction, manifest, latent	Means of production, facade of legitimacy	Social interaction, shared, meanings, interpretation
Vision of Society:	System of interrelated parts	Dominant and subordinate groups in conflicts over scarce and valued resources	Web of social interactions
Central question(s):	Why does a part exist? What are the anticipated and unintended consequences of a part?	Who benefits from a particular pattern or arrangement, and at whose expense?	How do involved parties experience, interpret, influence, and respond to what they and others are doing in the course of interacting?
Strength:	Balanced analysis of positive and negative effects	Encourages analysis beyond popular explanations	Encourages direct, firsthand, and extensive analysis
Weakness:	Defends existing social arrangements Difficult to determine overall effect	Presents simplistic view of dominant-subordinate groups or relationships	Generalizability of observation is difficult to determine

▲ **Figure 2.3**   Overview of the Three Theoretical Perspectives

Methods of Social Research

■ **CORE CONCEPT 5:** Sociologists adhere to the scientific method; that is, they acquire data through observation and leave it open to verification by others. **Research** is a data-gathering and data-explaining enterprise governed by strict rules (Hagan 1989). **Research methods** are the various techniques that sociologists and other investigators use to formulate or answer meaningful research questions and to collect, analyze, and interpret data in ways that allow other researchers to check the results. Theory and research are interdependent, because (1) theories inspire research, whose results can be used to support, disprove, or modify those theories; (2) the results of social research can inspire theories; and (3) theories are used to interpret facts generated through research. No matter which of the three perspectives they favor, all sociologists are guided by the scientific method when they investigate human behavior. The **scientific method** is an approach to data collection that relies on two assumptions: (1) knowledge about the world is acquired through observation, and (2) the truth of that knowledge is confirmed by verification—that is, by others making the same observations.

Researchers collect data that they and others can see, hear, taste, touch, and smell; that is, they focus on what they observe through the senses. They report the process they used to make their observations so that interested parties can duplicate, or at least critique, that process. If observations cannot be duplicated, or if repeating the study yields results that differ substantially from those of the original study, we consider the study suspect. Findings endure as long as they can withstand continued reexamination and duplication by the scientific community. "Duplication is the heart of good research" (Dye 1995, p. D5). No finding can be taken seriously unless other researchers can repeat the process and obtain the same results.

U.S. Fish and Wildlife Service photo by Pedro Ramirez, Jr.

Department of Defense photo by PH3 Bernardo Fuller, USN

Like all scientists, sociologists adhere to the scientific method. Just as this biologist observes the size of a mallard's egg through his sight and touch, sociologists use their senses to observe human behavior.

research A data-gathering and data-explaining enterprise governed by strict rules.

research methods Techniques that sociologists and other investigators use to formulate or answer meaningful research questions and to collect, analyze, and interpret data in ways that allow other researchers to verify the results.

scientific method An approach to data collection in which knowledge is gained through observation and its truth is confirmed through verification.

objectivity A stance in which researchers' personal, or subjective, views do not influence their observations or the outcomes of their research.

When researchers know that others are critiquing and checking their work, the process serves to reinforce careful, thoughtful, honest, and conscientious behavior. Moreover, this "checking" encourages researchers to maintain **objectivity**; that is, it encourages them not to let their personal, or subjective, views about the topic influence their observations or the outcome of the research.

This description of the scientific method is an ideal one, because it outlines how researchers and reviewers should behave. Ideally, researchers should be guided by the core values of honesty, skepticism, fairness, collegiality, and openness (National Academy of Sciences 1995). In practice, though, some research is dismissed as unimportant and unworthy of examination simply because the topic or the researcher is controversial or because the findings depart from mainstream thinking. Moreover, some researchers fabricate data to support a personal, an economic, or a political agenda. The extent to which researchers actually adhere to core values remains unknown.

Research should be carefully planned; the enterprise of gathering and explaining facts involves a number of interdependent steps (Rossi 1988):

1. Choosing the topic for investigation or deciding on the research question
2. Reviewing the literature
3. Identifying core concepts
4. Choosing a research design, forming hypotheses, and collecting data
5. Analyzing the data
6. Drawing conclusions

Researchers do not always follow these six steps in sequence, however. They may not define the topic (step 1) until they have familiarized themselves with the literature (step 2). Sometimes an opportunity arises to gather information (step 4), and a project is defined to fit that opportunity (step 1).

Although the six steps need not be followed exactly in sequence, all must be completed at some point to ensure the quality of the project. In the sections that follow, we will examine each stage. Along the way, we will refer to sociological and other research that focuses on unauthorized immigration and the fences' effect on deterring entry. We will emphasize two sociological studies: *Patrolling Chaos: The U.S. Border in Deep South Texas* by Robert Lee Maril (2004) and "The Social Process of Undocumented Border Crossing among Mexican Migrants" by Audrey Singer and Douglas S. Massey (1998). Singer and Massey's research is part of a larger project known as the Mexican Migration Project, a multidisciplinary effort involving researchers in the United States and Mexico (Mexican Migration Project 2007).

Step 1: Choosing the Topic for Investigation

■ **CORE CONCEPT 6:** Sociologists explain why their research topic is important, tie their research in with existing research, and specify the core concepts guiding investigation. The first step of a research project involves choosing a topic or deciding on a research question. It would be impossible to compile a comprehensive list of the topics that sociologists study, because almost any subject involving humans represents a potential target for investigation. Sociology is distinguished from other disciplines not by the topics it investigates, but by the perspectives it uses to study topics. Researchers choose their topics for a number of reasons. Personal interest is a common and often understated motive. It is perhaps the most significant reason that someone picks a specific topic to study, especially if we consider how a researcher eventually chooses one topic from a virtually infinite set of possibilities.

Good researchers explain to their readers why their topic or research question is significant. This explanation is vital because it clarifies the purpose and importance of the project and the researcher's motivation for doing the work. Sociologists Audrey Singer and Douglas S. Massey (1998) studied "the social process of undocumented border crossings among Mexican migrants" (p. 561). The two researchers chose this topic for several reasons: (1) it is a politically divisive topic in the United States; (2) little is known about how migrants evade borders guarded by agents, maneuver around or bypass fences (some triple-deep), and escape detection by surveillance equipment; and (3) knowing the extent of undocumented entries helps us judge whether fences or other barriers are effective deterrents.

In this step, researchers often announce the perspective guiding their investigation. Sometimes they announce it directly by indicating that they are writing from one or more theoretical traditions or perspectives. Sometimes they announce their guiding perspective(s) indirectly; that is, the reader surmises the perspective from the way researchers frame the question or analysis. That Singer and Massey (1998) focused on undocumented immigrants' social ties to others who have crossed the border successfully without authorization suggests that the two researchers are drawing upon the symbolic interaction perspective to frame their analysis. One also notices that Singer and Massey take a conflict perspective when they suggest that constructing fences and implementing other border control strategies "sit well with the public," since the U.S. government "appears to be defending the United States against alien invaders while not antagonizing U.S. business interests, since it does not really stop the entry of Mexican workers, most of whom simply try until they get in" (pp. 563–564).

Step 2: Reviewing the Literature

All good researchers consider existing research. They read what knowledgeable authorities have written on the chosen topic, if only to avoid repeating earlier work. More importantly, reading the relevant literature can generate insights that researchers may not have considered. Even if researchers believe that they have a revolutionary idea, they must still consider the works of other thinkers and show how their new research verifies, advances, and corrects past research.

At the end of most research papers, authors cite the literature that has influenced that work. This list can include dozens to hundreds of citations. For their research on undocumented border crossings, Singer and Massey (1998) cited 44 references. They used the existing literature to identify factors that help them predict how many times an undocumented migrant will be apprehensions trying to evade detection. Those factors: (1) the nature and intensity of U.S. enforcement efforts, (2) characteristics of the migrant that enhance the chance of success (such as previous success at crossing borders), and (3) ties to other

Department of Defense photo by Mass Communication Specialist Seaman Josue Leopoldo Escobosa, U.S. Navy

All good researchers place their research in the context of existing research. They review what other researchers have written on the chosen topic, if only to avoid repeating earlier work.

undocumented migrants (such as a parent, other relative, or friend) who succeeded in unauthorized entry.

Step 3: Identifying and Defining Core Concepts

After deciding on a topic and reading the relevant literature (albeit not necessarily in that order), researchers typically state their core concepts. **Concepts** are powerful thinking and communication tools that enable researchers to give and receive complex information efficiently. The mention of a concept triggers in the minds of people who know its meaning a definition and a range of important associations that help frame and focus observations. One core concept for sociologists studying illegal immigration is the *unauthorized immigrant* (also called an undocumented or illegal immigrant or migrant), who is defined as a non-citizen residing in the United States whom the American government has not admitted for permanent residence or

concepts Thinking and communication tools used to give and receive complex information efficiently and to frame and focus observations.

research design A plan for gathering data that specifies who or what will be studied and the methods of data collection.

methods of data collection The procedures a researcher follows to gather relevant data.

traces Materials or other forms of physical evidence that yield information about human activity.

for specific authorized temporary work or stays. Another core concept is *interpersonal ties*, as they relate to successful undocumented entry into the United States. These ties are defined as connections to a parent, sibling, or other relative or to friends or acquaintances who guide or otherwise help migrants enter the United States undetected.

Step 4: Choosing a Research Design and Data-Gathering Strategies

■ **CORE CONCEPT 7:** Sociologists decide on a plan for gathering data, identifying whom or what they will study and how they will select (sample) subjects for study. Once researchers have clarified core concepts, they decide on a **research design**, a plan for gathering data on the topic they have chosen. A research design specifies the population to be studied and the **methods of data collection**, or the procedures used to gather relevant data. One research design is not inherently better than another; researchers choose the design that best enables them to address the research question at hand (Smith 1991).

Researchers must decide whom or what they are going to study. The most common "thing" sociologists study is individuals, but they may also decide to study traces, documents, territories, households, small groups, or individuals (Rossi 1988).

- **Traces** are materials or other evidence that yields information about human activity, such as the items that people throw away, the number of lights turned on in a house, or changes in water pressure. Researchers who study undocumented border crossings might learn about paths undocumented immigrants take into the

CBP photo by James R. Tourtellotte

Researchers who study unauthorized border crossings may look for footprints as a way to document paths migrants take into the United States. Here we see U.S. Border Patrol agents "erasing" footprints and other signs that people have crossed. Later the Border Patrol will check for signs of subsequent crossings.

Sociologists who study activity at a border station, such as that pictured above, have chosen to study a territory. They may choose to determine the average time it takes a vehicle, once it pulls into a line, to cross from the Mexican side to the U.S. side.

CBP photo by James R. Tourtellotte

United States by observing the litter they leave behind, including "one-gallon plastic bottles, jeans, t-shirts, candy wrappers, socks, underwear, discarded purses, and inexpensive tennis shoes" (Maril 2004, pl 166).

- **Documents** are written or printed materials, such as magazines, advertisements, graffiti, birth certificates, death certificates, prescription forms, and traffic tickets. Researchers have studied letters undocumented workers have sent home to family members.

- **Territories** are places that have known boundaries or that are set aside for particular activities. They include countries, states, counties, cities, neighborhoods, streets, buildings, and classrooms. Sociologists who study territories focus on activity within the territory they select. Territories related to undocumented immigration include specific border stations, such as the Tijuana–San Diego, El Paso–Ciudad Juárez, and Nogales–Tucson stations.

- **Households** include all related and unrelated persons who share the same dwelling. When studying a household, researchers collect information about the household itself. They might want to determine the number of people living in the household and the household income (that is, the combined income of all people living in the same dwelling). For their research on undocumented border crossings, Singer and Massey (1998) drew upon interviews of households in 34 Mexican communities. Among other kinds of information, the researchers collected data on household size and on the number of household members who have made successful unauthorized entries into the United States.

- **Small groups** are defined as 2 to about 20 people who interact with one another in meaningful ways (Shotola 1992). Examples include father-child pairs, doctor-patient pairs, families, sports teams, circles of friends,

and committees. Concerning a doctor-patient interaction, a sociologist might study the length of time the doctor spends with each patient. Keep in mind that the focus is on the doctor-patient *relationship*, rather than on the doctor or the patient *per se*. For his book *Patrolling Chaos*, Robert Maril accompanied 12 Border Patrol agents on 60 ten-hour shifts along the border.

Because of time constraints, researchers cannot study entire **populations**—the total number of individuals, traces, documents, territories, households, or groups that exist. Instead, they study **samples,** or portions of the cases from a larger population.

Sampling. Ideally, a sociologist should study a **random sample**, in which every case in the population has an equal chance of being selected. The classic, if inefficient, way of selecting a random sample is to assign a number to every case, place cards on which the numbers are written into a container, thoroughly mix the cards, and pull out one card at a time until the desired sample size is achieved. Rather than employ this tedious system to generate their samples, most of today's researchers use computer programs. If every case has an equal chance of becoming part of a sample, then theoretically the sample should be a **representative sample**—that is, one with the same distribution of characteristics (such as age, sex, and ethnic composition) as the population from which it is selected. For example, if 56.4 percent of the population from which a sample is drawn is at least 30 years old, then approximately 56.4 percent of a representative sample should be that age. In theory, if the sample is representative, then whatever holds true for the sample should also hold true for the larger population.

documents Written or printed materials used in research.

territories Settings that have borders or that are set aside for particular activities.

households All related and unrelated persons who share the same dwelling.

small groups Groups of 2 to about 20 people who interact with one another in meaningful ways.

populations The total number of individuals, traces, documents, territories, households, or groups that could be studied.

samples Portions of the cases from a larger population.

random sample A type of sample in which every case in the population has an equal chance of being selected.

representative sample A type of sample in which those selected for study have the same distribution of characteristics as the population from which it is selected.

Obtaining a random sample is not as easy as it might appear. For one thing, researchers must begin with a **sampling frame**—a complete list of every case in the population—and each member of the population must have an equal chance of being selected. Securing such a complete list can be difficult. Campus and city telephone directories are easy to acquire, but lists of, say, U.S. citizens, adopted children in the United States, or American-owned companies with operations in Mexico are more difficult to obtain. Almost all lists omit some people (such as individuals with unlisted telephone numbers, members too new to be listed, or between-semester transfer students) and include some people who no longer belong (such as individuals who have moved, died, or dropped out). What is important is that the researcher considers the extent to which the list is incomplete and updates it before drawing a sample. Even if the list is complete, the researcher must also think of the cost and time required to take random samples and consider the problems of inducing all sampled persons to participate.

Researchers sometimes select nonrandom samples to study people who they know are not representative of the larger population but who are easily accessible. For example, they often use high school and college students as a sample because they are a captive audience. Researchers may choose unrepresentative samples for other important reasons: (1) little is known about members of the sample, (2) they have special characteristics, or (3) their experiences clarify important social issues.

Singer and Massey (1998) used data collected from a random sample of 6,341 households in 34 communities in Mexico. This data was supplemented by a nonrandom sample of 484 U.S. households in which undocumented immigrants from each of the 34 communities in Mexico were now living. For his study *Patrolling Chaos*, Maril chose a nonrandom sample of one Border Patrol station among nine in the McAllen, Texas, sector of the Southwest border. The border station, which he observed for two years, employed 300 men and women to guard a 45-mile stretch of the Rio Grande, which marks the Texas-Mexico border. In particular, Maril observed 12 agents as they worked 10-hour patrol shifts.

sampling frame A complete list of every case in a population.

self-administered questionnaire A set of questions given to respondents who read the instructions and fill in the answers themselves.

interviews Face-to-face or telephone conversations between an interviewer and a respondent, in which the interviewer asks questions and records the respondent's answers.

structured interview An interview in which the wording and sequence of questions are set in advance and cannot be changed during the interview.

Methods of Data Collection

■ **CORE CONCEPT 8:** Sociologists use a variety of data-collection methods, including self-administered questionnaires, interviews, observation, and secondary sources. Besides identifying whom or what to study, the design must include a plan for collecting information. Researchers can choose from a variety of data-gathering methods, including self-administered questionnaires, interviews, observation, and secondary sources.

Self-Administered Questionnaire. A **self-administered questionnaire** is a set of questions given to respondents, who read the instructions and fill in the answers themselves. The questions may require respondents to write out answers (open ended) or to select from a list of responses the one that best reflects their answer (forced choice). Self-administered questionnaires are one of the most common methods of data collection. The questionnaires found in magazines or books, displayed on tables or racks in service-oriented establishments (such as hospitals, garages, restaurants, grocery stores, and physicians' offices), and mailed to households are all self-administered questionnaires.

This method of data collection offers a number of advantages. No interviewers are needed to ask respondents questions, and the questionnaires can be given to large numbers of people at one time. Also, an interviewer's facial expressions or body language cannot influence respondents, so the respondents feel freer than they otherwise might to give unpopular or controversial responses.

Self-administered questionnaires pose some problems, too. Respondents can misunderstand or skip over questions. When questionnaires are mailed, set out on a table, or published in a magazine or newspaper, researchers must wonder whether the people who choose to fill them out have opinions that differ from those of people who ignore the questionnaires. The results of a questionnaire depend not only on respondents' decisions to fill it out, answer questions conscientiously and honestly, and return it, but also on the quality of the questions and on a host of other considerations.

Interviews. Compared with self-administered questionnaires, **interviews** are more personal. In these face-to-face or telephone conversations between an interviewer and a respondent, the interviewer asks questions and records the respondent's answers. As respondents give answers, interviewers must avoid pauses, expressions of surprise, or body language that reflects value judgments. Refraining from such conduct helps respondents feel comfortable and encourages them to give honest answers.

Interviews can be structured or unstructured, or some combination of the two. In a **structured interview**, the wording and sequence of questions are set in advance and cannot be altered during the interview. In one kind of structured interview, respondents choose answers from a response list that the interviewer reads to them. In another kind of structured interview, respondents are free to answer the questions as they see fit, although the interviewer may ask them to clarify answers or explain them in more detail. For their study on undocumented border crossings, Singer and Massey (1998) relied on data collected from structured interviews with heads of households in Mexico and the United States. Among other questions, interviewers asked the following: How many people have ever lived in the household, including those who no longer live there? How many people from the household have experience in migrating to the United States? For each person who has such experience, how many trips have they made to the United States, including the year of the first and last trip? How long did they stay in the United States? What occupation did they hold, and what did it pay? How many of those trips were made without documentation? How many trips resulted in deportation? What was the place of crossing and with whom was the crossing made? Did they use a coyote, and if yes, how much was the coyote paid? Interviewers also asked about each immigrant's ability to speak and understand English.

In contrast to the structured interview, an **unstructured interview** is flexible and open-ended. The question-and-answer sequence is spontaneous and resembles a normal conversation in that the questions are not worded in advance and are not asked in a set order. The interviewer allows respondents to take the conversation in directions they define as crucial. The interviewer's role is to give focus to the interview, ask for further explanation or clarification, and probe and follow up on interesting ideas expressed by respondents. The interviewer appraises

Interviews are face-to-face conversations in which the interviewer asks questions and records the respondent's answers. An interview can be structured or unstructured.

Department of Defense photo by PH3 Kristopher Wilson, U.S. Navy

the meaning of respondents' answers and uses the information learned to ask follow-up questions. Talk show hosts, for instance, often use an unstructured format to interview their guests. Sociologists, however, have goals much different from those of talk show hosts. For one thing, sociologists do not formulate questions with the goal of entertaining an audience. In addition, they strive to ask questions in a neutral way, and no audience reaction influences how respondents answer. Robert Maril (2004) used unstructured interviews as he rode with border agents during their 10-hour shifts and at other times. As he describes it, "under the scorching sun and in the dead of night along the banks of the Rio Grande, I asked these men and women what they knew, what they had seen, and what they thought. In no uncertain terms and with direct, sometimes alarming honesty, they told me" (p. 16).

Observation. As the term implies, **observation** involves watching, listening to, and recording behavior and conversations as they happen. This research technique may sound easy, but it entails more than just watching and listening. The challenge of observation lies in knowing what to look for while remaining open to other considerations; success results from identifying what is worth observing. "It is a crucial choice, often determining the success or failure of months of work, often differentiating the brilliant observer from the . . . plodder" (Gregg 1989, p. 53).

When observing, researchers must be careful not to misinterpret or misrepresent what is happening. How would you describe what is going on in the photo? Did you know that these are young, unranked sumo wrestlers ending their daily workout routine with a ritualized dance that emphasizes teamwork?

Department of Defense photo by PH1 (AW) M. Clayton Farrington

unstructured interview An interview in which the question-and-answer sequence is spontaneous, open-ended, and flexible.

observation A research technique in which the researcher watches, listens to, and records behavior and conversations as they happen.

INTERSECTION OF BIOGRAPHY AND SOCIETY

The Life of a Citrus Picker

FOR HIS BOOK *Coyotes: A Journey through the Secret World of America's Illegal Aliens,* Ted Conover (1987) used a participant observation research design to capture the intersection of biography and society as it related to undocumented workers. Conover decided that, to learn truly meaningful information about undocumented immigrants, he had to go beyond formal interviews, government statistics, and news reports. He had to live with his research subjects. Conover's book includes a chapter on the art of citrus fruit picking. The following excerpt stresses that by participating in, and not just observing, the life of a fruit picker, Conover learned that picking oranges is not something just anyone can do; certain skills are needed if one is to work all day at a fast enough pace to keep from getting fired:

> It required a vast store of special knowledge and dexterity. Not only did you need to know the optimum position of the ladder against a given tree, for example, you also had to be

able to get it there fast and deftly. Men half my size could manipulate the twenty-foot ladders as though they were balsa wood, but in my hands the ladder was a heavy, deadly weapon. I had bruised a friend's arm with my ladder one day and had nearly broken my own when, with sixty pounds of oranges in the bag around my neck, I had slipped from its fifth rung and been grazed as it followed me to the muddy ground . . . Handling the fruit itself was a further challenge. To be successful, you had to pick with both hands simultaneously. This meant your balance had to be good—you couldn't hold on to a branch or the ladder for support. And you had to twist the fruit off: pull hard on a Valencia without twisting and you're likely to get the whole branch in your hand. Planning was also important: ideally, you would start with your sack empty at the top of the tree and work your way down. Oranges would just be peeking out the top of the sack when your feet touched the soil, if you did it right; you topped off the bag by grabbing low-hanging fruit on your way to the tractor. (pp. 42–43)

Good observation techniques must be developed through practice; observers must learn to recognize what is worth observing, be alert to unusual features, take detailed notes, and make associations between observed behaviors.

If observers come from a culture different from the one under study, they must be careful not to misinterpret or misrepresent what is happening. Imagine for a moment how an uninformed, naive observer might describe a sumo wrestling match: "One big, fat guy tries to ground another big, fat guy or force him out of the ring in a match that can last as little as three seconds" (Schonberg 1981, p. B9). Actually, for those who understand it, sumo wrestling is "a sport rich with tradition, pageantry, and elegance and filled with action, excitement, and heroes dedicated to an almost impossible standard of excellence down to the last detail" (Thayer 1983, p. 271).

nonparticipant observation A research technique in which the researcher observes study participants without interacting with them.

participant observation A research technique in which the researcher observes study participants while directly interacting with them.

Observational techniques are especially useful for three purposes: (1) studying behavior as it occurs, (2) learning information that cannot be surveyed easily, and (3) acquiring the viewpoint of the persons under observation. Observation can take two forms: participant and nonparticipant. **Nonparticipant observation** consists of detached watching and listening; the researcher merely observes but does not interact with the study subjects or become involved in their daily life. In contrast, researchers engage in **participant observation** when they join a group and assume the role of a group member, interact directly with individuals whom they are studying, assume a position critical to the outcome of the study, or live in a community under study (see Intersection of Biography and Society: "The Life of a Citrus Picker"). Maril's research qualifies as participant observation, because, for all practical purposes, he became involved in the daily life of the people he studied, even agreeing to work on a "specific project that involved a presentation designed to inform and persuade the public of the need for new facilities" (p. 17). In addition, for safety reasons, Maril "actively participated in some of the policing techniques and procedures [he] was observing" (p. 18).

In both participant and nonparticipant observation, researchers must decide whether to hide their identity and purpose or to announce them. One major reason for choosing concealment is to avoid the **Hawthorne effect**, a phenomenon in which research subjects alter their behavior when they learn they are being observed. The term *Hawthorne effect* originated from a series of worker productivity studies conducted in the 1920s and 1930s and involving female employees of the Hawthorne, Illinois, plant of Western Electric. Researchers found that no matter how they varied working conditions—bright versus dim lighting, long versus short breaks, frequent versus no breaks, piecework pay versus fixed salary—workers' productivity increased. One explanation for this finding was that workers were responding positively to having been singled out for study (Roethlisberger and Dickson 1939).

If researchers choose to announce their identity and purpose, they must give participants adequate time to adjust to their presence. Usually, if researchers are present for a long enough time, their subjects will eventually display natural, uninhibited behaviors. Maril (2004) chose to announce his identity, and agents often referred to him as the "professor from the university." He noted that the majority of agents who worked at the station seemed to pay little attention to him after a couple of months. Maril believed that the agents were, with a few exceptions, "open and honest with me not just because they came to trust me, but because few, if any, had ever been asked about their work as agents. . . . They were anxious to talk and show me what they knew" (p. 16).

Secondary Sources (Archival Data). Another data-gathering strategy relies on **secondary sources (archival data)**—that is, data that have been collected by other researchers for some other purpose. Government researchers, for example, collect and publish data on many areas of life, including births, deaths, marriages, divorces, crime, education, travel, and trade. "Every researcher who uses an existing data set or who does a literature review in which published research findings are taken out of their original context and applied to a different issue or question is involved in 'archival research'" (Horan 1995, p. 423).

Another kind of secondary data source consists of materials that people have written, recorded, or created for reasons other than research (Singleton, Straits, and Straits 1993). Examples include television commercials and other advertisements, letters, diaries, home videos, poems, photographs, artwork, graffiti, movies, and song lyrics.

Identifying Variables and Specifying Hypotheses

■ **CORE CONCEPT 9:** Sociologists may choose to test hypotheses specifying the relationship between independent and dependent variables. As researchers acquire a conceptual focus, identify a population, and determine a method of data collection, they also identify the variables they want

CBP photo by Gerald L. Nino

Sociologists Singer and Massey worked to identify the variables that help us predict the probability that undocumented immigrants will be apprehended.

to study. A **variable** is any characteristic that consists of more than one category. The variable "sex," for example, is generally divided into two categories: male and female. The variable "mode of crossing the border" can be divided into three categories: alone, with family or friends, and with coyote.

Sometimes researchers strive to find associations between variables to explain or predict behavior. The behavior to be explained or predicted is the **dependent variable**. The variable that explains or predicts the dependent variable is the **independent variable**. The relationship between independent and dependent variables is described in a **hypothesis**, or trial explanation put forward as the

Hawthorne effect A phenomenon in which research subjects alter their behavior when they learn they are being observed.

secondary sources (archival data) Data that have been collected by other researchers for some other purpose.

variable Any trait or characteristic that can change under different conditions or that consists of more than one category.

dependent variable The variable to be explained or predicted.

independent variable The variable that explains or predicts the dependent variable.

hypothesis A trial explanation put forward as the focus of research; it predicts how independent and dependent variables are related and how a dependent variable will change when an independent variable changes.

focus of research, which predicts the relationship between independent and dependent variables. Specifically, it predicts how a change in an independent variable brings about a change in a dependent variable. Hypotheses that could be tested using Singer and Massey's data include the following:

Hypothesis 1: The more proficient in English undocumented immigrants are, the less likely they are to be apprehended by Border Patrol.

Hypothesis 2: The more times an undocumented immigrant crosses the border into the United States without being apprehended, the more likely the immigrant will be to cross alone.

In hypothesis 1, the independent variable is English-language proficiency and the dependent variable likelihood of apprehension. In hypothesis 2, the independent variable is number of successful border crossings and the dependent variable is likelihood of crossing alone.

A major reason that researchers collect data is to test hypotheses. If their findings are to matter, other researchers must be able to replicate their study. So, researchers need to give clear, precise definitions and instructions about how to observe and/or measure the variables under study. In the language of research, such definitions and accompanying instructions are called **operational definitions**.

An analogy can be drawn between an operational definition and a recipe. Just as anyone with basic cooking skills should be able to follow a recipe to achieve a desired end, anyone with basic research skills should be able to replicate a researcher's observations if he or she knows the operational definitions (Katzer, Cook, and Crouch 1991). Operational definitions include clear, precise definitions and instructions about how to observe or measure variables. They help researchers determine whether a behavior of interest has occurred.

Suppose a researcher is interested in the question of who washes hands after using a public toilet. An operational definition of hand washing would include an account of what must take place for a researcher to count someone as a hand washer. If people simply run water over their fingertips or rinse their hands quickly without using

soap, should the behavior count as hand washing? What if people use soap but wash only their fingertips? Should a behavior count as hand washing only if it satisfies the guidelines issued by the American Society of Microbiology? Those guidelines specify using warm or hot running water and soap while washing for 10 to 15 seconds "all surfaces thoroughly, including wrists, palms, back of hands, fingers and under fingernails" (American Society of Microbiology 1996). Researchers must address these kinds of questions in creating operational definitions.

Consider the operational definition of English-language proficiency used in the Singer and Massey study:

Do you speak and understand English?
____ Does not speak or understand.
____ Does not speak but understands a little.
____ Does not speak but understands well.
____ Both speaks and understands a little.
____ Both speaks and understands well.

Other questions related to English-language proficiency include asking the immigrants in which settings they speak English when living in the United States. The settings of interest include at home, at work, with friends, and while shopping; the frequency of using English in each setting is recorded as never, sometimes, often, or always.

Operational definitions do not have to take the form of questions. They may be precise accounts or descriptions of what a researcher observed and what the context of the observations was. If operational definitions are not clear or do not indicate accurately the behaviors they were designed to represent, they have questionable value. Good operational definitions are both reliable and valid. **Reliability** is the extent to which an operational definition gives consistent results. For example, the question How many magazines do you read each month? may not yield reliable answers, because respondents may forget some magazines. Thus, if you asked a respondent the question at two different times, he or she might give two different answers. One way to increase the reliability of this question would be to ask respondents to list the magazines they have read in the past week. The act of listing forces respondents to think harder about the question, and shortening the amount of time to one week makes it easier for them to remember what they have read.

Validity is the degree to which an operational definition measures what it claims to measure. Professors, for example, give tests to measure students' knowledge of a particular subject as covered in class lectures, discussions, reading assignments, and other activities. Students may question the validity of this measure if the questions on a test reflect only the material covered in lectures. They may argue that the test does not measure knowledge of all the material covered or assigned. Remember, when assessing validity, always ask, Is the operational definition really measuring what it claims to measure? For example, is the

---

operational definitions Clear, precise definitions and instructions about how to observe and/or measure the variables under study.

reliability The extent to which an operational definition gives consistent results.

validity The degree to which an operational definition measures what it claims to measure.

Is number of apprehensions the best way to measure the effectiveness of border fences? Increased apprehensions could mean one of two things: the fences are effective in forcing undocumented immigrants into territories where the Border Patrol has a strategic advantage, or the fences are ineffective, since for every one undocumented immigrant caught, four manage to cross.

question Do you speak and understand English? likely to yield valid responses about English-language proficiency? One problem with this operational definition is disagreement as to what constitutes speaking *a little* English and speaking English *well*. What if a person's proficiency falls somewhere between *a little* and *well*? Every person who

answers this question probably has his or her own understanding of what speaking *a little* or *well* means.

Likewise, is the number of apprehensions a valid measure of the effectiveness of a border control initiative such as fence construction? Should effectiveness be measured by increased apprehensions or decreased apprehensions? Increased apprehensions could mean fences are effective in forcing undocumented immigrants into territories where the Border Patrol has a strategic advantage; they could also mean that the fences are ineffective, since for every one undocumented immigrant caught, four manage to cross (see Figure 2.4).

Steps 5 and 6: Analyzing the Data and Drawing Conclusions

■ **CORE CONCEPT 10:** In presenting their findings, sociologists identify common themes and, if applicable, specify whether hypotheses are supported by the data. When researchers reach the stage of analyzing collected data, they search for common themes, meaningful patterns, and links. Researchers must "pick and choose among the available numbers [and observations] and then fashion a format" (Hacker 1997, p. 478). In presenting their findings, researchers may use graphs, frequency tables, photos, statistical data, and so on. The choice of presentation depends on which results are significant and how they might be best shown (see Table 2.1).

Table 2.1 Basic Statistics Researchers Use to Convey Findings

The table below shows the number of unauthorized immigrants apprehended each year from 1993 through 2005. The data can be presented in the following way:

Year	Unauthorized Immigrants Apprehended on U.S.-Mexico Border (in millions)
2005	1.17
2004	1.14
2003	.91
2002	.93
2001	1.24
2000	1.64
1999	1.54
1998	1.52
1997	1.37
1996	1.51
1995	1.27
1994	.98
1993	1.20

n: number of cases (13 years of data)

Mean: The sum of all apprehensions from 1993 through 2005 divided by the total number of years considered (13). Over the 13-year period the average number of apprehensions was 1.26 million.

Standard deviation (s.d.): How far the data are spread from the mean. Most of the data (at least 95 percent) will fall within two standard deviations of the mean, and almost all of the data (99.7 percent) will fall within three standard deviations of the mean. To calculate the spread of data around the mean, multiply the standard deviation by 2 or 3, and then add (1.26 million + (241,743 × 2) = 1.74 million) and subtract (1.26 million − (241,743 × 2) = .777 million) that number to and from the mean.

Minimum: The lowest number of apprehensions. (910,000 apprehensions in 2003)

Maximum: The highest number of apprehensions. (1.64 million apprehensions in 2005)

Range: The numbers separating the highest value from the lowest value. In this case, the range is 1.64 million − .910 million, which equals .730 million.

Mode: The value that occurs most often. In this case there is no mode.

n: A symbol that stands for the number of respondents or cases. In this case, *n* is 13.

Median: The number that 50 percent of the cases fall above and 50 percent fall below. Half of all the years listed had fewer than 1.24 million apprehensions, and half had more than 1.24 million.

Presenting Research Findings

CONSIDER ONE CONCLUSION that Singer and Massey drew from their research on border apprehensions. "Despite the apparent buildup of enforcement resources along the Mexico-U.S. border and highly publicized initiatives such as operations Hold-the-Line and Gatekeeper, the probability of apprehensions fell in the late 1980s . . . and continued through the early 1990s" (p. 585). Singer and Massey calculated probabilities over a 30-year period (1964–1994). The year 1974 is associated with the greatest chance of apprehension: two in five. The chances of apprehension steadily fell to a one-in-five chance in 1994. Data can be presented in many ways, including graphs, tables, and bar charts.

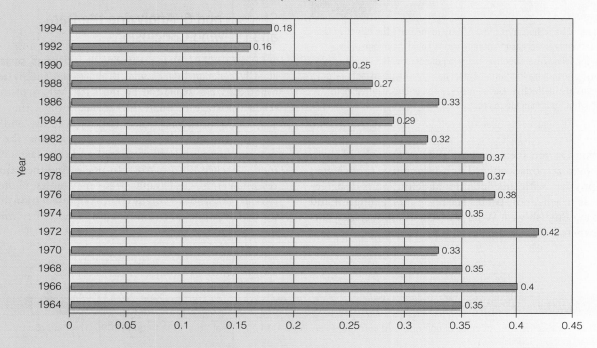

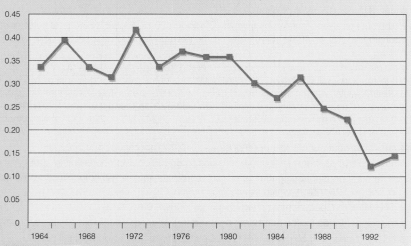

Year	Probability of Apprehension
1964	0.35
1966	0.40
1968	0.35
1970	0.33
1972	0.42
1974	0.35
1976	0.38
1978	0.37
1980	0.37
1982	0.32
1984	0.29
1986	0.33
1988	0.27
1990	0.25
1992	0.16
1994	0.18

▲ **Figure 2.4** Three Methods of Presenting Research Findings: Bar Graph, Line Graph, and Table.

Generalizability. Besides choosing a format in which to present data, sociologists comment on the **generalizability** of findings, the extent to which the findings can be applied to the larger population from which the sample was drawn. Both the sample used and the response rate are important in determining generalizability. If a sample is randomly selected, if almost all subjects agree to participate, and if the response rate for every question is high, we can say that the sample is representative of the population and that the findings are theoretically generalizable to that population. Recall that Singer and Massey used data derived from structured interviews with a sample of 6,341 randomly selected households in 34 Mexican communities. The response rate varied by community (from a 100-percent response rate in two rural communities to an 84-percent response rate in a large urban community); the average response rate was 94 percent. Such a high response rate suggests that Singer and Massey's findings are generalizable to the undocumented population. In particular, Singer and Massey found that, despite the apparent buildup of enforcement resources and the implementation of Operations Hold-the-Line and Gatekeeper, the probability of apprehensions fell in the late 1980s through early to mid-1990s.

If a sample is chosen for some other reason—perhaps because it is especially accessible or interesting—then the findings cannot be generalized to the larger population. Keep in mind that even though one goal of drawing conclusions is to make generalizations about the larger population, generalizations are not statements of certainty that apply to everyone. While Maril's findings probably cannot be generalized to all Border Patrol stations, his research on the border from the perspective of the Border Patrol agents supports Singer and Massey's findings: Highly visible border control strategies such as fences, increases in the number of agents, and increased surveillance are in reality "a grand pretense, because the majority of illegal immigration and opportunities for international terrorists existed far from the public eye" (p. 287).

Relationship between Independent and Dependent Variables

In the final step of the research process, researchers describe the relationship between independent and dependent variables. Instead of claiming that one variable causes another, researchers search for independent variables that contribute significantly toward explaining a dependent variable. At least three conditions must be met before a researcher can validly claim that an independent variable contributes significantly toward explaining a dependent variable.

First, the independent variable must precede the dependent variable in time. Time sequence can be established easily when the independent variable is a predetermined factor, such as sex or birth date. Such factors are fixed before a person becomes capable of any kind of behavior. Usually, however, time order cannot be established so easily.

Second, the two variables must be correlated. The strength of this correlation is often represented by a **correlation coefficient**, a mathematical representation of the extent to which a change in one variable is associated with a change in another variable (Cameron 1963). Correlation coefficients are numbers that fall between 0.00 and 1.00 and that are preceded by either a negative sign or a positive sign (for example, −0.23, +0.54). Researchers use these numbers to indicate two things about the relationship between two variables: the relationship's strength and its direction.

The closer the correlation coefficient is to 1.0, the stronger the relationship or association between variables is. A value of 1.0 represents a perfect association, meaning that if researchers know the value of the independent variable, they can predict the value of the dependent variable with 100-percent certainty. A value of 0.00 indicates that no relationship exists between variables.

The sign in front of the number shows the direction of the relationship. A positive sign means that as one variable increases or decreases, the second variable likewise increases or decreases. A negative sign in front of the number means that as one variable increases or decreases, the other variable moves in the opposite direction. A correlation of +0.90 between grade-point average (GPA) and income, for example, means the higher the GPA, the higher the income. Conversely, a correlation of −0.90 between GPA and income means the higher the GPA, the lower the income.

Third, establishing a correlation is a necessary step but is not in itself sufficient to prove causation. A correlation shows only that the variables are related; it does not mean that one variable causes the other. For one thing, a correlation can be spurious. A **spurious correlation** is one that is coincidental or accidental; the independent and dependent variables are not actually related, but some third variable related to both of them makes it seem as though they are.

For a researcher to claim that an independent variable causes a dependent variable, then, no evidence must indicate that another variable is responsible for a spurious

---

generalizability The extent to which findings can be applied to the larger population from which a sample is drawn.

correlation coefficient A mathematical representation that quantifies the extent to which a change in one variable is associated with a change in another variable.

spurious correlation A correlation that is coincidental or accidental because the independent and dependent variables are not actually related; rather, some third variable related to both of them makes it seem as though they are.

correlation between the independent and dependent variables. To check for this possibility, sociologists identify **control variables**, which are variables suspected of causing spurious correlations.

A good example of a significant relationship between two variables that disappears when we control for a third variable involves the strong correlation between the number of firefighters at a fire scene and the amount of damage expressed in dollars: the more firefighters, the greater the dollar amount of fire damage. Common sense, however, tells us that this relationship is a spurious correlation. The number of firefighters at the scene could not possibly be responsible for the dollar amount of fire damage. A third variable, the size of the fire, accounts for both the number of firefighters and the amount of damage. Although the number of firefighters called to the scene does help us predict the amount of damage, it is not the variable of cause. This logic can be applied to the relationship between mode of border crossing (alone, with family or friends, or with coyote) and apprehension rate (percentage of times caught relative to number of crossing attempts). If data shows that the apprehension rate increases when crossing with a coyote, one might question the role of English-language proficiency. It could be that those proficient in English cross alone or with family or friends, and it is their proficiency in English that helps them avoid detection. Likewise, it could be that the less proficient in English use coyotes, but it is their lack of proficiency that draws the attention of the Border Patrol.

control variables Variables suspected of causing spurious correlations.

ascribed characteristic Any physical trait that is biological in origin and/or cannot be changed, to which people assign overwhelming significance.

Department of Defense photo by TSGT Dave Ahlschwede, U.S. Air Force

A strong correlation exists between the number of firefighters at a fire scene and the amount of damage expressed in dollars: the more firefighters, the greater the dollar amount of fire damage. Common sense, however, tells us that this relationship is a spurious correlation. A third variable, the size of the fire, accounts for both the number of firefighters and the amount of damage.

Thinking about the possibility of a spurious correlation is especially important when the independent variable is an **ascribed characteristic**—any physical trait that is biological in origin and/or cannot be changed, to which people assign overwhelming significance. Ascribed characteristics include hair texture and color, eye shape, skin color, age, and country of birth. Findings that announce an ascribed characteristic as a "cause" can be used to stereotype and stigmatize groups of people. The hypothesis that an athlete's race determines athletic success at basketball is based on a spurious correlation (see Chapter 9 to learn why).

VISUAL SUMMARY OF CORE CONCEPTS

■ **CORE CONCEPT 1:** Sociological theories offer a set of guiding questions and key concepts that address how societies operate and how people relate to one another.

A sociological theory is a set of core assumptions and core concepts that speak to how societies operate and how people in them relate to one another and respond to their environment. Three major theories dominate the discipline of sociology: functionalist, conflict, and symbolic interactionist. The three perspectives can be used to analyze border fences and other significant issues.

Department of Defense photo by PH2 M. C. Farrington, U.S. Army

■ **CORE CONCEPT 2:** Functionalists focus on how the "parts" of society contribute in expected and unexpected ways to social order and stability and to social disorder and instability.

Functionalists view society as a system of interrelated, interdependent parts. Each part, whether it be a sports team or border fence, has a function; that is, it contributes to order and stability within society. Two types of functions are manifest (anticipated) and latent (unanticipated or unintended). Parts can also be disruptive to order and stability. When a disruption is anticipated, the dysfunction is manifest. When a disruption is unanticipated or unintended, the dysfunction is latent. When analyzing an issue, functionalists use these concepts to answer such questions as Why does a part (such as a border fence) exist? and What are the part's anticipated and unintended consequences?

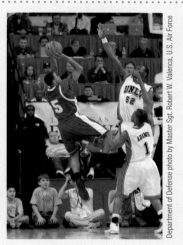
Department of Defense photo by Master Sgt. Robert W. Valenca, U.S. Air Force

■ **CORE CONCEPT 3:** The conflict perspective focuses on conflict over scarce and valued resources and the strategies dominant groups use to create and protect social arrangements that give them an advantage over subordinate groups.

Conflict theorists view conflict as an inevitable fact of social life and as the most important agent for social change. Conflict can take many forms, from physical confrontations to emotional manipulation. In any society, dominant and subordinate groups compete for scarce and valued resources.

National Archives and Records Administration

Conflict theorists ask this basic question: Who benefits from a particular social pattern or arrangement and at whose expense? Exploitation is disguised by a facade of legitimacy—an explanation that members of dominant groups give to justify their actions. On close analysis, however, this explanation falls apart.

■ **CORE CONCEPT 4:** Symbolic interactionists focus on social inter-action and related concepts of self-awareness/reflexive thinking, symbols, and negotiated order.

Symbolic interactionists focus on social interaction and ask, How do involved parties experience, interpret, influence, and respond to what they and others are doing while interacting? Symbolic interactionists draw upon the following concepts: (1) reflexive thinking, the process of stepping outside the self and observing and evaluating it from another's viewpoint; (2) symbol, any kind of physical phenomenon to which people assign a name, meaning, or value; and (3) negotiated order, the sum of existing and newly negotiated expectations, rules, policies, agreements, understandings, pacts, contracts, and other working arrangements.

Department of Defense photo by TSGT Scott Ree, U.S. Air Force

■ **CORE CONCEPT 5:** Sociologists adhere to the scientific method; that is, they acquire data through observation and leave it open to verification by others.

The scientific method is an approach to data collection that relies on two assumptions: (1) knowledge about the world is acquired through observation, and (2) the truth of that knowledge is confirmed by verification—that is, by others making the same observations.

U.S. Fish and Wildlife Service photo by Pedro Ramirez, Jr.

■ **CORE CONCEPT 6:** Sociologists explain why their research topic is important, tie their research in with existing research, and specify the core concepts guiding investigation.

The first step of a research project involves choosing a topic or deciding on a research question. Good researchers explain to their readers why their topic or research question is significant. In addition, researchers take existing research into account by showing how their new research verifies, advances, and corrects past research. Finally, researchers typically state the core concepts driving their research.

Department of Defense photo by Mass Communication Specialist Seaman Josue Leopoldo Escobosa, U.S. Navy

■ **CORE CONCEPT 7:** Sociologists decide on a plan for gathering data, identifying whom or what they will study and how they will select (sample) subjects for study.

The research design specifies the population to be studied and the methods of data collection. The population to be studied can be individuals, traces, documents, territories (such as a specific border crossing), households, or small groups. Researchers must decide whether to use a random or nonrandom approach to selecting subjects for study.

CBP photo by James R. Tourtellotte

■ **CORE CONCEPT 8:** Sociologists use a variety of data-collection methods, including self-administered questionnaires, interviews, observation, and secondary sources.

The research design must include a plan for collecting information. Researchers can choose from a variety of data-gathering methods, including self-administered questionnaires, interviews, observation, and secondary sources.

Department of Defense photo by PH3 Kristopher Wilson, U.S. Navy

■ **CORE CONCEPT 9:** Sociologists may choose to test hypotheses specifying the relationship between independent and dependent variables.

A variable is any characteristic that consists of more than one category. Examples of variables include *number of times apprehended* or *ability to speak English.* Sometimes researchers strive to find associations between variables to explain or predict behavior. The behavior to be explained or predicted is the dependent variable. The variable that explains or predicts the dependent variable is the independent variable. The relationship between independent and dependent variables is described in an hypothesis, or trial explanation put forward as the focus of research, which predicts the

CBP photo by Gerald L. Nino

relationship between independent and dependent variables. Researchers need to give clear, precise definitions and instructions (operational definitions) about how to observe and/or measure the variables under study. Operational definitions should be assessed for their reliability and validity.

■ **CORE CONCEPT 10:** In presenting findings, sociologists identify common themes and, if applicable, specify whether hypotheses are supported by the data.

When researchers reach the stage of analyzing collected data, they search for common themes, meaningful patterns, and links. They also comment on the generalizability of their findings—the extent to which the findings can be applied to the larger population from which the sample was drawn. If the study has involved testing hypotheses, researchers describe the relationship between independent and dependent variables. Keep in mind that it is very difficult to validly claim that an independent variable

Department of Defense photo by Staff Sgt. Dan Heaton, U.S. Air Force

(presence of a fence) causes a dependent variable (decline in the number of border crossings by undocumented workers). For such a claim to be valid, the independent variable must precede the dependent variable in time, the two variables must be correlated, and the correlation must not be spurious (coincidental or accidental).

Resources on the Internet

 Sociology: A Global Perspective Book Companion Web Site
www.cengage.com/sociology/ferrante

Visit your book companion Web site, where you will find flash cards, practice quizzes, Internet links, and more to help you study.

CENGAGENOW™

Just what you need to know NOW!

Spend time on what you need to master rather than on information you have already learned. Take a pre-test for this chapter, and CengageNOW will generate a personalized study plan based on your results. The study plan will identify the topics you need to review and direct you to online resources to help you master those topics. You can then take a post-test to help you determine the concepts you have mastered and what you will need to work on. Try it out! Go to www.cengage.com/login to sign in with an access code or to purchase access to this product.

Key Terms

ascribed characteristic   52
bourgeoisie   32
concepts   42
control variables   52
correlation coefficient   51
dependent variable   47
documents   43
dysfunctions   29
facade of legitimacy   32
function   28
generalizability   51
Hawthorne effect   47
households   43
hypothesis   47
independent variable   47
interviews   44
latent dysfunctions   30
latent functions   29

manifest dysfunctions   29
manifest functions   29
means of production   32
methods of data collection   42
nonparticipant observation   46
objectivity   40
observation   45
operational definitions   48
participant observation   46
populations   43
proletariat   32
random sample   43
reliability   48
representative sample   43
research   40
research design   42
research methods   40
samples   43

sampling frame   44
scientific method   40
secondary sources (archival data)   47
self-administered questionnaire   44
small groups   43
social interaction   36
sociological theory   28
spurious correlation   51
structured interview   44
symbol   37
territories   43
theory   28
traces   42
unstructured interview   45
validity   48
variable   47

3

Culture

With Emphasis on North and South Korea

Sociologists see people as products of their cultural experiences. At the same time, sociologists point out that people from the same culture are not replicas of one another.

Department of Defense photo by LCPL Adaecus G. Brooks, USMC

▲ Two U.S. military men take a break from field training in South Korea to visit children who live in the Hinsing Orphanage.

North and South Korea?

U.S. MILITARY INVOLVEMENT on the Korean Peninsula dates back to the end of World War II, when Premier Joseph Stalin of the Soviet Union, Prime Minister Winston Churchill of Great Britain, and President Franklin Roosevelt of the United States met in 1945 and, "without consulting even one Korean," agreed to chop Korea in half (Kang 1995, p. 75). The Korean War began in 1950, after the North Korean government invaded South Korea. Both sides endured heavy casualties as they fought to control the peninsula. In 1953, the war ended in a stalemate, with the 1945 boundary still in place.

Over the course of the U.S. military's 60-plus years of involvement, more than 7.5 million U.S. servicemen and servicewomen have fought, died, and otherwise served in the Koreas to maintain the division between the two countries. Today North Korea possesses a communist-style government and has one of the most isolated and centrally planned economies in the world (U.S. Central Intelligence Agency 2004). South Korea, on the other hand, is a republic, and its economy ranks among the top 20 in the world. The Korean War and the subsequent division of the Korean Peninsula into North and South have had a profound effect on Korean culture, on the meaning of being Korean, and on the relationship between the United States and the two Koreas. This division has affected the life of every North and South Korean resident who lived through the event and who has been born since. It has also affected the lives of millions of U.S. servicemen and servicewomen, their families, and significant others.

FACTS TO CONSIDER

- Some 29,000 U.S. military personnel are stationed in South Korea. There are no troops stationed in North Korea.

- U.S. military involvement in the Koreas dates back 60-plus years, to 1945.

- A 2.5-mile-wide border of barbed wire and land mines known as the demilitarized zone (DMZ) separates North Korea from South Korea. Just north of this border is the 1.2 million-strong North Korean army. In addition to the U.S. troops, there are 740,000 South Korean troops on the south side of the DMZ.

The Challenge of Defining Culture

■ **CORE CONCEPT 1: In the most general sense, culture is the way of life of a people.** Sociologists define **culture** as the way of life of a people; more specifically, culture includes the human-created strategies for adjusting to the environment and to those creatures (including humans) that are part of that environment. In responding to the environment and its creatures, humans draw upon strategies already in place or create or re-create their own. As we will learn, the list of human-created strategies is endless. These strategies include the automobile as a strategy for transporting people (and sometimes their pets) from one point to another; language as a strategy for communicating with others; and the online community MySpace.com as a strategy for presenting the self to others, learning about others, and building social networks. Culture cannot exist without a **society,** a group of interacting people who share, perpetuate, and create culture.

We use the word *culture* in ways that emphasize differences: "The cultures of X and Y are very different"; "There is a culture gap between X and Y"; "It is culture shock to come from X and live in Y." Our use of the word suggests that we think of culture as having clear boundaries, as an explanation for differences, and as a source of misunderstandings. In light of the seemingly clear way in which we use the word, we may be surprised to learn that sociologists face at least three conceptual challenges:

- *Describing a culture.* Is it possible to find words to define something so vast as the way of life of a people?
- *Determining who belongs to a group designated as a culture.* Does a person who "looks Korean" and who has lived in the United States most of his or her life belong to Korean or American culture?
- *Identifying the distinguishing characteristics that set one culture apart from others.* Is eating rice for breakfast a behavior that makes someone Korean? Is an ability to speak Korean a characteristic that makes someone Korean? Are ethnic Koreans who speak English or Spanish not Korean?

Such questions speak to the problem of being able to name a culture but not being able to define its geographic boundaries and of specifying the traits that mark some

people as members of a particular culture but finding that those traits do not apply to everyone designated as such. This chapter offers a framework for thinking about culture that considers both its elusiveness and its importance in shaping human life. This framework includes nine essential principles that define culture.

A few words of caution are in order before delving into this subject. Although this chapter compares cultures of South Korea, North Korea, and the United States, we can apply the concepts discussed here to understand any culture and frame other cross-cultural comparisons. As you read about culture, remember that North and South Korea are broadly referred to as countries possessing an Eastern (or Asian) culture and that the United States is regarded as a country possessing a Western culture. Therefore, many of the patterns described here are not necessarily unique to the Koreas or the United States; rather, they are shared with other Eastern or Western societies. At the same time, do not overstate the similarities among countries that share a broad cultural tradition.

Do not assume, for example, that South Korea and North Korea share the same culture. While there are undoubtedly many similarities, the two nations have very different economic and political systems. "North Korea is a place that is shrouded in mystery and conjecture; . . . for so long it has chosen to close itself off from the rest of the world that little information flows in or out of the place" (Sharp 2005a). Likewise, do not assume that South Korea is just like Japan. As we will see, much of South Korean identity is intricately linked with the idea of being "not Japanese" (Fallows 1988). To assume that South Korea is like Japan is equivalent to assuming that the United States is just like a Western European country, such as Germany. As we know, however, the United States is a country that celebrates its independence from European influence.

Since 1945, millions of U.S. soldiers of all races and ethnic groups, including Korean Americans, have served in South Korea. This photograph dates back to the 1960s.

Culture The way of life of a people; more specifically, the human-created strategies for adjusting to the environment and to those creatures (including humans) that are part of that environment.

Society A group of interacting people who share, perpetuate, and create culture.

GLOBAL COMPARISONS

U.S. Military Presence around the World

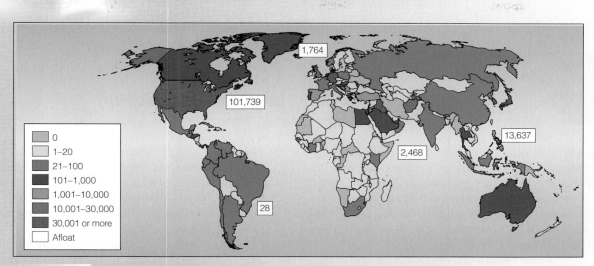

Legend:
- 0
- 1–20
- 21–100
- 101–1,000
- 1,001–10,000
- 10,001–30,000
- 30,001 or more
- Afloat

Map values: 1,764 · 101,739 · 13,637 · 2,468 · 28

▲ **Figure 3.1**   **U.S. Military Presence around the World**
The U.S. military presence in South Korea has shaped the cultures of North and South Korea. The Korean Peninsula is not the only place where the United States has such a presence. Figure 3.1 shows that U.S. military personnel are stationed in 140 countries. How might the presence of the U.S. military affect a country's culture?

Source: Department of Defense (2006).

Material and Nonmaterial Components

■ **CORE CONCEPT 2:** Culture consists of material and non-material components. **Material culture** consists of all the natural and human-created objects to which people have attached meaning. Material culture includes plants, trees, minerals or ores, dogs, cars, trucks, microwave ovens, computers, video cameras, and iPods. When sociologists think about material culture, they consider its most obvious and practical uses *and* the meanings assigned by the people who use it (Rohner 1984; See Working for Change: "Protecting Material Culture Considered Masterpieces of Human Creative Genius").

Learning the meanings that people assign to material culture helps sociologists grasp the significance of those objects in people's lives. Faucets, showers, tubs, soap, and towels are examples of material culture that people use to cleanse their bodies. Most Americans associate these items with bathrooms, relatively small rooms that offer a private space for washing the body. While private bathrooms exist in Korea, most Koreans also associate these items with

Department of Defense photo by CPL Robert J. Thayer, USMC

The Korean War Memorial in Washington, D.C., includes 19 statues each approximately 7 feet, 3 inches tall: 14 Army members, 3 Marines, 1 Navy member, and 1 Air Force member. The statues represent an ethnic cross section of the men who fought in the Korean War: 12 Caucasians, 3 African Americans, 2 Hispanic Americans, 1 Asian American, and 1 Native American.

Material culture All the natural and human-created objects to which people have attached meaning.

public bathhouses. An American woman visiting Korea described her experience with a bathhouse:

> Looking around, I noticed that all the women were completely naked—at a Korean bath, you check your modesty at the door, and the towel is for scrubbing, not drying or draping. After stripping down, I tentatively stepped through steamy glass doors, into the world of the baths—a large, noisy, cheerful area where about 100 women of all ages and small children of both sexes were scrubbing, chatting and soaking. To one side were rows of washing stations, with faucets, hand showers and mirrors set low to the ground. (Koreans, like Japanese, sit while washing). (McClane 2000)

WORKING FOR CHANGE

Protecting Material Culture Considered Masterpieces of Human Creative Genius

THE IDEA OF creating a system of international cooperation for protecting cultural and natural wonders emerged after World War I. But a particular event in 1959 spurred the international community to take action. That event involved the Egyptian government's decision to build what would become the world-famous Aswan High Dam. The dam is 11,811 feet long (the length of 3,937 football fields); 3,215 feet wide; and 364 feet high. To build it, 30,000 Egyptians worked day and night for 11 years. The dam addressed the country's need to control the Nile River (which flooded once a year). By controlling the Nile River, the Egyptian government could create an irrigation system, generate electricity, and provide clean water for drinking, bathing, and washing clothes. The dam created Lake Nasser, a 500-mile-long lake, considered to be one of the world's largest artificial lakes (Egypt Travel 1998).

The problem was that, upon completion, the valley containing the Abu Simbel and Philae Temples (more popularly known as Egyptian pyramids) would be flooded. The United Nations Educational, Scientific, and Cultural Organization (UNESCO) launched an international campaign to raise funds to dismantle, move, and reassemble the temples. Because the temples were carved into the sandstone cliffs above the Nile River, workers had to cut temples and monuments out and move them block by block to higher ground. Fifty countries contributed $40 million toward the $80 million preservation campaign. Other safeguarding campaigns followed, and UNESCO eventually institutionalized the effort in 1972 with the World Heritage program.

The Great Wall of China is one of 582 sites that UNESCO has designated as cultural treasures, masterpieces of human creative genius that offer exceptional testimony to cultural traditions that still exist or that have disappeared.

DoD photo by: TSGT JAMES PEARSON

To date 175 governments have signed the treaty known as the Convention Concerning the Protection of the World Cultural and Natural Heritage, agreeing that it is their duty to protect and preserve World Heritage Sites—sites designated as having such extraordinary environmental, cultural, or historic value that they should be preserved for the entire world citizenry and for future generations.

582 sites have been designated as cultural treasures, masterpieces of human creative genius that offer exceptional testimony to cultural traditions that still exist or that have disappeared (UNESCO 2005). Examples of cultural treasures:

- **Island of Gorée (Senegal)**, a slave-trading center from the 15th to 19th centuries on the West African coast.
- **Rock Drawings of Alta (Norway)**, petroglyphs made by hunting and gathering peoples 2,000 to 6,200 years ago.
- **Statue of Liberty (United States of America)**, a sculpture that has stood at the entrance to New York Harbor since 1886 as a symbol of welcome and liberty to millions of immigrants. The statue was a gift from France on the hundredth anniversary of American independence.
- **The Great Wall (China)**, a structure that stretches 4,163 miles (6,700 kilometers) from east to west, over deserts, grasslands, mountains, and plateaus (Smithsonian 1998).

For a list and description of all World Heritage Sites, see the UNESCO Web site, whc.unesco.org.

In addition to examining assigned meanings, sociologists consider the ways material culture shapes social relationships. American sociologists studying Korean bathhouses would be struck by the public nature of the bath, the relaxed and casual relationships among nude children and adult women, the lack of self-consciousness, and acceptance of one's own body and others' bodies. As one Western woman who went with her sister-in-law to a bathhouse explained, "She just stripped . . . and doing likewise to her son, didn't notice my very hesitant moves to do the same. . . . I felt so weird and exposed, but at the same time tried not to show it, as everyone seemed to be quite comfortable like that" (Chung 2003). This analysis suggests that the material component is shaped in some way by **nonmaterial culture**, the nonphysical creations which people cannot hold or see. In this case, the Western woman did not use the towel to cover herself (material culture) in the bathhouse, because the Korean cultural practice is to bathe in the nude (nonmaterial culture).

Some of the most important types of nonmaterial culture are beliefs, values, norms, symbols, and language.

Beliefs

Beliefs are conceptions that people accept as true, concerning how the world operates and where the individual fits in relationship to others. Beliefs can be rooted in blind faith, experience, tradition, or the scientific method. Whatever their accuracy or origins, beliefs can exert powerful influences on actions as they are used to justify behavior, ranging from the most generous to the most violent. For example, some people share the following beliefs:

- Anyone who wants to can grow up to become president of the United States.
- The whole world worshipped Kim Il Sung, the founding president of North Korea.
- Continuous conversation, rather than silence, validates a relationship.
- Athletic talent is essentially inherited.
- Athletic talent is essentially a product of hard work, practice, and persistence.
- It is acceptable for young children of both sexes to bathe with their mothers and other women in a public setting.

Values

A second component of nonmaterial culture is **values**: general, shared conceptions of what is good, right, appropriate, worthwhile, and important with regard to conduct, appearance, and states of being. One important study on values identified 36 values that people everywhere share to differing degrees, including the values of freedom, happiness, true friendship, broad-mindedness, cleanliness,

obedience, and national security. The study suggested that societies are distinguished from one another not according to which values are present in one society and absent in another, but rather, according to which values are the most cherished and dominant (Rokeach 1973). Americans, for example, place high value on the individual, whereas Koreans place high value on the group. These values manifest themselves in the American preference to bathe alone and the Korean preference to share the experience with others. Keep in mind that these value differences permeate many areas of life.

In sports, for example, the value Americans place on the individual is demonstrated when they single out the most valuable player of a game, a season, a league, or a tournament. Furthermore, when Americans view an outstanding athletic feat, they tend to give more credit to an athlete's innate talent or will to win than to hard work and discipline. In addition, American athletes work to find the style that is right for them and are willing to change that style if it does not bring success; Korean athletes work to find a style that has worked for others and tend to stay with that style because it has stood the test of time.

Koreans' emphasis on the group is demonstrated through their respect for tradition and form (time-tested and efficient methods of accomplishing goals). From a Korean point of view, athletic achievement does not occur simply because a person wants to excel or because he or she possesses raw talent. Instead, athletic competence develops over time, after the individual masters and appreciates the steps that combine to produce the intended result. Compared with the American system, the Korean system minimizes individual achievement because the achiever owes success to the mastery of technique and practice. These values are reflected in the comments of South Korea's top ice climber, Choi Won II: "I am not so special; anyone can do it" (*Korea News* 2004).

Norms

A third component of nonmaterial culture is **norms**, written and unwritten rules that specify behaviors appropriate and inappropriate to a particular social situation.

Nonmaterial culture Intangible human creations, which we cannot identify directly through the senses.

Beliefs Conceptions that people accept as true, concerning how the world operates and where the individual fits in relationship to others.

Values General, shared conceptions of what is good, right, appropriate, worthwhile, and important with regard to conduct, appearance, and states of being.

Norms Written and unwritten rules that specify behaviors appropriate and inappropriate to a particular social situation.

South Korea's top ice climber, Choi Won Il, claims that his athletic talents are "not so special; anyone can do it." Choi's statement reflects a belief in the power of perseverance and attention to technique rather than individual talent *per se.*

Examples of written norms are rules that appear in college student handbooks, on signs in restaurants (No Smoking Section), and on garage doors of automobile repair centers (Honk Horn to Open). Unwritten norms exist for virtually every kind of situation: wash your hands before preparing food; do not hold hands with a friend of the same sex in public; leave at least a 20 percent tip for waiters; remove your shoes before entering the house. Some norms are considered more important than others, and so the penalties for their violation are more severe. Depending on the importance of a norm, punishment can range from a frown to death. In this regard, we can distinguish between folkways and mores.

Folkways are norms that apply to the mundane aspects or details of daily life: when and what to eat, how to greet someone, how long the workday should be, how many times each day caregivers should change babies' diapers. As sociologist William Graham Sumner (1907) noted,

Folkways Norms that apply to the mundane aspects or details of daily life.

Mores Norms that people define as critical to the well-being of a group. Violation of mores can result in severe forms of punishment.

"Folkways give us discipline and support of routine and habit"; if we were forced constantly to make decisions about these details, "the burden would be unbearable" (p. 92). Generally, we go about everyday life without asking why until something reminds us or forces us to see that other ways are possible.

Consider the folkways that govern how a meal is typically eaten at Korean and American dinner tables. In Korea, diners do not pass items to one another, except to small children. Instead, they reach and stretch across one another and use their chopsticks to lift small portions from serving bowls to individual rice bowls or directly to their mouths. The Korean norms of table etiquette—reaching across instead of passing, having no clear place settings, and using the same utensils to eat and serve oneself food from platters and bowls—deemphasize the individual and reinforce the greater importance of the group.

Americans follow different dining folkways. They have individual place settings, marked clearly by place mats or blocked off by eating utensils. It is considered impolite to reach across another person's space and to use personal utensils to take food from the communal serving bowls. Instead, diners pass items around the table and use special serving utensils. That Americans have clearly marked eating spaces, do not trespass into other diners' spaces, and use separate utensils to take food reinforces values about the importance of the individual.

Often travel guides and international business guides list folkways that foreign travelers should follow when visiting a particular country. One guide advised business executives traveling to South Korea or hosting South Koreans to observe the following folkways:

- Don't write comments on others' business cards in their presence.
- Wait for the eldest person at the table to begin eating before you begin.
- Do not pour your own drink (Oregon Economic and Community Development Department (2003).

Mores are norms that people define as essential to the well-being of a group. People who violate mores are usually punished severely: they may be ostracized, institutionalized in prisons or mental hospitals, sentenced to physical punishment, or condemned to die. In contrast to folkways, mores are regarded as "the only way" or "the truth" and as thus unchangeable. Most Americans, for example, have strong mores against public nudity, especially when adults are in the presence of children. They believe that children should be shielded from seeing adults without clothes and that children are more vulnerable when naked in the presence of adults who are also nude. Koreans, on the other hand, do not view public nakedness as morally wrong or as a danger to children. Instead, they view the body as something to be accepted for what it is. Americans who visit Korean bathhouses report, to their surprise, that

The Korean table has no clear place settings. In addition, Korean diners use the same utensils for serving and eating. Most people living in the United States prefer individual place settings marked clearly by place mats or blocked off by eating utensils. Based on these two images, what conclusions might we draw about the relationship of the individual to the group?

they adjust quickly to social nudity and come to see being naked among others as unremarkable.

Symbols

We learned in Chapter 2 that **symbols** are any kind of physical or conceptual phenomenon—a word, an object, a sound, a feeling, an odor, a gesture or bodily movement, or a concept of time—to which people assign a name and a meaning. The meaning assigned is not evident from the physical phenomenon or idea alone. For example, what do the numbers 2-0-0-8 and 9-7, taken on their own, mean? Many societies, Christian and non-Christian, locate themselves in time by referencing the birth of Jesus Christ. Thus, "AD 2008" symbolizes 2,008 years since the traditionally recognized birth of Christ. AD is the abbreviation for *anno Domini* (Latin for "in the year of the Lord"). In North Korea, people locate themselves in time by referencing the year Kim Il Sung, the country's founding and "eternal" president, was born. The numbers 9-7 symbolize 97 years since his birth in 1912, with 1912 being year 1.

Language

In the broadest sense of the word, **language** is a symbol system involving the use of sounds, gestures (signing), and/or characters (such as letters or pictures) to convey meaning. When people learn language, they learn a symbol system. Those learning spoken languages must learn the agreed upon sounds that convey words, and they must learn the rules that specify relationships among the chosen words. That is, we cannot convey ideas by vocalizing

the relevant words in any order we choose. As the author of this textbook, I often ask my students, "Are you reading the book?" I cannot expect students to understand this question if I say the words in some random order like "Are book reading you?" English-language speakers follow a subject-verb-object order (We are reading the book). Koreans, on the other hand, follow a subject-object-verb format (We book are reading). Rules governing word order apply to stating first and last names. Consider that Koreans tend to identify themselves by stating their family name first and then their given name. In effect, the family name is given precedence over the individual's first name.

Learning a language and its rules is the key to human development and interaction. The level and complexity of human language sets people apart from the other animals. In addition, language is among the most important social institutions humans have created. That is, language is a predictable social arrangement among people that has emerged over time to facilitate human interaction and communication.

Symbols Any kind of physical or conceptual phenomenon—a word, an object, a sound, a feeling, an odor, a gesture or bodily movement, or a concept of time—to which people assign a name and a meaning or value.

Language A symbol system involving the use of sounds, gestures (signing), and/or characters (such as letters or pictures) to convey meaning.

The Role of Geographic and Historical Forces

■ **CORE CONCEPT 3: Geographic and historical forces shape culture.** Sociologists operate under the assumption that culture acts as a buffer between people and their surroundings (Herskovits 1948). Thus, material and nonmaterial aspects of culture represent responses to historical and geographic challenges and circumstances. Note that the responses are not always constructive and may have unintended consequences. The division of the Korean Peninsula into North and South Korea was a key geographic and historical event affecting the personal lives and culture of Koreans on both sides of the DMZ. It was a geographic event in that the division confined some Koreans to the north of the line and others to the south. The division was also a historic event in that it has affected all Koreans on the peninsula long after the fact. For the most part, people in the two Koreas have not communicated or interacted with each other since 1950, effectively evolving into two separate cultures. In 2005, fewer than 90,000 North and South Koreans traveled to one another's country. For the most part, the travelers were South Koreans visiting newly established resorts isolated from the North Korean population. That number, however, was greater than the total number of exchanges in the previous 15 years combined (Korean Overseas Information Service 2006).

Studies of North Korean defectors attending South Korean universities alert us to the consequences this divi-

The Dorasan Station once served a train line that connected North and South Korea. The station has been restored and the sign to the North Korean capital of Pyongyang has been put up in anticipation of resumed service between the two countries. The plans have yet to be finalized.

sion has had on Koreans. Unlike their South Korean counterparts, the defectors did not learn English as a second language and they tended to pursue careers as physical laborers despite achieving a college education. One defector commented that "I have lost my educational curiosity and was so accustomed to the instillation technique of education in North Korea. So the free-style and diverse education is very unfamiliar and strange for many North Korean defectors" (Chung 2006).

One important historical event that shaped American culture—especially the way Americans think about and

North Korea

Size of military. 1.2 million (active); 5 million (reserve)

Total military expenditures. $5.2 billion

Per capita military spending. $231.11

Military expenditures (percentage of GDP). 33.9%

South Korea

Size of military. 740,000 (active)

Total military expenditures. $21.06 billion (fiscal year 2006)

Per capita military spending. $289

Military expenditures (percentage of GDP). 3.0% (fiscal year 2006)

United States

Size of forces stationed in Korea. 29,600

Total annual cost of stationing troops in South Korea = $2+ billion

▲ **Figure 3.2** The Cost of Maintaining the Demilitarized Zone
The most heavily armed border in the world is the demilitarized zone that divides North and South Korea. Consider as one indicator of the costs of maintaining this division that North Korea spends about 33.9 percent of its gross domestic product on its military, whereas South Korea spends $21.06 billion, representing 3.0 percent of GDP. The United States spends about $2.0 billion each year to station 29,600 troops in South Korea.
Source: U.S. Central Intelligence Agency (2007)

The Korean War and its aftermath—a stalemate that left the Korean Peninsula divided into North and South Korea—has shaped life on both sides of the DMZ.

A relative abundance of domestic oil helped shape American culture. This Los Angeles oil field was one of three major oil fields in California that produced 77 million barrels of oil in 1910 (The Paleontological Research Institution 2006).

use energy—was the Spindletop, a Texas oil gusher of 1901. That discovery made the United States the largest producer of oil at the time and the most powerful nation in the world. "Oil was the new currency of the industrialized world, and America was rich. . . . Few Americans realized that their country was different or particularly fortunate . . . . They soon began to take their subterranean wealth for granted. . . . People in other industrialized nations were more aware of America's blessing. Being less sure of their sources of energy, they were warier about its dispensation. America quickly turned its industrial plant and its electrical grids over to oil" (Halberstam 1986, p. 87).

In contrast to the United States, North and South Korea produce no oil and consequently must import all their oil. Relative to the Koreas, the United States possesses abundant supplies of oil. As a result, South Koreans, North Koreans, and Americans use appliances such as refrigera-

tors differently. Consider that South Koreans tend to work harder than Americans to minimize the amount of electricity the refrigerator uses to keep food cold. South Koreans open the refrigerator door only as wide as necessary to remove an item, blocking the opening with their bodies to minimize the amount of cold air that escapes. Americans tend to open the refrigerator door wide and leave it open until they decide what they want or until after they move the desired item to a stove or countertop. Most North Koreans do not have refrigerators, and they follow a folkway of shopping for food daily (Winzig 1999).

Because Koreans depend entirely on other nations for oil and other resources, they are especially vulnerable to any world event that might disrupt the flow of resources into their country. This vulnerability reinforces the need to use resources sparingly and not to take them for granted. The relative lack of natural resources may

The photo on the left shows an American cemetery. The photo on the right shows a Korean cemetery. Think about how Korean and American beliefs and values about the relationship between individuals and the environment might have helped shape the appearance of the cemeteries.

ultimately explain Koreans' conservation-oriented behavior (see Table 3.1).

In summary, conservation- and consumption-oriented values and behaviors are rooted in circumstances of shortage and abundance. To understand this connection, recall a time when your electricity or water was turned off. Think about the inconvenience you experienced after a few minutes and how it increased after a few hours. The idea that one must conserve available resources takes root. People take care to minimize the number of times they open the refrigerator door. Now imagine how a permanent resource shortage or almost total dependence on other countries for resources can affect people's lives. In contrast, you can imagine how a greater abundance of resources breaks down conservation-oriented behaviors.

This child, born in Korea but adopted by parents living in the United States, is destined to learn the culture of the parents who raise him.

The Transmission of Culture

For the most part, people do not question the origin of the objects around them, the beliefs they hold, the values they follow, the norms to which they conform, the symbols they use, or the words they use to communicate and think about the world "any more than a baby analyzes the atmosphere before it begins to breathe it" (Sumner 1907, p. 76). Nor are people aware of alternative ways of thinking and behaving, because much of the material and nonmaterial components of their culture was in place before they were born. Thus, people think and behave as they do simply because they know no other way. And, because these behaviors and thoughts seem natural, we lose sight of the fact that culture is learned.

■ **CORE CONCEPT 4: Culture is learned.** Parents transmit to their offspring via their genes a biological heritage that is common to all humans yet uniquely individual. The genetic heritage that we share with all humans gives us a capacity for language development, an upright stance, four movable fingers and an opposable thumb on each hand, and other characteristics. If these traits seem overly obvious, consider that they allow humans to speak innumer-

able languages, perform countless movements, and devise and use many inventions and objects (Benedict 1976).

Regardless of their physical traits (for example, eye shape and color, hair texture and color, skin color), babies are destined to learn the ways of the culture into which they are raised. That is, our genes endow us with our human physical characteristics, but not our cultural characteristics. We cannot assume that someone comes from a particular culture simply because he or she looks like a person whom we expect to come from that culture. This fact becomes obvious to Korean American youth who participate in cultural immersion programs that involve study in South Korea. "Many say they have never felt so American as when they are slurping noodles in Korea. Even their slurps have an American accent" (Kristof 1995, p. 47).

The Role of Language

Human genetic endowment gives us a brain that is flexible enough to allow us to learn the languages that we hear spoken or see signed by the people around us. As children learn words and the meanings of words, they learn about

Table 3.1 Oil Consumption in United States, South Korea, and North Korea	United States	South Korea	North Korea
Proven Oil Reserves (in barrels)	21.4 billion	0.0	0.0
Barrels of Oil Consumed per Day	20.8 million	2.17 million	Unknown
Population Size	301 million	48.8 million	23.1 million
Per Capita Consumption of Oil per Year (in barrels)	25.2	16.5	Unknown

Note: A barrel of crude oil provides about 44 gallons of petroleum products.

Sources: U.S. Energy Information Administration (2007); *World Factbook* (2007)

their culture and what is important to it. They also acquire a tool that enables them to think about the world, interpret their experiences, establish and maintain relationships, and convey information.

For example, in Korean society it is nearly impossible to carry on a conversation, even among siblings, without considering age. This is because age is an exceedingly important measure of status: the older a person is, the more status or recognition he or she has in the society. Korean language acknowledges the importance of age by its use of special age-based hierarchical titles for everyone. In fact, words used to refer to a brother or sister acknowledge his or her age in relation to the speaker. Even twins are not equal, because one twin was born first. Furthermore, norms that guide Korean forms of address do not allow the speaker to refer to elder brothers or sisters by their first names. A boy addresses his elder brother as *hyung* and his elder sister as *muna*; a girl addresses her elder brother as *oppa* and her elder sister as *unni*. Regardless of gender, however, people always address their younger siblings by their first names (Kim and Kirby 1996).

Consider, as a final example of how language channels thinking, that Americans use the word *my* to express "ownership" of persons or things over which they do not have exclusive rights: my mother, my school, my country. The use of *my* reflects the American preoccupation with the needs of the individual over those of the group. In contrast, Koreans express possession as shared: our mother, our school, our country. The use of the plural possessive *our* reflects the Korean preoccupation with the group's needs over the individual's interests.

The language differences described above suggest that people see the world through the language(s) they have learned. The mind—or more precisely, the linguistic systems in our minds—gives order to a kaleidoscope of images, sounds, and impressions bombarding us. The words we have at our disposal allow us to organize the world, to notice some things and not others, and to ascribe significance to what we do notice. Keep in mind that when we learn a language we become parties to an agreement to communicate and organize our thoughts in a particular way—to "an agreement that holds throughout our speech community and is codified in the patterns of our language" (Whorf 1956, pp. 212–214).

Linguists Edward Sapir and Benjamin Whorf advanced the **linguistic relativity hypothesis**, which states that "No two languages are ever sufficiently similar to be considered as representing the same social reality. The worlds in which different societies live are distinct worlds, not merely the same world with different labels attached" (Sapir 1949, p. 162). Sapir and Whorf (1956) argue that unless people's linguistic backgrounds are similar, the same physical evidence does not lead to the same picture of the universe. So for example, the sound coming from a bird leads a speaker of English to think that the bird is *singing*, while it leads a speaker of Korean to think that the bird is *weeping*. Although the Korean and English languages channel thinking in different ways, do not assume that communication between the two speakers is impossible. It may take some work, but it is possible to translate one language into the other.

The Importance of Individual Experiences

■ **CORE CONCEPT 5:** People are products of cultural experiences but are not cultural replicas of one another. The information presented thus far may suggest that culture is simply a blueprint that guides, and even determines, thought and behavior. If that were true, then everyone would be cultural replicas of one another. Of course, they are not. Although culture is a blueprint of sorts, it also functions as a "toolkit" that allows people to select from and add to a menu of cultural options (Schudson 1989, p. 155).

How does this selection process work? A baby enters the world and becomes part of an already established set of human relationships. Virtually every event that the child experiences—being born, being fed, being cleaned, being

DPRK (North Korea) Archives

Kim Il Sung in 1944, four years before he assumed presidency of North Korea, with his son Kim Jong Il, who would eventually become the second, and current, president of North Korea.

Linguistic relativity hypothesis The idea that "no two languages are ever sufficiently similar to be considered as representing the same social reality. The worlds in which different societies live are distinct worlds, not merely the same world with different labels attached."

INTERSECTION OF BIOGRAPHY AND SOCIETY

Adding to the Menu of Cultural Options

RECALL THAT ALTHOUGH culture is a blueprint of sorts, it also functions as a "toolkit" that allows people to select from, and add to, a menu of cultural options. Here we meet two people whose lives were dramatically shaped by the division of North and South Korea: American servicewoman Andrea "Simone" Bowers and Korean-born adoptee Wayne Berry. Both expanded their "menu of cultural options" when they took steps to interact with Korean people and culture. Bowers joined the military and found herself stationed in South Korea because U.S. and Soviet leaders decided as long ago as 1945 that Korea was important to each country's national interests. Bowers, who lived in South Korea for 13 months, sought out new cultural experiences by befriending a Korean civilian who worked on her base. He introduced her to his country. Bowers climbed mountains every weekend with him. On these weekends, she met the "real" Korean people—not those who lived in and around the tourist towns next to the military bases, catering to the U.S. military. She sampled Korean cooking and slept at Korean houses on weekends. In the meantime, she picked up enough of the Korean language to communicate basic information.

Bowers visited mountain villages so remote that the inhabitants did not have access to television and had never seen dark-skinned people of African ancestry. She recalls that some Koreans were so fascinated with her skin color that they rubbed her skin just to feel it (Cherni 1998).

Korean-born Wayne Berry was adopted by U.S. parents 17 years after the Korean War ended in a stalemate. Holt International Children's Services, the agency that pioneered intercountry adoption in 1955, arranged the adoption. When Berry turned 26 he began looking for his birth parents—two people among 45 million in South Korea. He did not know their names, and he did not even speak Korean.

With the help of a translator, Berry sent one hundred letters to agencies and people in Korea. He made a two-minute video and sent it along with his baby picture to three TV stations in Seoul. Fortunately, his Korean aunt was watching the day the video aired, and she recognized him. Berry learned that his birth parents had never married but that they still wanted to see him. So he flew to Korea to meet his overjoyed birth parents. He is planning to stay in touch with them. His aunt is seeking a son living in the United States that she gave up for adoption (Smith 1996).

---

talked to, toilet training, talking, playing, and so on—involves people. The people present in the child's life at any one time may include various combinations of father, mother, grandparents, brothers, sisters, playmates, other adult relatives, neighbors, babysitters, and others (Wallace 1952). All of these people expose and pass on to the child their own "versions" of culture. In addition, caretakers make conscious or unconscious decisions about which aspects of the surrounding culture the child will be exposed to. The child, especially as he or she ages, may accept, reject, or modify others' versions; he or she may even seek new cultural experiences. Consider the case of Kim Il Sung, the founding president of North Korea. Kim Il Sung's father raised him as Christian. As a youth Kim attended church regularly, even playing the organ. However, when he took power in 1948, he abolished Christianity in the country, "keeping a couple of churches for show but staffing them with actors and actresses to impress foreign visitors with his tolerance" (Kristof 2005, p. 25). The case of Kim Il Sung suggests that individuals cannot be viewed as passive agents who simply absorb the culture around them (see Intersection of Biography and Society: "Adding to the Menu of Cultural Options").

Culture as a Tool for the Problems of Living

■ CORE CONCEPT 6: Culture provides formulas that enable the individual to adjust to the problems of living. Although our biological heritage is flexible, it presents all of us with a number of challenges. For example, everyone feels emotions and experiences hunger, thirst, and sexual desire; all humans age and eventually die. In turn, all cultures have developed "formulas" to help their members respond to these biological inevitabilities. Formulas exist for eliminating human waste; caring for children; satisfying the need for food, drink, and sex; channeling and displaying emotions; and eventually dying. In this section, we focus on the differing cultural formulas for dealing with two biological needs: relieving hunger and expressing social emotions.

Cultural Formulas for Relieving Hunger

All people become hungry, but the formulas for stimulating and satisfying appetite vary considerably across cultures. One indicator of a culture's influence is how people define only a portion of the potential food available to them as edible. For example, insects such as grasshoppers, locusts, and ants are edible and are an excellent source of protein, but not everyone chooses to eat them. Dogs and snakes are among the foods defined by many Koreans and other Asian peoples as edible. On the other hand, most Americans find it appalling that someone would eat dog meat, but they have no trouble eating lamb, beef, or pork. Cultural formulas for relieving hunger not only help people to "decide" what is edible, but to "decide" who should prepare the food, how the food should be served and eaten, how many meals should be consumed in a day, at what times meals should be eaten, and with whom one should eat.

South Korean formulas for satisfying hunger center around rice, whereas American and North Korean formulas center around corn. In fact, rice is the staple of the South Korean diet, whereas corn is the staple of the U.S. and North Korean diets. Of course, these preferences are influenced by a host of historical and social factors. In the case of North Korea, the country has relied on foreign aid to feed an estimated 16 million people (of 23 million). Corn, not rice, is one of the most donated food items. Even South Korea sent 100,000 tons of corn to North Korea (*Asia News* 2005). Furthermore, the average North Korean cannot afford rice, as 2.2 pounds cost the equivalent of 30 percent of the average monthly salary (BBC News 2005).

Corn as a formula for satisfying hunger in the United States extends beyond simply viewing it as a vegetable. Most of the American diet is affected by corn, although few Americans fully recognize its pervasiveness. Corn (in one form or another) appears in soft drinks, canned foods, candy, condensed milk, baby food, jams, instant coffee, instant potatoes, and soup, among other things (Vissar 1986). Like corn, rice and the by-products of rice plants have many uses: to feed livestock; to make soap, margarine, beer, wine, cosmetics, paper, and laundry starch; to warm houses; to provide inexpensive fuel for steam engines; to make bricks, plaster, hats, sandals, and raincoats; and to use as packing material to prevent items from breaking in shipping.

Cultural Formulas for Social Emotions

Culture also provides formulas for expressing **social emotions**, internal bodily sensations that we experience in relationships with other people. Empathy, grief, love, guilt, jealousy, and embarrassment are a few examples of social emotions. Grief, for instance, is felt at the loss of a relationship; love reflects the strong attachment that one person feels for another person; jealousy can arise from fear of losing the affection of another (Gordon 1981). People do not simply express social emotions directly, however. Rather, they interpret, evaluate, and modify their internal bodily sensations upon considering "feeling rules" (Hochschild 1976, 1979).

Feeling rules are norms that specify appropriate ways to express the internal sensations. They define sensations that one should feel toward another person. In the dominant culture of the United States, for example, same-sex friends are supposed to like one another but not feel anything resembling romantic love. It is also generally unacceptable for same-sex people to hold one another or to "celebrate" their friendship by holding hands in public. In this regard, the U.S. Army publishes a list of "must-know items" about South Korea for American soldiers stationed there. It informs them that their feeling rules do not apply in South Korea. One item says, "Don't be surprised to see two Korean women or men walking arm in arm. They are just good friends and there is nothing sexual implied" (U.S. Army 1998).

The process by which we come to learn feeling rules is complex; it evolves through observing others' actions and in interactions with others. In her novel *Rubyfruit Jungle,* Rita Mae Brown (1988) describes a situation in which feeling rules shape the way the central character, Molly, evaluates an encounter between her father, Carl, and his friend, Ep. Ep's wife has just died and Carl is comforting his friend. In this passage, Molly reflects on the feeling rules that apply to men:

> I was planning to hotfoot it out on the porch and watch the stars but I never made it because Ep and Carl were in the living room and Carl was holding Ep. He had both arms around him and every now and then he'd smooth down Ep's hair or put his cheek next to his head. Ep was crying just like Leroy. I couldn't make out what they were saying to each other. A couple of times I could hear Carl telling Ep he had to hang on, that's all anybody can do is hang on. I was afraid they were going to get up and see me so I hurried back to my room. I'd never seen men hold each other. I thought the only things they were allowed to do was shake hands or fight. But if Carl was holding Ep maybe it wasn't against the rules. Since I wasn't sure, I thought I'd keep it to myself and never tell. (p. 28)

This example shows that people learn norms that specify how, when, where, and to whom to display emotions. Somehow Molly learned that men do not hold one another—perhaps because she had never encountered such images. That is, her culture provided her with no images of men comforting one another in the way that Carl was comforting Ep.

Social emotions Internal bodily sensations that we experience in relationships with other people.

Feeling rules Norms that specify appropriate ways to express internal sensations.

Feeling rules can also apply to the social emotions people feel and display toward political leaders. In the case of North Korea, it is difficult for Americans and South Koreans to imagine how much Kim Il Sung, the country's founding and "eternal" president, and his son Kim Jong Il dominate North Korea's emotional life, culture, and landscape (McGeown 2003; Winzig 1999). Even defectors and outside observers maintain that most North Koreans feel genuine emotion for their leaders, especially for Kim Il Sung. One defector recalled the emotions he felt when he took ideology classes in college: "I cried often. I was so touched by the consideration Kim Il Sung showed for his people" (Kristof 2005). Why do North Koreans feel such emotion for leaders who, by many accounts, have mismanaged the country? Programs and activities in North Korea offer some answers:

- In North Korea students from nursery school through college take hundreds of hours of coursework that focuses on the lives and accomplishments of the two Kims, but especially of the father.
- Persons, places, and objects connected to Kim Il Sung are treated as sacred. "His parents, grandparents, wife and oldest son are still worshipped as an extension of Kim. Objects that he touched on his visits to collective farms or universities are covered with glass or draped with a veil" (Hunter 1999).
- Images of Kim Il Sung and Kim Jong Il are everywhere. Their portraits hang on the walls in all public buildings and households. All adults wear small badges showing Kim Il Sung's photo on their lapels or shirts (Sharp 2005b).
- An estimated 80 percent of the titles in a given bookshop are about the Kims or are written by one or both Kims. There are no dissident authors in North Korea who challenge the Kims' writings (Sharp 2005b; Winzig 1999).
- Major buildings and institutions are named after the Kims, and there are more than 40,000 Kim Il Sung Revolutionary Thought Study Rooms in the country (Koreascope 1998; Winzig 1999).

Cultural Diffusion

■ **CORE CONCEPT 7: People borrow material and nonmaterial culture from other societies.** Most people tend to think that the material and nonmaterial culture that surrounds them is "home-grown"—that it originated in their society. Journalist Nicholas D. Kristof (1998) offers an example:

Diffusion The process by which an idea, an invention, or some other cultural item is borrowed from a foreign source.

I once asked my 5-year-old son, who has grown up largely in Tokyo, about his favorite Japanese foods. Gregory thought for a moment and decided on rice balls and McDonald's. It makes perfect sense to think of Big Macs as Japanese food, since McDonald's is a much more important part of his life in Japan than it is when we are on vacation in America. In particular, given the cramped homes in which most Japanese live, when his Japanese friends have birthdays the most common place to hold parties is McDonald's. (p. 18)

The point is that most people tend to underestimate, ignore, or distort the extent to which familiar ideas, materials, products, and other inventions are connected in some way to outside sources or are borrowed outright from those sources (Liu 1994). The process by which an idea, an invention, or some other cultural item is borrowed from a foreign source is called **diffusion**. The term *borrow* is used in the broadest sense: it can mean to steal, imitate, plagiarize, learn, purchase, or copy. The opportunity to borrow occurs whenever people from different cultures make contact, whether face to face, by phone or fax, through televised broadcasts, or via the Internet (see No Borders, No Boundaries: "The Oprah Winfrey Show in 122 Countries").

Basketball, a U.S. invention, has been borrowed by people in 75 countries, including those in South Korea, where 21 clubs are registered with the *Federation Internationale de Basketball*. Baseball, another U.S. invention, has been borrowed by people in more than 90 countries, including South Korea, which won the world baseball championship in 1982 after it upset the United States and Japan (An 1997; *World Monitor* 1992, 1993). Likewise, 85,983 South Koreans have "borrowed" religion of the Jehovah's Witnesses, not just in name but as proselytizing members (Adherents .com 2002). Jehovah's Witnesses trace their beginning to a small group of Bible students who met near Pittsburgh, Pennsylvania, and eventually published the *Watchtower* magazine in 1879. Today the magazine is published in more than 140 languages and the religion counts 6.4 million practicing members in 240 lands (Watchtower Bible and Tract Society of Pennsylvania 2004).

Instances of opportunities for cultural diffusion are endless and can easily be found by skimming the newspaper headlines. Consider the following examples:

- Cross-listed Business Course to Send Indiana University Students to South Korea (University Wire 2006)
- World's Top Donut Chains Roll into South Korea (Young 2007).
- South Korea May Send Troops to Lebanon (UPI 2006)
- China Cultural Festival Opens in South Korea (*China View* 2006)
- South Korea Bulldozes Village to Expand US Military Base (Associated Press 2006)

People of one society do not borrow ideas, materials, or inventions indiscriminately from another society. Instead,

NO BORDERS, NO BOUNDARIES

The Oprah Winfrey Show in 122 Countries

THE UNITED STATES is known around the world as the largest exporter of popular culture. Sociologists define *popular culture* as any material or nonmaterial component of a society's culture that is embraced by the masses within and outside of that society. Examples include a sandwich (such as the Big Mac), a doll (such as Barbie), a television show (such as *Desperate Housewives*), a book (*Seven Habits of Highly Successful People*), a movie (such as *Harry Potter and the Goblet of Fire*), an item of clothing (such as blue jeans), and a phrase (such as "She's so fly"). Any analysis of American popular culture must consider the industries that sell popular culture and the avenues by which they reach the masses: commercials, television programs, radio, the Internet, and newspapers—to name a few. Australian social critic Peter Carey (1995) assesses American popular culture in this way:

> Americans have no understanding of the power that their culture has on everybody else; so even when there are people who are your political enemies, they are people who have taken on and internalized your popular music, your sitcoms, and there's a part in their heart which loves [the United States] and its culture. I don't know whether it's Mary Tyler

Moore or Bruce Springsteen or what it is, but these parts of this country are out there in the most unexpected places. (pp. 124–125)

The map below shows the global distribution list for the Oprah Winfrey Show. Which of the 122 countries did you least expect to carry her show? At the time of this writing, Oprah advertised the upcoming shows: "Wives Confess They Are Gay," "Back from the Brink of Suicide," and "Great Women Getting Older." What ideas are being diffused to foreign viewers?

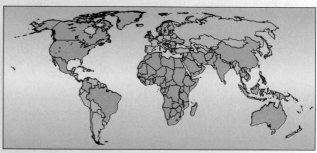

Orange-shaded countries air the Oprah Winfrey Show.

borrowing is usually selective. Even if people in one society accept a foreign idea or invention, they are nevertheless choosy about which features of the item they adopt. Even the simplest invention is really a complex of elements, including various associations and ideas of how it should be used. Not surprisingly, people borrow the most concrete and most tangible elements and then develop new associations and shape the item to serve new ends (Linton 1936). One might be surprised to learn that male circumcision in South Korea can be traced to contact with the U.S. military during the Korean War. Koreans, however, depart from the American practice of circumcising male babies at birth. In fact, only 1 percent of South Korean babies are believed to be circumcised at birth; most circumcisions occur during the elementary and middle school years (Ku, Kim, Lee, and Park 2003).

In contrast to South Korea, the North Korean government limits cultural diffusion opportunities by restricting access to information from the outside world. With rare exceptions, the 22 million people of North Korea cannot receive mail or telephone calls from outside the country. Nor can they travel beyond their country's borders (Brooke 2003a). One North Korean defector claims that "all the tape recorders and radios have to be registered. At registration, they cut off and solder the tuning dial to make sure you don't have a 'free' radio. If you have a cassette player, sometimes the police come to your apartment to check your cassette library" (Brooke 2003b, p. A8). In addition, North Korean officials remove brand-name labels, price tags, and images from all imported items or donated products that enter the country. Even something as simple as product labels provides clues about life elsewhere that might alert North Koreans to the impoverished state of their nation (*Korea Herald* 2004). While the North Korean government restricts cultural diffusion, North Korean people, especially those living along the border with China, are finding ways to acquire illicit radios, mobile phones, CD players, stereos, and televisions, which provide windows to a world beyond North Korea (MacKinnon 2005; Caryl and Lee 2006).

The Home Culture as the Standard

■ **CORE CONCEPT 8:** The home culture is usually the standard that people use to make judgments about the material and nonmaterial cultures of another society. Most people come to learn and accept the ways of their culture as natural. When they encounter foreign cultures, therefore, they can experience mental and physical strain. Sociologists use the term **culture shock** to describe the strain that people from one culture experience when they must reorient themselves to the ways of a new culture. In particular, they must adjust to a new language and to the idea that the behaviors and responses they learned in their home culture and now take for granted do not apply in the foreign setting. The intensity of culture shock depends on several factors: (1) the extent to which the home and foreign cultures differ, (2) the level of preparation for or knowledge about the new culture, and (3) the circumstances (such as vacation, job transfer, or war) surrounding the encounter. Some

Culture shock The strain that people from one culture experience when they must reorient themselves to the ways of a new culture.

© Kim Bo-Kyung Kirby and Kevin Kirby

Opportunities for cultural diffusion exist whenever people from different cultural backgrounds interact. The American-born boy behind the larger snowman built an "American" snowman, made of three balls of snow. The Korean-born girls built a Korean "two-ball" snowman. The simple act of playing together gave the children new perspectives on how a snowman might look.

cases of culture shock are so intense and unsettling that people become ill. Among the symptoms are "obsessive concern with cleanliness, depression, compulsive eating and drinking, excessive sleeping, irritability, lack of self-confidence, fits of weeping, nausea" (Lamb 1987, p. 270).

South Korean students in U.S. colleges, 2006	58,847
U.S. students in South Korea, 2006	631
Korean brides to U.S. soldiers since 1950	100,000
American citizens living in South Korea, 1999	30,000
Airline passengers from South Korea to the United States, 2003	618,000
Residents of the United States with Korean ancestry, 2003	1,190,353
U.S. military personnel in South Korea, 2007	29,600
Total Korean immigration, 1999–2002	27,696
Immigrant visas issued to Korean orphans adopted in the United States, 2003–2005	5,136
U.S. tourists to South Korea, 2004	547,020

▲ **Figure 3.3** **Opportunities for Cultural Diffusion between Americans and South Koreans**
Opportunities for cultural diffusion between the United States and South Korea exist because of the 60-year presence of U.S. troops in South Korea, international trade, and the departure and return of many Koreans who have attended colleges and universities in the United States. Notice, for example, that almost 59,000 South Koreans enroll in U.S. colleges and universities each year.

Sources: U.S. Embassy–Seoul (2003); U.S. Department of State (2002); U.S. Department of Defense (2007); French (2003); Institute for International Education (2006); Quinlan (2004); International Tourism Association (2004a, 2004b).

North Korea is one of the world's most isolated countries. Most of what we know about daily life inside this country comes from refugees who have fled to China, Russia, and Japan.

In his book *Communication Styles in Two Different Cultures*, Myung-Seok Park, a Korean-born professor of communication, describes an example of the kinds of stress encountered when someone enters a foreign society. This experience is typical of the kinds of adjustments that Koreans must make when they come to the United States. Taken by itself, the encounter does not seem especially stressful. Rather, the cumulative effect of a series of such encounters causes culture shock.

> When I studied at the University of Hawaii, my academic advisor was an old, retired English professor, Dr. Elizabeth Carr. All of the participants were struck by her enthusiasm, deep devotion, and her unfailing health. So at the end of the fall semester I said to her, "I would like to extend my sincere thanks to you for the enormous help and enlightening guidance you gave us in spite of your great age." Suddenly she put on a serious look, and I saw a portion of her mouth twisting. I had an inkling that she seemed unhappy about the way I expressed my thanks to her. Understandably enough, I was not a little embarrassed. A few hours later she told me that my remark "in spite of your great age" had reminded her suddenly that she was very old. I felt as if I had committed a big crime. I restrained myself from commenting on age anymore. (Park 1979, p. 66)

Do not assume that culture shock is limited to experiences with foreign cultures. People can also experience **reentry shock**, or culture shock in reverse, upon returning home after living in another culture (Koehler 1986). In fact, some researchers have discovered that many people find it surprisingly difficult to readjust to the return "home" after spending a significant amount of time elsewhere. As in the experience of culture shock, they face a situation in which differences jump to the forefront.

The intensity of reentry shock depends on an array of factors, including (1) the reason for being in the host culture, (2) the length of time lived in the host culture, and (3) the extent of the returnee's immersion in the everyday lives of people in the host culture. Symptoms of reentry shock are essentially the mirror image of those associated with culture shock. They include panic attacks ("I thought I was going crazy"), glorification of the host culture, nostalgia for the foreign ways, panic, a sense of isolation or estrangement, and a feeling of being misunderstood by people in the home culture. This comment by one American returning from abroad illustrates:

> America was a smorgasbord. But within two weeks, I had indigestion. Then things began to make me angry. Why did Americans have such big gas-guzzling cars? Why were all the commercials telling me I had to buy this product in order to be liked? Material possessions and dressing for success were not top priorities in the highlands. And American TV? I missed the BBC. (Sobie 1986, p. 96)

U.S. soldiers returning from their tour of duty in Iraq experienced reentry shock as well. One staff sergeant interviewed by the New York *Times* indicated that he was now "less tolerant of stupid people, . . . stupid people doing stupid things." He was particularly irritated by the questions "Did you kill anyone?" and "How did it feel?" (Myers 2003, p. A1).

Although many people expect to have problems adjusting to a stay in a foreign culture and even prepare for such difficulties, most do not expect to have trouble adjusting upon return to their home culture. Because reentry shock is unexpected, many people become anxious and confused and feel guilty about having problems with readjustment ("How could I possibly think the American way was anything but the biggest and the best?"). In addition, they may worry about how family, friends, and other acquaintances will react to their critical views of the home culture; they may be afraid that others will view them as unpatriotic.

The experience of reentry shock points to the transforming effect of an encounter with another culture (Sobie 1986). That the returnees go through reentry shock means that they have experienced up close another way of life and that they have come to accept the host culture's norms, values, and beliefs. Consequently, when they come home, they see things in a new light.

One reason people experience culture and reentry shock is that they hold the viewpoint of **ethnocentrism**. That is, they use one culture as the standard for judging the worth of foreign ways. From this viewpoint, one way is the center of everything, and all other ways are "scaled and rated with reference to it" (Sumner 1907, p. 13). Thus,

Reentry shock Culture shock in reverse; it is experienced upon returning home after living in another culture.

Ethnocentrism A viewpoint that uses one culture, usually the home culture, as the standard for judging the worth of foreign ways.

INTERSECTION OF BIOGRAPHY AND SOCIETY

Opposing Viewpoints on Same-Sex Adults Holding Hands

THE TWO VERY different reactions—an American in Ghana and a Moroccan in the United States—to hand holding between same-sex adults illustrate the process by which people use their home culture as a standard for judging behavior. The reactions also remind us of the many subtle ways our relationships with others are shaped by cultural forces we had no hand in creating.

> In the United States it is typically unacceptable for two men to hold hands; immediately the men would be labeled homosexual. I spent some time in Ghana, Africa, several years ago and one of the first cultural differences I noticed was that men, including the men I was with, hold hands. This cultural difference definitely hit home when one day one of the men I was with took my hand as we walked. In order not to offend him, I followed through with this until an appropriate opportunity allowed me to disengage our hands. Even though I was in a country where this was perfectly acceptable, I still felt extremely uneasy with this tradition. (An American in Ghana)

> In Morocco, it is common to see two men who are good friends walk down the street hand-in-hand, or even kissing each other on the cheek. In the U.S., it is seen as abnormal for people to act this way. The two friends are either seen as odd, or homosexuals. It took me a while to understand the American frame of mind, and to not act the way I would in Morocco. (A Moroccan in the United States)

In the United States, feeling rules dictate that adult men cannot hold hands. President Bush adjusted to another culture's norms by holding hands with Crown Prince Abdullah of Saudi Arabia.

© Rod Aydelotte/Bloomberg News/Landov

---

other cultures are seen as "strange," or worse, as "inferior" (see Intersection of Biography and Society: "Opposing Viewpoints on Same-Sex Adults Holding Hands").

Ethnocentrism

Several levels of ethnocentrism exist. Perhaps the most harmless type of ethnocentrism is simply defining foreign ways as peculiar, as did some Americans who attended the 1988 Summer Olympic Games in Seoul, South Korea. Upon learning that some Koreans eat dog meat, some visitors made jokes about it. People speculated about the

consequences of asking for a doggy bag, and they made puns about dog-oriented dishes: Great Danish, fettuccine Alfido, and Greyhound as the favorite fast food (Henry 1988). Keep in mind that Koreans don't eat their pet dogs; rather, they eat a "special breed of large tan-colored dogs raised especially for canine cuisine" (Kang 1995, p. 267). In fact, Koreans who eat dogs would argue that Americans who eat pigs, cows, chickens, and lambs are in no position to judge them (Kang 1995).

The most extreme and destructive form of ethnocentrism is **cultural genocide,** in which the people of one society define the culture of another society not as merely offensive, but as so intolerable that they attempt to destroy it. There is overwhelming evidence, for example, that Japanese tried to exterminate Korean culture between 1910 and 1945. After Japan annexed Korea in 1910, Japanese became the official language, Koreans were given Japanese names, Korean children were taught by Japanese teachers, Korean literature and history were abandoned, ancient

Cultural genocide An extreme form of ethnocentrism in which the people of one society define the culture of another society not as merely offensive, but as so intolerable that they attempt to destroy it.

Genocide involves a coordinated and systematic plan to destroy a people and their culture.

DoD photo by MSGT Rose Reynolds

temples—important symbols of Korean heritage—were razed, the Korean national anthem was banned, and the Korean flag could not be flown. Even Korean flowers were banned, as Japanese officials forced Koreans to dig up their national flowers and plant cherry trees (Kang 1995). Japanese brutally suppressed all resistance on the part of Korean people. When Koreans tried to declare their right to self-determination in March 1919, thousands of people were injured or killed in clashes with the Japanese military.

Sociologist Everett Hughes (1984) identifies yet another type of ethnocentrism:

> One can think so exclusively in terms of his own social world that he simply has no set of concepts for comparing one social world with another. He can believe so deeply in the ways and the ideas of his own world that he has no point of reference for discussing those of other peoples, times, and places. Or he can be so engrossed in his own world that he lacks curiosity about any other; others simply do not concern him. (p. 474)

A conversation between author Simon Winchester (1988) and the Korean owner of a bar frequented by U.S. service personnel shows that from a Korean viewpoint, many American military people display the kind of ethnocentrism that Everett Hughes describes. In particular, the bar owner pointed to soldiers who wore shirts with words like "Kill 'Em All: Let God Sort the Bastards Out." The bar owner remarked,

> They treat us like we're some backward Third World country, and you know we're not. . . . I've worked for 25 years trying to bring the two communities together. I organize them to go out to meet families. I try to persuade them to learn a bit of Korean, to eat some of the food, to understand why they're here. But they don't want to know. And it's the way some of them treat our women, and our men too. Some of them just have no respect for us. The way they see it, they're top of the pile, and everyone else is nothing. It makes me mad. (p. 144)

Another type of ethnocentrism is **reverse ethnocentrism**, in which the home culture is regarded as inferior to a foreign culture. People who engage in this kind of thinking often idealize other cultures as utopias. For example, they might label Japanese culture as a model of harmony, the United States as the model of self-actualization, and Native American culture as a model of environmental sustainability (Hannerz 1992).

People who engage in reverse ethnocentrism not only idealize other cultures, but also reject any information disproving their view. Journalist K. Connie Kang (1995) offers an excellent example of reverse ethnocentrism among the Korean students she taught while a visiting professor between 1967 and 1970 at Hankuk University of Foreign Studies:

> Going to the United States was a preoccupation with Koreans. America was akin to paradise, Koreans thought, and the image of America they carried in their heads was exaggerated. They really believed everyone was rich and lived in big houses with winding staircases. I tried to dispel their notions of what America was by saying that most people worked in uninteresting jobs to pay their bills and put their kids through school, but I do not think I succeeded. People want to hear only what they want to hear; it was hard to compete with the images of American life created by Hollywood. (p. 205)

Cultural Relativism

A perspective that runs counter to ethnocentrism is cultural relativism. **Cultural relativism** means two things: (1) that a foreign culture should not be judged by the standards of a home culture and (2) that a behavior or way of thinking must be examined in its cultural context—that

Reverse ethnocentrism A type of ethnocentrism in which the home culture is regarded as inferior to a foreign culture.

Cultural relativism The perspective that a foreign culture should not be judged by the standards of a home culture and that a behavior or way of thinking must be examined in its cultural context.

is, in terms of the society's values, norms, beliefs, environmental challenges, and history. Cultural relativism is a perspective that aims to understand—not condone or discredit—foreign behavior and thinking.

For example, whereas most Americans cannot understand why some Koreans eat dog meat, most Koreans are equally appalled that Americans often let dogs live in their homes, allow them to lick their faces, and spend so much money on them when the U.S. population includes many poor and homeless people. When we consider the historical and environmental challenges surrounding the Korean decision to eat dog meat, this practice might not seem so unreasonable. Whereas the United States has an abundance of fertile, flat land for grazing cattle, many Asian countries with limited space, such as North and South Korea, employ available land to grow crops, not to graze cattle. Likewise, in light of American feeling rules that limit touch and physical displays of affection between same-sex adults, Koreans might not be surprised at the close relationships many Americans have with their pets. (The results of an American Animal Hospital Association survey showed that almost 62 percent of respondents said that they celebrate their pet's birthday and 31 percent said they believed that, after their spouse, their pet understood them the best (Dole 2004)).

Critics argue that cultural relativism supports one or more of the following attitudes: "Whatever they do is fine"; "It's none of my business what others do"; or "Everything is relative." Such positions would allow every cultural trait—even some of the most harmful and violent ones (including infanticide, human sacrifice, foot binding, and witch hunts)—to escape judgment or criticism. Taking a position of cultural relativism does not mean that one has no values of one's own. Actually, the practice of understanding cultural differences teaches us that morality is both relative and universal (Redfield 1962). Morality is relative in that ideas about rightness and wrongness vary across time and place. Morality is also universal in that every culture has its own conceptions of morality. Although mores do exist that can make virtually any idea or behavior seem right or wrong, "some mores have a harder time making some things right than others" (Redfield 1962, p. 451).

Subcultures

■ **CORE CONCEPT 9: In every society, some groups possess distinctive traits that set them apart from the main culture.** Groups that share in certain parts of the domi-

Subcultures Groups that share in some parts of the dominant culture but have their own distinctive values, norms, beliefs, symbols, language, or material culture.

nant culture but have their own distinctive values, norms, beliefs, symbols, language, or material culture are called **subcultures.** Examples of subcultures in the United States:

- Retirement communities populated by people who have retired or reached a specific age.
- Military bases ranging from small outposts to cities that house members of the noncivilian population. The 29,000 American men and women stationed in South Korea constitute a subculture, as they spend most of their time at one of 46 bases that occupy approximately 100 square miles of the country (Brooke 2003).
- Sorority and fraternity houses, in which members live together as "brothers," or "sisters" in dwellings adjacent to a college campus.

Often we think we can identify subcultures based on physical traits, ethnicity, religious background, geographic region, age, gender, socioeconomic status, dress, or behaviors that society defines as deviant. Determining who belongs to a particular subculture, however, is actually a complex task that must go beyond simply including everyone who shares a particular trait. For example, using ethnic or racial categories as a criterion for identifying the various subcultures within the United States makes little sense. The broad racial category "Native American" ignores the fact that people lumped into this category practice a variety of customs and lifestyles, hold a variety of values and beliefs, speak many languages, and do not necessarily conceive of themselves as a single people (Berkhofer 1978).

U.S. Navy photo by JOC Al Fontenot, CNFK public affairs

The military is an example of a subculture. It uses, among other things, military uniforms to distinguish its members from the larger civilian population.

Sociologists determine whether a group of people constitute a subculture by learning whether they share such things as language, a symbol system, values, norms, or territory—and whether they interact with one another more than with people outside the group. One characteristic central to all subcultures is that their members are separated or cut off in some way from other people in the larger culture. This separation may be complete or it may be limited to selected aspects of life, such as work, school, recreation and leisure, dating and marriage, friendships, religion, medical care, or housing. It may be voluntary, result from an accident of geography, or be imposed consciously or unconsciously by a dominant group. It could also result from a combination of these three factors.

Subcultures within the United States experience separation in different ways or to varying degrees. Some integrate themselves into certain areas of mainstream culture when possible but remain excluded from other areas of life. In general, African Americans who work or attend school primarily with whites are often excluded or feel excluded from personal and social relationships with them. This exclusion forces them to form their own fraternities, study groups, support groups, and other organizations. Other subcultures are **institutionally complete** (Breton 1967)— that is, their members do not interact with anyone outside their subculture to shop for food, attend school, receive medical care, or find companionship, because the subculture satisfies these needs. Often we find a clear association between institutional completeness and language differences. Persons who cannot speak the language of the dominant culture are very likely to live in institutionally complete subcultures (such as Little Italy, Chinatown, Koreatown, or Mexican barrios).

Of the 750,000 Korean immigrants living in the United States, approximately 300,000 live in Southern California. A large portion of the California group resides in an institutionally complete subculture known as Koreatown, west of downtown Los Angeles. Still, the Korean experience in the United States varies and cannot be described by a few generalizations—if only because Korean immigrants to the United States are not always from Korea. They may have emigrated from the large Korean communities in Siberia, Canada, Japan, or Brazil (Lee 1994).

Sociologists use the term **countercultures** to describe subcultures in which the norms, values, beliefs, symbols, and language the members share emphasize conflict or opposition to the dominant culture. In fact, rejection of the dominant culture's values, norms, symbols, and beliefs

U.S. Navy photo by Mass Communication Specialist 3rd Class Jhoan Montolio

For the most part the U.S. military is an institutionally complete culture whose members tend to interact with Korean civilians only on specially arranged occasions, such as this ceremony in which Korean children welcome U.S. forces to the Republic of Korea for the start of military exercises.

is central to understanding a counterculture. In the United States the Older Order Amish (one of four Amish subcultures) constitute a counterculture in that they remain separate from "the rest of the world," organizing their life so that they do not even draw power from electrical grids. In South Korea, Buddhist monks constitute a counterculture. Their rejection of the material trappings of capitalistic society and their devotion to simple living, modest dress, and vegetarian diet are just some of the characteristics that distinguish the monks from the dominant South Korean culture. Of course, the Buddhist monks who live in the United States also qualify as a counterculture.

Institutionally complete subcultures Subcultures whose members do not interact with anyone outside their subculture to shop for food, attend school, receive medical care, or find companionship, because the subculture satisfies these needs.

Countercultures Subcultures in which the norms, values, beliefs, symbols, and language the members share emphasize conflict or opposition to the larger culture. In fact, rejection of the dominant culture's values, norms, symbols, and beliefs is central to understanding a counterculture.

■ VISUAL SUMMARY OF CORE CONCEPTS

■ **CORE CONCEPT 1:** In the most general sense, culture is the way of life of a people.

Sociologists define culture as the way of life of a people, specifically the human-created strategies for adjusting to the environment and to humans and other creatures that are part of the environment.

U.S. Marine Corps photo by Cpl. Michael S. Cifuentes, USMC

■ **CORE CONCEPT 2:** Culture consists of material and nonmaterial components.

Material culture is the physical creations (natural and man-made) to which people attach meaning. Nonmaterial culture includes the nonphysical creations. Sociologists are interested in meanings people assign to material culture, the ways in which material and nonmaterial culture shapes social relationships, and the ways material culture shapes and is shaped by values, norms, beliefs, symbols, and language.

Department of Defense photo by JO3 Monica Miles, USN

■ **CORE CONCEPT 3:** Geographic and historical forces shape culture.

Sociologists operate under the assumption that culture acts as a buffer between people and their surroundings. Thus, material and nonmaterial aspects of culture represent responses to historical and geographic challenges and circumstances. The division of the Korean Peninsula into North and South Korea was a key geographic and historical event affecting the personal lives and culture of all Koreans. Where people live—north or south of the DMZ—affects whether they will learn English as a second language, eat primarily corn or rice, have access to more than one radio station, or be free to travel outside their country.

National Archives and Records Administration

■ **CORE CONCEPT 4:** Culture is learned.

Regardless of their physical traits (for example, eye shape and color, hair texture and color, and skin color), people are destined to learn the ways of the culture into which they are raised. The Korean-born boy, adopted by an American family, will be American in culture.

Gregory A. Caldeira

■ **CORE CONCEPT 5: People are products of cultural experiences but are not cultural replicas of one another.**

No one can possibly experience all the material and nonmaterial components of their culture. Thus, we do not become replicas of one another simply because we experience different slices of the same culture. Sociologists view culture as a "toolkit" that allows people to select from an almost limitless menu of cultural options. How does this selection process work? One answer is that the people in a child's life expose the child to the parts of the culture they define as most important, and they emphasize some parts more than others.

Department of Defense photo by TSGT Curt Eddings

■ **CORE CONCEPT 6: Culture provides formulas that enable the individual to adjust to the problems of living.**

Our biological heritage presents us with a number of challenges. Cultures have developed formulas to help their members respond to these challenges. Formulas exist for eliminating human waste; caring for children; satisfying the need for food, drink, and sex; channeling and displaying emotions; and eventually dying. The photograph shows one formula for eliminating human waste. The United States formula involves sitting; another formula involves squatting.

Dr. Terry Pence

■ **CORE CONCEPT 7: People borrow material and nonmaterial culture from other societies.**

Cultural diffusion is the process by which an idea, an invention, or some other cultural item is borrowed from a foreign source. The borrowing may include imitating, stealing, purchasing, copying, or learning about something. The opportunity to borrow occurs whenever two people from different cultures make contact. These Americans are involved with tae kwon do, the national sport of South Korea, which has become one of the world's most widely practiced martial arts.

© Index Open

■ **CORE CONCEPT 8: The home culture is usually the standard that people use to make judgments about the material and nonmaterial cultures of another society.**

When people encounter a foreign culture they can experience mental and physical strain known as culture shock. One reason people experience culture shock is that they hold the viewpoint of ethnocentrism. That is, they use their home culture as the standard for judging the worth of foreign ways. Sociologists take a position of cultural relativity when they evaluate a foreign culture. That is, they analyze a cultural trait in the context of the society's values, norms, beliefs, environmental challenges, and history. When we consider the historical and environmental challenges surrounding the Korean decision to eat dog meat, this practice might not seem unreasonable. Whereas the United States has an abundance of fertile, flat land for grazing cattle, many Asian countries with limited space, such as North and South Korea, employ available land to grow crops, not to graze cattle.

© Claro Cortes IV/Reuters/Landov

■ **CORE CONCEPT 9:** In every society, some groups possess distinctive traits that set them apart from the main culture. Groups that share in certain parts of the dominant culture but have their own distinctive values, norms, beliefs, symbols, language, or material culture are called subcultures. One characteristic central to all subcultures is that their members are separated or cut off in some way from other people in the larger culture. Some subcultures are known as countercultures when their norms, values, beliefs, symbols, and language emphasize conflict with or opposition to the dominant culture. In fact, rejection of some aspect of the dominant culture is central to understanding a counterculture.

U.S. Navy photo by JOC Al Fontenot, CNFK public affairs

Resources on the Internet

Sociology: A Global Perspective Book Companion Website

www.cengage.com/sociology/ferrante

Visit your book companion Web site, where you will find flash cards, practice quizzes, internet links, and more to help you study.

CENGAGENOW™

Just what you need to know NOW!

Spend time on what you need to master rather than on information you have already learned. Take a pre-test for this chapter, and CengageNOW will generate a personalized study plan based on your results. The study plan will identify the topics you need to review and direct you to online resources to help you master those topics. You can then take a post-test to help you determine the concepts you have mastered and what you will need to work on. Try it out! Go to www.cengage.com/login to sign in with an access code or to purchase access to this product.

Key Terms

Beliefs   63
Countercultures   79
Cultural genocide   76
Cultural relativism   77
Culture   60
Culture shock   74
Diffusion   72
Ethnocentrism   75
Feeling Rules   71

Folkways   64
Institutionally complete
   subcultures   79
Language   65
Linguistic relativity
   hypothesis   69
Material culture   61
Mores   64
Nonmaterial culture   63

Norms   63
Reentry shock   75
Reverse ethnocentrism   77
Social emotions   71
Society   60
Subcultures   78
Symbols   65
Values   63

4 Socialization

With Emphasis on Israel and the Palestinian Territories

Socialization is a learning process that begins immediately after birth and continues throughout life. Through this process, people learn about and come to terms with the culture and behavior patterns of the society in which they live.

© Justin C. McIntosh

▲ A Palestinian child and Israeli soldier gaze at one another in front of the Wall/Barrier that separates the West Bank from Israel.

WHY FOCUS ON

Israel and the Palestinian Territories?

OUR EMPHASIS IN this chapter is on the fierce century-long conflict between Jews and Palestinians over land that both groups call home. The conflict has involved six wars (1948, 1956, 1967, 1968–1971, 1982, and 2006) and two major *intifadas* (uprisings). The conflict has been going on since Jews began their return "home" around 1900 in response to widespread persecution throughout Europe. Palestinians have used guns, knives, Molotov cocktails, stones, graffiti, strikes, boycotts, barricades, and suicide attacks to resist Israeli Jews' control of Palestinian land. Israelis have used gunfire, deportations, imprisonment, curfews, school closures, blockades, land seizures, targeted assassinations, and bulldozers to maintain their "right to return."

Over the years, the United States has acted as peace broker in this region, bringing Palestinian and Israeli Jewish leaders together in an effort to persuade them to work through their differences. Achieving peace has not been (and will not be) easy, because everyone on both sides has been significantly affected by the conflict.

In this chapter, we use sociological concepts and theories to help us understand how the conflict has been passed down through the generations and sustained. Keep in mind that socialization, or the process by which people come to learn the culture and behavior patterns of the society in which they live, is just one factor that helps us understand why the Israeli-Palestinian conflict has lasted so long. Although we concentrate on the role socialization plays in perpetuating this conflict, the concepts and theories presented apply to socialization in general.

FACTS TO CONSIDER

- *Sesame Street*, an educational TV series targeting children ages 4–7, is televised in 120 countries.
- In Israel and the Palestinian Territories the show airs as *Sesame Stories* (not *Sesame Street*) because the idea that a neutral street exists

where Palestinians and Israeli Jews might gather together is not a believable scenario in that part of the world.

- *Sesame Stories* concentrates on (1) teaching young viewers respect and understanding for their own

and other cultures, (2) promoting a peaceful resolution to the hundred-year conflict between Palestinians and Israeli Jews, and (3) presenting positive images of Palestinian and Israeli Jewish children.

Socialization

By the time children are two years old, most are biologically ready to show concern for what adults regard as the "rules of life." They are bothered when things do not match their expectations: Paint peeling from a table, broken toys, small holes in clothing, and persons in distress all raise troubling questions. From a young child's point of view, when something is "broken," then someone somewhere has done something very wrong (Kagan 1988, 1989).

To show this kind of concern with standards, two-year-olds must first be exposed to information that leads them to expect behavior, people, and objects to be a certain way (Kagan 1989). They develop these expectations as they interact with others in their world. For example, children go to adults with their questions and needs. Adults respond in different ways—perhaps by offering explana-

These Israeli children are too young to understand the full meaning of the celebration in which they are participating. The parents looking on and the rabbi directing them are introducing them to select aspects of Jewish culture.

socialization The process by which people develop a sense of self and learn the ways of the society in which they live.

internalization The process in which people take as their own and accept as binding the norms, values, beliefs, and language that their socializers are attempting to pass on.

nature Human genetic makeup or biological inheritance.

nurture The social environment, or the interaction experiences that make up every individual's life.

tions, expressing concern, trying to help, showing no concern, or paying no attention. These kinds of exchanges constitute socialization.

■ **CORE CONCEPT 1: In the broadest sense of the word, socialization is the process by which people develop a sense of self and learn the ways of the society in which they live.** **Socialization** is the process by which people develop a sense of self and learn the ways of the society in which they live. More specifically, it is the process by which humans (1) acquire a sense of self or social identity, (2) learn about the social groups to which they belong and do not belong, (3) develop their human capacities, and (4) learn to negotiate the social and physical environment they have inherited. Socialization is a lifelong process, beginning at birth and ending at death. It takes hold through **internalization**, the process in which people take as their own and accept as binding the norms, values, beliefs, and language that their socializers are attempting to pass on.

■ **CORE CONCEPT 2: Socialization involves nature and nurture.** No discussion of socialization can ignore the importance of two factors: nature and nurture. **Nature** comprises one's human genetic makeup or biological inheritance. **Nurture** refers to the social environment, or the interaction experiences that make up every individual's life. Some scientists debate the relative importance of genes and the social experiences, arguing that one is ultimately more important than the other. Such a debate is futile, because it is impossible to separate the influence of the two factors or to say that one is more forceful. *Both* nature and nurture are essential to socialization (Ornstein and Thompson 1984).

The relationship between the cerebral cortex and spoken language illustrates rather dramatically the inseparable qualities of nature and nurture. As part of our human genetic makeup (nature), we possess a cerebral cortex, which allows us to organize, remember, communicate, understand, and create. Scientists believe that humans inherit a cerebral cortex "set up" to learn any of the more than 6,000 known human languages. In the first months of life, all babies are biologically capable of babbling the essential sounds needed to speak any language. As children grow, this enormous potential is reduced, however, by the language (or languages) that the baby hears and eventually learns (nurture). For the most part, Palestinian babies hear standard Arabic spoken at home and learn Hebrew in school. Israeli babies, for the most part, hear modern Hebrew spoken at home. Since Israel is a land of immigrants, many Jewish children are exposed to Russian, Yiddish, Ladino, or Romanian as babies, but they all must eventually learn Hebrew. Most Israeli Jews do not learn Arabic.

While humans have a biological makeup that allows them to speak a language, the language itself is learned

The genes parents transmit to their offspring reach back over an indefinite period through four biological grandparents, eight biological great-grandparents, and beyond.

through interactions with others. If babies are not exposed to language, they will not acquire that communication tool. While social interaction is essential to language development, it is also essential to human development in general. If babies are deprived of social contact with others, they cannot become normally functioning human beings.

The Importance of Social Contact

■ **CORE CONCEPT 3: Socialization depends on meaningful interaction experiences with others.** Cases of children raised in extreme isolation or in restrictive and sterile environments show the importance of social contact (nurture) to normal development. Some of the earliest and most systematic work in this area was done by sociologist Kingsley Davis. His work on the consequences of extreme isolation demonstrates how neglect and lack of social contact influence emotional, mental, and even physical development.

Cases of Extreme Isolation

Davis (1940, 1947) documented and compared the separate yet similar lives of two girls: Anna and Isabelle. During the first six years of their lives, the girls received only the minimum of human care. Both children, living in the United States in the 1940s, were classified as illegitimate. Because of their status, both were rejected and forced into seclusion; each was living in a dark, attic-like room. Anna was shut off from her family and their daily activities. Isabelle was shut off in a dark room with her deaf-mute mother. Both girls were six years old when authorities intervened. At that time, they exhibited behavior comparable to that of six-month-old children.

Anna "had no glimmering of speech, absolutely no ability to walk, no sense of gesture, not the least capacity to feed herself even when food was put in front of her, and no comprehension of cleanliness. She was so apathetic that it was hard to tell whether or not she could hear" (Davis 1947, p. 434). Anna was placed in a private home for mentally disabled children until she died four years later. At the time of her death, she behaved and thought at the level of a two-year-old child.

Like Anna, Isabelle had not developed speech, but she did use gestures and croaks to communicate. Because of a lack of sunshine and a poor diet, she had developed rickets: "Her legs in particular were affected; they 'were so bowed that as she stood erect the soles of her shoes came nearly flat together, and she got about with a skittering gait'" (Davis 1947, p. 436). She also exhibited extreme fear of and hostility toward strangers. Isabelle entered into an intensive and systematic program designed to help her master speech, reading, and other important skills. After two years in the program, she had achieved a level of thought and behavior normal for someone her age. Isabelle's success may be partly attributed to her establishing an important bond with her deaf-mute mother, who taught her how to communicate through gestures and croaks. Although the bond was less than ideal, it gave her an advantage over Anna.

A person's overall well-being depends on meaningful interaction experiences with others. Social interaction is essential to developing and maintaining a sense of self.

Why Can't Palestinians and Jews Get Along?

THE UNITED STATES and other parties involved in the peace process have proposed a two-state solution to end the conflict. Specifically, the Palestinian Territories of the West Bank and Gaza—two geographically disconnected lands—would become the new Palestinian state. Before the two-state solution can be implemented, a number of critical issues must be resolved:

1. **Israeli settlements in the Palestinian Territories** Settlements are Jewish-populated communities in the Palestinian Territories. They are diverse in structure ranging from outposts composed of trailers, campers, and tents to self-contained towns and cities with populations of 10,000 or more. An estimated 325 such settlements house 400,000 Jewish residents (U.S. Central Intelligence Agency 2007). In 2005 the Israeli government evacuated settlers from 21 settlements in the Gaza Strip and from 4 West Bank settlements. Notwithstanding the recent evacuation, critics argue that the settlements are attempts to establish a significant Jewish presence on Palestinian land so that a permanent solution regarding the land cannot be achieved.

2. **Safe Passage between Gaza and the West Bank** If Gaza and the West Bank are eventually to be regarded as one state, how can they be geographically linked? Who will control the access roads from one territory to the other—Palestinians? Israelis? A joint force? Currently, Israelis control every access route for moving goods and people into and out of the Palestinian Territories.

3. **Right of Return** The creation of the state of Israel and the subsequent 1948 and 1967 Arab-Israeli wars resulted in Palestinian diasporas, forced scatterings of an ethnic population to various locations around the world. Several million Palestinians immigrated to surrounding countries and now live with their descendants in Jordan (est. 2,225,000); the Persian Gulf countries of Kuwait, Saudi Arabia, United Arab Emirates, Qatar (3,711,000), and Iraq (est. 450,000); Lebanon (est. 350,000); Syria (est. 340,000); and elsewhere. Approximately 3.9 million Palestinians are registered with the UN as refugees. Refugees seek the right to return to the land within Israel from which they fled or were evicted (Bennet 2003b).

4. **Status of Jerusalem** Both Palestinians and Israeli Jews claim Jerusalem, which is divided into East and West Jerusalem, as their capital. The Israeli government claims all of the city as its capital, and Palestinians claim East Jerusalem as the capital of their future state.

5. **The Wall** In June 2002 the Israeli government, with the support of 83 percent of Israelis, began construction on the West Bank Barrier—a 350-kilometer-long obstacle comprising electrified fencing, razor wire, trenches, concrete walls, and guard towers that winds through the West Bank. The wall puts 14 percent of the West Bank on the Israeli side; at one point the wall extends some 13 miles into the West Bank (Farnsworth 2004). The wall separates the West Bank from Israel and channels Palestinian movement from the West Bank into Israel through checkpoints. Israel claims that the wall is not a political border but rather a security border

Jerusalem, a city that both Palestinians and Israelis claim as their capital, is the home of many religious sites, such as the Western Wall (or Wailing Wall). Directly behind the Western Wall is the Dome of the Rock—a mosque built in the seventh century.

© Keith Levit Photography/Index Open

Cases of Less Extreme Isolation

Other evidence of the importance of social contact comes from less extreme cases of neglect. Psychiatrist Rene Spitz (1951) studied 91 infants who were raised by their parents during their first three to four months of life but who were later placed in orphanages. When they were admitted to the orphanages, the infants were physically and emotionally normal. Orphanage staff provided adequate care for their bodily needs—good food, clothing, diaper changes, clean

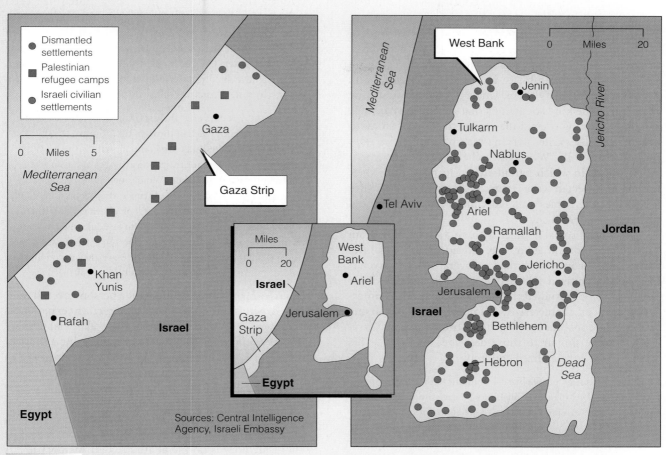

▲ Figure 4.1   Map of the West Bank and Gaza, Including Existing and Dismantled Israeli Settlements

designed to keep suicide bombers and other would-be attackers out of Israel and Jewish settlements. The UN estimates that the barrier has disrupted 600,000 Palestinian lives. For example, the wall is constructed so that it surrounds 12 Palestinian communities, allowing residents to leave only through Israeli-controlled checkpoints (Myre 2003).

6. **The Fatah-Hamas Divide** The political party Fatah, founded in 1958, renounced terrorism against Israel in 1988 and acknowledged Israel's right to exist in 1993. Hamas, founded in 1987, is labeled a terrorist organization by Israel, the United States, and the European Union. In 2006 Hamas won 74 of 132 elected seats in democratically held elections, giving it the majority in Palestinian Legislative Council. The victory reflected the Palestinian people's deep dissatisfaction with Fatah, which has been widely acknowledged as corrupt. The election results prompted Western leaders to shut off aid to the Hamas-dominated unity government. At the time of this writing, Hamas (backed by Syria and Iran) had taken control of the Gaza Strip after five days of civil war. Fatah (backed by the United States and Jordan) remains in control of the West Bank. This Fatah-Hamas division has jeopardized the two-state solution and further restricted Palestinian movement between the West Bank and Gaza.

nurseries—but gave the children little personal attention. Because only one nurse was available for every 8 to 12 children, the children were starved emotionally. The emotional starvation caused by the lack of social contact resulted in such rapid physical and developmental deterioration that a significant number of the children died. Others became completely passive, lying on their backs in their cots. Many were unable to stand, walk, or talk (Spitz 1951).

These cases and the cases of Anna and Isabelle teach us that children need close contact with and stimulation from others if they are to develop normally. Adequate stimulation means the existence of strong ties with a caring adult. These ties must be characterized by a bond of mutual expectation between caregiver and baby. In other words, there must be at least one person who knows the baby well enough to understand his or her needs and feelings and who will act to satisfy them. Under such conditions, the child learns the kinds of actions that generate predictable responses: getting excited may cause the child's father to become equally excited; crying may prompt the mother to soothe the child. When researchers set up experimental situations in which parents failed to respond to their infants in expected ways (even for a few moments), they found that the babies suffered considerable tension and distress (*Nova* 1986).

Meaningful social contact with and stimulation from others are important at any age. Indeed, strong social ties with caring people are linked to overall social, psychological, and physical well-being. British sociologist Peter Townsend (1962) studied the effects of the minimal interaction that can characterize life for the elderly in nursing homes. The consequences for the institutionalized elderly are strikingly similar to those described by Spitz in his studies of institutionalized children:

> In the institution, people live communally with a minimum of privacy, and yet their relationships with each other are slender. Many subsist in a kind of defensive shell of isolation. Their mobility is restricted, and they have little access to general society. Their social experiences are limited, and the staff leads a rather separate existence from them. They are subtly oriented toward a system in which they submit to orderly routine and lack creative occupation, and cannot exercise much self-determination. They are deprived of intimate family relationships. . . . The result for the individual seems to be a gradual process of depersonalization. He may become resigned and depressed and may display no interest in the future or things not immediately personal. He sometimes becomes apathetic, talks little, and lacks initiative. His personal and toilet habits may deteriorate. (Townsend 1962, pp. 146–147)

Children of the Holocaust

Anna Freud and Sophie Dann (1958) studied six German Jewish children whose parents had been killed in the gas chambers of Nazi Germany. The children were shuttled from one foster home to another for a year before being sent to the ward for motherless children at the Tereszin concentration camp. The ward was staffed by malnourished and overworked nurses, who were themselves concentration camp inmates. After the war, the six children were housed in three different institution-like environments. Eventually they were sent to a country cottage, where they received intensive social and emotional care.

During their short lives, these six children had been deprived of stable emotional ties and relationships with caring adults. Freud and Dann found that the children were ignorant of the meaning of family and grew excessively upset when they were separated from one another, even for a few seconds. In addition, they "behaved in a wild, restless, and uncontrollably noisy manner":

> During the first days after their arrival, they destroyed all the toys and damaged much of the furniture. Toward the staff they behaved either with cold indifference or with active hostility, making no exception for the young assistant Maureen who had accompanied them from Windermere and was their only link with the immediate past. At times, they ignored the adults so completely that they would not look up when one of them entered the room. They would turn to an adult when in some immediate need, but treat the same person as nonexistent once more when the need was fulfilled. In anger, they would hit the adults, bite or spit. (Freud and Dann 1958, p. 130)

It is clear from these studies that a person's overall well-being depends on meaningful interaction experiences with others. On a more fundamental level, social interaction is essential to developing and maintaining a sense of self. Yet, if the biological mechanisms (nature) involved in learning and then recalling names, faces, and the meaning of words and significant symbols were not present, people could not interact with one another in meaningful ways: "You have to begin to lose your memory, if only in bits and pieces, to realize that memory is what makes our lives. Life without memory is no life at all. . . . Our memory is our coherence, our reason, our feeling, even our action. Without it we are nothing" (Bunuel 1985, p. 22).

During the Holocaust many Jewish children were separated from their parents and confined to concentration camps.

Department of Defense photo by Sgt. Robert Holliway, US Army

Individual and Collective Memory

■ **CORE CONCEPT 4:** Socialization is impossible without memory; memories passed on from one generation to the next preserve and sustain culture. Memory, the capacity to retain and recall past experiences, is easily overlooked in exploring socialization. On an individual level, memory allows people to know others and remember interacting with them. On a societal level, memory preserves the cultural past. Without memory whole societies would be cut off from the past and left reinventing the wheel.

The latest neurological evidence on memory suggests that some physical trace remains in the brain after new learning takes place, stored in an anatomical entity called an *engram*. Engrams, or *memory traces*, are formed by chemicals produced in the brain. They store in physical form the recollections of experiences—a mass of information, impressions, and images unique to each person.

> It may have been a time of listening to music, a time of looking in at the door of a dance hall, a time of imagining the action of robbers from a comic strip, a time of waking from a vivid dream, a time of laughing conversation with friends, a time of listening to a little son to make sure he was safe, a time of watching illuminated signs, a time of lying in the delivery room at childbirth, a time of being frightened by a menacing man, a time of watching people enter the room with snow on their clothes. (Penfield and Perot 1963, p. 687)

Scientists do not believe that engrams store actual records of past events, in the same way that films are stored on DVDs or videocassettes. More likely, engrams store edited or consolidated versions of experiences and events, which are edited further each time they are recalled.

As we noted previously, memory has more than an individual quality; it is strongly social. First, no one can participate in society without the ability to recall such things as names, faces, places, words, symbols, and norms. Second, most "newcomers" easily learn the language, norms, values, and beliefs of the surrounding culture. Indeed, we take it for granted that people have this information stored in memory. Third, people born at approximately the same time and place have likely lived through many of the same events. These experiences—each uniquely personal and yet similar to one another—are remembered long after the event has passed. Sociologists use the term **collective memory** to describe the experiences shared and recalled by significant numbers of people (Coser 1992; Halbwachs 1980). Such memories are revived, preserved, shared, passed on, and recast in many forms, such as stories, holidays, rituals, and monuments (see Intersection of Biography and Society: "Collective Memory").

Virtually every Israeli Jew has memories of war and persecution. These memories may involve hiding in a shelter, saying good-bye to someone called up for military service, waiting for a loved one to return from war, or fleeing places where Jews were deemed unfit to exist.

Writer David Grossman (1998) explains that Israel is alive with memories and reminders of the past:

> There are nine hundred memorials to the war dead in this small country. . . . There is no week on the Israeli calendar in which there is not a memorial day of some sort for a traumatic event.
>
> A person walking through downtown Tel Aviv can, in the space of five minutes, set out from Dizengoff Street (where a suicide bomber murdered thirteen people two years ago), glance apprehensively at the No. 5 bus (on which another suicide bomber killed twenty-two civilians four years ago), and try to regain his composure at the Apropos Cafe (where a year ago yet another suicide bomber killed three women; none of us can forget the images of a blood-spattered, newly orphaned baby).
>
> When my daughter took her first field trip with her kindergarten classmates, they went to Yitzhak Rabin's grave and to the adjacent military cemetery. (p. 56)

Although Palestinians were part of many of the same historical events as Jews, the memories of the two groups differ because they witnessed these events from different vantage points. Displaced Palestinians retain memories of their homeland, and like the Jews, pass these memories down to their descendants, who did not experience the displacements firsthand. One way Palestinians pass on the memory is by naming their children after the cities and towns in which they lived before the 1948 war (Al-Batrawi and Rabbani 1991). Most tell their children about the places where they used to live; they teach their offspring to call those places home. They show their children keys and deeds to the houses in which they once lived (Kifner 2000). When author David Grossman (1988) asked a group of Palestinian children in a West Bank refugee camp to tell him their birthplace, each replied with the name of a formerly Arab town:

> Everyone I spoke to in the camp is trained—almost from birth—to live this double life; they sit here, very much here . . . but they are also there. . . . I ask a five-year-old boy where he is from, and he immediately answers, "Jaffa," which is today part of Tel Aviv.
>
> "Have you ever seen Jaffa?" "No, but my grandfather saw it." His father, apparently, was born here, but his grandfather came from Jaffa.

collective memory The experiences shared and recalled by significant numbers of people. Such memories are revived, preserved, shared, passed on, and recast in many forms, such as stories, holidays, rituals, and monuments.

INTERSECTION OF BIOGRAPHY AND SOCIETY

Collective Memory

THE FIVE PHOTOS show the faces of individuals who live in different parts of the world. Yet their individual biographies have been shaped by an event that occurred in 70 BC, the year the Romans conquered the land between the Jordan River and the Mediterranean Sea and expelled the Hebrew people (Israelites) who had lived their since 1200 BC. The dispersed Jews never forgot their "home."

In the late 19th century, a growing climate of anti-Semitism arose throughout Europe and Russia. Theodor Herzl, the founder of the modern Zionist movement, believed the only way to combat European anti-Semitism was for Jews to return to their homeland and establish the state of Israel. The Nazi Holocaust, which claimed 6 million lives, including one-third of the European Jewish population, gave the Jewish return movement a desperate urgency. Photo A shows the faces of Jews who fled Europe in the 1940s to establish their own state, with its own armed forces to protect them from future aggression.

But the land to which the Jews returned was now home to approximately 1.2 million Palestinians. The UN voted to partition Palestine into two independent states—one Jewish, the other Palestinian. The Palestinians could not tolerate this arrangement, arguing that they should not have to pay for European transgressions. Photo B shows the faces of Palestinians fleeing to neighboring countries in 1948, the year Jewish leaders declared Israel an independent state and defeated Arab armies from Egypt, Syria, Jordan, Lebanon, and Iraq.

Photo C shows the collapsed towers of the World Trade Center where thousands died on September 11, 2001. In the days following this "act of war," Americans learned that the attackers belonged to a global network of terrorists who resented American influence over their lives and especially resented America's long-standing support of Israel at the expense of Palestinians. The United States responded to the attack by taking military action against Taliban and al-Qaida bases in Afghanistan. The U.S. invasion sent Afghan refugees fleeing into neighboring countries (Photo D). In March 2003 the United States and its coalition partners (most notably the United Kingdom) attacked Iraq, under the pretext that a connection existed between Iraq and al-Qaida. This invasion forced people to flee from there as well (Photo E).

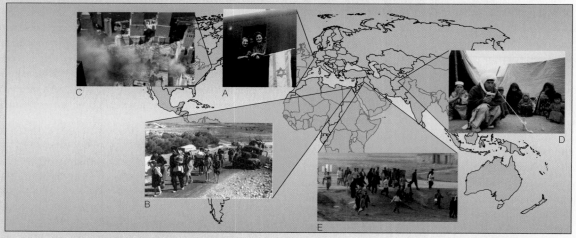

Sources: (a) National Archives and Records Administration. (b) © AP Photo. (c) DoD photo by JO1 Mark D. Faram, USN. (d) © IOM/USAID. (e) DoD Photo by Sgt. Paul L. Anstine II. USMC.

"And is it beautiful, Jaffa?" "Yes. It has orchards and vineyards and the sea." And farther down . . . I meet a young girl sitting on a cement wall, reading an illustrated magazine. . . . She is from Lod, not far from Ben-Gurion International Airport, forty years ago an Arab town. She is sixteen.

She tells me, giggling, of the beauty of Lod. Of its houses, which were big as palaces. "And in every room a hand-painted carpet. And the land was wonderful, and the sky was always blue. . . .

"And the tomatoes there were red and big, and everything came to us from the earth, and the earth gave us and gave us more." "Have you visited there, Lod?" "Of course not." "Aren't you curious to see it now?" "Only when we return." (Grossman 1988, pp. 6–7)

As another strategy for creating and reinforcing collective memory, a West Bank cultural center has established a memorial honoring the lives of the first hundred Palestinians killed since September 2000, when the second *intifada* (uprising) began. Each dead person is represented by one or two items, donated by family members, bundled into a package and displayed below a photograph of the deceased. Items include slingshots, sunglasses, shoes, book bags, soccer balls, and prayer shawls. One life is represented by a

The painting, by Palestinian children, reflects their internalization of significant symbols and gestures. Note that the Arab men are waving a significant symbol, the Jordanian flag (many Palestinians live in Jordan, and the expected Palestinian state was once part of Jordan). In addition, the Palestinian men are holding hands, an important gesture for expressing affection.

birdcage—that of a youth who had released his bird before heading off to confront Israeli troops (Sontag 2001).

Development of the Social Self

Significant Symbols

■ **CORE CONCEPT 5: Meaningful social interaction depends on the involved parties sharing significant symbols.** Meaningful interaction depends on the involved parties sharing significant symbols and gestures. A **significant symbol** is a word, gesture, or other learned sign that conveys "the same meaning for the person transmitting it as for the person receiving it" (Theodorson and Theodorson 1979, p. 430). Language is a particularly important significant symbol, because the shared meanings attached to words allow us to communicate with others.

Significant symbols also include **symbolic gestures** or signs—that is, nonverbal cues, such as tone of voice, inflection, facial expression, posture, and other body

The Miami Beach Holocaust Memorial, a copper green hand reaching to the heavens with human figures climbing up the forearm, calls out to those who visit it to remember the mass slaughter of Jews and other groups by the Nazis during World War II.

significant symbol A word, gesture, or other learned sign used to convey a meaning from one person to another.

symbolic gestures Nonverbal cues, such as tone of voice and body movements, that convey meaning from one person to another.

Palestinian children flash the V-sign, an important symbolic gesture representing victory over Israel. To the uninformed observer this symbolic gesture might represent peace.

movements or positions that convey meaning from one person to another.

As people learn significant symbols, they also acquire the ability to (1) step outside the self to observe and evaluate it from another's point of view and (2) adjust and direct their behavior in ways that meet others' expectations. People come to know how they are being evaluated by reading the signals others send through significant symbols and symbolic gestures. Likewise, people use significant symbols and gestures to respond to others' evaluations and expectations. Sociologist George Herbert Mead believed, however, that humans do not mechanically strive to meet others' expectations. Instead, two aspects of the self—the *me* and the *I*—engage in a continuous dialogue.

I and Me

The *me* is Mead's term for the social self—that part of the self that has learned and internalized society's expectations about what constitutes appropriate behavior and appearances. In other words, the *me* is the part of the self that knows the norms that govern behavior in specific situations. Before an individual acts or speaks, the *me* anticipates how others will respond. The individual then proceeds to act or speak accordingly.

role taking The process of stepping outside the self and imagining how others view its appearance and behavior from an outsider's perspective.

The *I* is the spontaneous, autonomous, creative self, capable of rejecting expectations and acting in unconventional, inappropriate, or unexpected ways. The *I* takes chances and violates expectations. Sometimes taking a chance pays off, and people react by defining the individual as unique, exceptional, or one of a kind. At other times, taking a chance backfires and the individual is punished or ostracized. Mead does not specify how the *I* emerges, but we know that a spontaneous, creative self must exist; otherwise, human behavior would be quite predictable. Mead is clear, however, about how the *me* develops—through a process known as role taking.

According to Mead (1934), a sense of self cannot arise outside of social interaction. Mead assumed that the self is a product of interaction experiences (see Working For Change: "Learning to Invest in the Community"). He maintained that children acquire a sense of self when they become objects to themselves. That occurs when children are able to imagine the effects of their words and actions on other people. According to Mead, a person can see himself or herself as an object after learning to role-take.

Role Taking

■ **CORE CONCEPT 6: People acquire a sense of self when they can role-take.** **Role taking** involves stepping outside the self and imagining how others view its appearance and behavior from an outsider's perspective. Researchers have devised an ingenious method for determining when a child is developmentally capable of role taking. A researcher puts a spot of rouge on the child's nose and then places the child in front of a mirror. If the child ignores the rouge, he or she presumably has not yet acquired a set of standards about how he or she ought to look; that is, the child cannot role-take or see himself or herself from another person's viewpoint. If the child shows concern over the rouge, however, then he or she presumably has formed some notion of self-appearance and therefore can role-take (Kagan 1989).

Mead hypothesized that children learn to take the role of others through three stages: (1) preparatory, (2) play, and (3) games. Each of these stages involves a progressively more sophisticated level of role taking.

The Preparatory Stage. In this stage, children have not yet developed the mental capabilities that allow them to role-take. They may mimic or imitate people in their environment but have almost no understanding of the behaviors that they are imitating. Children may imitate spontaneously (by mimicking a parent writing, cooking, reading, and so on), or they may repeat things that adults encourage them to say and then reward them for saying. In the process of imitating, children learn to function symbolically; that is, they learn that particular actions and words arouse predictable responses from others.

WORKING FOR CHANGE

Learning to Invest in the Community

SOCIOLOGISTS MAINTAIN THAT self-identity is fundamentally social in nature. This means that our sense of self (our sense of who we are) revolves around relationships with others. My thinking on self-identity has been influenced by my working relationship with the social entrepreneur Neal Mayerson. On behalf of the Manuel D. and Rhoda Mayerson Foundation (one of 18,367 family foundations in the United States), Mayerson funded a program at Northern Kentucky University known as the Mayerson Student Philanthropy Project. Among other things, this project encourages students to expand their definitions of self to include relationships with nonprofit leaders and with targeted populations of their choosing (such as homeless people, grieving children, Latino preschoolers, frail elderly people, and elementary art students). Students enroll in courses that challenge them to collaborate with each other and with community leaders to do the following:

- Identify unmet community needs
- Interact with people in the nonprofit sector to learn about unmet needs through interviews, site visits, and in-class presentations by nonprofit organizations
- Solicit, assess, and select proposals from nonprofit organizations that can address those needs
- Award a total of $4,000 to one or more nonprofit organizations
- Visit the nonprofit organization to observe the outcomes of the class investment

How did students come to identify a community need, learn about nonprofit organizations that could address that need, and select projects worth funding? They made a commitment to pay attention to things going on around them in their communities and to share their observations with classmates. For example, one student noticed

that bicyclists have no safe place to ride on high-traffic roads. Another noticed that when he picked up his little sister from school, an unexpected number of children who appeared to be Latino were greeted by parents who seemed unfamiliar with the English language.

Once students settled on a need, they searched the Internet and local newspaper archives to learn about nonprofit organizations with missions tied to that need. More importantly, they used their eyes to observe previously unnoticed signs and buildings associated with nonprofit agencies in their community. Eventually, students generated a list of relevant nonprofit groups and called their directors to determine interest and solicit proposals. Students reviewed the proposals, visited the nonprofit organizations, shared observations, argued, debated, and eventually reached an agreement about which proposals to fund. Funded projects included the following:

- A $1,700 investment to Cancer Family Care supported the Strength for the Journey Retreat—an overnight respite offering 16 cancer patients tools to improve their coping skills, gain inner strength, and share with others who understand what it means to have cancer. One participant described the retreat's effect on her this way: "I had the chance to acknowledge that I have cancer in the presence of others who know what I am feeling without passing judgment on me."
- A $1,000 investment to Licking River Watershed Watch supported the collection, interpretation, and presentation of data as they related to the river's water quality. According to the director of the project, this data system was needed to ensure that the Licking River met the vision of "swimable and fishable waters" as defined by the 1972 Clean Water Act.

For example, Israeli children may be taught early on that undivided Jerusalem is the capital of Israel. Similarly, Palestinian children may be taught that their "real" home is in a former Palestinian city, even before the children learn notions of geography and can understand the historical circumstances of their living arrangements.

Both Palestinian and Israeli children, like children in nearly every culture, learn to sing patriotic songs and say

prayers before they can understand the words. Observers to the West Bank and Gaza are struck by the fact that as soon as some two-year-old Palestinian children "saw the cameras come out, they were up and alert, hands outstretched as taut fingers made in unison the victory sign for our photos. No [Palestinian] child we met anywhere wanted to be photographed without that sign" (Bourne 1990, p. 70).

The Play Stage. Mead saw children's play as the mechanism by which they practice role taking. **Play** is a voluntary and often spontaneous activity with few or no formal rules that is not subject to constraints of time (for example, 20-minute halves or 15-minute quarters) or place (for example, a gymnasium or a regulation-size field).

Children in particular play whenever and wherever the urge strikes. If rules exist, they are developed by the children on their own, not imposed on participants by higher authorities (such as rulebooks and officials). Participants undertake play for amusement, entertainment, or relaxation. These characteristics make play less socially complicated than organized games, such as Little League baseball (Corsaro 1985; Figler and Whitaker 1991).

In the play stage, children pretend to be **significant others**—people or characters (such as cartoon characters or the family pet) who are important in a child's life, in that they greatly influence the child's self-evaluation or encourage the child to behave in a particular manner. Children recognize behavior patterns characteristic of these significant others and incorporate them into their play. When a little girl plays with a doll and pretends to be the doll's mother, she talks and acts toward the doll the same way her mother talks and acts toward her. By pretending to be the mother, she gains a sense of the mother's expectations and perspective and learns to see herself as an object. Similarly, two children playing doctor and patient are learning to see the world from viewpoints other than their own and to understand how a patient acts in relation to a doctor, and vice versa.

In the play stage, children's role taking comes from what they see and hear going on around them. Most Palestinian children in the West Bank and Gaza have never seen an adult male Israeli without a gun. Palestinian children's play reflects their experiences: The children pretend to be Israeli soldiers arresting and beating other Palestinian children, who are pretending to be stone throwers. They use sticks and cola cans as if they were guns and tear-gas canisters (Usher 1991). One evening ABC News featured a segment on Palestinian children who engaged in this type

Christopher Brown

The Palestinian child with the head scarf lives in the Ramallah refugee camp. He is playing at being a Palestinian militant.

of play. When asked by the reporter which they preferred to be, soldiers or stone throwers, the children replied, "Soldiers, because they have more power and can kill." For their part, Israeli children have had little experience with Palestinians except as manual laborers or "terrorists." Thus, it is hardly surprising that some Israeli kindergartners pretend that Israelis are Smurfs (the good guys in a TV program) and Palestinians portray Gargamel (the enemy of the Smurfs in the program). Israeli children pretend to be soldiers because both men and women must serve in the Israeli military, beginning at age 18.

The Game Stage. In Mead's theory, the play stage is followed by the game stage. **Games** are structured, organized activities that usually involve more than one person. They are characterized by a number of constraints, such as established roles and rules and an outcome toward which all activity is directed. Through games, children learn to (1) follow established rules, (2) take simultaneously the roles of all participants, and (3) see how their position fits in relation to all other positions.

When children first take part in games such as organized sports, their efforts seem chaotic. Instead of making an organized response to a baseball hit to the infield, for example, everyone tries to retrieve the ball, leaving no one at the base to catch the throw needed to put the runner out. This chaos exists because each child has not developed to the point at which he or she can see how the individual's role fits with the roles of everyone else in the game. Without such knowledge, a game cannot have order.

Through playing games, children learn to organize their behavior around the **generalized other**—a system of expected behaviors, meanings, and viewpoints that transcend those of the people participating. We can say that the expected behaviors, meanings, and viewpoints transcend

play A voluntary and often spontaneous activity with few or no formal rules that is not subject to constraints of time or place.

significant others People or characters who are important in an individual's life, in that they greatly influence that person's self-evaluation or motivate him or her to behave in a particular manner.

games Structured, organized activities that usually involve more than one person and a number of constraints, such as established roles, rules, time, place, and outcome.

generalized other A system of expected behaviors, meanings, and viewpoints that transcend those of the people participating.

those participating because the participants had no hand in creating them. When children play games such as baseball, they practice fitting their behavior into an already established behavior system that governs the game.

Not surprisingly, games are the tools used in programs designed to break down barriers between Palestinian and Israeli Jewish children and adolescents. The games involve activities like these:

> Throwing an orange into the air, calling a person's name to catch it, throwing it again with another's name, and again and again as the whoops of laughter fill the room. Then they all crowd together, take each other's hands, and turn around until they are enmeshed in a tangle of arms. Intertwined with each other, they try to unravel themselves without letting go. They talk to each other, giving advice, crouching so another can step over an arm, stooping so others can swing arms over heads, spinning around, trying to turn the snarled mess of Arab and Jewish bodies into a clean circle. (Shipler 1986, p. 537)

Although these games seem merely fun, sociologists contend that participants are learning to see things from another perspective and to play their parts successfully in a shared activity. The participants cannot be effective unless they understand their own roles in relation to everyone else's. Although the children trying to untangle themselves may not be fully aware of it, they are learning that a Palestinian (or an Israeli) can be in a position like their own. To untangle the knot, participants must be able to understand everyone else's situation.

Mead assumed that the self develops through interaction with others, and he identified the interaction that occurs in imitation, play and games as important to children's self-development. When children participate in play and games, they practice seeing the world from the view-

point of others and gain a sense of how others expect them to behave. Sociologist Charles Horton Cooley offered a more general theory about how the self develops.

The Looking-Glass Self

■ **CORE CONCEPT 7: The ability to imagine others' reactions to the self is critical to self-awareness.** Like Mead, Cooley assumed that the self is a product of interaction experiences. Cooley coined the term **looking-glass self** to describe the way in which a sense of self develops: People act as mirrors for one another. We see ourselves reflected in others' real or imagined reactions to our appearance and behaviors. We acquire a sense of self by being sensitive to the appraisals of ourselves that we perceive others to have: "Each to each a looking glass, / Reflects the other that [does] pass" (Cooley 1961, p. 824).

As we interact, we visualize how we appear to others, we imagine a judgment of that appearance, and we develop a feeling somewhere between pride and shame: "The thing that moves us to pride or shame is not the mere mechanical reflection of ourselves but . . . the imagined effect of this reflection upon another's mind" (Cooley 1961, p. 824). Cooley went so far as to argue that "the solid facts of social life are the facts of the imagination." Because Cooley defined the imagining or interpreting of others' reactions as critical to self-awareness, he believed that people are affected deeply even when the image they see reflected is exaggerated or distorted. The individual responds to the perceived reaction rather than to the actual reaction.

Nevertheless, we cannot overlook the fact that our imaginations of how other people will react and behave usually rest on experiences with others. In the case of Palestinians and Israelis, each group aims a number of powerful images at the other. For example, many Palestinians call Israeli Jews "nazis" and equate the Israeli military presence in the West Bank and Gaza—and the accompanying system of identification cards, checks, and imprisonments—with the concentration camps of World War II. These labels are quite painful to Jews, who see little similarity between the Holocaust and the Palestinian situation. One Israeli soldier assigned to a Palestinian refugee camp imagined how he appeared to Palestinian youth: "The Palestinian boys would look at you with hatred, such hatred that it reminded me of how the Jews in concentration camps in the Holocaust looked at [the guards]. You are the most evil thing on earth right now, like it was during the Holocaust. You are the persecutor" (Nelsen 2006, p. 56).

During the game stage, children practice fitting their behavior into an already established system of rules and take on a position with already defined relationships to other positions.

© Bill Romerhaus/Index Open

looking-glass self A process in which a sense of self develops, enabling one to see oneself reflected in others' real or imagined reactions to one's appearance and behaviors.

Many Israeli Jews retaliate by calling Palestinians "Arabs"—in effect telling Palestinians that the "Palestinian" people do not exist. By using the label "Arabs," Jews are declaring that "Arabs" do not need a state; they can simply live in one of the surrounding Arab countries. Many Jews also treat Palestinians as culturally primitive and incapable of managing their own affairs. They remind the Palestinians that Jews are responsible for turning the "worthless" desert land occupied previously by a "backward" Palestinian people into a modern, high-technology state.

Both Mead's and Cooley's theories suggest that self-awareness and self-identity derive from social relationships and from an ability to step outside the self and imagine it from another's perspective. Although Cooley and Mead described the mechanisms (imitation, play, games, looking-glass self) by which people learn to role-take, neither theorist addressed how a person acquires this level of cognitive sophistication. To answer this question, we turn to the work of Swiss psychologist Jean Piaget, whose work reaches across many disciplines, including sociology, biology, and education.

Cognitive Development

■ **CORE CONCEPT 8:** The ability to role-take emerges according to a gradually unfolding genetic timetable that must be accompanied by direct experiences with persons and objects. Piaget was the author of many influential and provocative books about how children think, reason, and learn. His ideas about how children develop increasingly sophisticated levels of reasoning stemmed from his study of water snails, which spend their early life in stagnant waters. When transferred to tidal water, these snails engage in motor activity that develops the size and shape of the shell to help them remain on the rocks and avoid being swept away (Satterly 1987). Building on this observation, Piaget arrived at the concept of active adaptation, a biologically based tendency to adjust to and resolve environmental challenges.

Piaget believed that learning and reasoning are important adaptive tools that help people meet and resolve environmental challenges. Logical thought emerges according to a gradually unfolding genetic timetable. This unfolding must be accompanied by direct experiences with persons and objects; otherwise, a child will not realize his or her potential ability.

Piaget's model of cognitive development includes four broad stages: sensorimotor, preoperational, concrete operational, and formal operational. Children cannot proceed from one stage to another until they master the reasoning challenges of earlier stages. Piaget maintained that reasoning abilities cannot be hurried; a more sophisticated level of understanding will not show itself until the brain is ready. Children construct and reconstruct their conceptions of the world as they experience the realities of living.

In the *sensorimotor stage* (from birth to about age 2), children explore the world with their senses (taste, touch, sight, hearing, and smell). The cognitive accomplishments of this stage include an understanding of the self as separate from other persons and the realization that objects and persons exist even when they are out of sight. Before this notion takes hold (at about 7 months of age), children act as if an object does not exist when they can no longer see it.

Children in the *preoperational stage* (ages 2 to 7) think *anthropomorphically*; that is, they assign human feelings to inanimate objects. They believe that objects such as the sun, the moon, nails, marbles, trees, and clouds have motives, feelings, and intentions (for example, dark clouds are angry; a nail that sinks to the bottom of a glass filled with water is tired). Children in the preoperational stage cannot appreciate that matter can change form but still

© Chris Lowe/Index Open

In the sensorimotor stage of development, children explore the world through tasting and touching. This baby is too young to recognize himself, which means he has not developed the ability to step outside the self and see it from another's viewpoint.

remain the same in quantity (for example, they believe a 12-ounce cup that is tall and narrow holds more than a 12-ounce cup that is short and wide; height is the variable by which they judge the amount, failing to consider diameter). In addition, children in this second stage cannot conceive how the world looks from another's point of view. Thus, if a child facing a roomful of people (all of whom are looking in his or her direction) is asked to draw a picture of how someone in the back of the room sees the audience, the child will draw the picture as he or she sees the people (showing their faces rather than the backs of their heads). Finally, children in the preoperational stage tend to center their attention on one detail and fail to process information that challenges that detail.

By the time children reach the *concrete operational stage* (from about ages 7 to 12) they have mastered the conceptual challenges of stage 2 but they have difficulty in thinking hypothetically or abstractly without reference to a concrete event or image. For example, children in this stage have difficulty envisioning life without them in it. One 12-year-old struggling to grasp this idea said to me, "I am the beginning and the end; the world begins with and ends with me."

In the *formal operational stage* (from the onset of adolescence onward), people can think abstractly. For example, people can conceptualize their existence as a part of a much larger historical continuum and a larger context. They no longer see the world in black-and-white terms; rather, they see it in shades of gray.

Apparently, this progression by stages toward increasingly sophisticated levels of reasoning is universal, although the content of people's thinking varies across cultures. For example, all children in the preoperational stage (stage 2) focus their attention on a single detail to the exclusion of information that might challenge it. But the details they learn to focus on vary by culture. Israeli children are taught to be suspicious not only of unattended packages and bags in airports but also of empty cups, bottles, and cans anywhere. In *Arab and Jew* David Shipler (1983) describes his frustrations in explaining to his young children that they did not have to follow this rule now that they were in the United States. When he set some bags of newspapers on the curb to be picked up, his children reported suspicious packages outside; and when he asked his seven-year-old son to pick up a plastic cup in the street and throw it away, "He adamantly refused to go near it, and he remained solidly unmoved by my extravagant assurances that we didn't have to worry about bombs on a quiet, tree-lined suburban street in America" (Shipler 1986, p. 83).

Agents of Socialization

■ **CORE CONCEPT 9:** Agents of socialization—significant others, primary groups, ingroups and outgroups, and institutions (such as mass media)—shape our sense of self and teach us about the groups to which we do and do not belong. Sociologists

Israeli soldiers carry guns and have access to other equipment such as armored vehicles. These Palestinian boys carry slingshots for launching stones. Guns and slingshots are symbols of the power Israelis possess relative to Palestinians.

Department of Defense Photo by Cherie A. Thurlby

© Justin C. McIntosh

have identified a number of **agents of socialization**: significant others, primary groups, ingroups and outgroups, and institutions that (1) shape our sense of self or social identity, (2) teach us about the groups to which we do and do not belong, (3) help us realize our human capacities, and (4) help us negotiate the social and physical environment we have inherited. It is impossible to list all the agents of socialization. Specific agents include the family, peers, military units, school, religion, and mass media. In the next section, we pay special attention to groups and mass media (a social institution). Note that earlier in this chapter, we considered significant others (see play stage).

Groups as Agents of Socialization

In the most general sense, a **group** is two or more persons who do the following:

1. Share a distinct identity (the ability to speak a specific language, biological descent from a specific couple, membership in a team or a military unit)
2. Feel a sense of belonging
3. Interact directly or indirectly with one another

Groups vary according to a host of characteristics, including size, cohesion, purpose, and duration. Sociologists identify primary groups, ingroups, and outgroups as particularly important agents of socialization.

Primary Groups

A **primary group**—such as a family, military unit, or peer group—is a social group that has face-to-face contact and strong emotional ties among its members. Primary groups are not always united by harmony and love; for example, they can be united by hatred for another group. In either case, the ties are emotional. The members of a primary group strive to achieve "some desired place in the thoughts of [the] others" and feel allegiance to the other members (Cooley 1909, p. 24). Although a person may never achieve the desired place, he or she may remain preoccupied with that goal. In this sense, primary groups are "fundamental

Michelle Frankfurther for the U.S. Census Bureau

Grandparents are important agents of socialization. In the United States an estimated 5.7 million children live with grandparents.

in forming the social nature and ideals of the individual" (Cooley 1909, p. 23).

The family is an important agent of socialization because it gives individuals their deepest and earliest experiences with relationships and their first exposure to the "rules of life." In addition, the family teaches its members about the social and physical environment and ways to respond to that environment. When that environment is stressful, the family can buffer its members against the effects of that stress; alternatively, it can exacerbate the effects. Sociologists Amith Ben-David and Yoav Lavee (1992) offer a specific example of this buffering function. The two sociologists interviewed 64 Israelis to learn how their families responded to the 1990 SCUD missile attacks Iraq launched on Israel during the Persian Gulf War. During these attacks, families gathered in sealed rooms and put on gas masks. The researchers found that families varied in their responses to this life-threatening situation.

Some respondents reported the interaction during that time as positive and supportive: "We laughed and we took pictures of each other with the gas masks on" or "We talked about different things, about the war, we told jokes, we heard the announcements on the radio" (p. 39).

Other respondents reported that little interaction among family members occurred but a feeling of togetherness prevailed: "I was quiet, immersed in my thoughts. We were all around the radio. . . . Nobody talked much. We all sat there and we were trying to listen to what was happening outside" (p. 40). Some respondents reported that interaction was tense: "We fought with the kids about putting on their masks, and also between us about whether the kids should put on their masks. There was much shouting and noise" (p. 39).

agents of socialization Significant others, primary groups, ingroups and outgroups, and institutions that (1) shape our sense of self or social identity, (2) teach us about the groups to which we do and do not belong, (3) help us to realize our human capacities, and (4) help us negotiate the social and physical environment we have inherited.

group Two or more people who share a distinct identity, feel a sense of belonging, and interact directly or indirectly with one another.

primary group A social group that has face-to-face contact and strong emotional ties among its members.

As this case illustrates, even under extremely stressful circumstances, such as war, the family can teach responses that increase or decrease that stress. Clearly, children in families that emphasize constructive responses to stressful events have an advantage over children whose parents respond in destructive ways.

Like the family, a military unit is a primary group. A unit's success in battle depends on the existence of strong ties among its members. Soldiers in this primary group become so close that they fight for one another, rather than for victory *per se*, in the heat of battle (Dyer 1985). In Israel the military represents a place where immigrants from almost 100 national and cultural backgrounds become Israelis, bonding with one another and with native Israelis (Rowley 1998). Military units train their recruits always to think of the group before themselves. In fact, the paramount goal of military training is to make individuals feel inseparable from their unit. Some common strategies employed to achieve this goal include ordering recruits to wear uniforms, to shave their heads, to march in unison, to sleep and eat together, to live in isolation from the larger society, and to perform tasks that require the successful participation of all unit members. If one member fails, the entire unit fails. Another key strategy is to focus the unit's attention on fighting together against a common enemy. An external enemy gives a group a singular direction, thereby increasing its internal cohesiveness.

The Israeli military is an important agent of socialization. Almost every Israeli can claim membership in this type of primary group because virtually every Israeli citizen—male and female—serves in the military (three years for men and two years for women). Men must serve on active duty for at least one month every year until they are 51 years old (Rodgers 1998).

Similarly, military training is an important experience for many Palestinians, although it takes place less formally. Palestinian youth—especially those living in Syrian, Lebanese, Egyptian, and Jordanian refugee camps (which are outside Israeli control)—join youth clubs and train to protect the camps from attack. The focus on defeating a common enemy helps establish and maintain the boundaries of the military unit. All types of primary groups have boundaries—a sense of who is in the group and who is outside it. The concepts of ingroup and outgroup help us understand these dynamics.

Ingroups and Outgroups

Sociologists use the term **ingroup** to describe a group with which people identify and to which they feel closely attached, particularly based on hatred or opposition toward another group, known as an outgroup. Ingroups cannot exist without an outgroup. An **outgroup** is a group of individuals toward which members of an ingroup feel a sense of separateness, opposition, or even hatred. Obviously, one person's ingroup is another person's outgroup.

The very existence of an outgroup heightens loyalty among ingroup members and magnifies the characteristics that distinguish the ingroup from the outgroup. The presence of an outgroup can unify an ingroup even when the ingroup members are extremely different from one another. For example, one could argue that the presence of Palestinians functions to unify Israeli Jews, who are culturally, linguistically, religiously, and politically diverse. Israel's diversity reflects the fact that, since 1948, Jews from 102 different countries, speaking 80 different languages, have settled there (Peres 1998). In fact, only about 20 percent of Israel's total population consists of Jews native to that land. About 32 percent consists of Jews born in Europe or America, 13 percent born in Asia, and 15 percent born in Africa; the remaining 20 percent of Israel's population consists of Palestinians (U.S. Central Intelligence Agency 2005a). To ease the inevitable communication problems caused by diversity, Israeli law requires that everyone learn Hebrew. In addition to their common language, unifying threads for Jews in Israel include the desire for a homeland free of persecution, and the conflict with the following

A military unit is a primary group. Because of military training, soldiers become so close that they fight for one another, rather than for victory *per se*, in the heat of battle.

DoD photo by Staff Sgt. Michael R. Holzworth, U.S. Air Force

ingroup A group with which people identify and to which they feel closely attached, particularly when that attachment is founded on hatred or opposition toward an outgroup.

outgroup A group toward which members of an ingroup feel a sense of separateness, opposition, or even hatred.

outgroups: Palestinians in the West Bank and Gaza, Palestinians in Israel, and Arabs in surrounding states. "Israeli Arab" is the label the Israeli government applies to the Palestinians who did not leave in 1948 or 1967 and to their descendants. These Palestinians number about 1.3 million people—about 20 percent of Israel's population. They live in 116 Arab-only communities and seven so-called "mixed" cities, in which Arabs live in separate communities adjacent to Jewish communities (Nathan 2006). The Palestinians who live in Israel prefer the label "Israeli Palestinians." Similarly, the presence of Jews acts to unite an equally diverse Palestinian society, which includes West Bank Palestinians, Gaza Palestinians, and Israeli Palestinians, who also come from different ethnic and religious groups, clans, and political orientations.

Loyalty to an ingroup is accompanied by an us-versus-them consciousness. This consciousness can be traced to the fact that the two groups live segregated lives. Palestinians and Israelis, for example, have little in common. For the most part, they do not share a language, religion, schools, residence, or military service. Before 1999 the two

groups did share an economic relationship, in that almost 20 percent of all Palestinian workers in the West Bank and 12 percent in Gaza commuted to jobs in Israel or Jewish settlements (Palestinian Central Bureau of Statistics 1998). Today only 12,000 Palestinians from the West Bank and Gaza are permitted to enter Israel each day to work; an estimated 20,000 enter illegally (Haaretz 2005). The Israeli government plans to ban Palestinian labor in Israel by 2008, and it has been gradually replacing such workers with workers from Thailand and Romania (Haaretz 2006).

Because little interaction occurs between ingroup and outgroup members, they know little about one another. This lack of firsthand experience deepens and reinforces misrepresentations, mistrust, and misunderstandings between members of the two groups. Thus, members of one group tend to view members of the other in the most stereotypical of terms. Yorum Bilu at Hebrew University of Jerusalem designed and conducted a particularly creative study to examine the consequences of ingroup–outgroup relations in the West Bank. Bilu and two of his students asked youths aged 11 to 13 from Palestinian refugee camps and Israeli settlements in the West Bank to keep journals of their dreams over a specified period. Seventeen percent of Israeli youths wrote that they dreamed about encounters with Arabs, whereas 30 percent of the Palestinian youths dreamed about meeting Jews. Among the 32 dreams of meetings between Jews and Arabs not one character was identified by name. Not a single figure was defined by a personal, individual appearance. All the descriptions were completely stereotyped—the characters defined only by their ethnic identification (such as "Jew," "Arab," or "Zionist") or by value-laden terms with negative connotations (such as "the terrorists" or "the oppressors"). The majority of the interactions in the dreams indicate a hard and threatening reality, a fragile world with no defense:

> An Arab child dreams: "The Zionist Army surrounds our house and breaks in. My big brother is taken to prison and is tortured there. The soldiers continue to search the house. They throw everything around, but do not find the person they want [the dreamer himself].
>
> "They leave the house, but return, helped by a treacherous neighbor. This time they find me, and my relatives, after we have all hidden in the closet in fright."
>
> A Jewish child dreams: ". . . suddenly someone grabs me, and I see that it is happening in my house, but my family went away, and Arab children are walking through our rooms, and their father holds me, he has a kaffiyeh and his face is cruel, and I am not surprised that it is happening, that these Arabs now live in my house." (Grossman 1988, pp. 30, 32–33)

There are at least 12 youth camps in the United States and Canada designed to bring Palestinians and Israeli Jews together with the hope of bridging the ingroup-outgroup divide with specially designed activities such as the following:

In 1945 these survivors of the Buchenwald concentration camp, like many other Jewish survivors of the Holocaust, resettled in Palestine. International outrage over Hitler's effort to exterminate the Jewish people was one of the factors that led the United Nations in 1947 to partition Palestine into two states, Jewish and Palestinian. The state of Israel was established in 1948. The Palestinians rejected a two state solution to a European-caused situation.

National Archives and Records Administration

The Israeli and Palestinian youth probably have never physically touched each other. They may arrive with images of the other as inhuman, even aliens or monsters. In one program, they are asked to find someone they don't know well in the group, someone they consider "other." It may be someone they were afraid of, or still are afraid of.

They are told to look at one another. And then, they are told to ask permission, and then find the pulse on one another's wrist. Then they do the same with the arteries in their necks. Then they place their hands on the others' breastbone to feel their hearts, while they continue to look at each other. "This can be very powerful, because they've never touched their enemy, and they realize that 'she feels like me,'" says the program director. (Fetzer Institute 2005)

Often an ingroup and an outgroup clash over symbols—objects or gestures that are clearly associated with and valued by one group. These objects can be defined by members of the outgroup as so threatening that they seek to eliminate the objects: destroying the symbol becomes a way of destroying the group. The wall the Israeli government built to divide the West Bank from Israel is a significant symbol for both Palestinians and Israeli Jews. Jews see it as a protection against suicide attacks. Palestinians see it as a severe restriction of their movement to and from and within the West Bank, impeding their ability to cross into Israel to work and shifting the boundaries between Israel and a future Palestinian state.

It is important to know that ingroups and outgroups also exist within Palestinian and Israeli society. Palestinians are divided according to which political party they support: the relatively secular Fatah or Islamic Hamas; in addition, about 10 percent of Palestinians are Christian. Among Israelis clear divisions exist between Sephardic Jews, with North African and Middle Eastern roots, and Ashkenazi Jews, with European roots. An estimated 50 percent of Israelis do not actively practice a religion, 30 percent consider themselves practicing Jews, and 18 per-

© Justin C. McIntosh

Another graffiti artist uses the wall to pay tribute to Ghassan Kanafani, the Palestinian writer on themes of exile, uprootedness, and national struggle.

cent are ultra-orthodox. The ultra-orthodox are exempt from military service, a policy that creates resentment among Jews who must serve in the military and then the reserves until they are 51 years old (BBC News 1998b).

Our discussion of ingroups and outgroups shows that one's self-identity revolves around group membership. In the section that follows we examine how the self-identity of Palestinian suicide bombers/martyrs is tied to the ingroup to which they belong, and by extension, to the outgroup to which they do not belong.

Suicide: The Severing of Relationships. We learned in Chapter 1 that Durkheim offered a definition of suicide that goes beyond its popular meaning (the act of intentionally killing oneself)—taking the spotlight off the victim and pointing it outward, toward the ties that bind (or fail to bind) people to others in the society. That is, he defined suicide as "the severing of relationships." Recall that Durkheim identified four types of problematic social relationships: (1) egoistic (weak ties to the group), (2) altruistic (strong ties to the group), (3) anomic (dramatic change in which the individual is cast out of the group into a higher or lower status), and (4) fatalistic (a state in which individuals see their future as hopelessly blocked).

Nancy Hemminger

Palestinian graffiti artists draw a scene that speaks to what the wall dividing Israel and the West Bank is not: a comfortable, home-like environment with a great view.

According to the Israeli security service, known as Shin Bet, between June 2002 and June 2004, 145 suicide bombers/martyrs carried out, or tried but failed to carry out, an attack (Harel 2004a). This represents 0.5 percent of the 29,000 attacks (ranging from stone throwing to planting bombs) Palestinians carried out against Israelis. Here are some things we know about suicide bombers that suggest *altruistic motives:*

- Palestinians who put their lives on the line are typically motivated by "revenge for acts committed by Israelis." In other words, the suicide and the publicity it receives let the Israelis know that the Palestinians will avenge acts of injustice and the bomber's response will be celebrated in the Islamic world. One suicide bomber/martyr explained his motives with these words: "I want to avenge the blood of the Palestinians, especially the blood of the women, of the elderly, and of the children, and in particular the blood of the baby girl Iman Hejjo, whose death shook me to the core" (Margalit 2003). Another suicide bomber/martyr's mother noted that her daughter was filled with pain after five Israeli soldiers killed her brother, Fadi. "She was full of pain about that. Some nights she woke screaming, saying she had nightmares about Fadi" (Burns 2003).
- Approximately 75 percent of Palestinians support suicide attacks (Lahoud 2001). Suicide attackers/martyrs are treated as celebrities in the sense that they are instantly "propelled from being no one to being someone who is glorified, and lionized with poems and [they] live on in this historical chain of heroic martyrs, being remembered and saluted for longer than if [they] had not undertaken this kind of operation" (Natta 2003). Even small children know the names of suicide bombers (Margalit 2003).

Several other characteristics distinguish suicide bombers/martyrs as well. While Durkheim would classify the conditions that support the individual Palestinian's decision to commit a suicide attack as altruistic, he would find it noteworthy that two-thirds of all suicide attackers were between the ages of 17 and 23 (and most were younger than 30) and that 86 percent were single. These characteristics suggest that, while suicide bombers/martyrs may be acting in the name of the larger Palestinian society and in the name of someone wronged by Israelis, they are also relatively detached from others (such as a spouse and children), and because of that detachment, they will encounter less resistance to carrying through on the sui-

mass media Forms of communication designed to reach large audiences without face-to-face contact between those conveying and those receiving the messages.

Suicide bombers/martyrs pose before their death for pictures that will be shown on posters honoring or celebrating their death.

cide than those attached to others through marriage or parenthood.

Mass Media

Mass media are forms of communication designed to reach large audiences without face-to-face contact between those conveying and those receiving the messages. Examples of mass media include magazines, newspapers, commercials, books, television programs, Web sites, radio broadcasts, video games, and music CDs. The tools of mass media—such as television, radio, the Internet, and MP3 players—introduce their audiences to a variety of people, including sports figures, news makers, cartoon characters, politicians, government leaders, actors, disc jockeys, and musicians. Some audience segments become so emotionally attached to these figures that they count themselves as fans following celebrities' lives, desiring the products celebrities endorse, and otherwise seeking to emulate them or live vicariously through them.

For the first time in human history, average people have, in the Internet, a tool for potentially reaching mass audiences (Internet World Stats 2006; see No Borders, No Boundaries: "Reaching Mass Audiences through the Internet"). At the time of this writing, two Web sites that allow users to present the self to millions of people—YouTube.com and MySpace.com—ranked fourth and sixth on the Global 500 (Alexa 2007b). The two Web sites were among the ten most popular sites among Internet users in Israel and the United States. YouTube ranked fourth among users in the Palestinian Territories (Alexa 2007a). YouTube contains hundreds of videos chronicling various viewpoints on the Palestinian-Jewish conflict. Titles include "Torture of Palestinians" (58 seconds), "Israel and the Palestinians" (4 minutes, 21 seconds), "Palestinian Propaganda about the Israeli Security Fence" (1 minute, 51 seconds), and "Israel's Oppression of the Palestinians" (5 minutes, 23 seconds).

NO BORDERS, NO BOUNDARIES

Reaching Mass Audiences through the Internet

THE MAP SHOWS the Internet penetration rate (or the percentage of the total population in each area who have Internet access) for seven regions of the world. Almost 75 percent of the North American population has access to the Internet. Only 5.6 percent of people across the continent of Africa have access. The map also highlights the 47 countries with the highest Internet penetration rates. Anyone with an Internet connection and the ability to create a Web site or blog has the potential to connect with the 1.6 billion people who have Internet access (about 24 percent of the world's population).

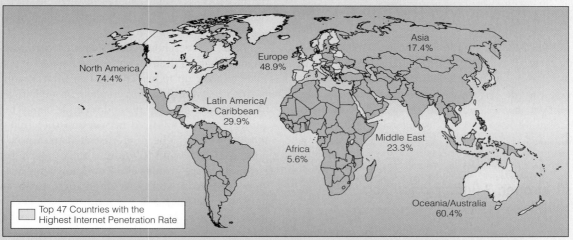

Asia
17.4%

Europe
48.9%

North America
74.4%

Latin America/
Caribbean
29.9%

Africa
5.6%

Middle East
23.3%

Oceania/Australia
60.4%

☐ Top 47 Countries with the
Highest Internet Penetration Rate

Source: Miniwatts Marketing Group (2009) "Internet Usage Statistics" http://www.internetworldstats.com/

Arguably, the most influential medium of mass communication in the United States is still television; 98 percent of American households have at least one television, and 82 percent have a cable connection or satellite dish antenna (U.S. Census Bureau 2006). Each day, the average American watches about 4 hours of TV. That translates into 28 hours per week, or two months of nonstop TV watching per year (A. C. Nielsen Co., 2006). It would be impossible to make definitive statements about the mass media's effect on people's sense of self and on their relationships with others. Studies have been done on virtually every type of mass media, showing both positive and negative consequences. For example, studies show that television can function as an important educational tool but it can also impede development of critical academic skills, such as reading.

One example of the power of television to reach mass audiences is *Sesame Street*. The most watched children's television show in history, it airs in at least 120 countries. It first aired in the United States in 1969, with a groundbreaking multiracial cast. *Sesame Street*'s mission is to level the educational playing field for all preschool children. Joan Gantz Cooney, the originator of *Sesame Street*, maintains that "our producers are like old-fashioned missionaries. . . . It's not religion they're spreading, but it's learning and tolerance and love and mutual respect" (Independent Lens 2006).

In Israel and the Palestinian Territories, *Sesame Street* airs as *Sesame Stories*, because the idea that a neutral street exists where Palestinians and Jews might gather together is not a believable scenario in that part of the world. *Sesame Stories* concentrates on (1) teaching young viewers respect and understanding for their own and other cultures, (2) promoting a peaceful resolution to the hundred-year conflict between Palestinians and Jews, and (3) presenting positive images of Palestinian and Jewish children. In one episode, Israeli and Arab friends stage a peaceful protest to stop one of the Muppets from banging his drum too loudly. In another episode, a Palestinian girl living in a refugee camp finds a water bottle on the street and takes it upon herself to plant something in it. She gets no encouragement from those around her, who claim that nothing can grow in a refugee camp. The producers place great emphasis on the girl's finding a *clear* container rather than, say, a tin can, as Jewish children are taught not to pick up items if they cannot see what is in them (Salamon 2002). These producers believe that shows emphasizing themes of hope, integration, and friendship will have lasting effects on Palestinian and Jewish viewers (Deutsche Welle 2003). A pre-test and post-test study of the impact of *Sesame Stories* programming suggests an increase in Jewish and Palestinian viewers' use of prosocial strategies to resolve conflicts and positive attributes to describe each other (Cole et al. 2003).

Socialization across the Life Cycle

■ **CORE CONCEPT 10:** Socialization takes place throughout the human life cycle. It is important to realize that socialization takes place throughout the human life cycle. Sociologists divide the life cycle into eight stages, emphasizing the interaction themes and social challenges peculiar to each. Because we have devoted considerable attention to early childhood socialization, we now focus on stages 4 through 8.

Stages 1 through 3 (Infancy, Toddler, Preschool)

During stage 1 (Infancy) it is important that caretakers give consistent, predictable care. Inadequate, unpredict-

able care leaves infants uncertain of their ability to elicit care and makes them feel that the world is not reliable (Erikson 1950). In stage 2 (Toddler) the child's nervous and muscular systems mature, and abilities in one area are frustrated by inabilities in another. Toddlers possess enough motor skills to move around and explore but are not yet aware of the consequences of their actions. Growing abilities clash with inabilities, and this stage becomes a battle for autonomy. Caregivers must protect children from danger, be tolerant yet firm, and support children in their efforts to be independent. If caregivers are critical, overprotective, or discouraging of children's attempts to master the environment, the children may feel shame and doubt, especially if parents resent their children's inabilities for interfering with the parents' independence (Erikson 1950).

Stage 3 (Preschool) corresponds with Mead's play stage, in that children play at being the kind of person they hope to grow up to be. They "hitch their wagons" to the persons they admire most in their immediate lives; they identify with these figures and imagine being them. If the gap between an admired person and the child's skills is too large, children become frustrated and think, "I am no good compared with X." Adults must show approval and encouragement and let children know they are equal in worth, if not yet ability. Without such reassurance, "the danger is that they will feel guilt over the goals contemplated and the acts initiated" (Erikson 1950, p. 84).

Stage 4 (Ages 6–12)

Systematic instruction is central to this stage. Recognition is won by doing things. If their teachers are secure and respected, children develop a positive identification with those who know how to do things. If assignments are meaningful, children learn the pleasure that comes with

Sesame Street, airing in at least 120 countries, is the most watched children's television show in history—an example of the power of television to reach and socialize mass audiences.

During stage 1 it is essential that children receive predictable and consistent care from caregivers, as such care sends the message that the world is reliable.

INTERSECTION OF BIOGRAPHY AND SOCIETY

Israeli Jews as Aggressors and Victims

IN *THE OTHER SIDE OF ISRAEL* Susan Nathan interviews an Israeli soldier who imagines how he would feel if Palestinians were in power and conducting home searches the way his Israeli unit had conducted them:

> I try to imagine the reverse situation: if they had entered my home, not a police force with a warrant but a unit of solders, if they had burst into my home and shoved my mother and little sister into my bedroom and forced my father and my younger brother and me into the living room, pointing their guns at us, laughing, smiling, and we didn't always understand what the soldiers were saying while they emptied drawers and searched through things. Oops, it fell, broken—all kinds of photos of my grandmother and grandfather, all kinds of sentimental things that you wouldn't want anyone else to see. There is no justification for this. If there is a suspicion that a terrorist has entered a house, so be it. But just to enter a home, any home: here I've chosen one, look, what fun. We go in, we check it out, we cause a bit of injustice, we have asserted our military presence and then we move on. (Nathan 2006, pp. 213–214)

In *Occupied Minds* Arthur Nelsen profiles Roni Hirschenson, a 62-year-old Jew whose sons Amir and Elad were Israeli soldiers. Hirschenson is a member of the group Families Forum, an organization of bereaved Palestinians and Israeli Jews working toward reconciliation. An excerpt from the profile follows:

> Amir volunteered for the paratroops, three months after he joined the army. He was a good, generous person, very close to his brother. He also believed in two states for two peoples but he was proud of the army. He wanted to be an architect. I was at my workshop when I heard about the attack that killed him. . . . It was on the news that a suicide bomber blew himself up among a group of soldiers at Beit Lid junction. I knew Amir was staying at another base so I thought he was safe. I didn't know that he'd been sent to the junction to guard against terrorists. He wasn't hurt by the first explosion but when he went to help his friends, another suicide bomber blew himself up, and he found his death. . . . Elad [who subsequently committed suicide] was also a soldier. The last words he wrote were that he couldn't stand the pain of the loss of his brother and that life was useless when you have lost your closest one like this. (Nelsen 2006, pp. 194–195)

---

steady attention and perseverance. In this setting, they come to enjoy performing tasks and take pride in doing things well. Industry involves doing things beside and with others (Erikson 1950).

The danger at this stage is developing a sense of inadequacy and inferiority—the feeling that one will never contribute anything or be any good. Palestinian children face the industry-inferiority tension when Israeli authorities close their schools for extended periods. The loss of meaningful, systematic instruction stimulates children to search for their own meaningful activity. Many Palestinian schoolchildren have chosen to participate by throwing stones, an action defined as a meaningful contribution toward securing a homeland.

Stage 5 (Adolescence)

This stage is characterized by rapid body growth and genital maturation. At this time, adolescents also develop ideas about what they will do with their lives. In other words, they begin to search for an identity. As they search, they are often preoccupied with how they appear in the eyes of others. They struggle with being a part of a group and being themselves. As they work to establish an identity, adolescents may overidentify with unrealistic cultural or group heroes and exclude people they deem "different."

Stage 6 (Young Adulthood)

During this stage, the young adult forms close and intimate bonds with others. A mark of a healthy personality is the ability to love and to work. People with this ability can work productively without losing the capacity to be sexual and loving. The opposite of intimacy is self-absorption, which involves "the readiness to isolate, if necessary, to destroy those forces and people whose essence seems dangerous to one's own" (Erikson 1950, p. 264).

The work lives and personal lives of most Israelis are interrupted repeatedly by military obligations. Israeli men must leave their jobs and families at least once a year to serve in the army. Palestinians frequently sacrifice work and family when they protest the occupation with strikes, work stoppages, and business closings. Israeli officials also disrupt Palestinian work and family life when they close access roads and Palestinian businesses. Each group holds the other responsible for the state of affairs, and each seeks to control or resist the other (see Intersection of Biography and Society: "Israeli Jews as Aggressors and Victims").

Stage 7 (Middle Age)

Ideally, those in middle age make an effort to guide and help establish the next generation and to pass on what they have contributed to life. This stage is characterized by a strengthened commitment to the care of cherished persons and objects. Its counterpart is stagnation and interpersonal impoverishment (Erikson 1988, p. 16).

Someone who chose a path of commitment to future generations is Israeli prime minister Yitzhak Rabin, who was assassinated by a Jewish right-wing extremist for seeking a peaceful solution to the Israeli-Palestinian conflict. Rabin "took the road of peace with the Palestinians, not because he possessed great affection for them or their leaders. . . . Rabin decided to act, because he discerned very wisely that Israeli society would not be able to sustain itself endlessly in a state of an unresolved conflict. He realized long before many others that life in a climate of violence, occupation, terror, anxiety and hopelessness, extracts a price Israel cannot afford" (Grossman 2006). In contrast to Rabin's response are the responses of Palestinians and Israeli Jews who see no end to the hundred-year conflict, "having been born into war and raised in it, and in a certain sense indoctrinated by it." They see this madness as "the only life," with "no option or even the right to aspire [to] a different life" (Grossman 2006).

Stage 8 (Old Age)

In old age "one faces the totality of life. Old people see their lives as a whole. At last, life hangs together," and one comes to accept the life one has lived and to acknowledge the people significant to it (Erikson 1982, p. 102). Old age

Joe Mabel

There are more than 60 Raging Granny Action Groups worldwide who believe they still have time to change the world for the better. According to their website their "purpose is to create a better world for our children and grandchildren." They "operate with a sense of outrage, a sense of humor, and a commitment to non-violence."

involves the realization that one's biography is the "accidental coincidence of but one life cycle in but one segment of history." With this realization comes a sense of comradeship with all human beings. If acceptance is not achieved, one feels despair about oneself and the world and believes that the remaining time is too short to try again or change things. Despair is often expressed as disgust or displeasure with institutions and people. On the other hand, if all goes well, wisdom is the fruit of the struggle to accept the life one has lived in the face of physical disintegration (see Table 4.1).

Table 4.1 Selected Demographic Characteristics of Gaza, the West Bank, and Israel

The table shows the population characteristics of Israel and the Palestinian Territories. Note that almost 50 percent of Gaza's population is age 14 or younger and its fertility rate is 5.8, which means the average Gazan woman over her lifetime has almost six children. Each year, Gaza has 40 births for every 1,000 people. To place these figures in context, consider that the highest recorded birth rate is 50 births per 1,000 people. A birth rate of 50 means that a population is reproducing as fast as humanly possible. If we consider that at least 63 percent of the Gaza population lives in poverty, how will the age distribution of that population affect prospects for peace? What insights do Erikson's life stages offer for thinking about the developmental challenges Palestinians and Israeli Jews face?

	Gaza	West Bank	Israel
Birth rate	40.0	31.2	18.0
Fertility rate	5.8	4.3	2.4
Infant mortality (per 1,000 live births)	22.4	19.5	6.89
Percent of population 14 or younger	48.1	42.9	26.3
Percent of population 65 or older	2.6	3.4	9.8
Percent of population living in poverty	63.1	45.7	21.6

Source: U.S. Central Intelligence Agency (2007)

© Justin C. McIntosh

GLOBAL COMPARISONS

The Jewish Population of the World

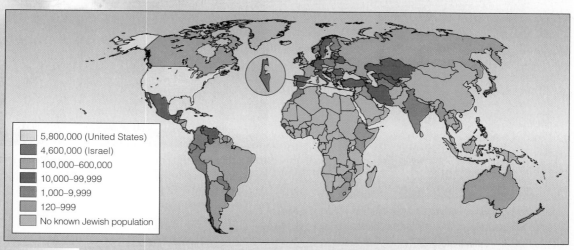

▲ Figure 4.4 **Estimated Number of Jews Living in Each of the World's Countries**
The map shows the estimated number of Jews living in each country. Keep in mind that the European pogroms (government-instigated massacres and persecution) and severe economic crises between 1880 and 1914 pushed Jews to leave Europe for many countries around the world, especially the United States. In addition, the Nazi Holocaust during World War II increased the flow of Jewish refugees out of Europe to other countries, especially to Israel. The Jewish immigrants had to adjust to their new homes through resocialization. According to the World Jewish Congress, an estimated 13 million Jews now live in 120 countries. The largest number of Jews live in the United States (5.8 million) and Israel (4.6 million). The eight cities with the largest Jewish populations are New York (1,750,000), Miami (535,000), Los Angeles (490,000), Paris (350,000), Philadelphia (254,000), Chicago (248,000), San Francisco (210,000), and Boston (208,000).

Source: Jewish Communities of the World, Institute of World Jewish Congress (2006).

Resocialization

■ **CORE CONCEPT 11: Over a lifetime, people make social transitions that entail acquiring new roles, shedding old roles, and integrating new roles with current roles. In making these transitions, people undergo resocialization.** Resocialization is the process of becoming socialized over again. In particular, this process involves discarding values and behaviors unsuited to new circumstances and replacing them with new, more-appropriate values and norms (standards of appearance and behavior). Much resocialization happens naturally over a lifetime and involves no formal training; people simply learn as they go. For example, people relocate to foreign countries (see Global Comparisons: "The Jewish Population of the World"), marry, change jobs, become parents, change religions, and retire without formal preparation or training. Conversely, some resocialization requires that, to occupy new positions, people must

undergo formal, systematic training and demonstrate that they have internalized appropriate knowledge, suitable values, and correct codes of conduct.

Such systematic resocialization can be voluntary or imposed (Rose, Glazer, and Glazer 1979). Voluntary resocialization occurs when people choose to participate in a process or program designed to "remake" them. Examples of voluntary resocialization are wide-ranging: the unemployed youth who enlists in the army to acquire a technical skill, the college graduate who pursues medical education, the drug addict who seeks treatment, the alcoholic who joins Alcoholics Anonymous.

resocialization The process of discarding values and behaviors unsuited to new circumstances and replacing them with new, more-appropriate values and norms.

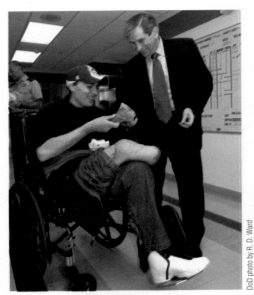

Soldiers recovering from serious physical injuries sustained while in Iraq and Afghanistan must learn to renegotiate life without limbs.

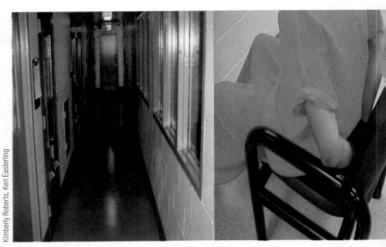

Total institutions are settings in which people voluntarily or involuntarily surrender control of their lives to an administrative staff that supervises their behavior.

Imposed resocialization occurs when people are forced to undergo a program designed to train them, rehabilitate them, or correct some supposed deficiency in their earlier socialization. Military boot camp (when a draft exists), prisons, mental institutions, and schools (when the law forces citizens to attend school for a specified length of time) are examples of environments designed to resocialize individuals whether or not they enter voluntarily.

In *Asylums: Essays on the Social Situation of Mental Patients and Other Inmates*, sociologist Erving Goffman writes about a setting—total institutions (with particular focus on mental institutions)—where people undergo systematic socialization. In **total institutions**, people surrender control of their lives, voluntarily or involuntarily, to an administrative staff.

As "inmates," they then carry out daily activities (eating, sleeping, recreation) in the "immediate company of a large batch of others, all of whom are [theoretically] treated alike and required to do the same thing together" (Goffman 1961, p. 6). Total institutions include homes for the blind, the elderly, the orphaned, and the indigent; mental hospitals; jails and penitentiaries; prisoner-of-war camps and concentration camps; army barracks; boarding schools; and monasteries and convents. Their total character is symbolized by barriers to social interaction, "such as locked doors, high walls, barbed wire, cliffs, water, forests, or moors" (p. 4).

Goffman (1961) identified wide-ranging general and standard mechanisms that the staffs of all total institutions employ to resocialize inmates. When the inmates arrive, the staff strips them of their possessions and their usual appearances (and the equipment and services by which their appearances are maintained). In addition, the staff sharply limits interactions with people outside the institution, to establish a "deep initial break with past roles" (p. 14). The staff typically employs "admission procedures," such as taking a life history, photographing, weighing, fingerprinting, assigning numbers, searching, listing personal possessions for storage, undressing, bathing, disinfecting, haircutting, issuing institutional clothing, instructing as to rules, and assigning to quarters. New arrivals allow themselves to be shaped and coded into objects that can be fed into the administrative machinery of the establishment (p. 16).

Goffman maintained that the admission procedures function to prepare inmates to shed past roles and to assume new ones. Inmates participate in enforced activities that staff members have designed to meet institutional goals. Those goals may be to care for the incapable, to keep inmates out of the community, or to teach people new roles (for example, to be a soldier, priest, or nun).

In general, it is easier to resocialize people when they want to be resocialized than when they are being forced to abandon old values and behaviors. Furthermore, resocial-

total institutions Institutions in which people surrender control of their lives, voluntarily or involuntarily, to an administrative staff and carry out daily activities with others required to do the same thing.

ization occurs more readily when acquiring new values and behaviors requires competence rather than subservience (Rose, Glazer, and Glazer 1979). Herein lies the dilemma in finding a resolution to the Palestinian-Israeli conflict. Factions on both sides attempt to forcibly resocialize the other side to change its position about land rights. Israelis deport, imprison, impose curfews, close schools, level houses, and kill. Palestinians throw stones, strike, boycott Israeli products, and employ suicide bombers/martyrs to kill. A significant percentage of Palestinians and Israelis seem to believe that if they make life miserable enough for the other group, their group will gain a homeland. The problem is that if one side wins through intimidation, the other side by definition assumes a subservient position and, in the long run, is likely to challenge the legitimacy of the other's victory (Elbedour, Bastien, and Center 1997).

■ VISUAL SUMMARY OF CORE CONCEPTS

■ CORE CONCEPT 1: In the broadest sense of the word, socialization is the process by which people develop a sense of self and learn the ways of the society in which they live.

Socialization is the process by which humans (1) acquire a sense of self or social identity; (2) learn about the social groups to which they primarily belong and about the groups to which they do not belong; (3) realize their human capacities, including the ability to talk, walk upright, speak a language, and reason; and (4) learn to negotiate the social and physical environment they have inherited.

Department of Defense photo by Scott H. Spitzer, CIV USAF

■ CORE CONCEPT 2: Socialization involves nature and nurture.

No discussion of socialization can ignore the importance of two factors: nature and nurture. Nature comprises one's human genetic makeup or biological inheritance. Nurture refers to the social environment or the interaction experiences that make up every individual's life. Children inherit genes transmitted through parents that reach back to include four biological grandparents, eight biological great-grandparents, and beyond. Parents and other caretakers are among the first to introduce children to the culture and behavior patterns of the society in which they live.

Amy Burke

■ **CORE CONCEPT 3: Socialization depends on meaningful interaction experiences with others.**

Meaningful social contact with and stimulation from others are important at any age. Indeed, strong social ties with caring people are linked to overall social, psychological, and physical well-being. More fundamentally, social interaction is essential to developing and maintaining a sense of self.

U.S. Census Bureau, Public Information Office (PIO)

■ **CORE CONCEPT 4: Socialization is impossible without memory; memories passed on from one generation to the next preserve and sustain culture.**

No one can participate in society without the ability to remember such things as names, faces, places, words, symbols, and norms. We take it for granted that people have language, norms, values, and beliefs of the surrounding culture stored in memory. People born at approximately the same time and place have likely lived through many of the same events. These experiences —each unique to one person and yet similar to experiences of other persons—are remembered long after an event has passed. Such collective memories are revived, preserved, shared, passed on, and recast in many forms, such as stories, holidays, rituals, and monuments. All societies support rituals that teach people about the past and encourage them to remember. The charred remains of an Israeli bus, in what is known as the Coastal Road Massacre, calls out to Israelis, born 40 or more years after that date, to remember.

Zachi Evenor, Israel

■ **CORE CONCEPT 5: Meaningful social interaction depends on the involved parties sharing significant symbols.**

A significant symbol is a word, gesture, or other learned sign that conveys "the same meaning for the person transmitting it as for the person receiving it" (Theodorson and Theodorson 1979, p. 430). Language is a particularly important significant symbol, because the shared meanings attached to words allow us to communicate with others. Significant symbols also include symbolic gestures or signs—that is, nonverbal cues, such as tone of voice, inflection, facial expression, posture, and other body movements or positions that convey meaning from one person to another.

Nancy Hemminger

■ **CORE CONCEPT 6: People acquire a sense of self when they can role-take.**

Role taking involves stepping outside the self and imagining how others view its appearance and behavior from an outsider's perspective. Children learn to role-take through a three-stage process involving imitation (preparatory stage), play (play stage), and games (game stage). In the play stage, children pretend to be significant others, or people or characters important in their lives; they recognize behavior patterns of significant others and incorporate them into play.

Elizabeth Mander

■ **CORE CONCEPT 7: The ability to imagine others' reactions to the self is critical to self-awareness.**
As we interact, we visualize how we appear to others, we imagine a judgment of that appearance, and we develop a feeling somewhere between pride and shame. The imagining or interpreting of others' reactions is critical to self-awareness. People are affected deeply even when the image they see reflected is exaggerated or distorted. The individual responds to the perceived reaction rather than to the actual reaction. Suicide bombers/martyrs pose for photos before their death, knowing that their faces will appear on posters and that many will salute them for undertaking that kind of operation.

Christopher Brown

■ **CORE CONCEPT 8: The ability to role-take emerges according to a gradually unfolding genetic timetable that must be accompanied by direct experiences with persons and objects.**
Self-awareness and self-identity derive from social relationships and from an ability to step outside the self and imagine it from another's perspective. The theory of cognitive development addresses how people come to acquire this level of cognitive sophistication. Piaget's model of cognitive development includes four broad stages: sensorimotor, preoperational, concrete operational, and formal operational. Children cannot proceed from one stage to another until they master the reasoning challenges of earlier stages. Piaget maintained that reasoning abilities cannot be hurried; a more sophisticated level of understanding will not show itself until the brain is ready. Children construct and reconstruct their conceptions of the world as they experience the realities of living. This progression by stages toward increasingly sophisticated levels of reasoning is apparently universal, although the content of people's thinking varies across cultures.

© Chris Lowe/Index Open

■ **CORE CONCEPT 9: Agents of socialization—significant others, primary groups, ingroups and outgroups, and institutions (such as mass media)—shape our sense of self, teach us about the groups to which we do and do not belong.**
Sociologists have identified a number of agents of socialization: significant others, primary groups, ingroups and outgroups, and institutions that (1) shape our sense of self or social identity, (2) teach us about the social groups to which we do and do not belong, (3) help us to realize our human capacities, and (4) help us negotiate the social and physical environment we have inherited.

Missy Gish

■ **CORE CONCEPT 10: Socialization takes place throughout the human life cycle.**

It is important to realize that socialization takes place throughout the human life cycle. Sociologists are impressed with Erikson's model because each of its eight stages addresses the importance of social interaction and forming constructive relationships with people. Each stage is marked by a crisis, a struggle with the challenges or the developmental tasks that a particular life stage presents. The struggle and its outcome shape one's thinking about the self and others.

Joe Mabel

■ **CORE CONCEPT 11: Over a lifetime, people make social transitions that entail acquiring new roles, shedding old roles, and integrating new roles with current roles. In making these transitions, people undergo resocialization.**

Resocialization is the process of becoming socialized over again. In particular, this process involves discarding values and behaviors unsuited to new circumstances and replacing them with new, more-appropriate values and norms (standards of appearance and behavior). People marry, change jobs, enlist in the military, pursue higher education, become parents, change religions, and retire. Resocialization can be voluntary or involuntary. The young man pictured is beginning a resocialization process that

Kimberly Roberts, Ken Easterling

begins with the staff stripping him of his possessions and of his usual appearance. In this case, street clothes are replaced with an orange jumpsuit and objects such as jewelry that give someone status on the "outside" are removed.

Resources on the Internet

Sociology: A Global Perspective Book Companion Web Site

www.cengage.com/sociology/ferrante

Visit your book companion Web site, where you will find flash cards, practice quizzes, Internet links, and more to help you study.

CENGAGENOW™

Just what you need to know NOW!

Spend time on what you need to master rather than on information you have already learned. Take a pre-test for this chapter, and CengageNOW will generate a personalized study plan based on your results. The study plan will identify the topics you need to review and direct you to online resources to help you master those topics. You can then take a post-test to help you determine the concepts you have mastered and what you will need to work on. Try it out! Go to www.cengage.com/login to sign in with an access code or to purchase access to this product.

Key Terms

agents of socialization   100
collective memory   91
games   96
generalized other   96
group   100
ingroup   101
internalization   86

looking-glass self   97
mass media   104
nature   86
nurture   86
outgroup   101
play   96
primary group   100

resocialization   109
role taking   94
significant others   96
significant symbol   93
socialization   86
symbolic gestures   93
total institutions   110

5

Social Interaction

With Emphasis on the Democratic Republic of the Congo

When sociologists study interaction, they seek to understand the social forces that bring people together. And once people come together, sociologists seek to identify the factors that shape the content and direction of interaction.

King Leopold's Soliloquy: A Defense of His Congo Rule, By Mark Twain, Boston: The P. R. Warren Co., 1905, Second Edition.

▲ Leopold II, king of Belgium, calculated that his "yearly income from the Congo is millions."

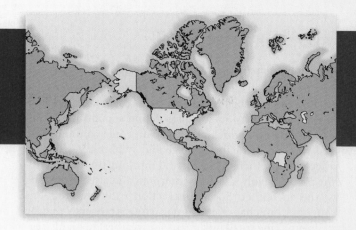

WHY FOCUS ON

The Democratic Republic of the Congo?

IN THIS CHAPTER, we consider social interaction related to the transmission of the human immunodeficiency virus (HIV) and the treatment of acquired immune deficiency syndrome (AIDS). In doing so, we focus on the Democratic Republic of the Congo (DRC)—formerly Congo Free State (1884–1908), Belgian Congo (1908–1960), and Zaire (1960–1997). Focusing on the DRC, the site of the earliest confirmed HIV-infected blood sample does not mean that African people started the global epidemic. Rather, focusing on the country in which this sample was stored forces us to consider how one case of HIV infection grew to 55 million cases worldwide. Of course, intimate interaction between the noninfected and the HIV-infected explains how HIV/AIDS became a global epidemic. But that is only part of the story. It is also important to know the circumstances that drew people from different countries and conti-

nents to the Congo, and the larger social forces that shaped interactions among them.

AIDS researchers believe that the story of this global epidemic started in southeastern Cameroon around 1930, when the virus "jumped" from a chimpanzee to a human host. For the virus to survive and lead to an epidemic, it had to jump from human to human and make its way to a big city, where human hosts were many; the closest such city was Léopoldville (later renamed Kinshasa), in what was then the Belgian Congo (Abraham 2006). Obviously, the story of HIV/AIDS involves the study of interaction at the intimate interpersonal level—at least two people engaging in sexual activity, sharing needles, or exchanging blood or other body fluids. As we will see, it also involves the study of such interaction within a larger social and globally organized context.

FACTS TO CONSIDER

- In 1980, AIDS researchers identified the first known sample of HIV-infected blood in a Congolese blood bank. The frozen stored blood sample was taken in 1959 from a volunteer participating in a medical study.

- According to United Nations estimates, 55 million people are HIV-infected or living with AIDS worldwide.

- Millions of intimate interactions—specifically an exchange of blood or other body fluids—between HIV-infected and noninfected people explain how HIV/AIDS became a global epidemic.

- HIV's origin cannot be understood apart from European colonial rule of Africa. HIV's spread was connected to colonial practices that shaped interactions between African peoples and colonial rulers (Moore 2006).

Social Interaction

■ **CORE CONCEPT 1: When sociologists study interactions, they seek to understand the larger social forces that bring people together in interaction and that shape the content and direction of that interaction.** Sociologists define **social interaction** as a situation in which at least two people communicate and respond through language and symbolic gestures to affect one another's behavior and thinking. In the process, the parties involved define, interpret, and attach meaning to the encounter. Social interaction includes situations ranging from chance encounters (two strangers standing in line making small talk about the weather) to highly regulated encounters in which the parties are required, even forced, to communicate and respond to others in specified ways (an army recruit addressing a drill sergeant).

When sociologists study social interaction, they seek to understand and explain the larger social forces that (1) bring people together in interaction and (2) shape the content and course of interaction. With regard to HIV/AIDS, sociologists ask, "What forces increase the likelihood that HIV-infected and noninfected people interact? What meanings do the involved parties assign to their encounter? How do the parties interpret each other's actions and reactions? How do these meanings and interpretations shape the interaction's direction, course, and outcome?"

We begin by exploring an important social force—the division of labor—that drew people from all over the world to Africa, especially to the DRC. If we can understand such larger, seemingly impersonal forces, we can understand more about the interaction in general and the transmission of viruses such as HIV in particular.

Division of Labor

■ **CORE CONCEPT 2: The division of labor is an important social force that draws people into interaction with one another and shapes their relationships.** Emile Durkheim was one of the first sociologists to provide insights into the social forces that contributed to the rise of a "global village." In *The Division of Labor in Society* ([1933] 1964) Durkheim provides a general framework for understanding the forces underlying global-scale interactions. According to Durkheim, an increase in population size and density increases the demand for resources (such as food, clothing, and shelter). This demand, in turn, stimulates people to find more-efficient methods for producing goods and services, thus advancing society "steadily towards powerful machines, towards great concentrations of [labor] forces and capital, and consequently to the extreme division of labor" (Durkheim [1933] 1964, p. 39). As Durkheim describes it, **division of labor** refers to work that is broken down into specialized tasks, each performed by a different set of workers specifically trained to do that task. The workers do not have to live near each other; they often live in different parts of a country or different parts of the world. Not only are the tasks geographically dispersed, but the parts and materials needed to manufacture products also come from many locations around the world.

In their search for natural resources and for low-cost (even free) labor, European governments vigorously colonized much of Asia, Africa, and the Pacific in the late 19th and early 20th centuries. In 1883 King Leopold II of Belgium claimed as his private property one million square miles of land that encompassed 15 million people. Leopold's personal hold over the land was formally legitimized in 1885 by the leaders of 13 European countries and the United States who were attending the Berlin West Africa Conference (see Figure 5.1). The purpose of that conference was to carve Africa into colonies and divide the continent's natural resources among competing colonial

social interaction An everyday event in which at least two people communicate and respond through language and symbolic gestures to affect one another's behavior and thinking.

division of labor Work that is broken down into specialized tasks, each performed by a different set of persons trained to do that task. The persons doing each task often live in different parts of the world. Not only are the tasks specialized, but the parts and materials needed to manufacture products also come from many different regions of the world.

Meeting the Rear Column at Banalya, from *In Darkest Africa* by H.M. Stanley, published 1890.

As early as 1871 the British-born journalist and explorer Henry Morton Stanley led expeditions along the Congo River, documenting the commercial possibilities of the land and making "deals" to acquire land for Leopold. In his diary he wrote, "Every cordial-faced aborigine whom I meet . . . I look upon . . . as a future recruit to the ranks of soldier-laborer" (Hochschild 1998, p. 68).

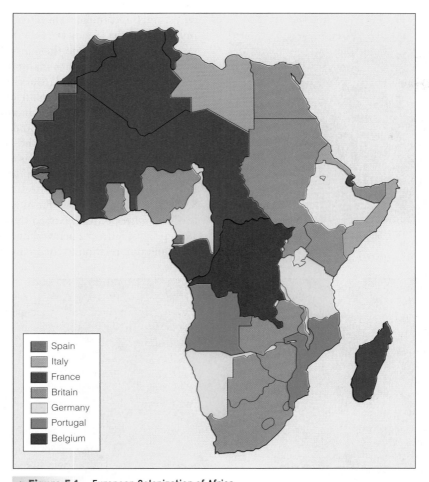

▲ **Figure 5.1**   European Colonization of Africa
By 1913 all of Africa had been divided into European colonies.

Spain
Italy
France
Britain
Germany
Portugal
Belgium

powers (Witte 1992). The vast amount of land awarded to Leopold contained as many as 450 ethnic groups, ranging "from citizens of large, organizationally sophisticated kingdoms to the Pygmies of the Ituri Rain Forest, who lived in small bands with no chiefs and no formal structure or government" (Hochschild 1998, p. 72).

For 23 years Leopold capitalized on the world's demand for rubber, ivory (to be used in making piano keys, billiard balls, snuff boxes), palm oil (a machine-oil lubricant and an ingredient in soaps such as Palmolive), coffee, cocoa, lumber, and diamonds. His reign over the Congo was the "vilest scramble for loot that ever disfigured the history of human conscience and geographical location" (Conrad 1971, p. 118). The methods he used to extract rubber for his own personal gain involved atrocities so ghastly that in 1908 international outrage forced the Belgian government to assume administration of the country. Both Leopold and the Belgian government forced the indigenous Africans to leave their villages to work the mines, cultivate and harvest

The Belgian colonists viewed and treated the Congolese "like animals, as beasts of burden" forcing them to carry or push loads long distances (Hochschild 1998).

A Léopoldville shipyard in the 1930s. Shipyard workers are pushing barges into the Congo River.

crops, and build and maintain roads and train tracks. Keep in mind that the "roads were not built for Africans. Their purpose was to provide the Europeans with easy access into the interior in order to maintain order, increase trade, and get raw materials out" (Mark 1995, p. 220).

The Belgian Congo's capital city, Léopoldville (now Kinshasa), became the major inland port connecting all of Central Africa to the West and beyond. The capital—located on the Congo River, the "river that swallows all rivers" running through what is now the DRC, Zambia, Angola, Tanzania, Cameroon, and Gabon—attracted Europeans and Africans. All trade originating within 8,000 miles of waterways "fanning out, like the veins of a leaf" from Léopoldville passed through that city (Forbath 1977, p. 12). Figure 5.2 below shows the location of the former capital in relation to the many surrounding rivers.

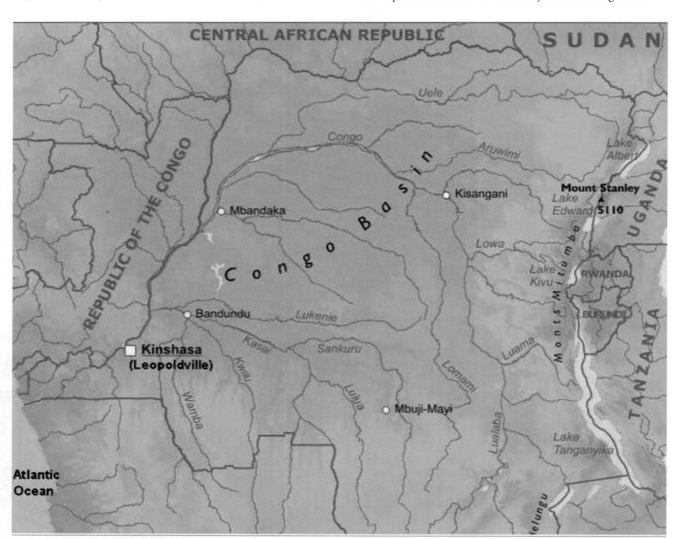

▲ **Figure 5.2** **The Rivers of Central Africa**
The map shows Léopoldville, the capital of the Belgian Congo. It was located on the Congo River—the "river that swallows all rivers" running through what is now the DRC, Zambia, Angola, Tanzania, Cameroon, and Gabon. The capital city attracted Europeans and Africans, who traded along the 8,000 miles of waterways which passed through the city and then out to sea.

As industrialization proceeded in Europe and the United States, the demand for various raw materials grew. The world came to depend on the Congo as a source of copper, cobalt (needed to manufacture jet engines), industrial diamonds, zinc, silver, gold, manganese (needed to make steel and aluminum dry-cell batteries), coltan (a heat-resistant mineral used in cell phones, laptops, and Playstations), and uranium (needed to generate atomic energy and fuel the atomic bomb) (Oliver 2006).

Solidarity: The Ties That Bind

■ **CORE CONCEPT 3:** Solidarity—the ties that bind people to one another and shape their interactions—can be mechanical or organic. Durkheim noted that as the division of labor becomes more specialized and as the sources of materials for products become more geographically diverse, a new kind of solidarity emerges. He used the term **solidarity** to describe the ties that bind people to one another in a society and shape their interactions. Durkheim observed that *mechanical* solidarity is characteristic of preindustrial societies and that *organic* solidarity is characteristic of industrial societies.

Mechanical Solidarity

Mechanical solidarity characterizes a social order based on a common conscience, or uniform thinking and behavior. In this situation a person's "first duty is to resemble everybody else"—that is, "not to have anything personal about

In a 24-hours-a-day, seven-days-a-week operation, European and Asian logging companies remove an estimated 10 million cubic meters of wood each year from forests in Central African countries, including the DRC (Peterson and Ammann 2003).

Brian Smithson

one's [core] beliefs and actions" (Durkheim [1933] 1964, p. 396). Durkheim believed that this similarity derived from the simple division of labor and the corresponding lack of specialization. In other words, a simple division of labor causes people to be more alike than different, because they do the same kind of tasks to maintain their livelihood. This similarity gives rise to common experiences, skills, core beliefs, attitudes, and thoughts. In societies characterized by mechanical solidarity, the ties that bind individuals to one another are based primarily on kinship, religion, and a shared way of life. To understand how society is organized around mechanical solidarity, consider the lifestyle of the Mbuti pygmies, a hunting-and-gathering people who, before colonization, lived in the Ituri Forest (an equatorial rain forest) in what is now the northeastern DRC. Their society represents one of the many ways of life that colonization forced people to abandon:

> The Mbuti share a forest-oriented value system. Their common conscience derived from the fact that the forest gives them food, firewood, and materials for shelter. It is not surprising that the Mbuti recognize their dependence upon the forest and refer to it as "Father" or "Mother" because as they say it gives them food, warmth, shelter, and clothing just like their parents. Mbuti say that the forest also, like their parents, gives them affection. . . . The forest is more than mere environment to the Mbuti. It is a living, conscious thing, both natural and supernatural, something that has to be depended upon, respected, trusted, obeyed, and loved. The love demanded of the Mbuti is no romanticism, and perhaps it might be better included under "respect." It is their world, and in return for their affection and trust it supplies them with all their needs. (1965, p. 19)

Mbuti are aware of the ongoing destruction of the rain forest by companies that push them farther into the forest's interior. The pygmies agree that if they leave the forest and become part of the modern world, their way of life will die: "The forest is our home; when we leave the forest, or when the forest dies, we shall die. We are the people of the forest" (Turnball 1961, p. 260).

Organic Solidarity

A society with a complex division of labor is characterized by **organic solidarity**—a social order based on interdependence and cooperation among people performing

solidarity The ties that bind people to one another in a society.

mechanical solidarity Social order and cohesion based on a common conscience, or uniform thinking and behavior.

organic solidarity Social order based on interdependence and cooperation among people performing a wide range of diverse and specialized tasks.

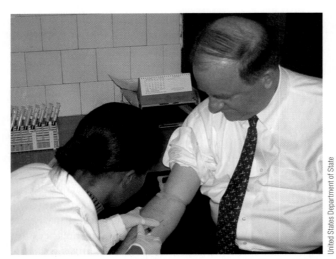

United States Department of State

In societies characterized by organic solidarity, most day-to-day interactions are largely transitory, limited, impersonal, and instrumental. We interact with people for a specific reason, not to get to know them. For example, we can interact with the the person who draws blood for medical tests without knowing them personally.

a wide range of diverse and specialized tasks. A complex division of labor increases differences among people, in turn leading to a decrease in common conscience. Nevertheless, Durkheim argued, the ties that bind people to one another, based on specialization and interdependence, can be very strong.

Specialization means that the tasks needed to make a product or to deliver a service are performed by workers who have been trained to do a specified task or tasks. The final product depends on contributions from all relevant occupational categories. Relationships reflect this specialization. People relate to one another in terms of their specialized roles in the division of labor. When we interact in this manner, we ignore personal differences and treat individuals who perform the same tasks as interchangeable. Thus, most day-to-day interactions are largely transitory, limited, impersonal, and instrumental (that is, we interact with people for a specific reason, not to get to know them). We buy tires from a dealer; we interact with a customer service representative by telephone, computer, or fax; we travel by airplane from city to city in a matter of hours and are served by flight attendants; we pay a supermarket cashier for coffee; and we deal with a lab technician when we give or sell blood. We do not need to know these people personally to interact with them. Likewise, we do not need to know the people behind the scenes: the rubber gatherer, the ivory hunter, the coffee grower, or the logger working in the DRC who contributed to making the products we purchase. When the division of labor is complex and when the materials for products are geographically scattered, few individuals possess the knowledge, skills, and mate-

rials to be self-sufficient. Consequently, people find that they must depend on others. Thus, social ties are strong because people need one another to survive.

Disruptions to the Division of Labor

■ **CORE CONCEPT 4:** Disruptions to the division of labor profoundly affect interaction and an individual's ability to connect with others meaningfully. Durkheim hypothesized that people become more vulnerable as the division of labor becomes more complex and more specialized. He was particularly concerned with the kinds of events that break down individuals' ability to connect with others meaningfully through their labor. Such events include (1) industrial and commercial crises caused by plant closings, massive layoffs, epidemics, technological revolutions, or war; (2) workers' strikes; (3) job specialization, insofar as workers are so isolated that few people grasp the workings and consequences of the overall enterprise; (4) forced division of labor, to the extent that occupations are filled according to inherited traits (such as nationality, age, race or sex) rather than ability; and (5) inefficient management and development of workers' talents and abilities, so that work for them is nonexistent, irregular, intermittent, or subject to high turnover. See Intersection of Biography and Society: "Disruptions to the Division of Labor" for two examples of how disruptions to the division of labor disrupt relationships and shape interactions.

Since at least 1883, disruptions to the division of labor have plagued the Congo region. Figure 5.3 puts these disruptions into historical context. In all cases, the disruptions severed established connections and interactions among people, threw people unfamiliar with one another together, and made them vulnerable to HIV/AIDS. While it is difficult to pinpoint the exact disruption and resulting interactions responsible for triggering the HIV/AIDS pandemic, one thing is clear: HIV's origin must be placed in the context of European colonial rule of Africa and the ways colonial practices shaped interactions among outsiders and the hundreds of Congolese ethnic groups (Moore 2006).

AIDS researchers believe that the story of this global epidemic started in southeastern Cameroon around 1930, when the virus "jumped" from a chimp to a human host. Researchers speculate that the host was a male from the Bantu ethnic group who hunted and butchered chimpanzees for meat. For the virus to survive and lead to an epidemic, it had to jump from human to human and make its way to a big city, where human hosts were many. The closest such city was Léopoldville (later renamed Kinshasa), in the Belgian Congo (Abraham 2006). But there must be more to the story than this, because this hunting practice had been going on for thousands of years. The key socio-

INTERSECTION OF BIOGRAPHY AND SOCIETY

Disruptions to the Division of Labor

IN *THE RECKONING* David Halberstam (1986) profiles the life of Joel Goddard after he was laid off by Ford Motor Company. Goddard was married with two children. After he lost his job, his daily interactions shifted from colleagues at work to contacts with people at the unemployment office. Moreover, he lost the structure in his life that came with the routine of his job. Instead of working, he watched TV, fished, or read want ads. His ties were further disrupted when friends from work moved out of state to find employment. Goddard eventually took a job selling insurance and found himself selling to his acquaintances. He later quit, and then his wife decided to go to work. However, her success at work strained their marriage because it reminded Goddard of his failures.

In "The Puzzling Origins of AIDS" biological anthropologist Jim Moore (2004) describes a series of interactions involving an African fisherman who experiences forced division of labor in the 1920s and 1930s:

[The] fisherman flees his small village to escape a colonial patrol demanding its rubber quota [which he cannot meet]; as he runs, he grabs one of the unfamiliar shotguns that recently arrived in the area. While hiding for several days, he shoots a chimpanzee [for food] and, unfamiliar with the process of butchering it, is infected with Simian immunodeficiency virus (SIV), a retrovirus that is found in primates. On return to the village he finds his family massacred and the village disbanded. He wanders for miles, dodging patrols, until arriving at a distant village. The next day he is seized by a railroad press gang and marched for days to the labor site, where he (along with several hundred others) receives several injections for reasons he does not understand. During his months working on the railroad, he has little to eat and is continually stressed, susceptible to any infection. He finds some solace in one of the camp prostitutes (themselves imported by those in charge), but eventually dies of an undiagnosed wasting—the fate of hundreds in that camp alone.

logical question is this: What is significant about the year 1930 (give or take 10 years)?

By 1930, Belgian-imposed division of labor had "forced" enough people together to sustain the virus.

> Forced labor camps of thousands had poor sanitation, poor diet and exhausting labor demands. It is hard to imagine better conditions for the establishment of an immune-deficiency disease. To care for the health of the laborers, well-meaning, but undersupplied doctors, routinely inoculated workers against small pox and dysentery, and they treated sleeping sickness (an infectious disease transmitted by the tsetse fly with symptoms that include severe headache, fever, and lymph node swelling, extreme weakness, sleepiness, and deep coma) with serial injections. The problem was that multiple injections given to arriving gangs of tens or hundreds were administered with only a handful of syringes. The importance of sterile technique was known but not regularly practiced. Transfer of pathogens would have been inevitable. And to appease the laborers, in some of the camps sex workers were officially encouraged. (Moore 2006, p. 545)

Furthermore, working conditions were such that they left the Congolese people vulnerable to illnesses and sent thousands of Congolese into the unfamiliar jungle in search of food, where they encountered chimpanzees. To complicate matters, HIV is exceptionally efficient, in that it does not kill its human host within a few days, a week, or even a month. The human host stays alive for years and in most cases is unaware of the infection, increasing the chances of inadvertently passing it on to others (Abraham 2006). How did HIV make its way out of the Congo region, leading to a global epidemic? Here we turn to the forces that resulted in global-scale interactions.

The Global Scale of Social Interaction

The demand for the Congo region's resources stimulated Western business interests to find more-efficient means for producing and transporting goods and services. The effort went beyond simply dividing labor into specialized tasks. Since the effort was profit driven, it also involved finding ways to increase the volume of goods and services produced, the reach of markets, and the speed at which goods, services, and labor were delivered. Railroads, freighters, and airplanes (by 1920 there was a Belgium–Congo flight) increased opportunities for more and more people to interact with people they would otherwise never have met (Wrong 2002).

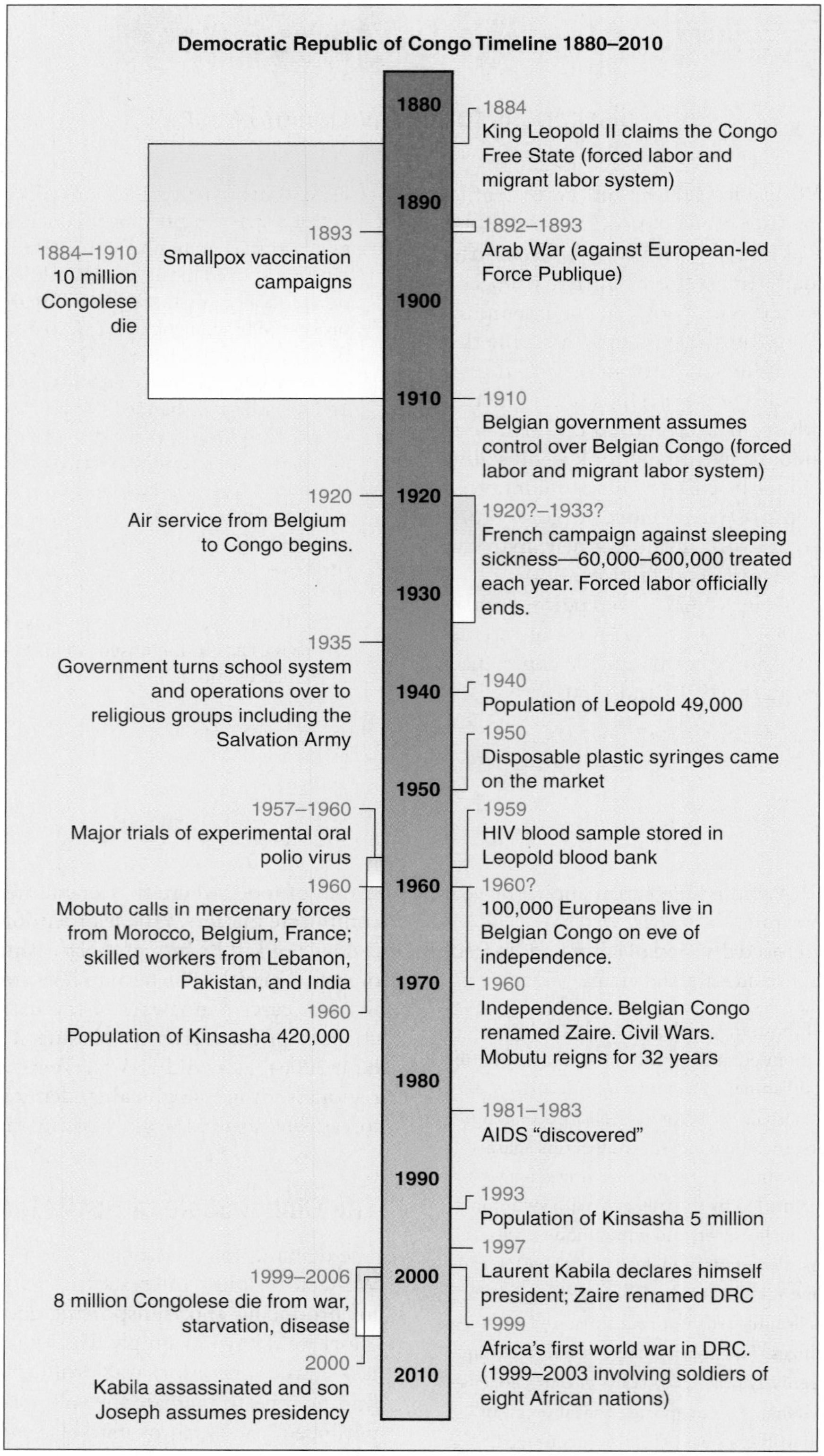

Democratic Republic of Congo Timeline 1880–2010

1880

1884
King Leopold II claims the Congo Free State (forced labor and migrant labor system)

1890

1892–1893
Arab War (against European-led Force Publique)

1884–1910
10 million Congolese die

1893
Smallpox vaccination campaigns

1900

1910

1910
Belgian government assumes control over Belgian Congo (forced labor and migrant labor system)

1920
Air service from Belgium to Congo begins.

1920

1920?–1933?
French campaign against sleeping sickness—60,000–600,000 treated each year. Forced labor officially ends.

1930

1935
Government turns school system and operations over to religious groups including the Salvation Army

1940

1940
Population of Leopold 49,000

1950
Disposable plastic syringes came on the market

1950

1957–1960
Major trials of experimental oral polio virus

1959
HIV blood sample stored in Leopold blood bank

1960
Mobuto calls in mercenary forces from Morocco, Belgium, France; skilled workers from Lebanon, Pakistan, and India

1960

1960?
100,000 Europeans live in Belgian Congo on eve of independence.

1970

1960
Independence. Belgian Congo renamed Zaire. Civil Wars. Mobutu reigns for 32 years

1960
Population of Kinsasha 420,000

1980

1981–1983
AIDS "discovered"

1990

1993
Population of Kinsasha 5 million

1997
Laurent Kabila declares himself president; Zaire renamed DRC

1999–2006
8 million Congolese die from war, starvation, disease

2000

1999
Africa's first world war in DRC. (1999–2003 involving soldiers of eight African nations)

2000
Kabila assassinated and son Joseph assumes presidency

2010

▲ **Figure 5.3**   Time Line of Major Events in the DRC

Global-scale interaction is not new. Industrialization and colonization drew people from the most remote regions of the world into a process that produced unprecedented quantities of material goods, primarily for the benefit of the colonizing powers. In this context, AIDS (and other infectious diseases) cannot be viewed simply as a biological event; it is a social event as well. The importance of considering interaction on a global scale is supported by the fact that disease patterns have historically been affected by changes in population density and changes in transportation. Physician Mary E. Wilson (1996) describes the relationship between travel and the emergence of infectious diseases:

> Travel is a potent force in the emergence of disease. Migration of humans has been the pathway for disseminating infectious diseases throughout recorded history and will continue to shape the emergence, frequency, and spread of infections in geographic areas and populations.
>
> The current volume, speed, and reach of travel are unprecedented. The consequences of travel extend beyond the traveler to the population visited and the ecosystem. When they travel, humans carry their genetic makeup, immunologic sequelae of past infections, cultural preferences, customs, and behavioral patterns. Microbes, animals, and other biologic life also accompany them. Concomitant changes in the environment, climate, technology, land use, human behavior, and demographics converge to favor the emergence of infectious diseases caused by a broad range of organisms in humans, as well as in plants and animals.

Keep in mind that it is impossible to state conclusively that the first case of HIV originated in what is now the DRC. Even if a particular Mbuti hunter were identified as the person to begin the chain reaction of interactions that resulted in the HIV pandemic, could we in good conscience define that hunter as the source?

In the second half of this chapter, we focus on face-to-face interaction, by which the interacting parties communicate and respond through language and symbolic gestures to affect another's behavior and thinking. While interacting, the parties define, interpret, and attach meaning to all that goes on around them and between them.

Social Status

■ **CORE CONCEPT 5:** When analyzing any social interaction, sociologists identify the social statuses of the participating parties. Sociology offers a rich conceptual vocabulary for analyzing the dialogue, actions, and reactions that take place when people interact, as well as for analyzing how the parties involved work to define, interpret, and attach meaning to everything that goes on during the encounter. When analyzing any interaction, sociologists identify the

Department of Defense photo by SPC Preston E. Cheeks

As one indicator of the massive movement of humans, consider that approximately 703 million people take international trips each year (World Tourism Organization 2007). Approximately 42 million foreign visitors enter the United States (excluding day-trippers), and 25.2 million U.S. residents travel to foreign countries (U.S. International Trade Administration 2007).

social statuses of the participating parties. **Social status** is a position in a social structure. A **social structure** consists of two or more people interacting in expected ways. Two-status social structures include relationships between a doctor and a patient, a husband and a wife, a professor and a student, a sister and a brother, and an employer and an employee. Multiple-status social structures include a family, an athletic team, a school, a large corporation, and a national government. Statuses are embedded in groups and institutions, both of which are also key components of social structures.

Individuals usually occupy more than one social status. Sociologists use the term **status set** to capture all the statuses a person assumes. Recall that Joel Goddard (see Intersection of Biography and Society: "Disruptions to the Division of Labor"), the laid-off Ford worker, was male, and unemployed, and he was a father, a husband, and a friend. Note that some of Joel Goddard's statuses, such as male, resulted from chance, and others, such as father, were achieved. At least one of Goddard's statuses, unemployed, seemed the most important of the statuses he occupied.

social status A position in a social structure.

social structure Two or more people occupying social statuses and interacting in expected ways.

status set All the statuses an individual assumes

GLOBAL COMPARISONS

Percentage of Adult Population with HIV/AIDS by Country

THE TABLE BELOW shows the years that HIV/AIDS was first noticed in selected regions of the world and the main modes of transmission. Why do you think MSM (men having sex with men) is not considered a main mode of transmission in some regions of the world? Notice on the map, which gives country-by-country breakdowns, that in the Democratic Republic of the Congo between 5 and 15 percent of the population is HIV infected or has AIDS. At a minimum, that represents 1 in every 20 people (U.S. Central Intelligence Agency 2007).

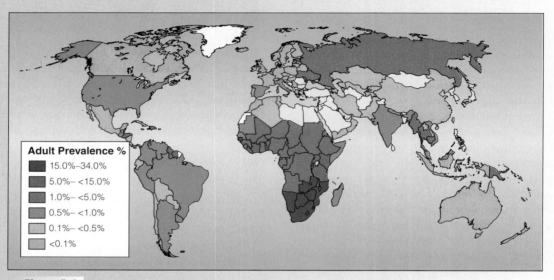

▲ Figure 5.4

Table 5.1

Region	When Epidemic Started	Main Mode(s) of HIV/AIDS Transmission	Number of Cases
Sub-Saharan Africa	Late 1970s–early 1980s	Hetero*	24.7 million
Southern and Southeast Asia	Late 1980s	Hetero; IDU**	7.8 million
Latin America	Late 1970s–early 1980s	MSM***; IDU; Hetero	1.7 million
North America	Late 1970s–early 1980s	MSM; IDU; Hetero	1.4 million
Western and Central Europe	Late 1970s–early 1980s	MSM; IDU; Hetero	740,000
Australia and New Zealand	Late 1970s–early 1980s	MSM; IDU; Hetero	81,000
Caribbean	Late 1970s–early 1980s	Hetero	250,000
Eastern Europe and Central Asia	Early 1990s	IDU; MSM	1.7 million
East Asia and Pacific	Late 1980s	IDU; Hetero; MSM	750,000
North Africa and Middle East	Late 1980s	IDU; Hetero	460,000

Notes: *Hetero: heterosexual transmission; **IDU: transmission through injecting drug use; ***MSM: transmission among men who have sex with men.

Source: UNAIDS (2006).

Ascribed, Achieved, and Master Statuses

Ascribed statuses result from chance; that is, the individual exerts no effort to obtain them. A person's birth order, race, national origin, sex, and age qualify as ascribed statuses. Of course, one can always think of cases in which people take extreme measures to achieve a status typically thought of as ascribed; they undergo a sex change operation, lighten their skin, or hire a plastic surgeon.

Achieved statuses are acquired through some combination of personal choice, effort, and ability. A person's marital status (excluding widowhood), occupation (quarterback, physician, miner), and educational attainment (high school dropout, college graduate) are considered achieved statuses. Ascribed statuses can play a role in determining achieved statuses. Simply consider that in the United States 99 percent of airplane pilots are male and 98 percent of child-care providers are female. Sometimes one status is so important to a person's social identity that it overshadows all other statuses that person occupies. It shapes every aspect of life and dominates social interactions. Such a status is known as a **master status**. Unemployed, retired, ex-convict, HIV infected, homosexual, and disabled can qualify as master statuses.

Master statuses, most notably a person's sex and racial classification, profoundly shaped interactions between European and African peoples between 1884 and 1960, the years the DRC was known as the Congo Free State and the Belgian Congo. During this time, white European males (predominantly Belgian) occupied the following kinds of positions: commissioned military officers, steamboat captains, district commissioners, and station chiefs. Belgian men and women also traveled to the Congo as physicians and missionaries. The black Congolese and other Africans who migrated to the Belgian colony worked as porters, canoe paddlers, miners, rubber gatherers, and low- to mid-level administrative and technical staff. Congolese women worked as maids, house servants, and prostitutes, and since there were always labor shortages, women were pressed to work on roads and plantations (Lyons 2002). Most Congolese were forced into these labor-intensive positions out of desperation or the threat of death from the Force Publique, a white-led repressive military force comprising African males from many ethnic groups both inside and outside of the Congo.

Until independence in 1960 the Congo maintained a rigid system of racial segregation in all areas of life. The segregated school system was such that the Congolese were denied access to education beyond primary school. At the time of independence, only 17 Congolese had earned a university degree (Wrong 2000, Edgerton 2002).

■ **CORE CONCEPT 6: People occupy statuses and perform roles.** A social status has meaning only in relation to other social statuses. For instance, the status of a physician takes

The status of disabled can overshadow all other statuses a person occupies. On the most observable level, this person also occupies the statuses of male and tennis player.

on quite different meanings depending on whether the physician is interacting with another physician or with a patient, spouse, or nurse. Thus, a physician's behavior varies, depending on the social status of the person with whom he or she is interacting.

Roles

All social structures allow sociologists to broadly anticipate the behavior of people occupying each status, no

ascribed statuses Social statuses that result from chance; that is, the individual exerts no effort to obtain them. A person's birth order, race, national origin, sex, and age qualify as ascribed characteristics.

achieved statuses Social statuses acquired through some combination of personal choice, effort, and ability. A person's marital status, occupation, and educational attainment are considered examples of achieved statuses.

master status One status in a status set that is so important to a person's social identity it overshadows all other statuses a person occupies—shaping every aspect of life and dominating social interactions.

matter the personality of the occupant. Just as the behavior of a person occupying the status of a football quarterback is broadly predictable, so too is the behavior of a person occupying the status of nurse, secretary, mechanic, patient, or physician. Sociologists use the term **role** to describe the behavior, obligations, and rights expected of a status in relation to another status (for example, professor to student or physician to patient).

The distinction between role and status is this: people *occupy* statuses and *enact* or *perform* roles. Thus, physician is a social status, a fixed position in a social structure. Attached to that social status are roles specifying how physicians are expected to behave toward people depending on the status (nurse, patient, colleague) they occupy. The role of physician in relation to a patient specifies that a physician has an obligation to establish a diagnosis, not overtreat the patient, respect patient privacy, work to prevent disease, "call on colleagues when the skills of another are needed for a patient's recovery," and avoid sexual relations with the patient (Hippocratic Oath 2006). In addition to their **role obligations**, physicians also have the **right** to demand that patients answer questions honestly and cooperate with a treatment plan.

Although roles set general limits on behavior, we should not expect behavior to be completely predictable. Quite often people do not meet their role obligations—as when physicians knowingly perform unnecessary surgery, overmedicate, engage in sexual relations with patients, or break confidentiality. Likewise, some patients do not give honest answers to questions their physicians ask, and they fail to comply with treatment plans. Two sources of such failures to meet role expectations are role strain and role conflict.

Role strain is a predicament in which contradictory or conflicting expectations are associated with a single role. For example, military doctors, as physicians, have an obligation to preserve life and to "first do no harm." At the same time, they are part of a system that deliberately places soldiers in situations that threaten their physical

Military doctors can experience role strain because they have an obligation to preserve life; yet they are part of a system that deliberately places soldiers in situations that threaten their physical and psychological well-being.

and psychological well-being. **Role conflict** is a predicament in which the expectations associated with two or more distinct roles contradict one another. For example, a person in a sick role has an obligation to want to get better and comply with treatment plans. This obligation, however, can interfere with a second role the person holds—as when a woman finds that the side effects of a prescribed drug (such as drowsiness) prevent her from carrying out her obligations in her role as mother if she is not alert enough to care for her children properly. In this situation, the woman cannot be a good mother and a good patient at the same time.

Despite variations in how people perform roles, role is still a useful concept because, for the most part, a predictability exists "sufficient to enable most of the people, most of the time to go about [the] business of social life without having to improvise judgments anew in each newly confronted situation" (Merton 1957, p. 370). Moreover, if people deviate too far from the expected range of behaviors, negative sanctions—ranging in severity from a frown to imprisonment and even death—may be applied. Serious punishment was meted out in the Congo when rubber gatherers and other forced laborers failed to meet imposed role obligations. To ensure that rubber gatherers met quotas, family members were held hostage until the quotas were delivered. Failing to gather enough rubber could mean death for a spouse, parents, or children. If a village refused to participate in forced labor, company troops shot everyone in sight so that nearby villagers were clear about expectations (Hochschild 1998, p. 165).

Cultural Variations in Role Expectations

Role expectations are intertwined with the larger cultural context. Recall from Chapter 3 that culture consists of

role   The behavior, obligations, and rights expected of a social status in relation to another social status.

role obligations   The relationship and behavior a person enacting a role must assume toward others occupying a particular social status.

right   A behavior that a person assuming a role can demand or expect from another.

role strain   A predicament in which the social role a person is enacting involves contradictory or conflicting expectations.

role conflict   A predicament in which the expectations associated with two or more roles in a role set contradict one another.

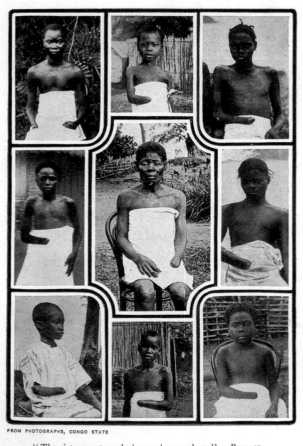

FROM PHOTOGRAPHS, CONGO STATE

"The pictures get sneaked around everywhere."— *Page 40.*

King Leopold's Soliloquy: A Defense of His Congo Rule, By Mark Twain, Boston: The P. R. Warren Co., 1905, Second Edition.

One of the most grisly policies of Leopold's rule was to sever the right hands of Congolese who refused to gather rubber. Soldiers were under orders to turn in a right hand for every bullet fired, as proof that they were using bullets to kill people (and not animals or for target practice). Often soldiers would sever the right hands of people to cover for other uses of bullets.

material and nonmaterial culture (norms, values, beliefs, language, and symbols)—both of which shape behavior and thinking. In light of culture's importance we should not be surprised to learn that the behavior expected of physicians varies across cultures. In the United States the main objective of the patient-physician interaction is to give a diagnosis and to use available technology to treat it. This objective is shaped by a core cultural belief in the ability of science to solve problems. Western medicine uses all available tools of science to establish the cause of a disease, combat it for as long as possible, and ideally return the body to a healthy state. In view of this cultural orientation, it is not surprising that U.S. physicians rely heavily on technological elements, such as X-rays and CAT scans, to diagnose conditions, and on vaccines, drugs, and surgery to cure them. This reliance is reflected in the

Department of Defense

The United States, with less than 5 percent of the world's population, consumes an estimated 44 percent of the world's pharmaceutical supply.

fact that the United States, with less than 5 percent of the world's population, consumes an estimated 44 percent of the world's pharmaceutical supply (*Online Newshour* 2004). This is equivalent to 2.8 billion prescriptions per year, or 10 prescriptions for every man, woman, and child (National Center for Health Statistics 2002).

We can contrast the U.S. physician-patient relationship with the traditional African healer-patient relationship. The social interaction between the African healer and patient is very different from that between the U.S. physician and patient. Like American physicians, traditional healers recognize and treat the organic and physical aspects of disease. But they also attach considerable importance to other factors: supernatural causes, social relationships (hostilities, stress, family strain), and psychological distress. This holistic perspective allows for a more personal relationship between healer and patient.

Sociologist Ruth Kornfield observed Western-trained physicians working in the Congo's urban hospitals and found that success in treating patients was linked to the physicians' ability to tolerate and respect other models of illness and include them in a treatment plan. Among some ethnic groups in the Congo, when a person becomes ill, the patient's kin form a therapy management group and make decisions about treatment. Because many people in the Congo believe that illnesses result from disturbances in social relationships, the cure must involve a "reorganization of the social relations of the sick person that [is] satisfactory for those involved" (Kornfield 1986, p. 369; also see Kaptchuk and Croucher 1986, pp. 106–108).

Given these culturally-rooted differences regarding the cause and treatment of disease, it is not surprising to learn that Europeans and Africans clashed over how to handle a sleeping sickness epidemic in the late 1920s and early 1930s. Europeans, confident in the superiority of their approach, assumed that sleeping sickness and other African diseases could be controlled, even eliminated, through European-developed techniques and technologies. With regard to sleeping sickness, those techniques and technologies revolved around needles, which were used to screen for and treat the condition.

Most Congolese were suspicious of the European campaign against sleeping sickness. They believed that Europeans were responsible for the dramatic increase in cases of sleeping sickness and other diseases. Specifically, the Congolese made a causal link between the European-imposed disruptions to their economic and social relations and the illness. They believed that slow, chronic illnesses like sleeping sickness were man made. For them diagnosis required

People line up for sleeping sickness screening. In light of the labor needs at the time, the Belgian government and other Western players in Congo ventures considered diseased and dead Africans as economic losses. Thus sleeping sickness posed a grave threat in need of control. Sleeping sickness campaigns were institutionalized through a network of rural clinics, screening and injection centers, and hospitals. Between 1923 and 1938 an estimated 26 million exams were performed in the Belgian Congo (Lyons 2002).

careful investigation to identify the ultimate cause behind physical symptoms; treatment focused on reintegrating the sick person into the social fabric (Lyons 2002). In *The Colonial Disease*, Maryinez Lyons (2002) argues that it "is vital to understand the profound importance and depth of this Congolese belief regarding cause and treatment" (p.183).

The Dramaturgical Model of Social Interaction

■ **CORE CONCEPT 7:** Social interaction can be viewed as if it were a theater, people as if they were actors, and roles as if they were performances before an audience. A number of sociologists have compared social roles with the dramatic roles played by actors. Sociologist Erving Goffman offered a **dramaturgical model** of social interaction. According to this model, social interaction is viewed as if it were theater, people as if they were actors, and roles as if they were performances before an audience in a particular setting. People in social situations resemble actors in that they must be convincing to others and must demonstrate who they are and what they want through verbal and nonverbal cues. In social situations, as on a stage, people manage the setting, their dress, their words, and their gestures to correspond to the impression they are trying to make or the image they are trying to project. This process is called **impression management**.

Impression Management

On the surface, impression management may strike us as manipulative and deceitful. Most of the time, however,

Europeans' approach to diagnosing sleeping sickness involved taking a blood sample through a finger prick or taking spinal fluid, shown here, through a lumbar puncture (spinal tap). In Africa, sleeping sickness was considered the AIDS of the early 20th century.

dramaturgical model A model in which social interaction is viewed as if it were a theater, people as if they were actors, and roles as if they were performances before an audience in a particular setting.

impression management The process by which people in social situations manage the setting, their dress, their words, and their gestures to correspond to the impression they are trying to make or the image they are trying to project.

The dramaturgical model portrays social interaction as theater, equating participants with actors performing their roles before an audience.

people are not even aware that they are engaged in impression management; they are simply behaving in ways they regard as natural. Women engage in impression management when they remove hair from their faces, legs, armpits, and other areas of their bodies and present themselves as hairless in these areas. From Goffman's perspective, even if people are aware that they are manipulating reality, impression management can be a constructive feature of social interaction. Social order depends on people behaving in broadly predictable ways. If people spoke and behaved entirely as they pleased, civilization would break down.

Goffman (1959) also recognized the dark side of impression management, which occurs when people manipulate their audience in deliberately deceitful and hurtful ways. Such was the case when King Leopold presented his interests in the Congo as purely philanthropic. He used the International African Association as a front for his profit-making ventures. His stated aim was to bring a humanizing influence to the Congo—an effort that included establishing a chain of medic posts and scientific centers across the region and abolishing Afro-Arab slave trade. In reality, Leopold (through Henry Morton Stanley) signed treaties with 450 African chiefs, who were unfamiliar with the written word and the concept of treaties beyond those made in friendship with neighboring chiefs and villages. The treaties read,

> In return for one piece of cloth per month to each of the undersigned chiefs, besides present cloth in hand, to freely and of their own accord, for themselves and their heirs and successors for ever, . . . give up to the said Association the sovereignty and all sovereign and governing rights to all their territories . . . and to assist by labour or otherwise any works, improvements or expeditions which the said Association shall cause at any time to be carried out in any part of these territories. . . . All roads and waterways running through this country, the right of collecting tolls on the same, and all mining, fishing, and

forest rights are to be the absolute right of the said Association. (Hochschild 1998, p. 72)

Impression management often presents people with a dilemma. If we reveal inappropriate, unacceptable, or unpleasant information, we risk losing or offending our audience. If we conceal the same information, we may feel that we are being deceitful, insincere, or dishonest. According to Goffman, in most social interactions, people weigh the costs of losing their audience against the costs of losing their integrity. If keeping our audience seems more important, deliberately concealing information is necessary; if being honest seems more important, we may risk losing the audience.

The tension between revealing and concealing information comes into play when people test positive for HIV. If they disclose the test results, they risk discrimination, including loss of their jobs, insurance coverage, friends, and family (Markel 2003). This risk explains why many HIV-infected people fail to disclose their HIV status, even to sexual partners or when giving blood. In a study of 203 HIV-infected patients treated at two East Coast hospitals, Michael Stein (1998) and his colleagues found that more than half the group claimed to be sexually active. Of these 129 patients 40 percent had not disclosed their HIV-positive status to partners with whom they had had sex in the past six months. In addition, only 42 percent reported that they always used a condom. In another study, almost one-third of a group of 304 HIV-positive blood donors indicated that they had donated blood because their colleagues had pressured them to do so (see Working For Change: "Uganda, a Success Story in Addressing HIV/AIDS").

Front- and Back-Stage Behavior

Goffman uses other theater analogies—front stage and back stage—to identify situations in which people are most likely to engage in impression management. Just as the theater has a front stage and a back stage, so too does everyday life. The **front stage** is the area visible to the audience, where people take care to create and maintain expected images and behavior. The **back stage** is the area out of the audience's sight, where individuals let their guard down and do things that would be inappropriate or

front stage The area of everyday life visible to an audience, where people take care to create and maintain the images and behavior the audience has come to expect.

back stage The area of everyday life out of an audience's sight, where individuals can do things that would be inappropriate or unexpected on the front stage.

WORKING FOR CHANGE

Uganda, a Success Story in Addressing HIV/AIDS

UGANDA IS CONSIDERED a success story in overcoming the HIV epidemic. It has experienced substantial declines in prevalence of HIV infection. Prevalence peaked at around 15 percent in 1991 and fell to 5 percent in 2001. The estimated cost of this success was $1.80 per adult per year over a 10-year period. The dramatic decline in prevalence appears to be unique. Not surprisingly, it has been the subject of curiosity and scientific scrutiny since the mid-1990s. Uganda's falling HIV prevalence can be traced to a number of behavioral changes:

1. **High-level political leaders supported the fight against HIV/AIDS.** Uganda's president, Yoweri Museveni, responded to evidence of the emerging epidemic with a proactive commitment to prevention, which has continued. In face-to-face interactions with Ugandans at all levels, he emphasized that fighting AIDS was a patriotic duty requiring openness, communication, and strong leadership from the village level to the national capitol.

2. **An aggressive media campaign informed the public.** The campaign involved using print materials, radio messages, billboards, and community mobilization for a grassroots offensive against HIV. The word was spread through thousands of community-based AIDS counselors, health educators, peer educators, and other specialists.

3. **Intervention campaigns addressed women and youth, stigmatization and discrimination.** Since 1989, Ugandan teachers have been trained to integrate HIV education and messages about changing sexual behavior into school curricula. Youth-friendly approaches promote reducing the risk of HIV infection by limiting the number of one's sexual partners, delaying the beginning of sexual activity, abstaining from sex until marriage, remaining faithful to one uninfected person if married, and using condoms if "you're going to move around." Of particular note has been the decline in the percentage of youth aged 13–16 who reported being sexually active. In 1994 nearly 60 percent of boys and girls aged 13–16 reported having already "played sex." In 2001 the figure declined to less than 5 percent.

© K. Burns/USAID

Uganda's AIDS Support Organization staff use motorbikes to take services to people. The services include "HIV prevention education; counseling and support activities; basic medical care for opportunistic infections, sexually transmitted infections, and family planning services; and social support, including skills training for income generating activities and orphan support."

4. **Religious leaders and faith-based organizations have been active on the front lines of the response to the epidemic.** Mainstream faith-based organizations wield enormous influence in Africa. Early and significant mobilization of Ugandan religious leaders and organizations resulted in their active participation in AIDS education and prevention activities. Mission hospitals were among the first to develop AIDS care and support programs in Uganda. The Roman Catholic church and mission hospitals provided leadership in designing AIDS mobile home care projects and special programs for AIDS widows and orphans.

5. **Uganda established Africa's first confidential voluntary counseling and testing (VCT) services.** In 1990 the first AIDS Information Center (AIC) for anonymous VCT opened in Kampala. By 1993, centers had opened in four major urban areas. AICs pioneered producing same-day results, using rapid HIV tests, as well as the concept of Post-Test Clubs to provide long-term support for behavior change to anyone tested, regardless of test results.

6. **Condom marketing has played a key role.** Condom promotion was not central to Uganda's early response to AIDS. Surveys show the rate of having ever used condoms, as reported by women, increased from 1 percent in 1989 to 16 percent in 2000. Rates of males having ever used condoms were 16 percent in 1995 and 40 percent in 2000.

7. **Uganda has experienced a decrease in multiple sexual partnerships and networks.** In general, Ugandans now have considerably fewer casual sex partners across all ages. Ugandan males in 1995 were less likely to have ever had sex (in the 15- to 19-year-old range), more likely to be married and keep sex within the marriage, and less likely to have multiple partners, particularly if never married. Such behavioral changes in Uganda appear to be related to more-open personal communication networks for acquiring AIDS knowledge.

Source: Adapted from "What Happened in Uganda?" U.S. Agency for International Development (September 2002).

unexpected on the front stage. Because back-stage behavior frequently contradicts front-stage behavior, we take great care to conceal it from the audience.

Goffman uses a restaurant as an example of a social setting that has clear boundaries between the back stage and front stage. Restaurant employees do things in the kitchen and pantry (back stage) that they would never do in the dining areas (front stage), such as eating from customers' plates, dropping food on the floor and putting it back on a plate, and yelling at one another. Once they enter the dining area, however, such behavior stops. If you have ever worked in a restaurant, you know that such actions are fairly common back-stage behavior.

The division between front stage and back stage can be found in nearly every social setting. In relation to the AIDS crisis, we can identify many settings that have a front stage and a back stage, including hospitals, doctors' offices, and blood banks. An awareness of how front-stage–back-stage dynamics shape behavior helps us anticipate problems such as those described in a U.S. government General Accounting Office (GAO 1997a) study.

The GAO found that 20 percent of blood donors claimed they would have answered screening questions differently if a more private setting had been provided. It appears that one in five donors did not answer screening questions honestly, because they were on the "front stage"—that is, others were within hearing distance and the donors feared disclosing answers that might be judged harshly. In fact, between 14 and 30 percent of blood donors believed that blood bank screening areas provided inadequate privacy, enabling others to hear answers to the questions technicians asked. Such questions included "Are you in good general health?" "Are you a male who has had sex with another male even once since 1977?" "Have you ever taken street drugs by needle, even once?" "Since 1977, have you ever exchanged sex for drugs or money?" "Have you had sexual contact with anyone who was born in or lived in Cameroon, Central African Republic, Chad, Congo, Equatorial Guinea, Gabon, Niger, or Nigeria since 1977?"

Attribution Theory

■ **CORE CONCEPT 8:** People assign cause to their own and others' behaviors. That is, they propose explanations for their own and other's behaviors, successes, and failures and then they respond accordingly. Social life is complex. People need a great deal of historical, cultural, and biographical information if they are to truly understand the causes of even the most routine behaviors. Unfortunately, it is nearly impossible for people to have this information at hand every time they seek to explain a behavior. Yet, despite this limitation, most people attempt to determine a cause, even if they rarely stop to examine critically the accuracy of their explanations. As most of us know very

well, ill-defined, incorrect, and inaccurate perceptions of cause do not keep people from forming opinions and taking action. The **Thomas theorem** describes this process very simply: "If [people] define situations as real they are real in their consequences" (Thomas and Thomas [1928] 1970, p. 572).

Attribution theory relies on the assumption that people make sense of their own and others' behavior by assigning a cause. In doing so, they often focus on one of two potential causes: dispositional (internal) or situational (external). **Dispositional causes** are forces over which individuals are supposed to have control—including personal qualities or traits, such as motivation level, mood, and effort. **Situational causes** are forces outside an individual's control—such as the weather, bad luck, and others' incompetence. Usually people stress situational factors to explain their own failures and dispositional factors to explain their own successes (for example, "I failed the exam because the teacher can't explain the subject" versus "I passed the exam because I studied hard"). With regard to other people's failures or shortcomings, people tend to stress dispositional factors ("She doesn't care about school"; "He didn't try"). With regard to others' successes, people tend to emphasize situational factors ("She passed the test because it was easy"). Right or wrong, the attributions people make affect how they respond.

Many people stress dispositional traits to explain the cause of HIV/AIDS, arguing that certain groups of people behaved in careless, irresponsible, bizarre, or immoral ways. They hypothesize that the disease originated in Africa and then spread to the United States and beyond through homosexual interactions. When I ask my students—many of whom were born after 1981 (the year AIDS was recognized as a disease)—about the origin of AIDS, they repeat these popular beliefs. With regard to Africa, students point to the evidence that the virus jumped from chimpanzees to humans—specifically from chimps to African hunters who were bitten or scratched by chimps. With regard to homosexuals, students point to high-risk and promiscuous behavior.

When evaluating the chimp-hunter and homosexual hypotheses, some questions must be asked. With regard to the chimp-hunter hypothesis, we must ask, If chimps

Thomas theorem An assumption focusing on how people construct reality: If people define situations as real, their definitions have real consequences.

dispositional causes Forces over which individuals are supposed to have control—including personal qualities or traits, such as motivation level, mood, and effort.

situational causes Forces outside an individual's immediate control—such as weather, chance, and others' incompetence.

Almost every country in the world produces AIDS-prevention messages tailored to the values and beliefs of people in that country.

AIDSCOM
Education is not enough . . .

National Library of Medicine

have been hunted for thousands of years, why did the leap between species take place around 1930 but HIV go unnoticed until the early 1980s? The answer lies in this simple equation: Suppose a single hunter is infected by an chimp and then transmits the virus to 2 people over 5 years, for a total of 3 people. In turn, each of those 2 transmits the virus to another 2 people over the next 5 years. We now have a total of 7 people infected over the course of 10 years. Now imagine a series of 5-year progressions. The number of cases climbs from 7 to 21 to 63 to 169 to 517 to 1,751 to 5,253, and so on. Now imagine the virus being "helped along" by forcing tens of thousands of unskilled hunters into the forests to hunt chimps for food, by administering blood transfusions, by conducting inoculation campaigns with unsterilized needles, and by rounding up large labor pools and then abandoning the weakest and sickest workers to scavenge for food in the forests (as their "unproductive state" merits no rations). The point is that, even with help, it took time for the virus to burst onto the global scene 50 years later (Moore 2000).

With regard to homosexual origins, we must ask, Are the categories male homosexual/bisexual and heterosexual mutually exclusive? To ask the question another way,

scapegoat A person or group blamed for conditions that (a) cannot be controlled, (b) threaten a community's sense of well-being, or (c) shake the foundations of an important institution.

Is heterosexual contact always vaginal? According to the researchers who conducted the most comprehensive study on sexual behavior in the United States, between 13.2 and 37.5 percent of respondents said they had engaged in anal sex with an opposite-sex partner at some time. Between 55.6 and 83.6 percent said they had had oral sex with an opposite-sex partner in the past year (Michael et al. 1994). An Urban Institute Study financed by the U.S. government surveyed a representative sample of 1,297 males between the ages of 15 and 19 and found that two-thirds had had experience "with non-coital behaviors like oral sex, anal intercourse, or masturbation by a female" (Lewin 2000, p. 18A). The point is that we could probably learn more about how HIV spreads if we classified HIV cases by type of sex act (oral, anal, vaginal) rather than identifying the categories of people having sex (homosexual/bisexual or heterosexual).

On the surface, attributing cause to dispositional factors—sexual practices, preferences, and appetite—seems to reduce uncertainty about the source and spread of the disease. The rules are clear: If we do not interact with or behave like members of the group to whom we attribute those dispositional factors, then we can avoid the disease. At best, such rules provide people with a false sense of security: The disease cannot affect *us* because it affects *them* (Grover 1987).

Dispositional attributions imply that these Africans and gays "earned" their disease as a penalty for perverse, indulgent, and illegal behaviors (Sontag 1989). Such logic supports the naming of a scapegoat. A **scapegoat** is a per-

What Is HIV/AIDS?

HIV (HUMAN IMMUNODEFICIENCY virus) is the virus that causes AIDS. This virus is passed from one person to another through exchanging infected blood or other body fluids, including semen, vaginal fluid, breast milk, cerebrospinal fluid surrounding the brain and spinal cord, synovial fluid surrounding bone joints, and amniotic fluid surrounding a fetus.

AIDS stands for **a**cquired **i**mmuno**d**eficiency **s**yndrome.

Acquired means that the disease is not hereditary but develops after birth from contact with a disease-causing agent (in this case, HIV).

Immunodeficiency means that the disease is characterized by a weakening of the immune system.

Syndrome refers to a group of symptoms that collectively indicate or characterize a disease. In the case of AIDS, symptoms can include the development of certain infections or cancers, as well as a decrease in the number of certain cells in a person's immune system.

A diagnosis of AIDS is made by a physician using specific clinical or laboratory standards. The symptoms of AIDS are not unique to that condition. An HIV-infected person receives a diagnosis of AIDS after developing one of the 25 CDC-defined AIDS indicator illnesses. These illnesses include invasive cervical cancer, recurrent pneumonia, Kaposi's sarcoma, toxoplasmosis of the brain, and wasting syndrome due to HIV. An HIV-positive person who has not experienced one of the 25 indicator illnesses can receive an AIDS diagnosis based on tests showing that the number of CD4+ T cells per cubic millimeter of blood has dropped below 200.

Keep in mind that a diagnosis of AIDS depends on how the condition is defined. Until 1993 the official definition of AIDS did not include HIV-related gynecological disorders, such as cervical cancer, or HIV-related pulmonary tuberculosis or recurrent pneumonia as conditions that indicated AIDS. Before 1993 many indicator illnesses were unique to the gay population with AIDS. Under the revised definition, HIV-positive women with cervical cancer are officially diagnosed as having AIDS. Before this definition change, physicians did not advise women with cervical cancer to be tested for HIV and did not treat HIV-positive women with this condition as if they had AIDS (Barr 1990; Stolberg 1996). The revised definition added 40,000 people to the total number diagnosed with AIDS.

Source: Adapted from Centers for Disease Control (2001a, 2004b).

---

son or group blamed for conditions that (a) cannot be controlled, (b) threaten a community's sense of well-being, or (c) shake the foundations of an important institution. Often the scapegoat belongs to a group that is vulnerable, hated, powerless, or viewed as different. The public identification of scapegoats gives the appearance that something is being done to protect the so-called general public; at the same time, it diverts public attention from those who have the power to assign labels. In the United States the early identification of AIDS as the "gay plague" diverted attention from blood banks and the risks associated with medical treatments involving blood and blood products. Blood bank officials could maintain that the supply was safe as long as homosexuals abstained from giving blood (See No Borders, No Boundaries: "Imports/Exports of Blood and Blood Products").

From a sociological perspective, dispositional explanations of a problem that point to a group—or characteristics supposedly inherent in members of that group—are simplistic and potentially destructive, not only to the group but also to the search for solutions. When a specific group and its behaviors emerge as targets, the solution is framed in terms of controlling that group. In the meantime, the problem can spread to members of other groups who believe that they are not at risk because they do not share the "problematic" attribute. This kind of misguided thinking about risk applies even to physicians and medical researchers. "HIV-positive patients older than 50 years tend to be more immunosuppressed and more often have AIDS at first evaluation than their younger counterparts" (Reuters Health Information 1999, p. 1). Attributions about who "should" have AIDS affect the way in which AIDS is defined and diagnosed. The definition and diagnosis, in turn, influence the statistics about who has AIDS. Such attributions also affect the content of the physician-patient interaction. In other words, if physicians do not believe a patient could be HIV-positive because he or she does not possess well-publicized attributes associated with risk (homosexual, young, male), they will not recommend testing (Henderson 1998; Villarosa 2003). Research shows, for example, that physicians do not suspect HIV/AIDS in older patients until their condition has reached advanced stages (see "What Is HIV/AIDS?")

In evaluating arguments about the origins of HIV/AIDS, we must recognize that we simply do not know how many people are infected with HIV in the United States

NO BORDERS, NO BOUNDARIES

Imports/Exports of Blood and Blood Products

U.S. CORPORATIONS SUPPLY about 60 percent of the world's plasma and blood products. The map shows the countries from which the United States imports blood and blood products and to which it exports them. For at least four years after HIV/AIDS was identified, U.S. blood bank officials continued to publicly affirm their faith in the safety of the country's blood supply, insisting that screening donors was unnecessary. Yet, these officials never revealed to the public the many shortcomings in production methods that could jeopardize the safety of blood. (By *shortcomings*, we mean deficiencies in medical knowledge and the level of technology rather than simply negligence.) Blood bank officials later argued that they practiced this concealment to prevent a worldwide panic. (In Goffman's terminology, blood bank officials did not want to "lose their audience.") Such a panic would have brought chaos to the medical system, which depends on blood products. This delay in implementing screening ultimately exposed many people to infection, especially hemophiliacs. Hemophiliacs were at higher risk of contracting HIV because

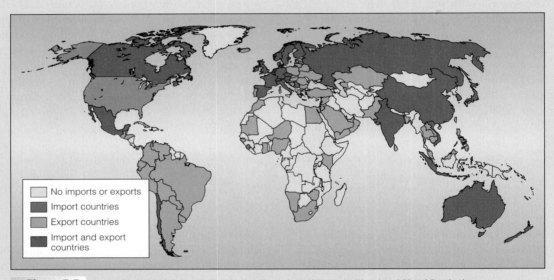

Legend:
- No imports or exports
- Import countries
- Export countries
- Import and export countries

▲ **Figure 5.5**   Countries to Which the United States Exports and Imports Blood and Blood Products

or worldwide. To obtain information on who is actually infected, every country in the world would have to administer blood tests to a random sample of its population. Unfortunately (but perhaps not surprisingly), people resist being tested. A planned random sampling of the U.S. population sponsored by the CDC was abandoned after 31 percent of the people in the pilot study refused to participate, despite assurances of confidentiality (Johnson and Murray 1988). Two researchers involved with this project concluded that "it does not seem likely that studies using data on HIV risk or infection status, even with complete protection of individual identity, will be practical until the stigma of AIDS diminishes" (Hurley and Pinder 1992, p. 625).

The United States is not the only country whose people do not want to be tested. Indeed, this resistance seems to be universal. For example, at one time Congolese officials were reluctant to disclose the number of AIDS cases and infection rates to UN officials or to allow foreign medical researchers to test their citizens because of sensitivity to the unsubstantiated but widely held belief that the Congo was the cradle of AIDS (Noble 1989).

their plasma lacks Factor VIII, a blood product that aids in clotting, or because their plasma contains an excess of anticlotting material (U.S. Department of Health and Human Services 1990). In fact, we now know that 50 percent of hemophiliacs became infected with HIV from contaminated Factor VIII treatments before the first case of AIDS appeared in this group (*Frontline* 1993).

To truly understand the global AIDS epidemic, we need to consider the role of exported blood products in transmitting HIV. The amount of blood products exported from the United States to other selected countries as a percentage of the receiving country's total need is graphed below. The data are from 1981, the year that HIV was "discovered" and scientists first learned that it was in the blood supply. Notice that at the time, Japan imported 98 percent of its blood products from the United States—46 million units of concentrated blood products and 3.14 million liters of blood plasma (Yasuda 1994).

Source: U.S. International Trade Administration (2006); International Federation of Pharmaceutical Manufacturers Associations (1981).

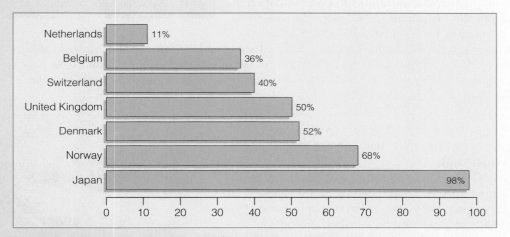

Amount of Blood Products Imported from the United States as a Percentage of Total Need

■ VISUAL SUMMARY OF CORE CONCEPTS

When sociologists study interaction, they seek to understand the social forces that bring people together. And once people come together, sociologists seek to identify the factors that shape the content and direction of interaction.

■ **CORE CONCEPT 1: When sociologists study interactions, they seek to understand the larger social forces that bring people together in interaction and that shape the content and direction of that interaction.**

IMTSSA Pharo Marseille

Social interaction occurs when two or more people communicate and respond through language and symbolic gestures to affect one another's behavior and thinking. In the process, the parties involved define, interpret, and attach meaning to the encounter. When sociologists study social interaction, they seek to understand and explain the larger social forces that (1) bring people together in interaction and (2) shape the content and course of interaction. With regard to HIV/AIDS, sociologists ask questions like these: What forces increase the likelihood that HIV-infected and noninfected people will interact? What meanings do the involved parties assign to their encounter? How do the parties interpret each other's actions and reactions? How do these meanings and interpretations shape the interaction's direction, course, and outcome?

■ **CORE CONCEPT 2: The division of labor is an important social force that draws people into interaction with one another and shapes their relationships.**

Urbain Ureel

In *The Division of Labor in Society* ([1933] 1964) Durkheim provides a general framework for understanding the forces underlying global-scale interactions. As Durkheim describes it, *division of labor* refers to a product or service that is broken down into specialized tasks, each performed by workers specifically trained to do that task. The workers do not have to live near each other; they often live in different parts of the country or different parts of the world. Not only are the tasks geographically dispersed but the parts and materials needed to manufacture products also come from many locations around the world. The Congolese workers shown here pushing a barge out to sea perform one of the tasks related to transporting products to European and other consumers.

■ **CORE CONCEPT 3:** Solidarity—the ties that bind people to one another and shape their interactions—can be mechanical or organic.

Solidarity describes the ties that bind people to one another in a society and shape their interaction. Mechanical solidarity is characteristic of pre-industrial societies, and *organic* solidarity is characteristic of industrial societies. Mechanical solidarity characterizes a social order based on a common conscience, or uniform thinking and behavior. The similarity derives from the simple division of labor and the corresponding lack of specialization. A society with a complex division of labor is characterized by organic solidarity—social order based on interdependence and cooperation among people performing a wide range of diverse and specialized tasks. The man getting his blood drawn does not need to know the woman personally to interact with her; the interaction is strictly instrumental.

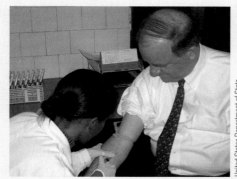
United States Department of State

■ **CORE CONCEPT 4:** Disruptions to the division of labor profoundly affect interaction and an individual's ability to connect with others meaningfully.

Durkheim maintained that people become more vulnerable as the division of labor becomes more complex and more specialized. He was particularly concerned with the kinds of events that break down individuals' ability to connect with others in meaningful ways through their labor. Such events include (1) industrial and commercial crises caused by plant closings, massive layoffs, crop failures, severe weather (such as Hurricane Katrina, which disrupted the work lives of thousands of people), technological revolutions, and war; (2) workers' strikes; (3) job specialization; (4) forced division of labor; and (5) inefficient management and development of workers' talents and abilities, so that work for them is nonexistent, irregular, intermittent, or subject to high turnover.

© FEMA photo/Andrea Booher

■ **CORE CONCEPT 5:** When analyzing any social interaction, sociologists identify the social statuses of the participating parties.

Social status is a position in a social structure. A social structure consists of two or more people interacting in expected ways. Individuals usually occupy more than one social status. Sociologists use the term *status set* to capture all the statuses a person assumes. Sociologists distinguish among ascribed, achieved, and master statuses. Ascribed statuses result from chance—that is, the individual exerts no effort to obtain them (the man pictured occupies the ascribed statuses of male and disabled). Achieved statuses are acquired through some combination of personal choice, effort, or ability, such as occupation and educational attainment. Ascribed statuses can play a role in determining achieved statuses. Sometimes one status, such as disability, is so important to a person's social identity that it overshadows all other statuses that that person occupies—shaping every aspect of life and dominating the course of social interaction. Such statuses are known as master statuses.

© Adam Clark/stock.xchng/HAAP Media Ltd.

■ CORE CONCEPT 6: People occupy statuses and perform roles.

A social status has meaning only in relation to other social statuses. The common characteristic of all social structures is that it is possible to broadly anticipate the behavior of people occupying each status, no matter the personality of the occupant. Just as the behavior of a person occupying the status of a football quarterback is broadly predictable, so too is the behavior of a person occupying the status of nurse, secretary, mechanic, patient, or physician. Sociologists use the term *role* to describe the behavior, obligations, and rights expected of a status in relation to another status. The two men occupying the status of surgeon are performing the role of operating on a person occupying the status of patient.

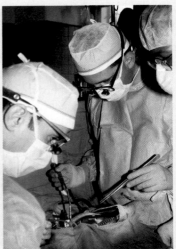

Department of Defense

■ CORE CONCEPT 7: Social interaction can be viewed as if it were a theater, people as if they were actors, and roles as if they were performances before an audience.

Social roles can be equated with the dramatic roles played by actors. In this model, social interaction is viewed as though it were theater, people as though they were actors, and roles as though they were performances before an audience in a particular setting. People in social situations resemble actors in that they must be convincing to others and must demonstrate who they are and what they want through verbal and nonverbal cues. In social situations, as on a stage, people manage the setting, their dress, their words, and their gestures to correspond to the impression they are trying to make or to the image they are trying to project. People alter their behavior depending on whether they are on the front stage (area visible to the audience, where people take care to create and maintain expected images and behavior) or back stage (area out of the audience's sight, where individuals let their guard down and do things that would be inappropriate or unexpected on the front stage).

Nic McPhee

■ CORE CONCEPT 8: People assign causes to their own and others' behaviors. That is, they propose explanations for their own and other's behaviors, successes, and failures and then they respond accordingly.

Attribution theory relies on the assumption that people make sense of their own and others' behavior by assigning a cause. In doing so they often focus on dispositional (internal) or situational (external) causes. Dispositional causes are forces over which individuals are supposed to have control—including personal qualities or traits, such as motivation level, mood, and effort. Situational causes are forces outside an individual's control, such as the weather, bad luck, and others' incompetence. Usually people stress situational factors to explain their own failures and dispositional factors to explain their own successes. With regard to other people's failures or shortcomings, people tend to stress dispositional factors. With regard to others' successes, people tend to emphasize situational factors. In the aftermath of Hurricane Katrina some observers believed the Gulf Coast residents did not heed warnings to leave (dispositional factor), while others argued that many people lacked transportation and money to evacuate (situational).

© Jocelyn Augustino/FEMA

Resources on the Internet

 Sociology: A Global Perspective Book Companion Web Site

www.cengage.com/sociology/ferrante

Visit your book companion Web site, where you will find flash cards, practice quizzes, Internet links, and more to help you study.

CENGAGENOW™

Just what you need to know NOW!

Spend time on what you need to master rather than on information you have already learned. Take a pre-test for this chapter, and CengageNOW will generate a personalized study plan based on your results. The study plan will identify the topics you need to review and direct you to online resources to help you master those topics. You can then take a post-test to help you determine the concepts you have mastered and what you will need to work on. Try it out! Go to www.cengage.com/login to sign in with an access code or to purchase access to this product.

Key Terms

achieved statuses 127
ascribed statuses 127
back stage 131
dispositional causes 133
division of labor 118
dramaturgical model 130
front stage 131
impression management 130

master status 127
mechanical solidarity 121
organic solidarity 121
right 128
role 128
role conflict 128
role obligations 128
role strain 128

scapegoat 134
situational causes 133
social interaction 118
social status 125
social structure 125
solidarity 121
status set 141
Thomas theorem 133

6 Formal Organizations

With Emphasis on McDonald's Operations

Sociologists view formal organizations as having a life that extends beyond the people who make them run. This idea is supported by the fact that most organizations continue to exist even as their members or employees die, quit, retire, or get fired.

© Bill Greenblatt/UPI/Landov

▲ Little two-year-old Gregory Chew tries to get the attention of the Ronald McDonald that used to adorn the "Big Mac Land" at the old Busch Stadium, during Fredbirds Garage Sale at America's Center in St. Louis, on November 27, 2005.

WHY FOCUS ON

McDonald's?

THIS CHAPTER FOCUSES on McDonald's, a fast-food service corporation headquartered in the United States. At this writing, McDonald's has 31,667 units in 118 countries (McDonald's 2007). On any given day, approximately 50 million people eat at a McDonald's restaurant, and the fast-food chain estimates that it handles 18.3 billion customer orders per year (McDonald's 2007). Sixty percent of McDonald's sales and profits are derived from its international units. As one measure of its global presence, McDonald's ranks 318th out of the 500 largest global corporations (*Fortune* 2006).

Our emphasis on the McDonald's organization in this chapter does not mean that McDonald's is better or worse than any other organization. Focusing on this well-known corporation allows us to apply the concepts that sociologists use to analyze formal organizations—in particular, to understand and appreciate their power, the way they function, their internal dynamics, and their impact on society (Perrow 2000).

FACTS TO CONSIDER

- McDonald's is the world's largest fast-food service chain and one of the world's largest purchasers of beef, chicken, and pork (buying 2.5 billion pounds per year) and of eggs (buying a billion per year) (Barboza 2003a, 2003b).

- Each year, McDonald's handles more than 18 billion customer visits worldwide.
- After Santa Claus, Ronald McDonald is the world's most recognized personality (McDonald's 2006).

- When McDonald's opened its first restaurant in Kuwait City, cars lined up for seven miles to experience drive-through service (McDonald's 2006).

Formal Organizations

■ **CORE CONCEPT 1:** The formal organization, a type of secondary group, is a coordinating mechanism people create to achieve a planned outcome. In this chapter, we examine the concepts that sociologists use to analyze **formal organizations**, coordinating mechanisms people create to achieve a planned outcome. Formal organizations are viewed as coordinating mechanisms because they bring together people, resources, and technology and then channel human activity toward achieving a specific outcome. That outcome may be to maintain order in a community (as does a police department); to challenge an established order (as does People for the Ethical Treatment of Animals); to keep track of people (as does a census bureau); to grow, harvest, or process food (as does PepsiCo); to sell goods (as does Wal-Mart); to produce oil (as does Exxon Mobil); or to provide a service (as does a hospital) (Aldrich and Marsden 1988).

Sociologists classify formal organizations as secondary groups, thereby distinguishing them from primary groups. Recall from Chapter 4 that primary groups are "fundamental in forming the social nature and ideals of the individual" and fuse members into a common whole (Cooley 1909, p. 23). Primary groups are characterized by face-to-face contact and by strong, emotional ties among members, who feel considerable allegiance to one another and strive to achieve "some desired place in the thoughts of [the] others" (Cooley 1909, p. 24). Examples of primary groups are the family, military units, and peer groups.

Secondary groups are impersonal associations among people who interact with a specific purpose. Thus, by definition, secondary-group relationships are narrow in scope (confined to a particular setting and specific tasks) and are seen as a means to achieve some agreed-upon end. Members relate to each other in terms of specific roles. One key to distinguishing primary and secondary groups is to examine the breadth of the relationships among members. If relationships are limited to a specific activity and setting and if members only know others by what they do and not by who they are outside that setting, the group would certainly be considered secondary.

Secondary groups can range from small to extremely large. Examples of small secondary groups include a work unit, college students assigned to work together on a class

formal organization Coordinating mechanisms that bring together people, resources, and technology and then channel human activity toward achieving a specific outcome.

secondary groups Impersonal associations among people who interact for a specific purpose.

This grandfather (right) and his son and grandson, who all share the same first and last names, constitute a primary group.

The 4,300 students enrolled in the United States Naval Academy constitute a secondary group.

project, and parent-teacher associations. Formal organizations, many of which are large, also qualify as secondary groups. Specific examples include McDonald's, the United States Naval Academy, and the Peace Corps. Keep in mind that members of secondary groups can form primary groups if they expand their relationships beyond the task at hand. Taken as a whole, the United States Naval Academy which enrolls 4,300 students is also a secondary group, as it is impossible for that number of people to interact with everyone in the group on a personal basis. Formal organizations are a taken-for-granted aspect of life. Simply consider that the human biography can be chronicled by encounters with them: "born in a hospital, educated in a school system, licensed to drive by a state agency, loaned money by a financial institution, employed by a corporation, cared for by a hospital or nursing home, and at death served by as many as five organizations—a law firm, a probate court, a religious organization, a mortician, and a florist (Aldrich and Marsden 1988, p. 362). Formal organizations can be voluntary, coercive, or utilitarian, depending on the reason that people participate (Etzioni 1975).

Voluntary organizations (also known as voluntary associations) draw in people who give time, talent, or treasure to support mutual interests, meet important human needs, or achieve a not-for-profit goal. Alexis de Tocqueville (1882) was the first to observe that "Americans of all ages, all conditions, and all dispositions constantly form associations" and that those "associations are created, extended, and worked in the United States more quickly and effectively than in any other country" (p. 129). Voluntary organizations include community service centers, politically oriented groups, religious organizations (such as ministries, churches, mosques, and synagogues), historical societies, and sports associations. Families and individuals who join the Ray and Joan Kroc Community Centers (named after the founder of McDonald's and his wife, who left $1.5 billion to fund such centers nationwide) qualify as members of a voluntary organization (Allen 2006).

Coercive organizations draw in people who have no choice but to participate. Organizations dedicated to compulsory socialization, such as elementary schools and the military (when there is a draft), qualify as coercive organizations. Other such organizations include resocialization or treatment facilities that work with individuals labeled as deviant because they are (or are seen as) not moving

along normal pathways or not adequately performing their social roles (Spreitzer 1971).

Utilitarian organizations draw people seeking material gain in the form of pay, health benefits, or a new status (as conferred through a college degree, certification, or voluntarily undergoing some treatment). McDonald's qualifies as a utilitarian organization as does any organization that employs people.

From a sociological perspective, formal organizations can be studied apart from the people who designed and created them. Indeed, even as their members or employees die, quit, retire, or get fired, formal organizations have a life that extends beyond the people who constitute them (unless a catastrophe hits and the organization is completely destroyed).

The Concept of Bureaucracy

■ **CORE CONCEPT 2: Most formal organizations can be classified as bureaucracies, organizational structures that strive to use the most efficient means to achieve a valued goal.** McDonald's Corporation coordinates the activities of 447,000 employees in more than 31,600 restaurants located in 118 countries to feed 50 million people each day. The concept of bureaucracy can help us appreciate the organizational structure behind this monumental feat.

Sociologist Max Weber defined a **bureaucracy**, in theory, as a completely rational organization—one that uses the most efficient means to achieve a valued goal, whether that goal is feeding people, making money, recruiting soldiers, counting people, or collecting taxes. A bureaucracy has seven major characteristics that allow it to coordinate people so their actions focus on achieving the organization's goals. After identifying each characteristic, we provide an example in parentheses, applying the characteristic to the McDonald's Corporation's organizational structure.

Department of Defense photo by PH2 Andrea Simmons, USN

Habitat for Humanity qualifies as a voluntary organization because people volunteer their time and talent to build affordable housing for low-income families. The houses are affordable because the labor costs and profit are not included in the sale price of these homes.

voluntary organizations Formal organizations that draw together people who give time, talent, or treasure to support mutual interests, meet important human needs, or achieve a not-for-profit goal.

coercive organizations Formal organizations that draw in people who have no choice but to participate; such organizations include those dedicated to compulsory socialization or to resocialization or treatment of individuals labeled as deviant.

utilitarian organizations Formal organizations that draw together people seeking material gain in the form of pay, health benefits, or a new status.

bureaucracy An organization that strives to use the most efficient means to achieve a valued goal.

1. A clear-cut division of labor exists: Each office or position is assigned a specific task geared toward accomplishing the organizational goals. (One of McDonald's Corporation's organizational goals is to deliver a meal 90 seconds after it is ordered. When a customer places an order for, say, a Big Mac, an order taker types it into a computer and the order appears on a video screen in the kitchen. An employee known as the initiator reads the screen. Other employees who work in assembly warm the bun, put on the main ingredients, including condiments and other toppings, and then wrap the sandwich and send it to the order taker, who hands it to the customer [*CNN Newstand Fortune* 1998].)

2. Authority is hierarchical: Each lower office or position is under the control and supervision of a higher one. (Individual McDonald's franchises are under the control of the corporate office; employees at each franchise are under the control of managers and assistant managers.)

3. Written rules specify the exact nature of relationships among personnel and describe the way tasks should be carried out. (McDonald's issues a 600-page *Operations and Training Manual* that specifies everything from where sauces should be placed on buns to how thick pickle slices should be [Watson 1997].)

4. Positions are filled based on objective criteria (such as academic degree, seniority, merit points, or test results) and not based on emotional considerations, such as family ties or friendship. (Each year, McDonald's Corporation receives 36,000 franchise applications, of which 360 will result in new franchises. One criterion for buying a McDonald's franchise is that an applicant must have $200,000 of unborrowed cash on hand [Williams 2006].)

5. Administrative decisions, rules, procedures, and activities are recorded in a standardized format and preserved in permanent files (such as McDonald's *Operations and Training Manual*).

6. Authority belongs to the position, not to the particular person who fills that position. This characteristic implies that one has authority over subordinates who are on the job because one holds a higher position, but one has little to no authority over subordinates who are not on the job (unless, for example, the company requires workers to lead a drug-free life). (A McDonald's manager has authority over other employees only when they are on the time clock, working for McDonald's. The manager cannot demand that other employees do nonwork-related tasks, such as washing the manager's car or babysitting his or her children.)

Organizations treat customers as "cases." If interaction between staff and customers were personalized, the speed of service would decline.

7. Organizational personnel treat clients as "cases" and "without hatred or passion, and hence without affection or enthusiasm" (Weber 1947, p. 340). To put it another way, no one receives special treatment. This utilitarian approach is believed necessary because emotion and special circumstances can interfere with the efficient delivery of goods and services. (According to standard operating procedures, every customer should be greeted with the words "Welcome to McDonald's. May I take your order?")

Taken together, these seven characteristics describe a bureaucracy as an **ideal type**—"ideal" not in the sense of being desirable, but as a standard against which real cases can be compared. An ideal type of a bureaucracy is a deliberate simplification or caricature, in that it exaggerates the defining characteristics of a bureaucracy (Sadri 1996). Anyone involved with an organization realizes that actual behavior departs from this ideal.

Formal and Informal Dimensions

■ **CORE CONCEPT 3:** Organizations include both formal and informal dimensions: The formal dimension is the official side of the organization governed by written guidelines, rules, and policies; the informal dimension includes employee-generated norms that depart from or otherwise bypass the formal dimension. Ideally organizations such as McDonald's have well-defined and predictable job descriptions, rules governing relationships among personnel, and procedures for carrying out work-related tasks. The actual workings of organizations are not always clear and predictable, however, because the people in organizations vary in their ability or willingness to follow through—as when man-

ideal type A deliberate simplification or caricature that exaggerates defining characteristics, thus establishing a standard against which real cases can be compared.

This sign reminds food service workers to wash their hands. Many food service workers fail to follow the policy before handling food. This informal practice explains, in part, why each year 76 million Americans get sick, 300,000 are hospitalized, and 5,000 die from food-borne illnesses (Centers for Disease Control and Prevention 2006).

agers hire family and friends or when employees do not always treat customers the same.

Sociologists distinguish between formal and informal dimensions of organizations. The **formal dimension** is the official aspect of the organization; it consists of job descriptions and written rules, guidelines, and procedures established to achieve valued goals. The **informal dimension** includes behaviors that depart from the formal dimension, such as employee-generated norms that evade, bypass, or ignore official policies and regulations (Sekulic 1978). For example, a manager may expect employees to work off the clock to meet goals related to labor costs; employees may give their friends free food and soft drinks when the manager is not looking; and servers may spit in a rude customer's drinks. In a British court case involving McDonald's, many former employees testified about employees and managers who routinely violated formal corporate policies to meet profit-related goals—squeezing French fry boxes before filling them to make them look fuller; watering down soft drinks, syrups, and shake mix; and failing to throw away food that had dropped on the floor (Beech 1994; Brett 1993; Coton 1995).

Another aspect of the informal organization is worker-generated norms that govern output or physical effort. They include informal norms against working too quickly (workers who do so are often called "rate busters"), working too slowly, or slacking off, as well as unofficial norms about the number and length of breaks. According to the Second Annual Survey of Restaurants and Fast Food Employees, worker-generated norms may govern how workers respond to managers in the industry who workers think have treated them unfairly. On average, entry-level fast-food service employees claim they each steal $238.72

in cash and merchandise per year, and they are more likely to do so when they believe they have been treated unfairly (Lowe 1997).

In light of these deviations from the ideal, one might ask, Why use the seven essential characteristics of bureaucracy to analyze an organization? These characteristics are useful tools because they focus our attention on important organizational features. By comparing an actual operation against the seven ideals, sociologists can evaluate how much a formal organization departs from them or adheres to them. As we will learn, many organizational problems result from not following official policies, but many also result from following them too rigidly. One process that can push organizational behavior to extremes is rationalization.

Rationalization

■ **CORE CONCEPT 4:** The concept of rationalization—a process in which thought and action rooted in custom, emotion, or respect for mysterious forces is replaced by instrumental-rational thought and action—helps us understand how striving to achieve valued goals can have undesirable, even disastrous, consequences. Formal organizations such as McDonald's strive to find the most efficient (time-saving and other cost-cutting) means to achieve its most valued goal, which is turning a profit. Rationalization describes the dynamics underlying the never-ending quest for efficiency. While that quest can result in amazing feats, such as filling 50 million orders per day, it can also have unintended, destructive consequences for workers, the public, and the environment.

The growth and dominance of formal organizations goes hand in hand with a process known as rationalization. Weber defined **rationalization** as a process in which thought and action rooted in emotion (such as love, hatred, revenge, or joy), superstition, respect for mysterious forces, or tradition is replaced by instrumental-rational thought and action. Through instrumental-rational thought and

---

formal dimension The official aspect of an organization, including job descriptions and written rules, guidelines, and procedures established to achieve valued goals.

informal dimension The unofficial aspect of an organization, including behaviors that depart from the formal dimension, such as employee-generated norms that evade, bypass, or ignore official rules, guidelines, and procedures.

rationalization A process in which thought and action rooted in custom, emotion, or respect for mysterious forces is replaced by instrumental-rational thought and action.

action, people strive to find the most efficient way (means) to achieve a valued goal (see Chapter 1) (Freund 1968).

One way to show how thought and action guided by emotion, tradition, superstition, or respect for mysterious forces differ from value-rational thought and action is to compare two distinct thought patterns that drive human behavior toward farm animals. One thought pattern is articulated by Matthew Scully (2003) in his book *Dominion: The Power of Man, the Suffering of Animals, and the Call to Mercy*. Scully portrays farm animals (all animals, for that matter) as possessing complex emotions, including love and sorrow. He argues that humans have a "moral responsibility to treat the animals in our care with kindness, empathy, and thoughtfulness. . . . It is wrong to be cruel to animals, . . . and when our cruelty expands and mutates to the point where we no longer recognize the animals in a factory farm as living creatures capable of feeling pain and fear, . . . we debase ourselves" (Angier 2002, p. 9). Clearly Scully believes that emotion must guide treatment of animals.

Instrumental-rational, or by-any-means-necessary to achieve the desired end thinking, drives the treatment of animals raised on factory farms. Obviously, the more chickens a factory farm can house and the faster it can raise them to egg-laying maturity, the more meat and eggs it can produce and sell. Egg and chicken suppliers raise chickens in crowded conditions that give each chicken an average of 49 square inches or 7 by 7 inches of space. Because chickens live in such close quarters, factory farm workers clip the wings and trim the beaks to prevent hens from injuring one another. Egg suppliers regularly practice "forced mating" by depriving hens of food and water for as long as two weeks (*Food Institute Report* 2000b; Yablen 2000). Apparently this practice increases egg production: Facing starvation, desperate hens will lay eggs so they can eat them. Of course, the eggs are taken before that can happen.

Weber made several important qualifications regarding instrumental-rational thought and action. First, he used the term *rationalization* to refer to the way daily life is organized socially to accommodate large numbers of people, but not necessarily to the way individuals actually think (Freund 1968). Individuals actually think that home-cooked meals are preferable to fast food, but they find it difficult to resist the efficiency of fast-food service, which can feed millions of people per day:

> In a highly mobile society in which people are rushing, usually by car, from one spot to another, the efficiency of a fast-food meal, perhaps without leaving one's car while passing by the drive-through window, often proves impossible to resist. The fast-food model offers us, or at least appears to offer us, an efficient method for satisfying many of our needs. (Ritzer 1993, p. 9)

Second, rationalization does not assume better understanding or greater knowledge. In fact, Weber argues that

People imagine that the chickens they eat come from farms where the animals are free to move around in the open air.

Instrumental-rational action guides human behavior toward chickens raised on factory farms. Under this system, chickens have little space to move around and will likely never run or see the light of day.

"'primitive' man in the bush knows infinitely more about the conditions under which he lives, the tools he uses, and the food he consumes" (Freund 1968, p. 20). People who live in an instrumental-rational environment typically know little about their surroundings (nature, technology, the economy). The consumer buys any number of products in the grocery store without knowing how they were made or what they consist of. Likewise, people who eat fast-food products regularly have no concept of how these foods were produced or how the foods can affect the body. In other words, they have no understanding of the processes by which the digestive system "ingests, absorbs, transports, utilizes, and excretes food substances" (National Library of Medicine 2004). People are not troubled by such ignorance; rather, they are content to let specialists or experts make corrections when something goes wrong with their digestive system or their general health.

Finally, when people identify a desired goal and decide on the means (actions) to achieve it, they seldom consider or dismiss less profitable or slower ways to achieve it. For example, people often turn to technology as the means to solve a problem they define as important. The problem may be as seemingly simple as growing potatoes that can be processed into French fries or potato chips that look and taste the same. In creating such a potato, people fail to consider that the demand for uniformity limits the varietal range of potatoes to those that are high-yielding, high in dry matter, low in sugar content, long and oval in shape, and uniform in color and flavor (International Potato Center 1998). With the help of science, such potatoes can be produced, but they require heavy doses of chemicals and threaten the longer-term viability of domestic potato production to meet commercial processing requirements (International Potato Center 1998).

The McDonaldization of Society

■ **CORE CONCEPT 5:** One organizational trend guided by value-rational action is the McDonaldization of society, a process in which the principles governing fast-food restaurants come to dominate other sectors of society. Of course, McDonald's and other fast-food service organizations are not the only formal organizations that strive to deliver a product or service by the most efficient means possible. Sociologist George Ritzer (1993) describes a larger organizational trend guided by value-rational action: the "McDonaldization" of society.

Ritzer sees the **McDonaldization of society** as "the process by which the principles of the fast-food restaurant are coming to dominate more and more sectors of American society as well as the rest of the world" (p. 1). Those principles are (1) efficiency, (2) quantification and calculation, (3) predictability, and (4) control. **Efficiency** is an organization's claim to offer the "best" products and services, which allow consumers to move quickly from one state of being to another (for example, from hungry to full, from fat to thin, from uneducated to educated, or from sleeplessness to sleep). **Quantification and calculation** are numerical indicators that enable customers to evaluate a product or service easily (for example, get delivery within 30 minutes, lose ten pounds in 10 days, earn a college degree in 24 months, limit menstrual periods to four times a year, or obtain eyeglasses in an hour). **Predictability** is the expectation that a service or product will be the same no matter where or when it is purchased. With regard to food products, this kind of predictability requires that they be modified. If the consumer expects cheese, for example, to have melted and to taste the same each time they eat it, it cannot be a naturally made cheese

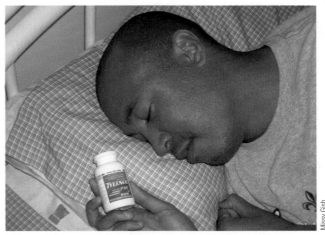

This man takes a pill to help in sleep at night. He chose the brand because it claims to help people sleep and because the pills' effect generally lasts between four to six hours.

Missy Gish

(Barrionuevo 2007). **Control**, "especially through the substitution of nonhuman for human technology" (Ritzer 1993), is the guiding or regulating, by detailed planning, of the production and delivery of a service or product (for example, by assigning a limited task to each worker, by filling soft drinks from dispensers that shut off automatically, or by having customers stand in line).

Ritzer's model allows for amazing organizational feats: each year, McDonald's handles 18.3 billion customer visits worldwide; pharmaceutical companies fill 42 million prescriptions for sleeping pills, "rescuing" consumers from an estimated 1.4 billion sleepless nights; and manufacturers of prefabricated houses assemble 1.3 million of them, taking hours to a few days to assemble each (Saul 2006). But the McDonaldization model also has drawbacks. Max

McDonaldization of society "The process by which the principles of the fast-food restaurant are coming to dominate more and more sectors of American society as well as the rest of the world" (Ritzer 1993, p. 1).

efficiency An organization's claim of offering the "best" products and services, which allow consumers to move quickly from one state of being to another (for example, from hungry to full, from fat to thin, or from uneducated to educated).

quantification and calculation Numerical indicators that enable customers to evaluate a product or service easily.

predictability The expectation that a service or product will be the same no matter where or when it is purchased.

control The guiding or regulating, by planning out in detail, the production or delivery of a service or product.

Weber used the phrase **iron cage of rationality** to describe the set of irrationalities generated by rational systems. "Ultimately, we must ask whether the creations of these rationalized systems create an even greater number of irrationalities" (Ritzer 1993). For example, it may seem efficient to divide up production so that "even a moron could learn to do the job." At the same time, a work setting that requires so little skill generates high employee turnover: in the fast-food industry, the annual turnover rate can be as high as 100 percent (Ritzer 1993). Likewise, it may seem rational to develop, produce, and modify potatoes that allow fries to be made uniform, but the overuse of pesticides and other chemicals the practice requires is irrational.

Value-Rational Action and Expanding Market Share

■ **CORE CONCEPT 6:** Value-rational action, or behavior guided by means-to-end thinking aimed at increasing market share and profits, helps explain how an organization expands its reach beyond local markets to regional, national, and global markets. One way to understand how organizations grow from local operations (McDonald's began as one restaurant in Illinois in 1955) to global giants is to consider the actions they have taken to increase profits. Organizations employ five major strategies to reach the valued goal of turning a profit: (1) lower production costs, (2) create new products that consumers "need" to buy, (3) improve existing products to make previous versions obsolete, (4) identify ways for people to purchase more products, and (5) create new markets.

Lower Production Costs. Hiring employees who will work for lower wages, introducing labor-saving technologies, reducing the number of employees, and moving production facilities out of high-wage zones are major ways corporations reduce production costs. McDonald's introduced a major labor-saving technology in the mid-1960s, when it replaced on-site potato peeling and slicing appliances with a system of flash-freezing half-cooked potatoes. This change not only lowered the labor cost associated with making fries, but also allowed the company to offer uniform-looking fries (Johnson 1997).

Create New Products That Consumers "Need" to Buy. Obviously, every new menu item represents a "new product." McDonald's and other fast-food chains have been particularly successful at marketing their products to chil-

iron cage of rationality The set of irrationalities that rational systems generate.

Headquarters: Oak Brook, Illinois
First restaurant: Des Plaines, Illinois, 1955
First foreign franchise: Canada, 1967
First foreign franchise outside North America: Costa Rica
Latest foreign franchise: French Guiana
Number of customer visits to McDonald's worldwide: 50 million/day; 18.3 billion/year
Number of U.S. restaurants: 13,774
Number of foreign-located restaurants: 17,893
Percentage of corporation profits derived from foreign sales: 66%
Total revenue, 2006: $20.4 billion
Total number of employees worldwide: 465,000
Most well-known charitable activity: Ronald McDonald House
U.S. Fortune 500 rank: 109
Global Fortune 500 rank: 318

▲ **Figure 6.1**   McDonald's at a Glance

Sources: Fortune (2007); McDonald's Corporation (2007)

dren. The company offers Happy Meals, which come with toys that many children (and some adults) "need to have," such as Teenie Beanie Babies and toys tied in with Disney movies, such as *Finding Nemo* and *The Haunted Mansion.*

Improve Existing Products to Make Previous Versions Obsolete. This strategy includes making existing food items bigger and better. Fast-food companies compete, for example, to make better (hotter or crispier) French fries than their competitors do.

Identify Ways for People to Purchase More Products. Obviously, one way for a fast-food company to increase profits is to persuade existing customers to eat more (Langley 2003). In fact, one might argue, at least from a purely business perspective, that the "fattening of America" and other places in the world "may well have been a necessity" (Critser 2003). Suggestive selling is one way to increase the amount of food customers order. Order takers ask customers if they would like to "super-size" their value meal—meaning that, for a small additional charge (of between 39 and 50 cents), customers can receive a large French fry order and drink instead of medium-sized ones with their meal. Market researchers have found that when consumers use credit cards to defer payments until a later date, they are likely to spend more, and thus buy more, at the time of purchase. Fast-food restaurants did not use to accept credit card payments, but that has changed. In 2004, McDonald's reached agreements with Visa, Mastercard, American Express, Discover Card, and Star that

had franchises for more than five years but have room for more franchises. The company has also identified more than three dozen countries (such as Bolivia, Egypt, India, and the Ukraine) where it is working to introduce the fast-food concept.

In their drive to create new markets, executives of fast-food corporations have come to realize that customers make choices about where to eat on the spur of the moment, based on speed of service and convenience (the closer the restaurant, the better the chance customers will choose to eat there). These assumptions inspired McDonald's to open its first drive-through window in 1975 (Tanner 2006). They also inspired McDonald's and other fast-food corporations to locate restaurants in Wal-Mart and Home Depot stores, gas stations, malls, airports, and hospitals, as well as on college campuses and military bases (Crecca 1997; Mannix 1996).

Multinational and Global Corporations

Multinational corporations (or just "multinationals") are enterprises that own, control, or license production or service facilities in countries other than where the corporations are headquartered. It is difficult to estimate the number of multinationals in the world, because Internet technologies allow as few as two people in different locations to form a corporation. The last estimate made

Staying open 24 hours has helped create the concept of a fourth meal—a meal between dinner and breakfast. This is one strategy that restaurants and other businesses use to sell more products.

© Daniel T. Yara/MorgueFile

allow use of those cards to buy its products (Associated Press 2004). A strategy Taco Bell uses to sell more food is to stay open until 1:00 a.m.; in conjunction with longer operating hours, they invite customers in for a fourth meal—a meal between dinner and breakfast. Similarly, many McDonald's franchises offer 24-hour service. A final strategy involves giving away free items to boost overall product sales. McDonald's gives away free coffee to boost breakfast sales; Wendy's sponsored a 25-city promotional taste tour, giving away 1,000 free hamburgers and 2,000 gift cards worth $5 at each stop (MacArthur 2007).

Create New Markets. McDonald's began its expansion outside the United States in 1967, when it opened a unit in Canada. It now operates units in more than 100 foreign countries (see No Borders, No Boundaries: "Locations of McDonald's by First Year a Franchise Opened"). In 2006 alone the corporation opened about 800 new restaurants worldwide (McDonald's 2006). In an annual report to stockholders, McDonald's distinguishes between established, well-penetrated markets (such as Australia, Brazil, Japan, and the United States) and emerging markets (such as Argentina, China, Italy, and Spain), which have

Ray Elfers

As a strategy for creating new markets, McDonald's opened its first drive-through window in 1975. Now drive-through windows are a standard feature of most fast-food restaurants.

multinational corporations Enterprises that own, control, or license production or service facilities in countries other than the one where the corporations are headquartered.

NO BORDERS, NO BOUNDARIES

Locations of McDonald's by First Year a Franchise Opened

THE FIRST MCDONALD'S restaurant opened in 1955 in Des Plaines, Illinois. Since then, operations have spread to more than 100 other countries, the most recent being French Guiana. Franchise holders are McDonald's employees and independent businesses. That is, they operate their own restaurants but must adhere to strict operating guidelines (Waters 1998). One way the corporation builds its global identity across units is by requiring all franchise holders to attend a two-week course on quality control and management at one of its four Hamburger Universities or other international training centers (Watson 1997).

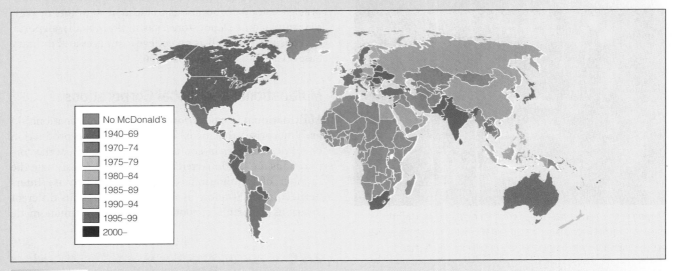

▲ **Figure 6.2**

Source: Wikipedia (2009).

by the United Nations put the number at 65,000 multinationals, with 820,000 foreign affiliates (Chanda 2003). Multinationals are headquartered disproportionately in the United States, Japan, and Western Europe (see Global Comparisons: "The Size of the Top 10 Global Corporations Relative to National Economies"). A multinational corporation can range in size from fewer than 10 employees to millions. In fact, most multinationals employ 250 people or fewer (Gabel and Bruner 2003). Regardless of their size, multinationals compete, plan, produce, sell, recruit, extract resources, acquire capital, and/or borrow technology on a multicountry scale (Kennedy 1993; Khan 1986; U.S. General Accounting Office 1978).

Multinationals establish operations in foreign countries for many reasons, including to obtain raw materials (such as oil and diamonds); to avoid paying taxes; to employ an inexpensive labor force; to provide low-cost services, such as call centers and help desks; and to manufacture goods, provide services, or sell products to consumers in a host country (as does Toyota, North America, Inc.). The world's largest multinational corporations are often referred to as "global corporations." Theoretically, a truly global corporation should have some kind of economic relationship with every country in the world. Probably no corporation is yet global in that sense. Still, because of their size and reach, many corporations are classified as global—most notably, those on the Fortune Global 500 list.

Critics of multinational corporations maintain that they are engines of destruction. That is, they exploit people and natural resources to manufacture products inexpensively. They take advantage of desperately poor labor forces, lenient environmental regulations, and sometimes nonexistent worker safety standards. Supporters of multinational corporations, by contrast, maintain that these companies are agents of progress. They praise the multinationals' ability to raise standards of living, increase

GLOBAL COMPARISONS

The Size of the Top 10 Global Corporations Relative to National Economies, 2008

ONE REASON THAT the largest multinational organizations have great influence on the societies in which they operate is related to their size. Taken together, the annual revenues of the top 10 global corporations are about $2.6 trillion. Only five countries in the world—the United States, China, Japan, India, and Germany—possess a gross national product that exceeds that amount. The annual revenue of the world's largest corporation, Wal-Mart, exceeds $378 billion. Only 30 countries have a gross national product that exceeds that amount: the United States, China, Japan, India, Germany, the United Kingdom, Russia, France, Brazil, Italy, Mexico, Spain, Canada, South Korea, Indonesia, Turkey, Iran, Australia, Taiwan, the Netherlands, Poland, Saudi Arabia, Argentina, Thailand, South Africa, Pakistan, Egypt, Colombia, Belgium, and Malaysia. Notice that petroleum industries dominate (Total is a French-based oil, gas, and chemical company). Since this list was created, GM declared bankruptcy and restructured its organization. GM will likely drop off this list when *Fortune* Magazine next releases it.

The World's Largest Global Corporations, 2008

Corporation	Revenues (in $ millions)	Profits (in $ millions)	Headquarters
Wal-Mart Stores	378,799	12,731	Bentonville, AR
Exxon Mobil	372,824	40,610	Irving, TX
Royal Dutch Shell	355,782	31,331	The Hague, Netherlands
BP	291,438	20,845	London, UK
Toyota Motor	230,201	15,042	Aichi, Japan
Chevron	210,783	18,688	San Francisco, CA
ING Group	201,516	12,649	Amsterdam, Netherlands
Total	187,280	18,042	Courbevoie, France
General Motors	182,347	−38,732	Detroit, MI
ConocoPhillips	178,558	11,891	Odessa, TX

Source: Fortune (2008)

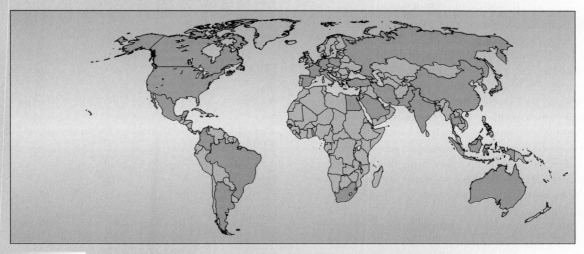

▲ **Figure 6.3** Countries (in orange) with Gross National Products that Exceed the Value of the Annual Revenue of Wal-Mart

Sources: U.S. Central Intelligence Agency (2008); *Fortune* (2008)

In annual revenues Exxon Mobil is the world's largest global corporation.

Missy Gish

employment opportunities, transcend political hostilities, transfer technology, and promote cultural understanding. In this regard, the president of McDonald's international operations notes that his corporation does not "force" itself on foreign countries. Rather, he argues, governments around the world actively recruit the company:

> I feel these countries want McDonald's as a symbol of something—an economic maturity and that they are open to foreign investments. I don't think there is a country out there we haven't gotten inquiries from. I have a parade of ambassadors and trade representatives in here regularly to tell us about their country and why McDonald's would be good for the country. (Freidman 1996, p. E15)

In reality, we cannot make a simple evaluation that would apply to all multinationals. Obviously, at some level, multinational corporations "do spread goods, capital, and technology around the globe. They do contribute to a rise in overall economic activity. They do employ hundreds of thousands of workers around the world, often paying more than the prevailing wage" (Barnet and Müller 1974, p. 151). Even so, the means that multinational companies use to achieve maximum profits for owners and stockholders (the valued goal) do not alleviate a host country's problems of poverty, hunger, mass unemployment, and gross inequality. Critics argue that, if anything, multinationals aggravate these problems, because the pursuit of profits is closely related to social inequalities and ecological imbalances.

One can also argue that multinationals are not responsible for such problems as inequality or obesity. Consider this statement made by a U.S. federal court judge: "If a person knows or should know that eating copious orders of super-sized McDonald's products is unhealthy, or may result in weight gain, it is not the place of the law to protect them from their own excesses. . . . Nobody forced them to eat at McDonald's" (Weiser 2003).

McDonald's former CEO Jack Greenburg (2001) elaborated on the issue of healthy foods by arguing that "we're selling meat and potatoes and bread and milk and Coca-Cola and lettuce and everything else you can buy in a grocery store. What you choose to eat is a personal issue. Every nutritionist I've talked to says a balanced diet is the key to health. You can get a balanced diet at McDonald's. It's a question of how you use McDonald's. Nobody's mad at the grocery store because you can buy potato chips and pastries there. Nobody wants a full diet of that either."

To complicate matters, corporations claim that they merely respond to consumer tastes. For example, virtually all of the major fast-food companies have introduced low-fat foods on their menus, and most have proven unpopular with consumers. McDonald's introduced the reduced-fat McLean hamburger, and Hardee's responded with the Lean One. Taco Bell even gave away 8 million Border Lights products (low-fat tacos and burritos) and spent $20 million on advertising to launch lower-fat menu

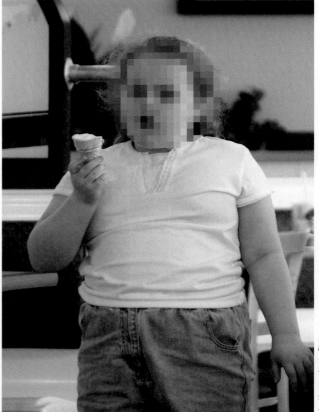

Robert Lawton, LawtonPhotos.com

Can McDonald's and other food service corporations be held responsible for the poor dietary choices many of its customers make? In one court case a judge ruled no on that question, but he did find "that the company ran ads aimed at getting children to pester their parents into going to McDonald's" (Koenig 1997: 20).

items that the company eventually discontinued for lack of demand.

Despite marketing campaigns promoting healthier choices, consumers clearly prefer menu items higher in calories and fat (Stone 1997). In fact, McDonald's dollar menu, which features the least healthy fast-food items (double cheeseburger and fried chicken sandwich), is credited with increasing profits by 33 percent since its introduction. The typical McDonald's restaurant sells 50 to 60 salads per day versus 300 to 400 double cheeseburgers (Warner 2006).

Nevertheless, many people question whether corporations should have the right to ignore the larger long-term effects of their products and business practices on people and the environment, even as they respond to consumer demand. The most profitable product for a corporation may prove costly for a society due to **externality costs**—hidden costs of using, making, or disposing of a product that are not figured into the price of the product or paid for by the producer. Yet, these costs must eventually be paid by someone (Lepkowski 1985). Such costs include those for cleaning up the environment and for medical treatment of injured workers, consumers, or other groups.

While multinational organizations are very powerful, consumers and "watchdog" groups have demonstrated that they can hold corporations in check. In many cases, informed consumers have gathered enough support from other consumers to change organizational behavior and

structure. When Hindu activists in India learned that McDonald's was cooking its fries in beef tallow, for example, they spread the word. The result: Customer lines that had been up to a mile long evaporated. The consumer

Table 6.1   Benefits and Drawbacks of Multinational Corporations

Common Multinational Corporation Claims of Benefits for Host Nations

Provide new products

Introduce and develop new technical skills

Introduce new managerial and organizational techniques

Promote higher employment

Yield higher productivity

Provide greater access to international markets

Provide for greater accumulation of foreign exchange

Supplement foreign aid objectives and programs of home countries directed toward the host

Serve as a point of contact for host-country businesspeople and officials in the home country

Encourage the development of new ancillary or spin-off industries

Assume investment risks that might not have been undertaken by others

Mobilize capital for productive purposes that might have gone to other, less fruitful uses

Common Criticisms of Multinational Corporations by Host Nations

Lead to a loss of cultural identity and traditions by creating new consumer tastes and demands

Are used as channels for foreign political influence

Possess a competitive advantage over local industries

Create inflationary pressures

Misapply host-country resources

Exploit host-country wealth primarily to benefit citizens of other nations

Lead to loss of control by host nations over their own economies

Possess neither sufficient understanding nor concern for the local economy, labor conditions, and national security requirements

Dominate key industries

Divert local savings from investment by nationals

Restrict access to modern technology by centralizing research and development facilities in the home country and by employing home-country nationals in key management positions

Source: Adapted from U.S. General Accounting Office (1978).

Thue Janus Kristensen

Should corporations ignore the environmental effects of the 500 billion to 1 trillion plastic bags that consumers use each year to cart home groceries and other products? These bags take months to hundreds of years to decompose. An estimated 3 percent of plastic bags end up in creeks, rivers, lakes, and oceans—where ducks, turtles, and other wildlife choke or strangle on plastic debris that they mistake for food (Roach 2003).

externality costs Hidden costs of using, making, or disposing of a product that are not figured into the price of the product or paid for by the producer.

WORKING FOR CHANGE

Margie Eugene Richard Takes on Shell Chemical

MARGIE RICHARD GREW up in the historically African American neighborhood of Old Diamond in Norco, Louisiana, in a house just 25 feet away from a Shell Chemical plant's fence line. Years later, she would stand on the front lines of a long, hard-fought battle to hold Shell accountable for the devastating health problems in her community. Four generations of Richard's family have lived in Old Diamond, which has a less attractive nickname: Cancer Alley. It was so named because of the high rates of cancer, birth defects, and other serious health ailments that plague the 1,500 residents who live on the four square blocks sandwiched between the Shell plant and a Motiva oil refinery owned by a Shell subsidiary. More than one-third of Norco's children suffer from asthma or bronchitis. Richard's sister Naomi died at age 43 from sarcoidosis, a rare bacterial infection. The disease typically strikes one in 1,000 people, yet Richard knows of at least three other neighbors who suffer from the same sickness.

Shell has been a fixture in Norco, which lies 25 miles west of New Orleans, since 1929. Over the years, the plant, with its looming tanks and belching vapor stacks, has grown to the size of nine football fields. The corporation has steadily bought the property of neighboring residents, many of whom were descendants of slaves and sharecroppers who farmed the land before the Civil War. Other families simply fled to escape the health hazards associated with living in a toxic "ghost town." Most of the residents who stayed were trapped by socioeconomic conditions and family responsibilities.

A Fatal Gas Blast

According to Richard, the defining event in her decision to become an activist occurred in 1973, when a Shell pipeline exploded, knocking one house off its foundation and killing an elderly woman and a teenage boy who was mowing the lawn. Richard recalls dashing out of her mother's house one block away and spotting a body lying beneath a sheet and the 16-year-old boy, then still alive, covered with raw burns and bubbling blisters. In 1988 another major industrial accident killed seven workers and released 159 million pounds of toxins into the air. Richard got into the habit of sleeping in her clothes so she could be ready to jump out of bed and run for her life if she needed to. In 1989 Richard, then a middle school teacher, founded Concerned Citizens of Norco to seek justice from Shell in the form of fair coverage of resettlement costs for her family and neighbors.

Fighting Back

Over the next 13 years, Richard led a community campaign that consisted of equal parts hard science, grassroots organizing, and media savvy. She joined forces with environmentalists and researchers to release a report revealing that the Shell refinery in Norco was releasing more than 2 million pounds of toxic chemicals into the air each year. In addition to serving as a community representative on a high-level EPA regulatory committee, Richard took her battle to the courts, acting as the plaintiff in a high-profile but ultimately unsuccessful class-action suit against Shell. She has organized press conferences and local "bucket brigades," empowering her neighbors by providing them with specially outfitted plastic buckets so they can monitor hazardous air pollutants on their own. She has also participated in educational workshops sponsored by Xavier University's Deep South Center for Environmental Justice.

Richard has a sharp eye for political theater. At one point, she had a Web camera installed on her trailer home to broadcast live feeds of the refinery spewing petro-

boycott succeeded in getting McDonald's to honor Indian food traditions, such as its taboo against eating beef and beef products (Rai 2003). Unfortunately, consumers do not always use the power they have to push corporations to make lasting and meaningful changes in the way they do business. For example, the European Union plans to require that pregnant pigs be allocated enough space in their pens to turn around, although this law will not take effect until 2012. While we should applaud a change that improves the quality of life of factory farm animals, we might also wonder whether this is enough space to live a healthy life and wonder why it should take until 2012 to implement such a change.

In sum, we have a responsibility to be informed consumers. When corporate executives feel pressure from consumers, they act; if only a small number of consumers speak out, however, their claims are often dismissed. Consider the comments from McDonald's CEO after four days

(a) Shell Chemical's Norco, Louisiana, petrochemical plant, viewed from the fence line that separates it from the Old Diamond neighborhood. (b) Margie Eugene Richard, 2004 Goldman Environmental Prize winner, North America, grew up in a house just 25 yards away.

chemical by-products. While speaking at an international environmental conference in the Netherlands, Richard approached Shell officials and invited them to take a sniff from a bag of Norco air.

Her untiring efforts have attracted powerful allies, including U.S. congresswoman Maxine Waters, and spurred a criminal investigation by the EPA that faulted Shell for failing to ensure plant safety and for falsifying its emissions reporting, a practice confirmed by a company whistleblower. Throughout the campaign, Richard, whom the local news media call Norco's "chief rabble-rouser," faced stiff resistance from Shell officials, who aggressively discouraged her group from seeking outside counsel and refused to hold open meetings with community members.

"There were times I thought it was an impossible task," Richard recalls. "I remember standing in my yard thinking,

'Lord, will there ever be hope?' But a little voice within me kept saying, 'If we don't tell them, how will they know?'"

No More Shell Games

In 2000, thanks largely to Richard's efforts, Shell agreed to reduce its emissions by 30 percent and to improve its emergency evacuation routes. The company also agreed to pay voluntary relocation costs for residents who lived on the two streets closest to the plant. But Richard and Concerned Citizens turned up the heat, leading to a meeting at the Royal Dutch Shell offices in London, where they secured a $5 million community development fund and full relocation coverage for residents of all four Old Diamond streets. Since the agreement was brokered in 2002, Shell has bought about 200 of the 225 lots for at least $80,000 per lot.

Adapted from Goldman Environmental Prize (2004). Reprinted with permission.

of protest in Seattle against the World Trade Organization, where as many as 2,000 protesters trashed McDonald's restaurants and other businesses. The CEO noted that while 2,000 people protested, 17.5 million other people visited a McDonald's restaurant to eat (Greenberg 2001). The point is that there is no need to be concerned about the voices of 2,000 activists when 17.5 million consumers are voting with their feet, or mouths (see Working for Change: "Margie Eugene Richard Takes on Shell Chemical").

The Extremes of Value-Rational Action

To this point, we have focused on rationalization and value-rational action, two concepts that help us understand how means-to-end thinking applied to the goal of turning a profit has affected organizational structure, the McDonaldization of society, the iron cage of rationality, the rise of multinational and global corporations, and externality costs. In the pages to come we will consider how means-to-end thinking can be taken to extremes, resulting in trained incapacity, tunnel vision concerning statistical measures of performance, oligarchy, and alienation.

Trained Incapacity

■ **CORE CONCEPT 7:** To be efficient, organizations sometimes train employees to respond mechanically or mindlessly to the dictates of the job, leaving them unable to respond creatively to new or changing circumstances. Formal organizations such as McDonald's train workers to perform their jobs a certain way and reward them for good performance. When workers are trained to respond mechanically or mindlessly to the dictates of the job, however, they risk developing what economist and social critic Thorstein Veblen (1933) called **trained incapacity**—the inability, because of limited training, to respond to new or unusual circumstances or to recognize when official rules or procedures are outmoded or no longer applicable. In other words, workers are trained to do their jobs only under normal circumstances or in a certain way; they are not prepared to respond creatively to what-if scenarios so that they can perform under a variety of changing circumstances.

Social psychologist Shoshana Zuboff distinguishes between work environments that promote trained incapacity and those that promote empowering behavior. In her research, Zuboff (1988) found that management can choose to use computers as tools to automate or to "informate." To *automate* means to use computers to increase workers' speed and consistency or to monitor workers' performance (for example, by keeping precise records of the number of keystrokes a worker makes per minute). "Smart" equipment, such as automatic timers, cash registers that calculate change, and time clocks that monitor the speed at which orders are filled—any device that does the thinking for employees, "watches" them, or pushes them to produce—represents the use of computers as automating tools.

Alternatively, management can choose to use computers as "informating" tools. To *informate* means to empower workers with decision-making tools, such as employee-scheduling software, which ensures that enough employees are scheduled for the busiest times and shifts. Other decision-making tools include software to keep track of payroll, sales, inventory, and purchasing. On the surface it may seem as if the software is doing the work for managers. Keep in mind, however, that managers must interpret the results and use this information to make decisions. Workers who use computers as informating tools experience work very differently from those whose work is monitored by computers.

Statistical Records of Performance

■ **CORE CONCEPT 8:** Statistical measures of performance are used to measure how well an organization and its members or employees are performing; these measures can be problematic when they encourage employees to concentrate on achieving good scores and to ignore problems generated by their drive to score well. McDonald's performance evaluators posing as customers conduct more than 500,000 unannounced visits each year to the company's 31,000 restaurants. The company also uses a checklist of 500 performance measures or questions to evaluate each of its restaurants. Items include (1) Are restrooms clean, in good repair, and fully stocked with paper towels, soap, and toilet paper? and (2) Are current promotional materials properly displayed? (McDonald's 2003). These criteria exemplify **statistical measures of performance**, quantitative (and sometimes qualitative) measures of how well an organization and its members or employees are performing.

In an industry such as fast-food service, where profits depend on being cost-conscious about every item used, statistical measures of performance exist for everything,

> whether it be the amount of milk-shakes sold per gallon of shake mix used, or the amount of cola drinks per liter of cola syrup, the amount of burgers sold per box of burgers used, the number or portions of chips per kilo used, the monetary amount of cleaning materials used as a percentage of the taking, even small things like the amount of sauces used per portion of Chicken McNuggets sold, or the amount of ketchup used per burger sold, and so on. (Gibney 1993)

Executives and other managers often compile statistics on profits, losses, market share, customer satisfaction, total sales, production quotas, and employee turnover as a way to measure individual, departmental, and overall organizational performance. Such measures can be convenient and useful management tools for two reasons: They are considered to be objective and precise, and they permit comparison across individuals, time, or departments.

Based on these measures, management can reward good performance through pay increases, profit sharing, and promotions and can act to correct poor performance. But statistical measures of performance have inherent shortcomings. One problem is that a chosen measure may not be a valid indicator of the performance it is intended to measure. For example, a corporation's occupational

---

trained incapacity The inability, because of limited training, to respond to new or unusual circumstances or to recognize when official rules or procedures are outmoded or no longer applicable.

statistical measures of performance Quantitative (and sometimes qualitative) measures of how well an organization and its members or employees are performing.

safety record is often gauged by counting the number of accidents that occur on the job. Based on this indicator, chemical corporations have one of the lowest accident rates of any industry. This measure, however, has been criticized as too narrow and thus lacking validity: Chemical workers are more likely than other workers to suffer from exposure-related illnesses, which can take years to produce symptoms and cannot be directly connected to their work the way burn injuries from cooking hamburgers can.

A second problem with statistical measures of performance is that they can encourage employees to concentrate on achieving good scores and to ignore problems generated by their drive to score well. In other words, people tend to pay attention only to the areas of performance being measured and to overlook the areas not being measured. For example, sales increases are a common statistical mea-

McDonald's claims to employ more than 2,000 statistical measures of performance to monitor food as it moves from the farm to the restaurant. For instance, pork used to make the McRib sandwich is inspected for joint enlargement, which can be a sign of infectious arthritis, the cause of 14 percent of condemned carcasses (Goetzinger 2006). How might meat suppliers' goal of making a profit affect willingness to notice this and other unhealthy conditions?

sure of performance. Employees are asked to meet hourly, weekly, monthly, or annual sales goals. In addition, sales goals often increase from one evaluation period to the next. Even after achieving record sales, employees are expected to achieve higher levels in the future. This attitude—"We can do better; there is no limit to profit increases"—is reflected in a McDonald's press release, which quotes the president and CEO: "Our U.S. business had an outstanding month with comparable sales up 13.5%. . . . I remain confident in our plans and the opportunities that exist to add even more customers to our restaurants in the future" (Bell 2004). (See Intersection of Biography and Society: "Statistical Measures of Performance.")

If sales increases and profits are the main criteria by which employees (and especially managers) are evaluated, then problems are inevitable. Managers, under pressure to make and increase profits may force employees to work unpaid overtime, as happened at 62 Seattle-area Taco Bells. A jury found that the managers of these restaurants had illegally required at least 12,000 workers to "prepare food, pick up trash, or do other chores" while off the clock (Rousseau 1997). Moreover, when managers are preoccupied with meeting the goal of filling customer orders within 90 seconds and keeping labor costs low, worker safety may suffer. One reason that the food service industry rates first in total recordable injuries and illnesses, with burns as the largest injury category, could be related to pressure on employees to fill orders quickly and to chronic understaffing (Personick 1991).

Oligarchy

■ **CORE CONCEPT 9:** Large formal organizations inevitably tend to become oligarchical: Power becomes concentrated in the hands of a few people, who hold the top positions, and these people often draw upon the advice of experts. **Oligarchy** is rule by the few, or the concentration of decision-making power in the hands of a few persons, who hold the top positions in a hierarchy.

> One of the most bizarre features of any advanced industrial society in our time is that the cardinal choices have to be made by a handful of men . . . who cannot have firsthand knowledge of what those choices depend upon or what their results may be. . . . [By] "cardinal choices," I mean those which determine in the crudest sense whether we live or die. For instance, the choice in England and the United States in 1940 and 1941, to go ahead with work on the fission bomb: the choice in 1945 to use that bomb when it was made (Snow 1961, p. 1).

oligarchy Rule by the few, or the concentration of decision-making power in the hands of a few persons, who hold the top positions in a hierarchy.

INTERSECTION OF BIOGRAPHY AND SOCIETY

Statistical Measures of Performance

STATISTICAL MEASURES OF performance are a part of daily life. Three Northern Kentucky University students describe how such measures were used to evaluate and motivate employees:

- At my place of work, sales staff are rewarded according to a point system. When a customer applies for and is accepted for a store credit card, the salesperson receives 1,000 points. Employees are also awarded points according to the number of items per sale, the dollar amount per sale, and total hourly sales. When a salesperson accumulates a certain number of points, he or she can redeem them for prizes such as CD players, TVs, and gift certificates.
- At my workplace, to earn a raise, employees must be in uniform every day. That is, employees' shoes must be the right color and they must wear the hat, pants, and shirt issued by the store. In addition, they must have the proper nametag on and shirts tucked in. Employees must rarely call in sick (a maximum of two or three times per quarter), and they cannot be late. It also helps to rarely request days off. The employees who meet these standards earn a 25-cent raise every quarter.
- I work at a nursing home where employee absenteeism and tardiness for work were such big problems that management began to give bimonthly bonuses to correct the problem. If over the course of a two-month period an employee was never late for work and did not miss a day of work, he or she earned a $100 bonus. If the employee was late only one time and did not miss a day of work, he or she received $75. If an employee missed one day of work but had a medical excuse, he or she earned $50. Attendance improved dramatically under this system.

Political analyst Robert Michels (1962) believed that large formal organizations inevitably tend to become oligarchical, for the following reasons: First, democratic participation is virtually impossible in large organizations. Size alone makes it "impossible for the collectivity to undertake the direct settlement of all the controversies that may arise" (p. 66). For example, McDonald's employs about 500,000 people and has franchises located in more than 100 countries (see Figure 6.1, "McDonald's at a Glance"). Obviously, "such a gigantic number of persons . . . cannot do any practical work upon a system of direct discussion" (Michels 1962, p. 65). Second, as countries become more interdependent and technology grows increasingly complex, many organizational features inevitably become incomprehensible to workers. As a result, many employees work toward achieving organizational goals they did not define, cannot control, may not share, or may not understand. This lack of knowledge prevents workers from participating in or evaluating decisions made by executives.

A danger of oligarchy is that key decision makers may become so preoccupied with preserving their own leadership that they do not consider the greater good. In addition, they may not have the background information necessary to understand the full implications of their choices. In such situations, they draw upon experts to provide the information.

In his writings about bureaucracy, Weber emphasizes that power lies not in the person but rather in the position that person occupies in the division of labor. The kind of power Weber describes is clear-cut and familiar: A superior gives orders to subordinates, who are required to carry out those orders. The superior's power is supported by the threat of sanctions, such as demotions, layoffs, or firings. Sociologists Peter Blau and Richard Schoenherr (1973) recognize the importance of this form of power but identify a second, more ambiguous type—expert power—that they believe is "more dangerous than seems evident for democracy and . . . is not readily identifiable as power" (p. 19).

Expert Knowledge and Responsibility

According to Blau and Schoenherr (1973), expert power is connected to the fact that organizations are becoming increasingly professionalized. **Professionalization** is a trend in which organizations hire experts (such as

---

professionalization A trend in which organizations hire experts with formal training in a particular subject or activity—training needed to achieve organizational goals.

BP (formerly British Petroleum) is the fourth largest global corporation in the world, employing 96,200 employees and running operations in 26 countries. BP's size dictates that many decisions affecting its employees, millions of consumers, and the environment will be made in the boardroom.

Dr. Kaihsu Tai

chemists, physicists, accountants, lawyers, engineers, psychologists, or sociologists) as consultants or full-time employees. Experts have formal training in a particular subject or activity—training that an organization must draw upon to achieve valued goals. Experts receive their training not from the organization, but from colleges and universities. Theoretically, they are self-directed and not subject to narrow job descriptions or direct supervision.

Experts use the framework of their profession to analyze situations, solve problems, or invent new technologies. From the experts' viewpoint, the information, service, or innovation they provide to the organization is technical and neutral. Experts do not have, nor do they seek, control over how corporations use the information, service, or invention they provide. They may, however, feel pressure to deliver a product or to present a position that executives "need" or want to hear. For instance, when McDonald's issued recommendations specifying the proper treatment of egg-laying hens, the company highlighted the expertise of the Scientific Advisory Committee on Animal Welfare for United Egg Producers. Among other things, the recommendations called for "a minimum of 72 square inches [of space] per bird" and beak breaking only when necessary to prevent widespread feather pecking and cannibalism, and only when carried out by properly trained and monitored personnel (United Egg Producers 2006). Note that even though McDonald's drew upon the expertise of an advisory council concerned with animal welfare, that council raised no questions about the morality of confining hens to eight by nine inches of space or about the fact that crowded conditions lead to epidemics of pecking and cannibalism.

Blau and Schoenherr regard this arrangement between experts and organizations as problematic because it leaves no one accountable for the actions of powerful executives and because it complicates attempts to identify individuals "whose judgments [are] the ultimate source of a given action" (pp. 20–21). Who will take responsibility when something goes wrong?

In addition, decision making in large organizations is complex because no single person provides all the input that goes into a decision. Rather, a decision is a joint product of the decision makers and the experts who provide requested information and judgments. Often a decision maker does not fully understand the principles underlying an expert's recommendations and judgments. The problem with the specialization of knowledge is that when something goes wrong, experts claim that they provided only the patent, information, or recommendations and thus could not control the ultimate implementation. Executives, on the other hand, claim that they relied on the experts' advice to make decisions about processes they might not have understood.

Keep in mind that, as Blau and Schoenherr emphasize, the men and women who give expert advice may be decent people, but their position as consultants and their specialized training and point of view often make them unable to anticipate, plan, or control unintended consequences. Another drawback to organizations' reliance on expertise, not named by Blau and Shoenherr, is that experts may offer the "right" advice or product to ensure that their consulting contracts are extended.

Alienation

■ **CORE CONCEPT 10:** The growth of bureaucracies to coordinate the efforts of humans as well as machines and other technology is accompanied by alienation, a state of being in which human life is dominated by the forces of its inventions. Human control over nature increased with the development of more and more sophisticated tools and with the growth of bureaucracies to coordinate the efforts of both people and machines. Machines and bureaucratic organizations, in turn, combined to extract raw materials from the earth more quickly and more efficiently and to increase the speed with which necessities such as food, clothing, and shelter could be produced and distributed. Karl Marx believed that this increased control over nature is accompanied by **alienation**, a state of being in which human life is dominated by the forces of its inventions.

Chemical substances represent one such invention; they have reduced the physical demands and risk of failure involved in producing goods such as potatoes. Syn-

alienation A state of being in which human life is dominated by the forces of its inventions.

Karl Marx believed that alienation occurs when the production process is divided up so that workers are treated like parts of a machine rather than as active, creative, social beings.

thetic fertilizers, herbicides, pesticides, and chemically treated seeds give people control over nature, because they eliminate the need to fight weeds with hoes, prevent pests from destroying crops, and help people produce unprecedented amounts of standard and uniform-looking fruits and vegetables.

These gains also have a dark side, however. In the long run, people are dominated by the effects of this invention. For example, heavy reliance on chemical technologies can cause the soil to erode and become less productive; it can also prompt insects and disease-causing agents to develop resistance to the chemicals. In addition, chemical technologies have altered the ways farmers plant crops. Planting patterns have changed from many crop varieties to a single standard cash crop, planted in rows. Consequently, some farmers have lost knowledge of how to control insects and diseases without chemicals, by interplanting a variety of flowers, herbs, and vegetables. Today many farmers are economically dependent on a single crop and the chemical industry.

Although Marx discussed alienation in general, he wrote more specifically about alienation in the workplace. Marx maintained that workers are alienated on four levels: (1) from the process of production, (2) from the product, (3) from the family and the community of fellow workers, and (4) from the self. Workers are alienated from the *process* because they produce not for themselves or for known consumers but rather for an abstract, impersonal market. In addition, they do not own the tools of production.

Workers are alienated from the product because their roles are rote and limited. Each fast-food worker, for instance, performs a specialized task, such as warming buns, adding condiments, or wrapping an order. Thus, many workers are treated as being replaceable or interchangeable, like machine parts. That is, they are treated as economic components rather than as active, creative, social beings (Young 1975). Marx believed that the conditions of work usually impair an individual's "capacity to become a multidimensional, authentic being with human qualities of compassion, reflection, judgment, and action" (Young 1975, p. 27).

Workers are alienated from the *family and the community of fellow workers* because households and work environments remain separate from one another. In the fast-food industry, for instance, workers can lose touch with their families when they work shifts (for example, late at night, early in the morning, or on weekends) that keep them from participating in other family members' lives. Workers are alienated from the community of fellow workers because they compete for jobs, business, advancement, and awards. As they compete, they fail to consider how they might unite as a force and control their working conditions.

Often workers are alienated from more than the family and the community of fellow workers; they are alienated from other communities, such as the school. McDonald's and other fast-food companies aggressively recruit high-school-aged employees, who make up 65 percent of the labor force (Waters 1998). While McDonald's considers homework schedules and other school demands, we must question the extent to which after-school jobs curtail involvement in academic and social activities (Crispell 1995).

"Alienation from self" can occur in service industries when management standardizes or routinizes virtually every aspect of the employee-customer relationship, including the employees' appearance and the words they must say when greeting customers.

Finally, workers are alienated from the self because "one's genius, one's skills, one's talent is used or disused at the convenience of management in the quest of private profit" (Young 1975, p. 28). When Karl Marx developed his ideas about alienation in the late 1800s, he was describing "alienation from self" as it relates to industrial society. More recently, sociologist Robin Leidner (1993) has described the alienation from self that can occur in service industries when management standardizes or routinizes virtually every aspect of the service provider–customer relationship, so that neither party feels authentic, autonomous, or sincere:

> Employers may try to specify exactly how workers look, exactly what they say, their demeanors, their gestures, even their thoughts. The means available for standardizing interactions include scripting; uniforms or detailed dress codes; rules and guidelines for dealing with service-recipients and sometimes with co-workers. . . . Surveillance and a range of incentives and disincentives can be used to encourage or enforce compliance. (pp. 8–9)

■ VISUAL SUMMARY OF CORE CONCEPTS

■ CORE CONCEPT 1: The formal organization, a type of secondary group, is a coordinating mechanism people create to achieve a planned outcome.

Formal organizations are viewed as coordinating mechanisms because they bring together people, resources, and technology and then channel human activity toward achieving a specific outcome. Formal organizations can be voluntary, coercive, or utilitarian, depending on the reason that people participate. Sociologists classify formal organizations as secondary groups, impersonal associations between people who interact for a specific purpose. Secondary group relationships are confined to a particular setting and specific tasks and are seen as a means to achieve some agreed-upon end. Members relate to each other in terms of specific roles. The 4,300 students enrolled in the United States Naval Academy constitute a secondary group.

Department of Defense photo by Marine Corps
Master Sgt. Robert Carr, USMC

■ CORE CONCEPT 2: Most formal organizations can be classified as bureaucracies, organizational structures that strive to use the most efficient means to achieve a valued goal.

In theory, a bureaucracy is a completely rational organization—one that uses the most efficient means to achieve a valued goal, whether that goal is feeding people, making money, recruiting soldiers, counting people, or collecting taxes. A bureaucracy has seven major characteristics that allow it to coordinate people so that their actions focus on achieving the organization's goals: (1) a clear-cut division of labor; (2) a hierarchical authority structure; (3) written rules and procedures; (4) positions filled on the basis of qualifications, not emotional considerations; (5) recorded and preserved administrative decisions, rules, guidelines, procedures, and activities; (6) authority tied to the position, not the person in the position; and (7) clients treated as "cases."

Missy Gish

■ **CORE CONCEPT 3:** Organizations include both formal and informal dimensions: The formal dimension is the official side of the organization governed by written guidelines, rules, and policies; the informal dimension includes employee-generated norms that depart from or otherwise bypass the formal dimension.

The formal dimension or official side of an organization consists of job descriptions and written rules, guidelines, and procedures established to achieve valued goals. The informal dimension includes behaviors that depart from the formal dimension, including employee-generated norms that evade, bypass, or ignore official policies, rules, and procedures. Despite an official policy requiring food service workers to wash their hands before handling food, many workers fail to follow the policy. This informal practice explains, in part, one important source of food-borne illnesses.

Center for Disease Control, Public Health Image Library

■ **CORE CONCEPT 4:** The concept of rationalization—a process in which thought and action rooted in custom, emotion, or respect for mysterious forces is replaced by value-rational thought and action—helps us understand how striving to achieve valued goals can have undesirable, even disastrous, consequences.

The growth and dominance of formal organizations goes hand in hand with rationalization, a process in which thought and action rooted in emotion, superstition, respect for mysterious forces, or tradition are replaced by value-rational thought and action. Value-rational thought and action mean that people strive to find the most efficient way to achieve a valued goal. One way to show how value-rational thought and action differs from thought and action driven by emotion, tradition, superstition, or respect for mysterious forces is to consider two distinct thought patterns that drive human behavior toward farm animals. One thought pattern views animals as having complex emotions, including the ability to feel love and sorrow, and being "capable of feeling pain and fear." Value-rational thinking sees animals as profit-generating creatures: The more animals owners can raise in the space available and the faster they can raise them to maturity, the greater the profit.

USDA Photo by Bill Tarpening

■ **CORE CONCEPT 5:** One organizational trend guided by value-rational action is the McDonaldization of society, a process in which the principles governing fast-food restaurants come to dominate other sectors of society.

The McDonaldization of society is a process by which the organizing principles driving the fast-food industry come to dominate other sectors of society. Those principles are (1) efficiency, which allows consumers to move quickly from one state of being to another (for example, from hungry to full); (2) quantification and calculation, which are numerical indicators that allow customers to evaluate a product or service easily; (3) predictability, which allows customers to expect the same product or service no matter where or when it is purchased; and (4) control, which allows the organization to manage in detail the way a product or service is produced and delivered. Weber used the phrase "iron cage of rationality" to describe the set of irrationalities generated by so-called rational systems, the drawbacks to the McDonaldization model.

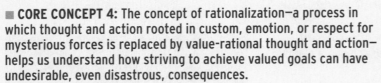

Jean-noël Lafargue

■ **CORE CONCEPT 6:** Value-rational action, or behavior guided by means-to-end thinking aimed at increasing market share and profits, helps explain how an organization expands its reach beyond local markets to regional, national, and global markets.

One way to understand how organizations grow from local operations (McDonald's began as one restaurant in Illinois in 1955) to global giants is to consider the actions they have taken to increase profits. Organizations employ five major profit-generating strategies: (1) lower production costs, (2) create new products that consumers "need" to buy, (3) improve existing products to make previous versions obsolete, (4) identify ways for people to purchase more products, and (5) create new markets. Regardless of their size, multinational corporations compete, plan, produce, sell, recruit, extract resources, acquire capital, and/or borrow technology on a multicountry scale. Taken together, the annual revenues of the top 10 global corporations is about $2.3 trillion. Only five countries in the world have a gross national product that exceeds that amount.

■ **CORE CONCEPT 7:** To be efficient, organizations sometimes train employees to respond mechanically or mindlessly to the dictates of the job, leaving them unable to respond creatively to new or changing circumstances.

Formal organizations such as McDonald's train workers to perform their jobs a certain way and reward them for good performance. When workers are trained to respond mechanically or mindlessly to the dictates of the job, however, they risk developing what economist and social critic Thorstein Veblen (1933) called trained incapacity, the inability, because of limited training, to respond to new or unusual circumstances or to recognize when official rules or procedures are outmoded or no longer applicable.

■ **CORE CONCEPT 8:** Statistical measures of performance are used to measure how well an organization and its members or employees are performing; these measures can be problematic when they encourage employees to concentrate on achieving good scores and to ignore problems generated by their drive to score well.

Statistical measures of performance are quantitative (and sometimes qualitative) measures of how well an organization and its members or employees are performing. Executives and other managers often compile statistics on profits, losses, market share, customer satisfaction, total sales, production quotas, and employee turnover as a way to measure individual, departmental, and overall organizational performance. Such measures can be convenient and useful management tools, but they can also encourage employees to concentrate on achieving good scores and to ignore problems generated by their drive to score well. Considering that statistical measures of performance govern meat inspection, we must ask how meat suppliers' goal of turning a profit might affect willingness to condemn carcasses.

■ **CORE CONCEPT 9: Large formal organizations inevitably tend to become oligarchical: Power becomes concentrated in the hands of a few people, who hold the top positions, and these people often draw upon the advice of experts.**

Oligarchy is rule by the few, or the concentration of decision-making power in the hands of a few persons who hold the top positions in a hierarchy. A danger of oligarchy is that key decision makers may become so preoccupied with preserving their own leadership that they do not consider the greater good. In addition, these decision makers may not have the necessary background knowledge to understand the full implications of their choices. Instead, they draw upon experts to provide background information. The experts may feel pressure to deliver a product or to present a position that executives "need" or want to hear. If something goes wrong, the experts can claim that they provided only the patent, information, or recommendations and thus could not control the ultimate implementation. Executives can claim that they relied on the experts' advice to make decisions about processes they might not have understood.

■ **CORE CONCEPT 10: The growth of bureaucracies to coordinate the efforts of humans as well as machines and other technology is accompanied by alienation, a state of being in which human life is dominated by the forces of its inventions.**

Karl Marx believed that humans' increased control over nature is accompanied by alienation, so that people are dominated by the forces of their inventions. Marx was specifically concerned about alienation in the workplace. He maintained that workers are alienated on four levels: (1) from the process of production, (2) from the product, (3) from the family and the community of fellow workers, and (4) from the self.

Resources on the Internet

 Sociology: A Global Perspective Book Companion Web Site

www.cengage.com/sociology/ferrante

Visit your book companion Web site, where you will find flash cards, practice quizzes, Internet links, and more to help you study.

CENGAGENOW™

Just what you need to know NOW!

Spend time on what you need to master rather than on information you already have learned. Take a pre-test for this chapter, and CengageNOW will generate a personalized study plan based on your results. The study plan will identify the topics you need to review and direct you to online resources to help you master those topics. You can then take a post-test to help you determine the concepts you have mastered and what you still need to work on. Try it out! Go to www.cengage.com/login to sign in with an access code or to purchase access to this product.

Key Terms

alienation   161
bureaucracy   145
coercive organizations   145
control   149
efficiency   149
externality costs   155
formal dimension   147
formal organizations   144

ideal type   146
informal dimension   147
iron cage of rationality   150
McDonaldization of society   149
multinational corporations   151
oligarchy   159
predictability   149
professionalization   160

quantification and calculation   149
rationalization   147
secondary groups   144
statistical measures of
   performance   158
trained incapacity   158
utilitarian organizations   145
voluntary organizations   145

7 Deviance, Conformity, and Social Control

With Emphasis on the People's Republic of China

The sociological contribution to understanding deviant and conforming behaviors goes beyond studying individual character or motives. Instead, sociologists emphasize the social context in which deviant and conforming behaviors occur. In fact, depending on the social context, any behavior can qualify as deviant or conforming.

Department of Defense photo by Chief Mass Communication Specialist David Rush, U.S. Navy

▲ The concept of deviance cannot be discussed without also considering the concepts of conformity and social control.

WHY FOCUS ON

The People's Republic of China?

IN THIS CHAPTER, we pay special attention to the People's Republic of China because it represents an interesting case for studying issues of deviance, conformity, and social control. Many behaviors that constituted deviance in China from 1966 to 1976, the period known as the Cultural Revolution, are no longer judged that way. During that time any person who held a position of authority, worked to earn a profit, showed the slightest leaning toward foreign ways, or expressed academic interests was subject to interrogation, arrest, and punishment. Included in this group were scientists, teachers, athletes, performers, artists, writers, and owners of private businesses. The list of suspicious characters also included people who wore glasses, wore makeup, spoke a foreign language, owned a camera or a radio, had traveled abroad, or had relatives living outside China (Mathews and Mathews 1983). Understanding the Cultural Revolution is especially relevant today, if only because all of China's current leaders were in their teens or early 20s during that time, "and they would not be human if the scars did not run deep" (Spence 2006).

Contrast the events of the Cultural Revolution with Asia scholar Robert Oxman's (1993a) description of contemporary China: "All over China people are jumping into the sea. That's the new Chinese expression for going into business, making money privately in a country that used to forbid any form of capitalism." Consider that in the late 1970s, when the people of Datang, a rice-farming village of 1,000 residents, took it upon themselves to stitch socks and sell them along a highway, the Chinese government ordered them to stop, branding the money-making venture as capitalist. Today, Datang is the socks capital of the world, producing nine billion per year (Barboza 2004).

The changes described above speak to an unspoken deal, evolving since 1990, between the Chinese Communist party and the people: Stay out of politics and the government will allow the people to "get on with the business of making money" and retreat from managing their everyday life. Prior to this "deal," the Chinese lived in tiny birdcages; now they live in an aviary. The people of China "cannot yet fly up to the clear blue sky" but there is more room in which to fly around (Gifford 2007, p. 15).

FACTS TO CONSIDER

- From 1966 to 1976, the Chinese government declared profit-making activities to be criminal and closed the country to foreign investment and influences. During this time any person who worked to earn a profit, showed the slightest leanings toward foreign ways, traveled abroad, wore makeup, or owned foreign-made items was considered suspect and subject to interrogation, arrest, and punishment.

- The newly constructed South China Mall is four times larger than the Mall of America in Minnesota. In fact, four other shopping malls in China are also larger than the Minnesota mall (Barboza 2005).

- On July 20, 2001, the International Olympic Committee announced that the 2008 Summer Olympics would be held in Beijing. Millions of foreign visitors from every country in the world are expected to attend the event.

- Since 1976 approximately 700,000 students from China have studied at foreign universities, with U.S. universities being the number one destination.

Deviance, Conformity, and Social Control

■ **CORE CONCEPT 1:** The only characteristic common to all forms of deviance is that some social audience challenges or condemns a behavior or an appearance because it departs from established norms. The topics of this chapter are deviance, conformity, and social control. **Deviance** is any behavior or physical appearance that is socially challenged or condemned because it departs from the norms and expectations of a group. **Conformity** comprises behaviors and appearances that follow and maintain the standards of a group. All groups employ mechanisms of **social control**—methods used to teach, persuade, or force their members, and even nonmembers, to comply with and not deviate from its norms and expectations.

These topics are complex, because almost every behavior has at some time qualified as deviant or conforming. When sociologist J. L. Simmons (1965) asked 180 men and women in the United States from a variety of age, educational, occupational, and religious groups to "list those things or types of persons whom you regard as deviant," they identified a total of 1,154 items.

> Even with a certain amount of grouping and collapsing, these included no less than 252 different acts and persons as "deviant." The sheer range of responses included such expected items as homosexuals, prostitutes, drug addicts, beatniks, and murderers; it also included liars, democrats, reckless drivers, atheists, self-pitiers, the retired, career women, divorcées, movie stars, perpetual bridge players, prudes, pacifists, psychiatrists, priests, liberals, conservatives, junior executives, girls who wear makeup, and know-it-all professors." (Simmons 1965, pp. 223–224)

Although Simmons conducted this study more than 40 years ago, his conclusion remains relevant today: The only characteristic common to all forms of deviance is "the fact

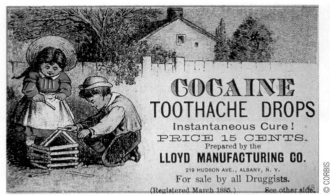

At one time in the United States, cocaine was an ingredient in many over-the-counter medications.

that some social audience regards them and treats them as deviant" (p. 225). Consequently, it is difficult to generate a precise list of deviant behaviors and appearances, because something that some people consider deviant may not be considered deviant by others. Likewise, something considered deviant at one time and place may not be considered deviant at another. For example, wearing makeup is no longer considered a deviant behavior in China, as evidenced by the fact that the U.S.-based cosmetics company Avon has more than 6,300 stand-alone stores and more than 1,000 Avon counters inside department stores within the country (ConsumerAffairs.com 2005). As another example, cocaine and other now-illegal drugs once were legal substances in the United States. In fact, "[w]e unwittingly acknowledge the previous legality of cocaine every time we ask for the world's most popular cola by brand name" (Gould 1990, p. 74). Originally Coca-Cola was marketed as a medicine that could cure various ailments. One ingredient used to make the drink came from the coca leaf, which is also used to produce cocaine (Henriques 1993). These examples alert us to the fact that deviance exists only in relation to norms in effect at a particular time and place.

Deviance: The Violation of Norms

■ **CORE CONCEPT 2:** Most people abide by their society's norms, because they have been socialized to accept those norms as "good, proper, appropriate, and worthy" (Sumner 1907). In Chapter 3 we learned that norms give order and predictability to life. We also learned that some norms are considered more important than others (Field 2002). In that chapter, we highlighted two kinds of norms—folkways and mores—distinguished by sociologist William Graham Sumner. We learned that **folkways** are customary ways of handling the routine matters of everyday life—for example, how one should look, eat, greet another person,

deviance Any behavior or physical appearance that is socially challenged or condemned because it departs from the norms and expectations of a group.

conformity Behavior and appearances that follow and maintain the standards of a group. Also, the acceptance of culturally valued goals and the pursuit of those goals through means defined as legitimate.

social control Methods used to teach, persuade, or force a group's members, and even nonmembers, to comply with and not deviate from its norms and expectations.

folkways Customary ways of handling the routine matters of everyday life.

or express affection toward same-sex and opposite-sex persons. **Mores** are norms that people define as essential to the well-being of their group. People who violate mores are usually punished severely: They may be ostracized, institutionalized in prisons or mental hospitals, sentenced to physical punishment, or condemned to die. Unlike folkways, mores are considered unchangeable and regarded as "the only way" and "the truth" (Sumner 1907).

During the Cultural Revolution, the dominant mores rejected special social status and the accumulation of worldly possessions. Any person in a position of authority or with the slightest leaning toward foreign ways, including a farmer who planted extra crops, was considered suspect. If someone simply remarked that a foreign-made product, such as a can opener, was better than its Chinese counterpart, or if someone wrapped food or garbage in a piece of newspaper with then–Communist Party Chairman Mao Zedong's picture, he or she was regarded with distrust (Mathews and Mathews 1983). As Jung Chang writes in *Wild Swans: Three Daughters of China* (1991), such conditions reduced many people "to a state where they did not dare even to think, in case their thoughts came out invol-

untarily" (1992, p. 6). The slightest misstep could make one a target of intense criticism.

While mores rejecting the accumulation of material possessions have all but disappeared in China, mores rejecting special social status still exist on some level, as illustrated by the following assessment of how best to market products to Chinese youth. While ads may appear to "stress Western individualism" a close analysis suggests that the individual is never outside the group.

> Individualism in China—it's the illusion of being different, of having a better, smarter, cleverer way, but it always has to remain understated within the context of social barriers of a Confucian society. One thing that is interesting about Chinese youth is they're very ambitious. They want to show off in front of their peers, but because it's not truly an individual culture, they show off but in an understated way so the tone and the manner . . . [are] very quiet. . . . So Chinese don't want to define themselves independent of society. (Doctoroff 2006)

Most people abide by established folkways and mores because they accept them as "good and proper, appropriate and worthy" (Sumner 1907, p. 60). For the great majority of people, "the rule to do as all do suffices." Recall from Chapter 4 that *socialization* is the process by which most people come to learn and accept the ways of their culture as natural.

Socialization as a Means of Social Control

Because socialization begins as soon as a person enters the world, one has little opportunity to avoid exposure to a culture's folkways and mores. If we compare the ways preschoolers are socialized in China and the United States, we can see that different but important cultural lessons are incorporated into their daily activities. Even though it is impossible to generalize about preschools in countries as large and diverse as the United States and China, we can identify some broad differences. For the most part, Chinese preschoolers are taught to suppress individual impulses, play cooperatively with other children, and attune themselves to group enterprises. In contrast, American preschoolers are taught to cultivate individual interests and to compete with other children for success and recognition by the teacher.

Three researchers from the University of Hawaii filmed daily life in Chinese and U.S. preschools to learn how teachers in each system socialize children to participate

Ted and Mary Ann Weiss

This shopping center–located in Beijing, China–could not have existed during the Cultural Revolution, when working to earn a profit was considered deviant.

mores Norms that people define as essential to the well-being of their group or nation. People who violate mores are usually punished severely.

Schools are agents of socialization that transmit important cultural lessons. Broadly speaking, school in China emphasize discipline, cooperation, and social-mindedness. U.S. schools emphasize individuality, self-direction, and freedom of choice.

effectively in their respective societies. (It is worth noting that the great majority of preschool children whom they filmed in both countries seemed happy and productive.) While this study was published in 1989, its broad themes and findings remain relevant today. The researchers found that, compared with U.S. preschools, Chinese preschools are highly structured and socially minded: Chinese teachers discipline their four-year-old students "by stopping them from misbehaving before they even know they are about to misbehave" (Tobin, Wu, and Davidson 1989, p. 94), and they promote loyalty to the group. The following bathroom scene exemplifies the extent to which Chinese children are taught to follow instructions and attune themselves to group enterprises:

It is now 10 o'clock, time for children to go to the bathroom. Following Ms. Wang, the twenty-six children walk in single file across the courtyard to a small cement building toward the back of the school grounds. Inside there is only a long ditch running along three walls. Under Ms. Wang's direction and, in a few cases, with her assistance, all twenty-six children pull down their pants and squat over the ditch, boys on one side of the room, girls on the other. After five minutes Ms. Wang distributes toilet paper, and the children wipe themselves. Leaving the toilet, again in single file, the children line up in front of a pump, where two daily monitors are kept busy filling and refilling a bucket with water that the children use to wash their hands. (pp. 78–79)

In the United States, preschoolers are also taught discipline, but they are more likely to be corrected *after* they do something wrong or get out of hand. U.S. teachers are more likely than their Chinese counterparts to encourage self-direction, freedom of choice, independence, and individuality—qualities that often depend on a supply of material items. For example, American children typically use as much paper as they want; they start a drawing, decide they do not like it, and crumple up the paper. When they play house, store, or firefighter, they use costumes, plastic dishes, children's versions of household appliances, plastic food items, and so on. American teachers also encourage children to choose from a number of activities. A typical exchange between a preschool teacher and students follows. It is difficult to imagine such an exchange in a Chinese classroom:

Who would like to paint? Michelle. Mayumi. Nicole. Okay, you three get your smocks from your cubbies and you can paint. [The teacher holds up a wooden block.] Who wants to do this? Mike? Okay, that's one. Stu, that makes two. Billy is three.... Here's a puzzle piece. You want to start on the puzzles? Okay? [The teacher holds up a toy frying pan.] Who wants to start in the house? Lisa, Rose, Derek. Go ahead to the housekeeping corner. Kerry, what do you want to do? The Legos? You're going to work on the radio, Carl? That's fine. Who is going to come over to the book corner to read this book? It's called *Stone Soup*. Okay, come on with me. (p. 130)

Reaction to Socialization in Another Culture

Both the Chinese and the American audiences who watched such preschool exchanges on film were disturbed by their counterparts' system of socializing preschoolers. Comments by Chinese viewers showed their clear preference for their own way of structuring early education. They maintained that their form of discipline expresses care and concern, and they regarded U.S. preschools as chaotic, undisciplined, and promoting self-centeredness. As one Chinese viewer remarked, "There are so many toys in the classroom that children must get spoiled. When they have so much, children don't appreciate what they have" (p. 88). On the other side, U.S. viewers criticized the Chinese preschools for being rigid, totalitarian, too group-oriented, and over-restrictive, "making children drab, colorless, and robot-like" (p. 138). Americans were particularly disturbed by the bathroom scene and asked why children were forced to go to the bathroom in this manner. One Chinese educator replied:

> Why not? Why have small children go to the bathroom separately? It is much easier to have everyone go at the same time. Of course, if a child cannot wait, he is allowed to go to the bathroom when he needs to. But, as a matter of routine, it's good for children to learn to regulate their bodies and attune their rhythms to those of their classmates. (p. 105)

The people in each country were uncomfortable with the other country's system because the lessons it taught clashed with the prevailing mores about which behaviors are essential to their own country's well-being. Such early socialization experiences are intended to help children fit into the existing system. Each society tries to prepare its people to mesh with and accept their environment. Even so, primary socialization experiences such as those that take place during preschool are uneven at best. Not all preschools are alike, and some children do not attend preschool. Even among those who do attend, some children do not internalize (take as their own and accept as binding) the values, norms, and expectations to which they are exposed. Therefore, all societies establish other mechanisms of social control to ensure conformity.

Ideally, conformity should be voluntary. That is, people should be internally motivated to maintain group standards and to feel guilty if they deviate from them. As noted earlier, during the Cultural Revolution it was considered deviant to wear glasses, use makeup, speak a foreign language, or break or destroy items that displayed Mao Zedong's picture. Many Chinese conformed to these rules and felt guilty if they broke the rules, even if only by accident. The memories of one Chinese man illustrate this point:

> As a boy, I did not know what a god looked like, but I knew that Mao was the god of our lives. When I was six, I accidentally broke a large porcelain Mao badge. Fear gripped me. In my life until that moment, the breaking of the badge seemed the worst thing I had ever done. Desperate to hide my crime, I took the pieces and threw them down a public toilet. For months I felt guilty. (Author X 1992, p. 22)

In this case, the guilt was a sign of voluntary conformity. If conformity cannot be achieved voluntarily, however, people may employ various means to teach, persuade, or force others to conform.

Mechanisms of Social Control

■ **CORE CONCEPT 3:** When socialization fails to produce conformity, other mechanisms of social control—sanctions, censorship, or surveillance—may be used to convey and enforce norms. Ideally, socialization brings about conformity and conformity is voluntary. When conformity cannot be achieved voluntarily, other mechanisms of social control may be used to convey and enforce norms. One such mechanism is **sanctions**—reactions of approval or disapproval to others' behavior or appearance. Sanctions can be positive or negative, formal or informal.

A **positive sanction** is an expression of approval and a reward for compliance; it may take the form of applause, a smile, or a pat on the back. In contrast, a **negative sanction** is an expression of disapproval for noncompliance; the punishment may consist of withdrawal of affection, ridicule, ostracism, banishment, physical harm, imprisonment, solitary confinement, or even death.

Informal sanctions are spontaneous, unofficial expressions of approval or disapproval that are not backed by the force of law. An example of an informal sanction is people making fun of a woman wearing a skirt when her legs are unshaven. Clearly no law requires women to shave their

Mao Zedong, the architect of the Cultural Revolution, died in 1976. Still, his image hangs in many prominent locations throughout China. The Chinese People's Armed Police Force is a symbol of social control. When police are in view, people are more conscious of their behavior.

sanctions Reactions of approval or disapproval to others' behavior or appearance.

positive sanction An expression of approval and a reward for compliance.

negative sanction An expression of disapproval for noncompliance.

informal sanctions Spontaneous, unofficial expressions of approval or disapproval that are not backed by the force of law.

GLOBAL COMPARISONS

Sentenced Prisoners in the United States and Other Nations

THESE MAPS SHOW the annual incarceration rate per 100,000 population across the United States and around the world. Louisiana has a higher incarceration rate (816 per 100,000) than the United States as a whole, and higher than Russia. Notice that China and the European Union have very low incarceration rates. How might one explain these differences?

Source: U.S. Department of Justice (2006).

The United States has the highest incarceration rate in the world, at 715 incarcerated for every 100,000 people.

Ten Countries with Highest Incarceration Rates (per 100,000 population)

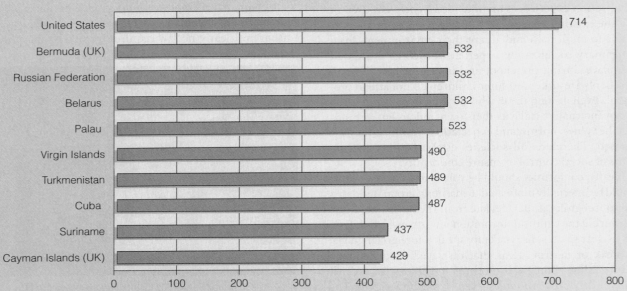

Country	Rate
United States	714
Bermuda (UK)	532
Russian Federation	532
Belarus	532
Palau	523
Virgin Islands	490
Turkmenistan	489
Cuba	487
Suriname	437
Cayman Islands (UK)	429

legs, but they are still penalized. **Formal sanctions**, by comparison, are expressions of approval or disapproval backed by laws, rules, or policies that specify (usually in writing) the conditions under which people should be rewarded or punished and the procedures for allocating

formal sanctions Expressions of approval or disapproval backed by laws, rules, or policies that specify (usually in writing) the conditions under which people should be rewarded or punished and the procedures for allocating rewards and administering punishments.

rewards and administering punishments. Examples of formal positive sanctions include awarding medals, cash bonuses, and diplomas. Formal negative sanctions may take the form of fines, prison sentences, the death penalty, corporal punishment, or the firing of tear gas to disperse demonstrators (see Global Comparisons: "Sentenced Prisoners in the United States and Other Nations").

During the Cultural Revolution, a paramilitary youth organization called the Red Guards applied negative sanctions to people targeted as holding a special social status and as accumulating worldly goods:

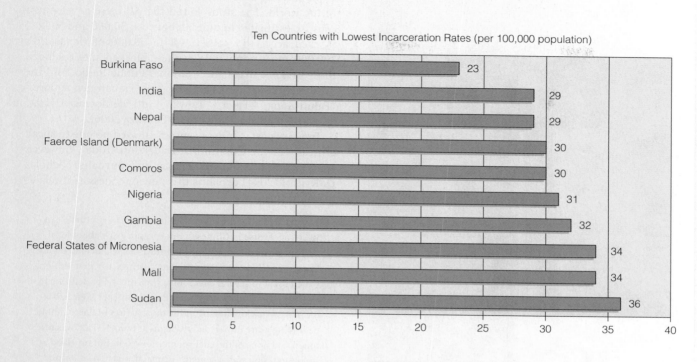

Ten Countries with Lowest Incarceration Rates (per 100,000 population)

Country	Rate
Burkina Faso	23
India	29
Nepal	29
Faeroe Island (Denmark)	30
Comoros	30
Nigeria	31
Gambia	32
Federal States of Micronesia	34
Mali	34
Sudan	36

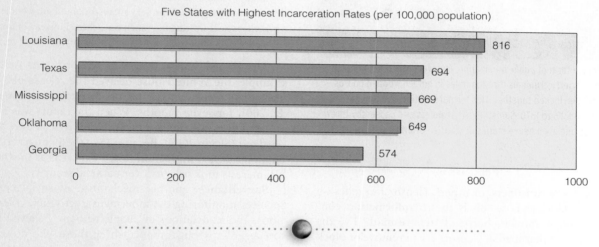

Five States with Highest Incarceration Rates (per 100,000 population)

Louisiana	816
Texas	694
Mississippi	669
Oklahoma	649
Georgia	574

Targets might be required to stand on a platform, heads bowed respectfully to the masses, while acknowledging and repeating their crimes. Typically they had to "airplane," stretching their arms out behind them like the wings of a jet. In the audience tears of sympathy might be in a friend's eyes, but from his mouth would come only curses and derisive jeering, especially if the victim after an hour or two fell over from muscular collapse. To Chinese people, who were especially sensitive to peer-group esteem, to be beaten and humiliated in public before a jeering crowd including colleagues and old friends was like having one's skin taken off. (Fairbank 1987, p. 336)

Censorship and Surveillance

In addition to sanctions, other mechanisms of social control include censorship and surveillance. **Censorship** is a method of preventing information from reaching an audience. That audience may consist of children, vot-

censorship A method of preventing information from reaching an audience.

The pillory, a form of public humiliation, involved inserting the head and arms of petty criminals through holes in hinged wooden boards that were then locked together. This formal sanction was popular in Europe from 1275 to 1870. Sometimes bystanders and passersby threw rotten food and even heavy stones at pilloried offenders.

Book of Days, 1st Edition by Robert Chambers

ers, employees, prisoners, or others. Censorship relies on **censors**—people whose job is to sift information conveyed through movies, books, letters, e-mail, TV, the Internet, and other media. Censors try to remove or block any material that those in power consider unsuitable or

censors People whose job is to sift information conveyed through movies, books, letters, TV, the Internet, and other media and to remove or block any material that those in power consider unsuitable or threatening.

surveillance A mechanism of social control that involves monitoring the movements, activities, conversations, and associations of people who are believed likely to engage in wrongdoing; catching those who do engage in it; preventing people from engaging in it; and ensuring that the public is protected from wrongdoers.

threatening. A Harvard Law School study identified China as the country with the most extensive Internet censorship in the world. The study found that Chinese government censors, estimated to number between 30,000 and 50,000, blocked access on some level to 50,000 of the world's 200,000 most popular Web sites. Blocked sites included those of *Time* magazine, National Public Radio, and the *Washington Post*. Links to the topics "Tiananmen Square," "Falun Gong," "Tibet," "Taiwan," and "democracy China" were also blocked (Kahn 2002; Barboza 2006).

In years past, Chinese censors allowed no one to speak critically of the Communist leadership. Today, however, some criticism is allowed. Scholar and journalist Nicholas D. Kristof (2004) explains this new openness as it relates to online forums:

> You can't go online and say that President Hu is a turtle's egg (it sounds worse in Chinese), but you can gripe about local corruption or poor highway planning. I experimented on my last trip to China and tried various postings. My first version, which I sent to several chat rooms (in Chinese, pretending to be Chinese myself) was "Why is Prime Minister Wen Jiabao off in America kowtowing to the imperialists when he should be solving more important problems at home!" That was too tough and none of the chat rooms allowed it. But my third and mildest version was accepted: "Prime Minister Wen Jiabao's visit to America has been very successful, but I am wondering if perhaps he is wasting too much time abroad instead of focusing on our own important problems like unemployment." (p. 58)

The United States engages in censorship as well. For example, the White House delayed the release of a congressional report on the terrorist attacks of September 11, 2001, for seven months. When the report was finally released, some sections related to Saudi Arabia were blackened out (Slavin 2004). Because 15 of the 19 hijackers were Saudi nationals, this action raised concerns as to whether oil interests took priority over national security.

Surveillance, another mechanism of social control, involves monitoring the movements, activities, conversations, and associations of people who are believed likely to engage in wrongdoing; catching those who do engage in it; preventing people from engaging in it; and ensuring that the public is protected from wrongdoers. Surveillance activities include telephone tapping, interception of letters and e-mail, observation via closed-circuit television, and electronic monitoring. The U.S. Justice Department's post–September 11 appeal to "couriers, meter readers, cable installers, and telephone repairmen" to report suspicious activities they might notice while servicing customers qualifies as a form of surveillance (Kayal 2002, p. A21). A Gallup poll conducted shortly after the attacks showed that 60 percent of Americans favored special surveillance checks on any person of Arab descent (including U.S. citizens) traveling by air. In addition, 50 percent believed that all persons of Arab descent should be required to carry

These surveillance devices, aimed at passengers riding in cars approaching terminals, have facial recognition and body temperature-reading capabilities.

Steve Jurvetson, www.DFJ.com/J

special identification. One-third of the Americans surveyed agreed that Arabs living in the United States should be put under special surveillance, similar to that applied to Japanese Americans during World War II (Jones 2001; Zogby 2001).

In July 2004 the Chinese government launched a surveillance campaign to monitor the estimated 220 billion (and climbing) text messages that 300 million Chinese send to one another each year. The government is taking this step to prevent text-messaging technology from spreading unauthorized information that might undermine its one-party rule (Kahn 2004d).

The Functionalist Perspective

■ **CORE CONCEPT 4: It is impossible for a society to exist without deviance. Always and everywhere there will be some behaviors or appearances that offend collective sentiments.** Émile Durkheim (1901) argued that although definitions of what constitutes deviance varies by place, it is present in all societies. He defined *deviance* as acts that offend collective norms and expectations. The fact that always and everywhere some people will offend collective sentiments led him to conclude that deviance is normal as long as it is not excessive and that "it is completely impossible for any society entirely free of it to exist" (p. 99). According to Durkheim, deviance will be present even in a "community of saints" living in an "exemplary and perfect monastery" (p. 100). Durkheim used the analogy of the "perfect and upright" person who judges his or her smallest failings with a severity that others reserve for the most serious offenses. Likewise even among the exemplary, some seemingly insignificant act or appearance will be greeted as deviant, even criminal, if only because "it is impossible for everyone to be alike if only because each of us cannot stand in the same spot" (p. 100).

Durkheim believed that what makes an act or appearance deviant is not so much its character or consequences, but that a group has defined it as dangerous or threatening to its well-being. Wearing eyeglasses, for example, clearly does not harm others. As we have learned, however, this behavior was identified in China during the Cultural Revolution as a clear indicator of other threatening behaviors, such as the crimes of "special status" or abandonment of revolutionary spirit, that were not so easily observable.

According to Durkheim, deviance has an important function in society, for at least two reasons. First, the ritual of identifying and exposing the wrongdoing, determining a punishment, and carrying it out is an emotional experience that binds together the members of a group and establishes a sense of community. Durkheim argued that a group that went too long without noticing deviance or doing something about it would lose its identity as a group. In evaluating Durkheim's argument, consider that when U.S. officials criticize China for some alleged wrongdoing, they are simultaneously arguing that the United States is a better place to live. Likewise, when Chinese officials criticize the United States, they are indirectly sending the message "Be glad you live here and not there!" Second, deviance is functional because it helps bring about necessary change and prepares people for change. Nothing would change if someone did not step forward and introduce a new

Department of Defense photo by PH1 Steven Batiz

Durkheim believed that it is impossible for a society to exist without deviance. Some difference, however slight, is bound to offend. If Durkheim were analyzing this photograph, what might he point out as something that might offend?

perspective or a new way of doing things. By definition this "step forward" is considered deviant, as it departs from the normal way of thinking or of doing things.

Durkheim's theory offers an intriguing explanation for why almost anything can be defined as deviant. Yet, Durkheim did not address an important question: Who decides that a particular activity or appearance is deviant? Labeling theory provides one answer to this question.

Labeling Theory

■ **CORE CONCEPT 5:** Labeling theorists maintain that an act is deviant when people notice it and then take action to label it as a violation and apply appropriate sanctions. In *Outsiders: Studies in the Sociology of Deviance*, sociologist Howard Becker states the central thesis of labeling theory:

> All social groups make rules and attempt, at some times and under some circumstances, to enforce them. When a rule is enforced, the person who is supposed to have broken it may be seen as a special kind of person, one who cannot be trusted to live by the rules agreed on by the group. He is regarded as an outsider. (1963, p. 1)

As Becker's statement suggests, labeling theorists are guided by two assumptions: (1) rules are socially constructed, and (2) these rules are not enforced uniformly or consistently. Support for the first assumption comes from the fact that because definitions of deviant behavior vary across time and place, people must decide what is deviant. The second assumption is supported by the fact that some people break rules and escape detection, whereas others are treated as offenders even though they have broken no rules. Labeling theorists maintain that whether an act is deviant depends on whether people notice it and, if they do notice, on whether they label it as a violation of a rule and subsequently apply sanctions.

Such contingencies suggest that violating a rule does not automatically make a person deviant. That is, from a

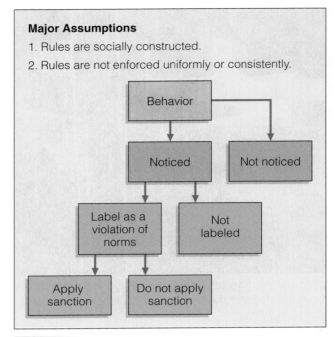

Major Assumptions

1. Rules are socially constructed.
2. Rules are not enforced uniformly or consistently.

▲ **Figure 7.2** Flow Chart Illustrating Labeling Theory

sociological point of view, a rule breaker is not deviant (in the strict sense of the word) unless someone *notices* the violation and decides to take corrective action (see Figure 7.2). "The critical variable in the study of deviance, then, is the social audience rather than the individual actor, since it is the social audience [that] eventually determines whether or not any episode or behavior . . . is labeled deviant" (Erikson 1966, p. 11).

Labeling theorists suggest that for every rule a social group creates, four categories of people exist: conformists, pure deviants, secret deviants, and the falsely accused. The category to which one belongs depends on whether a rule has been violated and on whether sanctions are applied. **Conformists** are people who have not violated the rules of a group and are treated accordingly. **Pure deviants** are people who have broken the rules and are caught, punished, and labeled as outsiders. As a result, these rule breakers take on the **master status of deviant**, an identification that "proves to be more important than most others. One will be identified as a deviant first, before other identifications are made" (Becker 1963, p. 33). We must remember that although pure deviants undeniably violate rules, rule enforcers play a role in "noticing and deciding to punish" those they apprehend.

Secret deviants are people who have broken the rules but whose violation goes unnoticed or, if it is noticed, prompts those who notice to look the other way rather than reporting the violation. Becker maintains that "no one really knows how much of this phenomenon exists,"

conformists People who have not violated the rules of a group and are treated accordingly.

pure deviants People who have broken the rules of a group and are caught, punished, and labeled as outsiders.

master status of deviant An identification marking a rule breaker first and foremost as a deviant.

secret deviants People who have broken the rules of a group but whose violation goes unnoticed or, if it is noticed, prompts no one to enforce the law.

Table 7.1 Number of Victimizations by Crime Type and Percentage Not Reported to Police, 2005

Crime	Number (millions)	Percentage Not Reported to Police
All crimes	23.4	57.4
All personal crimes	5.4	51.3
Violent crimes	5.1	50.7
Property crimes	18.0	59.3

Source: U.S. Department of Justice, Bureau of Justice Statistics (2005).

but he is convinced that the "amount is very sizable, much more so than we are apt to think" (1963, p. 20). A survey of crime victims in the United States found that 23.4 million crimes were committed against U.S. residents 12 years old and older. Of the victims, 57.4 percent did not report the crimes to police (U.S. Department of Justice 2005; see Table 7.1).

The **falsely accused** are people who have not broken the rules but are treated as if they have. The ranks of the falsely accused include victims of eyewitness errors and police cover-ups; they also include innocent suspects who make false confessions under the pressure of interrogation. For the book *In Spite of Innocence*, sociologist Michael L. Radelet and his co-author reviewed more than 400 cases of innocent people convicted of capital crimes and found that 56 had made false confessions. Apparently, some innocent suspects admitted guilt, even regarding heinous crimes, to escape the stress of interrogation (Jerome 1995). As with secret deviance, no one knows how often false accusation occurs, but it probably occurs more often than we think. Moreover, the taint of guilt often lingers even after the falsely accused is cleared of all charges. Such cases lead us to ask a larger question: Under what circumstances are people most likely to be falsely accused?

The Falsely Accused

Sociologist Kai Erikson (1966) identified a particular situation in which people are likely to be falsely accused of a crime: when the well-being of a country or a group is threatened. The threat can take the form of an economic crisis (such as a depression or recession), a moral crisis (such as family breakdown), a health crisis (such as AIDS), or a national security crisis (such as war). At times like these, people need to identify a clear source of the threat. Thus, whenever a catastrophe occurs, it is common to blame someone for it. Identifying a threat gives an illusion of control. In a crisis, the person blamed is likely to be someone who is at best indirectly responsible, someone

in the wrong place at the wrong time, or someone who is viewed as different.

This defining activity can take the form of a **witch hunt**—a campaign to identify, investigate, and correct behavior that has been defined as undermining a group or country. In actuality, a witch hunt rarely accomplishes these goals, because the real cause of a problem is often complex, extending far beyond the behavior of a targeted person or category. Often people who are identified as the cause of a problem are simply being used to make the problem appear as if it is being managed. These dynamics help explain how, during the Cultural Revolution, the most seemingly insignificant acts—such as wearing makeup or eyeglasses—were classified as crimes against the country. The Cultural Revolution was Chairman Mao Zedong's response to a specific crisis: the failure of the Great Leap Forward—Mao's plan to mobilize the masses to transform China from a land of poverty into one of agricultural abundance in five short years (1958 through 1962).

Under this plan, Mao enlisted hundreds of thousands of people for projects ranging from killing insects to building giant dams with shovels and wheelbarrows (Butterfield 1976). The Great Leap Forward was an ill-conceived, hastily planned, sweeping reorganization of Chinese society that created economic and environmental disruption on a massive scale, leading to the deaths of 30 to 50 million Chinese from human-induced famine. In one region, peasants stopped harvesting crops and dug tunnels in their fields to seek coal, which Communist Party officials believed was plentiful there. No coal was found, however, and the crops rotted.

In light of this crisis, Mao was particularly vulnerable to political attack. He blamed the failure of his plan on entrenched authority, which he loosely defined as the "Four Olds": old ideas, old culture, old customs, and old habits. Mao also blamed the failure on the abandonment of revolutionary spirit and on the "evils of special status and special accumulation of worldly goods" (Fairbank 1987, p. 319). He used the Cultural Revolution as an attempt to eliminate anyone in the Communist Party and in the masses who opposed his policies. Of course, such targeted behaviors as wearing eyeglasses or makeup could

falsely accused People who have not broken the rules of a group but are treated as if they have.

witch hunt A campaign to identify, investigate, and correct behavior that is believed to be undermining a group or country. Usually this behavior is not the real cause of a problem but is used to distract people's attention from the real cause or to make the problem seem manageable.

National Park Service, U.S. Department of the Interior

The U.S. government's internment of more than 110,000 people of Japanese descent (80 percent of whom were U.S. citizens) during World War II is an example of a witch hunt. Japanese Americans living on the West Coast were forced from their homes and taken to desert prisons surrounded by barbed wire and guarded with machine guns. None had been found guilty of anti-American activity. Nevertheless, wartime hysteria, combined with long-standing prejudice, led to this internment (Kometani 1987). Here we see Japanese American children tagged like luggage to be sent to various detention centers.

not possibly have been responsible for the failure of the Great Leap Forward. Nevertheless, targeting such behaviors diverted the public's attention from this disruptive event and from the shortcomings of those in power, and it united the public behind a cause.

Of course, witch hunts are not confined to China. After the terrorist attacks of September 11, 2001, Muslim and Arab Americans were "caught up in the biggest criminal investigation in U.S. history" (Kaye 2001). Although the FBI did not keep statistics on the ethnicity or religious affiliation of the people questioned about the attacks, it is believed that most people interrogated were or appeared to be Muslim or Middle Eastern. In southern California alone, the FBI pursued more than 22,000 leads in the month following the attacks. While many Arab and Muslim Americans expressed understanding about the real and perceived needs for such increased scrutiny, an unknown number

white-collar crime "Crimes committed by persons of respectability and high social status in the course of their occupations" (Sutherland and Cressey 1978, p. 44).

corporate crime Crime committed by a corporation as it competes with other companies for market share and profits.

of FBI and police interrogations went beyond questioning to include demands that women remove headscarves. We also know that many federal agents, haunted by the September 11 attacks, acted on "information from tipsters with questionable backgrounds and motives, touching off needless scares and upending the lives of innocent suspects" (Moss 2003, p. A1).

The existence of the falsely accused underscores the fact that the study of deviance must look beyond people identified or labeled as rule breakers. After all, the falsely accused are innocent, but they have been labeled as deviant by those with the power to do so. (See Working For Change: "The Falsely Accused on Death Row.")

Rule Makers and Rule Enforcers

■ **CORE CONCEPT 6: Sociologists are concerned less with rule violators than with rule makers and enforcers.** Sociologist Howard Becker (1973) recommends that, when studying deviance, researchers pay particular attention to who the rule makers and rule enforcers are and to how they achieve power and then use it to define how others "will be regarded, understood, and treated" (p. 204). This topic, of course, draws on conflict theorists' emphasis on dominant and subordinate groups. According to conflict theorists, the members of a society with the most wealth, power, and authority have the ability to create laws and establish crime-stopping and crime-monitoring institutions. Consequently, we should not be surprised that law enforcement efforts tend to focus disproportionately on the poor and other powerless groups rather than on the wealthy and politically powerful.

This uneven focus gives the widespread impression that the poor, the uneducated, and members of minority groups are more prone to criminal behavior than are people in the middle and upper classes, the educated, and members of the majority group. In fact, crime exists in all social classes, but the type of crime, the extent to which the laws are enforced, access to legal aid, and the power to shape laws to one's advantage vary across classes (Chambliss 1974). In the United States, for example, police efforts are largely directed at controlling crimes against individual life and property (crimes such as drug offenses, robbery, assault, homicide, and rape) rather than controlling white-collar and corporate crimes.

White-collar crime consists of "crimes committed by persons of respectability and high social status in the course of their occupations" (Sutherland and Cressey 1978, p. 44). **Corporate crime** is crime committed by a corporation as it competes with other companies for market share and profits. White-collar and corporate crimes—such as the manufacturing and marketing of unsafe products, unlawful disposal of hazardous waste, tax evasion, and money laundering—are usually handled not by the police but by regulatory agencies, such as the Environmental

WORKING FOR CHANGE

The Falsely Accused on Death Row

IN SEPTEMBER 1998 Anthony Porter was within 50 hours of his scheduled execution for a double homicide when the Illinois Supreme Court granted him a stay of execution. Porter had been defended by a lawyer who fell asleep in court, assigned a judge who later left the bench over a financial scandal, and convicted by a jury prejudiced by a witness's false testimony. After his appeals had failed, a lawyer volunteered his services and had Porter's IQ tested. When the lawyer learned that Porter was borderline mentally retarded, a team of four other pro bono lawyers and journalism students agreed to take the case. The team interviewed two crime-scene witnesses.

One eyewitness, Inez Jackson, told the team that she had seen her then-husband, Alstory Simon, shoot both victims. Simon admitted his guilt, claiming that he killed one victim in self-defense while fighting over a drug deal and that the other victim's death was an accident (Center on Wrongful Convictions 2003).

George Ryan had just been inaugurated governor of Illinois in early 1999 when he saw Simon's videotaped confession. A proponent of capital punishment when elected, Ryan said this case left him feeling "jolted into reexamining everything I believed in" (Shapiro 2001). Weeks later, Andrew Kokoraleis, another Illinois prisoner, was scheduled to die. After agonizing over the decision, Ryan chose to sign off on the execution.

Later that year, Ryan learned that 13 death-row inmates, some convicted 25 years before, had been found innocent after DNA evidence was discovered. Among them was an inmate who was within days of a scheduled execution. Ryan's views on capital punishment changed drastically. While he never questioned the state's right to take a life, he argued that the sentencing process was so flawed that it must be shut down until it was repaired (*Christian Century* 2003). In January 2000 he ordered a moratorium on executions in Illinois.

Ryan then appointed a panel to investigate Illinois's capital punishment system. The panel recommended 85 changes, including videotaping all police questioning of capital suspects and revising the procedures for conducting lineups. Because the panel found that the death penalty was unevenly applied, they recommended that it should be applied only when the defendant had murdered two or more people, a police officer or firefighter, a prison officer or inmate, or a crime-scene witness. The panel also suggested that the death penalty should not be a sentencing option when only eyewitness evidence existed (Governor's Commission on Capital Punishment 2002).

In October 2002 Ryan pardoned four black men who, after serving 15 years on death row for the 1986 slaying and rape of a medical student, were exonerated by DNA evidence. Then in January 2003, after a three-year battle to reform the Illinois capital punishment system, Ryan announced that he was commuting the death sentences of another four black men tortured by Chicago police officers into confessing to crimes they did not commit. Ryan stated that these four men were "perfect examples of what is so terribly broken about our system" (Kelly 2003).

Two days before leaving office in January 2003, George Ryan commuted the death sentences of 167 inmates to life without parole. Because Illinois citizens were evenly divided on the issue of sparing prisoners and commuting death sentences to life (*The Economist* 2002), it should come as no surprise that Ryan's actions were both praised as courageous and scorned as irresponsible (Johnson 2003, p. 34). At a press conference, Ryan stated, "Our capital system is haunted by the demon of error—error in determining guilt, and error in determining who among the guilty deserve to die" (Ryan 2003).

Ryan's actions have raised public awareness about death penalty misuse and the plight of the wrongfully convicted. He has forced us to ask hard questions: How many of the 3,557 prisoners on death row in the United States were falsely convicted? How many of the 4,744 prisoners put to death since 1930 were innocent? (U.S. Department of Justice 2004). Ryan has won many awards recognizing his courage and conviction, and he was nominated for the 2003 Nobel Peace Prize. Ryan's efforts run parallel to those of the Innocence Project, a not-for-profit legal clinic that "only handles cases where post-conviction DNA testing of evidence can yield conclusive proof of innocence" (Innocence Project 2004). Students, supervised by a team of attorneys and clinic staff, investigate and handle the cases. Of the 144 inmates the clinic has exonerated, 23 (16 percent) of them were convicted in Illinois.

Source: Missy Gish, Class of 2005, Northern Kentucky University, Updated June 2007.

In 2006 two top executives at Enron Corporation—Ken Lay and Jeffrey Skilling—were found guilty of conspiracy, fraud, and making false statements. Their actions caused 4,000 Enron employees to lose their jobs and life savings and caused investors to lose billions of dollars.

lation of all efforts toward overture and modernization—none of these could ever enter into consideration once the Party's authority was at stake. (Leys 1989, p. 17)

When studying rule makers and rule enforcers, sociologists question how the powerful influence the public to accept their definitions of what is deviant and then apply the recommended sanctions. The work of social psychologist Stanley Milgram provides some insights.

Obedience to Authority

■ **CORE CONCEPT 7: The firm commands of a person holding a position of authority over a person hearing those commands can elicit obedient responses.** When Stanley Milgram (1974) conducted the research for his book *Obedience to Authority*, he wanted to learn how people in positions of authority persuade other people to accept the authorities'

Protection Agency and the Food and Drug Administration, which have minimal staff to monitor compliance.

Escaping punishment is easier for white-collar and corporate criminals than for other criminals. In the case of white-collar crime, offenders are part of the system: They occupy positions in the organization that permit them to carry out illegal activities discreetly. In the case of corporate crime, everyone in the organization contributes to illegal activities simply by doing their jobs. In addition, both white-collar and corporate crimes are "directed against impersonal—and often vaguely defined—entities such as the tax system, the physical environment, competitive conditions in the market economy, etc." (National Council for Crime Prevention in Sweden 1985, p. 13). These crimes are without victims in the usual sense, because they are "seldom directed against a particular person who goes to the police and reports an offense" (p. 13). In 2004, less than 1 percent (1,108 persons) of all people sentenced to U.S. federal prisons were classified as white-collar criminals, whereas 54.3 percent (56,291 persons) were classified as drug offenders (Federal Bureau of Prisons 2004).

The focus on rule makers and rule enforcers suggests that sociologists are just as concerned with those who make and apply rules as with rule violators. So, for example, sociologists could not study deviance in China without considering that, since 1949, those in power have adhered to the following code:

In any circumstance and at any cost, political power must be retained in its totality. This rule is absolute, it tolerates no exception and must take precedence over any other consideration. The bankruptcy of the entire country, the ruin of its credit abroad, the destruction of national prestige, the annihi-

When sociologists study deviance, they place at least as much emphasis on the rule makers and rule enforcers as they do on the rule violators.

definitions of deviance and to conform to orders about how to treat people classified as deviant. His study gives us insights into how events such as the Holocaust, the Cultural Revolution, and more recently, prisoner abuse at Abu Ghraib and other U.S.-run prisons, could have taken place. That such atrocities required the cooperation of many people raises important questions about people's capacity to obey authority.

> The person who, with inner conviction, loathes stealing, killing, and assault may find himself performing these acts with relative ease when commanded by authority. Behavior that is unthinkable in an individual who is acting on his own may be executed without hesitation when carried out under orders. (Milgram 1974, p. xi)

Milgram designed an experiment to see how far people would go before they would refuse to conform to an authority's orders. The findings of his experiment have considerable relevance for understanding the conditions under which rules handed down by authorities are enforced by the masses.

The participants in Milgram's experiment were volunteers who answered an ad he had placed in a local paper. When participants arrived at the study site, they were greeted by a man in a laboratory jacket who explained to them and another apparent volunteer that the study's purpose was to learn whether the use of punishment improves the ability to learn. (Unknown to each subject, the other "volunteer" was actually a confederate—someone working in cooperation with the investigator conducting the study.) The participant and the confederate drew lots to determine who would be the teacher and who would be the learner. The draw was fixed, however, so that the confederate was always the learner and the real volunteer was always the teacher.

The learner was strapped to a chair, and electrodes were placed on his or her wrists. The teacher, who could not see the learner, was placed in front of an instrument panel containing a line of shock-generating switches. The switches ranged from 15 to 450 volts and were labeled accordingly, from "slight shock" to "danger, severe shock." The researcher explained that when the learner made a first mistake, the teacher was to administer a 15-volt shock; the teacher would then increase the voltage with each subsequent mistake. In each case, as the strength of the shock increased, the learner expressed greater discomfort. One learner even said that his heart was bothering him and went silent.

When the volunteers expressed concern, the researcher firmly told them to continue administering shocks. Although many of the volunteers protested, a substantial number obeyed and continued "no matter how vehement the pleading of the person being shocked, no matter how painful the shocks seemed to be, and no matter how much the victim pleaded to be let out" (Milgram 1987, p. 567).

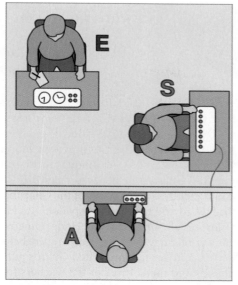

In Milgram's classic study, the experimenter (E) commands the subject (S) to increase the intensity of electric shocks to a confederate posing as a learner (A) when the learner makes mistakes, even after the learner begs the subject to stop.

The results of Milgram's experiments are especially significant when one considers that the participants received no penalty if they refused to administer shocks. Obedience in this situation was founded simply on the firm command of a person with a status that gave minimal authority over the subject. If this level of obedience is possible under the circumstances of Milgram's experiments, one can imagine the level that is possible when disobedience brings severe penalties or negative consequences.

For the Cultural Revolution to have taken place, millions of Chinese must have cooperated to carry out Mao Zedong's mission of finding and purging those deemed responsible for the failure of the Great Leap Forward. Mao defined those loyal to the past and undevoted to his thoughts and words as the culprits. Chinese author Wang Shuo (1997) described the Cultural Revolution as a time when "we were out of control. Everything was turned upside down. The teachers who used to do the educating were sent away to be reeducated. Children were allowed to correct their parents" (p. 51). Between 1966 and 1971, more than 36 million Chinese were persecuted in some way, with the most extreme forms of persecution being "enforced suicide" and being "hounded to death" (Spence 2006; MacFarquhar and Schoenhals 2006).

Mao initially assigned the Red Guards (his name for the youths of China between the ages of 9 and 18) the tasks of finding these culprits. In fact, Mao ordered classes suspended nationwide to free up 500,000 college students

INTERSECTION OF BIOGRAPHY AND SOCIETY

Obedience to Authority during the Cultural Revolution and at Abu Ghraib Prison

THE REFLECTIONS OF a former Red Guard, some 15 years after the Cultural Revolution, show the intensity with which people were hunted down and persecuted: "I was very young when the Cultural Revolution began. . . . My schoolmates and I were among the first in Peking to become Red Guards; we believed deeply in Chairman Mao. I could recite the entire book of the Chairman's quotations backward and forward; we spent hours just shouting the slogans at our teachers." Hong remembered in particular a winter day, with the temperature below freezing, when she and her faction of Red Guards put on their red armbands and made three of the teachers from their high school kneel on the ground outside without coats or gloves. "We had gone to their houses to conduct an investigation, to search them, and we found some English-language books. They were probably old textbooks, but to us it was proof they were worshipping foreign things and were slaves to the foreigners. We held a bonfire and burned everything we had found." After that, she recalled, the leader of her group—a tall, charismatic eighteen-year-old boy, the son of an army general, whose nickname was "Old Dog"—ordered them to beat the teachers. He produced some wooden boards, and the students started hitting the teachers on their bodies. "We kept on till one of the teachers start[ed] coughing blood," Hong said. . . ."We felt very proud of ourselves. It seemed very revolutionary" (Butterfield 1982, p. 183).

Staff Sergeant Ivan L. Frederick II was caught up in the Abu Ghraib prison scandal. Abu Ghraib, one of the world's most notorious prisons under Saddam Hussein, was converted to a U.S. military prison after Hussein was removed from power. Frederick faced charges of conspiracy, dereliction of duty, cruelty toward prisoners, maltreatment, assault, and indecent acts. He was sentenced to eight years in prison. Letters and e-mail messages that Frederick wrote to family members were presented in court to bolster his defense lawyer's assertion that he was simply carrying out orders (Hersh 2004). In January 2004, Frederick wrote "I questioned some of the things that I saw, . . . such things as leaving inmates in their cell with no clothes or in female underpants, handcuffing them to the door of their cell—and the answer I got was, 'This is how military intelligence [MI] wants it done.' . . . MI has also instructed us to place a prisoner in an isolation cell with little or no clothes, no toilet or running water, no ventilation or window, for as much as three days." Frederick also wrote that when he questioned his superior officer about this mistreatment, the officer said, "Don't worry about it." Frederick wrote about the situation of an Iraqi prisoner under CIA control: "They stressed him out so bad that the man passed away. They put his body in a body bag and packed him in ice for approximately twenty-four hours in the shower. . . . The next day the medics came and put his body on a stretcher, placed a fake IV in his arm and took him away" (Hersh 2004).

---

and 113 million primary and middle school students for political action (Spence 2006; MacFarquhar and Schoenhals 2006). (See Intersection of Biography and Society: "Obedience to Authority during the Cultural Revolution and at Abu Ghraib Prison.")

constructionist approach A sociological approach that focuses on the way specific groups, activities, conditions, or artifacts become defined as problems.

claims makers People who articulate and promote claims and who tend to gain in some way if the targeted audience accepts their claims as true.

The Constructionist Approach

■ **CORE CONCEPT 8:** In an effort to define deviance, rule makers and rule enforcers assume the role of claims maker and engage in various claims-making activities. Regarding deviance, the **constructionist approach** focuses on the way specific groups (such as illegal immigrants and homosexuals), activities (such as child abuse), conditions (such as teenage pregnancy, infertility, and pollution), or artifacts (such as song lyrics, guns, art, and eyeglasses) become defined as problems. In particular, constructionists examine claims makers and claims-making activities. **Claims makers** are people who articulate and promote claims and who tend to gain in some way if the targeted audience accepts their claims as true. Claims makers include government officials, marketers, scientists, professors, and other special-interest groups. **Claims-making activities** are actions taken to draw attention to a claim—actions such as "demanding services, filling out forms, lodging complaints, filing lawsuits, calling press conferences, writing letters of protest, passing resolutions, publishing exposés, placing ads in newspapers, . . . setting up picket lines or boycotts" (Spector and Kitsuse 1977, p. 79).

An example of a claims-making activity is the annual human rights reports that the U.S. Department of State releases on 190 countries, including one on the People's Republic of China. In that report the United States makes a number of claims about China, including this one:

> The government's human rights record remained poor, and the government continued to commit numerous and serious abuses. There was a trend towards increased harassment, detention, and imprisonment by government and security authorities of those perceived as threatening to government authority. The government also adopted measures to control more tightly print, broadcast and electronic media, and censored online content. Protests by those seeking to redress grievances increased significantly and were suppressed, at times violently, by security forces.

Each year the Chinese government responds by issuing a human rights report on the United States. It began its 2006 report with this claim:

> The U.S. Department of State, posing once again as "the world's judge of human rights," released its Country Reports on Human Rights Practices for 2005. As in previous years, the State Department pointed the finger at human rights situations in more than 190 countries and regions, including China, but kept silent on the serious violations of human rights in the United States. To help people realize the true features of this self-styled "guardian of human rights," it is necessary to probe into the human rights abuses in the United States.

The success of a claims-making campaign depends on a number of factors, including access to the media, available resources, and the claims maker's social status and skill at fund-raising, promotion, and organization (Best 1989). According to sociologist Joel Best, when constructionists study the process through which a group or behavior is defined as a problem to society, they focus on who makes the claims, whose claims are heard, and how audiences respond to them. Constructionists are guided by one or more of the following questions: What kinds of claims are made about the problem? Who makes the claims? Which claims are heard? Why is a claim made at a particular time? What are the responses to the claim? Is there evidence that the claims maker has misrepresented or inaccurately characterized the situation? To answer the last question, constructionists examine how claims makers characterize a condition. Specifically, constructionists pay attention to any labels that claims makers attach to a condition, the examples they use to illustrate the nature of the problem, and their orientation toward the problem (describing it, for example, as medical, moral, genetic, or educational).

Labels, examples, and orientation are important because they tend to evoke a particular cause of a problem and a particular solution to it (Best 1989). For example, to label AIDS as a moral problem is to locate its cause in the goodness or badness of human action and to suggest that the solution depends on the victims changing their evil ways. To call it a medical problem is to locate the cause in the biological workings of the body or mind and to suggest that the solution rests with a drug, a vaccine, or surgery.

When Mao died in 1976, the Cultural Revolution ended, and China's new leaders faced a great many problems, including how to undo the effects of this movement, which had taken place at the expense of China's economic, technological, scientific, cultural, and agricultural development and had drained the Chinese physically and mentally. It also created a 10-year gap in training, such that one leading Chinese surgeon claimed, "I can't honestly let any of the young doctors in my hospital operate on a patient. . . . They went to medical school. But they studied Mao's thought, planting rice or making tractor parts. They never had to take exams and a lot of them don't know basic anatomy" (Butterfield 1980, p. 32).

To solve the problems resulting from the Cultural Revolution, the Communist Party, under the leadership of Deng Xiaoping, rallied the teachers, technicians, artists, and scientists whom the revolution had struck down. Deng Xiaoping claimed that the government's new policies were in the best interest of the country and that they did not promote individual self-interest, special status, or accumulation of wealth. These claims allowed Chinese leaders to send tens of thousands of students overseas to study in the capitalist West. They allowed farmers and factory workers to keep profits from surplus crops after meeting government quotas, and they permitted the government to establish Special Economic Zones (SEZs), designated areas within China that could enjoy capitalist privileges.

To this point, we have examined several sociological concepts—socialization, norms (folkways and mores), and mechanisms of social control—and discussed how they relate to deviance. In addition, we have examined the

claims-making activities Actions taken to draw attention to a claim, such as "demanding services, filling out forms, lodging complaints, filing lawsuits, calling press conferences, writing letters of protest, passing resolutions, publishing exposés, placing ads in newspapers, . . . setting up picket lines or boycotts" (Spector and Kitsuse 1977, p. 79).

functionalist perspective to gain insights about how any behavior or appearance can come to be defined as deviant. We have looked at labeling theory and the constructionist approach to gain insights into the role that rule makers play in shaping deviance. Next, we turn to the theory of structural strain, which helps us answer the following question: Under what conditions do people engage in behavior defined as deviant?

Structural Strain Theory

■ **CORE CONCEPT 9: Deviant behavior is a response to structural strain, a situation in which a disjuncture exists between culturally valued goals and legitimate means for achieving those goals.** Robert K. Merton's theory of structural strain takes three factors into account: (1) goals defined as valuable and legitimate for all members of society, (2) norms that specify the legitimate means of achieving those goals, and (3) the actual number of legitimate opportunities available to people to achieve the goals. According to Merton, **structural strain** is any situation in which (1) the valued goals have unclear limits (that is, people are unsure whether they have achieved them), (2) people are unsure whether the legitimate means will allow them to achieve the goals, or (3) legitimate opportunities for reaching the goals remain closed to a significant portion of the population. The rate of deviance is likely to be high in any one of these situations. Merton uses the United States, a country where all three conditions exist, to show the relationship between structural strain and deviance.

Structural Strain in the United States

In the United States most people place a high value on the culturally valued goal of economic success and social mobility. Americans tend to believe that anyone can achieve such success; that is, they believe that all people, regardless of the circumstances into which they were born, can achieve affluence and social stature. Such a viewpoint suggests that persons who fail to do so have only themselves to blame.

---

structural strain Any situation in which (1) the goals defined as valuable and legitimate for a society have unclear limits, (2) people are unsure whether the legitimate means that the society provides will allow them to achieve the goals, and (3) legitimate opportunities for reaching the goals remain closed to a significant portion of the population.

conformity The acceptance of the cultural goals and the pursuit of those goals through legitimate means.

innovation The acceptance of cultural goals but the rejection of the legitimate means to achieve them.

Merton argues that "Americans are bombarded on every side" with the message that this goal is achievable, "even in the face of repeated frustration" (1957, p. 137). Even when people do achieve monetary success, at no point can they say they feel secure about having achieved their ultimate goal. "At each income level . . . Americans want just about 25 percent more (but of course this 'just a bit more' continues to operate once it is obtained)" (p. 136).

Merton also maintains that structural strain exists in the United States because the legitimate means of achieving wealth do not always lead to its achievement. The individual's task is to choose a path that leads to success. That path might involve education, hard work, or development of a talent. The problem is that schooling, hard work, and talent do not always guarantee economic success. Regarding education, for example, many Americans believe that a diploma (and especially a college diploma) in itself entitles a person to a high-paying job. In reality, a diploma is merely one of many components needed to achieve success.

Finally, structural strain exists in the United States because too few legitimate opportunities exist to achieve the goal of financial success. Although Americans are supposed to seek it, all cannot expect to achieve it legitimately. The opportunities for achieving success remain closed to many people, especially those in the lower classes. For example, many young black males living in poverty believe that one seemingly sure way to achieve success is through sports. The opportunities dwindle rapidly, however, as an athlete advances. Fewer than 2 percent of the athletes who play at the college level (in basketball, for instance) even have a chance at joining the professional ranks.

Merton believed that people respond in identifiable ways to structural strain and that their response involves some combination of acceptance and rejection of the valued goals and means (see Figure 7.3). He identified the following five responses—only one of which is not deviant:

- **Conformity** is the acceptance of the cultural goals and the pursuit of those goals through legitimate means.
- **Innovation** is the acceptance of the cultural goals but the rejection of legitimate means to achieve them. For the innovator, success is equated with winning the game rather than with playing by the rules (achieving a desired end by whatever means). After all, money may be used to purchase the same goods and services, whether it was acquired legally or illegally. According to Merton, when the life circumstances of the middle and upper classes are compared with those of the lower classes, the lower classes clearly face the greatest pressure to innovate, although no evidence suggests that they do so more than the middle and upper classes (for example, white-collar and corporate crimes are forms of innovation, often committed by middle- and upper-class people).

Robert K. Merton's theory of structural strain applies to the situation of black athletes who seek to achieve financial success through sports, as fewer than 2 percent of college basketball players have a chance to join the NBA.

- **Ritualism** involves the rejection of the cultural goals but a rigid adherence to the legitimate means of achieving them. This response is the opposite of innovation: The game is played according to the rules despite defeat. Merton maintains that this response can be a reaction to the status anxiety that accompanies the ceaseless competitive struggle to stay on top or to get ahead. Ritualism finds expression in clichés such as "Don't aim high and you won't be disappointed" and "I'm not sticking my neck out." It can also be the response of people who have few employment opportunities open to them. To see how a ritualist might be defined as deviant, consider the case of a college graduate who can find only a job bagging groceries at minimum wage. Most people would probably consider this person a failure, even though he or she may be working full-time.
- **Retreatism** involves the rejection of both culturally valued goals and the means of achieving them. People who respond this way have not succeeded by either legitimate or illegitimate means and thus have resigned from society. According to Merton, retreatists are the true aliens or the socially disinherited—the outcasts, vagrants, vagabonds, tramps, drunks, and addicts. They are "in the society but not of it."

- **Rebellion** involves the full or partial rejection of both the goals and the means of attaining them and the introduction of a new set of goals and means. When this response is confined to a small segment of society, it provides the potential for the emergence of subgroups as diverse as street gangs and the Old Order Amish. When rebellion is the response of a large number of people who wish to reshape the entire structure of society, a great potential for a revolution exists.

Structural Strain in China

In China one source of structural strain involves the number of legitimate opportunities for married couples to achieve the culturally valued goal of producing children, especially a son. Since 1979 China has worked to impose a limit of one child per couple in urban areas and a limit of two children per couple in rural areas in hopes of slowing its population growth. This family-planning policy poses

According to Merton, the response to structural strain that he calls "rebellion" applies to the Old Order Amish, who reject many of the dominant society's valued goals related to self-promotion and a false sense of self-reliance. Instead, they place great value on such characteristics as calmness, composure, and cooperation.

ritualism The rejection of cultural goals but a rigid adherence to the legitimate means of achieving them.

retreatism The rejection of both cultural goals and the means of achieving them.

rebellion The full or partial rejection of both cultural goals and the means of achieving them and the introduction of a new set of goals and means.

Mode of Adaptation	Goals	Means
Conformity	+	+
Innovation	+	−
Ritualism	−	+
Retreatism	−	−
Rebellion	+/−	+/−

+ Acceptance/achievement of valued goals or means

− Rejection/failure to achieve valued goals or means

▲ **Figure 7.3** Merton's Typology of Responses to Structural Strain

Merton believed that people respond to structural strain and that their responses involve some combination of acceptance and rejection of valued goals and means.

Source: Adapted from Merton (1957, p. 140), "A Typology of Modes of Individual Adaptations."

a problem, because a cultural preference for boys exists in China, especially among the people living in the countryside. One reason that sons are valued more than daughters is that sons and their families are expected to care for parents in old age. Thus, couples who produce a son can be confident that someone will care for them later in life. The legitimate means of starting a family include the following steps: Obtain permission to have a child, accept the sex of the child (that is, do not abort female fetuses or kill a daughter), report the birth and sex of the child to appropriate agencies, and practice birth control to avoid conceiving other children.

We can apply Merton's typology of responses to structural strain to describe the reactions of Chinese couples to this one-child policy (Figure 7.3). Couples most likely to be conformists are those whose first child is a healthy son. Conformists would also include couples who have no preference as to the sex of their child and couples who are firmly committed to upholding the laws related to birth control because they see them as critical to China's quality of life (Bolido 1993, p. 6). Most people in urban China can be classified as conformists.

Innovators accept the culturally valued goal of one child per couple but reject the package of legitimate means to obtain this goal. Upon learning that she is expecting a child, a woman may undergo an ultrasound exam to learn the sex of the fetus. If it is female, the couple may decide

to abort the fetus. Alternatively, upon the birth of a girl baby, the parents may kill her or have a midwife kill her. Such practices are blamed for the so-called "missing girls" problem. Many rural governments in China have launched "respect girls" campaigns backed by special incentives, such as free tuition to poor families with one or more girls (Yardley 2005).

Government statistics show that for every 129 boys born in China, there are 100 girls (French 2005; Kahn 2004c; Yardley 2005). A second category of innovators includes couples who abandon an unhealthy or female child or abandon a second child born in violation of family planning regulations. These children fill the country's 67 state-run orphanages (Tyler 1996b). From 2005 to 2006, Americans adopted 14,000 babies from China (U.S. Department of State 2007). (See No Borders, No Boundaries: "The Foreign Adoption Process.")

Ritualists reject the one-child-per-couple goal of population control, but they adhere to the rules. They do not agree with the government policies, but they are afraid they will be punished if they do not follow them (Remez 1991).

Retreatists reject the one-child goal as well as the legitimate means open to them. This category of deviants include couples who continue having children until a son is born and who hide the births of girls from party officials. In fact, Sterling Scruggs of the United Nations Population Fund argues that many of the so-called missing girls have just never been officially registered as having been born (Rosenthal 2000). Because of the high number of unreported births, some experts claim that it is impossible to know the actual size of China's population (Rosenthal

For every 129 boys born in China, there are 100 girls. For second and third births, the rate is 147 boys for every 100 girls. In some rural areas the ratio is 134 to 100.

Dereck and Elizabeth Bradley

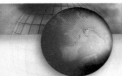

No Borders, No Boundaries

The Foreign Adoption Process

SINCE 1990, AMERICANS have adopted about 250,000 babies from more than 100 countries. The map shows the countries from which Americans adopted babies in 2006.

In 2006 American adopted 20,679 babies, 6,500 of which were from China. Some of China's eligibility requirements follow:

Age of Children: Chinese law allows for the adoption of children up to and including age 13; children ages 14 and up may not be adopted.

Civil Status of Prospective Adoptive Parents: Chinese law permits adoption by married couples (one man, one woman) and single heterosexual persons. Chinese law prohibits homosexual individuals or couples from adopting Chinese children.

Residency: China does not require that prospective adoptive parents reside in China for a specified period prior to completing an adoption. However, in order to finalize the adoption, at least one adopting parent must travel to China to execute the required documents in person before the appropriate Chinese authorities. If the prospective adoptive parents are married, they must adopt the child jointly. If only one member of an adopting married couple travels to China, that person must have in his/her possession a power of attorney from the other spouse, notarized and authenticated by the Chinese Embassy in Washington or one of the Chinese Consulates General elsewhere in the United States.http://www.china-embassy.org/eng/hzqz/t84229.htm

Time Frame: It is hard to predict with certainty how much time is required to complete an adoption in China. The time frames provided in this flyer are intended as guidelines only, and the specific circumstances of each case could affect significantly how long it takes.

As of February 2006, adoptions were taking approximately ten to twelve months from the time the U.S. adoption agency submitted the paperwork of the prospective adopter to CCAA to the time the CCAA gave the prospective adoptive parent(s) their initial referral. Cases involving children with special needs may take longer.

After the referral is sent and the prospective parent(s) accept the child (see the step-by-step description of the Chinese adoption process, below), four to eight more weeks are likely to elapse before the CCAA gives the prospective adoptive parents final approval to travel to China.

With regard to time required in China, the CCAA has advised local officials to try to complete the process within 15 days after the arrival of the prospective parent(s) in China. The Chinese passport, exit permits, and U.S. visa process can take another 7-10 days after the adoption is finalized. Some U.S. families have been able to complete the in-country process, including obtaining the U.S. immigrant visa for the adoptive child, in approximately two weeks.

Source: Adapted from U.S. Department of State 2007

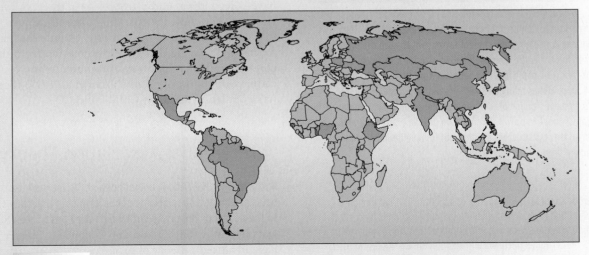

▲ **Figure 7.4**

Culturally valued goal: One child per couple
Culturally valued means: One-child limit
Sources of structural strain: Preference for boys

Response	Goal Population control	Means One-child policy	Examples
Conformists	+	+	• Couples with no preference • "Ideologically sound" couples
Innovators	+	−	• Couples who abort female or unhealthy babies • Couples who abandon babies
Ritualists	−	+	• Couples who follow the rules but reject policies
Retreatists	−	−	• Couples who reject the idea of population control and hide the "extra" babies
Rebels	+/−	+/−	• Couples who replace official goals and means with new goals and means

+ Acceptance/achievement of valued goals or means
− Rejection/failure to achieve valued goals or means

▲ **Figure 7.5**   Merton's Typology Applied to China's One-Child Policy

Julius Zielinski

Students in Wrocław, Poland, created this memorial to the June 5, 1989, Tiananmen Square demonstration. It shows a damaged bicycle and broken bricks caused by government tanks sent to stop the protests. Tiananmen Square is the most well-known site for a series of demonstrations that took place in China between April 15 and June 4, 1989. In all, 84 cities and as many as 3 million students—joined by workers, teachers, and even police officers and soldiers—took part in demonstrations against those in power. Although no official figures are available, hundreds of these protesters are thought to have been killed and thousands injured.

differential association A theory of socialization that explains how deviant behavior, especially delinquent behavior, is learned. It states that "when persons become criminal, they do so because of contacts with criminal patterns and also because of isolation from anticriminal patterns" (Sutherland and Cressey 1978, p. 78).

2000). In a sense, these retreatist parents are in society but, because of their secret, "not of it."

Rebels reject the culturally-valued goal of population control as well as the legitimate means of achieving it; instead, they introduce new goals and new means. One could argue that this response applies to couples who belong to any of China's 55 ethnic minority populations, which make up 8.1 percent (107 million) of the total population (U.S. Central Intelligence Agency). Most of these ethnic groups are exempt from the one-child policy. The government permits couples from such a group to have—depending on the group—two, three, or even four children. For these groups, the culturally valued goal is to increase the size of the minority populations, and the means of achieving that goal is exemption from the one-child policy. This option has prompted many couples to claim ethnic minority status (Tien et al. 1992) (See "China's Population and One-Child Policy").

Differential Association Theory

■ **CORE CONCEPT 10:** Criminal behavior is learned; thus, criminals constitute a special type of conformist in that they conform to the norms of the group with which they associate. Sociologists Edwin H. Sutherland and Donald R. Cressey (1978) have advanced a theory of socialization called **differential association** to explain how deviant behavior, especially delinquent behavior, is learned. This theory states that

China's Population and One-Child Policy

THE SIZE OF the Chinese population is one major reason for the country's rigid system of population control. More than 1.3 billion Chinese—one of every five people alive in the world—live in a space roughly the same size as the United States. After subtracting deserts and uninhabitable mountain ranges, China's habitable land area is about half that of the United States. Although almost 21 percent of the world's population lives in China, this country has only 6.3 percent of the world's agricultural land (U.S. Central Intelligence Agency 2007). Even though China manages to feed most of its population well, approximately 65 million people live in a state of absolute poverty; that is, they lack the resources to satisfy the basic needs of food and shelter. "Most Americans could not begin to comprehend what 65 million Chinese have to endure and what that level of deprivation is all about" (Piazza 1996, p. 47).

Another 700 million rural poor live in a state of substantial deprivation as reflected in the 80,000 incidents of rural unrest in 2005. The unrest may involve a few older women demanding their pensions be paid or thousands

The hundreds of high-rise apartment buildings in the Chaoyang District of Beijing are one indicator of the housing pressures facing China's large population. Beijing's population is 15.8 million.

protesting the confiscation of their land for development projects (Gifford 2007). It is believed that between 150 to 200 million people from rural villages have migrated to the cities in search of work. "It is the largest migration in human history" (Gifford 2007, p. xvii).

On average, 75,616 births and 22,027 deaths occur every day in China. If these patterns persist, the total population could increase by another 168 million people over the next 10 years. To complicate the situation, the population has expanded rapidly in the past 45 years, growing from approximately 500 million in 1949 to 1.32 billion in 2007 (U.S. Central Intelligence Agency 2007). Such rapid growth strains China's ability to house, clothe, educate, employ, and feed its people and has overshadowed its industrial and agricultural advancements. When one considers that approximately 1 billion people live under approximately 50 independent governments in Europe, North America, and South America (Fairbank 1989), the magnitude of China's population problem appears even more overwhelming.

"when persons become criminal, they do so because of contacts with criminal patterns and also because of isolation from anticriminal patterns" (1978, p. 78). These contacts take place within **deviant subcultures**—groups that are part of the larger society but whose members adhere to norms and values that favor violation of the larger society's laws. That is, people learn crime techniques by closely interacting with those who engage in and approve of criminal behaviors.

Sutherland and Cressey maintain that impersonal forms of communication, such as television, movies, and newspapers, play only a relatively small role in the genesis of criminal behavior. If we accept the premise that criminal behavior is learned, then criminals constitute a special type of conformist. That is, they conform to the norms of the group with which they associate. The theory of differential association does not seek to explain how a person makes the initial contact with a deviant subculture (unless, of course, the person is born into such a subculture). Once contact is made, however, the individual

learns the subculture's rules for behavior the same way that all behavior is learned.

Sociologist Terry Williams studied a group of teenagers, some as young as 14, who sold cocaine in the Washington Heights section of New York City. These youths were recruited by major drug suppliers because, as minors, they could not be sent to prison. Williams (1989) argues that the teenagers were susceptible to recruitment for two reasons: They saw little chance of finding high-paying jobs, and they perceived drug dealing as a way to earn money that would enable them to pursue a new life.

Williams's findings suggest that once teenagers become involved in drug networks, they learn the skills to perform

deviant subcultures Groups that are part of the larger society but whose members adhere to norms and values that favor violation of the larger society's laws.

their jobs the same way everyone learns to do a job. Indeed, success in a "deviant" job is measured in much the same way that success is measured in mainstream jobs: pleasing the boss, meeting goals, and getting along with associates.

In China, notions of differential association are the basis of the philosophy underlying rehabilitation: A deviant individual becomes deviant because of "bad" education or associations with "bad" influences (see Figure 7.5). To correct these influences, society must reeducate deviant individuals politically, inspire them to support the Communist Party, and teach them a love of labor. These principles guided Communist Party leaders in their handling of the Chinese students who participated in a series of pro-democracy demonstrations in Tiananmen Square in 1989. The government eventually suppressed these demonstrations violently, accusing the students of "counterrevolutionary revolt" and "bourgeois liberalization." While these terms have never been clearly defined, they suggest wanton expressions of individual freedom, which threatened the country's stability and unity. Such expres-sions had to be curbed. In the government's view, students who demonstrated led a sheltered life. As a result they ignored larger collective interests and concerns and failed to grasp the complexities of government reforms (Kwong 1988, pp. 983–984; Rahn 2004b).

The government undertook a number of measures to persuade the students to learn more about the complexities of life and resist subversive ideas. These measures included sending them "to rural areas to teach them to endure hardship, work hard and appreciate the daily difficulties faced by China's mostly rural population" (Kristof 1989, p. Y1). The rationale was that proper ideological commitment could be instilled through association with the masses and through manual labor. Other measures included limiting the number of students entering the humanities and social sciences. In the year following the Tiananmen Square incident, almost no students were admitted to study academic subjects that government officials considered "ideologically suspect," such as history, political science, sociology, and international studies (Goldman 1989).

Theories of Deviance: Summary

Theory	Central Question	Answer
Functionalist	How does deviance contribute to order and stability?	Deviance—especially the ritual of identifying and exposing wrongdoing, determining a punishment, and carrying it out—is an emotional experience that binds together members of groups and establishes a sense of community.
Labeling theory	What is deviance?	Deviance depends on whether people notice it and, if they do notice it, on whether they label it as such and subsequently apply sanctions/punishment.
Obedience to authority	How do rule makers get people to accept their definition of deviance and to act on those definitions?	Behavior that is unthinkable in an individual acting on his or her own may be executed without hesitation when authority figures command such behavior.
Constructionist	How do specific groups, activities, conditions, or artifacts come to be defined as problems?	Claims makers with the ability to define something as a problem, to have those claims heard, and to shape public responses play important roles in defining deviance and responses to it.
Structural strain	Under what conditions do people engage in behavior defined as deviant?	People engage in behavior defined as deviant when valued goals have no limits or clear boundaries, when legitimate means do not guarantee valued goals, and when the number of legitimate opportunities are in short supply.
Differential association theory	How is behavior defined as deviant learned?	People learn deviant behavior through close associations and interactions with those who engage in and approve of criminal behavior.

■ VISUAL SUMMARY OF CORE CONCEPTS

■ CORE CONCEPT 1: The only characteristic common to all forms of deviance is that some social audience challenges or condemns a behavior or an appearance because it departs from established norms.

Deviance is any behavior or physical appearance that is socially challenged or condemned because it departs from the norms and expectations of a group. A norm might be as simple as "girls wear pink and boys wear blue." Conformity comprises behaviors and appearances that follow and maintain the standards of a group. All groups employ mechanisms of social control—methods used to teach, persuade, or force their members, and even nonmembers, to comply with and not deviate from norms and expectations.

Department of Defense photo Juan Carlos Lopez, USM

■ CORE CONCEPT 2: Most people abide by their society's norms because they have been socialized to accept those norms as "good, proper, appropriate, and worthy" (Sumner 1907).

Because socialization begins as soon as a person enters the world, one has little opportunity to avoid exposure to the culture's folkways and mores. That exposure may take place in schools. For example, Chinese preschools and primary schools emphasize discipline, cooperation, and social-mindedness. U.S. preschools and primary schools emphasize individuality, self-direction, and freedom of choice. When Chinese and Americans were shown each other's system of socializing preschoolers, each group expressed preference for their own culture's system. Of course this principle applied to more than Chinese and American preschools. These Amish children have come to see their way of life as good, proper and worthy.

Scott Bookman

■ CORE CONCEPT 3: When socialization fails to produce conformity, other mechanisms of social control—sanctions, censorship, or surveillance—may be used to convey and enforce norms.

Ideally, conformity is voluntary. That is, people feel internally motivated to maintain group standards and to feel guilty if they deviate from them. When conformity cannot be achieved voluntarily, other mechanisms of social control may be used to convey and enforce norms. These mechanisms include sanctions, censorship, and surveillance. Sanctions can be positive (reward oriented) or negative (punishment oriented), informal (spontaneous and unofficial) or formal (official, backed by force of law).

Steve Jurvetson, www.DFJ.com/J

■ CORE CONCEPT 4: It is impossible for a society to exist without deviance. Always and everywhere there will be some behaviors or appearances that offend collective sentiments.

Durkheim ([1901] 1982) argued that although deviance does not take the same form everywhere, it is present in all societies. He defined *deviance* as acts that offend collective norms and expectations. The fact that always and everywhere some people will offend collective sentiments led him to conclude that deviance is normal as long as it is not excessive and that "it is completely impossible for any society entirely free of it to exist" (p. 99).

Department of Defense photo by PH1 Steven Batiz

According to Durkheim, deviance has an important function in society, for at least two reasons. First, the ritual of identifying and exposing wrongdoing, determining a punishment, and carrying it out is an emotional experience that binds together the members of a group and establishes a sense of community. Second, deviance is functional because it helps bring about necessary change and prepares people for change. It is the first step toward the future. Nothing would change if someone did not step forward and introduce a new perspective or a new way of doing things.

■ CORE CONCEPT 5: Labeling theorists maintain that an act is deviant when people notice it and then take action to label it as a violation and apply appropriate sanctions.

Labeling theorists maintain that violating a rule does not automatically make a person deviant. That is, a rule breaker is not deviant (in the strict sense of the word) unless someone *notices* the violation and decides to take corrective action. Labeling theorists suggest that for every rule a social group creates, four categories of people exist: conformists, pure deviants, secret deviants, and the falsely accused. Sociologists are particularly interested in the circumstances of the falsely accused. People are likely to be falsely accused of a crime when the well-being of a country or a group is threatened. The threat can

National Park Service, U.S. Department of the Interior

take the form of an economic crisis or a moral, health, or a national security crisis. The response to the crisis can take the form of a witch hunt—a campaign to identify, investigate, and correct behavior that has been defined as undermining a group or a country.

■ CORE CONCEPT 6: Sociologists are concerned less with rule violators than with rule makers and enforcers.

Sociologists pay particular attention to rule makers and rule enforcers—to who they are and to how they achieve power and then use it to define how others are regarded, understood, and treated. The members of a society with the most wealth, power, and authority have the ability to create laws and establish crime-stopping and crime-monitoring institutions. Consequently, we should not be surprised that law enforcement efforts tend to focus disproportionately on crimes committed by the poor and other powerless groups rather than on crimes committed by the wealthy, well connected, and politically powerful (such as white-collar and corporate crimes). In reality, crime exists in all social classes, but the type of crime, the extent to which the laws are enforced, access to legal aid, and the power to shape laws to one's advantage vary across classes.

CBP Photo by Gerald L. Nino

■ **CORE CONCEPT 7:** The firm commands of a person holding a position of authority over a person hearing those commands can elicit obedient responses.

When Stanley Milgram (1974) conducted the research for his book *Obedience to Authority*, he wanted to explain how large numbers of people cooperate in and carry out atrocities. He learned that obedience can be founded simply on the firm command of a person with a social status that gives him or her minimal authority over a subject.

■ **CORE CONCEPT 8:** In an effort to define deviance, rule makers and rule enforcers assume the role of claims maker and engage in various claims-making activities.

Claims makers are people who articulate and promote claims and who tend to gain in some way if the targeted audience accepts their claims as true. Claims makers include government officials, marketers, scientists, professors, and other special-interest groups. Claims-making activities draw attention to a claim and involve a range of activities, from filing lawsuits to holding signs. The success of a claims-making campaign depends on a number of factors, including access to the media, available resources, and the claims maker's social status and skill at fund-raising, promotion, and organization. When constructionists study the claims-making process, they ask who makes the claims, whose claims are heard, how audiences respond, and so on.

■ **CORE CONCEPT 9:** Deviant behavior is a response to structural strain, a situation in which a disjuncture exists between culturally valued goals and legitimate means for achieving those goals.

Robert K. Merton's theory of structural strain takes three factors into account: (1) culturally valued goals defined as legitimate for all members of society, (2) norms that specify the legitimate means of achieving those goals, and (3) the actual number of legitimate opportunities available to people to achieve the goals. Structural strain occurs when the valued goals have unclear limits, people are unsure whether the legitimate means will allow them to achieve the goals, or legitimate opportunities for meeting the goals remain closed to a significant portion of the population.

Merton believed that people respond in identifiable ways to structural strain and that their response involves some combination of acceptance and rejection of the valued goals and means. He identified the following five responses—only one of which is not deviant: conformity, innovation (such as taking illegal performance-enhancing drugs to achieve success in sports), ritualism, retreatism, and rebellion.

■ **CORE CONCEPT 10:** Criminal behavior is learned; thus, criminals constitute a special type of conformist in that they conform to the norms of the group with which they associate.

The theory of differential association explains how deviant behavior, especially delinquent behavior, is learned. This theory states that "when persons become criminal, they do so because of contacts with criminal patterns and also because of isolation from anticriminal patterns" (1978, p. 78). These contacts take place within deviant subcultures—groups that are part of the larger society but whose members adhere to norms and values that favor violation of the larger society's laws. That is, people learn crime techniques by closely interacting with those who engage in and approve of criminal behaviors. Youth Centers try to counteract the forces of differential association by offering youth activities, friendships, and mentors.

Resources on the Internet

Sociology: A Global Perspective Book Companion Web Site

www.cengage.com/sociology/ferrante

Visit your book companion Web site, where you will find flash cards, practice quizzes, Internet links, and more to help you study.

CENGAGE**NOW**™

Just what you need to know NOW!

Spend time on what you need to master rather than on information you already have already learned. Take a pre-test for this chapter, and CengageNOW will generate a personalized study plan based on your results. The study plan will identify the topics you need to review and direct you to online resources to help you master those topics. You can then take a post-test to help you determine the concepts you have mastered and what you will need to work on. Try it out! Go to www.cengage.com/login to sign in with an access code or to purchase access to this product.

Key Terms

censors 176
censorship 175
claims makers 184
claims-making activities 185
conformists 178
conformity 170, 186
constructionist approach 184
corporate crime 180
deviance 170
deviant subcultures 191
differential association 190

falsely accused 179
folkways 170
formal sanctions 174
informal sanctions 173
innovation 186
master status of deviant 178
mores 171
negative sanction 173
positive sanction 173
pure deviants 178

rebellion 187
retreatism 187
ritualism 187
sanctions 173
secret deviants 178
social control 170
structural strain 186
surveillance 176
white-collar crime 180
witch hunt 179

Social Stratification

With Emphasis on the World's Richest and Poorest

When sociologists study social stratification, they focus on the connection between the various social statuses people occupy and their life chances, including the chance to stay alive past the first year of life, the chance to live beyond age 75, and the chance to experience an infinite number of potential events in between those points.

T. Scott McLaren

▲ What do you think it would be like to have grown up in the United States—one of the world's richest countries—and, after graduating from college, volunteered to go to Mauritania, which is among the world's poorest countries? That is what T. Scott McLaren, a student in one of my sociology classes, did. Scott is holding the child wearing the baseball cap.

The World's Richest and Poorest?

ON SEPTEMBER 27, 2008, I received a letter from Scott McLaren, a former student. He wrote:

I want to take a few minutes and share with you some thoughts regarding my first few months in Mauritania. Before I begin, I would like to say how appreciative I am of the fate that my last semester held within it the basics of sociology. That discipline has given me yet another set of eyes to view this experience. I have found that culture shock is but a mild symptom of travel and service in the Peace Corps. What is the most painful symptom is what I call self-death awareness—the self, being my understanding of my own personage as revealed and shaped through my home culture and interactions with others in my society of origin. This awareness has hit me so very hard this week and I believe that it is what has been at the epicenter of the repeated tidal wave of emotion within me . . . I am now the barely-able foreigner who has to ask where to pee. I depend completely on others to sustain my needs and it is humiliating. The self-death shock has been debilitating this week and limited me to self-loathing and pity for my situation . . . Self-death is painful; I am now beginning to deconstruct the person I had worked to be known as. I will be reconstructed into a person, not of my own making or choosing. My new self will be molded to the needs of survival.

This chapter focuses on the world's richest and poorest people. We begin by describing how wealth, income, and other valued resources are unequally distributed among the 6.6 billion people living in the 243 or so countries on the planet. Sociologists seek to understand the patterns of inequality. For example, sociologists are interested in which countries the richest and poorest tend to live and why some countries have larger percentages of poor than others. Likewise, sociologists are interested in the patterns of inequality that exist *within* countries. They are obligated to ask the kinds of questions Scott faces head-on each day of his service in the Peace Corps:

- How do we explain the extremes of wealth and poverty in the world?
- Why should a very small percentage of the world's population enjoy an inordinate share of the income, wealth, and other valued resources, while so many others struggle to survive?
- Can we assume capitalism and globalization will correct these dramatic inequalities, or should we rethink the way wealth and other valued resources are distributed?

FACTS TO CONSIDER

- T. Scott McLaren grew up and lived in a solid middle- to upper-middle-class community in which the median household income was $62,000 and the median home/condo value was $241,472 (U.S. Bureau of the Census 2008).

- As a Peace Corps volunteer, Scott traveled to Mauritania and was assigned to a village in which people live on the equivalent of $1.00 per day.

The Extremes of Poverty and Wealth in the World

It is difficult to define the condition of poverty except to say that it is a situation in which people have great difficulty meeting basic needs for food, shelter, and clothing. Poverty can be thought of in absolute or relative terms. **Absolute poverty** is a situation in which people lack the resources to satisfy the basic needs no person should be without. Absolute poverty is often expressed as a state of being that falls below a certain threshold or a minimum. In this regard, the United Nations has set the absolute poverty threshold in developing countries at the equivalent of US$1.00 per day. The World Bank (2009), on the other hand, believes that threshold should be set at US$1.25 per day. According to the UN threshold, there are 1.1 billion people who live in a state of absolute poverty. Based on the World Bank (2008) threshold, that number is 1.4 billion people. There are other criteria by which to determine the number of people living in absolute poverty. For example, if we use the lack of access to a toilet as a measure of absolute poverty, one-third of the world's people—2.6 billion—do not have a decent place to go to the bathroom. Some use plastic bags and then throw them into ditches and streets; others squat in fields, yards, and streams (Dugger 2006). Scott McLaren described the absolute poverty he observed in Mauritania in this way: "Poverty is

- eating the same meal every day.
- living in the same quarters as goats and cows.
- not having access to clean drinking water.
- sleeping on a plastic mat between your body and the sand.
- no longer caring about improving your life."

Relative poverty is measured not by some objective standard, but rather by comparing the situation of those at the bottom against an average situation or against the situation of others who are more advantaged. When thinking of poverty in relative terms, one thinks not just about an inability to meet basic needs, but about a relative lack of access to goods and services that people living in a particular time and place have come to expect as necessities. Today, in the United States, examples of such goods and services that people have come to "need" to survive in the society in which they live might be an automobile, a cell phone, satellite television service, and the Internet.

Extreme wealth, on the other hand, is the most excessive form of wealth, in which a very small proportion of people in the world have money, material possessions, and other assets (minus liabilities) in such abundance that a small fraction of it (if spent appropriately) could provide adequate food, safe water, sanitation, and basic health care for the 1 billion poorest people on the planet. About how many people in the world are excessively wealthy? The World Institute for Development Economics Research of the United Nations University (2006) released a groundbreaking study estimating the world's total household wealth and describing how it was distributed. The study found that 1 percent of the world's adult population (or 37 million people) own more than 40 percent of the global household wealth (estimated to be $124 trillion). Contrast that with the poorest 50 percent of the world's adult population, who own less than 1 percent of global wealth (see Table 8.1).

■ **CORE CONCEPT 1:** When sociologists study systems of social stratification, they seek to understand how people are ranked on a scale of social worth and how that ranking affects life chances. **Social stratification** is the systematic process of ranking people on a scale of social worth such that the ranking affects life chances in unequal ways. Sociologists define **life chances** as the probability that an individual's life will follow a certain path and will turn out a certain way. Life chances apply to virtually every aspect of life—the chances that someone will survive the first year of life after birth, complete high school and go on to college, see a dentist twice a year, work while going to school, travel abroad, be an airline pilot, play T-ball, major in elementary education, own 50 or more pairs of shoes, or live a long life. **Social inequality** describes a situation in which these valued resources and desired outcomes (i.e., a col-

---

absolute poverty A situation in which people lack the resources to satisfy the basic needs no person should be without.

relative poverty Measured not by some objective standard, but rather by comparing the situation of those at the bottom against an average situation or against the situation of others who are more advantaged.

extreme wealth The most excessive form of wealth, in which a very small proportion of people in the world have money, material possessions, and other assets (minus liabilities) in such abundance that a small fraction of it (if spent appropriately) could provide adequate food, safe water, sanitation, and basic health care for the 1 billion poorest people on the planet.

social stratification The systematic process of ranking people on a scale of social worth such that the ranking affects life chances in unequal ways.

life chances The probability that an individual's life will follow a certain path and will turn out a certain way.

social inequality A situation in which these valued resources and desired outcomes (i.e., a college education, long life) are distributed in such a way that people have unequal amounts and/or access to them.

Table 8.1   The Distribution of Global Household Wealth

Researchers at the United Nations University have estimated how wealth is distributed among the adult population of the world. Of course, a significant percentage of the adult population have children or will have children who share in their wealth or poverty. We know that the world's poorest peoples have the greatest number of children. The table shows various income categories and the number of adults in each category. To be classified as "extremely wealthy," a person must have a minimum wealth (assets minus liabilities) of $1 billion. Worldwide, only about 800 adults are categorized as "extremely wealthy." To be among the richest 1 percent, a person must have between $500,000 and $1 million (excluding the value of their home) in wealth. The richest 1 percent holds 40 percent of the world's wealth. Note that the poorest 50 percent of the world's adult population (1.9 billion people) possess less than 1 percent of the world's wealth.

Category	Estimated Number of Adults in Category	Minimum Wealth (Assets − Liabilities)*	Percent of Total Household Wealth	Estimated Combined Wealth
Extremely wealthy	800	$1 billion	Not known	Not known
Ultra rich	85,400	$30 million	Not known	Not known
Richest 0.025%	8.3 million	$1 million	Not known	Not known
Richest 1%	37 million	$500,000	40%	$49.6 trillion
Richest 10%	370 million	$61,000	85%	$105 trillion
Middle 40%	1.4 billion	$2,200	14%	$17 trillion
Poorest 50%	1.9 billion	<$2,200	<1%	$1.2 trillion

* Does not include the value of the house in which the person resides.

Source: United Nations University (2006)

lege education, long life) are distributed in such a way that people have unequal amounts and/or access to them.

Every society in the world stratifies (ranks and categorizes) its people. Almost any criterion can be used (and, at one time or another, probably has been used) to categorize people and assign them a status, whether it be based on hair color and texture, eye color, physical attractiveness, weight, height, occupation, sexual preference, age, grade point average, test scores, or many others. People's status in society can be ascribed or achieved.

Miley Cyrus is among the richest 85,000 people on the planet. *People* magazine reported that Miley Cyrus was on track to be a billionaire by 2009.

DoD photo by Mass Communication Specialist 1st Class Mark O'Donald, U.S. Navy/Released

Ascribed versus Achieved Statuses

Ascribed statuses are social positions assigned on the basis of attributes people possess through no fault of their own—those attributes are acquired at birth (such as skin shade, sex, or hair color), develop over time (such as height, weight, baldness, wrinkles, or reproductive capacity), or are possessed through no effort or fault of their own (such as the country into which one is born and religious affiliation "inherited" from parents). **Achieved statuses** are attained through some combination of personal choice, effort, and ability. In other words, people must act in some way to acquire an achieved status. Achieved statuses include a person's wealth, income, occupation, and educational attainment.

ascribed statuses Social positions assigned on the basis of attributes people possess through no fault of their own–those attributes are acquired at birth (such as skin shade, sex, or hair color), develop over time (such as height, weight, baldness, wrinkles, or reproductive capacity), or possess through no effort or fault of their own (such as the country into which one is born and religious affiliation "inherited" from parents).

achieved statuses Attained through some combination of personal choice, effort, and ability.

Ascribed and achieved statuses may seem clearly distinguishable, but that is not the case. One can always think of cases where people take extreme measures to "achieve" a status typically thought of as ascribed; they undergo a sex change operation, lighten their skin, or hire a plastic surgeon to look a different age. Likewise, one's ascribed statuses affect one's opportunities to "achieve" wealth, a college education, certain occupations, or high income. We can also raise questions as to what statuses children are able to achieve independent of their parents.

The various achieved and ascribed statuses hold **social prestige**, a level of respect or admiration for a status apart from any person who happens to occupy it. There is social prestige associated with a person's occupation, level of education, and income, race, sex, and age, and so on. The social prestige accompanying each of these characteristics can complicate the overall experience of prestige. As a case in point, the occupation of physician—in the abstract—is a prestigious one. But that prestige can be complicated by the prestige assigned to race, sex, and age. Social prestige is further complicated by **esteem**, the reputation that someone occupying an ascribed or achieved status has earned from people who know and observe the person.

In a related vein, sociologists are also interested in the status value ascribed and achieved statuses take on. **Status value** is the social value assigned to a status such that people who possess one status (white skin versus brown skin, blonde hair versus dark hair, low income versus high income, single versus married, professional athlete versus high school teacher) are regarded and treated as more valuable or worthy than people who possess another status. The compensation guidelines for the September 11, 2001, attacks on the United States show how many factors—age, annual income, occupation, marital status, potential earnings, and family status (with children versus childless)—affected victims' "worth" and the money awarded to survivors. The actual awards ranged from $250,000 (least valued life) to $7.1 million (most valued life). One of the "least valued" categories was single, childless persons age

65 and older with an annual income of $10,000. Under the guidelines, their survivors were to be awarded a one-time payment of $300,000. One of the "most valued" categories was married persons age 30 and younger with two children and an annual income of $225,000. Their survivors were to be awarded a one-time payment of $3,805,087. But the survivors of a person of the same age and marital status with two children but with an annual income of $101,000 would be awarded only $694,588 (Chen 2004; September 11 Victim Compensation Fund 2001).

Sociologists give special attention to stratification systems in which

- ascribed status is used to explain certain abilities (such as athletic talent or intelligence and personality traits, such as propensity for violence or honesty);
- some ascribed statuses are considered inferior to others;
- ascribed statuses are associated with inequalities in wealth, income, and other socially valued items.

Life Chances across and within Countries

Broadly speaking, what does it mean for a baby to be born in the United States as opposed to, say, Mauritania? For one thing, the country into which one is born has an incalculable effect on a person's aspirations about what they can accomplish in life. Scott McLaren has observed that Mauritanian boys, especially those who live in rural villages, "talk about herding large flocks of goats or owning lots of cattle. Cattle here are a sign of wealth and privilege. Many also talk about going to Nouakchott (the capital city) to open boutiques or become cab drivers. Very few really even mention going to university." By contrast, American children are likely to believe they can be anything they want, however unrealistic that sentiment. And if they do not accomplish their dreams, it is largely their own fault.

■ **CORE CONCEPT 2: Dramatic inequalities exist across countries and within countries.** We have no control over which of the world's 243-odd countries we are born in; in that sense, a person's country of birth is an ascribed status. Still, that country has important effects on our life chances. The chances that a baby will survive the first five years of life depend largely on the country where he or she is born. Babies born in Iceland, Sweden, and Singapore have some of the best chances of surviving their first five years, as fewer than 3 of every 1,000 babies born there die before reaching the age of 5. Babies born in Sierra Leone and Afghanistan have some of the worst chances of surviving that first year; at least 250 of every 1,000 babies born—25 percent—die within the first five years of life (see Table 8.2).

One startling example of how life chances vary by country relates to consumption patterns of the 56 highest-income countries, where approximately 20 percent of the

social prestige A level of respect or admiration for a status apart from any person who happens to occupy it.

esteem The reputation that someone occupying an ascribed or achieved status has earned from people who know and observe them.

status value The social value assigned to a status such that people who possess one status (white skin versus brown skin, blonde hair versus dark hair, low income versus high income, single versus married, professional athlete versus high school teacher) are regarded and treated as more valuable or worthy than people who possess another status.

Table 8.2   Selected Life Chances in the Country with the Lowest and the Highest Under Age 5 Mortality Rates

In the African country of Sierra Leone, 26.2 percent of babies—262 of every 1,000 born—die before reaching age 5. In Sweden 0.03 percent—3 of every 1,000 babies—die before reaching age 5. In Sierra Leone, the average woman has 6.5 children, compared to the average Swedish woman, who has 1.7 children. Largely because of the high mortality rate among children, the life expectancy is 41.8 years in Sierra Leone—some 38 years shorter than the life expectancy in Sweden.

Life Chance	Sierra Leone	Sweden
Under-5 mortality rate (number of babies per every 1,000 born who die before reaching age 5)	262.0	3.0
Fertility rate (average number of babies born to woman over lifetime)	6.5	1.7
Life expectancy at birth	41.8	80.5
Percentage of population living below national poverty line	70.2	2.0

Sources: United Nations (2008); UNICEF (2009)

In the United States, there is considerable status associated with being a college graduate. Many people feel inferior if they do not have a college degree; it is thought of as a necessary, but not sufficient, credential for success. The photo above shows Scott on the day he graduated from college. The photo on the left shows Scott wearing a bubu. The quality of the fabric or, in some cases, the color can indicate one's status in Mauritania. The average Mauritanian wears a white or light blue bubu (particularly in Moor culture) but will wear other colors to distinguish themselves during social events where position in the community is on display.

world's people live, compared to the patterns of the rest of the world, especially of the lowest-income countries. The 20 percent in the highest-income countries account for 86 percent of total private consumption expenditures, while the 20 percent in the lowest-income countries account for only 1.3 percent of this spending. And consider these inequalities in consumption (as percentages of total resources) between the richest fifth and the poorest fifth of the world's people:

	Richest Fifth	Poorest Fifth
Meat and fish	45%	5%
Energy	58%	<4%
Paper	84%	1%
Vehicles	87%	<1%

Source: United Nations (1998)

Ranking countries according to specific kinds of life chances captures only part of the story. Life chances vary within countries as well. In the United States, on average, 30.9 babies per 1,000 live births die before reaching age 5. But the chances of survival vary by racial and ethnic classification, with 49.8 babies classified as Native American and 47.4 babies classified as black dying before reaching age 5. Babies classified as Asian have the best chances of surviving the first five years (see Table 8.3).

One way of gauging the inequalities that exist within individual countries is to compare the incomes of their richest and poorest residents. From a global perspective, the African country of Sierra Leone, a former British colony, has the greatest inequality between the richest and poorest 20 percent of its population, with the richest segment receiving 58 times more income than the poorest. Azerbaijan, a largely socialist country that was once part of the Soviet Union, has the greatest equality, with the richest 20 percent receiving 2.5 times the income of the poorest 20 percent (World Bank 2006).

Within the United States, the richest 20 percent earn 11 times that of the poorest 20 percent. The average after-tax income of the richest 20 percent living in the United States is $184,400, or 11 times that of the poorest 20 percent, who average $16,500. That is, for every $1,000 of taxed income earned by the poorest one-fifth, the top one-fifth earns $11,000. When we compare the after-tax income of the top 1 percent with that of the bottom 20 percent, the inequality is even greater. That 1 percent's after-tax income is $1.2 million, or 73 times greater than the bottom 20 percent (see Table 8.4).

To this point, we have shown that valued resources are distributed unequally both across countries and within countries. Next we will focus on the processes by which those resources are unevenly distributed.

Caste and Class Systems

Real-world stratification systems fall somewhere on a continuum between two extremes: a **caste system** (or "closed" system)—in which people's ascribed statuses (over which they have no control) figure most prominently in determining their life chances—and a **class system** (or "open" system)—in which merit, talent, ability, or past performance figure most prominently in determining their life chances.

caste system Any form of stratification in which people are categorized and ranked by characteristics over which they have no control and that they usually cannot change.

class system A system of social stratification in which people are ranked on the basis of achieved characteristics, such as merit, talent, ability, or past performance.

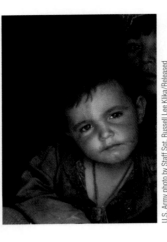

Official U.S. Marine Corps photo by Cpl. Ruben D. Maestre

U.S. Army photo by Staff Sgt. Russell Lee Klika/Released

These children have dramatically different life chances of surviving until age 5. The average American-born child has a 97 percent chance of surviving beyond the first five years of life. The average Afghan child has a 75 percent chance of surviving beyond age 5.

Three characteristics distinguish a caste system from a class system:

1. The rigidity of the system, or how difficult it is for people to change their status.
2. The relative importance of ascribed and achieved statuses in determining opportunities in life.
3. The extent to which there are restrictions on interaction between people of unequal status.

■ **CORE CONCEPT 3:** Caste and class systems of stratification are opposite, extreme points on a continuum. The two systems differ in the ease of social mobility, the relative importance of achieved and ascribed statuses, and the extent to which each restricts interaction among people considered unequal. Most sociologists use the term caste to refer to any form of stratification in which people are categorized and ranked by characteristics over which they have no control and that they usually cannot change. Whenever people are ranked and rewarded on the basis of such traits, they are part of a caste system of stratification. A direct connection exists between caste rank and the opportunities to exercise power and acquire wealth and access to other valued resources. People in lower castes are seen and portrayed as innately inferior in intelligence, morality, ambition, and many other traits. Conversely, people in higher castes consider themselves to be superior in such traits. Moreover, caste distinctions are treated as if they are absolute; that is, their significance is never doubted or questioned (especially by those who occupy higher castes), and the categories into which people are classified are viewed as unalterable and clear-cut. Finally, heavy restrictions constrain

Table 8.3   Infant and Under Age 5 Mortality Rates in U.S. by Race/Ethnicity

In the United States, an average of 6.7 babies per 1,000 live births die before reaching age 1. But the chances of survival vary by racial and ethnic classification, with 13.7 babies classified as black and 8.6 babies classified as Native American dying in the first year of life. Within the first five years of life, we can expect almost 50 American Indian/Alaska Native babies per 1,000 to die before reaching age 5. That translates to 5 babies for every 100 (or 1 in 20 babies).

Race/Ethnicity	Under Age 1 Infant Mortality (per 1,000 live births)	Under Age 5 Mortality Rate (per 1,000 live births)
American Indian/Alaska Native	8.6	49.8
Asian/Pacific Islander	4.8	22.6
Black/African American	13.7	47.4
White	5.6	29.1
Hispanic/Latino	5.7	27.2
Average, all groups	6.7	30.9

Sources: CDC (2005); U.S. Bureau of the Census (2005)

Table 8.4   Average After-Tax Income in the United States by Income Group, 1979 and 2006 (in 2006 dollars)

The table compares after-tax income in the United States across six income groups at two points in time—1979 and 2006. Perhaps most striking is that the greatest financial gains within this time period have been made by those in the top fifth, including the top 1 percent. The financial gain made by those in the top 1 percent is 540 percent greater than that of the lowest one-fifth. (The 540 percent figure is obtained by dividing the dollar change of the top 1 percent ($863,200) by the dollar change of the lowest fifth ($1,600).)

Income Category	1979 Average After-Tax Income	2006 Average After-Tax Income	Percent Change	Dollar Change
Lowest fifth	$14,900	$16,500	11%	$1,600
Second fifth	$30,100	$35,400	18%	$5,300
Middle fifth	$42,900	$52,100	21%	$9,200
Fourth Fifth	$56,100	$73,800	32%	$17,700
Top Fifth	$98,000	$184,400	87%	$85,500
Top 1 Percent	$337,100	$1,200,300	256%	$863,200

Source: Congressional Budget Office (2009)

interactions among people in higher and lower castes. For example, marriage between people of different castes is forbidden. An excerpt from Scott McLaren's journal helps to clarify how caste systems in which gender is central operate in the rural Mauritanian village to which he has been assigned:

> There are very few girls in my village who can go to school. Our nearest school, which just opened this past year, is the first chance all of the kids in my area will have at getting an education. I'm praying that 2% literacy rate rises very soon. The girls are hard to get a handle on. Because of the cultural restrictions on men and women interacting, I can't really get close enough to any of them to really get to know them. The sisters in my family don't really talk much about what they'd like to do; they just talk about what they are doing in their daily chores.

Theoretically, in class systems, "people rise and fall on the strength of their abilities" (Yeutter 1992, p. A13). While class systems contain inequality, that inequality is not centered on ascribed statuses; that is, a person's sex, skin shade, hair texture, age, or other ascribed statuses are not associated with the inequalities that exist. An ideal class system differs from a caste system in that the inequality results from differences in talent, ability, and past performance and not from attributes over which people have no control. Thus, in class systems, people assume that they can achieve a desired level of education, income, and standard of living through hard work; that they can raise their class position during their own lifetime; and that their children's class position can differ from (and ideally be higher) than their own.

This fresh fish market in the United States illustrates consumption by the richest fifth of the world's people, who consume 45 percent of all meat and fish products. By contrast, the poorest fifth consume 5 percent.

Social Mobility

Within a class system, movement from one social class to another is termed **social mobility** (see Chart 8.1).

When thinking about mobility in the United States, we often think about the possibility that children's economic status will exceed that of their parents. Many Americans believe that we live in a country in which it is possible to move from rags to riches. But how many people born poor become wealthy? While data to answer this question are hard to come by, one study conducted by economist Paul Hertz (2006) followed 4,004 children into adulthood to calculate the odds of changing or maintaining economic status. The researcher averaged the income of each of the 4,004 children's households for the years 1967 through 1971 and compared it to their household income (averaged over four years, 1996 through 2000) when they became 40-year-old adults. The study found that the chances of moving from the lowest-income category to the highest are actually quite low but that the chances of remaining in the highest 20 percent income category are quite high (see Table 8.5). Hertz also found that children from high-income households (top fifth of household income) receive more education and are healthier as adults, factors that surely contributed to them maintaining their economic status. Hertz's study also found that children classified as white have advantages over children classified as black

social mobility Movement from one social class to another.

In Mauritania, girls who live in rural areas are married off at relatively young ages, in some cases as early as 15. Over the past few decades, girls have begun to choose their partners, but it is the parents who hold the final say. Theoretically, girls can move away from home if they choose, but their support network becomes very limited, and the challenges of finding ways to support themselves are often very great.

with regard to financial mobility. Specifically, 63 percent of black children who grow up in low-income households retain that low-income status as adults, compared to the 32.3 percent of poor whites. Black children from the lowest-income households have a 3.6 percent chance of moving into the highest 20 percent income category as adults, compared to a 14.2 percent for poor white children.

Conceptualizing Inequality

Sociologists draw upon the theoretical traditions of functionalist, conflict, and symbolic interaction to think about inequality—why it exists, who benefits from inequality, and how it is enacted in interaction with others.

Chart 8.1   Types of Social Mobility

Many kinds of social mobility exist in class systems of social stratification. This chart offers a list of the various types along with specific examples.

Type of Social Mobility	Definition	Example
1. Horizontal	A change in social situation that does not involve a change in social status.	A waitress moves into a customer service position for an insurance agency.
2. Vertical	A change in a person's social situation that involves a gain or loss of social status.	See following for examples of various types of vertical mobility.
2A. Upward	A change in a person's social situation that involves a gain in social status.	A medical student moves up in rank and becomes a physician.
2B. Downward	A change in a person's social situation that involves a loss of social status.	A wage earner loses a job, goes on unemployment, or otherwise moves down in rank.
2C. Intragenerational	A change in social situation that involves a loss or gain of social status over the course of an individual's lifetime.	A laid-off factory worker takes a job with a lower salary/a bank teller is promoted to a bank manager.
2D. Intergenerational	A change in social situation that involves a loss or gain of social status relative to a previous generation.	A son or daughter goes into an occupation that is higher or lower in rank and prestige than a parent's occupation.

Table 8.5   Probability of Children Attaining Each Income Quintile Based on the Income Quintile of the Household in Which They Were Born

The table shows the relationship between a person's household income as a child (Parents' Quintile) and their household income as an adult (Adult Children's Quintile). The highlighted numbers represent the percentage of adult children who remained in the same income grouping as their parents. To interpret the data in this table, ask the following kinds of questions: What are the chances that someone born in the lowest household income category (lowest fifth) will achieve a household income that is among the top 5 percent? What are the chances that someone born in the top 5 percent will retain that income status into adulthood? What are the chances that a person born in the top 5 percent income category will be among the lowest fifth as an adult?

Childhood Home (4-Year Average Household Income, 1967–1971)	Adult Children's Quintile (4-Year Average Household Income, 1997–2001)					
	Lowest Fifth	Second Fifth	Third Fifth	Fourth Fifth	Top Fifth	Top 5 Percent
Lowest fifth	**41.5**	24.0	15.5	13.2	5.9	1.1
Second fifth	22.6	**25.8**	23.1	18.5	10.0	1.5
Middle fifth	18.7	25.8	**24.1**	19.6	16.9	1.8
Fourth fifth	11.1	19.0	20.7	**25.1**	24.0	5.6
Top fifth	6.1	11.1	17.2	23.7	**41.9**	14.2
Top 5 percent	2.9	9.0	15.5	21.5	51.1	**21.7**

Source: Hertz (2006)

■ **CORE CONCEPT 4:** Functionalists maintain that poverty exists because it contributes to overall order and stability in society; it is the mechanism by which societies attract the most qualified people to the most functionally important occupations. Sociologists Kingsley Davis and Wilbert Moore (1945) wrote what is now considered the classic functionalist argument about why social inequality exists. The authors argue that social inequality—the unequal distribution of social rewards—is the device by which societies ensure that the most functionally important occupations are filled by the best-qualified people. So, in the United States, garbage collectors earn an average of $40,000 per year and heart surgeons earn an average of $350,000 per year (U.S. Bureau of Labor Statistics 2008). The $310,000 difference in average salary represents the greater functional importance of the physician relative to the garbage collector.

How do Davis and Moore define functional importance? They offer two indicators:

1. The degree to which the occupation is functionally unique (i.e., whether few other people can perform the same function adequately).
2. The degree to which other occupations depend on the one in question.

With this definition in mind, Davis and Moore argue that because people need little training and talent to be a garbage collector, that position does not need to be highly rewarded. However, with regard to the second indicator—the degree to which other occupations depend on garbage collectors—we must acknowledge that virtually every occupation and area of life depends on garbage collectors to maintain sanitary environments. Regardless of this fact, Davis and Moore argue that society does not have to offer extra incentives to attract people to the occupation of garbage collector because many can do that job with little training. Society does have to offer extra incentives to attract the most talented people to occupations such as physician that require long and arduous training and a high level of skills. Davis and Moore concede that the stratification system's ability to attract the most talented and qualified people is weakened when:

1. capable people are overlooked or not granted access to the needed training;
2. elite groups control the avenues of training (by limiting admissions);
3. parents' influence and wealth (rather than the ability of their offspring) determine the status that their children attain.

Davis and Moore maintain, however, that society eventually adjusts to such shortcomings. Medical schools, once dominated by white males, eventually began admitting people from groups once denied access. When there are shortages of people to fill functionally important occupations (such as teachers, nurses, and doctors), society acts

From a functionalist point of view, sanitation workers are essential to the smooth operation of society, and yet they need not be rewarded highly because their job requires little training and skill.

Department of Defense photo by Sgt. James R. Richardson, USMC

to increase opportunities for people to enter those occupations and to admit those previously denied entry. If such a step is not taken, the society as a whole will suffer and will be unable to meet its needs.

In another, now-classic essay, "The Functions of Poverty," written from a functionalist perspective, sociologist Herbert Gans (1972) asked, "Why does poverty exist?" He answered that poverty performs at least 15 functions, several of which are described here.

Fill Unskilled and Dangerous Occupations. First, the poor have no choice except to take on the unskilled, dangerous, temporary, dead-end, undignified, menial work of society at low pay. Obviously, the lower the wage, the lower the labor cost needed to make goods and services and the greater the employer's profits. Hospitals, hotels, restaurants, factories, and farms draw employees from a large pool of workers who are forced to work, who are willing to work, or who appear willing to work at minimum wage or below. In the United States, for example, there are 1.3 million residents in 17,000 certified nursing homes across the United States. Workers at nursing homes must "lift and turn patients [who often weigh more than they do], help them in and out of baths, make beds, and take residents to and from the toilet." They have the highest rate of occupational-related injuries, especially to the back. They earn an average salary of $15,000 per year (SUIE 2008).

Provide Low-Cost Labor for Many Industries. The U.S. and many other economies depend on cheap labor from around the world and within its borders. According to Pew Hispanic Center (2009) estimates, at least 12 million undocumented adults and children are living in the United States. Of these 12 million, 8.3 million are employed. Twenty-five percent of all farm workers, 17 percent of all construction workers, and 12 percent of all food preparation workers are believed to be undocumented. That is a "whole lot of cheap labor. Without it, fruits and vegetables would rot in the fields. Toddlers in Manhattan would be without nannies. Towels at hotels in states like Florida, Texas, and California would go unlaundered . . . bedpans and lunch trays at nursing homes would go uncollected" (Murphy 2004).

Serve the Affluent. Affluent persons contract out and pay low wages for many time-consuming activities, such as housecleaning, yard work, and child care. On a global scale, millions of poor women work outside their home countries as maids in middle- and upper-class homes. Consider that an estimated 460,000 Indonesian women from poor villages work as maids in Malaysia and Saudi Arabia. Many of these women must be trained in how to use toasters, vacuum cleaners, microwave ovens, refrigerators, and other appliances (Perlez 2004). Even the U.S. military depends on low-wage workers from the Philippines. While only 51 Filipino troops served as part of the U.S.-led coalition in the Iraq war, more than 4,000 "serve food, clean toilets and form the backbone of the support staff for American forces. The military would be hard pressed to operate in Iraq without them" (Kirka 2004, p. 1A).

Volunteer for Drug Trial Tests. The poor often volunteer for over-the-counter and prescription drug tests. Most new drugs, ranging from AIDS vaccines to allergy medicine, must eventually be tried on healthy human subjects to determine their potential side effects (such as rashes, headaches, vomiting, constipation, or drowsiness) and appropriate dosages. Money motivates people to volunteer as subjects for these clinical trials. Because payment is relatively low, however, the tests attract a disproportionate share of low-income, unemployed, or underemployed people (Morrow 1996).

Sustain Organizations and Employees Serving the Poor. Many businesses, governmental agencies, and nonprofit organizations exist to serve poor people or to monitor their behavior, and, of course, the employees of these entities draw salaries for performing such work. The United States allocates about $1.6 billion toward food aid worldwide each year, but that money does not go directly to the poor; instead, it goes to corporations, agencies, and individuals who serve the poor. U.S. law mandates that American farmers must grow all government-donated food and that the food must be shipped by U.S.-flagged vessels. Four agricultural corporations, five shipping companies, and seven food-aid charities receive a disproportionate share of food-aid monies. For example, $341 million of the food-aid budget went toward paying packing, shipping, and storage costs, and one-quarter of the operating budgets of seven participating food-aid charities comes from government food-aid allocations. Rising transportation costs consume an even larger share of money the United States allocates to food aid. By one estimate, the United States is feeding 20 million fewer people a year because of costs lost to transportation (Dugger 2005).

Purchase Products That Would Otherwise Be Discarded. Poor people use goods and services that would otherwise go unused and be discarded. Day-old bread, used cars, and secondhand clothes are purchased by or donated to the poor. In the realm of services, the labor of many professionals (such as teachers, doctors, and lawyers) who lack the competency to be hired in more-affluent areas, is purchased by low-income communities. Where do most of the estimated 63 million computers deemed obsolete each year just in the United States go? Most are donated in the name of "bridging the digital divide" to poor countries. Each month, Nigeria alone receives 400,000 such computers, 75 percent of which are obsolete or nonfunctional and end up in landfills, where toxic materials pollute the surrounding environment, including groundwater (Dugger 2005).

Gans (1972) outlines the functions of poverty to show how a part of society that everyone agrees is problematic and should be eliminated remains intact: it contributes to the supposed stability of the overall system. Based on this reasoning, the economic system as we know it would

T. Scott McLaren

These boys live in the African country of Mauritania. Notice that the little boy on the left is wearing a John Deere cap. Both are wearing clothes likely given away by Americans to a Goodwill store. The clothes come from what are called "dead *toubab* stores" (donated white people clothes). Given the racial and ethnic diversity in the United States, it is interesting that the clothes are seen as once belonging to white people.

be strained seriously if we completely eliminated poverty; industries, consumers, and occupational groups that benefit from poverty would be forced to adjust.

A Conflict View of Social Inequality

■ **CORE CONCEPT 5:** Conflict theorists take issue with the premise that social inequality is the mechanism by which the most important positions in society are filled. As you might imagine, conflict theorists challenge the fundamental assumption underlying the functionalist theory of stratification—that social inequality is a necessary device used by societies to ensure that the most functionally important occupations attract the best-qualified people. Sociologists Melvin M. Tumin (1953) and Richard L. Simpson (1956) point out that some positions command large salaries and bring other valued rewards even though their contributions to society are questionable. Consider the salaries of athletes. For the 2008–2009 season, the average NBA athlete earned $5.2 million, with the highest-paid athlete earning $24.7 million (Berkshire 2008). We might argue that professional athletes deserve such enormous salaries because they generate income for owners, cities, advertisers, and media giants. The functionalist argument becomes less convincing, however, when we consider that schoolteachers are paid between $29,000 and $70,000, depending on years of service (U.S. Bureau of Labor Statistics 2007). In spite of a shortage of teachers and the suspicion that a significant percentage of teachers are unqualified to teach, we do not significantly raise their salaries to attract the most qualified.

Tumin and Simpson maintain that the functionalist argument cannot explain why some workers receive a lower salary than other workers for doing the same job, just because they are of a different race, age, sex, or national origin. After all, the workers are performing the same job, so functional importance is not the issue. This question relates to issues of pay equity. For example, why do women working full-time as registered nurses in the United States earn a median weekly wage of $1,011 while their male counterparts earn $1,168? In fact, the U.S. Bureau of Labor Statistics (2009) data show only 3 of 300 occupational categories in which median weekly earnings for females exceed those of their male counterparts: "other life, physical, and social sciences technicians" ($752 versus $751); construction and extraction occupations ($747 versus $688); and installation and repair occupations ($779 versus $774).

In addition to the issue of pay equity is the question of comparable worth. Advocates of comparable pay for comparable work ask whether women who work in predominantly female occupations (such as registered nurse, secretary, and day-care worker) should receive salaries comparable to those earned by men who work in predominantly male occupations that are judged to be of roughly comparable worth (such as those of a housepainter, a carpenter, and an automotive mechanic). For example, assuming comparable worth, why should full-time workers at a child day-care center (performing a traditionally female occupation) receive a median weekly salary of $393 while auto mechanics (performing a traditionally male occupation) earn $678 (U.S. Department of Labor 2009)?

Even if physicians and chief executive officers (CEOs) occupy the most functionally important positions, how much inequality in salary is necessary to ensure that qualified people choose these positions over unskilled ones? In the United States, the average base salary of the CEO of a large public (top 500) corporation is $10.5 million (Jones 2009). This average base salary (excluding bonuses, stock options, and other benefits) is 250 times the median household income. If we compare such CEO compensation with the wages of workers outside the United States, the inequality is even more dramatic.

Consider that the Wal-Mart CEO's total compensation package in fiscal year 2009 was $31.6 million (Reuters 2009). Notwithstanding the CEO's skills, the company's success can be partly attributed to the fact that it imports about $18 billion in manufactured goods from the People's Republic of China (Jiang 2004). If the average Chinese factory worker earns about $1,920 per year, the Wal-Mart CEO earns about 16,000 times more than this worker. In contrast, the average full-time Wal-Mart hourly employee earns $10.11 per hour, or $21,000 per year. The CEO earns about 1,500 times more than the average clerk (*Business Week* 2006; UC Berkeley Labor Center 2007).

Conflict theorists ask if such high salaries are really necessary to make sure that someone takes the job of CEO over, say, the job of a factory worker. Probably not. Nevertheless, these high salaries are justified as necessary to recruit and retain the most able people to run a corporation in the context of a global economy. In *Wealth and Commonwealth*, William H. Gates Sr. (father of Bill Gates) and Chuck Collins (2003) critique one CEO who justified his "enormous compensation package" by stating, "I created about $37 billion in shareholder value." The CEO made no mention of the role the company's 180,000 employees played in creating shareholder value. Gates and Collins argue that "the problem with this individualistic way of assessing one's own contribution is that it is inaccurate and dishonest" (p. 113).

Finally, Tumin and Simpson argued that in societies characterized by a complex division of labor, it is very difficult to determine the functional importance of any occupation, because the accompanying specialization and interdependence make every position necessary to the smooth operation of society. Thus, to judge that physicians are functionally more important than garbage collectors fails to consider the historical importance of sanitation relative to medicine. Contrary to popular belief, advances in medical technology had little influence on death rates

For the book *Nickel and Dimed*, Barbara Ehrenreich (2001) worked at many different kinds of low-paying jobs as a way of learning about the work, family, and social lives of those who work in jobs with little status, benefits, and pay.

until the turn of the 20th century—well after improvements in nutrition and sanitation had caused dramatic decreases in deaths due to infectious diseases.

A Symbolic Interactionist View of Social Inequality

■ **CORE CONCEPT 6: Symbolic interactionists emphasize how social inequality is communicated and enacted.** When symbolic interactionists study social inequality, they seek to understand the experience of social inequality; specifically, they seek to understand how social inequality is communicated and how that inequality shapes social interactions—interactions that involve a self-awareness of one's superior or inferior position relative to others. Social inequality is also conveyed through symbols that have come to be associated with inferior, superior, and equal statuses. In addition, there is a negotiation process by which the involved parties reinforce that inequality in the course of interaction or they ignore, challenge, or change it.

In the tradition of the symbolic interaction, journalist Barbara Ehrenreich (2001) studied inequality in everyday life as it is experienced by those working in jobs that paid $8.00 or less per hour. Ehrenreich, a "white woman with unaccented English" and a professional writer with a Ph.D. in biology, decided to visit a world that many others—as many as 30 percent of the workforce—"inhabit full-time, often for most of their lives" (p. 6). Her aim was just to see if she could "match income to expenses, as the truly poor attempt to do everyday" (p. 6). In the process, Ehrenreich worked as a "waitress, a cleaning person, a nursing home aid, or a retail

clerk" (p. 9). The only "lie" she told in presenting herself to others was that she had completed three years of college. Yet no supervisor or co-worker ever indicated to Ehrenreich that they found her "special in some enviable way—more intelligent, for example, or clearly better educated than most" (p. 8). Ehrenreich reflected on her lack of specialness in this way: "Low-wage workers are no more homogeneous in personality or ability than people who write for a living, and no less funny or bright. Anyone in the educated classes who thinks otherwise ought to broaden their circle of friends" (p. 8).

Ehrenreich's on-the-job observations show the many ways inequality is enacted. Working as a retail clerk in ladies' wear, one of Ehrenreich's jobs is to put away the returns—"clothes that have been tried on and rejected . . . there are also the many items that have been scattered by customers, dropped on the floor, removed from the hangers and strewn over the racks, or secreted in locations far from their natural homes. Each of these items, too, must be returned to its precise place matched by color, pattern, price, and size" (p. 154).

Ehrenreich tells of a colleague who becomes "frantic about a painfully impacted wisdom tooth and keeps making calls from our houses (we are cleaning) to try and locate a source of free dental care" (p. 80). She tells of a colleague who would like to change jobs but the act of changing jobs mean "a week or possibly more without a paycheck" (p. 136); then there is the colleague making $7.00 per hour at KMart thinking about trying for a $9.00-per-hour job at a plastics factory. Ehrenreich also tells of "single mothers who live with their own mothers, or share apartments with a co-worker or boyfriend" and of a woman "who owns her own home, but she sleeps on a living room sofa, while her four grown children and three grandchildren fill up the bedrooms" (p. 79).

Explaining Inequalities across Countries

The United Nations (2009) uses three criteria to classify a country among the least economically developed: (1) annual gross domestic product (GDP) per capita (that is, per person) of $750 or less; (2) low quality of life measures related to life expectancy at birth, per capita calorie intake, primary and secondary school enrollment rates, and adult literacy; and (3) measures of economic vulnerability (instability of agricultural productions, heavy reliance on food aid, heavy dependence of a single or few commodities). Figure 8.1 in the Global Comparisons box shows the poorest and the richest countries in the world. How did the poorest countries become so poor? To answer this question, sociologists draw upon two views: modernization theory and dependency theory.

GLOBAL COMPARISONS

The World's 22 Richest and 50 Poorest Economies

THE COUNTRIES HIGHLIGHTED in orange represent the world's 50 poorest economies. Notice that most are concentrated in Africa. The world's 22 richest countries are highlighted in blue. In the sections that follow, we will try to understand possible reasons for the differences in wealth.

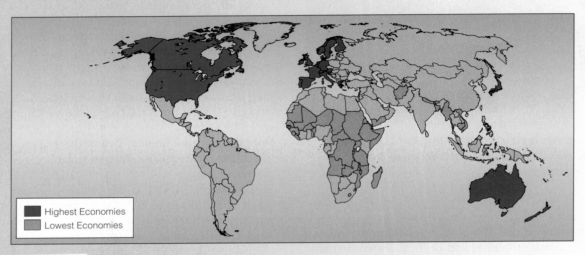

Highest Economies
Lowest Economies

▲ **Figure 8.1**

Source: United Nations (2008)

Modernization Theory

■ **CORE CONCEPT 7:** Modernization theory holds that poor countries are poor because they have yet to develop into modern economies and that their failure to do so is largely the result of internal factors such as a country's resistance to free-market principles or to the absence of cultural values that drive material success. **Modernization** is a process of economic, social, and cultural transformation in which a country "evolves" from preindustrial or underdeveloped status to a modern society in the image of the most developed countries (i.e., western European and North American countries). A country is considered modern when it possesses the following eight characteristics (after each characteristic are comments from Scott McLaren about how the country to which he has been assigned—Mauritania—stands):

1. A high proportion of the population lives in and around cities such that the society is urban-centered. (Scott notes that there has been a steady migration to the cities as infrastructure becomes more available. The number of new villages surrounding cities is growing. With the establishment of Nouakchott as the capital, rural people have migrated there looking to open boutiques or to find work so they can send money home.)
2. The energy to produce food, make goods, and provide services does not come primarily from physical exertion (human and animal muscle) but from inanimate sources of energy such as oil and electricity. (Scott notes that in Mauritania, there are a few areas with access to

modernization A process of economic, social, and cultural transformation in which a country "evolves" from preindustrial or underdeveloped status to a modern society in the image of the most developed countries.

heavy machinery; however, much of the farming/gardening is done by hand. Government agencies sometimes provide the large rice plantations with machines to help with the harvest.)

3. There is widespread access to goods and services, which are features of productive economies with high standards of living. (Scott observes that the capital city—Nouakchott—has some very wealthy families that can afford many of the nuances we in the United States enjoy. The regional capitals have steady access to a broad range of products and services, although the quality or quantity of those services is far lower than those found in Nouakchott.)

4. People have a voice in economic and political affairs. (Scott has found that the people have very little voice. The country is dominated by a small elite, and although ideals of democracy have been planted, democracy does not exist in practice.)

5. Literacy is widespread and there is a scientific, rather than secular, orientation to solving problems. (Scott notes that although some government statistics report relatively high literacy, the reality in rural areas is that literacy is not widespread. There is a continually expanding school system that is striving to eventually include all people, but an infrastructure needs to develop.)

6. A system of mass media and communication is in place that offsets the influence of the family and local cultures. (Scott notes that the Mauritanian government controls the media. Little unapproved news is seen or heard. There is some access to outside news sources via radio, satellite TV, and the Internet.)

7. There are large-scale, impersonally administered organizations such as government, businesses, schools, and hospitals that reduce dependence on family for child care, education, and social security. (Scott believes that there are some government-funded organizations set up to help, but their funding is limited and their reach is largely confined to the immediate areas around the capital.)

8. People feel a sense of loyalty to a country (a national identity), not to an extended family and/or tribe (Naofusa 1999). (Scott believes that in Mauritania, loyalty to family is much stronger than to country. If Scott had to rank people's loyalties, people's sense of loyalty and connection to family is first, followed by a loyalty to culture, then ethnicity, and lastly country).

Modernization theorists seek to identify the conditions that launch underdeveloped countries on the path to modernization and to identify the stages through which those countries must pass to reach modernization. The road to modernization begins with a tradition-oriented way of life (stage 1) that is characterized by kinship-related obli-

This village school in Afghanistan enrolls more than 600 students. It has no modern conveniences, and these girls are allowed to attend on the condition that their father allows it. Modernization theorists would argue that a country will not modernize until the decision is out of the fathers' hands and becomes a government mandate.

gations and loyalties that modernization theorists claim discourage change and personal mobility. Stage 1 is also characterized by a level of productivity limited by an inaccessibility to modern science, including its applications and frame of mind (Rostow 1960). W. W. Rostow (1960), an economic advisor to President John F. Kennedy and a proponent of modernization theory, described tradition-oriented cultures as possessing a long-run fatalism fueled by the "pervasive assumption that the range of possibilities open to one's grandchildren would be just about what it had been for one's grandparents."

According to Rostow's model of modernization, the next stages are the pre-take-off (stage 2), the take-off (stage 3), the drive to maturity (stage 4), and the age of high mass-consumption (stage 5). Western countries can set the preconditions for take-off by jump-starting modernization (stage 2) through foreign aid and investments that include technology transfers (e.g., fertilizers, pesticides), birth control programs, loans, cultural exchange, and medical interventions (e.g., inoculation programs). Ideally, these interventions "shock" the traditional society and hasten its undoing so that the country "takes off." Such interventions set into motion the ideas and sentiments by which a "modern alternative to the traditional society" evolves "out of the old culture" (Rostow 1960). The developing countries can hasten modernization through appropriate government reforms and policies. Eventually—perhaps as many as 60 years later—the developing country will reach a final state of modernization characterized by technological maturity and high mass consumption.

According to Rostow, modernization involves a transformation of cultural beliefs and values away from those that supposedly support fatalism and collective orientation to those that support a work ethic, deferred gratification, future-orientation, ambition, and individualism (important attitudes and traits essential to the development of a free market economy or capitalism). As the country modernizes, "the idea spreads, not merely that economic progress is possible, but that economic progress is a necessary condition for some other purpose . . . be it national dignity, private profit, the general welfare, or a better life for the children."

Dependency Theory

■ **CORE CONCEPT 8: Dependency theory holds that for the most part, poor countries are poor because they are products of a colonial past.** Dependency theorists challenge the basic tenet of modernization theory—that poor countries fail to modernize because they reject free-market principles and because they lack the cultural values that drive entrepreneurship. Rather, dependency theorists argue that poor countries are poor because they have been, and continue to be, exploited by the world's wealthiest governments and by the global and multinational corporations that are based in the wealthy countries. This exploitation began with colonialism.

Colonialism is a form of domination in which a foreign power uses superior military force to impose its political, economic, social, and cultural institutions on an indigenous population so it can control their resources, labor, and markets (Marger 1991). The age of European colonization began in 1492 with the voyage of Christopher Columbus.

By 1800 Europeans had learned of, conquered, and colonized much of North America, South America, Asia, and coastal parts of Africa, setting the tone of interna-

colonialism A form of domination in which a foreign power uses superior military force to impose its political, economic, social, and cultural institutions on an indigenous population so it can control their resources, labor, and markets.

decolonization A process of undoing colonialism such that the colonized achieves independence from the so-called mother country.

tional relations for centuries to come (see Figure 8.2 in No Borders, No Boundaries). During this time, European colonists forced local populations to cultivate and harvest crops and to extract minerals and other raw materials for export to the colonists' home countries. When the colonists' labor needs could not be met by indigenous populations, the colonists imported slaves from Africa or indentured workers from Asia and Europe. In fact, an estimated 11.7 million enslaved Africans survived their journey to the "New World" between the mid-15th century and 1870 (Chaliand and Rageau 1995; Conrad 1996; Holloway 1996).

The scale of social and economic interdependence changed dramatically with the Industrial Revolution, which gained momentum in Britain around 1850 and then spread to other European countries and the United States. The Industrial Revolution even drew people from the most remote parts of the world into a process that produced unprecedented quantities of material goods, primarily for the benefit of the colonizing countries. Between 1880 and 1914, pursuit of and demand for raw materials and labor increased dramatically. This period, known as the Age of Imperialism, saw the most rapid colonial expansion in history. During this time, rival European powers (such as Britain, France, Germany, Belgium, Portugal, the Netherlands, and Italy) competed to secure colonies and influence in Asia, the Pacific, and especially Africa (see, again, No Borders, No Boundaries).

Consider as one measure of the extent of colonization that during the 20th century, 130 countries and territories gained political independence from their colonial masters. That process of gaining political independence is known as decolonization. **Decolonization** is a process of undoing colonialism such that the colonized country achieves independence from the so-called mother country. Decolonization can be a peaceful process by which the two parties negotiate the terms of independence, or it can be a violent disengagement that involves civil disobedience, insurrection, or armed struggle (war of independence). Once independence is achieved, civil war between rival factions often takes place as each seeks to secure the power relinquished by the colonizer. Some scholars argue that the Americas (which include the United States, Canada, and Central and South America) are technically still colonized lands because the indigenous peoples were not the ones to revolt and declare independence; rather, it was the colonists and/or their descendants who revolted and declared their independence. Those who took power simply continued colonizing and exploiting the land and resources belonging to indigenous peoples (e.g., people now known as Native Americans) and others (i.e., the enslaved and indentured peoples forced to immigrate) (Cook-Lynn 2008; Mihesuah 2008).

Gaining political independence does not mean, however, that a former colony no longer depends on its colonizing

No Borders, No Boundaries

The Legacy of Colonization

THIS MAP OF Africa shows the decade in which each country gained independence from its "mother" country. The colonizing country exploited the labor and resources of the colony. After independence, the ties between the two countries do not end, however.

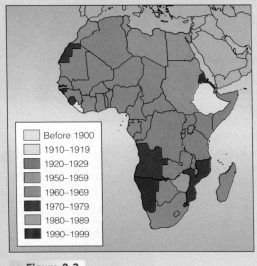

Before 1900
1910–1919
1920–1929
1950–1959
1960–1969
1970–1979
1980–1989
1990–1999

▲ **Figure 8.2**

country (see Chapter 5). In *How Europe Underdeveloped Africa*, Walter Rodney argues that in the end, the African continent—90.4 percent of which was once controlled by colonial powers—has been "consigned to the role of producer of primary products for processing in the West" (Obadina 2000). Examples of primary products include products that are mined (gold) or extracted (oil) from the earth, as well as fish and agricultural products. Mauritania, a former colony of France, gained independence in the 1960s. The country has rich deposits of iron ore that account for 40 percent of total exports, mostly to European Union countries and China. Its coastal waters are considered to be among the richest fishing areas in the world, but overexploitation by foreigners is threatening to deplete a major source of revenue (U.S. Central Intelligence Agency 2009).

This continuing economic dependence on former colonial powers is known as **neocolonialism**. In other words, neocolonialism is a new form of colonialism where more powerful foreign governments and foreign-owned busi-

nesses continue to exploit the resources and labor of the post-colonial peoples. Specifically, resources still flow from the former colonized countries to the wealthiest countries that once controlled them. Some critics would argue that the U.S. military presence in and around the continent of Africa is a form of neocolonialism.

In 2007 the U.S. Department of Defense put the continent of Africa under one military command, known as Africom. Prior to this, the command of the African continent was split between the European and Pacific Commands. The Pentagon's decision has faced resistance and criticism

neocolonialism A new form of colonialism where more powerful foreign governments and foreign-owned businesses continue to exploit the resources and labor of the post-colonial peoples.

because "many African leaders questioned the formation of the command, calling it a U.S. grab for African resources—while others felt the command represented the militarization of U.S. foreign policy." The official U.S position is that Africom "allows the U.S. military to help the Africans help themselves, provide security, and to support the far larger U.S. civilian agency programs on the continent" (Garacome 2008; U.S. African Command 2008).

A Response to Global Inequality: The Millennium Declaration

■ **CORE CONCEPT 9:** Structural responses to global inequality include transferring wealth from highest-income countries to lowest-income countries through foreign aid and fair-trade policies. Clearly one obvious way to reduce global inequality is to redistribute wealth by transferring it away from those with the most wealth and income to those with the least wealth and income. The United Nations has devised such a plan. In 2000, the 189 United Nations member countries endorsed the Millennium Declaration (2000), which states:

> As leaders we have a duty . . . to all the world's people, especially the most vulnerable and in particular the children of the world, to whom the future belongs. . . . We will spare no effort to free our fellow men, women, and children from the abject and dehumanizing conditions of extreme poverty, to which more than a billion are currently subjected. . . . We also undertake to address the special needs of the least developed countries including the small island developing states, the landlocked developing countries, and the countries of sub-Saharan Africa.

The Millennium Development Project set 20 targets and 60 measures of success to be reached by 2015, including these:

- Halve the proportion of people whose income is less than $1 a day.
- Halve the proportion of people who suffer from hunger.
- Reduce by three-quarters the maternal mortality ratio.

Clearly these are ambitious targets. Their success hinges on at least two major commitments from the world's 22 richest countries:

1. Increase current levels of foreign aid by investing 0.7 percent (seven-tenths of 1 percent) of their annual gross national product (GNP) in the Millennium Development Project.
2. Develop an open, nondiscriminatory trading system that addresses the needs of the lowest-income economies by eliminating the subsidies, tariffs, and quotas that put those economies' products at a disadvantage in the global marketplace.

A review of the $27.5 billion in foreign aid the United States allocated each year suggests that the bulk of the assistance goes not toward development but toward crisis intervention such as disaster and famine relief (bottom photo) and refugee programs, military training (top photo) and financing, and narcotics control. (U.S. Department of State 2005)

Increase Foreign Aid from Richest Countries

The United States is the largest donor of foreign aid in absolute dollars ($27.5 billion). But that number is deceiving when we consider that amount in light of the United States's size and wealth. The United States has more than 305 million people living within its borders. Its per-capita spending on foreign aid translates into $84 per person. The European Union, by contrast, spends $145 per person. If we consider foreign aid as a percentage of gross national income (GNI), the United States is tied with Japan as the least generous (see Figure 8.3).

Supporters of the U.S. level of giving argue that the United States gives assistance in other, less officially recognized forms, including private donations, wage remittances from immigrants working in the United States, and trade investments (see Working for Change: Reducing

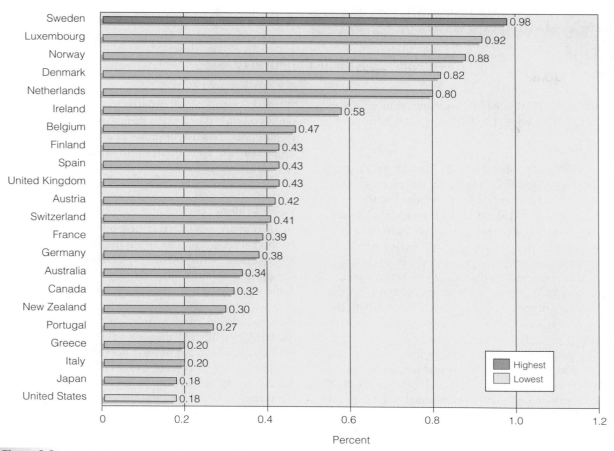

▲ **Figure 8.3**   Foreign Assistance as a Percentage of Gross National Income, 2008

The bar lines show that the United States falls very short of meeting the 0.7 percent UN-recommended donation target. So, for example, the U.S. GDP is $14.3 trillion; 0.7 percent of that is $101 billion. On the bright side, five countries—Denmark (0.82%), the Netherlands (0.80%), Luxembourg (0.92%), Sweden (0.98%), and Norway (0.88%)—have exceeded that 0.7 target.

Source: OECD (2009)

Poverty). Jeffrey D. Sacks (2005), the UN Millennium Project director and the author of *The End of Poverty*, argues that the biggest myth held by most Americans is that the United States already gives a large amount of money and that much, if not all, is wasted. Sacks maintains that the United States puts in "almost no funding, and it accomplishes almost nothing. And then we bemoan the waste. I don't know how to break through that misunderstanding. That's what I've been trying to do for many years, but it's very, very powerful in this country." Here is how Sacks assesses U.S. financial assistance to Africa:

> The U.S. aid to Africa is $3 billion this year. That $3 billion is roughly divided into three parts: The first is emergency food shipments. Of the billion or so in emergency food shipments, half of that, roughly $500 million, is just for transport costs. So the commodities are maybe half a billion dollars. That's not development assistance, that's emergency relief. The second bil-

lion is the AIDS program, now standing at about $1 billion. That, on the whole, is a good thing. I would call it a real program. It's providing commodities; it's providing relief. It started late and it's too small, but it's there. The third billion is everything else we do for child survival, maternal survival, family planning, roads, power, water and sanitation, malaria; everything is the third $1 billion. Most of that, approaching 80 percent, is actually for American consultant salaries. There's almost no delivery of commodities, for example. There's essentially zero financing to help a country build a school or build a clinic or dig a well.

Jeffrey Sacks's assessment contrasts starkly with the American public's belief that their government is contributing about 20 percent of its $2.472 trillion federal budget (or $494 billion) in financial assistance to other countries (Office of Management and Budget 2005). Could the UN and Jeffrey Sacks be mistaken? According to U.S. Department of State data, they are not. A review of the $27.5 billion

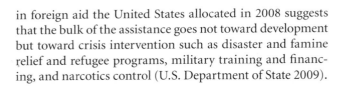

WORKING FOR CHANGE

Reducing Poverty

HUNDREDS OF CREATIVE and successful efforts to reduce poverty offer hope. Two of them are described here.

Positive Deviants

Save the Children staff member Jerry Sternin was charged with a seemingly impossible assignment: save starving children in Vietnam. He drew inspiration from the theory of positive deviants—individuals "whose exceptional behaviors and practices enable them to get better results than their neighbors with the exact same resources." Sternin identified the Vietnamese children whose weight (relative to their age) suggested they were well nourished. He observed the children's mothers and learned that they were behaving in ways that defied conventional wisdom. Among other things, the mothers were (1) using alternative food sources available to everyone (they were going to the rice paddies to harvest tiny shrimp and crabs, and they were picking sweet potato greens—considered low-class food—and mixing both food sources with rice); (2) feeding their children when the children had diarrhea, contrary to traditional practice; and

(3) making sure their children ate, rather than "hoping children would take it upon themselves to eat." The strategy of observing positive deviants as a problem-solving technique reached 2.2 million Vietnamese in 276 villages. In addition, this model was used in at least 20 other countries where malnutrition is widespread (Dorsey and Leon 2000).

Micro-Lending

In 1976 micro-lending was first piloted in Bangladesh, a country whose economy is among the poorest. The goal was to examine the possibility of extending tiny loans to the poorest of rural poor women. The anticipated outcome was to eliminate money lenders' exploitation of the poor and create opportunities for self-employment among the unemployed rural population of Bangladesh. Today, the bank that piloted micro-lending has 1,175 branches, with an estimated 2.4 million borrowers living in 41,000 villages (about 60 percent of all villages in Bangladesh). An estimated 90 percent of borrowers repay the loans (Grameen 2005).

---

in foreign aid the United States allocated in 2008 suggests that the bulk of the assistance goes not toward development but toward crisis intervention such as disaster and famine relief and refugee programs, military training and financing, and narcotics control (U.S. Department of State 2009).

End Subsidies, Tariffs, and Quotas

Notwithstanding an agreement to eliminate tariffs on products imported from the poorest economies in the world, the wealthiest economies are resisting the UN request to dismantle a trade system structured to their advantage and to the advantage of some of the wealthiest segments of their populations (Bradshear 2006). In particular, the United States, Japan, the European Union, and other high-income countries subsidize agriculture and other sectors, such as steel and textiles manufacturing, so that producers in these countries are paid more than world market value for their products. Considerable attention has been given to agricultural subsidies, which give farmers in high-income countries an estimated $372

billion in support (United Nations 2009). It is well documented that the farmers to which most such subsidies go are large agricultural corporations, not small farmers. For example, Riceland Foods (the top recipient) received $554 million in subsidies between 1995 and 2005 (Oxfam 2006; Environmental Working Group 2007).

In addition to subsidizing their own producers, the wealthiest economies put the poorest ones at a disadvantage by applying tariffs and quotas to imported goods that compete with protected products, making less expensive imported goods cost as much as or more than the domestic versions. Consider sugar. The European Union protects its sugar industry so that domestic producers earn double, sometimes triple, the world market price. Because of subsidies, 6 million tons of surplus sugar each year is dumped into the world market at artificially low prices. Of course, sugar is not the only protected commodity. Other protected commodities include rice, textiles, and steel. The United States, while not a major exporter of sugar, applies tariffs and quotas on imported sugar cane and other sugars grown in Brazil, Vietnam, and other countries to protect its domestic sugar producers (Thurow and Winestock 2005). Subsidies, tariffs, and quotas on sugar help to keep prices on the world market artificially low and prices within protected markets artificially high.

The United Nations estimates that this system of subsidies, tariffs, and quotas costs poor nations $50 billion annually in lost export revenue—in effect, negating the aid monies given to the poorest economies each year. In 2006 that aid totaled $104 billion. Such policies affect not just workers in the poorest countries, but also workers in the protected markets. For example, Brach and Kraft Foods closed their U.S.-based candy plants and outsourced more than 1,000 jobs to Argentina and Canada, where sugar can be purchased at lower, world-market prices (Kher 2002).

Moreover, in the United States, the highest tariffs are often placed on goods purchased by low-income consumers. That is, tariffs are usually lowest on luxury items and highest on essential items, such as shoes and clothing, especially the cheaper varieties. For example, the United States places a 2.4 percent tariff on women's silk underwear but a 16.2 percent tariff on the polyester variety; a 1.9 percent tariff on silk suits, a 12 percent tariff on wool suits, and a 29 percent tariff on polyester suits; a 5.3 percent tariff on snakeskin handbags but an 18 percent tariff on plastic-sided handbags; and no tariff on silver-handled forks but a 15 percent tariff on stainless-steel ones (Gresser 2002).

Like poor people in the United States, poor people in the lowest-income economies must pay more for essential goods than they otherwise would because their governments also place tariffs on such goods that come from other countries. In Vietnam, bicycles are essential for the rural poor, but the 50 to 60 percent tariff that the government slaps on imported bicycles makes the cost of even the least expensive model prohibitive, at about twice a poor peasant's monthly income. As another example, the 25 percent tariff that some sub-Saharan governments place on mosquito nets prevents the poor there from taking proven action to prevent malaria, a disease that kills almost 1 million people a year in that region of the world (*Bicycle Retailer* 2005).

Progress Toward Reaching Millennium Development Goals (at Midpoint)

Recall that the UN Millennium Development goals were set in 2000 with the goal of achieving them by 2015. In 2008—the halfway point—the UN released a report highlighting the areas in which progress has been made and those in which greater efforts must be focused. Some areas of progress include the following:

- The Millennium project is on track to reduce absolute poverty by half for the world as a whole.
- Deaths from measles have fallen from approximately 750,000 in 2000 to 250,000 in 2006, and an estimated

80 percent of children in developing countries receive measles vaccines.
- Since 1990, about 1.6 billion people have gained access to safe drinking water.
- The use of ozone-depleting substances has been virtually eliminated.
- The private sector has increased access to some critical and essential drugs.
- Mobile phone technology has increased dramatically throughout the developing world. (UN 2009, p. 4)

These successes notwithstanding, there are many goals the UN (and, by extension, the world community) has failed to meet. In particular, the world's richest countries have failed to eliminate agricultural and other subsidies, to dedicate 0.7 percent of their national budgets toward foreign aid, and to implement fair-trade practices. These failures, in conjunction with the global economic crisis, mean that among other things, the UN will not be able to:

- Reduce the proportion of people in sub-Saharan Africa living on less than $1 per day by 50 percent.
- Alleviate childhood undernourishment and its effects.
- Reduce maternal mortality.
- Improve sanitation of 2.5 billion people, or 50 percent of the developing world's population.
- Prevent carbon dioxide emissions from increasing. (United Nations 2009, p. 4)

Criticism of the Millennium Declaration

Critics of the UN recommendation to increase foreign aid and reduce tariffs argue that eliminating trade barriers is only a first step. There are other factors that make it difficult for poor economies to compete—one being that employers can always find desperate workers willing to work for less. For example, American retailers offered Laotian clothing manufacturers the opportunity to cut and sew synthetic-lined winter jackets for $2.65 per jacket—the same price that the retailers offered Chinese manufacturers. The problem was that it cost the Laotian manufacturers $4.00 to make each jacket. Laotian laborers simply could not compete against their Chinese counterparts, who earn between 36 cents and $1.00 an hour, depending on location (inland versus coastal) (Bradsher 2006).

A second reason critics argue for why increasing the amount of foreign aid and ending subsidies, tariffs, and quotas are not sufficient to substantially reduce global inequality relates to **brain drain**, the emigration from a country of the most-educated and most-talented people, including actual or potential hospital managers, nurses,

brain drain The emigration from a country of the most educated and most talented people.

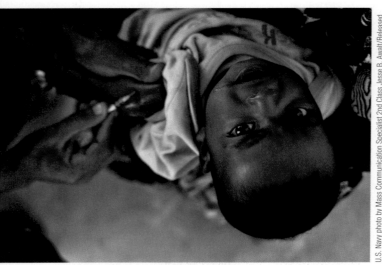

The United Nations has met its goal of vaccinating children against measles. It has failed to meet a number of other goals, such as alleviating childhood malnutrition.

accountants, teachers, engineers, political reformers, and other institution builders. One estimate of brain drain shows that 30 percent (Sri Lanka) to 84 percent (Haiti) of college-educated citizens from the poorest countries live abroad in high-income countries (Dugger 2005).

The rich economies have facilitated this movement by implementing immigration policies that give preference to educated, skilled foreigners. The British Medical Association has grown so concerned about the shortage of health-care workers around the world and the migration of such workers from poor to rich countries that it has called for the following measures:

1. All countries must strive to attain self-sufficiency in their health-care workforce without generating adverse consequences for other countries;
2. Developed countries must assist developing countries to expand their capacity to train and retain physicians and nurses, to enable them to become self-sufficient.

Analyzing Social Class

■ **CORE CONCEPT 10:** Social class is difficult to define. It depends on many factors, including people's relationship to the means of production, their sources of income (such as land, labor, or rent), their marketable abilities, their access

class A person's overall economic and social status in a system of social stratification.

finance aristocracy Bankers and stockholders seemingly detached from the world of "work."

to consumer goods and services, their status group, and their membership in political parties. Sociologists use the term **class** to designate a person's overall economic and social status in a system of social stratification. They see class as an important factor in determining life chances. We begin with the writings of Karl Marx and Max Weber, who represent the "two most important traditions of class analysis in sociological theory" (Wright 2003, p. 1).

Karl Marx and Social Class

Karl Marx believed that the most important engine of change is class struggle. In *The Communist Manifesto*, published in 1848 and written with Friedrich Engels, Marx describes how conflict between two distinct social classes propelled society from one historical epoch to another. He observed that the rise of factories and mechanization as a means of production created two fundamental classes: the owners of the means of production (the bourgeoisie) and the propertyless workers (proletariat) who must sell their labor to the bourgeoisie (Allen and Chung 2000). For Marx, then, the key variable in determining social class is source of income.

In *Das Kapital*, Marx names three classes, each of which is comprised of people whose revenue or income "flow from the same common sources" (p. 1032). For wage laborers, the source is wages; for capitalists, the source is profit; for landowners, the source is ground rent. In the *Class Struggles of France 1884–1850*, Marx named another class, the **finance aristocracy**, who lived in obvious luxury among masses of

This 1883 political cartoon captures Karl Marx's vision of social class. It shows four of the wealthiest people who have ever lived–Cyrus Field, Jay Gould, Cornelius Vanderbilt, and Russell Sage. These men were wealthier in real dollars than those considered the world's richest today. The four are seated on bags of money atop a large raft, which is being held afloat by millions of workers.

starving, low-paid, and unemployed workers. The finance aristocracy includes bankers and stockholders seemingly detached from the world of "work." Here is how Marx described this source of income: "it is a source of income created from nothing—without labor and without creating a product or service to sell in exchange for wealth." The finance aristocracy speculates or employs people who know how to speculate for them. "But while speculation has this power of inventiveness, it is at the same time also a gamble and a search for the 'easy life'; as such it is the art of getting rich without work." According to Marx, the financial aristocracy appropriates to themselves "public funds or private funds without giving anything equivalent in exchange; it is the cancer of production, the plague of society and of states" (Marx 1856, Bologna 2008, and Proudhon 1847).

Max Weber and Social Class

Karl Marx clearly states that social class is based on people's relationship to the means of production, a relationship that determines the sources of their income. For Max Weber, the basis extends beyond someone's relationship to the means of production. According to Weber (1947), people's class standing depends on their marketable abilities (work experience and qualifications), their access to consumer goods and services, their control over the means of production, and their ability to invest in property and other sources of income. According to Weber, people completely lacking in skills, property, or employment, or who depend on seasonal or sporadic employment, constitute the very bottom of the class system. They form the **negatively privileged property class**.

Individuals at the very top—the **positively privileged property class**—monopolize the purchase of the highest-priced consumer goods, have access to the most socially advantageous kinds of education, control the highest executive positions, own the means of production, and live on income from property and other investments. Weber viewed class as a series of rungs on a social ladder, with the top rung being the positively privileged property class and the bottom rung being the negatively privileged property class. Between the top and the bottom of this social-status ladder is a series of rungs. He argued that a "uniform class situation prevails only among the negatively privileged property class." We cannot speak of a uniform situation regarding the other classes, because people's class standing is complicated by their occupation, education, income, group affiliations, consumption patterns, and so on.

Weber states that class ranking is complicated by the status groups and political parties to which people belong. He defines a **status group** as an amorphous group of persons held together by virtue of a lifestyle that has come to be "expected of all those who wish to belong to the circle" (Weber 1948, p. 187) and "by the level of social

Bodybuilders are a status group in that they have developed a lifestyle around maximizing the size and appearance of their muscles. They are held together by a shared way of living, including eating high-protein foods and sleeping at least eight hours of the day to help muscles recuperate and build efficiently after intense training.

esteem and honor accorded to them by others" (Coser 1977, p. 229). This lifestyle may encompass leisure activities, eating, time devoted to sleeping, occupation held, and friendships.

Weber's definition suggests that wealth, income, and position are not the only factors that determine an individual's status group. Simply consider that some people possess equivalent amounts of wealth yet hold very different statuses due to their upbringing and education. In addition to status group, people can also belong to **political parties**—organizations "oriented toward the planned acquisition of social power [and] toward influencing social action no matter what its content may be" (Weber 1982, p. 68). Parties are organized to represent people of a certain class and status and with certain interests; they exist at all levels (within an organization, a city, a country). The

negatively privileged property class Weber's category for people completely lacking in skills, property, or employment or who depend on seasonal or sporadic employment; they constitute the very bottom of the class system.

positively privileged property class Weber's category for the people at the very top of the class system.

status group Weber's term for an amorphous group of people held together both by virtue of a lifestyle that has come to be expected of them and by the level of esteem in which other people hold them.

political parties According to Weber, "organizations oriented toward the planned acquisition of social power [and] toward influencing social action no matter what its content may be."

means to obtain power can include violence, vote canvassing, bribery, donations, the force of speech, suggestion, and fraud. Examples of political parties include NOW (National Organization for Women), Promise Keepers, the UAW (United Auto Workers), the NRA (National Rifle Association), and AARP (American Association of Retired Persons).

Erik Orin Wright and Social Class

Sociologist Erik O. Wright (2003), who has spent his professional career working to clarify the concept of social class and its historical, structural, and personal significance, offers a series of questions that capture some of the ways sociologists approach class.

- *Is there an objective way to assess the distribution of material inequality?* Once sociologists settle on the number of class categories relevant to their analysis, they work to identify objective criteria by which to classify people into those categories. If they decide, for example, to categorize class in terms of upper, middle, and lower, the next step is to decide how to objectively measure social class: should income or accumulated wealth be used to determine one's social class? **Income** refers to the money a person earns usually on an annual basis through salary or wages. **Wealth** refers to the combined value of a person's income and other material assets such as stocks, real estate, and savings minus debt. If we divide U.S. households into four groups of 28 million households each, the top quartile controls 87 percent of the wealth (the equivalent of $43.6 trillion) and the bottom 28 million households hold no wealth after debt is paid out (Di 2007).
- *How much income or wealth qualifies someone to be upper, middle, or lower class?* The Pew Research Center (2008) defines social class in this way: a person is considered middle-income class if he or she lives in a household with an annual income that falls within 75 percent to 150 percent of the median household income. A person with a household income above that range is upper-income class; a person whose household income is below that range is the low-income class (Figure 8.4). In 2008, the median household income in the United States was $59,494. Seventy-five percent of

$59,494 is $44,620; 150 percent of $59,494 is $89,241. So a middle-class household is one with an annual income between $44,620, and $89,241. Of course, this represents just one way to conceptualize social class—in the 2008 presidential elections then-candidate Barack Obama defined the upper limits of middle class status as $250,000 with no lower limit specified.

- *How do people see the class structure, and how do they go about locating themselves in that structure?* Objective measures of social class do not usually correspond with people's self-assessment of their class location. For example, Pew Research Center found that an objective measure of social class (using income cut-offs) puts 35 percent of households in the middle-class category, but a subjective measure of social class (asking people to place themselves in a social class) puts 53 percent of households in the middle class.

Pew researchers found that the "greater the income, the higher the estimate of what it takes to be middle class." Those with household incomes between "$100,000 and $150,000 a year believe, on average, that it takes $80,000 to live a middle-class life." Conversely, those with household incomes of "less than $30,000 a year believe that it takes about $50,000 a year to be middle class" (Pew Research Center 2008, p. 15).

The Disadvantaged in the United States

One in eight people, or 12.5 percent of the U.S. population (37.3 million people), is officially classified as living in poverty (see Figure 8.4). To determine who lives in poverty, the U.S. Bureau of the Census sets a dollar-value threshold that varies depending on household size and age (under 65 and 65 and over). If the total household annual income is less than the specified dollar value, then that household is considered as living in poverty. The poverty thresholds are also adjusted for geography (cost of living). The poverty threshold for a four-person family consisting of one adult and three children is $21,100. The poverty threshold for a single person under age 65 is $10,787. How does the U.S. government determine the various thresholds? The formula was set in 1963 and is based on the estimated daily cost per person of a nutritionally adequate diet; that estimate is then multiplied by 3. The resulting number is the amount of money (threshold) a person needs on a daily basis to live outside of poverty. That daily cost is multiplied by 365 to determine the yearly cost. In 2007, the government estimated that a person needed about $29 per day to live above poverty; $9.80 per day for food and $19.80 for shelter and clothes. For the year, a single person under 65 needs a total of $10,787 (U.S. Department of Health and

---

income The money a person earns, usually on an annual basis through salary or wages.

wealth The combined value of a person's income and other material assets such as stocks, real estate, and savings minus debt.

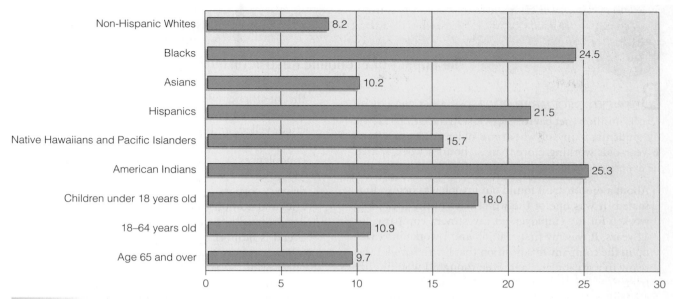

▲ **Figure 8.4**   Percentage of Population in Poverty by Selected Characteristics, 2007

Poverty rates vary according to race, ethnicity, age, and sex. What percentage of non-Hispanic whites fall below the poverty thresholds? What percentage of Native Americans fall below that threshold? What percentage of children under 18 do?

Source: U.S. Bureau of the Census (2008)

Human Services 2009). Those living in poverty can be classified into one of three groups: inner city, suburban, or rural poor.

Inner-City Poor

What happens to people when a factory closes? What happened to those who found work supporting and servicing the people who worked at that factory? Detroit automobile plants, built in 1907, could employ as many as 40,000 people, enough employees to sustain a department store, two schools, and grocery store on the premises. The Detroit plants began to close in 1957, laying off 130,000 auto workers by 1967 as car companies restructured their operations, relocating plants to the suburbs and automating production facilities (Sugrue 2007). What happened to those who worked in the stores and schools? Of course, there was a second wave of restructuring and layoffs beginning in the 1970s and continuing though today, as Detroit automakers steadily moved operations to overseas locations and downsized their operations in the United States. Beginning in 2008, the global economic crisis pushed Ford, General Motors, and Chrysler to restructure even further in response to foreign competition and declining car sales. GM plans to close 16 of 47 operating plants, lay off 23,000 production and 10,000 white-collar workers, and shut down 50

percent of its 6,200 dealerships (Goldstein 2009). This restructuring and downsizing was on the mind of sociologist Julius Wilson when he wrote *The Truly Disadvantaged in the United States.*

■ **CORE CONCEPT 11:** Economic or occupational restructuring can devastate people, leaving them without jobs and disrupting the networks of occupational contacts crucial to moving affected workers into and up job chains. In *The Truly Disadvantaged* and other related studies, sociologist William Julius Wilson (1983, 1987, 1991, 1993) describes how structural changes in the U.S. economy helped create what he termed the ghetto poor, who are now known as the inner-city poor. Since the 1970s, a number of economic transformations have taken place, including:

- The restructuring of the American economy from a manufacturing base to a service and information base.
- The rise of a labor surplus marked by the entry of women and the large "baby boom" segment of the population into the labor market.
- A massive exodus of jobs from the cities to the suburbs.
- The transfer of manufacturing jobs out of the United States.
- The transfer of customer service and information jobs out of the United States over the past decade (see Intersection of Biography and Society).

INTERSECTION OF BIOGRAPHY AND SOCIETY

The Impact of Economic Restructuring

BETWEEN DECEMBER 2007 and April 2009, there were 5.7 million (net payroll) jobs lost in the United States. My students, many of whom are older adults or 18- to 26-year-olds working more than 20 hours a week, shared the experience of losing their job at this time.

- About a month ago I found out my job was being eliminated. It was one of the worst days of my life. I had worked for my employer, an investment company, for 14 years. It was my first real job and I worked my way up in the company to a position that I truly loved. I was really in a great place, and even thought I would retire from that company! I was loyal, dedicated, and gave 100 percent each day. So needless to say, I was heart-broken when I found out I was part of the "RIF" (reduction in force). Not to mention the fact that my husband and I are in the process of trying to adopt two children. The timing couldn't have been worse.

- Today, before coming to class, I was called into the office and told I was to be laid off. I work for one of the largest banks in the United States. Already I have experienced a wide range of reactions from pretending to be totally confident about my future to what I can only describe as losing my mind. I even yelled at my husband, saying that I wished he would be more upset and stop telling me everything would be fine. My layoff wasn't totally unexpected, but to be told by my manager that the work I did is "above and beyond" that of other employees but that my services just aren't needed anymore is overwhelming. I have laughed, cried, yelled, and sat in silence. It feels like I am living through my own funeral—people stare or offer condolences.

- My boyfriend works as a driver for a major package delivery corporation that has cut out all domestic deliveries and now only deliver international. The company has dropped from 150 full-time drivers to 11. My boyfriend is number 12, which means he is on-call when other drivers are overloaded or when someone is off or on vacation. So far, he has been called in to work every day, but every day he worries about whether he will be needed. He is on call day or night, so the company might call at 6:00 a.m. or 5:00 p.m. His life completely revolves around his job. He is very cautious about spending money. He needs a new car but is afraid to take on the payment. I feel sorry for him, because he's in a terrible situation.

- During the start of the economic crisis, I was working for an acrylic manufacturer when orders really slowed down due to the declines in new-home sales. Fortunately, a job opportunity came up with the post office and I took it. Now my work hours have been reduced at the post office due to the decrease in mail volume. Within the past six months, I went from working 50-plus hours per work week down to about 40 hours. I feel fortunate that I haven't lost full-time status.

These changes, combined with other historical factors, are major forces behind the emergence of the inner-city poor or **urban underclass**—a "heterogeneous grouping of families and individuals in the inner city [who] are outside the mainstream of the American occupational system and [who] consequently represent the very bottom of the economic hierarchy" (Wilson 1983, p. 80).

Wilson (in collaboration with sociologist Loïc J. D. Wacquant) studied Chicago to illustrate this point.

urban underclass The group of families and individuals in inner cities who live "outside the mainstream of the American occupational system and [who] consequently represent the very bottom of the economic hierarchy" (Wilson 1983, p. 80)

The events surrounding Hurricane Katrina in 2005 alerted television viewers to the situation of the inner-city poor and provided visual evidence to a global audience of the incomprehensible level of poverty and disability within the city of New Orleans. (Ranjan 2000)

U.S. Navy photo by Petty Officer 1st Class Brien Aho

(The same point applies to every large city in the United States.) In 1954 Chicago was at the height of its industrial power. Between 1954 and 1982, however, the number of manufacturing establishments within the city limits dropped from more than 10,000 to 5,000, and the number of jobs declined from 616,000 to 277,000. This reduction, along with the out-migration of stably employed working-class and middle-class families, which was fueled by access to new housing opportunities outside the inner city, profoundly affected the daily life of people left behind.

According to Wacquant (1989), the single most significant consequence of these historical economic events was the "disruption of the networks of occupational contacts that are so crucial in moving individuals into and up job chains, . . . [because] ghetto residents lack parents, friends, and acquaintances who are stably employed and can therefore function as diverse ties to firms . . . by telling them about a possible opening and assisting them in applying [for] and retaining a job" (Wacquant 1989, pp. 515–516).

Suburban and Rural Poor

Because the inner-city poor are the most visible and most publicized underclass in the United States, many Americans associate poverty with minority groups or urban areas. In fact, the suburban poor outnumber the urban poor by 1.2 million. Many of the suburban poor were pushed out of the city when factories and other businesses closed; they headed to the suburbs in search of jobs and low-cost housing (Jones 2006).

The rural poor in the United States are another population that needs attention. Demographers William P. O'Hare and Kenneth M. Johnson (2004) estimate that approximately 2.6 million rural children can be classified as underclass. In fact, 48 of the 50 counties with the highest child poverty rates are rural. Like their urban counterparts, members of the rural underclass are concentrated in geographic areas with high poverty rates. They, too, have felt the effects of economic restructuring, including the decline of the farming, mining, and timber industries and the transfer of routine manufacturing out of the United States.

The rural, suburban, and urban poor represent three distinct segments of the U.S. population living below the poverty line. Together, the three underclasses number almost 35.9 million Americans (or 12.5 percent of the population). Included among the poor are 12 million people whose incomes are less than half the amount officially defined as the poverty level (U.S. Bureau of the Census 2004). Public perception of the level of poverty in the United States is not in line with reality, however. For example, a Catholic Campaign for Human Development poll found that most Americans surveyed underestimate the number of poor people in the United States by at least 13 million. In addition, most believed that a family of four needed an income of at least $40,000 a year to cover their daily expenses (O'Brien 2004). This estimate is almost double the figure set by the U.S. government as the poverty threshold for a family of four ($21,000).

The Indebted

Since the 1970s, credit has helped drive the U.S. and global economy, giving people money to spend that they did not have. Many people acquired unmanageable levels of debt, which created a division in society between the debt-free and the indebted. Simply put, *debt* is money owed to another party. People take on debt so they can consume something for which they do not have the money or because they do not want to use money they have on hand. Consumer debt is one way to fuel economic growth because credit puts money in the hands of consumers who purchase goods and services. Some of the most common "short-term" sources of consumer debt are credit cards, payday loans, and other financing arrangements (two years same as cash, no payments for two years). Typically, these short-term credit sources are financed at higher interest rates than are mortgage, car, and student loans, with those borrowers least able to afford credit and to pay off credit card debt each month subject to the highest interest rates. Debt becomes a problem when the borrower cannot make payments or does not have the money to make payments large enough to reduce the overall amount owed. While in the short term, debt temporarily frees borrowers from their financial constraints, it can severely constrain their life chances if it becomes unmanageable. The Pew Research Center conducted a nationally representative sample of 2,000 adults to learn the extent of debt problems in the United States. The survey found that one in seven adults(14 percent) have experienced a debt problem at some point in their lives that was so severe they used a debt consolidation service or declared bankruptcy. That percentage varies by income, with 8 percent of those with household incomes $100,000 or greater experiencing a debt problem and 19 percent of adults earning $30,000 or less experiencing such a problem (see Figure 8.5).

Americans who are late making credit card payments pay an estimated $15 billion in penalty fees a year. One in every five credit card holders carries over debt each month and pays interest rates of 20 percent or more (Baker 2009). President Barack Obama (2009) argues that Americans have a responsibility not to use credit cards to live beyond their means, but he also points out that "We're lured in by ads and mailings that hook us with the promise of low rates while keeping the right to raise those rates at any time for any reason—even on old purchases. You should not have to worry that when you sign up for a credit card you're signing away all your rights. You shouldn't need a magnifying glass or a law degree to read the fine print that sometimes doesn't even appear to be written in English."

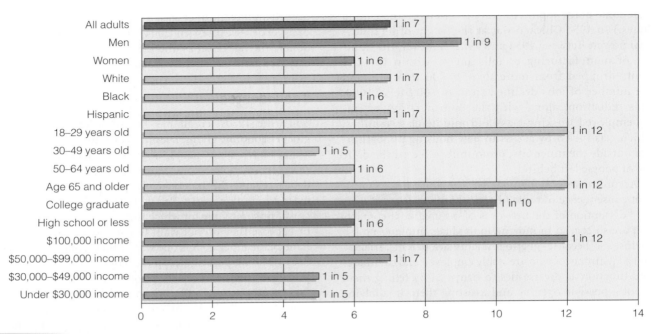

▲ **Figure 8.5**   Chances of Ever Having a Debt Problem

The figure shows the chances of someone having a debt problem by selected characteristics. For this study, a debt problem was defined as debt so severe that a person uses a debt consolidation service or files for bankruptcy. Which category has the lowest chance of having a debt problem? Which category has the highest chance of having a debt problem?

Source: Pew Research Center (2009)

Payday loans represent another type of lending practice that can trap its users in a cycle of debt. Payday loan companies offer credit in the form of cash advances to be repaid when borrowers receive their next paycheck (usually one to two weeks later). Typically, the interest charged is equivalent to an annual percentage rate exceeding 400 percent. Borrowers who fail to repay the loan in full at the designated time can renew the loan for an added fee (plus interest). While the research on payday loans is limited, existing data suggest that a large fraction of payday loan customers roll over their principal multiple times. Political scientist Robert Mayer (2009) sought to create a profile of the person who uses payday loan services. While Mayer could watch the people as they go into payday loan stores to learn something about them, outward appearances can be deceiving because many who use such services appear to be middle or upper class. Mayer thought of an ingenious way to gain insight into this industry, its clients, and lending practices. He realized that when debtors file for bankruptcy, they are required to list the names of creditors from whom they are seeking protection, including the amount of debt and the date the debt was incurred. He examined a sample of 500 bankruptcy petitions filed by residents of one U.S. county—Milwaukee County—in 2004. It is worth noting that there are 66 licensed payday

services in that county, the equivalent of one store for every 10,000 adults. Mayer found that 15.2 percent (*n* = 76) of petitions listed a debt owed to one payday loan company and 10.6 percent listed debts owed to two or more payday loan companies, with some petitions listing as many as nine loans. At least 50 percent of the petitioners were in a situation in which they owed the payday lender(s) more than the amount of their next paycheck. Mayer then created a profile of those filing for bankruptcy who list payday loans as a source of debt (see Table 8.6).

Table 8.6	Profile of a Typical Person Filing a Bankruptcy Petition in Milwaukee County Listing a Debt(s) Owed to a Payday Lender(s)
Gross household income (median)	$26,573
Net monthly income (median)	$1,823
Payday loan debt (median)	$928
Homeowner	25%
Married	30%
Single mother	45%
Resides in majority black neighborhood	53%

Source: Mayer (2009)

■ VISUAL SUMMARY OF CORE CONCEPTS

■ **CORE CONCEPT 1:** When sociologists study systems of social stratification, they seek to understand how people are ranked on a scale of social worth and how that ranking affects life chances.

Every society in the world stratifies its people according to ascribed and achieved statuses. Ascribed statuses are social positions assigned on the basis of attributes people develop over time or possess through no effort or fault of their own; achieved statuses are attained through some combination of personal choice, effort, and ability. Ascribed and achieved statuses may seem clearly distinguishable, but that is not the case. The various achieved and ascribed statuses hold status value and social prestige. Sociologists give special attention to stratification systems in which ascribed statuses shape people's life chances and access to achieved statuses.

T. Scott McLaren

■ **CORE CONCEPT 2:** Dramatic inequalities exist across countries and within countries.

The country into which we are born has important effects on our life chances. For example, the chances that a baby will survive the first five years of life depend largely on the country where it is born. Babies born in Iceland, Sweden, and Singapore have some of the best chances of surviving their first five years; fewer than 3 of every 1,000 babies born there die before reaching the age of five. Babies born in Sierra Leone and Afghanistan have some of the worst chances of surviving that first year; at least 250 of every 1,000 babies born die within the first five years of life. Life chances vary *within* countries as well. In the United States on average 30.9 babies per 1,000 live births die before reaching age five. But the chances of survival vary by racial and ethnic classification.

Official U.S. Marine Corps photo by Cpl. Ruben D. Maestre

U.S. Army photo by Staff Sgt. Russell Lee Klika/ Released

■ **CORE CONCEPT 3:** Caste and class systems of stratification are opposite, extreme points on a continuum. The two systems differ in the ease of social mobility, the relative importance of achieved and ascribed statuses, and the extent to which each restricts interaction among people considered unequal.

Real-world stratification systems fall somewhere on a continuum between two extremes: a caste system (or "closed" system), in which people are ranked by ascribed statuses (over which they have no control), and a class system (or "open" system), in which people are ranked by merit, talent, ability, or past performance.

T. Scott McLaren

■ **CORE CONCEPT 4:** Functionalists maintain that poverty exists because it contributes to overall order and stability in society; it is the mechanism by which societies attract the most qualified people to the most functionally important occupations.

According to functionalists, poverty performs many functions that contribute to overall social order and stability. As one example, the poor have no choice but to take the unskilled, dangerous, temporary, dead-end, undignified, menial work of society at low pay. In addition, functionalists see the unequal distribution of social rewards as the mechanism by which societies ensure that the most functionally important occupations are filled by the best qualified people.

Department of Defense photo by Sgt. James R. Richardson, USMC

■ **CORE CONCEPT 5:** Conflict theorists take issue with the premise that social inequality is the mechanism by which the most important positions in society are filled.

From a conflict point of view, social inequality works ineffectively as a mechanism for filling the most important positions. First, some occupations command large salaries and bring other valued rewards, even though their contributions to society are questionable. Second, it is common for one worker to receive a lower salary than other workers for doing the same job, just because he or she is of a certain race, age, sex, or national origin. Third, occupations that are roughly comparable in social importance are not rewarded accordingly. Fourth, conflict theorists question how much inequality is necessary to ensure that the most qualified and capable people will choose and stay with their occupation—say, that of doctor—instead of choosing some other less critical occupation.

■ **CORE CONCEPT 6:** Symbolic interactionists emphasize how social inequality is communicated and enacted.

Social inequality is also conveyed through symbols that have come to be associated with inferior, superior, and equal statuses. In addition there is a negotiation process by which the involved parties reinforce that inequality in the course of interaction or ignore, challenge, or change it. In the tradition of the symbolic interactionist, journalist Barbara Ehrenreich worked at many different kinds of low-paying jobs as a way of learning about the work, family, and social lives of those who work in jobs with little status, benefits, and pay ($8.00 or less per hour).

■ **CORE CONCEPT 7:** Modernization theory holds that poor countries are poor because they have yet to develop into modern economies and that their failure to do so is largely the result of internal factors such as a country's resistance to free-market principles or the absence of cultural values that drive material success.

Modernization is a process of economic, social, and cultural transformation in which a country "evolves" from preindustrial or underdeveloped status to a modern society in the image of the most developed countries (i.e., western European and North American countries). A country is considered modern when it possesses qualities that make it modern, such as widespread literacy, an urban-centered environment, mechanized farming, and a system of mass communication. Modernization theorists seek to identify the conditions that launch underdeveloped countries on the path to modernization and to identify the stages through which those countries must pass to reach modernization.

■ **CORE CONCEPT 8:** Dependency theory holds that for the most part, poor countries are poor because they are products of a colonial past.

Dependency theorists challenge the basic tenet of modernization theory—that poor countries fail to modernize because they reject free-market principles and because they lack the cultural values that drive entrepreneurship. Rather, dependency theorists argue that poor countries are poor because they have been, and continue to be, exploited by the world's wealthiest governments and by the global and multinational corporations that are based in the wealthy countries. This exploitation began with colonialism, a form of domination in which a foreign power uses superior military force to impose its political, economic, social, and cultural institutions on an indigenous population so it can control their resources, labor, and markets.

■ **CORE CONCEPT 9: Structural responses to global inequality include transferring wealth from highest-income countries to lowest-income countries through foreign aid and fair-trade policies.**

The United Nations plan to reduce global inequality hinges on the world's 22 richest countries (1) investing 0.7 percent (seven-tenths of 1 percent) of their annual gross national income in foreign aid and (2) implementing a trading system that eliminates the subsidies, tariffs, and quotas that put the poorest economies at a disadvantage in the global marketplace. To this point, the richest countries have failed to deliver in both areas. Critics of the UN plan argue that giving aid and eliminating trade barriers are only first steps in addressing inequality. Other problems that create inequality include intense competition with low-wage laborers working in other countries and brain drain.

Department of Defense photo by SSGT Maritza Fernandez, USAF

■ **CORE CONCEPT 10: Social class is difficult to define. It depends on many factors, including people's relationship to the means of production, their sources of income (such as land, labor, or rent), their marketable abilities, their access to consumer goods and services, their status group, and their membership in political parties.**

Lisa Southwick

Sociologists use the term *class* to designate a person's overall economic and social status. Marx alerts us that the key variable in determining social class is source of income. For wage laborers, the source is wages; for capitalists, the source is profit; for landowners, the source is ground rent. The finance aristocracy's source of income is created without labor and without creating a product or service. For Weber, the basis for a person's class also depends on marketable abilities (work experience and qualifications), access to consumer goods and services, control over the means of production, and ability to invest in property and other sources of income. Social class spans a range of categories, with the "negatively privileged" property class at the bottom and the "positively privileged" property class at the top. Class ranking is also affected by the status group and political parties to which people belong.

■ **CORE CONCEPT 11: Economic or occupational restructuring can devastate people, leaving them without jobs and disrupting the networks of occupational contacts crucial to moving affected workers into and up job chains.**

Lance Cpl. Bryan G. Carfrey/United States Marines

Major changes in a society's economic or occupational structure—such as factory closings or a decline in farming—can have life-devastating consequences for the affected groups. Such structural changes in the United States help explain the situation of the inner-city, suburban, and rural poor, whose networks of occupational contacts (so crucial to moving them into and up job chains) have collapsed. We can also think of lending practices in structural terms. Since the 1970s, credit has helped drive the U.S. and global economy, giving people money to spend that they did not have. Many people acquired unmanageable levels of debt, which created a division in society between the debt-free and the indebted.

Resources on the Internet

 Sociology: A Global Perspective Book Companion Web Site

www.cengage.com/sociology/ferrante

Visit your book companion Web site, where you will find flash cards, practice quizzes, Internet links, and more to help you study.

CENGAGENOW™

Just what you need to know NOW!

Spend time on what you need to master rather than on information you have already learned. Take a pre-test for this chapter, and CengageNOW will generate a personalized study plan based on your results. The study plan will identify the topics you need to review and direct you to online resources to help you master those topics. You can then take a post-test to help you determine the concepts you have mastered and what you will need to work on. Try it out! Go to www.cengage.com/login to sign in with an access code or to purchase access to this product.

Key Terms

absolute poverty   200
achieved statuses   201
ascribed statuses   201
brain drain   219
caste system (or "closed" system)   204
class   220
class system (or "open" system)   204
colonialism   214
decolonization   214
esteem   202
extreme wealth   200

finance aristocracy   220
income   222
life chances   200
modernization   212
negatively privileged property class   221
neocolonialism   215
political parties   221
positively privileged property class   221
relative poverty   200

social inequality   200
social mobility   206
social prestige   202
social stratification   200
status group   221
status value   202
urban underclass   224
wealth   222

9 Race and Ethnicity

With Emphasis on the Peopling of the United States (A Global Story)

Sociologists study systems of racial and ethnic classification, which divide people into racial and ethnic categories that are implicitly or explicitly ranked on a scale of social worth. They study the origins of these racial and ethnic categories and their effect on life chances.

DoD photo by Sgt. Andy Meissner, USA

▲ The peopling of the United States is one of the great dramas of human history. The 143 men and women pictured came to the United States legally from many countries but, instead of waiting five years to apply for citizenship, were immediately eligible to apply upon completing military training.

The Global Story of Race and Ethnicity in the United States?

THE PEOPLING OF the United States is a global story and "one of the great dramas of all human history" (Sowell 1981, p. 3). It involved the European conquest of Native American peoples; the annexation of Mexican territory, along with many of its inhabitants (who lived in what is now New Mexico, Utah, Nevada, Arizona, California, and Texas); and an influx of voluntary and involuntary immigrants from practically every part of the world. One-half of Mexico's territories and the peoples residing within them were ceded to the United States at the close of the war with Mexico in 1848. Indentured servants from Europe and Africa were part of a forced transfer of labor to what is now the United States beginning as early as 1619. By 1640 the first formal laws stated that an African could serve a master "for the rest of his natural life." Puerto Ricans, American Samoans, Hawaiians, and other peoples all became part of the United States through a form of domination known as conquest or colonialism.

And then there are the voluntary immigrants—the millions of people who chose to move (often in response to a disaster or crisis) to the United States. One well-known example is the thousands of people who moved from Ireland to the United States in the 1840s and 1850s in response to a potato famine and British discrimination. Another example is Chinese immigrants who came in the 1850s as contracted labor to help build the transcontinental railroad and to work in agriculture and in mines.

One of the most interesting, significant, and long-lasting aspects of this global story is the U.S. government's establishment of a racial and ethnic classification scheme that applied to all who lived in and immigrated to the United States. Perhaps as many as 2,000 distinct groups of Native American peoples, speaking seven different language families, were placed in a single category: "Indian." The millions of voluntary and involuntary immigrants from Europe eventually became "white." The peoples from all of Latin America became "Hispanic." Those of African descent became "black." Those from the Far East, Southeast Asia, and the Indian subcontinent were lumped into the category "Asian." The peoples from Hawaii and other Pacific Islands (such as American Samoa and Guam) were eventually lumped into the category of "Native Hawaiian and Pacific Islander." The categories to which people were assigned reflected and reinforced the prejudices and discrimination of the times and set the tone for race relations then and far into the future.

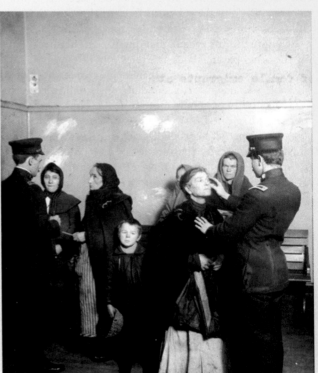

Immigration officers at Ellis Island, New York are shown here giving physical exams to newly arrived immigrants from Europe in 1911.

Library of Congress Prints and Photographs Division

- In the United States, parents and their biological offspring can be classified as belonging to different races.

- In the United States, *Hispanic* refers to an ethnic group, not a race; people of Arab ancestry are classified as white.

- The largest life expectancy gap in the United States is 33 years: 91 years for Asian American females living in Bergen County, New Jersey, versus 58 years for Native American males living in a five-county cluster in South Dakota (Murray et al. 2006).

- An estimated 75 percent of Arab Americans are Christian (U.S. Bureau of the Census 2003).

Race and Ethnicity

■ **CORE CONCEPT 1:** The concepts of race and ethnicity cannot be understood apart from systems of racial and ethnic classification. Most biologists and social scientists have come to agree that race is *not* a biological fact. The reason is that parents from different racial categories can produce offspring. The offspring, by definition, are mixtures of the two categories and therefore cannot be placed in just one category. Despite this obvious fact, in the United States, children of mixed parentage are almost always classified as belonging to the race of just one parent, in effect treating the other parent's race as irrelevant.

While race has no basis in biology, it does have considerable social significance. Sociologists define **race** as a vast collectivity of people more or less bound together by shared and selected history, ancestors, and physical features. These people are socialized to think of themselves as a distinct group, and they are regarded by others as such. In evaluating this definition the emphasis must be placed on *selected*. It is the social significance assigned to sharing certain physical features (such as coarse or straight hair; dark or light skin; and oval, round, or almond-shaped eyes) believed to belong to certain broad catego-

The photo shows a mother who appears white and her two daughters, who appear black. We have been socialized to believe that a few select biological markers are clear indicators of a person's race, even when the facts of ancestry tell us otherwise.

ries of ancestors, such as Africans, Europeans, Asians, and Native Americans. The social significance of race is also a product of emphasizing or feeling connected to a history shared by a certain broad category of ancestors, who were commonly forced by laws and other social practices to become socially distinct from other broad categories of ancestors.

Ethnicity refers to people who share, believe they share, or are believed by others to share a national origin; a common ancestry; a place of birth; distinctive concrete social traits (such as religious practices, style of dress, body adornments, or language); or socially important physical characteristics (such as skin color, hair texture, or body structure). Unlike race, which emphasizes physical features and geographic origin, ethnicity can be based on an almost infinite number of traits.

race A vast collectivity of people more or less bound together by shared and selected history, ancestors, and physical features; these people are socialized to think of themselves as a distinct group, and they are regarded by others as such.

ethnicity People who share, believe they share, or are believed by others to share a national origin; a common ancestry; a place of birth; distinctive concrete social traits (such as religious practices, style of dress, body adornments, or language); or socially important physical characteristics (such as skin color, hair texture, or body structure).

This "black" Iraqi soldier who have just returned from Iraq meets his "white" son for the first time. This image speaks to the product of chance in determining one's physical characteristics. Sociologists ask: given the role of chance, what are the consequences of dividing family members into racial categories based on hair texture and skin color?

No discussion of race and ethnicity can proceed without reference to **systems of racial and ethnic classification**, processes that divide people into racial or ethnic categories that are implicitly or explicitly ranked on a scale of social worth. Emphasis should be placed on *systems*, as most societies if not all, have established racial and ethnic categories and have established rules for placing people in those categories. Until just recently, the U.S. system of racial classification centered around the rule that an individual can belong to only one racial category. Despite evidence to the contrary, the United States has never created formal or legal categories for so-called mixed-race offspring, although many Americans have acknowledged such mixture informally by using (often derogatory) words like "mixed," "half-breed," "mongrel," and "biracial." Instead, such offspring were assigned to a single race category and that category was almost always nonwhite (unless a person could pass as "white").

The Brazilian system of racial classification offers an interesting contrast, as it emphasizes mixture and includes a range of categories—perhaps as many as 100—including "blue" (darkest black, bordering on purple) and "green" (lighter than white) and everything in between. Other categories include *moreno claro* (light-skinned, dark-haired person of primarily European ancestry) and *mulato claro* (light-skinned person of mixed African and European ancestry). Moreover, in Brazil a person's race can change, depending on the situation and with whom he or she is talking (Reeve 1977).

We must remember that racial and ethnic categories and identities are not static; they shift over time. Governments add and remove racial categories; over the past 200 years, the United States has used as few as 3 and as many as 14 racial categories or subcategories. In addition, governments change the labels assigned to racial categories (for instance, from "Negro" to "black").

The U.S. System of Racial Classification

The U.S. government recognizes five official racial categories (plus a sixth category, "Some Other Race"—a catch-all used as a last resort when people resist identifying with one of these five racial categories):

American Indian or Alaskan Native—A person having origins in any of the original peoples of North and South America (including Central America) and who maintain tribal affiliation or community attachment (by some estimates, this category covers more than 2,000 distinct groups).

Asian—A person having origins in any of the original peoples of the Far East, Southeast Asia, or the Indian subcontinent.

Black or African American—A person having origins in any of the Black racial groups of Africa.

Native Hawaiian or Other Pacific Islander—A person having origins in any of the original peoples of Hawaii, Guam, Samoa, or other Pacific Islands.

White—A person having origins in any of the original peoples of Europe, the Middle East, or North Africa.

It is significant that the definition for "Black or African American" omits the words "original peoples" and substitutes "Black racial groups of Africa." If original peoples were included in the definition of "Black or African American," every person in the United States would have to claim this racial category. In the view of evolutionary biologists, all people evolved from a common African ancestor. Moreover, how many people can trace their ancestors to the geographic region in which they originated?

The 2000 census represented the first time in U.S. history that the federal government allowed people to identify themselves as belonging to more than one racial category. Prior to that census, people could claim to be members of only one racial category. If they identified with two or more races, the government "forced" them into a single racial category. The government has yet to decide what to call people who identify with two or more races. One thing is clear, however: the government has stated that it will not classify them as *multiracial* (U.S. Bureau of the Census 1999). This decision seems odd in light of the fact that people can now self-identify with as many as six racial categories (see Table 9.1).

systems of racial and ethnic classification A systematic process that divides people into racial or ethnic categories that are implicitly or explicitly ranked on a scale of social worth.

Table 9.1 Official Racial Categories in the United States

The U.S. government now records data for 63 single- and multiple-race categories, including 6 categories for census respondents who marked one race only and 57 categories for those who marked two or more races. The 63 categories can be collapsed into seven broad categories: (1) "White" alone, (2) "Black or African American" alone, (3) "American Indian and Alaska Native" alone, (4) "Asian" alone, (5) "Native Hawaiian and Other Pacific Islander" alone, (6) "some other race" alone, and (7) "two or more races." There are 15 two-race categories (e.g., "Black or African American-Asian"), 24 three-race categories (e.g., "White-Black or African American-Asian"), 10 four-race categories (e.g., "White-Black or African American-Native Indian and Alaska Native-Asian"), 7 five-race categories, and 1 six-race category.

When given the chance to identify themselves with more than one race, only 6.8 million people (about 1 in 50) did so. Why, do you think, did almost everyone—97.6 percent—self-identify with one racial category?

Number of Racial Categories Used to Self-Identify	Number of People	Percentage of Total Population
One	274,595,678	97.6
Two or more	6,826,228	2.4
Two only	6,368,075	2.3
Three only	410,285	0.1
Four only	38,408	—
Five only	8,637	—
Six only	823	—
Total	281,595,678	100

Note: Dash indicates number too small to calculate a percentage from.

Source: U.S. Census Bureau (2000).

The Problem with Racial Categories

■ **CORE CONCEPT 2:** That governments and other powerful groups have created illogical racial and ethnic categories and have worked hard to present them as logical is one measure of their social importance. Trying to assign people to one racial category has many shortcomings, which immediately become evident when we imagine trying to classify the more than 6.6 billion people in the world.

First, many people do not fit easily into a single racial category, because no sharp dividing line distinguishes characteristics such as so-called black skin from white skin or tightly curled hair from wavy hair. The lack of an obvious line, however, has not discouraged people from trying to devise ways to make the dividing lines seem clear-cut. For example, a hundred years ago in the United States, some churches "had a pinewood slab on the outside door . . . and a fine tooth comb hanging on a string" (Angelou 1987, p. 2). People could go into the church if they were no darker than the pinewood and if they could run the comb through their hair without it snagging. At one time

in South Africa, the government board that oversaw racial classification used a pencil test to classify individuals as white or black. If a pencil placed in a person's hair fell out, the person was classified as white (Finnegan 1986).

A second problem with trying to assign people to one racial category is that boundaries between races can never be clear-cut if only because males and females of any alleged race can produce offspring together. Millions of people in the world have mixed ancestry and possess physical traits that make it impossible to assign them to just one racial category. The U.S. news media often depict mixed ancestry as a recent phenomenon connected to the dismantling of laws in 1967 forbidding interracial marriages and a subsequent societal openness to interracial marriages (which produce so-called mixed-race children). Since colonial days, however, intermixture has been common. Clearly, the offspring of such unions cannot fit into a single racial category, and yet most of these people became accepted as, or came to identify with, just one race (Pollitzer 1972).

Booker T. Washington, an African American leader and educator, was born into slavery in 1856 as the son of a white father and and enslaved black mother. In his autobiography Washington wrote of his father, "I have heard reports to the effect that he was a white man who lived on one of the near-by plantations" (Washington 1901).

Library of Congress Prints and Photographs Division

Table 9.2 Number and Percentage of U.S. Population Classified as Black and Mulatto by Census Year

Census Year	"Negro" Population			Percentage of Total "Negro" Population	
	Total	Black	Mulatto	Black	Mulatto
1910	9,827,763	7,777,077	2,050,688	79.1	20.9
1890	7,488,676	6,337,980	1,132,060	84.8	15.2
1870	4,880,009	4,295,960	584,049	88.0	12.0
1860	4,441,830	3,853,467	588,363	86.8	13.2
1850	3,638,808	3,233,057	405,751	88.6	11.2

Source: U.S. Census Bureau (1910).

Evidence of this intermixing before 1967 is reflected in U.S. census data from 1850 through 1910 (see Table 9.2). Notice that in 1910, 20.9 percent of the "Negro" population was subclassified as "mulatto," and yet, despite their mixed ancestry, "mulattos" were still part of the "Negro" racial category, not the "White" or "Native American" category. Note also that the percentage of the "Negro" population subclassified as "mulatto" increased from 11.2 percent in 1850 to 20.9 percent in 1910, the last census in which the U.S. government employed the term "mulatto."

A third shortcoming of single-race classification is that rules or guidelines for placing people in racial categories are often vague, contradictory, unevenly applied, and subject to change. Consider that for the 1990 U.S. census, coders were instructed to classify as "White" respondents who classified themselves as "White-Black or Negro" and to classify as "Black or Negro" respondents who classified themselves as "Black or Negro-White" (U.S. Census Bureau 1994). The National Center for Health Statistics (1993) has changed its guidelines for recording race on birth and death certificates. Before 1989 a child born in the United States was designated as "White" if both parents were white; if only one parent was white, the child was classified according to the race of the nonwhite parent; if the parents were of different nonwhite races, the child was assigned to the race of the father. If the race of only one parent was known, the child was assigned to the race of that parent. After 1989 the rules for classifying newborns changed, so that a child is now assigned to the race of the mother (Lock 1993)—as if identifying the mother's race presents no challenges.

While the U.S. government has tried to address the problem of forcing people into one racial category, it has not dealt with a fourth shortcoming of the approach: Most classification schemes force large aggregates of people who vary in ethnicity, language, generation, social class, and physical appearance under one umbrella category. People expected to identify themselves as "Asian" include those with ancestors from Cambodia, India, Japan, Korea, Malaysia, Pakistan, Siberia, the Philippine Islands, Thailand, Vietnam, and dozens of other Far Eastern or Southeast Asian countries. Similar heterogeneity exists within populations labeled as "Black or African American," "White," "American Indian and Alaska Native," and "Native Hawaiian and Other Pacific Islander."

The U.S. System of Ethnic Classification

In addition to being assigned to one or more racial categories, everyone in the United States is assigned to one of two ethnic categories: (1) "Hispanic or Latino" and (2) "Not Hispanic or Latino." The government defines a "Hispanic or Latino" as "a person of Mexican, Puerto Rican, Cuban, Central or South American culture or other Spanish culture or origin, regardless of race" (U.S. Office of Management and Budget 1977, 1997). The term applies to people from, or with ancestors from, 21 Latin American countries (excluding Brazil), each country consisting of peoples with distinct histories, cultures, and languages. While "Blacks or African Americans" from many ethnic backgrounds have been forced into one racial category, ethnically diverse peoples from 21 countries have been forced into one ethnic category, with the qualification that they may be of any race.

Most people still commonly known in the United States as "Hispanics" do not define themselves as such (Novas 1994). The label *Hispanic* is confusing, because it forces people to identify themselves with conquistadors and settlers from Spain, who imposed their culture, language, and religion on indigenous people and on the African peoples they enslaved. "For Latin Americans, who, like North Americans, fought hard to win their independence from European rule, identity is derived from their native lands and from the heterogeneous cultures that thrive within their borders" (Novas 1994, p. 2).

To complicate matters even further, the history of Latin America is intertwined with that of Asia, Europe, the Middle East, and Africa. Because of this interconnected past, the countries of Latin America are populated not by a homogeneous group known as "Hispanics," but by native- and foreign-born persons, immigrants, nonimmigrant residents, and persons from every conceivable ancestry, not just Spanish ancestry (Toro 1995).

According to the U.S. system of ethnic classification, "Hispanics or Latinos" can be of any race or combination of races (see Table 9.3). The government does not recognize multi-ethnic origins. In other words, people with one "Hispanic or Latino" parent and one who is not cannot officially claim both heritages. Rather, the government considers such people to share their mother's ethnicity only. If the mother happens to be both "Hispanic or Latino" and not Hispanic, the child is assigned the first ethnicity the mother names when asked her ethnic background. Thus, people who claim that their mother is both "not Hispanic or Latino" and "Hispanic or Latino" are classified differently from those who claim that their mother is both "Hispanic or Latino" and "not Hispanic or Latino." Just

as "Hispanic or Latino" is applied as an umbrella term to people of every possible race, ethnic ancestry, and culture, so is the term "not Hispanic or Latino," which includes anyone who is not Hispanic or Latino, regardless of their race, ethnic ancestry, and culture.

This critical examination of the racial and ethnic categories used by the U.S. government might lead you to conclude that racial and ethnic categories tell us nothing meaningful about the people assigned to them. If anything, the existence of these categories tells us that they are a product of at least three factors: chance, context, and choice.

Under the U.S. system of ethnic classification, everyone pictured is "Hispanic or Latino." What race would you assign to each person pictured?

Department of Defense photo by SSG Helen Miller, USA

© AbleStock/Index Open

© Justin C. McIntosh

Department of Defense photo by Rilet Hansom Jr., Civ.

Table 9.3   U.S. Hispanic or Latino Population Classified by Race

The U.S. government officially recognizes two ethnic categories: (1) "Hispanic or Latino" and (2) "not Hispanic or Latino." Hispanics or Latinos can be of any race. The table below shows the percentage and number of Hispanics or Latinos by racial category. Why, do you think, did 47.9 percent of these people classify themselves as "White" and 42.2 percent classify themselves as "Some Other Race"?

Racial Category Hispanics/Latinos Identified Themselves As	Number of Hispanics/Latinos	Percentage of Hispanic/Latino Population
White Only	16,907,852	47.9
Black Only	710,000	2.0
American Indian or Alaska Native Only	407,073	1.2
Asian	119,829	.3
Native Hawaiian or Other Pacific Islander	45,326	0.1
Some Other Race	14,891.303	42.2
More Than One Race	2,224,082	6.3
Total	35,305,818	100

Source: U.S. Census Bureau (2001a).

The Roles of Chance, Context, and Choice

■ **CORE CONCEPT 3:** The racial and ethnic categories to which people belong are a product of three interrelated factors: chance, context, and choice. **Chance** is something not subject to human will, choice, or effort. We do *not* choose our biological parents, nor can we control the physical characteristics we inherit from them. **Context** is the social setting in which racial and ethnic categories are recognized, created, and challenged. **Choice** is the act of choosing from a range of possible behaviors or appearances. The choices one makes may emphasize or reject the behaviors and appearances that have come to be associated with a racial or ethnic group. As we will see, individual choices are constrained by chance and context (Hanley-Lopez 1994).

In evaluating the relative importance of chance, context, and choice, consider the cases of several prominent "black" Americans. By chance, former Secretary of State Colin Powell is the son of Jamaican immigrants to the United States with "African, English, Irish, Scottish, Jewish, and probable Arawak Indian ancestry" (Gates 1995, p. 65). Similarly, by chance, the golfer Eldrick "Tiger" Woods is the son of a mother who is half Thai, a quarter Chinese, and a quarter "white." Woods's father is half "black," a quarter Chinese, and one-quarter American Indian (Page 1996, p. 285). Both Powell and Woods live in a country (the context) where the biological facts of ancestry have no bearing on race. By chance, both men inherited physical features that place them in the category "black."

Powell and Woods are also part of a society (context) that gives people in some racial categories more freedom

Study the physical features of this American family. What roles do chance, context, and choice play in determining how people perceive them and classify them?

Photographed by Marty Lueders for the U.S. Census Bureau

chance Something not subject to human will, choice, or effort; it helps determine a person's racial and ethnic classification.

context The social setting in which racial and ethnic categories are recognized, constructed, and challenged.

choice The act of choosing from a range of possible behaviors or appearances; a person's choices may evoke associations with a particular race or ethnic group.

Hawaii-born President Barack Obama is the son of a Kenyan immigrant father and Kansas-born mother. In his autobiography Obama described his father as "black as pitch" and his mother as "white as milk." Obama writes: "When people who don't know me well, black or white, discover my background (and it usually is a discovery, for I ceased to advertise my mother's race at the age of twelve or thirteen, when I began to suspect that by doing so I was ingratiating myself to whites), I see the split-second adjustments they have to make" (1995, p. xv).

James A. Baker III, a high-ranking Cabinet official in the Reagan, Ford, and Bush administrations, and most recently co-chair of the *Iraq Study Group*, also known as the Baker-Hamilton Commission, learned in 2004 that he has cousins classified as black. One cousin, also named James Baker, broke the news at a political event when he extended his hand, saying, "How do you do, sir? My name is James Baker." The "white" James Baker replied, "That's interesting; my name is James Baker, too." The "black" cousin replied, "I know, I have followed your career a long time. I'm your cousin" (Baker 2006, p. 420).

than those in other racial categories to claim an ethnic identity. Specifically, people racially classified as "white" have a great deal of freedom to identify with one or more ethnic identities (such as Italian, German, and Swedish). Those racially classified as other than "white" have far fewer choices. People classified as "black"—whether they were born in the United States or immigrated here from Haiti, Jamaica, Trinidad, Germany, Kenya, or elsewhere—are pressured in overt and subtle ways to identify as just "black," even when they know or believe they have biological connections to other races and ethnic groups (Waters 1994, 1990).

Powell's and Woods's racial classification camouflages the shared histories and interconnected experiences among the various "races" that came together to eventually produce the two. A system of racial classification that acknowledges only one of their many racial and ethnic ancestries supports an illusion of racial purity and of dis-

tinct racial histories and experiences. In the United States the origins of this illusion can be traced to slavery. In the context of slavery, sexual relationships (usually forced, sometimes willing) between enslaved "black" women and their "white" masters were commonplace. Those sexual relationships produced not only "mixed-race" offspring, but sons and daughters, half-brothers and half-sisters, grandsons and granddaughters (Parent and Wallace 2002). Yet, the system of racial classification assigned the offspring of these unions to the race and the enslaved status of the mother. This system allowed the master to own his offspring, thereby increasing the size of the enslaved population and his wealth.

In light of this context, can Colin Powell present himself as "white"? Can Tiger Woods choose to present himself as "Asian"? When Woods first came on the golf scene, he tried to present himself as "Cablinasian"—a mixture of Caucasian, black, American Indian, and Asian—but as he has pointed out, context trumps choice: "In this country I'm looked at as being black. When I go to Thailand, I'm considered Thai. It's very interesting. And when I go to Japan, I'm considered Asian. I don't know why it is, but it just is" (Woods 2006).

Classifying People of Arab or Middle Eastern Ancestry

It is interesting to note that the U.S. government classifies people of Middle Eastern and Arab ancestry as "white" (Samhan 1997). Yet, in light of the September 11, 2001, attacks (as well as previous attacks), one must question whether people of Arab and Middle Eastern ancestry are really viewed and treated as "white." The U.S. Census Bureau explored the idea of creating a separate racial or ethnic category for this group. In doing so, the bureau asked two questions:

- Should it create a *geographically* oriented racial category called "Middle Eastern" for persons having origins in the Middle East, North Africa, and West Asia? The category would include the Arab states, Israel, Turkey, Afghanistan, Iran, and possibly Pakistan and Asian Indians.
- Should the bureau take a linguistic and cultural approach and create an "Arab–Middle Eastern" *ethnic* category? Like the "Hispanic or Latino" category, the "Arab–Middle Eastern" category would consist of people of any race who share broad cultural traditions and a language.

The Census Bureau struggled with these questions because it recognized that many people of Arab–Middle Eastern descent do not think of themselves as "white"; nor are they treated as such. In addition, it recognized that collecting data on this specific category of people would be useful for civil rights monitoring and enforcement. In the end, however, it decided *against* creating an "Arab" or "Middle Eastern" racial or ethnic category, for several reasons.

First, the Census Bureau maintained that such a category would be too small—perhaps comprising only 1.2 million people—although Arab American groups projected a number as large as 3.5 million. Second, the bureau found little agreement about the geographic meaning of "the Middle East" or about which countries should be considered "Arab" (U.S. Office of Management and Budget 1995).

Despite confusion over the meaning of "Arab" and "Middle Eastern," the U.S. Census Bureau (2003) made history on December 3, 2003, when it released a report on the "population of Americans who indicated an Arab *ancestry* on the 2000 Census." The only source of data the bureau has on the ancestry comes from responses to this question asked every 10 years on the census survey: "What is this person's ancestry or ethnic origin?" Respondents can list as many as two ancestries. The bureau defined as "Arab"

anyone reporting their ancestry to be Algerian, Alhuceman, Arab, Bahraini, Bedouin, Berber, Egyptian, Emirati (United Arab Emirates), Iraqi, Jordanian, Kuwaiti, Kurdish, Lebanese, Libyan, Middle Eastern, Moroccan, North African, Omani, Qatari, Palestinian, Rio de Oro, Saudi Arabian, Syrian, Tunisian, or Yemeni. The Census Bureau pointed out that this definition of "Arab" includes some peoples, such as Kurds and Berbers, who do *not* necessarily identify themselves as Arab, while excluding other peoples who do identify themselves as Arab (such as Mauritanians, Somalis, Djiboutians, Sudanese, and Comoros Islanders) (U.S. Census Bureau 2003). According to the bureau's definition of "Arab," the following celebrities would be classified as such: Steve Jobs (cofounder of Apple, Inc.; Syrian father), Marlo Thomas (actress; father with Lebanese ancestry); Paula Abdul (*American Idol* judge and vocalist; father with Syrian ancestry); and George Allen (67th governor of Virginia; mother with Tunisian ancestry).

Here we might pause and ask, Why does the U.S. government classify people of Middle Eastern and Arab ancestry as "white"? While this question has no clear answer, some critics argue that the Middle East holds important symbolic value that whites (or at least whites who have the power to classify) hope to associate with their race. For example, the Middle East is the birthplace of Christianity and the site of many important biblical cities (such as Jerusalem, in Israel and the Palestinian territories, and Babylon, in Iraq). The Middle East boasts the Egyptian pyramids, considered to be among the wonders of the world. Finally, the Middle East has renowned geographic landmarks associated with the cradle of civilization, including ancient Mesopotamia (the land between the Tigris and Euphrates Rivers, in what is now Iraq).

We might also ask why the Census Bureau's report on the "Arab" population in the United States was historic. It was the first time the government agency published a brief on a subpopulation not officially designated as a minority (Arab Institute 2003). In the United States the officially designated minorities are "Blacks or African Americans," "Hispanics or Latinos," "American Indians and Alaska Natives," "Asians," and "Native Hawaiians and Other Pacific Islanders" (U.S. Census Bureau 2006). Some highlights from this report follow:

- Approximately 1.2 million people reported being of Arab ancestry. Arab American organizations estimated the number to be more than 3 million.
- About half of the Arab population was concentrated in five states: California, Florida, Michigan, New Jersey, and New York.
- Arabs represented 30 percent of the population in Dearborn, Michigan.
- People claiming Lebanese, Syrian, and Egyptian ancestries accounted for about three-fifths of the Arab American population.

Alan Light

Department of Defense photo by Spc. Jeffery Sandstrum, U.S. Army

After the 9/11 attacks, the U.S. Census Bureau released a first-of-a-kind report, *The Arab Population: 2000*. In the popular imagination the report was about people who looked like the woman wearing the headscarf, not like *American Idol* judge and vocalist Paula Abdul.

- An estimated 75 percent of Arab Americans are Christian. Many immigrants from Middle Eastern and Arab countries belonged to religious minorities in their home countries (Hertz 2003).

To this point, we have examined the U.S. system of racial and ethnic classification. In addition to placing U.S. residents in official racial and ethnic categories, the government classifies them as either native or foreign-born.

The Foreign-Born Population

■ **CORE CONCEPT 4:** The legal status of the foreign-born varies by country and is often connected to race and ethnicity. Every country in the world has residents considered **foreign-born**—people living within the political boundaries of

foreign-born People living within the political boundaries of a country who were born elsewhere.

the country who were born elsewhere. The foreign-born can include legally and illegally admitted immigrants, refugees, temporary residents, and temporary workers. The legal status of the foreign-born—particularly their right to pursue citizenship and permanent residence—varies by country. The oil-exporting countries of the Persian Gulf, whose labor force is 40 to 70 percent foreign-born, offer immigrants no possibility of citizenship. The United States, Canada, Australia, Israel, and New Zealand are countries that welcome immigrants as permanent residents and as citizens. Japan and Germany make obtaining such status difficult but not impossible. No government welcomes just anyone; often race, ethnicity, and skills figure into the probability that one will gain admission to a country legally or will have to enter illegally.

In the United States the foreign-born consist of U.S. residents born in a foreign country, to parents who were not U.S. citizens. A native of the United States is someone (1) born on American soil or on the soil of a U.S. Island Area, such as Puerto Rico, or (2) born abroad but to at least one parent who was an American citizen. How do foreign-born differ from natives? First, the foreign-born are immi-

Cities with the Largest Percentage of Foreign-Born Residents

Table 9.4

City	%
Miami, U.S.	59
Toronto, Canada	44
Los Angeles, U.S.	41
Vancouver, Canada	37
New York City, U.S.	36
Singapore City, Singapore	33
Sydney, Australia	31
Abidjan, Ivory Coast	30
London, U.K.	28
Paris, France	23

Source: United Nations (2004).

Lloyd Wolf for the U.S. Census Bureau

Cities, especially the largest cities in the world, represent locations that are disproportionately impacted by immigration. The cities on this list are well-known gateways for immigrants into the United States and elsewhere.

grants who have experienced life in another country and culture. They bring these experiences with them, and the experiences continue to influence them and their relationships with their offspring and others as they live their lives in the United States. Second, their life chances in the United States are shaped by a number of factors associated with their arrival, including the following:

- The nature of their migration: Did they come as political refugees, forced out of their home countries; as voluntary immigrants in search of a better life; or as involuntary immigrants forced into servitude?
- The social and economic status they occupied in their home countries: Did they live in poverty, or were they among the elite, or somewhere in the middle?
- The social atmosphere that greets them: Did they arrive during an economic boom or bust? Are there services to ease their adjustment to the new culture? Do locals resent their arrival? (Pedraza 1999).

More than 33 million (33.5 million) residents of the United States are counted as foreign-born. That number represents 11.7 percent of the total U.S. population and 15 percent of the civilian labor force. More than half (53%) of all foreign born residents were born in a Central American,

Caribbean, or South American country); another 25 percent were born in an Asian country (U.S. Census Bureau 2004; U.S. Department of Labor 2006). The seven states with the largest percentage of foreign-born residents were California (25 percent), New York (20 percent), Hawaii (18 percent), Florida (16 percent), New Jersey (15 percent), Arizona (14 percent), and Texas (11 percent). In some U.S. cities, foreign-born residents make up 36 percent or more of the population (see Global Comparisons: "Cities with the Largest Percentage of Foreign-Born Residents).

If we take a long view of U.S. history (since 1820, when the U.S. government began keeping such records), we see that of the top 10 countries from which people have immigrated to the United States, 6 are European (see Figure 9.1). Of the 70.8 million people who have immigrated to the United States and obtained permanent legal status since 1820, 38.9 million (55 percent) have come from these six European countries, whereas only 980 thousand (1.4 percent) have come from African countries. In evaluating the last number, keep in mind that by 1798 it was illegal to import slaves to the United States. The best estimate indicates that between 1619 and 1798 the British colonies and the United States imported 500,000 slaves. By 1860, according to that year's federal census, 4 million slaves

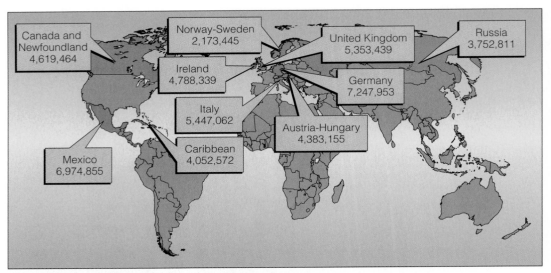

▲ **Figure 9.1** Number of Immigrants to the United States, 1820–2005: Top 10 Countries of Origin

Source: U.S. Department of Homeland Security (2007).

were living in the United States, which suggests that the slave population (with the help of masters and other people not enslaved) reproduced itself (U.S. Bureau of the Census 1860).

The Role of Race and Ethnicity in Immigration Policy

Historically, race and ethnicity considerations have played a major role in U.S. immigration policy. Legislation that focused on race and ethnicity to curb the numbers and types of immigrants entering the United States includes the Chinese Exclusion Act of 1882 and the Immigration Act of 1924. The Chinese Exclusion Act of 1882 prohibited the entry of Chinese laborers into the United States for 10 years; it was then renewed in 1892, requiring Chinese immigrants already in the United States to carry resident permits. In 1943, legislation opened the door a crack by allowing 105 Chinese to immigrate to the United States each year.

The Immigration Act of 1924 established a quota system that set numerical limits on immigration, based on national origin. The quota was set at 2 percent of the number of people from each nation already living in the United States at the time of the 1890 census. Note that the quota was not based on the most current census data at the time (from 1920). Because very few southern, central, and eastern Europeans—and virtually no Asians—had immigrated to the United States by 1890, people from these regions faced the greatest immigration restrictions. In 1965 Congress abolished the discriminatory national origin quotas. This act opened the door to immigrants from Latin America and Asia.

Other immigration legislation has focused on specific racial and ethnic groups. The Bracero Program, which began in 1942, allowed Mexicans to work legally in the United States to relieve labor shortages in rural areas and to bolster the American workforce during World War II. The Mexican laborers, known as braceros, replaced American workers who were needed to work in defense plants or

Farm Security Administration - Office of War Information Photograph Collection, Library of Congress Prints and Photographs Division

This photo from 1942 shows men standing in line outside a soccer stadium in Mexico City, seeking to sign up for work in the United States under the Bracero Program. Some stood in line for five days and nights.

NO BORDERS, NO BOUNDARIES

Foreign-Born by Region of Birth

IN THE UNITED States 33.5 million residents (11.7 percent of the total population) are classified as foreign-born. The following figure shows the percentage of foreign-born U.S. residents by region of birth. Which region of the world accounts for the greatest percentage of foreign-born? Which region contributes the lowest percentage?

Source: U.S. Census Bureau (2004).

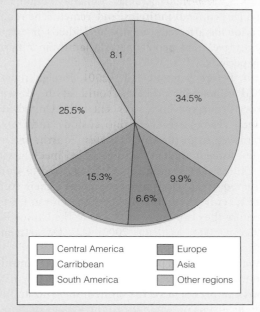

Central America   ■ Europe
Carribbean   ■ Asia
South America   ■ Other regions

▲ **Figure 9.2** Percentage of Foreign-Born Residents Living in the United States by Region of the Birth.

to serve in the armed forces. After World War II the Bracero Program was extended several times to supplement the U.S. workforce during the Korean War and to provide laborers for agricultural and other low-wage work. The program ended in 1965.

The Immigration Reform and Control Act of 1986 permitted illegal workers in the United States to apply for amnesty if they could prove they had worked here for at least 90 days between May 1, 1985, and May 1, 1986. Under this law, 1.3 million illegal workers were granted amnesty and legal status. Despite this amnesty program, an estimated 12 million undocumented immigrants live in the United States (Kochhar 2006).

After the terrorist attacks of September 11, 2001, the U.S. government made many changes to immigration law and procedures. First, the U.S. Immigration and Naturalization Service (INS) was renamed the Bureau of Citizenship and Immigration Services (BCIS) and assigned to the newly established Department of Homeland Security. Second, nationals from selected countries identified as presenting an elevated national security concern had to

Department of Defense photo by Sgt. 1st Class Doug Sample

A Jamaican-born servicewoman and a Dominican-born serviceman recite the pledge of allegiance during U.S. citizenship ceremonies.

register with the Department of Homeland Security to verify their location, activities, and expected departure date. Those countries are Afghanistan, Algeria, Bahrain, Bangladesh, Egypt, Eritrea, Indonesia, Iran, Iraq, Jordan, Kuwait, Lebanon, Libya, Morocco, North Korea, Oman, Pakistan, Qatar, Saudi Arabia, Somalia, Sudan, Syria, Tunisia, United Arab Emirates, and Yemen. The required registration has affected students, individuals on extended business travel, and people visiting family members for lengthy periods.

Third, before September 11, 2001, foreign nationals who had minor visa problems could, at the discretion of border or airport inspectors, enter the United States. Now such visitors are taken into custody and immediately removed from the United States. Fourth, under the old rules, foreign students who attended medical school in the United States could stay if they agreed to work in underserved areas. Now, after completing their studies, they must return to their home countries for at least two years before they can apply to return to the United States (About.com 2004; Madison 2002; U.S. Department of Homeland Security 2003).

A final example of a post-9/11 change in U.S. immigration policy is Executive Order 13269, signed on July 3, 2002. It allows the president of the United States to authorize expedited citizenship while the country is at war. Foreign-born members of the U.S. armed forces (including National Guard and reserve personnel) who have served any amount of time beginning on or after September 11, 2001, are immediately eligible to become U.S. citizens (Gilmore 2004).

The Long-Standing Debate over Immigration

In the United States, immigration policy has always inspired debate. Generally, the debate has centered on economic considerations—whether immigrants pose an economic threat to American workers, taking the workers' jobs away or depressing their wages (especially the wages of low-skill workers). The resolution to this debate is uncertain. Consider the case of California, a state that has attracted about 25 percent of all immigrants who have ever come to the United States:

- Between 1900 and 1970 California's population grew by attracting American-born migrants from across the United States, especially the East Coast.
- From 1970 onward, California's population growth came from an influx of immigrants.
- The influx of immigrants, combined with skyrocketing housing prices, discouraged native-born Americans from moving to California.
- To attract native-born Americans to California, corporations and other business would have had to raise wages to compensate for the higher housing costs.

- The large influx of immigrants willing to rent and live several to a room meant that many businesses did not have to raise wages to counter rising housing prices.
- After 1990, American-born workers moved out of California at about the same rate as foreign-born workers moved in.

One might argue that California's shift to foreign-born labor was good for business. After all, corporations and other businesses made money, and eight million immigrants were better off than they had been in their Asian and Latin American home countries. On the other hand, the low-wage American-born workers, who eventually had no choice except to move out of the state, would likely label the labor shift as devastating and disruptive (Jencks 2001).

Some states and metropolitan areas across the United States are working diligently to attract immigrants, hoping that the immigrants will replenish the population, jump-start economic activities in depressed neighborhoods, offset labor shortages, and inject greater ethnic diversity into specific areas (Schmitt 2001). As a case in point, Pittsburgh lost 9.5 percent of its population between 1990 and 1999. Other metropolitan areas, such as Louisville and Albuquerque, suffered substantial population losses as well. All are working to make their cities immigrant-friendly, to encourage foreign college students to stay after graduation, to integrate immigrants into the workplace, and to open resource centers. Critics of these efforts argue that the United States is overpopulated and that the real need is better distribution of the people already living here (Schmitt 2001).

The Consequences of Racial and Ethnic Classification

Sociologists are particularly interested in the ways that race and ethnic categories affect life chances. Table 9.5 shows dramatic differences between racial groups with regard to the chances of living a long life, surviving the first year of life, dropping out of high school, earning a living wage, and so on. The comparisons show that the most advantaged racial categories are Asian and white and the least advantaged are black and Native American.

In evaluating the information in Table 9.6, keep in mind that the data describe the nation as a whole; they do not account for local, state, and regional variations. For example, consider that on a national level, people classified as white have the lowest chance of dropping out of high school overall but that the rate varies by locality from 77 percent (Cleveland City School District) to 1 percent (Albuquerque Public Schools). For Hispanics or Latinos the drop-out rate ranges from 77 percent (Cleveland City School District) to 27 percent (Montgomery County Public Schools). The corresponding drop-out rates for blacks

Table 9.5 Differences in Life Chances by Race and Sex

Chance of ...	Highest Chance	Lowest Chance	Difference
living a long life (life expectancy)	Asian Female 86.7 years	Black Male 69.8 years	16.9 years
dropping out of high school*	Black 44%	White 22%	22 %
going to prison	Black Male 32.5%	White Female 1%	31.5%
earning a high median weekly income (working full-time)	Asian Male $825	Black Female $499	$326 per week
dying before reaching one year of age (per 100,000)	Black Male 1,410.2 per 100,000	Asian Female 427 per 100,000	983.2 babies per 100,000
living in poverty*	Native American 24.5%	White (nonHispanic) 8.7%	15.8%
having no health insurance*	Native American 35%	White 11.9%	23.1%

Note: *Data not available for sex categories.

Sources: U.S. Department of Labor (2006); Murray (2006); Manhattan Institute (2001); U.S. Department of Justice (2007); U.S. Bureau of the Census (2005).

are 74 percent (Cleveland City School District) to 15 percent (Boston School District) (Greene 2002).

We can extend this contrast beyond high school dropout rates. The largest life expectancy gap in the United States is 33 years (91 years versus 58 years), between Asian American females living in Bergen County, New Jersey, and Native American males living in a five-county cluster in South Dakota. Using this contrasting assessment of life chances, Harvard researchers divided the United States into population clusters based on geographic region, race, and population density (rural versus urban) and identified what they call "eight Americas" (see Table 9.6). Again we see that the most advantaged racial categories, as measured by average per capita income, are whites and Asians. The least advantaged are Native Americans and blacks.

Minority Groups

■ **CORE CONCEPT 5:** Minorities are populations that are systematically excluded (whether consciously or unconsciously) from full participation in society and denied equal access to positions of power, prestige, and wealth. **Minority groups** are subpopulations within a society that can be distinguished from members of the dominant group by visible identifying characteristics, including physical and cultural attributes. Members of minorities are regarded and treated as inherently different from those in the dominant group. They are systematically excluded (whether consciously or unconsciously) from full participation in society and denied equal access to positions of power, prestige, and wealth. Consequently, members of minority groups tend to be concentrated in inferior political and economic positions and to be isolated socially and spatially from members of the dominant group.

Based on these characteristics, many groups can be classified as minorities, including some racial, ethnic, and religious groups, women, the very old, the very young, and the physically different (for example, visually impaired people or overweight people). Although we focus on ethnic and racial minorities in this chapter, the concepts described here can be applied to any minority.

Sociologist Louis Wirth (1945) made a classic statement on minority groups, identifying a number of essential traits that are characteristic of all minority groups. First, membership is involuntary: As long as people are free to join or leave a group, no matter how unpopular it is, they do not constitute a minority. This trait is quite controversial, because the meaning of "free to join or leave" is unclear. For example, if very light-skinned people of African descent can pass as white, are they really "free" to leave the African connections in their life?

A second characteristic of minority groups is that their minority designation is not necessarily based on numbers; that is, a minority may be the numerical majority

minority groups Subgroups within a society that can be distinguished from members of the dominant group by visible identifying characteristics, including physical and cultural attributes. These subgroups are systematically excluded, whether consciously or unconsciously, from full participation in society and denied equal access to positions of power, privilege, and wealth.

Table 9.6 Life Chances for Eight Population Subgroups in the United States

Population Subgroup	Population (in millions)	Per Capita Income*	Percent Not Completing High School
Western Native Americans	1	$10,029	31
Southern Rural Low-Income Blacks	5.8	$10,463	49
High-Risk Urban Blacks	7.5	$14,800	28
Black Middle America	23.4	$15,412	25
Appalachian and Mississippi Valley Low-Income Whites	16.6	$16,390	28
Northland Rural Low-Income Whites	3.6	$17,758	17
Asians	10.4	$21,566	20
Middle America (Mostly Whites)	214	$24,640	16

Note: *Per capita income is the total income for the subgroup divided by its size. The resulting number is the average income for each man, woman, and child in a specific population subgroup.

Source: Murray et al. (2005)

in a society. The key to minority status, then, is not size but access to and control over valued resources. South Africa presents one obvious example: Even under the post-apartheid multiracial government, roughly 9 percent of the South African population—the people classified as white—control access to valued resources. As one measure of inequality, 32 percent of South Africans in the highest income category (the top 10 percent) are classified as black. Since blacks make up 80 percent of South Africa's population, one would expect 80 percent of the people in the highest income category to be black (Nduru 2007).

A third, and the most important, characteristic of minority groups is nonparticipation in the life of the larger society. That is, minorities do not enjoy the freedom or the privilege to move within the society the way the dominant group does. Sociologist Peggy McIntosh (1992) identifies a number of privileges that most members of the dominant group take for granted, including the following:

- I can, if I wish, arrange to be in the company of people of my race [or ethnic group] most of the time.
- If I should need to move, I can be pretty sure of renting or purchasing housing in an area which I can afford and in which I would want to live.
- I can go shopping alone most of the time, fairly well assured that I will not be followed or harassed by store detectives.
- I can be late to a meeting without having the lateness reflect on my race [or ethnicity].
- Whether I use checks, credit cards, or cash, I can count on my skin color not to work against the perception that I am financially reliable. (pp. 73–75)

The final and most troublesome characteristic of minority groups is that people who belong to them are "treated

This woman possesses ascribed characteristics that allow her to shop most, if not all the time, without feeling store security is following or watching her.

as members of a category, irrespective of their individual merits" (Wirth 1945, p. 349) and often irrespective of context. In other words, people outside a minority group focus on the visible characteristics that identify someone as belonging to a minority. These visible characteristics become the focus of interaction, as in the scene described here, in which a woman can focus on only the physical features of the people she encounters, to the neglect of what is happening around her:

> "It obsesses everybody," declaimed my impassioned friend, "even those who think they are not obsessed. My wife was driving down the street in a black neighborhood. The people at the corners were all gesticulating at her. She was very frightened, turned up the windows, and drove determinedly. She discovered, after several blocks, she was going the wrong way on a one way street and they were trying to help her. Her assumption was they were blacks and were out to get her. Mind you, she's a very enlightened person. You'd never associate her with racism, yet her first reaction was that they were dangerous" (Terkel 1992, p. 3)

The characteristics that Wirth identifies as associated with minority group status indicate that minorities stand apart from the dominant culture. Some people argue that minorities stand apart because they do not wish to be assimilated into mainstream culture. To assess this claim, we turn to the work of sociologist Milton M. Gordon, who has written extensively on assimilation.

Assimilation

■ **CORE CONCEPT 6:** Social and cultural differences between racial and ethnic groups "disappear" when one group is absorbed into another group's culture and social networks or when two groups merge to form a new, blended culture. **Assimilation** is a process by which ethnic and racial distinctions between groups disappear because one group is absorbed into another group's culture or because two cultures blend to form a new culture. Two main types of assimilation exist: absorption assimilation and melting pot assimilation.

Absorption Assimilation

In **absorption assimilation**, members of a minority group adapt to the ways of the dominant group, which sets the standards to which they must adjust (Gordon 1978). According to Gordon, absorption assimilation has at least seven levels. That is, an ethnic or racial minority is completely "absorbed" into the dominant group when it goes through all of the following levels:

1. The minority abandons its culture (language, dress, food, religion, and so on) for that of the dominant group.
2. The minority enters into the dominant group's social networks and institutions.
3. The minority intermarries and procreates with members of the dominant group.
4. The minority identifies with the dominant group.
5. The minority encounters no widespread prejudice from members of the dominant group.

These women attended a school in Bismarck, North Dakota, run by the U.S. Bureau of Indian Affairs. The purpose of such schools was to "Americanize," or assimilate, the Native American peoples of the United States. To further this goal, the schools required students to speak English and wear uniforms.

6. The minority encounters no widespread discrimination from members of the dominant group.
7. The minority has no value conflicts with members of the dominant group.

Gordon advances a number of hypotheses to explain how the various levels of assimilation relate to one another. First, he maintains that level 1 assimilation is likely to take place before the other six levels of assimilation are achieved. He also states, however, that even if an ethnic or racial group abandons its culture (level 1), it does not always lead to the other levels of assimilation.

Gordon proposes that if the dominant group permits people from ethnic and racial minorities to join its social cliques, clubs, and institutions on a large enough scale (level 2 assimilation), a substantial number of interracial or interethnic marriages are bound to occur (level 3 assimilation):

> If children of different ethnic backgrounds belong to the same play group, later the same adolescent cliques, and at college the same fraternities and sororities; if the parents belong to the same country club and invite each other to their homes for dinner; it is completely unrealistic not to expect these children, now grown, to love and to marry each other, blithely oblivious to previous ethnic extraction. (Gordon 1978, pp. 177–178)

assimilation A process by which ethnic or racial distinctions between groups disappear because one group is absorbed into another group's culture or because two cultures blend to form a new cultural system.

absorption assimilation A process by which members of a minority group adapt to the ways of the dominant culture.

Of the seven levels of assimilation, Gordon believes that gaining access to the dominant group's social networks and institutions is the most important; if that occurs, the other levels of assimilation inevitably follow. Yet, in practice, gaining such access is very difficult. Why? Because members of ethnic or racial minorities are kept apart from the dominant group through **segregation**—the physical or social separation of categories of people. In fact, all of the important and meaningful primary relationships that are close to what Gordon calls "the core of personality and selfhood" are confined largely to people of the same racial or ethnic group:

> From the cradle in the sectarian hospital to the child's play group, the social clique in high school, the fraternity and religious center in college, the dating group within which he [or she] searches for a spouse, the marriage partner, the neighborhood of his [or her] residence, the church affiliation and the church clubs, the men's and the women's social and service organizations, the adult clique of "marrieds," the vacation resort, and then, as the age cycle nears completion, the rest home for the elderly and, finally, the sectarian cemetery—in all these activities and relationships which are close to the core of personality and selfhood—the member of the ethnic group may if he wishes follow a path which never takes him [beyond] the boundaries of his [or her] ethnic structural network. (Gordon 1978, p. 204)

This scenario applies especially to the situation of **involuntary minorities**—ethnic or racial groups that did not at first choose to be a part of a country; rather, they were forced to become part of it by slavery, conquest, or colonization. Native Americans, African Americans, Mexican Americans, and Native Hawaiians are examples of involuntary minorities in the United States. Unlike **voluntary minorities**, whose members come to a country expecting to improve their way of life, members of involuntary minorities first immigrated with no such expectations. Their initial forced incorporation involves a loss of freedom and status (Ogbu 1990).

segregation The physical or social separation of categories of people.

involuntary minorities Ethnic or racial groups that were forced to become part of a country by slavery, conquest, or colonization.

voluntary minorities Racial or ethnic groups that come to a country expecting to improve their way of life.

melting pot assimilation Cultural blending in which groups accept many new behaviors and values from one another. The exchange produces a new cultural system, which is a blend of the previously separate systems.

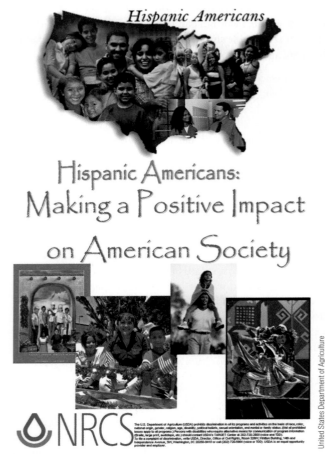

The U.S. Department of Agriculture issued this poster celebrating Hispanic Heritage Month. According to the U.S. Bureau of the Census, people of Hispanic or Latino origin constitute 15 percent of the total population.

Melting Pot Assimilation

Assimilation need not be a one-sided process in which a minority group disappears, or is absorbed, into the dominant group. Ethnic and racial distinctions can also disappear through **melting pot assimilation** (Gordon 1978). In this process, groups accept many new behaviors and values from one another. This exchange produces a new cultural system, which is a blend of the previously separate systems. Melting pot assimilation becomes total when significant numbers of people from different ethnic or racial groups take on each other's cultural patterns, enter each other's social network, intermarry and procreate, and identify with the blended culture.

The melting pot concept can be applied to the various African ethnic groups transported to the British colonies/ United States as slaves. They were "not one but many peoples," who spoke many languages and came from many cultures (Cornell 1990, p. 376). Slave traders capitalized on this diversity: "Advertisements of new slave cargoes frequently referred to ethnic origins, while slave owners often purchased slaves on the basis of national identities and the characteristics they supposedly indicated" (Cornell 1990, p. 376; see also Rawley 1981). Although slave owners and traders acknowledged ethnic differences among Africans, they treated Africans from various ethnic groups as belonging to the same category of people—the enslaved. Because slave traders sold and slave owners purchased individual human beings, not ethnic groups, this treatment broke down ethnic concentrations. In addition, slave owners tended to mix together slaves of different ethnic origins to decrease the likelihood that the enslaved would plot a rebellion. To communicate with one another, slaves invented pidgin and Creole languages.

In addition to developing new languages, the enslaved created a common and distinctive culture based on kinship, religion, food, songs, stories, and other features. The harsh conditions of enslavement, along with the mixing together of people from many ethnic groups, encouraged the enslaved to borrow aspects of one another's cultures and create a new, blended culture.

The melting pot concept can also be applied to interrelationships between various Native American populations and between Native Americans and other groups within the United States. The U.S. Census Bureau attempted to document this assimilation in the 1880 and 1890 censuses. Below is a list of its instructions for coding responses related to Native American identities:

- If this person is of full-blood of this tribe, enter "/."
- For mixture with another tribe, enter name of latter. For mixture with white, enter "W.;" with black, "B.;" with mulatto, "Mu."
- If this is a white person adopted into the tribe, enter "W.A.;" if a negro or mulatto, enter "B.A."
- If this person has been for any time habitually on the reservation, state the time in years or fractions.
- If this person wears citizen's dress, state the time in years or fractions since he or she has habitually so worn it.
- If other than native language is spoken by this person, enter for English, "E.;" Spanish, "S.;" French, "F."

The coding instructions show that the U.S. government, at least at that point in history, was motivated to document the interrelationships between Indians, blacks, and whites. It was also interested in which non-native languages Indians spoke, whether Indians were living on reservations, and even how they dressed (Ferrante and Brown 2001).

The examples of melting pot and absorption assimilation we have discussed are not meant to suggest that the dominant culture as a whole has not been shaped and influenced by minority cultures. Rather, the dominant group has usually failed to learn about or acknowledge

A woman sits between her son and his Native American wife. At this time in history U.S. census takers were interested in classifying Native Americans by their degree of assimilation.

others' contributions to the society in which they all live, thereby deeming minority accomplishments and contributions irrelevant or insignificant.

Stratification theory is one major approach to understanding the forces that work against assimilation (both

This cowboy lived on a South Dakota Indian reservation. He is believed to be of Native American and African descent.

the absorption and melting pot forms) between dominant and minority groups. This theory is guided by the assumption that racial and ethnic groups compete with one another for scarce valued resources. On the whole, the dominant group retains an advantage over competing groups, because its members are in a position to preserve the system that gives them the advantage. Stratification theorists focus on the mechanisms people employ to preserve difference and inequality (Alba 1992). These mechanisms include racist ideologies, prejudice and stereotyping, discrimination, and institutional discrimination.

Racist Ideologies

■ **CORE CONCEPT 7: Racist ideologies claim that biological factors explain and even justify inequalities between racial and ethnic groups.** An **ideology** is a set of beliefs that are not challenged or subjected to scrutiny by the people who hold them. Thus, ideologies are taken to be accurate accounts and explanations of the way things are. On closer analysis, however, ideologies are at best half-truths, based on misleading arguments, incomplete analyses, unsupported assertions, and implausible premises. They "cast a veil over clear thinking and allow inequalities to persist" (Carver 1987, pp. 89–90). One such ideology is racism.

People who embrace racist ideology believe that something in the biological makeup of an ethnic or racial group explains and justifies its subordinate or superior status. Racist ideologies are structured around three notions:

1. People can be classified according to physical characteristics.
2. A close correspondence exists between physical traits and characteristics such as language, dress, personality, intelligence, and athletic ability.
3. Physical attributes such as skin color and hair texture are so significant that they explain and determine behavior and inequalities.

Any racial or ethnic group may use racist ideologies to explain their own or another group's behavior. One indication of a racist ideology is the hypothesis offered by former Los Angeles police chief Daryl Gates to explain why so many blacks have died from restraining chokeholds: Their "veins and arteries do not open up as fast as they do on normal people" (Dunne 1991, p. 26). An alternative explanation is that many police officers, of all racial classifications, tend to handle black suspects more harshly

than they do white suspects. Another example of how racist ideology shapes thinking comes from students.

Many students in the author's class work as waiters and claim that "black" and "white" customers tip differently. In particular, they claim that "black" customers do not tip as well as "white" customers do. As one student wrote, "We talk about it at work all the time; it's that big of an issue. One coworker believes it's black people's way of paying us back for enslaving them in historical times. They must get off seeing us wait on them and they rub it in by stiffing us." To the author's knowledge, no study has ever examined "black" versus "white" tipping, but this difference, if it exists, has other, more plausible explanations. Think about this question: Who are the best tippers? When members of the author's class were asked this, they answered, practically in unison, "former or current waiters." According to the U.S. Census Bureau (2005), only 5.1 percent of waiters are classified as "black" and only 4.1 percent of bartenders are classified as such. One likely explanation for a difference in tipping is that customers classified as "black" have not learned the "norms" about what constitutes a good tip because, until very recently, they were not allowed to wait on tables or tend bar.

On the other hand, the claim that black customers leave smaller tips than white customers might have to do with the former's reaction to waiters' negative attitudes and prejudgments toward them. Waiters may be so sure that they are going to receive a bad tip that they give poor service. That poor service then becomes the reason for the poor tip.

The Origins of Race as a Concept

The premise of racial superiority lies at the heart of other rationalizations used by one group to dominate another. Sociologist Larry T. Reynolds (1992) observes that race as a concept for classifying humans is a product of the 1700s, a time of widespread European exploration, conquest, and colonization that did not begin to subside until the end of World War II. Racist ideology also supported Japan's annexation and domination of Korea, Taiwan, Karafuto (the southern half of the former Soviet island Sakhalin), and other Pacific islands prior to World War II. Both Japanese and Europeans used racial schemes to classify people they encountered; the idea of racial differences became the "cornerstone of self-righteous ideology," justifying their right by virtue of racial superiority to exploit, dominate, and even annihilate conquered peoples and their cultures (Lieberman 1968).

Racial classification was the cornerstone of the Indian Removal Act of 1830, in which Congress authorized the removal, by force if necessary, of Native Americans from their ancestral lands east of the Mississippi to lands to the west. Subsequent laws restricted most Native Americans to reservations. Because the reservations were set aside for

---

ideology A set of beliefs taken to be accurate accounts and explanations of why things are as they are. The beliefs are not challenged or subjected to scrutiny by the people who hold them.

USDA photo by Russell Lee

Jim Crow laws (1876-1965) in the United States enforced so-called separate but equal segregation to the point of even requiring separate water fountains for whites and nonwhites.

a particular category of people, the system of racial classification had to be clear about exactly who belonged in that category.

Racial classification was the cornerstone of the Jim Crow laws, which were enacted in the 1880s and overturned with the ratification of the Civil Rights Act of 1964, the Voting Rights Act of 1965, and the Fair Housing Act of 1968. The Jim Crow laws enforced segregation between blacks and whites and often between whites and other nonwhite groups (Nile Valley Solutions 2001). Examples of such laws from various states follow:

- No person or corporation shall require any white female nurse to nurse in wards or rooms in hospitals, either public or private, in which Negro men are placed. (Alabama)
- It shall be unlawful for colored people to frequent any park owned or maintained by the city for the benefit, use, and enjoyment of white persons . . . and unlawful for any white person to frequent any park owned or maintained by the city for the use and benefit of colored persons. (Georgia)
- All marriages of white persons with Negroes, Mulattos, Mongolians, or Malaya hereafter contracted in the State of Wyoming are and shall be illegal and void. (Wyoming)

Clear evidence of widespread sexual relations between whites, blacks, and Native Americans required a system of racial classification that specified how to classify the offspring of such relations; otherwise, the Jim Crow laws would become increasingly difficult to enforce. For hundreds of years the U.S. government insisted that all

its residents should be placed in single racial categories. This long-standing practice has shaped a common belief that racial categories are accurate reflections of biological reality.

Prejudice and Stereotyping

A **prejudice** is a rigid and usually unfavorable judgment about an outgroup (see Chapter 4) that does not change in the face of contradictory evidence and that applies to anyone who shares the distinguishing characteristics of the outgroup. Prejudices are based on **stereotypes**—inaccurate generalizations about people who belong to an outgroup. Stereotypes "give the illusion that one [group] knows the other" and has the right to control images of the other (Crapanzano 1985, pp. 271–272).

Stereotypes are supported and reinforced in a number of ways. In **selective perception**, prejudiced persons notice only the behaviors or events related to an outgroup that support their stereotypes about the outgroup. In other words, people experience "these beliefs, not as prejudices, not as prejudgments, but as irresistible products of their own observations. The facts of the case permit them no other conclusion" (Merton 1957, p. 424). But many of these supposed facts are unfounded. Many people believe that white men can't jump and are slow, based on the relatively small number of white men who play professional basketball and who are involved in short-distance track events. If one takes the time to look beyond these sports, it is easy to find white athletes who possess leaping ability and speed (see "Selective Perception and Stereotypes of Athletic Ability").

Stereotypes persist in another way: When a prejudiced person encounters a minority person who contradicts a stereotype, the former sees the latter as an exception. The encounter with a minority group member who is "different" merely reinforces the prejudiced person's stereotype. In addition, prejudiced people use "the facts" to support their stereotypes. A prejudiced person can point to the large number of "black" athletes in college and professional basketball as evidence of natural leaping ability and

prejudice A rigid and usually unfavorable judgment about an outgroup that does not change in the face of contradictory evidence and that applies to anyone who shares the distinguishing characteristics of the outgroup.

stereotypes Inaccurate generalizations about people who belong to an outgroup.

selective perception The process in which prejudiced persons notice only the behaviors or events related to an outgroup that support their stereotypes about the outgroup.

Selective Perception and Stereotypes of Athletic Ability

WHICH BASKETBALL PLAYER in this photo do you think is the better athlete? When posed with this question, many of us probably think to ourselves that "blacks" are naturally better athletes than other people, especially at basketball. It is no secret that black athletes dominate collegiate and pro basketball. Why? Over the years many of us have heard explanations for why blacks run faster and jump higher than whites. One popular belief is that plantation owners of the 1700s and 1800s bred their slaves to produce strong laborers. Although this practice might have seemed, from a slave owner's point of view, a logical way to produce strong laborers, it has been dismissed as an accurate description of how slave owners built their labor force. An unknown yet significant number of the enslaved were offspring of owners and enslaved women. These owners repeatedly impregnated their female slaves, thereby enlarging their pool of labor. They were, in essence, enslaving their own children.

Massimo Finizio

Racist folklore has long told of blacks having extra muscles in their legs and other genetic advantages, none having any basis in truth. Even famous "black" athletes have argued publicly that "black" success in sports is due to their physical superiority (Hoberman 2000). The track and field champion Carl Lewis was quoted as saying, "Blacks are just made better" (Malik 2000). Is this belief based on actual physiological evidence or just on the high number of "blacks" participating in the "money sports"— the most visible, best-paid, and most televised sports (basketball, football, baseball, and boxing)?

Athletes and entertainers have always been the most highly publicized "black" achievers, and they have arguably been just as influential in shaping "black" identity as Martin Luther King Jr., Malcolm X, and other "black" political leaders. Studies show that "black" families are four times more likely than "white" families to push their children toward the most visible sports careers (Hoberman 2000). Consequently, many "black" youths chan-

nel all their energies into becoming an athlete in a money sport. A Northeastern University study reported that 66 percent of African American youths between the ages of 13 and 18 believed that they could support themselves as professional athletes (Hoberman 2000).

What about hockey, tennis, gymnastics, swimming, soccer, cycling, sailing, rowing, archery, volleyball, skiing, and other less lucrative "white"-dominated sports? Several of these sports require speed, strength, and jumping ability, and yet "black" representation is almost nonexistent. While it may be true that many "blacks" have fewer opportunities than "whites" to participate in such sports, those who do do not generally rank among the top athletes. If "blacks" actually are superior athletes, then that generalization should hold for any sport. Can we make the case that "black" basketball players are better athletes than "white" gymnasts or that "black" sprinters are better athletes than "white" skiers?

In evaluating the argument that "blacks" are the superior athletes, we must also ask why "black" athletes from African countries do not dominate international competitions, such as the Olympic Games? Consider that African Americans originated from a variety of cultures in western and west-central Africa. Modern West Africans, however, do not run and jump their way to many Olympic medals. Most of the slaves brought to North America came from the areas that are now Ghana, Togo, Benin, and Nigeria, although significant numbers also came from the areas that are now Senegal, Gambia, and Angola (Microsoft 2001). Athletes from these countries have competed in the Olympics for more than 50 years, and during that time, they have won a total of 22 Olympic medals. American athletes classified as "black" earned more than twice that number during each Summer Olympics (Darmoni 2005).

Source: William Hatfield, Northern Kentucky University (class of 2001; revised July 2007 by Joan Ferrante).

Department of Defense photo by JO1 Preston Keres, USN

Mark Sadowski

Which of the athletes pictures are the fastest or more athletic? Is it accurate to claim that the bobsled runners are slower or less athletic than the athlete running the relay?

quickness. And yet, this person does not use the same kind of logic to explain why "white" athletes dominate sports such as gymnastics, rowing, and hockey.

Finally, prejudiced individuals keep stereotypes alive when they evaluate the same behavior differently at different times, depending on the race or ethnicity of the person exhibiting the behavior (Merton 1957). For example, incompetent behavior of racial and ethnic minority members is often attributed to innate flaws in their biological makeup; in contrast, incompetence exhibited by someone from the dominant group is almost always treated as an individual shortcoming.

Discrimination

■ **CORE CONCEPT 8: Both prejudiced and nonprejudiced people can discriminate.** In contrast to prejudice, **discrimination** is not an attitude but a behavior. It includes intentional or unintentional unequal treatment of individuals or groups based on attributes unrelated to merit, ability, or past per-

formance. Discrimination is behavior aimed at denying people equal opportunities to achieve socially valued goals (such as education, employment, health care, and long life) or blocking their access to valued goods and services.

Sociologist Robert K. Merton explored the relationship between prejudice (the attitude) and discrimination (the behavior). He distinguishes between the nonprejudiced (who believe in equal opportunity) and the prejudiced (who do not). Merton asserts that knowing whether someone is prejudiced does not necessarily predict discriminatory conduct, as he makes clear in his four-part typology (see Figure 9.3).

Nonprejudiced nondiscriminators (all-weather liberals) accept the creed of equal opportunity, and their conduct conforms to that creed. They represent a "reservoir of culturally legitimized goodwill" (Merton 1976, p. 193) because they not only believe in equal opportunity but take action against discrimination. One example involves a white employee who complained to a district manager about her supervisors' racist policies toward black customers. Apparently, this white employee's supervisors "told her that to prevent theft, she should follow around black customers as they browsed. They also told her to withhold large shopping bags from such customers and to refrain from inviting them to apply for credit cards or telling them about sales." After receiving no response to her complaints, the employee took them to the state attorney general's office. The complaints were investigated and found to be true (Goldberg 2000).

Nonprejudiced discriminators (fair-weather liberals) believe in equal opportunity but discriminate because doing so gives them an advantage or because they simply fail to consider the discriminatory consequences of their actions. For example, "white" people decide to move out of their neighborhood after a "black" family moves in— not because they are prejudiced against blacks *per se*, but because they are afraid that property values might start to decline and they want to "get out" while time remains. The act of moving out contributes to lower property values for the new neighbor and for other neighbors left behind.

discrimination Intentional or unintentional unequal treatment of individuals or groups because of attributes unrelated to merit, ability, or past performance—treatment that denies equal opportunities to achieve socially valued goals.

nonprejudiced nondiscriminators (all-weather liberals) Persons who accept the creed of equal opportunity and whose conduct conforms to that creed.

nonprejudiced discriminators (fair-weather liberals) Persons who believe in equal opportunity but discriminate because doing so gives them an advantage or because they fail to consider the discriminatory consequences of their actions.

Prejudiced nondiscriminators (timid bigots) reject the creed of equal opportunity but refrain from discrimination, primarily because they fear possible sanctions or being labeled as racists, Timid bigots rarely express their true opinions about racial and ethnic groups, and often use code words to camouflage their true feelings. An example of a prejudiced nondiscriminator is the parents of a minority college student who chooses a white professor as her mentor. The parents, who want to steer their child toward a same-race mentor, do not ask their daughter to change mentors but they do question the white professor's motives.

Prejudiced discriminators (active bigots) reject the notion of equal opportunity and profess a right, even a duty, to discriminate. They derive significant social and psychological gains from the conviction that anyone from the ingroup (including the village idiot) is superior to any members of the outgroup (Merton 1976). Active bigots are most likely to believe that they "have the moral right" to destroy the people whom they see as threatening their values and way of life. Of the four categories in Merton's typology, prejudiced discriminators are the most likely to commit hate crimes, actions aimed at humiliating minority-group people and destroying their property or lives (see Table 9.7).

Within a one-month period following the September 11 attacks on the United States, 326 different media reports told of hate-based acts in 38 states against people who were or who appeared to be of Arab or Middle Eastern descent. Several reported incidents resulted in death. Another case involved two men on a motorcycle "who pulled up next to a Sikh woman stopped at a red light, yanked her door open, shouting, 'This is what you get for what you have done to us!' then 'I'm going to slash your throat. . . .' She was slashed on her head at least twice before the men, hearing a car approach, sped off" (Hamad 2001). A less extreme case involved an Arab American real estate agent who received phone threats ordering him to "Leave the country or else" (Hamad 2001).

---

prejudiced nondiscriminators (timid bigots) Persons who reject the creed of equal opportunity but refrain from discrimination primarily because they fear the sanctions they may encounter if they are caught.

prejudiced discriminators (active bigots) Persons who reject the notion of equal opportunity and profess a right, even a duty, to discriminate.

individual discrimination Any overt action of an individual that depreciates someone from an outgroup, denies outgroup members opportunities to participate, or does violence to their lives and property.

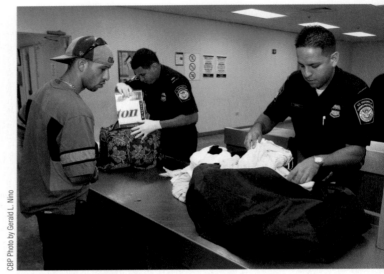

Sociologists distinguish between individual discrimination and institutional discrimination. Currently the United States has security procedures in place that give some groups "special attention" when they pass through security barriers. That discrimination is institutional in nature.

Individual Discrimination

Sociologists distinguish between individual discrimination and institutionalized discrimination. **Individual discrimination** is any overt action of an individual that depreciates someone from an outgroup, denies outgroup members opportunities to participate, or does violence to their lives and property. The U.S. Commission on Civil Rights (1981) describes ways individuals can take discriminatory action against each other:

- A personnel officer who holds stereotyped beliefs about members of outgroups may justify hiring them for low-level and low-paying jobs exclusively, regardless of their potential experience or qualifications for higher-level jobs.
- A teacher may interpret linguistic and cultural differences as indications of low potential or lack of academic interest.
- A guidance counselor and teacher may hold low expectations that lead them to steer members of outgroups away from "hard" subjects, such as mathematics and science, and toward subjects that do not prepare them for higher-paying jobs.
- An owner of an apartment complex may refuse to rent to members of a minority by falsely informing them that no apartments are available (U.S. Department of Justice 1996).

	Attitude Dimension Prejudice and Nonprejudice	**Behavior Dimension** Discrimination and Nondiscrimination	**Example**
Type I Nonprejudiced Nondiscriminator	+	+	All-weather liberal; nonprejudiced person notices discrimination and takes action to correct it.
Type II Nonprejudiced Discriminator	+	−	Fair-weather liberal; unprejudiced person moves out of neighborhood when someone from another racial and ethnic group moves in, not because he or she is prejudiced but because he or she believes housing values will drop.
Type III Prejudiced Nondiscriminator	−	+	Timid bigot; prejudiced person is afraid to discriminate because he or she fears sanctions.
Type IV Prejudiced Discriminator	−	−	Active bigot; prejudiced person who acts accordingly.

+ Attitude/behavior supports equal opportunity

− Attitude/behavior rejects equal opportunity

▲ **Figure 9.3** A Typology of Ethnic Prejudice and Discrimination

Source: Adapted from Merton (1976, p. 192).

Table 9.7 A Profile of Reported Hate Crimes in the United States

The data below show that people of all racial categories commit hate crimes, sometimes even against members of their own racial categories. How many hate crimes were classified as anti-white? Why, do you think, do people commit hate crimes against someone of their own race?

What do you make of the "race unknown" category? Under what circumstances would we not know the offender's race?

Bias Motivation		Known Offender's Race						
	Total Offenses	White	Black	American Indian	Asian	Multiple Races	Unknown Race	Unknown Offender
Total Offenses	8,380	3,585	933	68	44	215	751	2,784
Race:	4,691	2,205	484	45	22	128	387	1,420
Anti-White	935	204	368	19	4	22	85	233
Anti-Black	3,200	1,803	87	13	6	84	239	968
Anti–American Indian	95	40	6	12	0	1	11	25
Anti-Asian	231	88	11	1	8	11	29	83
Anti–Multiple Races	230	70	12	0	4	10	23	111
Religion:	1,314	306	42	6	5	11	132	812
Anti-Jewish	900	198	9	4	2	6	86	595
Anti-Islamic	146	52	20	1	1	2	14	56
Ethnicity/National Origin:	1,144	548	179	9	5	35	116	252
Anti-Hispanic	660	340	115	3	1	18	50	133
Anti–Other Ethnicity	484	208	64	6	4	17	66	119

Source: U.S. Department of Justice, 2004.

Institutionalized Discrimination

Institutionalized discrimination is the established, customary way of doing things in society—the unchallenged rules, policies, and day-to-day practices established by dominant groups that impede or limit minority members' achievement and keep them in subordinate and disadvantaged positions. It is "systematic discrimination through the regular operations of societal institutions" (Davis 1978, p. 30).

Laws and practices clearly designed to keep members of outgroups in subordinate positions fall under the category of institutionalized discrimination. Institutionalized discrimination is more difficult to identify, condemn, hold in check, and punish than individual discrimination, because it can exist in a society even if the society's members are not prejudiced. Institutionalized discrimination cannot be traced to the motives and actions of individuals; instead, discriminatory actions result from simply following established practices that seem on the surface to be impersonal and fair or part of the standard operating procedures.

One case of institutionalized discrimination involved 50 different Cracker Barrel restaurants in seven states. These restaurants violated Title II of the Civil Rights Act of 1964 by engaging in a pattern or practice of discrimination against African American customers and prospective customers based on their race or color. Specifically, a complaint filed by the U.S. Justice Department alleged that the restaurants allowed white servers to refuse to wait on African American customers; segregated customer seating by race; seated white customers before African American customers who arrived earlier; provided inferior service to African American customers after they were seated; and treated African Americans who complained about the quality of Cracker Barrel's food or service less favorably than white customers who lodged similar complaints. The Justice Department's investigation included interviews with approximately 150 persons, mostly former Cracker Barrel employees, of whom 80 percent stated that they had experienced or witnessed discriminatory treatment of customers at a Cracker Barrel restaurant. The investigation suggested that managers often directed, participated in, or condoned the discriminatory behavior.

To this point, we have looked at barriers that exist in society to prevent racial and ethnic groups from adapting or assimilating. These barriers include racist ideology, prejudice, and discrimination (individual and institutional). Although we have viewed these barriers in a general way, we have not examined how they operate in everyday interaction. In *Stigma: Notes on the Management of Spoiled Identity*, sociologist Erving Goffman (1963) offers us a framework for such an examination.

Social Identity and Stigma

■ **CORE CONCEPT 9:** Sociologists are interested in stigmas—attributes that are so deeply discrediting that they come to dominate interaction. A **stigma** is an attribute that is deeply discrediting. That is, when someone possesses a stigma, he or she is reduced in the eyes of others from a multifaceted person to a person with one tainted status. In addition, the attribute dominates both the stigmatized person's interaction with others and the way others think about the person. To illustrate these points, Goffman refers to a letter written by a 16-year-old girl born without a nose. Although she is a good student, has a good figure, and is a good dancer, no one she meets seems able to get past the fact that she has no nose.

Consider as one example of how a stigma reduces a person in the eyes of others the situation of noted historian John Hope Franklin (1990), where the element of race has affected the ways people describe and evaluate his accomplishments:

> It's often assumed that I'm a scholar of Afro-American history, but the fact is that I haven't taught a course in Afro-American history in 30 . . . years. They say I'm the author of 12 books on black history, when several of those books focus mainly on whites. I'm called a leading black historian, never minding the fact that I've served as president of the American Historical Association, the Organization of American Historians, the Southern Historical Association, Phi Beta Kappa, and on and on. The tragedy . . . is that black scholars so often have their specialties forced on them. My specialty is the history of the South, and that means I teach the history of blacks and whites. (1990, p. 13)

Goffman was particularly interested in social encounters known as **mixed contacts**—interactions between stigmatized persons and "normals" (see Intersection of Biography and Society: "The Dynamics of Mixed Contacts"). Note that Goffman did not use the term *normal*

---

institutionalized discrimination The established, customary way of doing things in society—the unchallenged rules, policies, and day-to-day practices that impede or limit minority members' achievement and keep them in subordinate and disadvantaged positions.

stigma An attribute defined as deeply discrediting because it overshadows all other attributes that a person might possess.

mixed contacts "The moments when stigmatized normals are in the same 'social situation,' that is, in one another's immediate physical presence, whether in a conversation-like encounter or in the mere co-presence of an unfocused gathering" (Goffman 1963, p. 12).

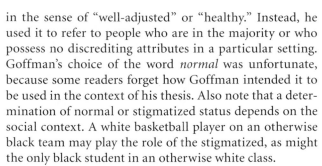

WORKING FOR CHANGE

Lester Ward (1841–1913)

THROUGHOUT HIS LIFE, Lester Ward, the founder of American sociology, worked diligently to improve ethnic and racial relations. He vehemently opposed slavery, fought to have it abolished, and was seriously wounded in the Civil War doing so. After the war, both he and his wife worked to help educate emancipated slaves, and subsequently he was involved with the founding of Howard University. During the 1870s, as editor of *The Iconoclast,* he published the writings of the noted African American reformer Frederick Douglass, and he regularly addressed groups seeking to recognize the rights of African Americans.

In the early 1900s, the eugenics movement became popular in American society. Proponents contended that some groups were naturally or innately superior to others due to genetic or hereditary differences. Furthermore, they believed that the best way to improve society was to increase the birth rates of genetically superior peoples and decrease the rates of others. They held that unless steps, such as mandated sterilizations, were taken to reduce the birth rates of African Americans, Southern Europeans, and other minority immigrant groups, there would be a steady deterioration of the fabric of American society.

Lester Ward worked tirelessly to resist the eugenics movement. He held that while there are indeed differences in ability *within* the respective groups, there are no significant differences among the groups. He held that most differences observed with regard to intelligence, crime, and deviant behavior were rooted in differences of opportunity. In short, he argued that all people are inherently equal in their potential regardless of the racial or ethnic group with which they are identified. In hindsight, one might say that Ward recognized that the proponents of eugenics were setting the stage for the emergence of racial superiority doctrines that were ultimately expressed in Nazism and genocide, that is, the elimination of groups on the basis of alleged hereditary traits.

In his final public lecture before his death, Ward called upon sociologists to vigorously challenge the *eugenics* movement and instead advocate for *euthenics*: a science seeking to improve society through the betterment of living conditions, such as the expansion of educational opportunity. He also challenged social scientists of his era who opposed ethnically and racially mixed marriages. Ward felt that such marriages should be encouraged and that they would ultimately lead to an improved and more tolerant society.

Source: Gale Largey, Professor Emeritus of Sociology, Mansfield University.

---

in the sense of "well-adjusted" or "healthy." Instead, he used it to refer to people who are in the majority or who possess no discrediting attributes in a particular setting. Goffman's choice of the word *normal* was unfortunate, because some readers forget how Goffman intended it to be used in the context of his thesis. Also note that a determination of normal or stigmatized status depends on the social context. A white basketball player on an otherwise black team may play the role of the stigmatized, as might the only black student in an otherwise white class.

Goffman wrote that mixed contacts are "the moments when stigmatized and normals are in the same 'social situation'—that is, in one another's immediate physical presence, whether in a conversation-like encounter or in the mere co-presence of an unfocused gathering" (Goffman 1963, p. 12). According to Goffman, when normals and stigmatized persons interact, the stigma attached to one characteristic—such as skin color, hair texture, or eye shape—comes to dominate the interaction. Such dynamics are especially evident in situations where "race matters."

Patterns of Mixed Contact

A stigma can come to dominate interaction in many ways. First, the very anticipation of contact can cause normals and stigmatized individuals to avoid one another. One reason they avoid contact is to escape one another's scrutiny. People may prefer interacting with others of the same race or ethnicity over interacting with people of a different race or ethnicity to avoid the discomfort, rejection, and suspicion they encounter from those who differ from them.

INTERSECTION OF BIOGRAPHY AND SOCIETY

The Dynamics of Mixed Contacts

ERVING GOFFMAN BELIEVED that sociologists should focus on mixed contacts, situations in which stigmatized persons and normals find themselves in each other's presence. The following excerpts from students in my introduction to sociology course describe such situations.

I grew up in an all-white neighborhood and went to an all-white elementary school. The middle school I went to was predominantly black. In middle school I was seen as an "oreo" (black on the outside, white on the inside). Another kid at this school was also called the same name. This kid's situation was a little bit different from mine, though. He was from Africa; his father was a doctor and his family was very well off. After a couple of months of middle school I was able to change the way I talked—less white, whatever that means—so that I was not made fun of every time I spoke. But it was not that simple for the African kid. No matter what he did, the other black students did not accept him. In fact both the white and black students teased him by saying "You act whiter than a white person." By the time I graduated from middle school, I was very popular with all students at school. The African guy could only make it with some of the white kids; he was never able to break into the social circles of the black students.

My mother is white and my father is Pakistani. I appear Pakistani. A friend and I were driving to Panama City Beach, Florida, and were passing through a small town just outside Birmingham, Alabama. . . . A police officer pulled us over for following too close and began to search our car [and] handcuffed me after finding vodka in a cooler in the trunk. . . . As they put me in the back of the car one cop said to the other, "We are going to make a million dollars turning in Bin Laden." . . . I was in jail nine hours, got a lawyer and charges were dropped.

Social psychologist Claude Steele (1995) offers this:

Imagine a black and a white man meeting for the first time. Because the black person knows the stereotypes of his group, he attempts to deflect those negative traits, finding ways of trying to communicate, in effect, "Don't think of me as incompetent." The white, for his part, is busy deflecting the stereotypes of his group: "Don't think of me as a racist." Every action becomes loaded with the potential of confirming the stereotype, and you end up with these phantoms they're only half aware of. The discomfort and tension [are] often mistaken for racial animosity. (p. A25)

Goffman (1963) suggests that the stigmatized individual has good reason to feel anxious about mixed social interaction and that "normals will find these situations shaky too" (p. 18). Normals may feel that if they show direct sympathy toward a stigmatized person, they will be calling attention to differences when they should be "color-blind." Normals may also feel that a stigmatized person is "too ready to read unintended meanings into our actions" (p. 18).

Another reason that stigmatized persons and normals make conscious efforts to avoid one another is because they believe that widespread social disapproval will undermine any relationship. One of my "white" students wrote about her experiences of trying to date and establish a relationship with a "black" classmate, against the disapproval of

Lisa Southwick

Sociologists are interested in studying mixed contacts—situations in which normals and stigmatized persons are in one another's immediate physical presence.

her father and many of her friends: "We tried to carry on a 'secret' relationship, but it just didn't work. We were always worried about who might see us out together. We both agreed it wasn't worth the trouble, so we called it off."

The response of avoidance is related to a second pattern that characterizes mixed contacts: Upon meeting, normals and stigmatized people are unsure how the other views them or will act toward them. For the stigmatized, the source of the uncertainty is not that everyone they meet views them negatively and treats them accordingly. Rather, it is the chance that they might encounter prejudice and discrimination that gives them reason to be cautious about all encounters. According to the *Kolts Report* (a report written in response to the Rodney King beating, one of the most publicized and replayed police encounters with a black man, which took place in 1992), there were 62 "problem" deputies out of a total of 8,000 members of the Los Angeles County Sheriff's Department. The report concluded that "nearly all deputies treat nearly all individuals, most of the time, with at least minimally acceptable levels of courtesy and dignity" (*Los Angeles Times* 1992, p. A18). Although only a small proportion of deputies (fewer than 1 percent) were identified as "problems," the cases of mistreatment were "outrageous enough and frequent enough to poison the well in some communities" (p. A18).

A third pattern characteristic of mixed contacts is that normals often define accomplishments by the stigmatized—even minor accomplishments—"as signs of remarkable and noteworthy capacities" (Goffman 1963, p. 14) or as evidence that they have met someone from a minority group who is an exception to the rule. In a fourth pattern, normals tend to interpret a stigmatized person's failings, both major and minor (such as being late for a meeting, cashing a bad check, or leaving a small tip), as related to the stigma.

A fifth pattern noted with mixed contacts is that the stigmatized are likely to experience invasion of privacy, especially when people stare. Questions such as What are *they* doing in *our* neighborhood? Why are *they* crashing *our* party? and Why is she dating that white man? represent an invasion into the personal matters of the stigmatized person.

Responses to Stigmatization

Goffman describes five ways that the stigmatized respond to people who fail to accord them respect or who treat them as members of a category. First, they may attempt to "correct" the source of stigma. This response includes changing the visible cultural and physical characteristics believed to represent barriers to status and belonging. For example, a person may undergo plastic surgery or do other things to alter the shape of the nose, eyes, or lips, or may straighten the hair.

Often such persons find these changes difficult to make, because they may be considered traitors to their own racial or ethnic group. In addition, they themselves may feel conflicting emotions in response to their own visible changes.

Lawrence Otis Graham (2001) experienced such emotions after he underwent plastic surgery on his nose:

> Did I have the operation in order to become less black—to have features that were more white? Had I bought into the white definition of beauty—the sharp nose, the thin lips, the straight hair? Did I think that my less Negroid-looking black friends were more attractive than me? (p. 36)

A second way people respond to being stigmatized involves devoting a great deal of time and effort to overcoming stereotypes or to appearing as if they are in full control of everything around them. The stigmatized may try to be perfect—to always be in a good mood, to be extra friendly, to outperform everyone else, or to master an activity ordinarily thought to be beyond the reach of or closed to people with their traits. This response is common among nonwhite plumbers, electricians, building contractors, and other service workers (especially males) who must make house calls in predominantly white neighborhoods. One black contractor interviewed in the *New York Times* maintained that when doing business in white neighborhoods, he tries to "Avoid working at night or showing up at a job early in the morning. Never linger inside houses or gaze at a resident's possessions. And always keep your tools at hand to allay suspicion" (p. A1).

As a third response to being stigmatized, people may use their subordinate status for secondary gains, including personal profit, or as "an excuse for ill success that has come [their] way for other reasons" (Goffman 1963, p. 10). If an employee accuses his or her employer of racism and

During World War II, Japanese American men joined the U.S. military (even as their families were being interned) to show that they were patriotic and not a danger to the United States and its way of life.

National Archives & Records Administration, Department of the Interior, War Relocation Authority.

threatens to file a lawsuit when he or she has been justly sanctioned for poor work, then that person is using stigma for secondary gains. Keep in mind, however, that the person can use stigma this way only because discrimination is commonplace in the larger society.

A fourth response to being stigmatized is to view discrimination as a blessing in disguise, especially for its ability to build character or for what it teaches about life and humanity. Finally, a stigmatized person can condemn all the normals and view them negatively.

■ VISUAL SUMMARY OF CORE CONCEPTS

■ CORE CONCEPT 1: The concepts of race and ethnicity cannot be understood apart from systems of racial and ethnic classification.

Race has no basis in biology. It does, however, have considerable social significance. Unlike race, which emphasizes physical features and geographic origin, ethnicity can be based on an almost infinite number of traits. No discussion of race and ethnicity can proceed without reference to systems of racial and ethnic classification.

Ray Elfers

■ CORE CONCEPT 2: That governments and other powerful groups have created illogical racial and ethnic categories and have worked so hard to present them as logical is one measure of their social importance.

The shortcomings of racial and ethnic classification schemes become evident when we imagine trying to classify the world's 6.3 billion people. This chapter identifies at least four such shortcomings: (1) Many people do not fit easily into a single racial category, because no sharp dividing line distinguishes characteristics such as so-called black skin from white skin. (2) Boundaries between races can never be fixed and definite, if only because males and females of any alleged race can produce offspring together. (3) Rules or guidelines for placing people in racial categories are often vague, contradictory, unevenly applied, and subject to change. (4) Most classification schemes force large aggregates of people who vary in ethnicity, language, generation, social class, or physical appearance under one umbrella category.

Photographed by Marty Lueders for the U.S. Census Bureau

■ CORE CONCEPT 3: The racial and ethnic categories to which people belong are a product of three interrelated factors: chance, context, and choice.

Chance is something not subject to human will, choice, or effort. We do *not* choose our biological parents, nor can we control the physical characteristics we inherit from them. Context is the social setting in which racial and ethnic categories are recognized, created, and challenged. Choice is the act of choosing from a range of possible behaviors or appearances. Individual choices are constrained by chance and context.

Department of Defense photo by Staff Sgt. Lanie McNeal, U.S. Air Force

■ **CORE CONCEPT 4: The legal status of the foreign-born varies by country and is often connected to race and ethnicity.**

Every country in the world has residents considered foreign-born—people living within the political boundaries of the country who were born elsewhere. The legal status of the foreign-born—particularly their right to pursue citizenship and permanent residence—varies by country. No government welcomes just anyone; often race, ethnicity, and skills figure into the probability that one will gain admission to a country legally or will have to enter illegally. Historically, race and ethnicity considerations have played a major role in U.S. immigration policy.

■ **CORE CONCEPT 5: Minorities are populations that are systematically excluded (whether consciously or unconsciously) from full participation in society and denied equal access to positions of power, prestige, and wealth.**

Members of minorities are regarded and treated as inherently different from those in the dominant group. They are systematically excluded (whether consciously or unconsciously) from full participation in society and denied equal access to positions of power, prestige, and wealth. Consequently, members of minorities tend to be concentrated in inferior political and economic positions and to be isolated socially and spatially from members of the dominant group.

■ **CORE CONCEPT 6: Social and cultural differences between racial and ethnic groups "disappear" when one group is absorbed into another group's culture and social networks or when two groups merge to form a new, blended culture.**

Assimilation is a process by which ethnic and racial distinctions between groups disappear because one group is absorbed into another group's culture or because two cultures blend to form a new culture. Two main types of assimilation exist: absorption assimilation and melting pot assimilation.

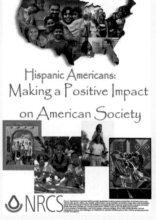

Hispanic Americans

Hispanic Americans:
Making a Positive Impact
on American Society

NRCS

■ **CORE CONCEPT 7: Racist ideologies claim that biological factors explain and even justify inequalities between racial and ethnic groups.**
An ideology is a set of beliefs that are not challenged or subjected to scrutiny by the people who hold them. Thus, ideologies are taken to be accurate accounts and explanations of the way things are. On closer analysis, however, ideologies are at best half-truths, based on misleading arguments, incomplete analyses, unsupported assertions, and implausible premises. People who embrace the ideology of racism believe that something in the biological makeup of an ethnic or racial group explains and justifies its subordinate or superior status.

Massimo Finizio

■ **CORE CONCEPT 8: Both prejudiced and nonprejudiced people can discriminate.**
In contrast to prejudice, discrimination is not an attitude but a behavior. It is aimed at denying people equal opportunities to achieve socially valued goals (such as education, employment, health care, and long life) or blocking their access to valued goods and services. Prejudice, does not necessarily predict discriminatory conduct.

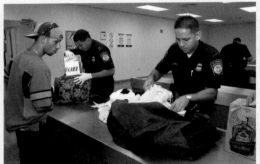

CBP Photo by Gerald L. Nino

■ **CORE CONCEPT 9: Sociologists are interested in stigmas—attributes that are so deeply discrediting that they come to dominate interaction.**
Sociologists are particularly interested in social encounters known as mixed contacts—interactions between stigmatized persons and "normals." When normals and stigmatized persons interact, the stigma can come to dominate the interaction in many ways.

Lisa Southwick

Resources on the Internet

 Sociology: A Global Perspective Book Companion Web Site
www.cengage.com/sociology/ferrante

Visit your book companion Web site, where you will find flash cards, practice quizzes, Internet links, and more to help you study.

CENGAGENOW™

Just what you need to know NOW!

Spend time on what you need to master rather than on information you have already learned. Take a pre-test for this chapter, and CengageNOW will generate a personalized study plan based on your results. The study plan will identify the topics you need to review and direct you to online resources to help you master those topics. You can then take a post-test to help you determine the concepts you have mastered and what you will need to work on. Try it out! Go to www.cengage.com/login to sign in with an access code or to purchase access to this product.

Key Terms

absorption assimilation   249
assimilation   249
chance   239
choice   239
context   239
discrimination   255
ethnicity   234
foreign-born   242
ideology   252
individual discrimination   256
institutionalized discrimination   258

involuntary minorities   250
melting pot assimilation   250
minority groups   247
mixed contacts   258
nonprejudiced discriminators (fair-weather liberals)   255
nonprejudiced nondiscriminators (all-weather liberals)   255
prejudice   253
prejudiced discriminators (active bigots)   256

prejudiced nondiscriminators (timid bigots)   256
race   234
segregation   250
selective perception   253
stereotypes   253
stigma   258
systems of racial and ethnic classification   235
voluntary minorities   250

10 Gender

With Emphasis on American Samoa

When sociologists study gender, they focus on male-female differences in behavior and appearance that have been socially created. Sociologists seek to understand the mechanisms by which people learn and perpetuate society's expectations about sex-appropriate behavior and appearance.

Department of Defense photo by Paul Adams, USA

▲ Sociologists are interested in how people learn and perpetuate their society's expectations about appropriate behaviors, dress, and appearances for males and females.

American Samoa?

IN THIS CHAPTER, the topic of gender is paired with American Samoa. Most Americans probably do not know that American Samoa is a territory of the United States. Many Americans probably do know an American Samoan who plays for their favorite NFL or Division I college football team, as 228 Samoans play at those levels. That number is truly amazing, given that the entire population of American Samoa is about 67,000, hardly large enough to fill a Super Bowl stadium.

In contrast to the rugged image of the football player, Samoans recognize a third gender which they call *fa'afafine*. They are men who dress as famous female celebrities such as Britney Spears, Madonna, or Kelly Clarkson. Their suc-

cess in Samoan society comes from giving the most stunningly accurate imitation of these women. There is even a Miss Island Queen contest where very high-profile men and women outside this transgender community are judges. Last year a second grade teacher won.

Sociologists are fascinated by a society where both football players and *fa'afafines* are commonplace. They look for social forces that support the existence of both in such a tiny population. Perhaps what is most fascinating about studying American Samoa is what we will learn about gender in the United States.

FACTS TO CONSIDER

- In American Samoa (and other Pacific island areas) people recognize three genders: males, females, and *fa'afafine*.

- More than 200 American Samoan men play division I college football in the United States; at least 28 American Samoans play for NFL teams. These numbers are amazing, given that the entire population of American Samoa (65,000) is not large enough to fill a Super Bowl stadium.

- At one time in Samoa, the transition from boyhood to manhood was marked by the long and painful process of tattooing the body from the waist to below the knee. Some present-day Samoan males choose to continue this tradition.

- In Samoa 90 percent of those who work in health service occupations are female; 99 percent of those who work in construction trades are male.

Distinguishing Sex and Gender

In this chapter, we explore the basic concepts that sociologists use to analyze the connection between gender and life chances. In outlining this connection, we examine how sociologists distinguish between sex (a biologically based classification scheme) and gender (a socially constructed phenomenon). Although many people use the words *sex* and *gender* interchangeably, the two terms do not have the same meaning. Sex is a biological concept, whereas gender is a social construct.

Sex as a Biological Concept

■ **CORE CONCEPT 1:** Sex is a biological distinction determined by the anatomical traits essential to reproduction. A person's **sex** is based on **primary sex characteristics**, the anatomical traits essential to reproduction. Most cultures classify people in two categories—male and female—based largely on what are considered to be clear anatomical distinctions. Biological sex is not a clear-cut category, however, if only because some babies are born **intersexed**. The medical profession uses this broad term to classify people with some mixture of male and female biological characteristics. Although we do not know how many intersexed babies are born each year, one physician who treats intersexed children estimates that number to be one thousand (Dreifus, 2005).

If some babies are born intersexed, why does society not recognize an intersexed category? In the United States and many other countries, no such category exists because parents of such children collaborate with physicians to assign their newborns to one of the two recognized sexes. Intersexed infants are treated with surgery and/or hormonal therapy. The rationale underlying medical intervention is the belief that the condition "is a tragic event" resulting in "a hopeless psychological misfit doomed to live always as a sexual freak in loneliness and frustration" (Dewhurst and Gordon 1993, p. A15).

Determining biological sex becomes even more complicated when we consider that a person's primary sex characteristics may not match his or her sex chromo-

sex A biological concept based on primary sex characteristics.

primary sex characteristics The anatomical traits essential to reproduction.

intersexed A broad term used by the medical profession to classify people with some mixture of male and female biological characteristics.

transsexuals People whose primary sex characteristics do not match the sex they perceive themselves to be.

Department of Defense photo by Petty Officer 1st Class Bobby R. McRill, U.S. Navy

Theoretically, this baby is a girl, because her mother contributed an X chromosome and her father contributed an X chromosome as well.

somes. Theoretically, one's sex is determined by two chromosomes. Each parent supposedly contributes one sex chromosome: The mother contributes an X chromosome, and the father contributes an X or a Y chromosome. If the chromosome carried by the sperm that fertilizes the egg is a Y, then the baby will be a male. In an unknown number of cases, however, sex chromosomes do not match anatomy. The results of mandatory "sex tests" of female participants in international athletic competitions before 1999 (the year such testing officially ended) have shown that such cases exist. Indeed, a few women were disqualified from Olympic competition and other major international competitions because they "failed" the tests (Grady 1992). That is, they had the physical appearance of females but the chromosomes of a male.

Perhaps the most highly publicized case involved Spanish hurdler María José Martínez Patiño, who, although "clearly a female anatomically, is, at a genetic level, just as clearly a man" (Lemonick 1992, p. 65). When a test before a track meet identified her as genetically male, meet officials forbade her to compete and advised her to fake an injury to deflect the media's attention from her unclear sexual identity (Grady 1992). After losing her right to compete in international athletic events, Martínez Patiño spent three years challenging the decision. The International Amateur Athletic Federation (IAAF) restored her status after deciding that her X and Y chromosomes gave her no advantage over female competitors with two X chromosomes (Kolata 1992; Lemonick 1992).

Consider, as further evidence that no clear line separates male from female, the existence of **transsexuals**—people whose primary sex characteristics do not match the sex they perceive themselves to be. Those motivated to undergo a sex change are labeled "high-intensity" transsexuals (Bloom 1994). The *Diagnostic and Statistical Manual of Mental Disorders*, a reference book used by mental health practitioners, estimates that 1 in 30,000 people born male and 1 in 100,000 people born female have what the manual calls a "gender identity disorder" (Barton 2005).

Why does no clear line exist to separate everyone into one of the two biological categories male and female? One answer lies in the biological mechanisms involved in creating males and females. In the first weeks after conception, the human embryo develops the potential to form a "female set of ovaries and a male set of testes." Approximately eight weeks into development, "a molecular chain of events orders one set to disintegrate." One week later, the embryo begins to develop an outer appearance that matches its external sex organs (Lehrman 1997, p. 49). This complex chain of events may not occur "perfectly"; instead, it may be affected by any number of factors, including exposure to medications taken by the mother.

In addition to primary sex characteristics and chromosomal sex, we use **secondary sex characteristics** to distinguish one sex from another. These physical traits (such as breast development, quality of voice, distribution of facial and body hair, and skeletal form) are not essential to reproduction but result from the action of so-called male (androgen) and female (estrogen) hormones. We use the term *so-called* because, although testes produce androgen and ovaries produce estrogen, the adrenal cortex produces androgen and estrogen in both sexes (Garb 1991). Like primary sex characteristics, none of the secondary ones represents a clear line by which to separate males from females.

Gender as a Social Construct

■ **CORE CONCEPT 2:** Gender is the socially created and learned distinctions that specify the ideal physical, behavioral, and mental and emotional traits characteristic of males and females. Whereas sex is a biological distinction, **gender** is a social distinction based on culturally conceived and learned ideals about appropriate appearance, behavior, and mental and emotional characteristics for males and females (Tierney 1991). *Ideal* in this context means a standard against which real cases can be compared. A gender ideal is at best a caricature, in that it exaggerates the characteristics that make someone the so-called perfect male or female. Keep in mind that very few people achieve a gender ideal; even those who do cannot usually sustain it, if only because they age. In fact, such ideals may be so difficult to achieve that they may not exist in reality. Consider that at one time the ideal foot length for Chinese women was 4–6 inches—an impossible standard that no females could achieve without enduring foot binding as young girls. This practice involved breaking the four smallest toes on each foot and binding them toward the heel so the feet could no longer grow. Likewise, few grown women have 18-inch waists, and yet, women in the United States and in other societies have worn corsets to achieve that impossible standard.

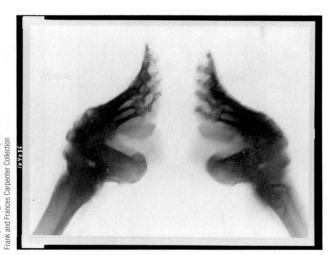

Library of Congress Prints and Photographs Division, Frank and Frances Carpenter Collection

Bob and Cathie Jung

Gender ideals may not exist in reality, but that does not stop people from trying to achieve them. To achieve the ideal foot length of 4 to 6 inches, young Chinese girls from the 10th to the mid-20th centuries endured foot binding. This practice involved breaking the four smallest toes on each foot and binding them toward the heel so the feet could no longer grow. To achieve an ideal waistline of 13-18 inches, women have worn corsets and even had lower ribs removed. Cathy Jung (pictured) is believed to hold the Guinness World Record for the smallest waist of a living adult: 15 inches.

secondary sex characteristics Physical traits not essential to reproduction (such as breast development, quality of voice, distribution of facial and body hair, and skeletal form) that result from the action of so-called male (androgen) and female (estrogen) hormones.

gender A social distinction based on culturally conceived and learned ideals about appropriate appearance, behavior, and mental and emotional characteristics for males and females.

Ideal

Reality

Ideal

Reality

Gender ideals are socially constructed standards against which real cases can be compared. Do not confuse this meaning of *ideal* with "the best way to be." Few people can achieve and sustain gender ideals. Still, people draw upon them to evaluate themselves and others. Even the most attractive men and women fall short. For example, their shoulders are not broad enough or their hair is brown instead of blond. Whenever we compare people against ideals they fall short.

The terms **masculinity** and **femininity** signify the physical, behavioral, and mental and emotional traits believed to be characteristic of males and females (Morawski 1991). The following saying captures both the significance and elusiveness of a gender ideal: "You cannot take hold of it. But you cannot lose it. In not being able to get it, you get it."

To grasp that gender is a culturally conceived "reality," we must note that no fixed line separates masculinity from femininity. The French painter Paul Gauguin pointed out this ambiguity in his observations about Tahitian men and women, which he recorded in a journal that he kept while painting in Tahiti in 1891. His observations were influenced by 19th-century Western norms of femininity:

masculinity The physical, behavioral, and mental and emotional traits believed to be characteristic of males.

femininity The physical, behavioral, and mental and emotional traits believed to be characteristic of females.

At the turn of the twentieth century, the French artist Paul Gauguin observed that differences between the sexes in Tahitian society were less accentuated than they were in Western society: there was "something virile in the women and something feminine in the men."

Among peoples that go naked, as among animals, the difference between the sexes is less accentuated than in our climates. Thanks to our cinctures and corsets we have succeeded in making an artificial being out of woman. . . . We carefully keep her in a state of nervous weakness and muscular inferiority, and in guarding her from fatigue, we take away from her possibilities of development. Thus modeled on a bizarre ideal of slenderness . . . our women have nothing in common with us [men], and this, perhaps, may not be without grave moral and social disadvantages.

On Tahiti, the breezes from forest and sea strengthen the lungs, they broaden the shoulders and hips. Neither men nor women are sheltered from the rays of the sun nor the pebbles of the sea-shore. Together they engage in the same tasks with the same activity. . . . There is something virile in the women and something feminine in the men. ([1919] 1985, pp. 19–20)

Often we mistakenly attribute masculinity and femininity to biology, when in fact, they are socially created. In the United States, for example, norms specify the amount and distribution of facial and body hair appropriate for females. It is deemed acceptable for women to have eyelashes and well-shaped eyebrows but certainly not to have hair above their lips, under their arms, on their inner thighs (outside the bikini line), or on their chin, shoulders, back, chest, breasts, abdomen, legs, or toes. Many men, and even some women, do not realize that women work to achieve these cultural standards and that the women's compliance makes males and females appear more physically distinct in terms of hair distribution than they are in reality. We lose sight of the fact that normal biological events—puberty, pregnancy, menopause, stress—contribute to the balance between two hormones, androgen and estrogen. Changes in the proportions of these hormones trigger hair growth that departs from societal norms about the appropriate amount and texture of hair for females. When women grow hair because of these events, they tend to think something is wrong with them instead of seeing this development as natural. A "female balance" between androgen and estrogen is seen as one in which a woman's hair fits these norms.

Just as women strive to meet norms for facial and body hair, both men and women work to achieve the ideal standards of masculine and feminine beauty, as portrayed in the media or as conveyed and reinforced elsewhere. On a personal level, these ideal standards are not viewed objectively—as something created. Despite all evidence to the contrary, for example, we believe that facial and body hair is a masculine quality.

Consider also gender ideals regarding hair length. Long hair on Samoan women (and American women, for that matter) is the ideal. In fact, long hair does not simply *signify* feminine sexual attractiveness; it *is* feminine sexual attractiveness. Jeanette Mageo (1996) argues that ideal standards of beauty affect us personally because

what has personal significance is at least in part a product of how we are regarded and treated by others. When a Samoan girl acts under constant threat of having her hair cut off or of being pulled home by her hair [because she is attracting male attentions], when her beauty is judged, at contests and elsewhere, by the length of her hair, the public symbol of hair cannot fail to touch her feelings. (p. 158)

While Samoan women's femininity is judged according to the length of their hair, some Samoan men seek to acquire tattoos as a sign of masculinity. Before the Christianization of Samoa, the transition from boyhood to manhood was accompanied by a "long and painful process of body tattooing, from the waist to below the knees" (Cote 1997, p. 2). Tattooing or *tatau* (ta-TAH-oo) did not merely *signify* manhood; it *was* manhood:

The man who was not tattooed . . . was not respected. . . . Until a young man was tattooed . . . he could not think of marriage,

© Charles & Josette Lenars/CORBIS

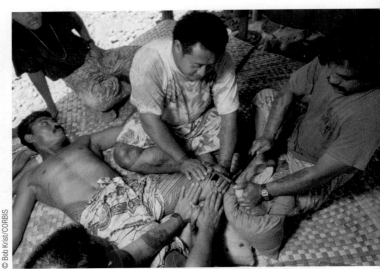

© Bob Krist/CORBIS

At one time in Samoa the transition from boyhood to manhood was marked by the long and painful process of tattooing the body from the waist to below the knee. Some present-day Samoan males choose to continue this tradition.

and he was constantly exposed to taunts and ridicule, as being poor and of low birth, and having no right to speak in the society of men. But as soon as he was tattooed, he passed in his majority, and considered himself entitled to the respect and privileges of mature years. When a youth, therefore, reached the age of sixteen, he and his friends were all anxious that they should be tattooed. . . . On these occasions, six or a dozen young men would be tattooed at one time. . . . In two or three months the whole is completed. The friends of the young men are all the while in attendance with food. (Turner [1861] 1986, pp. 87–89)

gender polarization The organizing of social life around male-female ideals, so that people's sex influences every aspect of their life, including how they dress, the time they get up in the morning, what they do before they go to bed a night, the social roles they take on, the things they worry about, and even the ways they express emotion and experience sexual attraction.

Many Samoan males still choose to tattoo their bodies in this fashion. In fact, some Los Angeles–based Samoans on visits to American Samoa will get the painful body *tatau* to show they are men and responsible to family (Channell 2002).

To this point, we have drawn a distinction between sex and gender. Although sociologists acknowledge that no clear biological markers exist to distinguish males from females, they would not argue that no biological differences exist. Sociologists are, however, interested in the extent to which those differences are socially induced. That is, they study the actions that men and women take to be different from one another and to accentuate biological differences. As we will see in the next section, these actions lead to gender polarization.

Gender Polarization

■ **CORE CONCEPT 3:** Gender ideals shape practically every aspect of life—influencing, among other things, how people dress, how they express emotions, and what occupations they choose. **Gender polarization** is the organizing of social life around male-female ideals, so that people's sex influences every aspect of their life, including how they dress, the time they wake up in the morning, what they do before they go to bed at night, the social roles they take on, the things they worry about, and even ways of expressing emotion and experiencing sexual attraction (Bem 1993). To understand how life becomes organized around gender, we consider research by Alice Baumgartner-Papageorgiou (1982) summarizing the results of a study of elementary and high school students. In the study, she asked the students how their lives would be different if they were members of the opposite sex. Their responses reflected culturally conceived and learned ideas about sex-appropriate behaviors and appearances and about the imagined and real advantages and disadvantages of being male or female (Vann 1995).

The boys generally believed that their lives would change in negative ways if they became girls. Among other things, they would become less active and more restricted in what they could do. In addition, they would become more conscious about tending to their appearance, finding a husband, and being alone and unprotected in the face of a violent attack:

- I would start to look for a husband as soon as I got into high school.
- I'd use a lot of makeup and look good and beautiful.
- I'd have to shave my whole body.
- I'd have to know how to handle drunk guys and rapists.
- I couldn't have a pocket knife.
- I would not be able to help my dad fix the car and truck and his two motorcycles. (pp. 2–9)

The girls, on the other hand, believed that if they became boys they would be less emotional, their lives would be more active and less restrictive, they would be closer to their fathers, and they would be treated as more than "sex objects":

- I would have to stay calm and cool whenever something happened.
- I could sleep later in the mornings, since it would not take [me] very long to get ready for school.
- My father would be closer, because I'd be the son he always wanted.
- People would take my decisions and beliefs more seriously. (pp. 5–13)

Although the Baumgartner-Papageorgiou study was published more than 25 years ago, these beliefs about how the character of one's life depends on one's sex seem to hold up across time, even among the college students enrolled in the author's introductory sociology classes. These students were asked to take a few minutes to write about how their lives would change in positive and negative ways as members of the other sex. The men generally believed that they would be more emotional and more conscious of their physical appearance and that their career options would narrow considerably. Some responses illustrate:

- I would be much more sensitive to others' needs and what I'm expected to do.
- I wouldn't always have to appear like I am in control of every situation. I would be comforted instead of always being the comforter.
- People would put me down for the way I look.
- I would be more emotional.

- I would worry more about losing weight instead of trying to gain weight.
- If I stayed in the construction program, I would have to fight the belief that men are the only real construction workers.
- My career options would narrow.
- I would have to be conscious of the way I sit.

The women in the class believed that as men they would have to worry about asking women out and about whether their major was appropriate. They also believed, however, that they would make more money, be less emotional, and be taken more seriously. Some of their responses follow:

- I would worry about whether a woman would say yes if I asked her out.
- I would earn more money than my female counterpart in my chosen profession.
- People would take me more seriously and not attribute my emotions to PMS.
- My dad would expect me to be an athlete.
- I'd have to remain cool when under stress and not show my emotions.
- I think that I would change my major from "undecided" to a major in construction technology.

These comments by high school and college students show the extent to which life is organized around male–female distinctions. Both men and women can feel constrained by their gender roles. They also reveal that students' decisions about how early to get up in the morning, which subjects to study, whether to show emotion, how to sit, and whether to encourage a child's athletic development are gender-schematic decisions. Decisions and viewpoints about any aspect of life are considered **gender-schematic** if they are influenced by a society's polarized definitions of masculinity and femininity rather than by criteria such as self-fulfillment, interest, ability, or personal comfort. For example, one study found that 90 percent of women routinely buy and wear shoes that are too narrow by one or two sizes (Harris 2003).

College students make gender-schematic decisions about possible majors if they ask, even subconsciously, about the "sex" of the major; if it matches their own sex, they consider the major to be a viable option, and if it does not match, they reject the major outright (Bem 1993). Consider that 75 percent of bachelor's degrees in computer and information sciences are awarded to men, whereas 94

gender-schematic A term describing decisions that are influenced by a society's polarized definitions of masculinity and femininity rather than by criteria such as self-fulfillment, interest, ability, and personal comfort.

Many women buy shoes that are too narrow and wear high heels in spite of foot, back, neck, and even jaw pain.

percent of bachelor's degrees in library sciences are awarded to females. Other majors dominated by women include education, health professions, and public administration/services (at least 80 percent of all bachelor's degrees awarded in these fields go to women) (*Chronicle of Higher Education* 2006).

Even sexual attraction between men and women is organized around characteristics unrelated to reproduction, such as height and age differences. Neither women nor men in American society tend "to appreciate heterosexual relationships in which the woman is bigger, taller, stronger, older, smarter, higher in status, more experienced, more educated, more talented, more confident, or more highly paid than the man; they do tend to appreci-

social emotions Internal bodily sensations experienced in relationships with other people.

ate heterosexual relationships in which the man is bigger, taller, stronger, and so forth, than the woman" (Bem 1993, p. 163).

The negative consequences of channeling sexual attraction according to age differences, so that the woman in a heterosexual relationship is usually younger than her partner, becomes evident when we consider that the median age at first marriage for women is 25.1 and for men is 26.7 (U.S. Census Bureau 2005). In the United States the average life expectancy for women is 5.8 years longer than that for men (U.S. Central Intelligence Agency 2007). The difference in life expectancy can be partly explained by the fact that men tend to hold the most hazardous jobs in society. The practice of men marrying women younger than themselves, combined with differences in life expectancy, means that women who marry can expect to live a significant portion of their lives as widows.

Social Emotions and Feeling Rules

Gender-polarized ideas strongly influence not only sexual attraction between men and women but also the expression of emotions toward persons of the same sex. In Chapter 3 we learned that **social emotions** are internal bodily sensations we experience in relationships with other people and that feeling rules are norms specifying appropriate ways to express those sensations. These other people may be boyfriends, spouses, parents, same-sex friends, teachers, and so on. Sometimes the rules are conveyed informally, as when a father tells his three-year-old son not to hold hands with another three-year-old boy; other times the rules are conveyed more formally. Such is the case in American Samoa, where feeling rules are formalized for young men who choose to get "tattooed from above the waist, to all the way around the back, sides and front, down the knees, and around the legs."

> The entire community is involved when a boy receives his tattoo. While he is getting his tattoo, or cycling into manhood, other men from the village surround him. They put their hands on his body and hold him steady while the artist tattoos. The men already have their tattoos and understand how it feels. They talk to him and sing songs to him. . . . The boy will have one partner, sometimes a brother or close friend, who will stay with him for about two weeks after the tattooing and help him bathe and make sure he does not get an infection. That relationship is very important, and helps build a sense of community. (Channell 2002, p. 18)

Compliance and Resistance to Gender Ideals

If we simply think about the men and women we encounter every day, we quickly realize that most people fall short of society's gender ideals. This fact does not stop most people from using these ideals to evaluate their own and

INTERSECTION OF BIOGRAPHY AND SOCIETY

Social Emotions and Feeling Rules

MOST OF THE author's students have acknowledged knowing a same-sex person for whom they felt strong affection and who made them feel "alive." They have also said, however, that it is important that their feelings not be interpreted as sexual:

- I have a set of friends that I absolutely love hanging out with, but I have one friend in particular that, when I'm around him, I feel really great. If I told him that I liked his company that much, he would most likely call me a "fag." When I read the question "Do you feel guilty [for your feelings]?" at first I thought that I did not. However, as I am getting further into writing this, I am beginning to feel weird. I have never said anything to him about how I feel, I guess because I'm a guy and guys just don't do that, generally speaking. What would I say, anyway? "Hey there buddy, I was just thinking about you and I realized that I really love your company. What do you say we have a couple beers tonight?" I doubt it.

- I do know a same-sex friend with whom I have such an intense relationship, and until today I thought I was just weird. I have had a very close friendship with this girl for about seven years, and I've always felt this way about her. Whenever we're together, I feel so energized and so "healthy." I feel an almost uncomfortable closeness to her sometimes. For example, the other day we were watching TV and for some reason I just felt the urge to curl up next to her and snuggle. There was no feeling of sexual arousal in any way at all; it was just a type of closeness. It's almost like the same closeness I feel when I snuggle up to my mom, almost a moth-

The man is hugging his so-called best buddy, but would he feel comfortable hugging a same-sex friend in this fashion if he were a real person? That would depend on the context. Certainly, if this man lived in the United States, he might be reluctant to do so.

erly bond. I have never, ever told her of these feelings because it would just come out so homosexual and it's nothing like that at all.

Why, do you think, does our society offer no vocabulary to describe strong same-sex friendships separate from the vocabulary we use to describe romantic relationships? Why have these students feared that their feelings might be interpreted as sexual? What does that fear suggest about American culture? Why have the students dismissed their affection for same-sex friends by suggesting that they do not want to be thought of as a "fag" or "so homosexual"?

others' behavior and appearance. For more than 10 years, the author has collected response papers from students at Northern Kentucky University in which they explain how gender ideals shape their lives. A reading of hundreds, even thousands, of such responses has shown that most of the students struggle with gender ideals. While many try to conform to these ideals, they do so reluctantly, or they resist them or even outright challenge them. A sample of individual responses illustrates five broad responses:

1. *Learning about and Coming to Accept Gender Ideals.* In high school I began to realize how important my appearance and behavior were to being accepted. In the bathrooms, the juniors and seniors showed freshmen how to throw up after lunch in order to stay thin. Anorexia and bulimia were widespread, and we were very ruthless and materialistic in our criticism of others. I began to practice walking with a sway in my hips and to flirt with my eyes lowered and head tilted. I started to wear makeup to accentuate my feminine features. I began to giggle, something I never really did growing up. Gender ideals have had their effects upon me. I work as an amateur model and actress, banking on the male-female distinction.

2. Attempting to Change the Behaviors That Deviate from Gender Ideals. There is something that drove me crazy. The feeling I experienced and the warmth I felt toward this woman were extraordinary, but so wrong. I didn't have anywhere to turn and no one to talk to. I did not want to tell anyone who or what I had done. Society said it was wrong to feel and act this way, but why did I feel so good and think about her so much? I nearly drove myself crazy and ran as far as I could. I joined the Army to punish myself. Every day I hated what I had done and I wanted to hurt myself to make it go away. I hoped that by surrounding myself with men, I would forget what I had done and find some man that would help make me into a "normal" heterosexual woman.

3. Giving in to Gender Ideals, with Regrets about Their Superficiality. I distinctly remember when I first felt freakish about my facial hair. I was a sophomore in high school. I had a huge crush on the star of our basketball team. Because I was painfully shy, I could never talk to him. My friends sensed my crush and decided to tell him. One day a friend told me that she had asked him whether he might be interested in me, and he said, "No way," and called me "Blackbeard." She asked me what he could be talking about. At first, I didn't have a clue. I had never paid attention to the new, dark hair growing on my face. It finally dawned on me that he was talking about the hair on my face. I was so embarrassed and ashamed that I didn't think I could ever face anyone again. Since that day, I have bleached the hair on my face. I am sad that our culture defines something natural as unnatural. I'm a slave to creme bleach every month, and for what? Acceptance from men? No one has ever commented about my facial hair since then, because I have conformed to the societal standards.

4. Challenging Those Who Do Not Conform to Gender Ideals/5. Refusing to Conform to Gender Ideals. I went on a blind date and met a girl who seemed too good to be true. Everything went well for the next three weeks. We hadn't been very intimate, so when she asked me to come over to her house and stay with her, I jumped at the opportunity. I entered her house to find candles lit and soft music playing. We began to kiss, and I noticed that she hadn't shaved her legs in a while. Disgusting, but under the circumstances I tried to block it out. As we got more involved I was surprised to find a lot of hair under her arms. I quickly pulled away. She asked what was wrong, and after I told her she laughed. Needless to say I felt like I was going to throw up. I put my clothes on and quickly left. On the way home I kept thinking how sick it made

me that she didn't shave. I decided that I would never talk to her again. Sometime later she called me to explain. She told me that she didn't think women should have to shave and that she hadn't shaved for at least a year. She believed I should respect and even like her more for making a stand. Making a stand toward what, I thought? Anyway, I told her that I thought it was disgusting and I told her I didn't want to see her again.

These accounts of resistance, compliance, and the accompanying inner conflicts and social strains speak to the importance of gender ideals in shaping our lives. To this point, we have shown that not everyone fits easily into two biological categories, that gender is a social construction, and that gender ideals are impossible to realize and maintain. So, it should not surprise us to learn that not every society divides people into so-called opposite sexes and genders. People in American Samoa and other Pacific island areas accept a third gender: *fa'afafine*.

A Third Option

■ **CORE CONCEPT 4: While all societies distinguish between male and female, some also recognize a third category.** Jeannette Mageo (1992) begins her article "Male Transvestitism and Cultural Change in Samoa" by describing the guests attending a wedding shower in Samoa. Of approximately 40 "women," 6 were *fa'afafine*—people who are not biologically female but who have taken on the "way of women" in dress, mannerism, appearance, and role. *Fa'afafine* means "in the way of the woman." The closest word we have to express this idea in American society is *transvestite*. Those who study *fa'afafine* maintain that to understand this third gender, we must set aside any cultural preconceptions we have about being male, female, gay, or transvestite (Fraser 2002).

During that wedding shower, the *fa'afafine* staged a beauty contest in which each sang and danced a love song. Such beauty contests are well known in Samoa, and the winner "is sometimes the 'girl' who gives the most stunningly accurate imitation of real girls, such that even Samoans would be at a loss to tell the difference; sometimes the winner is the most brilliant comic" (Mageo 1998, p. 213). Often the *fa'afafine* imitate popular foreign female vocalists, such as Whitney Houston, Britney Spears, Madonna, or Kelly Clarkson. For the past 23 years, the American Samoan Island Association has crowned a *fa'afafine* Miss Island Queen. In 2006 the winner was a second-grade school teacher. The judges were high-profile men and women outside the *fa'afafine* community, including a U.S. military officer, a writer, Miss American Samoa, and a musician (Matàafa 2006).

Mageo believes that "transvestitism" was not common in pre-Christian Samoa (before 1830). If it had been, early Christian missionaries to Samoa would have mentioned

A *fa'afafine* candidate for Miss Island Queen impersonates Tina Turner.

Economic opportunities are limited in American Samoa. Consequently, some men look to the U.S. military; others, to football. Here American Samoan Army Reserve soldiers demonstrate a traditional war dance while a visiting officer (with neck ornaments) looks on.

it in their written accounts of Samoan society, because they were preoccupied with documenting the sexual habits of the Samoans; they would most certainly have mentioned the *fa'afafine*. How, then, did "transvestitism" become commonplace among males, especially in urban areas? Mageo (1992, 1998) argues that *fa'afafine* could not have become commonplace unless something about Samoan society supported gender blurring. She notes that "on a personal level Samoans do not distinguish sharply between men and women, boys and girls." For example, "boys and girls take equal pride in their skills in fights; pre-Christian personal names are often not marked for gender, and outside school little boys and girls still wear much the same clothing" (1998, p. 451).

Another practice that encourages gender blurring relates to the separation of boys and girls and to norms that do not simply channel affection toward so-called opposite-sex persons. Once Samoan boys reach the age of five or six, they begin spending the majority of their time in the company of other boys; at this point, they are prohibited from flirting with girls. At the same time, "close and physically affectionate relations with same-sex people are established practices. In Samoa, as in much of the Pacific, boys may walk about hand-in-hand or with an arm draped around their comrade, and so may girls" (Mageo 1992, p. 452).

Samoans also make a clear distinction between situations in which they must show respect and those in which they may engage in highly sexualized entertainment, including joking, jesting, and imitating. At one time, such entertainment was institutionalized, in the form of ceremonies involving young girls who were part of a village. The Christian missionaries pressured the Samoans to change this practice and, as a result, girls abandoned this entertaining role. Mageo makes the case that *fa'afafine* became both "stand-ins for them and reminders of how girls are not to behave" (1998, p. 454).

Another factor that may account for the widespread emergence of the *fa'afafine* relates to changes in the positions of and opportunities open to men in Samoa. Specifically, these changes are connected to the gradual and ongoing decline of the *augama*, an organization of younger and older men without titles. At one time, the *augama* was considered the "strength of the village" (Mead 1928, p.34), serving "as a village police force or an army reserve" (Mageo 1992, p. 444). It took responsibility for the heavy work, whether that be "on the plantation, or fishing, [or] cooking for the chiefs" (Mead 1928, p. 34).

The system of mass education introduced by the missionaries, the shift away from an agriculture-based economy to a wage-based economy, and the introduction of new technology eventually transformed the *augama*. In the process, this transformation removed an important source of status for Samoan males. This loss of status has been compounded by an unemployment rate of 12 percent in American Samoa. Moreover, when we consider that the total number of paid employees there is 11,618 and that the two largest employers are the tuna canneries (employing

Table 10.1   Occupational Categories in Which Samoan Men and Women Were Overrepresented

In American Samoa, women make up 41.2 percent of the labor force and men make up 58.7 percent. If occupations were filled without regard for gender, 41.2 percent of the people in each category would be female and 58.7 percent would be male.

Top Five Categories in Which Samoan Women Were Overrepresented

Occupation	Female Workers (%)
Secretaries, stenographers, and typists	94
Health service occupations (excluding home care)	90
Personal service occupations	72
Finance, insurance, and real estate	71
Machine operators, assemblers, and inspectors	65

Top Five Categories in Which Samoan Men Were Overrepresented

Occupation	Male Workers (%)
Construction trades	99
Mechanics and repairers	98
Forestry and fisheries	98
Transportation and material moving	98
Precision production, craft, and repair	90

Source: U.S. Department of the Interior (1992).

4,282) and the Samoan government (employing 4,000), we can see the problems of status that many males might experience (Infonautics Corporation 1998; U.S. Department of the Interior 1998; U.S. Census Bureau 2005; see Table 10.1). This economic situation has left the average man without a clear sense of status in Samoan society. For some men, becoming a *fa'afafine* offers them an opportunity to step out of their lowered status and assume the status of well-known female impersonators. Other options for men include joining the military or becoming football players for the U.S., Canadian, or European league.

Mechanisms That Perpetuate Gender Ideals

We have learned that people vary in how much they conform to their society's gender expectations. This fact, however, does not prevent us from using gender expectations to evaluate our own and other people's behavior. For many people, nonconformity (whether deliberate or reluctant) serves as a source of intense confusion, pain, and pleasure.

Sociologists are therefore interested in identifying the mechanisms by which individuals learn and perpetuate a society's gender expectations. To address this issue, we examine four important factors: socialization, situational constraints, the commercialization of gender ideals, and ideologies.

Socialization

■ **CORE CONCEPT 5:** Gender expectations are learned and culturally imposed through a variety of social mechanisms, including socialization, situational constraints, and commercialization of gender ideals. In Chapter 4, we learned that socialization is a life-long learning process by which people develop a sense of self and learn the ways of the society in which they live. The socialization process may be direct or indirect. It is indirect when children learn gender expectations by observing others' behavior, such as the jokes, comments, and stories they hear about men and women or portrayals of men and women in magazines, books, and television (Raag and Rackliff 1998). Socialization is direct when significant others intentionally convey the societal expectations to children.

Socialization theorists argue that an undetermined yet significant portion of male-female differences are products of the ways in which males and females are socialized. Child development specialist Beverly Fagot and her colleagues observed how toddlers in a play group interacted and communicated with one another and how teachers responded to the children's attempts to communicate with them at ages 12 months and 24 months (Fagot et al. 1985). Fagot found no real sex differences in the interaction styles of 12-month-old boys and girls: All of the children communicated by gestures, gentle touches, whining, crying, and screaming.

The teachers, however, interacted with the toddlers in gender-polarized ways. They were more likely to respond to girls when the girls communicated in gentle, "feminine" ways and to boys when the boys communicated in assertive, "masculine" ways. That is, the teachers tended to ignore assertive acts by girls and to respond to assertive acts by boys. Thus, by the time these toddlers reached two years of age, their communication styles showed quite dramatic differences.

Fagot's research was conducted more than 20 years ago. A more recent study found that early childhood teachers are more accepting of girls' cross-gender behaviors and explorations than they are of such behaviors from boys. Apparently, teachers believe that boys who behave like "sissies" are at greater risk of growing up to be homosexual and psychologically ill-adjusted than are girls who behave like "tomboys." This finding suggests that while American society has expanded the range of behaviors and appearances deemed acceptable for girls, it has not extended the range for boys in the same way (Cahill and Adams 1997).

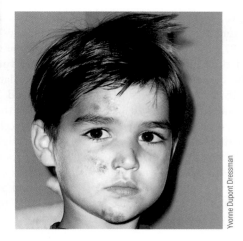

This four-year old boy has rolled his hair in curlers. Should someone tell him that curlers are only for girls, or should he be allowed to play? The same boy (with scrapes and bruises) is shown in a second photo. Doesn't he look more masculine?

This girl is interested in dance. She is also working with her dad on construction sites. Should someone encourage her to give up construction work for dance?

Children's toys and celebrated images of males and females figure prominently in the socialization process, along with the ways in which adults treat children. Barbie dolls, for example, have been marketed for more than 40 years and currently are available in 67 countries. Executives at Mattel present Barbie as an aspirational doll or as a role model to the child. An estimated 95 percent of girls between ages 3 and 11 in the United States have Barbie dolls, which come in several different skin colors and 45 nationalities (including African Barbie and Egyptian Barbie) (Mattel 2005). In fact, most American women have felt Barbie's presence in their life (Lee 2004). Market analysts attribute Mattel's success to the fact that the company has "correctly assessed what it means to a little girl to be grown-up" (Morgenson 1991, p. 66).

For boys, G.I. Joe became the first "action figure" toy on the market, being launched in 1964. It thrived until 1976, when the line was canceled. G.I. Joe was reintroduced in 1978 (Hasbro Toys 1998). Keep in mind that this toy is merely one in a long line of action figures, including Transformers, Micronauts, Star Wars, Power Rangers, X-Men, Street Fighter, Bronze Bombers, and Mortal Kombat. The popularity of G.I. Joe has generated several lines of comic books, 750 different action figures and vehicles, a motion picture, and cartoons. The G.I. Joe logo also appears on school supplies, video games, card games, lunch boxes, posters, and party supplies (Son 1998).

We might conclude that if we change socialization experiences, behavior will change accordingly. The Christian missionaries assigned to Samoa must have recognized this principle, as they sought to "destroy most of the social institutions that guided young Samoans through childhood to adulthood" (Cote 1997, p. 7). Among other things, these missionaries attempted to end the practice of tattooing, and they targeted the *aualuma*, a group of unmarried adolescent girls who "lived together," "supported one another emotionally" and carried out village work projects. As part of the missionaries' efforts, unmarried girls, instead of going to live with the *aualuma*, were brought to live with the pastors and their wives (Cote 1997, p. 8). By introducing mass education, the missionaries also changed the role of the *augama* (the organization of younger and older men without titles). Instead of learning skills as members of the *augama* boys studied inside the classroom, which prepared them to work for wages instead of for the village as a whole.

Male-Female Body Language

Socialization theory helps to explain the differing norms governing body language for males and females. According to women's studies professor Janet Lee Mills, norms governing male body language suggest power, dominance, and high status (see "Language of Body Language"), whereas norms governing female body language suggest submissiveness, subordination, vulnerability, and low status. Mills argues that these norms are learned and that people give them little thought until someone breaks them, at which point everyone focuses on the rule breaker. Such norms can prevent women from conveying a sense of security and control when they are in positions that demand these qualities, such as a lawyer, politician, or physician. Mills suggests that women face a dilemma: "To be successful in terms of femininity, a woman needs to be passive, accommodating, affiliative, subordinate, submissive, and vulnerable. To be successful in terms of the managerial or professional role, she needs to be active, dominant, aggressive, confident, competent, and tough" (Mills 1985, p. 9).

Socialization of Samoan Boys

There are about 4,600 high school students (grades 9–12) in American Samoa. Of this number, about 400 males graduate from high school each year. Each year about 46 graduates leave the Pacific island to play football in the United States at the collegiate level (Syken 2003). That figure translates into about 1 in every 9 high school gradu-

ates. What socialization mechanisms encourage young Samoan males' interest in this sport, which may take them 4,150 miles or more from home?

First, for young Samoan boys to play football, the sport must be available in their society. Football was introduced to the island in 1969, after a U.S. government official decided that the public schools should field football teams. Second, the celebrity status of successful Samoan football players highlights for young Samoan males the rewards of pursuing a football career. For example, 120 football players of Samoan descent played at the University of Hawaii and Arizona State between 1997 and 2000 under Coach Dick Tomey. Today 200 players of Samoan descent play on Division I college teams and a few play in the NFL—most notably Lofa Tatupu. Others play in the Canadian Football League and NFL Europe (U.S. House of Representatives 2004; Ferguson 2005). Samoan youth hear and read praise from college and NFL coaches about Samoan players, which fuels their interest in the game and channels their choices toward football and not some other sport, such as golf or tennis:

- They're so physical. Even in scrimmages they go all out. (Busch 2003)
- There are no athletes that are, in my estimation, more competitive, more athletic, or more family-oriented, or who fit into a team concept as well as Samoan athletes. The more we could get on our team, the better I felt. (Tomey 2003)
- Why not use what God gave you? You wouldn't mind putting a golf club in their hands, but you have to be realistic. I don't see a [Samoan] Tiger Woods out there. We're going to use what we know. (Malauulu 2003)

Third, high school football is very popular in American Samoa. The six high school football teams use the same 10,000-seat stadium to host all of their games—five games per weekend. The schools play each other twice each season and meet again for playoffs. In addition to the fans who attend games, other fans listen to the games on the radio or watch them on TV (Syken 2004; Ferguson 2005).

Finally, American Samoan males have relatively few career opportunities. Consider that one in every 50 persons moves away each year. Surely this high rate is tied in part to the limited opportunities. As a case in point, consider that one-third of all Samoan jobs are connected to tuna fishing and processing. The importance of this industry to the economy is such that tuna represents 93 percent of American Samoa's total exports (U.S. Central Intelligence Agency 2004).

Commercialization of Gender Ideals

In Chapter 6 we described strategies capitalists use to generate profits. Three of the six measures apply to the **commercialization of gender ideals**—the process of

commercialization of gender ideals The process of introducing products to the market by using advertising and sales campaigns that draw on socially constructed standards of masculinity and femininity.

Department of Defense photo by SSGT James M. Bowman

The celebrity status of Samoan NFL football players helps to channel the energies of American Samoan males in that sport's direction. Approximately one in every eight American Samoan males goes on to play football in the United States at the collegiate level.

introducing products into the market by using advertising and sales campaigns that promise consumers they will achieve masculine and feminine ideals if they buy and use the products. Those measures are creating new products that consumers "need" to buy, improving existing products, and creating new markets. Keep in mind that sales depend on buyers. One way to convince people to buy products is to play on their insecurities over whether they meet or maintain (as they age) appearances that conform to gender ideals. Of course, meeting gender ideals requires a great deal of effort.

Creating New and Improving Existing Products That Consumers "Need" to Buy. The list of "new" products is endless, especially for women. From hair dye to toenail polish, products are available to improve almost every female body part or body function (from the top of the head to the tip of the toe). Other such products include four-times-a-year menstrual period–control pills, vaginal moisturizers, chin gyms (a mouthpiece that includes a miniature weightlifting system to help those who use it avoid or lose a double chin), botox, wrinkle creams, hair removers, and artificial fingernails. Relatively new products on the market for men include the erectile dysfunction drugs Viagra, Levitra, and Cialis, which are being advertised even to men who do not have the medical condition for which these drugs should be prescribed. Millions of men have tried these drugs to date, and the manufacturers hope to attract millions more "by suggesting that if men cannot have an erection 'on demand,' if they 'fail' even once, they are candidates for these drugs" (Tuller 2004).

Virtually every product listed above will eventually be improved and then billed as a "new and improved" version of its original. When "old" products are thus made obsolete, their manufacturers encourage users to abandon or throw them away in favor of the new—thereby generating further profits for the manufacturers.

Creating New Markets. The female market is saturated with products. From a marketing perspective, the amount of money that female consumers have to spend on cosmetic products may have reached its limit. Thus, marketers must search for a new market—and that new market appears to be males. The problem for marketers is how to sell men products that have traditionally been viewed as "feminine." One strategy is to "masculinize" feminine products:

- A body wash ad tells men that this is a "body wash that's not for sissies." It's what "guys want. For less."
- An ad for a hair removal product offers men "five reasons to cut a rug," including "[Your] Willy will look more like a William."
- An ad for a revitalizing face cream maintains that the product is "more evolved," playing on a hierarchy that puts men at the top of the evolutionary chain.

Structural Constraints

Structural constraints are the established and customary rules, policies, and day-to-day practices that affect a person's life chances. One example is the structural con-

Nick Richards

Creating new markets is one way capitalists increase profits. In the area of cosmetics and toiletries, manufacturers are seeking to expand the market beyond females to males by "masculinizing" their products.

structural constraints The established and customary rules, policies, and day-to-day practices that affect a person's life chances.

The Language of Body Language

"Could you say no to this woman?" In it she violates many traditional female behaviors with relaxed posture: arms, legs positioned away from body; direct, confrontational eye contact; no affiliative smile.

Mills demonstrates the "power spread," another typical high-status male pose, with hands behind head, elbows thrust out, legs in a "brokenfour" position, and an unaffiliative facial expression. Women dressed for success appear shocking in this pose.

Mills recruited Richard Friedman, assistant to the president at the University of Cincinnati, to model for these photographs. Friedman does not appear shocking in the same pose, because many male executives conduct business from a similar position.

Ah, but doesn't Mills look feminine in this typically feminine pose—with canted head, affiliative smile, ankles crossed, and hands folded demurely?

And doesn't Friedman look ridiculous in the same pose?

Power is often wielded in postural align-ment, gesture, and use of objects. Mills holds power, along with papers on which attention is focused; Friedman signals deference with lowered gaze and constricted body.

Now Friedman holds power, along with papers, with lower limbs spread away from his body; Mills signals deference with an attentive pose, submissively bowed head, and a hand-to-mouth gesture of uncertainty.

In this typical office scene, Friedman holds power with an authoritative stance—one hand in pocket and the other at mid-chest, straight posture, and head high; Mills is submissive with canted head, smile, arms and hands close to her body. Note Friedman's wide, stable stance and Mills's unstable stance. Many women tend to slip into a posture similar to that of Mills when talking to a shorter male authority figure.

Now the tables are turned: Friedman defers to authority by assuming a feminine, subordi-nate posture—with scrunched-up spine, con-stricted placement of arms and legs, canted head, and smiling attentiveness.

In the United States 90 percent of registered nurses, including those who serve in the military, are female; 99 percent of electricians are men. Sociologists argue that one's position in the social structure can channel behavior in stereotypically male or female directions.

straints that push men and women into jobs that correspond with society's ideals for sex-appropriate work. Women are pushed into work roles that emphasize personal relationships and nurturing skills or that pertain to family-oriented and "feminine" products and services. Men are pushed into jobs that emphasize decision making and control and that pertain to machines and "masculine" products and services. The job descriptions demand that occupants behave in sex-appropriate ways. Table 10.2 shows the top 20 occupations for women who work full-time. Since 43.6 percent of all full-time jobs are filled by women, we can see that they are most overrepresented in the secretaries/administrative assistants category. Notice that 2.6 million women in the United States work as secretaries or administrative assistants and that 96.7 percent of such positions are filled by women. Other occupational categories in which women make up 90 percent or more of the workforce include registered nurses, home health aides, receptionists, bookkeepers, teacher assistants, preschool and kindergarten teachers, and childcare workers.

In Table 10.3, notice that 2.7 million men in the United States work as truck drivers and that 96 percent of such positions are filled by men. Other occupational categories dominated by men include construction laborers, carpenters, grounds maintenance workers, and electricians. Sociologist Renee R. Anspach's (1987) research observing nurses and physicians working in neonatal care units illustrates vividly how one's position in a social structure can channel behavior in stereotypically male or female directions.

The Case of Physicians and Nurses

Anspach spent 16 months observing and holding interviews in two neonatal intensive care units (NICUs). Among other things, she found that nurses (almost all of whom were female) and physicians (most of whom, at the time of her research in the mid-1980s, were male) used different criteria to answer the question, "How can you tell if an infant is doing well or poorly?" Physicians tended to draw on so-called objective (technical or measurable) information and immediate perceptual cues (skin color, activity level) noted during routine examination:

> Well, we have our numbers. If the electrolyte balance is OK and if the baby is able to move one respirator setting a day, then you can say he's probably doing well. If the baby looks gray and isn't gaining weight and isn't moving, then you can say he probably isn't doing well. (Anspach 1987, pp. 219–220)

Although technical and measurable signs were likewise important to the nurses, Anspach found that the nurses also considered interactional clues, such as the baby's level of alertness, ability to make eye contact, and responsiveness to touch:

> Basically emotionally if you pick them up, the baby should cuddle to you rather than being stiff and withdrawing. Do they quiet when held or do they continue to cry when you hold them? Do they lay in bed or cry continuously or do they quiet after they've been picked up and held and fed? . . . Do they have a normal sleep pattern? Do they just lay awake all the time really interacting with nothing or do they interact with toys you put out, the mobile or things like that, do they interact with the voice when you speak? (p. 222)

Anspach concluded that the differences between nurses' and physicians' responses to the question, "How can you tell if an infant is doing well or poorly?" could be traced to their daily work experiences. In the division of hospital labor, nurses interact more with patients than do physicians. Also, doctors and nurses have access to different types of knowledge about infants' condition—types of knowledge that correspond to our stereotypes of how

Table 10.2   20 Leading Occupations of Full-Time Female Workers, 2006

Occupations	Total Women (in 1,000s)	% Workers Who Are Women	Women's Median Weekly Income	Men's Median Weekly Income
All	46,269	43.6	$600	$743
Secretaries/administrative assistants	2,595	96.7	$584	$559
Elementary and middle school teachers	1,916	82.2	$824	$920
Registered nurses	1,713	90.3	$971	$1,074
Nursing, psychiatric, and home health aides	1,238	88.1	$395	$471
Customer service representatives	1,110	70	$533	$615
Cashiers	1,044	73.6	$327	$387
First-line supervisors/office managers	980	70.1	$658	$812
Bookkeeping, accounting, and auditing clerks	898	88.6	$582	$607
Accountants and auditors	886	61.2	$844	$1,160
Receptionists and information clerks	882	92.9	$467	$562
First-line supervisors/managers of retail sales workers	872	42.3	$526	$732
Retail salespersons	867	43	$405	$471
Maids and housekeepers	773	88	$348	$404
Preschool and kindergarten teachers	702	97.6	$520	—
Office clerks (general)	583	88.6	$534	$564
Waiters	580	68.8	$348	$407
Secondary school teachers	535	54.2	$890	$950
Teacher assistants	517	92	$405	—
Social workers	500	69.4	$728	$749
Child-care workers	425	95.3	$345	—

Source: U.S. Department of Labor, Bureau of Labor Statistics 2007.

Table 10.3   20 Leading Occupations of Full-Time Male Workers, 2006

Occupations	Total Men (in 1,000s)	% Workers Who Are Men	Men Median Weekly Income	Women's Median Weekly Income
All	59,747	56.3	$734	$600
Driver/sales workers and truck drivers	2,708	96	$651	$436
First-line supervisors or managers of retail sales workers	1,321	58	$732	$536
Construction laborers	1,292	96.5	$513	—
Laborers and freight, stock, and material movers (hand)	1,210	85.8	$484	$412
Carpenters	1,169	98	$598	—
Janitors and building cleaners	1,076	72.1	$447	$375
Sales representatives (wholesale and manufacturing)	905	75.1	$997	$739
Grounds maintenance workers	832	95.2	$403	—
Chief executive officers	782	75.1	$1,907	$1,422
Electricians	769	99	$754	—
Food preparation workers	738	60.0	$377	$340
Automotive service technicians and mechanics	687	98.7	$635	—
General and operations managers	655	72	$1,256	$957
First-line supervisors or production and operating workers	650	93	$832	$587
Store clerks and order fillers	642	61	$457	$425
First-line supervisors of construction trades and extraction workers	634	98.4	$866	—
Computer software engineers	623	80	$1,410	$1,272
Police and sheriff's patrol officers	568	88	$884	$758
Accountants and auditors	562	39.9	$1,160	$844
Security guards and gaming surveillance officers	531	76.1	$498	$416
Industrial truck and tractor operators	517	94.1	$514	—

Source: U.S. Department of Labor, Bureau of Labor Statistics 2007.

females and males manage and view the world. Because physicians have only limited amounts of daily interaction and contact with infants, they tend to rely on perceptual and technological (measurable) cues. By comparison, nurses remain in close contact with infants throughout the day; consequently, they are more likely to consider interactional cues as well as perceptual and technological ones. Anspach suggests that a person's position in the division of labor "serves as a sort of interpretive lens through which its members perceive their patients and predict their futures" (p. 217).

Sexism

■ **CORE CONCEPT 6:** The belief that one sex—and by extension, one gender—is superior to another, and that this superiority justifies inequalities between sexes, is sexism. **Sexism** is the belief that one sex—and by extension, one gender—is innately superior to another, justifying unequal treatment of the sexes. Sexism revolves around three notions:

1. People can be placed into two categories: male and female.
2. A close correspondence exists between a person's primary sex characteristics and characteristics such as emotional activity, body language, personality, intelligence, the expression of sexual desire, and athletic capability.
3. Primary sex characteristics are so significant that they explain and determine behavior and the social, economic, and political inequalities that exist between the sexes.

One example of sexism is the belief that men are prisoners of their hormones, making them powerless in the face of female nudity or sexually suggestive dress or behavior. Still another sexist ideology claims that men are not capable of forming relationships with other men that are as meaningful as those formed between women. While hundreds of books have been written by men to refute these stereotypes, they persist in popular culture (Shweder 1994).

We might also add a fourth notion underlying sexism: People who behave in ways that depart from ideals of masculinity or femininity are considered deviant, in need of fixing, and subject to negative sanctions ranging from ridicule to physical violence. This ideology is reflected in U.S. military policy toward gay men and lesbians, for example. According to a U.S. Department of Defense (1990) directive formulated in 1982,

sexism The belief that one sex—and by extension, one gender—is innately superior to another, justifying unequal treatment of the sexes.

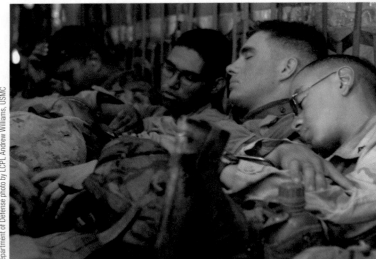

Department of Defense photo by Staff Sgt. Stacy L. Pearsall, U.S. Air Force

Department of Defense photo by LCPL Andrew Williams, USMC

Whether taking part in training exercises or relaxing afterward, soldiers often make physical contact with other soldiers. Would the presence of gay men and lesbians in the military disrupt these activities? Might the presence of two men or a man and a woman who share a strong friendship or attraction also be disruptive?

Homosexuality is incompatible with military service. The presence of such members adversely affects the ability of the Armed Forces to maintain discipline, good order, and morale; to foster mutual trust and confidence among the members; to ensure the integrity of the system of rank and command; to facilitate assignment and worldwide deployment of members who frequently must live and work under close conditions affording minimal privacy; to recruit and retain members of the military services; to maintain the public acceptability of military services; and, in certain circumstances, to prevent breaches of security. (p. 25)

GLOBAL COMPARISONS

Countries That Allow Gay Men and Lesbians to Serve in the Military Openly and Countries That Ban Them from Military Service

BETWEEN 1998 AND 2007, the Department of Defense discharged 11,000 servicemen and servicewomen for homosexuality under the "don't ask, don't tell" policy (Frank 2004). This number is significant in light of the fact that U.S. military forces are currently stretched to the limit. Gay men and lesbians in 161 different occupations, including foreign-language specialists and infantrymen, have been discharged (Online Newshour 2007). Yet little if any scientific evidence supports this directive. In fact, whenever Pentagon researchers (with no links to the gay and lesbian communities and with no ax to grind) have found evidence that runs contrary to this directive, high-ranking military officials have generally refused to release or directed researchers to rewrite their reports. A 2007 Zogby poll found that 75 percent of U.S. soldiers returning from Iraq and Afghanistan indicated that they were comfortable interacting with gay colleagues; only 5 percent indicated that they were extremely uncomfortable (Shalikashville 2007).

People who oppose the presence of gay men and lesbians in the military stereotype them as sexual predators just waiting to pounce on heterosexuals while they shower, undress, or sleep. Opponents seem to believe that *any* same-sex person is attractive to a gay man or lesbian. But as one gay ex-midshipman noted, "Heterosexual men have an annoying habit of overestimating their own attractiveness" (Schmalz 1993, p. B1). Supporters of gay men and lesbians in the military point out that almost 75 percent of troops say that they usually shower alone; only 8% say they usually take showers where others can see them (Zogby 2007). Apparently openly gay soldiers serve in the British military without any major problems; the British military even recruits soldiers at gay pride events (Lyall 2007)

Countries that allow openly gay men and lesbians in the military: Argentina, Australia, Austria, Bahamas, Belgium, Bulgaria, Canada, Colombia, Croatia, Czech Republic, Denmark, Estonia, Finland, France, Germany, Hungary, Ireland, Israel, Italy, Lithuania, Luxembourg, Netherlands, New Zealand, Norway, Peru, Poland, Portugal, Romania, Slovenia, South Africa, Spain, Sweden, Switzerland, Taiwan, Thailand, United Kingdom

Countries that ban openly gay men and lesbians from the military: Brazil, Cuba, Egypt, Iran, North Korea, the Philippines, Saudi Arabia, Syria, United States, Venezuela, Yemen

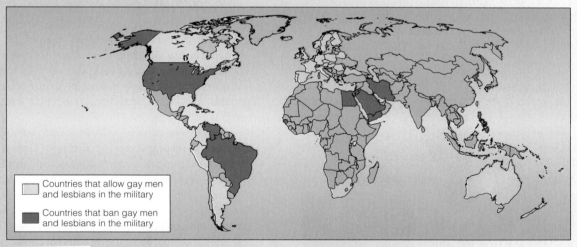

▲ **Figure 10.1**

Source: Wikipedia 2009.

Gender Inequality

■ **CORE CONCEPT 7:** When sociologists study inequality between males and females, they seek to identify the social factors that put one sex at a disadvantage relative to the other. Social inequality exists between men and women when one category relative to the other (1) faces greater risks to physical and emotional well-being, (2) possesses a disproportionate share of income and other valued resources, and/or (3) is accorded more opportunities to succeed. There are many areas in which men are disadvantaged relative to women. One example is life expectancy. The average life expectancy for males living in American Samoa is 72. That is eight years less than the life expectancy of females living there (World Health Organization 2007). The average man in the United States can expect to live 5.4 fewer years than the average woman. In this regard, some groups of men are more disadvantaged than others and some groups of females are more advantaged. For example, females classified as Asian can expect to live 16.9 years longer than males classified as black.

There are also many areas in which women are disadvantaged relative to men. Notably, women tend to earn less money than men. Table 10.4 shows a $10,850-per-year difference between the median incomes of males and females age 25 and older. It also shows that regardless of the level of education, men earned more money than women. Income differences varied according to age, with the greatest inequality between men and women at ages 45–55 and the least inequality between men and women at ages 16–24. Note that in 2005, women ages 16–24 earned 93.2 cents for every dollar men earned and that women ages 45–54 earned 75.4 cents for every dollar (see Figure 10.2).

Likewise, male-female income differences vary by occupation. Tables 10.2 and 10.3 show the top 20 male- and female-dominated occupations and the percentage of males and females in each occupation. Notice that for the 20 occupations in which women dominated, their median weekly earnings ranged from a low of $327 (for cashiers) to a high of $971 (for registered nurses). Likewise, in the top 20 male-dominated occupations, men's median weekly earnings ranged from a low of $377 (for food preparation workers) to a high of $1,907 (for chief executive officers). Note that in all male-dominated occupational categories for which data is available, men earned more than their female counterparts.

There are many possible explanations for this income difference. They include the following:

- Women are disproportionately employed in lower-paying, lower-status occupations.
- Women choose or are forced into lower-paying positions that are considered sex-appropriate, such as teacher, secretary, and caregiver.
- Women choose or are forced into positions that offer fewer, more-flexible hours to meet caregiving responsibilities.
- Women choose or are forced into lower-paying subspecialties within high-paying professions (for example, women tend to be divorce lawyers rather than corporate lawyers, and pediatricians rather than heart surgeons).
- Employers underinvest in the careers of childbearing-age women because they assume the women will eventually leave to raise children.
- Women leave the labor market to take care of children and elderly parents, and then they re-enter it.
- Women choose or are forced into occupations that will not require them to relocate, work in unpleasant environments (such as mines), or take on hazardous assignments—three activities associated with higher pay (Sahadi 2006).
- Employers view women's salary needs as less important than men's and pay women accordingly. Unfortunately, women's earnings are often considered supplemental to men's—earnings that can be used to buy "extras"—when in reality many women are heads of households.
- Female-dominated occupations (such as child-care workers and nursing aides) are valued less and thus rewarded less than male-dominated occupations (such as auto mechanics and construction laborers).
- When negotiating for salaries, women underestimate their worth to employers and ask for less than their male counterparts.
- Some employers steer males and females into different gender-appropriate assignments and offer them different training opportunities and chances to move into better-paying jobs. Lawyers representing 1.5 million current and former female Wal-Mart employees filed a

One reason that life expectancy is shorter for males than for females is that males are assigned the most hazardous jobs in society.

USDA Photo by Tim McCabe

Table 10.4 Median Income of Males and Females Age 25 and Older Working Year Round and Full-time, by Education Level

	Overall	Less Than High School	High School Graduate	4-Year Degree	Graduate/ Professional Degree
Male, age 25+	$32,850	$22,138	$31,683	$53,693	$71,918
Female, age 25+	$22,000	$13,076	$20,179	$36,250	$47,319
Difference	$10,850	$ 9,062	$11,504	$17,443	$24,599

Source: U.S. Census Bureau (2006).

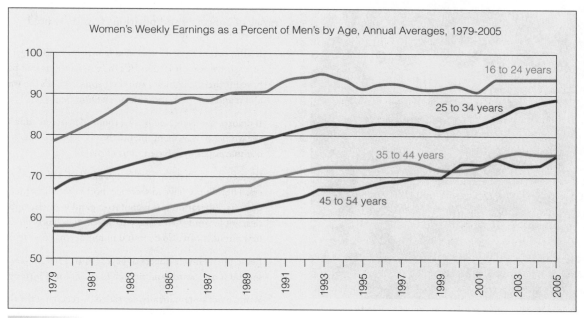

▲ **Figure 10.2**

This graph shows year-by-year variations in money women earned relative to men. If we take a longer view—total earnings over a 15-year period—we find that the average woman earned about 38 percent of what the average man did (Madrick 2004). Economists Stephen J. Rose and Heidi Hartman (2004) compared men's and women's total earnings between 1983 and 1998 and found that the average woman earned $273,592 while the average man earned $722,693.

Source: U.S. Bureau of Labor Statistics (2006).

class-action suit against that company, claiming that it pushed women into positions (such as sales clerk in the baby clothes department rather than in the hardware department) that were less likely to help them advance to the ranks of management (Love 2007). Wal-Mart, the world's largest retail company and employer of 1.2 million people at 3,500 stores, countered with the claim that women are not interested in managerial positions (O'Conner 2007; Greenhouse 2003, 2004).

Feminism

■ **CORE CONCEPT 8:** In its most basic sense, the feminist perspective advocates equality between men and women. When

women living in the United States are asked, "Do you consider yourself a feminist or not?" only one in four or 25 percent of respondents answers yes. When asked the same question accompanied by a definition of *feminist* ("A feminist is someone who believes in social, political, and economic equality of the sexes. Do you think of yourself as a feminist or not?"), 65 percent of women answer yes. One possible explanation for the difference is that only 12 percent of women and 10 percent of men consider the label of "feminist" to be a compliment (CBS News Polls 2005). People correctly or incorrectly associate feminists with many mainstream and controversial positions, including support for equal rights, equal pay, affordable daycare, abortion rights, opposition to sexual harassment, a lack of respect for "stay-at-home" moms, and a dislike or even

Department of Defense photo by Mass Communication Specialist Seaman Michelle Rhonehouse, U.S. Navy

Feminists believe that both men and women should be free to make choices that defy gender expectations. They would support the 98,000 "stay-at-home" dads, who have remained out of the labor force for at least one year to care for children while their wives worked outside the home.

hatred of men (Time.com 2007). It appears that respondents see themselves as feminist on some positions and not others.

A basic sociological definition of **feminism** is "a perspective that advocates equality between men and women." Questions about what that equality looks like and how equality should be achieved distinguish feminist camps from one another. Some feminist men, for example, believe that the dominant model of manhood or masculinity oppresses both women and men. Other feminists take a separatist position. For example, Mary Daly, a

feminism In its most basic sense, a perspective that advocates equality between men and women.

Boston College professor, refused to allow male students to attend her women's studies classes, arguing that class discussion was inhibited and "dumbed-down" by their presence. She did agree to meet individually with her male students. In an out-of-court settlement with the college, Daly agreed to retire (Daly 2006). Many feminists believe that any inequality between males and females, including that which gives females an advantage over males, needs to be addressed (see Working For Change: "Historic Events That Opened Opportunities for Women"). The following quotations from well-known feminists demonstrate the range of concerns and positions feminists hold:

> It's important to remember that feminism is no longer a group of organizations or leaders. It's the expectations that parents have for their daughters, and their sons, too. It's the way we talk about and treat one another. *Anna Quindlen*

> If divorce has increased by one thousand percent, don't blame the women's movement. Blame the obsolete sex roles on which our marriages were based. *Betty Friedan*

> Women do not have to sacrifice personhood if they are mothers. They do not have to sacrifice motherhood in order to be persons. Liberation was meant to expand women's opportunities, not to limit them. The self-esteem that has been found in new pursuits can also be found in mothering. *Elaine Heffner*

> Remember, Ginger Rogers did everything Fred Astaire did, but she did it backwards and in high heels. *Faith Whittlesey*

> Women are systematically degraded by receiving the trivial attentions which men think it manly to pay to the sex, when, in fact, men are insultingly supporting their own superiority. *Mary Wollstonecraft*

> No one sex can govern alone. I believe that one of the reasons why civilization has failed so lamentably is that it had one-sided government. *Nancy Astor*

> I myself have never been able to find out precisely what feminism is: I only know that people call me a feminist whenever I express sentiments that differentiate me from a doormat, or a prostitute. *Rebecca West*

Sociologists take a feminist perspective when they emphasize in their teaching and research the following kinds of themes: the right to bodily integrity and autonomy, access to safe contraceptives, the right to choose the terms of pregnancy, access to quality prenatal care, protection from violence inside and outside the home, freedom from sexual harassment, equal pay for equal work, workplace rights to maternity and other caregiving leaves, and the inescapable interconnections between sex, gender, social class, race, culture, and religion. Important feminist concepts include those covered in this chapter—such as sex, gender, and sexism—and others not covered here—including patriarchy, sexual objectification, and oppression.

WORKING FOR CHANGE

Historic Events That Opened Opportunities for Women

1848 A group of women and men assembled in Seneca Falls, New York, to discuss the status of women in American society. During this event, regarded as the beginning of the women's rights movement in the United States, Elizabeth Cady Stanton delivered the "Declaration of Rights and Sentiments" (a statement calling for women's rights, patterned after the Declaration of Independence).

1872 Susan B. Anthony and 15 other women were arrested and indicted for "knowingly, wrongfully, and unlawfully voting for a representative to the Congress of the United States." She appealed her conviction to the U.S. Supreme Court, but her conviction was upheld.

1920 The Nineteenth Amendment to the U.S. Constitution became law, guaranteeing women the right to vote.

1921 The American Birth Control League (which later became the Planned Parenthood Federation of America) was founded by Margaret Sanger and Mary Ware Dennett.

1938 Congress passed the Fair Labor Standards Act, prohibiting child labor in factories and mines. The law also established a minimum wage, a maximum work week, and overtime pay, which benefited many working women.

1943 As more and more men enlisted during World War II, more and more women entered the civilian workforce. Over 6 million women held factory jobs as welders, machinists, and mechanics. About 310,000 women worked in the U.S. aircraft industry alone.

1963 The Equal Pay Act, signed by President Kennedy, prohibited the practice of paying women less money than men for the same job.

1964 President Lyndon B. Johnson signed the Civil Rights Act, outlawing discrimination in unions, public schools, and the workplace on the basis of race, creed, national origin, or sex.

During World War II, over 6 million women held factory jobs as welders, machinists, and mechanics; 310,000 women were employed in the U.S. aircraft industry alone. When women demonstrated their ability to do jobs previously held by males it shattered stereotypes.

National Archives and Records Administration

1965 Under Title VII of the Civil Rights Act of 1964, the Equal Employment Opportunity Commission was established. Its function has been to enforce federal law prohibiting discrimination in the workplace on the basis of sex, religion, race, color, national origin, age, or disability.

1966 The National Organization for Women (NOW) was founded by Betty Friedan. It has challenged sex discrimination in the workplace through public demonstrations, lobbying, and litigation.

1972 The U.S. Senate approved the Equal Rights Amendment—49 years after it was first introduced in Congress. It was then sent to the state legislatures for ratification.

Title IX of the Education Amendments of 1972 made sex discrimination in schools that receive federal funding illegal. Title IX also requires that schools that receive federal funds give women and girls an equal opportunity to participate in sports and to receive millions of dollars in athletic scholarships.

1973 The U.S. Supreme Court ruled (7–2) that, according to a woman's right to privacy guaranteed in the Fourteenth Amendment, the Texas law restricting abortion in the first trimester was unconstitutional. As a result, anti-abortion laws in nearly two-thirds of the states were declared unconstitutional, legalizing abortion nationwide.

1975 President Gerald Ford signed a defense appropriations bill to allow women to be admitted into U.S. military academies.

1978 The Women's Army Corps (WAC) as a separate military entity was dismantled, and women began integrating into the U.S. Army.

1993 President Bill Clinton signed the Family Medical Leave Act. It allows eligible employees to take up to 12 weeks of leave for reasons of illness, maternity, adoption, or a child's serious health condition.

Source: Adapted from Barbara Boxer, U.S. senator from California, Historical Timeline for Women's History (2007), http://boxer.senate.gov/whm/time.cfm

Gender, Ethnicity, Race, and the State

■ **CORE CONCEPT 9: Sex and gender are complicated by racial classification and legal relationship to a state or country.** **Ethgender** refers to people who share (or are believed by themselves or others to share) the same sex, race, and ethnicity. This concept acknowledges the combined (but not additive) effects of gender, race, and ethnicity on life chances. Ethgender merges them into a single social category. In other words, a person is not just a resident of Samoa, but an ethnic Samoan or a white Samoan (Geschwender 1992). In addition to their sex and ethnic or racial classification, people have a legal relationship with a country or state, which puts them in such categories as citizen, refugee, temporary worker, or illegal migrant. In a legal sense, people born in American Samoa are classified as U.S. nationals, not citizens.

We use the term *country* or *state* here to mean a governing body organized to manage and control specified activities of people living in a given territory. The governing body is almost always male dominated. For example, in the 50-year history of the American Samoan legislature, only 5 women have been elected to office. In the November 1998 election, only 5 of 50 candidates competing to fill 20 seats in the Samoan House of Representatives were women (*Samoa Daily News* 1998c). The U.S. Congress is similarly male dominated. Of the 435 members that make up the U.S. House of Representatives, only 74 (17 percent) are women. Of the 100 voting members of the U.S. Senate, 16 (16 percent) are women (U.S. Senate 2007).

Sociologists Floya Anthias and Nira Yuval-Davis (1989) give special attention to women, their ethnicity, and the state. They argue that "women's link to the state is complex" and that while men are subject to state control, women "are a special focus of state control because of their role in human reproduction—women carry and give birth to children who become the state's citizens and future labor force." Anthias and Yuval-Davis identify five areas over which the state may choose to exercise control.

1. Women as Biological Reproducers of Babies of a Particular Ethnicity or Race.
As factors that can underlie a state's population control policies, Anthias and Yuval-Davis (1989) name "fear of being 'swamped' by different racial and ethnic groups" or fear of a "demographic

Being classified as male or female affects life chances. But so do other factors, such as race and citizenship. In other words, a person is not just a male or a female, but an American Samoan female or male classified as "Native Hawaiian or other Pacific Islander."

holocaust"—that is, fear that a particular racial or ethnic group will die out or become too small to hold its own against other ethnic groups (p. 7). Such policies can range from physically limiting numbers of a particular racial or ethnic group deemed undesirable to actively encouraging the "right kind" of women to produce more children. Policies that limit numbers include immigration control (limiting or excluding members of certain ethnic groups from entering a country and subsequently producing children), extermination, forced sterilization, and massive birth control or family planning campaigns.

2. Women as Reproducers of the Boundaries of Ethnic or National Groups.
In addition to implementing policies affecting the numbers and kinds of babies born, states may institute policies that define who the parents should be. Examples of such policies include laws prohibiting sexual relationships between men and women of different

---

ethgender A social category that combines sex, gender, race, and ethnicity.

GLOBAL COMPARISONS

Countries with Paid Maternity Leave

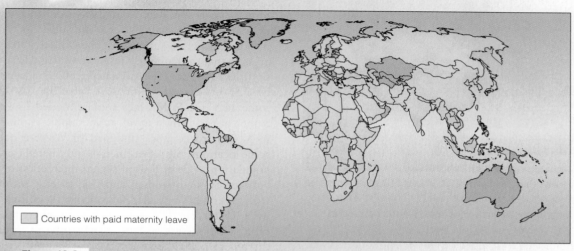

Countries with paid maternity leave

▲ **Figure 10.3**
Source: Wikipedia (2007)

races or ethnicities, laws specifying marriage for a couple if their child is to be recognized as legitimate, and laws connecting a child's race to the mother's or father's race (see Chapter 9).

3. Women as Transmitters of Social Values and Culture. The state can institute policies that either encourage women to be the main socializers of their offspring or leave socialization in the hands of others or the state. For example, it may tie welfare payments to nonemployment (or, under welfare reform, to employment) so that mothers are forced to stay home with their children or to enter the workforce. Some countries have generous maternity leave policies (see Global Comparisons: "Countries with Paid Maternity Leave"); others have no policies at all. Sometimes state leaders become concerned that children of particular ethnic or racial groups are not learning the cultural values or language that they need to succeed in the dominant culture. This concern may motivate them to fund programs such as preschools that expose particular kinds of children to the personal, social, and learning skills deemed necessary.

4. Women as Symbols of Urgent Issues. Political leaders often use images of women and men to symbolize what the leaders believe are the most urgent issues a country faces.

In wartime, for example, the country may be represented as "a loved woman in danger or as a mother who lost her sons in battle" (Anthias and Yuval-Davis 1989, pp. 9–10). Men are called to battle to protect the women and children. Often the leaders present the image of a woman who meets the culture's ideal of femininity and who belongs to the dominant ethnic group. Sometimes political leaders evoke images of women of a certain ethnic or racial group as the source of a country's problems, such as Hispanic women who produce many children, or African American "welfare mothers" with no economic incentives to practice birth control. Checking the facts often reveals that such images are unfounded. For instance, in "Fertility among Women on Welfare: Incidence and Determinants," sociologist Mark R. Rank (1989) maintains that "it is impossible to calculate with any precision the fertility rate of women on public assistance," (p. 296) because the data available have serious flaws. For one thing, women move in and out of welfare, and we simply do not keep track of their fertility during these periods. "There is no way of judging whether the fertility rate of women on welfare is high or low" (p. 296) relative to the fertility rate of other women.

5. Women as Participants in National, Economic, and Military Struggles. States may implement policies governing the roles that women and men can assume in crises, nota-

DoD photo by TSGT Curt Eddings

This photo shows children who live at St. Vincent's Home for Amerasian Children in South Korea. Amerasian is the name given to the offspring of U.S. servicemen and servicewomen from countries classified as Asian.

bly in war. Historically, women have played supportive and nurturing roles, even in situations in which they have been exposed to great risks. In most countries, women are not drafted; they volunteer to serve. If they are drafted, the state defines acceptable military roles for them. If women fight, they often do so as part of special units or in an unofficial capacity. In the United States, two milestone events affected women's military roles. In 1973 the draft came to an end and a volunteer military was instituted. Recruiters focused on women to fill the ranks. In 1994 the United States ended the risk rule, thereby giving women access to 90 percent of positions in the military, including positions that support male soldiers fighting in hostile zones (Wilgoren 2003).

Through its military institutions, the state even establishes policies that govern male soldiers' sexual access to women outside military bases, both in general and in times of war. In *Let the Good Times Roll: Prostitution and the U.S. Military in Asia* Saundra Pollock Sturdevant and Brenda Stoltzfus (1992) examine "the sale of women's sexual labor outside U.S. military bases" (p. vii). They present evidence that the U.S. military helps regulate prostitution;

that retired military officers own some of the clubs, massage parlors, brothels, discotheques, and hotels; and that the military provides the women with medical care to prevent the spread of sexually transmitted diseases between the soldiers and the women.

As an example of the complex relationship between military men and local women, consider that thousands of Filipino women who live near the Subic Bay Naval Base filed a class-action suit against the United States in 1993, arguing that the United States has moral and legal responsibilities to support the estimated 8,600 children fathered by U.S. servicemen stationed at Subic Bay. In the 1940s, some 14,371 U.S. Marines were stationed in American Samoa at the Tutuila naval station alone; another 25,000 to 30,000 troops were stationed in Western Samoa. In some locations, the servicemen outnumbered the Samoan population. A great deal of sexual interaction took place between Samoan women and U.S. troops. Although we will never know exactly how many illegitimate babies were fathered by U.S. soldiers, the numbers are significant. "One mission society reported that in Upolu alone there were 1,200 known instances of illegitimate children by American soldiers" (Stanner 1953, p. 327).

In evaluating the role of the military in the women's lives, we must consider that many poor women who live near the bases see a relationship with a U.S. recruit as their only way out of poverty and a desperate situation. The Subic Bay situation is therefore more complicated than the story presented by the media. To clarify this point, we will use an example from a student at Northern Kentucky University. A young male student, perhaps 23 or 24 years old, stopped the author after a class in which we had discussed the Subic Bay base closing. He explained that when he was in the navy and his ship was docked at Subic Bay, it seemed as if the entire town turned out to welcome the ship. Local women were everywhere. Whereas many recruits visited prostitutes for one-night stands, others fell in love with local women. Usually the commanding officers attempted to discourage permanent relationships. Although some might criticize the officers' actions, they might have been anticipating the difficulties ahead for a recruit who took a local woman as his wife and then continued life at sea, leaving her alone. Unfortunately, the complexity of the relationship between the military and local populations is often overlooked.

■ VISUAL SUMMARY OF CORE CONCEPTS

■ CORE CONCEPT 1: Sex is a biological distinction determined by the anatomical traits essential to reproduction.

Most cultures place people in two categories—male and female—based largely on what are considered to be clear anatomical distinctions. Biological sex is not a clear-cut category, however, if only because some babies are born intersexed. In addition, a person's primary sex characteristics may not match his or her sex chromosomes. The existence of transsexuals also challenges the practice of separating everyone into the two biological categories male and female.

Department of Defense photo by Petty Officer 1st Class Bobby R. McRill, U.S. Navy

■ CORE CONCEPT 2: Gender is the socially created and learned distinctions that specify the ideal physical, behavioral, and mental and emotional traits characteristic of males and females.

Gender is a social distinction based on culturally conceived and learned ideals about appropriate appearance, behavior, and mental and emotional characteristics for males and females. *Masculinity* and *femininity* signify the ideal physical, behavioral, and mental and emotional traits believed to be characteristic of males and females, respectively. Gender ideals are socially constructed standards against which real cases can be compared. Few people can realize such ideals; in fact, the ideals may not exist in reality, yet they are still regarded as the standard. Often we attribute differences between males and females to biology; in fact, they are more likely to be socially created.

DoD photo by Sgt. Adrian R. Pascual, USMC

© ImageDJ/Jupiter Images

■ CORE CONCEPT 3: Gender ideals shape practically every aspect of life—influencing, among other things, how people dress, how they express emotions, and what occupations they choose.

Gender polarization is the organizing of social life around male-female ideals, so that people's sex influences every aspect of their life, including how they dress, the way they pose for photographs, the time they wake up in the morning, what they do before they go to bed at night, the social roles they take on, the things they worry about, and even ways of expressing emotion and experiencing sexual attraction. Decisions and viewpoints about any aspect of life are considered gender-schematic if they are influenced by a society's polarized definitions of masculinity and femininity rather than by criteria such as self-fulfillment, interest, ability, or personal comfort.

DoD photo by Paul Adams, USA

■ CORE CONCEPT 4: While all societies distinguish between male and female, some also recognize a third gender.

The existence of *fa'afafine*—people who are not biologically female but who take on the "way of women" in dress, mannerism, appearance, and role in American Samoa and other Pacific island areas—challenges the two-gender classification scheme. Jeannette Mageo argues that *fa'afafine* could not have become commonplace unless something about Samoan society had supported gender blurring

Joey Cummings KKHJ-FM / WVUV-AM

■ **CORE CONCEPT 5:** Gender expectations are learned and culturally imposed through a variety of social mechanisms, including socialization, situational constraints, and commercialization of gender ideals.

Socialization theorists argue that an undetermined yet significant portion of male-female differences are products of the ways in which males and females are socialized. Another powerful mechanism for conveying gender expectations is the commercialization of gender ideals—the process of introducing products into the market by using advertising and sales campaigns that promise consumers they will achieve masculine and feminine ideals if they buy and use the products. Finally, structural constraints—the established and customary rules, policies, and day-to-day practices that affect a person's life chances—channel people's behavior in desired directions. Structural constraints push men and women into jobs that correspond with society's ideals for sex-appropriate work.

Department of Defense photo by Petty Officer 2nd Class Susan Cornell, U.S. Navy

■ **CORE CONCEPT 6:** The belief that one sex—and by extension, one gender—is superior to another, and that this superiority justifies inequalities between sexes, is sexism.

Sexism revolves around four notions: (1) people can be placed into two categories: male and female; (2) a close correspondence exists between a person's primary sex characteristics and characteristics such as emotional activity, body language, personality, intelligence, the expression of sexual desire, and athletic capability; (3) primary sex characteristics are so significant that they explain and determine behavior and the social, economic, and political inequalities that exist between the sexes; and (4) people who behave in ways that depart from ideals of masculinity or femininity are considered deviant, in need of fixing, and subject to negative sanctions ranging from ridicule to physical violence. This early 20th century political cartoon shows "sexist" women inspecting what appears to be a "bug" but is someone of the male species.

Library of Congress Prints and Photographs Division

■ **CORE CONCEPT 7:** When sociologists study inequality between males and females, they seek to identify the social factors that put one sex at a disadvantage relative to the other.

Social inequality exists between men and women when one category relative to the other (1) faces greater risks to physical and emotional well-being, (2) possesses a disproportionate share of income and other valued resources, and/or (3) is accorded more opportunities to succeed. There are many areas in which men are disadvantaged relative to women. For example, life expectancy is shorter for men than for women. There are also many areas in which women are disadvantaged relative to men. For example, women tend to earn less money than men. Sociologists seek to identify the social factors that explain such differences.

USDA Photo by Tim McCabe

■ **CORE CONCEPT 8:** In its most basic sense, the feminist perspective advocates equality between men and women.

A basic sociological definition of *feminism* is "a perspective that advocates equality between men and women." Questions about what that equality looks like and how equality should be achieved distinguish feminist camps from one another. Sociologists take a feminist perspective when they emphasize in their teaching and research the following kinds of themes: the right to bodily integrity and autonomy, access to safe contraceptives, the right to choose the terms of pregnancy, access to quality prenatal care, protection from violence inside and outside the home, freedom from sexual harassment, equal pay for equal work, workplace rights to maternity and other caregiving leaves, and the inescapable interconnections between sex, gender, social class, race, culture, and religion.

Department of Defense photo by Mass Communication Specialist Seaman Michelle Rhonehouse, U.S. Navy

■ **CORE CONCEPT 9: Sex and gender are complicated by racial classification and legal relationship to a state or country.**
In addition to their sex and ethnic or racial classification, people have a legal relationship with a country or state, which puts them in such categories as citizen, refugee, temporary worker, or illegal migrant. Both men and women are subject to state control, but women are a special focus of such control because of their role in human reproduction. The state may regulate the number and kind of babies born; prohibit sexual relationships between certain men and women and between same-sex partners; institute policies that either encourage or discourage caregiving (such as maternity leave policies); or use images of certain kinds of women and men to symbolize what political leaders believe are the most urgent issues a country faces. States may also implement policies governing the roles that women and men can assume in crises such as war.

Department of Defense photo by Chief Petty Officer Edward G. Martens, U.S. Navy

Resources on the Internet

 Sociology: A Global Perspective Book Companion Web Site

www.cengage.com/sociology/ferrante

Visit your book companion Web site, where you will find flash cards, practice quizzes, Internet links, and more to help you study.

CENGAGENOW™

Just what you need to know NOW!

Spend time on what you need to master rather than on information you have already learned. Take a pre-test for this chapter, and CengageNOW will generate a personalized study plan based on your results. The study plan will identify the topics you need to review and direct you to online resources to help you master those topics. You can then take a post-test to help you determine the concepts you have mastered and what you will need to work on. Try it out! Go to www.cengage.com/login to sign in with an access code or to purchase access to this product.

Key Terms

commercialization of gender
   ideals   280
ethgender   292
feminism   290
femininity   270
gender   269

gender polarization   272
gender-schematic   273
intersexed   268
masculinity   270
primary sex characteristics   268
secondary sex characteristics   269

sex   268
sexism   286
social emotions   274
structural constraints   281
transsexuals   268

11 Economics and Politics

With Emphasis on Iraq

When sociologists study economic institutions, they seek to understand how societies produce, distribute, and consume goods and services. When they study political institutions, they focus on the way power is distributed within and across societies.

U.S. Navy photo by Petty Officer 2nd Class James Wagner/Released

▲ These Iraqi boys were born after the U.S. military entered Iraq in March 2003. This American soldier, on patrol in their neighborhood, stops to talk. While the boys know life only after the fall of Saddam Hussein, they are still part of a country with an economy that revolves around oil exports. The soldier is from a country that depends on oil imports. Whether they are aware of it or not, they share a dependence on oil to sustain their ways of life.

WHY FOCUS ON

Iraq?

IN THIS CHAPTER, we emphasize Iraq for several reasons. First, in 2003, then President George W. Bush made it clear to the world that he intended to change the political system in Iraq from a dictatorship to a democracy. The United States also committed itself to changing Iraq's economic system from a centrally planned (socialist) one to a free-market (capitalist) one. After the U.S.-led invasion and occupation of Iraq, U.S. officials abolished Iraqi laws that had restricted foreign ownership and investment in the country, began selling at least 150 of Iraq's 200 state-owned enterprises (Eckholm 2004). One benchmark of success involves transforming Iraq's oil industry from a nationalized one (like that of its neighbors Iran, Kuwait, and Saudi Arabia) to a commercial one that is open to international oil companies (Juhasz 2007). In 2009 Iraq opened up six oil fields, which hold one-third of the country's oil reserves, and invited 32 of the world's biggest energy companies to submit bids to service them for the next 20 years. To date, BP and China's CNPC have been awarded contracts.

Shortly after President Barack Obama (2009) took office, he stated in a speech he made to an audience of U.S. soldiers that:

> We sent our troops to Iraq to do away with Saddam Hussein's regime—and you got the job done. We kept our troops in Iraq to help establish a sovereign government—and you got the job done. And we will leave the Iraqi people with a hard-earned opportunity to live a better life—that is your achievement; that is the prospect that you have made possible.

For better or worse, the Iraqi people's chances for a better life revolve around oil. Iraq lies in a region that possesses two-thirds of the world's known oil reserves. After Saudi Arabia, Iraq holds the largest proven oil reserves in the world. Iraq's economy is dominated by the oil sector, which provides 95 percent of its foreign exchange earnings. Oil accounts for 85 percent of all Iraq's total exports. Iraq's number one trading partner is the United States. The United States, a country with 4.6 percent of the world's population, consumes 25 percent of the world's annual oil production. This evolving partnership between Iraq and the United States is a story of two countries—one that is heavily dependent on oil imports; the other, highly dependent on oil revenue (U.S. Central Intelligence Agency 2009; U.S. Department of Energy 2009).

FACTS TO CONSIDER

- Since the 1930s, Iraq's primary export has been oil. Oil accounts for 85 percent of that country's total exports and 95 percent of the government's revenue. Iraq's economy is oil.

- The United States uses 21 million barrels of oil per day, 60 percent of which comes from outside the United States. Oil is the engine of the U.S. economy.

The Economy

■ **CORE CONCEPT 1:** The ongoing agricultural, industrial, and information revolutions have profoundly shaped the world's economic systems. A society's **economic system** is the social institution that coordinates human activity to produce, distribute, and consume goods and services. **Goods** include any product that is extracted from the earth, manufactured, or grown—such as food, clothing, petroleum, natural gas, automobiles, coal, and computers. **Services** include activities performed for others that result in no tangible product, such as entertainment, transportation, financial advice, medical care, spiritual counseling, and education. Three major, ongoing revolutions have shaped the world's economic systems: agricultural, industrial, and information revolutions.

Agricultural Revolutions

Many agricultural revolutions have taken place over the course of human history. The first important agricultural revolution occurred approximately 10,000 years ago, when humans domesticated plants and animals. **Domestication** is the process of bringing plants and animals under human control. Instead of hunting for and gathering wild grains and whatever other edible vegetation was available, people planned when they would plant and harvest crops. This planning allowed for a more predictable food source and made it possible to produce more grain than people needed just to survive. The availability of excess grain supported the domestication of cattle, oxen, and other animals, because enough food became available for animals as well as people. Domestication reduced the need to hunt for animals, and domesticated animals offered people a source of power in addition to human muscle to transport heavy loads, guard sheep, and so on.

Another major agricultural revolution occurred around 5000 BC with the invention of the scratch plow, "with a forward-curving wooden blade for cutting the soil, and a backward-curving pair of handles" by which farmers could

economic system A socially created institution that coordinates human activity in the effort to produce, distribute, and consume goods and services.

goods Any products manufactured, grown, or extracted from the earth, such as food, clothing, housing, automobiles, coal, computers, and so on.

services Activities performed for others that result in no tangible product, such as entertainment, transportation, financial advice, medical care, spiritual counseling, and education.

domestication The process by which plants and animals were brought under human control.

Oil is classified as goods because it is a product extracted from the earth. Half the oil imported by the United States each year–about 3 billion gallons–arrives by supertanker (Intertanko 2007).

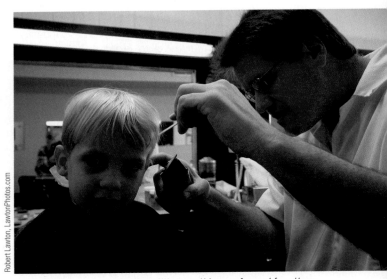

Cutting hair is a service because it is something performed for others but results in no tangible product.

direct oxen (Burke 1978, p. 9). The plow helped to create food surpluses, which enabled a society to support non-food occupations, such as potters, weavers, bakers, musicians, and craftsmen (Burke 1978). Figure 11.1 shows how food surpluses triggered a chain of events that changed society in fundamental ways. Other innovations—such as irrigation systems, the windmill, the tractor, pesticides, and genetic engineering—also triggered "agricultural revolutions," in that they dramatically increased crop yields, decreased the number of people needed to produce food, and further ensured predictable food supplies.

The rise of one of the earliest civilizations, Sumer, in the valley of the Tigris and Euphrates rivers (in what is

Department of Defense photo by Cpl. James P. Johnson, U.S. Army

As part of a humanitarian relief effort, a U.S. soldier offloads a sheep (a domesticated animal) near Baghdad, Iraq.

Once people gained control over the Tigris and Euphrates rivers more than 5,000 years ago, other developments, such as the Sumerian writing system, followed.

now Iraq), showed the power of agricultural revolutions to shape and organize human life. For at least 90,000 years, hunters and gatherers from the surrounding mountains and foothills were drawn to that valley because it was a source of fresh water, fish, and vegetation. The rivers were unpredictable, so the hunters and gatherers did not settle permanently next to them, but moved to avoid devastating floods and droughts. Archaeological evidence shows that by around 10,000 BC small stone hoes, millstones, and other agricultural implements were in use in the area.

The volatility of the rivers—attested to by ancient Mideastern flood stories, such as the biblical account of Noah, which scholars and scientists have linked to an actual flood in the region—necessitated the development of flood control and irrigation systems around 4800 BC (Roux 1965). Once the rivers were harnessed, the ability to grow surplus food followed. Consequently, people did not have to follow nomadic lives, wandering about in search of food or to escape floods and droughts. Soon urban centers emerged, supporting occupations other than agriculturally based ones. Many of the great legacies of this region—the invention of writing, irrigation, the wheel, and literature—"can be seen as adaptive responses to the great rivers" (Roux 1965).

The Industrial Revolution

As we learned in Chapter 1, a fundamental feature of the Industrial Revolution was **mechanization**, the addition of external sources of power, such as oil or coal, to hand tools and modes of transportation. The new forms of energy supported machines, factories, and mass production—thereby reducing the physical requirements for producing goods. As a result industrialization transformed individual workshops into factories, craftspeople into machine operators, and manual production into machine production.

The Industrial Revolution cannot be separated from **colonization**, a form of domination in which one country imposes its political, economic, social, and cultural institutions on an indigenous population and the land it occupies (see Chapters 1 and 8). The Industrial Revolution and colonization drew people from even the most remote parts of the world into a production process that manufactured unprecedented quantities of material goods, primarily for

mechanization The addition of external sources of power, such as oil or steam, to hand tools and modes of transportation.

colonization A form of domination in which one country imposes its political, economic, social, and cultural institutions on an indigenous population and the land it occupies.

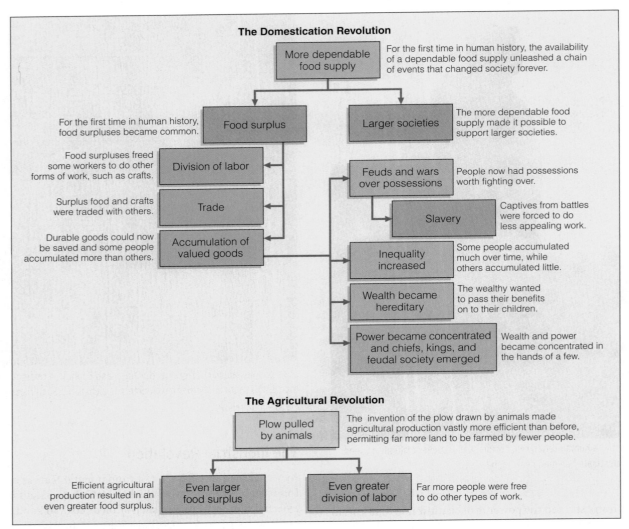

▲ **Figure 11.1**   **Plant and Animal Domestication and the Plow: Effects on Civilization**

Source: From *Sociology Timeline.* Copyright © 1995, IdeaWorks, Inc.™ Portions copyright © 1995, Board of Curators, University of Missouri.

the benefit of the colonizing countries. Consider as one measure of the extent of colonization that during the twentieth century 130 countries gained independence from their "mother countries." In light of this history, it should come as no surprise that 10 of the 11 members of the Organization of Petroleum Exporting Countries (OPEC) were once colonies of a European power (see Figure 11.2).

It should also come as no surprise that when European influence in the Middle East declined after World War II, the United States worked to exert more influence over the region. Consequently, "today virtually every regime in the region owes its survival to some combination of U.S. military and economic assistance" (Elamine 2004). Consider as one indicator of U.S. dependence on oil that the country possesses less than 3 percent of the world's proven oil reserves within its borders, and yet it consumes 25 percent of the world's annual oil production (U.S. Energy Information Administration 2007).

The Post-industrial Society and the Information Revolution

Sociologist Daniel Bell, who has been writing about the coming of a post-industrial society since the 1950s, describes the essential changes that are accompanying the emergence of a **post-industrial society**, one that

post-industrial society   A society that is dominated by intellectual technologies of telecommunications and computers, not just "large computers but computers on a chip." These intellectual technologies have had a revolutionary effect on virtually every aspect of social life.

relies on intellectual technologies of telecommunications and computers, not just "large computers but computers on a chip" (Bell 1999, p. xxxvii). According to Bell, this intellectual technology encompasses four interdependent revolutionary innovations: (1) electronics that allow for incredible speed of data transmission and calculations in the realm of nanoseconds or one billionth of a second or one trillionth of a second; (2) miniaturization, or the drastic reduction in the size of electronic devices, so that things like computers, cell phones, and iPods have become sleeker and smaller thanks to the development of the silicon chip, which can hold hundreds of millions of transistors connected by wires smaller in diameter than a 500th of a human hair; (3) digitalization that permits voice, text, images, and data to be carried with the same efficiency and also integrated with other technology; and (4) software, the programs that enable computers to operate, to perform a variety of tasks, and to generate a variety of simulated experiences.

In addition to the associated technology, post-industrial societies are distinguished by

- a substantial proportion of the working population employed in service, sales, and administrative support occupations (from 29 percent in 1950 to 41 percent in 2010 in the U.S.)
- an extraordinary rise in the percentage of workers in management, professional, and related occupations (from 18 percent in 1950 to 36 percent in 2010 in the U.S.)
- an increased emphasis on education as the avenue of social mobility, such that 6.2 percent of the U.S. population had at least a four-year college degree in 1950 and 28.7 percent in 2010. U.S. Census Bureau data (2009) shows that workers 18 and over with a bachelor's degree earn an average of $68,167 a year, while those with a high-school diploma earn $39,038. Workers with a doctoral degree make an average of $106,000, and those without a high school diploma average $18,734.
- recognition that capital is not only financial (cash and property) but also social, as measured by people's access to social networks that serve as sources of information and opportunity
- the dominance of intellectual technology (as opposed to machine technology) based on mathematics and linguistics in the form of algorithms (decision rules), programs (software), models, and simulations
- the creation of an electronically mediated global communication infrastructure, which includes broadband, cable, digital TV, optical fiber networks, fax, e-mail, and ISDN (integrated system digital networks)
- an economy defined not simply by the production of goods and labor-saving devices (automation and outsourcing), but by applied knowledge as the source of invention and innovation and by the manipulation of

A

B

These two photographs taken in the early 20th century show that Industrial Revolution and colonization were inseparable events. Photograph A shows Dodge trucks being assembled at a Chrysler factory. Of course trucks need tires which are made of rubber. Photo B shows rubber gatherers of French Equatorial Africa (now Gabon, the Republic of the Congo, the Central African Republic, and Chad) bringing rubber to a trading-weighing station.

numbers, words, images, and other symbols. The jobs associated with this knowledge-driven, information-based economy include computer programmers, technical writers, financial analysts, market analysts, and customer-service representatives.

The challenge of post-industrial society is interpersonal, as the "basic experience of each person's life is his relationship between himself and others." The communication

Country		Former Colony of		Year Independence Gained
Algeria		France		1962
Indonesia		Netherlands		1949
Iraq		United Kingdom		1932
Kuwait		United Kingdom		1961
Libya		Italy		1651
Nigeria		United Kingdom		1960
Qatar		United Kingdom		1971
Saudi Arabia		United Kingdom		1960
United Arab Emirates		United Kingdom		1971
Venezuela		Spain		1811
Iran		Iran was not a colony. Iran gained its independence during the Islamic Revolution, when the Shah Mohammad Reza Pahlavi was overthrown and exiled.		1979

▲ **Figure 11.2**   **Members of OPEC That Were Once Colonies of a European Power**
Eleven countries belong to the Organization of Petroleum Exporting Countries (OPEC). Ten of them are former colonies of a European power. Iran is the only member of OPEC that was not a colony. Keep in mind that industrialization has depended on petroleum products beyond just gasoline and heating oil. Nonfuel uses for petroleum include solvents (used in paints, lacquers, and printing ink), petroleum wax (used in candy making, packaging, candles, matches, and polishes), asphalt (used to pave roads and make roofing materials), and petroleum feed stocks (used to produce plastics, synthetic fibers, detergents, and even drugs) (U.S. Department of Energy 2003).

infrastructure and the nature of work multiply the interactions (however superficial and fleeting) between people, making interpersonal relationships a primary focus of one's thoughts and actions. That focus is complicated by the fact that we leave records of most transactions, whether it be related to banking transactions or to the time a text message was sent. The great difference in the post-industrial society is "the enlargement of an individual's world" or the "tremendous change of scale" in the number of people one knows or can know (Bell 1976, p. 48).

In an environment that emphasizes knowledge and interpersonal relationships, the institutions of science and education take center stage. Education becomes key to negotiating an information society and is viewed as something that takes place across the lifespan, not confined to a specific time or place. As a result, we can think of learning not just in classrooms of brick and mortar, but in other forms such as home schooling, adult education, continuing education, professional development, on-the-job training, and self-directed learning.

The New York Stock Exchange building is a symbol of capitalism, because the building is a place set aside for brokers to buy and sell stocks on behalf of private investors. Selling stock is one method corporations use to raise capital.

With regard to science, Bell described the rise and importance of science-based industries, which involve applications of theoretical knowledge. These industries are fundamentally different from the industries of the Industrial Revolution, such as steel, automobile, and telephone. For the most part, these industries were "founded or created by talented tinkers" who were not connected to the scientific establishment. As examples, Thomas Edison, who was "a great genius" credited with "creating the electric light, the motion picture and the gramophone, was literally mathematically illiterate; Alexander Graham Bell was an elocutionist who invented the telephone largely to amplify sound so that deaf people could hear, and he knew nothing of the work of James Maxwell on electromagnetism" (p. 46). Post-industrial industries, on the other hand, "derive directly from the investigations of scientists into the basic phenomena of nature and the applications of this research to technological problems" (p. 46). For example, "William B. Shockley's research on transistors is responsible for a huge transformation in modern electronics and the work of Felix Bloch on solid-state physics is the basis for much of computer technology" (p. 46).

Major Economic Systems

■ **CORE CONCEPT 2:** The world's economic systems fall along a continuum whose endpoints are ideal forms of capitalism and socialism. No economy fully realizes capitalist or socialist principles. The capitalist and socialist principles described here represent economic ideals, the standards against which real economic systems can be compared.

Capitalism

Capitalism is an economic system in which the raw materials and the means of producing and distributing goods and services remain privately owned. This economic system is profit driven and free of government interference. Private ownership means that individuals (rather than workers, the government, or communal groups) own the raw materials, machines, tools, labor, trucks, buildings, and other inputs needed to produce and distribute goods and services. "Profit-driven" is the most important characteristic of capitalist systems. In such systems, the owners of the means of production and distribution are driven to continually increase and maximize profits. Profit motivates owners, in their quest to maximize the return on their investment, to make the most efficient use of labor and resources. Theoretically, production and distribution are consumer-driven, because businesses depend on consumers "choosing" their products and services over a competitor's.

Capitalist systems are governed by the laws of supply and demand; that is, as the demand for a product or service increases, its price rises. Manufacturers and service providers respond to increased consumer demand by increasing production, which in turn "increases competition and drives the price down" (Hirsch, Kett, and Trefil 1993, p. 455). Although most economic systems in the world are classified as capitalist, in reality no system fully realizes capitalist principles. Simply consider that the U.S. and other governments ignored capitalist ideals when they intervened in 2008 and 2009 to bail out the auto industry, banks, and other financial institutions. In the case of banks, the government provided direct infusions of cash to banks. In addition, it insured deposits up to $250,000 (to encourage consumers to leave money in banks and prevent a run) and allowed banks to issue debt (bonds) backed by the FDIC to raise much-needed capital. So Citigroup received $50 billion in bailout money but was able to issue $27.6 billion in FDIC-insured bonds, thus effectively adding $77.6 billion in capital. The value of the assistance to the financial institutions is beyond calculation (Story 2009).

Karl Marx, who was considered a student of capitalism, believed that it was the first economic system that could maximize the immense productive potential of human labor and ingenuity. He also felt, however, that capitalism ignored too many human needs and was driven by greed. As one example of a way in which capitalism ignores human

capitalism An economic system in which the raw materials and the means of producing and distributing goods and services remain privately owned.

Department of Defense photo by CPL D.W. Yarnall, USMC

This sculpture, a symbol of socialist ideals, celebrates the common worker who ideally labors for the good of society. The following principle dominates: "From each according to his ability, to each according to his needs." That is, each person shall produce to the best of his or her ability and shall take in accordance with his or her needs. Guided by such principles, there will not be the wide disparities in income and wealth that separate, even divide, people in the professional and working classes.

Since the 1930s, Iraq's main export has been oil. Oil accounts for 85 percent of the country's total exports and 95 percent of the government's revenue. Iraq's and the world's extreme dependence on this one commodity explains why insurgents target oil pipelines.

need, consider that when the price of oil exceeded $80 to $100 per barrel, American farmers were motivated to grow corn for the corn-based fuel known as ethanol. This use of corn for fuel drove up prices not only of corn, but also of wheat, rice, and soy, because farmers were growing less of these crops. This shift in production affected food prices on a global scale and especially harmed the poorest peoples, who could not afford staple foods such as tortillas, which are made from corn.

As a second example, the largely unregulated pursuit of profit fueled a housing mortgage crisis that reached a tipping point in 2008. At that time about 1 in every 50 mortgage holders was late with house payments, and foreclosure proceedings were initiated on more than 1.5 million homes. The financial meltdown came about in part because so many, at all levels of the home-buying and lending process, made decisions based on profit and personal gain. What is the connection between the pursuit of profit and the housing crisis? The *Japan Times* (2008) described the connection in this way: "Banks gave mortgages to unworthy borrowers and failed to explain the terms of loans. Borrowers took loans that they knew they could not repay. Financial companies repackaged those loans and

sold them, knowing that their value was uncertain at best. Companies bought those securities, not understanding their value. And regulators adopted a hands-off approach that facilitated the spread of toxic debt."

Socialism

The term *socialism* was first used in the early 19th century in response to excesses of poverty and inequality that accompanied the capitalist-driven Industrial Revolution, and by extension, colonization (see Chapter 1). In contrast to capitalism, **socialism** is an economic system in which raw materials and the means of producing and distributing goods and services are collectively owned. That is, public ownership—rather than private ownership—is an essential characteristic of this system. Socialists reject the idea that what is good for the individual and for privately owned businesses is good for society. Instead, they believe the government or some worker- or community-oriented organization should play the central role in regulating economic activity on behalf of the people as a whole. Socialists maintain that things like water, oil, banks, medical

socialism An economic system in which raw materials and the means of producing and distributing goods and services are collectively owned.

care, transportation, and the media should be state owned. In socialism's most extreme form, the pursuit of personal profit is forbidden. In less extreme forms, profit-making activities are permitted as long as they do not interfere with larger collective goals. As with capitalism, no economic system fully realizes socialist principles. The People's Republic of China, Cuba, North Korea, and Vietnam are all officially classified as socialist economies, but they permit varying degrees of personal wealth-generating activities.

Americans often hear the word *socialist* in reference to government-run universal health care coverage. Critics claim that socialized medicine will lower the quality of health care and point out that Canada's socialist health care system is not able to meet demand. These critics fail to consider that the capitalist-driven health care system in the United States leaves the uninsured and those with preexisting conditions without health care. The charge of "socialism" was most recently leveled at government intervention in the economic crisis. U.S. and other governments intervened because the "free enterprise" system had failed. Critics of intervention argue that we will never know if the system would have failed without such intervention because the bailout disrupted a critical check and balance—that check and balance was risk. If people who made high-risk investments and who gave and took risky loans were allowed to fail, the checks would have triggered, people would have acted to change their financial situation, and the losses would have acted as a deterrent against future risk-taking. Instead, the bailouts actually encouraged risk-taking.

In reality, no economic system is purely capitalist or purely socialist. The **welfare state** is a term that applies to an economic system that is a hybrid of capitalism and socialism. In this economic model, the government (through taxes) assumes a key role in providing social and economic benefits to some or all of its citizens, including unemployment benefits, supplemental income, child care, social security, basic medical care, transportation, education (including college), and/or housing. Under the welfare state model followed by the United States (with the exception of Social Security and Medicare, to which everyone over 65 is entitled), such benefits are provided to those who fall below a set minimum standard such as a poverty line or a certain income level. Under a second model, the benefits are awarded in a more comprehensive way to everyone in the population (e.g., all families with children; all college-age students; universal health care). Most European countries follow the second model, as do oil-revenue-rich countries such as Saudi Arabia, Kuwait, Qatar, Bahrain, Oman, and the United Arab Emirates.

Iraq under Socialism

Under Saddam Hussein, Iraq followed a socialist model. After Saddam Hussein came to power in Iraq in 1979, his government pursued the socialist economic policies of

In light of Iraq's extreme dependence on oil revenue, one should not be surprised that insurgent groups target its oil industry, including its pipelines, oil facilities, and transportation routes. The U.S. military trains to handle such fires.

regimes that had been in power since 1958 (the year Iraq gained independence from British control). On the surface, Iraq's centrally planned economy seemed to work well, because the country had one of the most advanced economies among the Arab nations. Iraq boasted a large middle class, a well-developed transportation and health care system, and one of the best educational systems in the Arab world. Critics of Iraq's economy maintain that it depended heavily on oil revenues, so large that they allowed the Iraqi government to support social welfare and other subsidization programs. For example, Iraqi law guaranteed every adult lifetime employment. In return, Iraqi workers could not change jobs or careers. Under this system, labor costs accounted for 20 to 40 percent of output, far above the 10 percent costs found in most nonsocialist economies. By some estimates, half of all individuals who worked in Iraq's state-run enterprises were nonproductive administrative staff. Despite such inefficiencies, the country ran a budget surplus because of its large oil revenues. The Iran-Iraq War, an eight-year conflict that resulted in more than 1 million casualties between 1980 and 1988, left the Iraqi government $50 billion in debt. This debt forced the government to implement economic reforms, which included relaxing state control over the economy. These reforms ended with the Gulf War (1991), which was followed by 13 years of sanctions and the implementation of the Uncontrolled Oil-for-Food Program (OFFP), and then the Iraq War (Metz 1988; Sanford 2003). Prior to the

welfare state A term that applies to an economic system that is a hybrid of capitalism and socialism.

OFFP, virtually all oil-related transactions in Iraq were conducted through government contracts (Sanford 2003, p. 34). After OFFP was implemented in 1997, the United Nations functioned as a watchdog, reviewing all oil transactions. Then when the United States took control of the country with military force in March 2003, it abolished laws that had nationalized the oil industry.

World System Theory

■ **CORE CONCEPT 3:** Capitalists' responses to economic downturns and stagnation have driven a 500-year-plus economic expansion, which has facilitated interconnections between regional and national economies. World system theory focuses on the forces underlying the development of economic transactions that transcend national boundaries (see Chapter 8). In particular, world system theorists identify profit-generating strategies that have caused capitalism to dominate and facilitate a global network of economic relationships. These strategies, which have come in response to economic downturns and stagnation, include (1) finding ways to lower labor-associated production costs (moving production facilities out of high-wage zones and into lower-wage zones inside or outside a country); (2) securing at the lowest possible price the raw materials needed to make products; (3) creating new products that consumers "need" to buy; (4) improving existing products and thus making previous versions obsolete; and (5) "redistributing" wealth to enable more people to purchase products and services. Because of these strategies, capitalism has spread steadily throughout the globe. In addition, every country has come to play one of three different and unequal roles in the global economy: core, peripheral, or semiperipheral.

Peripheral economies rely on a few commodities or even a single commodity, such as coffee, peanuts, or tobacco, or on a natural resource, such as oil, tin, copper, or zinc (Van Evera 1990). Core economies include the wealthiest, most highly diversified economies with strong, stable governments. When economic activity weakens in core economies, the economies of other countries suffer, because exports to core economies decline and price levels fall. Semiperipheral economies are characterized by moderate wealth (but extreme inequality) and moderately diverse economies. Semiperipheral economies exploit peripheral economies and are in turn exploited by core economies. Iraq is classified as a peripheral economy, because of its extreme dependence on oil revenue.

Iraq: A Peripheral or One-Commodity Economy

■ **CORE CONCEPT 4:** Peripheral economies are built upon a few commodities or a single commodity; thus, they are extremely vulnerable to fluctuations in price and demand. Peripheral economies have a dependent relationship with core economies that traces its roots to colonialism. Peripheral economies operate on the so-called fringes of the world economy. Most of the jobs that connect their workers to the world economy pay little and require few skills. Amid widespread and chronic poverty islands of greater economic activity may exist, including manufacturing zones and tourist attractions.

Iraq lies in a region that holds two-thirds of the world's known oil reserves. In addition, about half of the world's total demand for oil is met by a handful of countries in the Middle East. After Saudi Arabia, Iraq possesses the largest amount of oil reserves in that region. Since the 1930s, its main export has been oil. In 1979 (the year Saddam Hussein came to power) oil trade accounted for 83 percent of the country's exports.

Of course, extreme dependence on one commodity causes many problems. First, fluctuations in the price of that commodity can leave a government flush with money when prices rise and poor when prices fall, leading to cycles of wild government spending followed by drastic spending cuts. Second, strong oil revenues (which are collected in U.S. dollars) increase the value of a country's currency, making it difficult for domestic manufacturers to produce goods at competitive prices for the world market. Consequently, the country has relatively few manufacturing jobs and "the whole country suffers, since it loses the benefits—such as technological innovation and good management—that a strong domestic manufacturing sector can provide" (Birdsall and Subramanian 2004, p. 81). Third, corruption and political rivalries inevitably arise when a government or a few powerful people control a country's natural resources, the distribution of those resources, and the revenues from them. Finally, governments presiding over economies that depend on one commodity do not have to tax their people to generate operating funds. While such a scenario might seem ideal to people who must pay taxes, it leaves citizens with "no effective mechanism by which to hold government accountable" (Birdsall and Subramanian 2004, p. 81).

Iraq's experience with oil illustrates the problems that come with extreme dependence on a single commodity. Iraq suffered a severe blow when the international community denied it access to worldwide oil markets in 1990, after Saddam Hussein invaded Kuwait. In August of that year, the UN Security Council issued Resolution 661, asking member and nonmember states not to import oil, other commodities, and products from Iraq and not to export any commodities and products to Iraq except for medical supplies and foodstuffs (Resolution 661, 1990). In 1991 the UN assessed the situation in Iraq and reported that the Iraqi people faced imminent catastrophe, including epidemic and famine, unless massive amounts of life-supporting aid were delivered to them (United Nations 2004). In response, the UN proposed the Oil-for-Food Program (OFFP). Not until December 1996 did the Iraqi government sign a memorandum of understanding, agreeing that the UN would oversee the program. At first, Iraq was permitted to sell $4 billion worth of oil per year. This amount then increased to $10.5 billion per year, and eventually the cap was lifted altogether.

The Iraq Survey Group, a U.S.-led fact-finding mission, issued a 1,000-page report revealing a global-scale collusion in which the Iraqi regime, UN officials, and contractors in 40 countries sought to circumvent the OFFP. Apparently, Iraqi officials asked suppliers of supposedly essential goods and services to inflate their bills by at least 10 percent; then, upon being paid, the suppliers deposited the excess amount in a bank account accessible to Saddam Hussein. One estimate put the amount of money siphoned from the OFFP at $10 billion (Raghavan 2004).

The United States: A Core Economy

■ **CORE CONCEPT 5:** Core economies are wealthy societies with highly diversified economies and secure, stable governments. According to the *World Factbook* (U.S. Central Intelligence Agency 2009), "the U.S. has the largest and most technologically powerful economy in the world." The *World Factbook* (2009) also claims that "U.S. firms are at or near the forefront in technological advances, especially in computers and in medical, aerospace, and military equipment." Since early 2008, the U.S. economy has been in a recession—a recession triggered by a variety of factors including the sub-prime mortgage crisis; investment, bank, and other financial services failures; falling home prices; and tight credit (see Chapter 1).

The United States is a consumption-oriented society. It represents 4.6 percent of the world's population, yet it consumes 25 percent of oil and 40 percent of pharmaceuticals. Per capita personal consumption (broadly defined as money spent on goods and services intended for individual/household consumption or use) exceeds that of the European Union countries and Japan (see Table 11.1).

While it is difficult to give an exhaustive analysis of the largest economy in the world, we can highlight a few notable characteristics.

The Size of the U.S. Economy and Debt

As measured by GDP, the United States has the largest economy in the world at $14 trillion (U.S. Central Intelligence Agency 2009). In per capita terms, the United States has the tenth highest per capita (per person) gross domestic product in the world—$47,000—after Liechtenstein ($118,000), Qatar ($103,500), Luxembourg ($81,100), Bermuda ($69,900), Kuwait ($57,400), Jersey ($57,000), Norway ($55,200), Brunei ($52,000), and Singapore ($52,000) (U.S. Central Intelligence Agency 2009).

The U.S. national debt is $11.6 trillion, which is 81 percent of the country's GDP. The debt is climbing at a rate of $3.9 billion per day. If we consider the national debt in relation to the annual federal budget, which is about $3.1 trillion a year, the national debt is almost four times that. About $4.3 trillion of the $11.3 trillion in debt is held by other U.S. government entities, including the Social Security Trust Fund. Foreign investors and governments hold an estimated $4 trillion of the debt (see Figure 11.3). Other holders of national debt include U.S. state and local governments ($467 billion), individual investors ($423 billion), and public and private pension funds ($319 billion) (MSNBC 2007; U.S. Department of Treasury 2009).

In addition to the national debt, total U.S. consumer debt (not counting mortgage debt) stands at $2.41 trillion, with $800 billion in credit card debt (Wyman 2009). If mortgage debt were counted, total consumer debt would be $9.3 trillion (U.S. Federal Reserve 2007). Each month, 1 in every 15 credit card holders (6.5 percent) misses a credit card payment (Streitfeld 2009). Consumer credit card debt was such in 2008 that Bank of America waived late fees, lowered interest charges, and/or even reduced loan balances of more than 700,000 credit card holders. Other credit card companies offered similar incentives to encourage debt repayment (Dash 2009).

The large consumer debt has been building since the mid-1980s and indicates that U.S. consumption fueled by debt has played a key role in driving the domestic and global economy. As American spending has slowed due to the economic crisis, many foreign economies have been affected, especially in countries whose economies have been built around exports and savings, most notably Asian economies. In fact, the International Monetary Fund estimated the losses of the global economic crisis to

Table 11.1	Total Consumption and Per Capita Consumption by United States Relative to the European Union, Japan, China, and India

U.S. personal consumption far exceeds that of any other economy. Note that per capita personal consumption averages $31,596 for every man, woman, and child in the country. Per capita personal consumption in the United States is much greater than that of the European Union and Japan and far exceeds that of China and India.

Country	Total Personal Consumption	Per Capita Personal Consumption
United States	$9.7 trillion	$31,596
European Union	$8.9 trillion	$18,126
Japan	$2.5 trillion	$19,685
China	$1.2 trillion	$923
India	$0.6 trillion	$521

Sources: New York Times (2008); U.S. Bureau of Economic Analysis (2009)

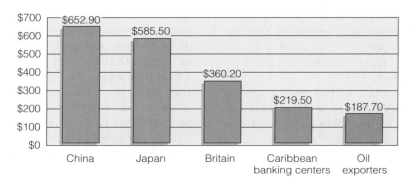

▲ **Figure 11.3**   The Top Five Foreign Holders of the U.S. National Debt (in billions, October 2008)
The top foreign holder of the U.S. national debt is China. After China, Japan holds the second largest share of the national debt, followed by Britain and various Caribbean banking centers and oil-exporting countries.

Sources: U.S. Department of Treasury (2009); *Washington Post* (2009)

be $4.05 trillion. About 65 percent of that debt originates in the United States (Landler 2009).

After losing weight, Americans name managing debt and saving money as their top New Year's resolutions (USA .gov 2009). There are three different types of consumer debt—safe debt, stupid debt, and survival debt. **Safe debt** is secured debt; that is, debt secured through collateral, such as a house, that is associated with the debt. **Stupid debt** results from using credit cards to finance spending sprees and impulse buying. **Survival debt** is debt acquired from using credit cards to pay living expenses associated with food, rent, and transportation (Hunt 2004). Although we cannot be sure how much of the total consumer debt can be classified as stupid or survival, we can hypothesize that it is a significant share.

The Trade Deficit and Dependence on Oil and Mineral Imports

The overall U.S. trade deficit—the difference between the dollar value of goods and services imported and exported—was $700.3 billion in 2008. That is, the United States imported $700 billion more than it exported. Figure 11.4 in No Borders, No Boundaries shows the U.S. trade balance with individual countries.

If we break the trade deficit into two parts—goods and services—we find that the United States runs a surplus (+$119.1 billion) in services and a deficit in goods (−$819.4 billion). Petroleum imports accounted for 33 percent of the overall trade deficit. The fact that 33 percent of the U.S. trade deficit is related to oil imports points to a major weakness or vulnerability—the U.S. dependence on foreign sources for raw materials. According to the U.S. Department of Energy (2009), the United States produces approximately 2.1 billion barrels of crude oil each year. This amount meets approximately 40 percent of the country's crude oil needs. The United States must import the remaining 60 percent—about 3.5 billion barrels (see Figure 11.5).

The extent of U.S. dependence on foreign oil becomes evident when we consider that the country has an estimated 20.2 billion barrels of proven oil reserves—less than 3 percent of the world's total proven reserves (see Figure 11.6). At the current rate of domestic production (40 percent of need per year), these reserves will last about nine

Each month in the United States, 1 in every 15 credit card holders (6.5 percent) misses a payment.

Lance Cpl. Bryan G. Carfrey/United States Marines

safe debt   Debt secured through collateral, such as a house.

stupid debt   Debt from using credit cards to finance spending sprees and impulse buying.

survival debt   Debt from using credit cards to pay living expenses associated with food, rent, and transportation.

NO BORDERS, NO BOUNDARIES

U.S. Trade Deficits and Surpluses with Countries of the World, 2008

THE COUNTRIES SHOWN in dark blue represent those with which the United States has a trade surplus. Its largest surplus is with the Netherlands (+$14.7 billion) and its smallest surplus is with Western Sahara (+$297 thousand). The countries in light blue are those with which the United States has a trade deficit. Its largest trade deficit is with China (−$256 billion) and its smallest is with Bosnia and Herzegovina (−$4.7 million). The total trade deficit represents 6.5 percent of U.S. GDP (total value of economic activity) (Lynch 2008).

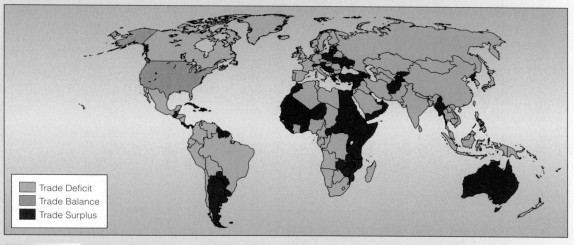

Trade Deficit
Trade Balance
Trade Surplus

▲ **Figure 11.4**

Source: U.S. Bureau of the Census, Foreign Trade Division (2009)

---

years. If the United States could produce and refine 100 percent of its crude oil needs, it would deplete the reserves in less than four years.

While most Americans are likely to know of their dependence on oil, they are not likely to know of the country's dependence on imported minerals. The United States depends on foreign sources for 21 of the 24 major nonfuel minerals/metals; its reliance is 100 percent for the minerals/metals listed in Table 11.2. Notice that many of the minerals/metals listed relate to electronic equipments such as computers and cell phones.

One might speculate that worldwide demand for oil and minerals/metals will continue to increase, if only because China and India, home to 2.5 billion people, have experienced dramatic economic growth. This growth has increased their demand for these commodities. As the world's demand for energy increases, the United States must address a critical question: How can a country with 4.6 percent of the world's population continue to consume 40 percent of the world's energy resources?

Table 11.2   Strategic and Critical Materials for Which the United States is 100 Percent Dependent on Foreign Sources

U.S. dependence on oil makes the headlines, but the United States also relies on other nonfuel mineral materials. This chart shows some of the critical materials for which the United States relies completely on foreign sources. The chart also shows uses of those materials.

Material/Metal	Uses
Arsenic	Semiconductors, pyrotechnics, insecticides
Columbium	Specialty steels
Indium	Semiconductors, metal organics, light-emitting diodes
Manganese	Specialty steels
Quartz crystals (high purity)	Electronic and photonic devices
Scandium	Refractory ceramics, aluminum alloys
Strontium	Medium- or high-temperature fuel cells

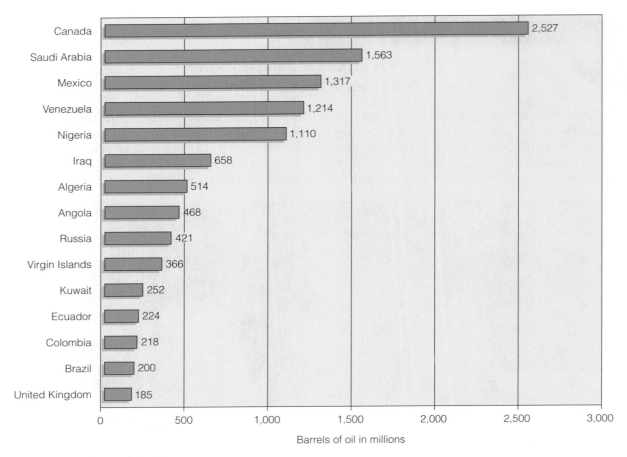

▲ Figure 11.5   Top 15 Countries from Which the United States Imports Petroleum, 2008

Source: U.S. Department of Energy (2009)

Decline in Union Membership

Of all workers in the United States, 12 percent (or 15.4 million) are union members. Union affiliation varies by state. In four states, 20 percent or more of the workers belong to unions: Hawaii (24.7 percent), New York (24.4 percent), Alaska (22.2 percent), and New Jersey (20.1 percent). Five states report union affiliation of less than 5 percent of the workforce: North Carolina (3.3 percent), South Carolina (3.3 percent), Virginia (4.0 percent), Georgia (4.4 percent), and Texas (4.9 percent) (U.S. Department of Labor 2007c). Unions, or organized labor, seek to maintain and sometimes improve workers' pay, benefits, and working conditions. Ideally, they also strive to ensure that the workplace remains secure for future generations of workers.

Union membership in the United States has declined from a high of 35 percent of the workforce in the 1950s to 12 percent in 2006 (U.S. Department of Labor 2007c). The steady drop in union membership over the past 30 years has been connected to many factors, including the following:

- The declining significance of the manufacturing sector (the traditional base of union membership) in the overall economy.
- Increasing percentages of females in the workforce (who have tended to work in nonunion positions).
- Increasing global competition (which has added pressure to keep wages down and minimize union influence).

As part of the 2008–2009 $17.4 billion federal bailout of the automobile industry, the United Automobile Workers (UAW) was forced to take a 20 percent cut in hourly wages. The rationale behind such a cut was that the higher wages and costly benefit packages, including health insurance and retirement benefits, were key factors in American automakers' inability to compete against foreign counterparts. Union leaders counter that unions are a necessary force against corporate desire to lower wages and reduce or eliminate benefits. Unions (as a group) are visible supporters of Democratic candidates—as evidenced by the $50 million they contributed during

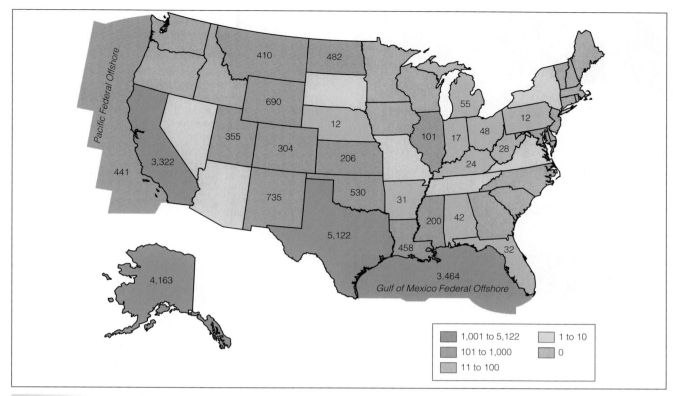

▲ **Figure 11.6**   **Amount of Proven Reserves of Crude Oil (in millions of barrels)**

The map shows that there are 4.2 million barrels of known crude oil in the state of Alaska. Texas has 5.1 billion barrels within its borders. How do we assess the amount of oil in Alaska? One way is to place this figure in the context of annual consumption. The United States consumes 20 million barrels of oil per day; that translates into 7.3 billion barrels per year. Alaska, with 4.2 billion barrels, contains enough oil to satisfy six months of oil demand. The Arctic National Wildlife Refuge (ANWR) is believed to contain about 1.9 billion barrels of known oil, the equivalent of about three months of oil consumption. Some government geologists believe that there could be an additional 10 billion barrels of oil waiting to be discovered, an amount equivalent to 1.5 years of consumption.

Source: United States Energy Information Agency (2008)

the 2008 election cycle, an amount greater than any other category of contributors (Center for Responsive Politics 2009). Unions pushed for the passage of the Employee Free Choice Act, which would significantly change the six-decade-old labor laws, making it easier for workers who want unions to form them and to negotiate contracts with employers (Library of Congress 2009).

A Tertiary Sector That Dominates the Economy

We can think of an economy as being comprised of three sectors: primary, secondary, and tertiary. The **primary sector** includes economic activities that generate or extract raw materials from the natural environment. Mining, fishing, growing crops, raising livestock, drilling for oil, and planting and harvesting forest products are examples. The **secondary sector** consists of economic activities that transform raw materials from the primary sector into manufactured goods.

The **tertiary sector** encompasses economic activities related to delivering services such as health care or entertainment and those activities related to creating and distributing information. One way to identify the relative importance of each sector of an economy is by determining how much it contributes to the **gross domestic product**

primary sector Economic activities that generate or extract raw materials from the natural environment.

secondary sector Economic activities that transform raw materials from the primary sector into manufactured goods.

tertiary sector Economic activities related to delivering services such as health care or entertainment and those activities related to creating and distributing information.

gross domestic product (GDP) The monetary value of the goods and services that a nation's workforce produces in a year or some other time period.

Strontium, an essential metal used to make fuel cells that power alternative fuel vehicles, cannot be obtained in the United States. China is the top producer of strontium; it is believed to have 66 percent of the world's strontium within its borders. Spain and Mexico also have significant amount within their borders (British Geological Survey 2008).

Cpl. Gabriela Gonzalez/United States Marine Corps

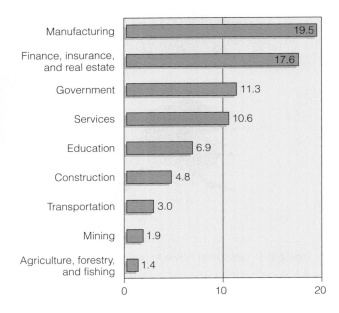

Chart 11.1 Gross Domestic Product by Industry in Current Dollars as a Percentage of Gross Domestic Product, 2007

In 2007, the GDP of the United States was $13.9 trillion. Which industry contributed the most to the GDP? Which contributed the least? Why do you think agriculture and mining are so low? Notice that finance, insurance, and real estate account for almost 18 percent of GDP. Services, government, education, and transportation are part of the tertiary sector.

Industry	Percentage
Manufacturing	19.5
Finance, insurance, and real estate	17.6
Government	11.3
Services	10.6
Education	6.9
Construction	4.8
Transportation	3.0
Mining	1.9
Agriculture, forestry, and fishing	1.4

(GDP), the monetary value of the goods and services that a nation's workforce produces in a year or some other time period (see Chart 11.1). Agriculture accounts for less than 1 percent of the United States GDP, and the manufacturing sector accounts for 20.9 percent. The tertiary sector accounts for 78.5 percent of the GDP (U.S. Central Intelligence Agency 2008).

A Sustained Period in Which Job Losses Outpaced Job Creation

Since the economic crisis began in December 2007, 6.5 million net jobs were lost in 16 months' time. The extent of the economic crisis cannot be measured only by job loss. By April 2009, 24 million Americans—16 percent of the workforce—were looking for work, working fewer hours than they would like, or so discouraged that they gave up looking for work. Putting this many people back to work will take years, even in the event of a recovery (Uchitelle 2009). The unemployment statistics offer some insights about occupational categories that have been most and least affected. According to the U.S. Bureau of Labor Statistics (2009), the five occupational categories most affected by job loss as measured by percentage of people unemployed are:

- Construction and extraction occupations   19.7
- Farming, fishing, forestry   15.2
- Production (manufacturing)   14.7
- Cleaning and maintenance   12.0
- Food preparation and serving-related occupations   9.6

The occupations least affected by job loss as measured by percentage of people unemployed are:

- Health care practitioner and technical occupations   1.8
- Education, training, and library   2.6
- Legal occupations   3.1
- Life, physical, and social services   2.4
- Community and social services   3.9

In assessing the least and most affected occupations, think about where those once employed in each of the five job categories with highest unemployment will find work. One answer is to look for occupations in which unemployment rates are low. How easy do you think it is for those in construction or food preparation to make a transition to an occupation in health care, law, or education? Some groups have been more affected by the economy than others. Table 11.3 shows that 1 in 10 males was unemployed in 2009 and that young males were especially vulnerable to job loss. Of course, these rates vary by race and ethnicity as well, with overall black unemployment standing at 15 percent and black teen unemployment at 34.7 percent.

Political Systems and Authority

A society's **political system** is the institution that regulates the use of and access to the power that is essential to articulating and realizing individual, local, regional, national, international, or global interests and agendas. **Power** is the probability that an individual can achieve his or her will, even against another individual's opposition (Weber 1947). That probability increases if the individual can force people to obey his or her commands or if the individual has authority over others.

■ **CORE CONCEPT 6:** When people believe that power differences are legitimate, those with power possess authority, which may be classified as charismatic, traditional, or legal-rational. **Authority** is legitimate power—power that people believe is just and proper. A leader has authority to the extent that people view him or her as being entitled to give orders. Max Weber identified three types of authority: traditional, charismatic, and legal-rational. **Traditional authority** relies on the sanctity of time-honored norms that govern the selection of someone to a powerful position (such as chief, king, or queen) and that specify responsibilities and conduct appropriate for the individual selected. People comply with a leader because they believe they are accountable to the past and have an obligation to perpetuate it. (Their rea-

soning is apt to be, "It has always been like that.") To give up past ways of doing things is to renounce a heritage and an identity (Boudon and Bourricaud 1989).

Charismatic authority derives from the exceptional and exemplary qualities of the person who issues commands. That is, charismatic leaders are obeyed, because their followers believe in and are attracted irresistibly to the leaders' vision. These leaders, by virtue of their special qualities, can unleash revolutionary changes; they can persuade their followers to behave in ways that depart from rules and traditions.

Charismatic leaders often emerge during times of profound crisis (such as economic depressions and wars), because in these situations, people are most likely to be drawn to someone with exceptional personal qualities who offers them a vision of a new order. A charismatic leader is more than popular, attractive, likable, or pleasant. A merely popular person, "even one who is continually in our thoughts" (Boudon and Bourricaud 1989, p. 70), is not someone for whom we would break all previous ties and give up our possessions. Charismatic leaders successfully persuade their followers to make extraordinary personal sacrifices, cut themselves off from ordinary worldly connections, or devote their lives to achieving a vision that the leaders have outlined.

The source of a charismatic leader's authority, however, does not rest with the ethical quality of the command or vision. Adolf Hitler, Franklin D. Roosevelt, Mao Zedong, and Winston Churchill were all charismatic leaders. Each assumed leadership of a country during turbulent times. Likewise, each conveyed a powerful vision (right or wrong) of his country's destiny.

Charismatic authority results from the intense relationships between leaders and followers. From a relational

| Table 11.3 | Unemployment Rates by Age and Sex, April 2009 |

Which age-sex category has the highest unemployment rate? Which has the lowest? Why do you think this is the case? Keep in mind that the unemployment rate considers job loss among those actively looking for work; it does not consider discouraged workers.

	Unemployment Rate	
	Males	Females
16 years and over	10.0	7.6
16 to 17 years	26.3	19.9
18 to 19 years	25.3	17.1
20 years and over	9.4	7.1
20 to 24 years	17.5	11.5
25 to 54 years	8.8	6.7
25 to 34 years	11.1	7.9
35 to 44 years	8.2	6.7
45 to 54 years	7.1	5.7
55 years and over	6.7	No data

Source: U.S. Bureau of Labor Statistics (2009)

political system A socially created institution that regulates the use of and access to power that is essential to articulating and realizing individual, local, regional, national, international, or global interests and agendas.

power The probability that an individual can achieve his or her will even against another individual's opposition.

authority Legitimate power in which people believe that the differences in power are just and proper—that is, people view a leader as being entitled to give orders.

traditional authority A type of authority that relies on the sanctity of time-honored norms that govern the selection of someone to a powerful position (chief, king, queen) and that specify responsibilities and appropriate conduct for the individual selected.

charismatic authority A type of authority that derives from the exceptional and exemplary qualities of the person who issues the commands.

From the U.S. government's point of view, one measure of success in Iraq has been the holding of elections. Here Iraqi soldiers line up to vote (left) and an Iraqi voter pulls out a finger soaked in indelible ink to show a vote was cast.

point of view, then, charisma is a highly unequal "power-relationship between an inspired guide and a cohort of followers" who believe wholeheartedly in the guide's promises and visions (Boudon and Bourricaud 1989, p. 70).

Over time, charismatic leaders and their followers come to constitute an "emotional community" devoted to achieving a goal and sustained by a belief in the leaders' special qualities. Weber argues, however, that eventually the followers must be able to return to a normal life and to develop relationships with one another based on something other than their connections to the leader. Attraction and devotion cannot sustain a community indefinitely, if only because the object of these emotions—the charismatic leader—is mortal.

Legal-rational authority derives from a system of impersonal rules that formally specify the qualifica-

legal-rational authority A type of authority that rests on a system of impersonal rules that formally specifies the qualifications for occupying a powerful position.

government The organizational structure that directs and coordinates people's involvement in the political activities of a country or other territory (city, county, state) within that country.

democracy A system of government in which power is vested in the citizen body, and in which members of that citizen body participate directly or indirectly in the decision-making process.

tions for occupying a powerful position. These rules also specify the scope of that power and appropriate conduct. In cases of legal-rational authority, people comply with commands, decisions, and directives because the power belongs to the position, and by extension, to the person occupying the position.

Forms of Government

■ **CORE CONCEPT 7:** Government is an organizational structure that directs and coordinates people acting in the name of a country or some other political entity. That structure may be democratic, authoritarian, totalitarian, or theocratic. **Government** is the organizational structure that directs and coordinates people's involvement in the political and economic activities of a country or some other territory, such as a city, county, or state. In this chapter, we consider four forms of government: democracy, totalitarianism, authoritarianism, and theocracy.

Democracy

Sir Winston Churchill, the prime minister of the United Kingdom during World War II, once said that "democracy is the worst form of government except for all others that have been tried" (wordiq.com 2004a). **Democracy** is a system of government in which power is vested in the citizens or "the people" and in which the citizenry participates directly or indirectly in making decisions. Usually, the size of the citizen population makes direct participation impossible. Therefore, decision making usually takes place indirectly, through elected representatives; this indirect form of governance is known as **representative democracy**. Representative democracies hold free elections and, theoretically, give every citizen the right to vote. In a democracy, political candidates and parties can campaign in opposition to the party holding power, and the choice of candidates is not limited to a single party. In addition, when a majority of citizens votes to change the party in power, an orderly and peaceful transition in government occurs. In democracies, elected representatives legislate; vote on taxes; control the budget; and debate, support, question, discuss, criticize, and oppose government policies.

Democratic forms of government extend basic rights to all of their citizens (and legal residents). These rights include freedom of speech, movement, religion, press, and assembly (that is, the right to form and belong to parties and other associations) as well as freedom from "arbitrary arrest and imprisonment" (Bullock 1977, pp. 210–211). Other characteristics of democracies include free and fair elections, a constitution that sets limits on executive and other powers, access to alternative information sources (free press), and an educated or informed citizenry.

Today Iraq is classified as a representative democracy (parliamentary in form). In June 2005, elections were held for a 275-member national Council of Representatives. The council elected Jalal Talabani as president of the Republic of Iraq.

Although the United States, a world superpower, is also classified as a representative democracy, not everyone who is eligible to vote does vote. In the 2008 presidential election, 64.9 percent of those eligible to vote actually voted. Barack Obama received 52.9 percent of all votes actually cast but only 33.4 percent of all votes Americans were eligible to cast. The percentages of voting-age Americans who reported being registered to vote varied dramatically by age and racial or ethnic classification. It is striking that only 54.9 percent of people classified as Hispanic or Latino and 58.5 percent of those aged 18–24 reported being registered. In contrast, 72.0 percent of voting-age people classified as white and 78.1 percent of those aged 65 and older reported being registered (U.S. Census Bureau 2009).

In light of these statistics, we might question whether elected officials represent everyone or just those groups who are likely to vote. We might also question whether elected officials represent all voters or whether they simply serve the powerful constituents who donate generously to their campaigns.

In assessing whether a form of government is a true democracy, it is important to consider who has the right to vote. At one time or another, many governments classified as democratic have—on the basis of race, sex, income, property, criminal status, mental health, religion, age, or other characteristics—excluded portions of their populations from decision making (Creighton 1992, pp. 430–431).

Totalitarianism

Totalitarianism is a system of government characterized by (1) a single ruling party led by a dictator, (2) an unchallenged official ideology that defines a vision of the "perfect" society and the means to achieve that vision, (3) a system of social control that suppresses dissent, and (4) centralized control over the media and the economy. Ideological goals vary but may include overthrowing capitalist and foreign influences (as in China under Mao Zedong), creating the perfect race (as in Germany under Hitler), or meeting certain economic and development goals (as during China's Great Leap Forward). Whatever the government's goals, the political leaders, the military, and the secret police intimidate and mobilize the masses to help the state meet the goals (see Chapter 7 for an example—China's Cultural Revolution).

Totalitarian governments are products of the 20th century, because by that time technologies existed that allowed a few people in power to control the behavior of the masses and the information the masses could hear. Many of the governments labeled as totalitarian have followed Communist principles. Traditionally, Communist governments instituted polices that outlawed private ownership of property, supported the equal distribution of wealth, and offered status and power to the working class (proletariat). Communist leaders mobilized the masses to help them realize Communist principles and to bring about economic changes that benefited everyone, not just the upper and middle classes.

The Soviet Union under the leadership of Joseph Stalin provides one example of how Communist ideals were put into practice and helps to explain why most Communist governments eventually collapsed. Stalin believed that if the Soviet Union did not become industrialized, it could not compete with European nations that were expanding

representative democracy A system of government in which decision making takes place indirectly through elected representatives.

totalitarianism A system of government characterized by (1) a single ruling party led by a dictator, (2) an unchallenged official ideology that defines a vision of the "perfect" society and the means to achieve that vision, (3) a system of social control that suppresses dissent and opposition, and (4) centralized control over the media and the economy.

After Saddam Hussein took control of Iraq in 1979, the United States both supported and opposed him. Here Hussein is shown addressing the Iraqi Special Tribunal at his trial for crimes against humanity. The tribunal sentenced him to death by hanging, and the sentence was carried out on December 30, 2006.

their territories and fighting to gain influence. He maintained that the Soviet Union was 50 years behind these countries and that the gap must be closed within 10 years. To accomplish this goal, Stalin conducted a reign of terror. He forced millions of peasants to work in factories, seized millions of private land holdings, and created massive state-owned agricultural collectives. In addition, he controlled a brutal system of repression (characterized by secret police, forced labor camps, mass deportations, and purges) as means of controlling or eliminating anyone who opposed him, or was even suspected of opposing him. This system was so relentless and brutal that it caused the death of more than 20 million people. Because Stalin controlled and censored all mass media, Soviet people learned only about his vision of socialism and heard only

authoritarian government A system of government in which there is no separation of power and a single person (dictator), group (family, military, single party), or social class holds all power.

good things about his policies. In addition, Stalin strengthened the government's control over the economy so that state bureaucrats decided what should be produced, how much should be produced, how much the products should cost, and where they should be distributed.

Authoritarianism

Under **authoritarian government**, no separation of powers exists; a single person (a dictator), a group (a family, the military, a single party), or a social class holds all power. No official ideology projects a vision of the "perfect" society or guides the government's political or economic policies. Indeed, authoritarian leaders do not seek to mobilize the masses to help realize a vision or meet ideological goals. Instead, the government functions to serve those in power, who may or may not be interested in the general welfare of the people. Common to all authoritarian systems is the "leader's freedom to exercise power without restraint, unencumbered by a commitment to law, ideology, or values" (Chehabi and Linz 1998).

How does a single person, a group, or a social class gain control of an entire country? The political culture must be amenable to personal leadership. Authoritarian leaders typically receive support from a foreign government that expects to benefit from their leadership (Buckley 1998). During the Cold War, for example, the United States supported anti-Communist dictators and the Soviet Union supported anti-capitalist dictators. Between 1945 and 1989, the foreign and domestic policies of the United States and the Soviet Union were shaped largely by Cold War dynamics. Virtually every U.S. policy—from the 1949 Marshall Plan to the covert aid given to the Contra guerrillas during the Reagan administration—was shaped in some way by America's professed desire to protect the world from Soviet influence and the spread of Communism, even to the point of supporting antidemocratic, brutally repressive authoritarian regimes (McNamara 1989).

Making a clear-cut distinction between totalitarian and authoritarian governments is difficult. In the early 1980s, the U.S. Ambassador to the United Nations, Jeanne Kirkpatrick, distinguished between totalitarian and authoritarian governments by using a "lesser of two evils" principle. That principle holds that authoritarian dictatorships are likely to be anti-Communist and not "ideology" driven, whereas Communist dictatorships are likely to be totalitarian. Both control and punish those who oppose them, but totalitarian dictatorships go even further, in that they seek to control thoughts and beliefs. Both use propaganda as a method of social control, totalitarian regimes use propaganda *and* brainwashing. Kirkpatrick maintained that authoritarian regimes are the lesser of two evils, because they are not ideologically driven and because they do not engage in thought control. Kirkpatrick also argued that, in its effort to defeat Communism, the United States could support authoritarian regimes because they posed less of a danger to the American way of life and were more capable of reform.

Saddam Hussein is one example of an authoritarian leader whom the U.S. government both supported and opposed. The U.S. government supported Hussein during the Iran-Iraq War—even though he invaded Iran, abused the human rights of Iraqi citizens, used chemical weapons on Iranians and his own people, and clearly hoped to produce nuclear weapons. Of course, the U.S. government opposed Hussein during and after the 1991 Gulf War. In 2003 it invaded Iraq and deposed Hussein—primarily, it claimed, because the dictator had been secretly developing and building "weapons of mass destruction." When no evidence of such weapons turned up, the U.S. government justified the invasion on the grounds that Hussein had abused the human rights of Iraqi citizens and used chemical weapons against his own people.

Theocracy

Theocracy, which means "rule of the deity," is a form of government in which political authority rests in the hands of religious leaders or a theologically trained elite group. The primary purpose of a theocracy is to uphold divine laws in its policies and practices. Thus, it recognizes no legal separation of church and state.

Government policies and laws correspond to religious principles and laws. Contemporary examples of theocracies include the Vatican under the pope, Afghanistan under the Taliban, and Iran under Ayatollah Ali Khamenei. In other forms of theocracy, power is shared by a secular ruler (such as a king) and a religious leader (such as a pope or an ayatollah), or by secular government leaders who are devoted to the principles of the dominant religion. At one time, England was dominated by the Anglican Church, France by the Roman Catholic Church, and Sweden by the Lutheran Church. Today Saudi Arabia is a monarchy whose head of state and government is a king, not a religiously trained leader; its constitution and legal system, however, are grounded in Islamic laws and principles.

Since the removal of Saddam Hussein from power in March 2003, "the scare word in American debates about Iraq is 'theocracy'—as in 'If we don't stay the course the result could be a theocracy' like that of Iran, Afghanistan, or Saudi Arabia" (Bulliet 2004). In assessing this claim, it is important to remember that many of Iraq's most prominent religious leaders reject the Iranian model of theocracy.

Power-Sharing Models

There are two models of power sharing: the power elite and pluralistic models. The two models help us to evaluate whether an elite few hold the power in society or whether power is dispersed among competing interest groups. In addition to elite and pluralist models, which help us to understand how power is distributed within societies, we will consider concepts such as imperialism and hegemony, which focus our attention on unequal power relations between countries.

The Power Elite

■ **CORE CONCEPT 8:** The power elite consists of people whose positions in the military, the government, and the largest corporations are so high that their decisions affect the lives of millions, even billions, of people. Sociologist C. Wright Mills wrote about the connection between government, industry, and the military in *The Power Elite* (1956). The **power elite** are those few people who occupy such lofty positions in the social structure of leading institutions that their decisions affect millions, even billions, of people worldwide. For the most part, the source of this power is legal-rational—residing not in the personal qualities of those in power, but rather in the positions that the power elite have come to occupy. "Were the person occupying the position the most important factor, the stock market would pay close attention to retirements, deaths, and replacements in the executive ranks" (Galbraith 1958, p. 146).

The power elite use their positions, and the tools of their positions, to rule, control, and influence others. These tools might include weapons, surveillance equipment, and specialized modes of communication. According to C. Wright Mills, since World War II, rapid advances in technology have allowed power to become concentrated in the hands of a few; those with access to such power exercise an extraordinary influence over not only their immediate environment, but also millions of people, tens of thousands of communities, entire countries, and the globe.

In writing about the power elite, Mills does not focus on any single individual, but rather on those who occupy the highest positions in the leading U.S. institutions. According to Mills, these leading institutions are the military, corporations (especially the 200 or so largest), and the government. "The power to make decisions of national and international consequence is now so clearly seated in political, military, and economic institutions that other areas of society seem off to the side and, on occasion, readily subordinated to these" (Mills 1963, p. 27).

theocracy A form of government in which political authority rests in the hands of religious leaders or a theologically trained elite. Under this system, there is no separation of church and state.

power elite Those few people who occupy such lofty positions in the social structure of leading institutions that their decisions have consequences affecting millions of people worldwide.

The origins of these institutions' power can be traced to World War II, when the political elite mobilized corporations to produce the supplies, weapons, and equipment needed to fight the war. U.S. corporations, which were left unscathed by the war, were virtually the only institutions in the world that could offer the services and products war-torn countries needed for rebuilding. The interests of the U.S. government, the military, and corporations became further intertwined when the political elite decided that a permanent war industry was needed to contain the spread of Communism. Thus, over the past 45 to 50 years, these three institutions have become deeply and intricately interrelated in hundreds of ways (see Working for Change).

Consider that each year the U.S. Department of Defense (2009) awards more than 60 million contracts to more than 2,000 businesses. The value of these contracts ranges from less than $25,000 to $21.9 billion. Figure 11.7 shows the top 10 military contractors for fiscal years 2006 and 2007. In evaluating this list, think about Lockheed Martin's dependence on the military-industrial complex. More than 92 percent of its sales are to the U.S. Department of Defense, the Department of Homeland Security, and other U.S. government agencies (Lockheed Martin, 2007b). For fiscal years 2006–2007, the company received $74.6 billion in Pentagon contracts, nearly three-quarters of which were won through less than full and open competition (Makinson 2004). Department of Defense and other government contracts support 140,000 Lockheed employees working in 939 facilities in 457 cities in 45 states and 56 countries and territories (Lockheed Martin 2009).

WORKING FOR CHANGE

President Eisenhower's Farewell Warning

WHEN PRESIDENT DWIGHT D. Eisenhower (a former five-star general who served in World War II) left office in 1961 after two terms, he issued a warning and a challenge to the American people. The warning concerned the "military-industrial complex," and the challenge was to contain it. As you read the following portion of Eisenhower's farewell speech, consider whether we have heeded his warning.

Until the latest of our world conflicts, the United States had no armaments industry. American makers of plowshares could, with time and as required, make swords as well. But now we can no longer risk emergency improvisation of national defense; we have been compelled to create a permanent armaments industry of vast proportions. Added to this, three and a half million men and women are directly engaged in the defense establishment. We annually spend on military security more than the net income of all United States corporations.

This conjunction of an immense military establishment and a large arms industry is new in the American experience. The total influence—economic, political, even spiritual—is felt in every city, every state house, every office of the Federal government. We recognize the imperative need for this development. Yet we must not fail to comprehend its grave implications. Our toil, resources and livelihood are all involved; so is the very structure of our society.

In the councils of government, we must guard against the acquisition of unwarranted influence, whether sought or unsought, by the military-industrial complex. The potential for the disastrous rise of misplaced power exists and will persist.

We must never let the weight of this combination endanger our liberties or democratic processes. We should take nothing for granted. Only an alert and knowledgeable citizenry can compel the proper meshing of the huge industrial and military machinery of defense with our peaceful methods and goals, so that security and liberty may prosper together.

Source: Eisenhower (1961).

The 34th president of the United States, Dwight D. Eisenhower, was also a five-star general.

Library of Congress Prints and Photographs Division

To get an idea of the extent to which government, military, and corporate interests became intertwined during World War II, consider these two images. The poster, issued by the U.S. Navy, shows the worker and the soldier as the combination needed for victory. The photograph shows female workers at a Long Beach, California, Douglas Aircraft Company plant installing fixtures and assemblies to a tail fuselage section of a B-17 bomber.

Lockheed Martin's board of directors includes a former chairman of the U.S. president's National Security Telecommunication Advisory Committee; a former undersecretary of defense; a former Pentagon assistant director of defense research and engineering; a former commissioner of the Social Security Administration; a former chair and permanent member of the board of directors of the Armed Forces Communication and Electronics Association; and a retired commander, U.S. European Command, and Supreme Allied Commander, Europe, NATO (Lockheed Martin 2007).

Because the military, government, and corporations are so interdependent and because decisions made by the elite of one sector affect the elite of the other two sectors, Mills believes that everyone has a vested interest in coop-

Company Name	Awards (Billion $)		Percentage of Revenue Derived from Defense Contracts
	2007	2006	
Lockheed Martin Corporation	38.5	36.1	92.0
Boeing Company	32.1	32.4	48.0
BAE Systems	29.8	25.1	95.0
Northrop Grumman Corporation	24.6	23.6	77.0
General Dynamics Corporation	21.5	18.8	79.0
Raytheon Company	19.8	19.5	93.0
EADS	12.2	13.2	21.3
L-3 Communications	11.2	10.0	81.0
Finmeccanica	10.6	9.1	53.6
United Technologies	8.8	7.7	16.0

▲ **Figure 11.7** Top 10 Military Contractors for Fiscal Years 2006 and 2007

The U.S. Department of Defense awards hundreds of billions of dollars in contracts each year. More than one-third of this amount goes to the 10 companies listed in the table. What companies depend almost exclusively on government contracts?

Source: U.S. Department of Defense (2009)

eration. Shared interests cause those who occupy the highest positions in each sector to interact with one another. Out of necessity, then, a triangle of power has emerged. We should not assume, however, that the alliance among the three sectors is untroubled, that the powerful in each realm share the same mindset, that they know the consequences of their decisions, or that they are joined in a conspiracy to shape the fate of a country or the globe.

> At the same time it is clear that they know what is on each other's minds. Whether they come together casually at their clubs or hunting lodges, or slightly more formally at the Business Advisory Council or the Committee for Economic Development or the Foreign Policy Association, they are definitely not isolated from each other. Informal conversation elicits plans, hopes, and expectations. There is a community of interest and sentiment among the elite. (Hacker 1971, p. 141)

Mills gives no detailed examples of the actual decision-making process at the power elite level, nor does he assess the players' motives. Instead, he focuses on understanding the consequences of this alliance. Mills acknowledges that the power elite are not free agents subject to no controls. A chief executive officer of a major corporation is answerable to unions, the Occupational Safety and Health Administration, the Food and Drug Administration, and perhaps other regulatory bodies. Pentagon officials are subject to congressional investigations and budget constraints. Defense contractors must be aware of the Federal False Claims Act, which gives a share in the settlement to any employee who can prove that the contractor has defrauded the government (Stevenson 1991). The president of the United States is constrained by bureaucratic "red tape" and by the judicial and legislative branches. A pluralist model of power acknowledges these constraints and offers an alternative vision of how power is distributed.

pluralist model A model that views politics as an arena of compromise, alliances, and negotiation among many competing and different special-interest groups, and power as something that is dispersed among those groups.

special-interest groups Groups composed of people who share an interest in a particular economic, political, and social issue and who form an organization or join an existing organization with the goal of influencing public opinion and government policy.

political action committees (PACs) Committees that raise money to be donated to the political candidates most likely to support their special interests.

527 group A tax-exempt advocacy organization that seeks to influence federal elections by running issue-related advertisements criticizing the record of a candidate or by mobilizing voters to register and vote.

Pluralist Models

■ **CORE CONCEPT 9:** Pluralist models view politics as a plurality of special-interest groups competing, compromising, forming alliances, and negotiating with each other for power. A **pluralist model** views politics as an arena of compromise, alliances, and negotiation among many competing special-interest groups, and it views power as something dispersed among those groups. **Special-interest groups** consist of people who share an interest in a particular economic, political, or social issue and who form an organization or join an existing organization to influence public opinion and government policy. Some special-interest groups form **political action committees (PACs)**, which raise money to be donated to the political candidates who seem most likely to support their special interests. Examples of the more than 4,500 registered PACs listed among the top 20 contributors to 2005–2006 U.S. political campaigns included the National Association of Realtors ($4.0 million), International Brotherhood of Electrical Workers ($3.3 million), AT&T ($3.1 million), American Bankers ($2.9 million), National Beer Wholesalers ($2.9 million), National Auto Dealers ($2.9 million), International Association of Firefighters ($2.7 million) (Center for Responsive Politics, 2009).

Another kind of special-interest group is the **527 group**. This is a tax-exempt advocacy organization that seeks to influence federal elections by running issue-related advertisements criticizing the record of a candidate or by mobilizing voters to register and vote. Prominent 527 groups in the 2008 U.S. presidential election included MoveOn.org and The Fund for America (Center for Responsive Politics 2009b). Together these two organizations spent $50 million.

According to the pluralist model, no single special-interest group dominates the U.S. political system. Rather, competing groups thrive and can express their views through opinion polls, unions, protests, e-mails, and PACs. "If a variety of interest groups are able to exercise power and influence in the system through a variety of means, then the system, with all its flaws, can be considered democratic" (Creighton 1992). One problem with the pluralist model is that, even though several thousand PACs operate in the U.S. political arena, we cannot conclude that every special-interest group has enough resources to organize and donate enough money to defend its interests.

The pluralist model helps us to think about the many groups in Iraq competing for representation: Shiites, Sunni Muslims, Kurds, ethnic Turks, and Assyrian Christians. While we may think of Iraq as consisting of ethnic and religious groups, some scholars maintain that the real divisions are tribal. Iraq is home to "100 major tribes, 25 tribal confederations and several hundred cohesive clans" (Sachs 2004). Strong family bonds hold tribes together, as evidenced by the fact the 50 percent of Iraqi marriages are between first and second cousins and the divorce rate among participants in these marriages is less than 2 percent. These extraordinarily strong family bonds complicate the road to democracy, because in theory, democracies depend on individuals who value public good over family obligation (Tierney 2003).

Imperialism and Related Concepts

■ **CORE CONCEPT 10:** Empire, imperialism, hegemony, and militarism are concepts that apply to political entities such as governments that can exercise their will over other political entities. An **empire** is a group of countries under the direct or indirect control of a foreign power or government, which shapes their political, economic, and cultural development. An **imperialistic power** exerts control and influence over foreign entities either through military force or through political policies and economic pressure. Imperialists believe that their cultural, political, or economic superiority justifies control over other entities, and they maintain that such control is for the greater good of all humankind.

Hegemony is a process by which a power maintains its dominance over foreign entities. Those in power use institutions (such as the World Bank and the United Nations) to formalize their power; work through bureaucratic structures "to make power seem abstract (and therefore, not attached to any one individual)"; use advertising, education, television, music, movies, and other forms of media to influence the foreign populace; and mobilize a military or police force to intimidate and subdue opposition forces (Felluga 2002).

A **militaristic power** believes that military strength, and the willingness to use it, is the source of national and even global security. Usually a "peace through strength" doctrine—peace depends on military strength and force—is cited to justify military buildups and interventions on foreign soil (see "The Personal Cost of War").

Is the United States an Imperialistic Power?

Throughout the history of mankind certainly no country has existed that so thoroughly dominates the world with its policies, its tanks, and its products as the United States does today. (*Der Spiegel* 2003)

Talk of "empire" makes Americans distinctly uneasy . . . and yet though the rest of the world is under no illusion, in the United States today there is a sort of wishful denial. We don't want an empire, we aren't an empire—or else if we are an empire, then it is one of a kind. (Judt 2004, p. 38)

"American empire" and the related phrases "American imperialism," "American militarism," and "American hegemony" yield more hits on the Google search engine than terms for any other empire (including the British and Roman). Is the United States an empire, an imperialistic power, militaristic, or hegemonic? Our goal in this section is *not* to assess whether the United States is an evil or a benev-

olent empire. Rather, we simply wish to assess the United States' world influence and power in light of these concepts.

Because Iraq is the focus of this chapter, we begin by noting that the United States intervened in Iraq despite having limited international support for this action. Note that although 29 countries sent troops to Iraq, the United States supplied seven-eighths of the ground troops (more than 130,000). After the United States, the largest contributors at peak were the United Kingdom (45,000 troops), followed by Italy (3,200) and South Korea (3,600). Of the 28 coalition partners, 14 sent 200 or fewer troops, and 3 (Norway, Moldova, and Kazakhstan) sent 30 or fewer (U.S. Department of Defense 2009). As of July 31, 2009, the United States remains the only country with combat troops in Iraq.

Of course, the ability to exercise power over foreign entities requires a strong military, the ability to put weapons into the hands of allies, and the ability to keep those same weapons out of the hands of adversaries (See No Borders, No Boundaries: "Countries in Which the United States Intervened to Support Regime Change, Support Dictators, or Oppose Reactionary Movements, 1902–2007"). Between 2000 and 2007, the United States delivered $134.8 billion worth of arms to foreign countries. This figure represents 38.0 percent of all arms delivered, whose value was $449 billion. After the United States, the top arms suppliers during that seven-year period were Russia ($65.5 billion), France ($32.1 billion), the United Kingdom ($26.4 billion) and China ($12.5 billion). Of all arms delivered, 66 percent went to developing nations, including India ($58.4 billion), Saudi Arabia ($49.4 billion), China ($31.4 billion), United Arab Emirates ($26.9 billion), Pakistan ($26.9 billion), Egypt ($22.5 billion), and Israel ($17.9 billion) (Congressional Research Service 2009).

In terms of its ability to invest in its military, the United States accounts for almost 57 percent of all military spending worldwide. In 2008, worldwide military expenditures totaled $1.1 trillion. The United States budgeted $657.3 billion, followed by the People's Republic of China ($65

empire A group of countries under the direct or indirect control of a foreign power or government such that the dominant power shapes the subordinate entities' political, economic, and cultural development.

imperialistic power A political entity that exerts control and influence over foreign entities through conquest or force and/or through policies and economic pressures.

hegemony A process by which a power maintains its dominance over other entities.

militaristic power One that believes military strength, and the willingness to use it, is the source of national—and even global—security.

INTERSECTION OF BIOGRAPHY AND SOCIETY

The Personal Cost of War

THIS ACCOUNT OF an Iraqi translator is from a news article published by the U.S. Department of Defense. It can be found in its entirety at www.defenselink.mil/news/newsarticle.aspx?id=25244. The stated purpose of this Web site is "to support the overall mission of the Department of Defense." Keep in mind that the information on this webpage, however factual, is to further the war cause. On the other hand, the story speaks to a major tension between Iraqis who have risked life and limb to help the U.S. and the seeming unwillingness of the United States to protect them from insurgent forces by allowing them (among other things) to immigrate to the United States (Guardian Unlimited 2007).

Sally's children were taken away from her more than six months ago. Her husband beat her. Her brother threatened her life while holding a gun to her head. Her own father contracted her death with a $5,000 reward.

Sally, an Iraqi translator working with Multinational Force Iraq, lost everything by working to help Americans rebuild Iraq. Still, she feels her service with Americans is the right thing for her country. . . .

Early one morning when the war started, she heard yelling outside her home. Americans in a Humvee were talking to one of her neighbors. "They were speaking English and trying to talk to a man," she said. "They were going to arrest him. So I went outside to help him and talked to the Americans for the man. The Americans were very

appreciative and [offered me] a job. I told them they know where I live if they ever need my help."

She thought being a translator would be a great way to help her country. She took an English test and was accepted to become a translator.

Sally's decision, though, was unpopular with her Iraqi neighbors.

"My neighbors found out that I was helping the Americans, and they beat my children," she said. "They threw rocks at my daughter and broke both arms on my son. They told me to watch out or I will be killed."

It wasn't just her neighbors who harbored hatred for Sally's assistance to the coalition. Her family was infuriated. . . .

Her husband threatened to tell her family where she was, sealing her death sentence.

She escaped only because of her oldest son.

"My older son, who is 13, opened the bathroom door and said, 'Mom you need to run away,'" she recalled. "You cannot stay here. They will kill you. Mom, they will kill you!"

Sally said she did not want to leave her children behind.

"He pushed me out the door and I ran," she said. "I don't know where, but I ran."

She left with nothing but the clothes she was wearing, a picture of her kids and a stuffed tiger her son slept with at night. It was the last time she saw her children.

Source: Tuskowski 2004.

---

billion), Russia ($50 billion), France ($45 billion), the United Kingdom ($42.8 billion), and Japan ($41.7 billion) (Global Security.org 2009).

U.S. military strength and power were displayed in the first days of the 2003 Iraq War, when U.S. forces engaged in a "shock and awe" campaign. Part of this campaign involved launching 800 cruise missiles in two days—more than two times the number of missiles launched during the entire 1991 Gulf War. The purpose of "shock and awe"

insurgents Groups who participate in armed rebellion against some established authority, government, or administration with the hope that those in power will retreat.

was to "shatter Iraq 'physically, emotionally, and psychologically'" so that Iraqis would lose their will to fight. One Pentagon official pointed out in a news briefing that "the sheer size of this has never been seen before, never been contemplated before" (West 2003).

President George W. Bush declared major military operations in Iraq to be over on May 2, 2003—43 days after they began. Since then, however, the most powerful military in the world has been battling Iraqi **insurgents**, groups who participate in armed rebellion against an established authority, government, or administration with the hope that those in power will retreat or pull out. From the occupier's or liberator's point of view, insurgents have no legitimate cause; from the insurgent's point of view, the authority of the occupier or liberator is illegitimate. Iraqi

NO BORDERS, NO BOUNDARIES

Countries in Which the United States Intervened to Support Regime Change, Support Dictators, or Oppose Reactionary Movements, 1902–2007

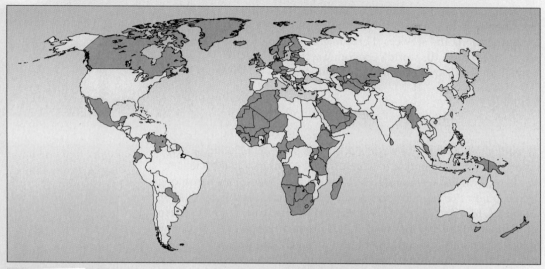

▲ **Figure 11.8**

Consider as one indicator of American power the countries (highlighted in yellow) in which the United States intervened to support regime change, support dictators, or oppose reactionary movements during the period 1902–2007. These interventions took a variety of forms: overt force, covert operations, and subverted elections. This map is not intended to judge a decision to intervene, as each intervention has a specific history. Before making such judgments, the reader must study the reasons for and events leading up to each intervention.

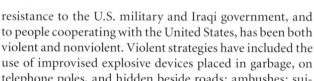

resistance to the U.S. military and Iraqi government, and to people cooperating with the United States, has been both violent and nonviolent. Violent strategies have included the use of improvised explosive devices placed in garbage, on telephone poles, and hidden beside roads; ambushes; suicide bombings; kidnappings; beheadings; and other attacks on military and civilian targets (Wordiq.com 2004c).

By the time this book went to press, more than 1 million U.S. troops had served in Iraq; 4,250 had died and 30,182 had been wounded (U.S. Department of Defense 2009). The number of Iraqi civilians reported killed because of U.S.-led military actions since January 2003 ranged from 92,871 to 101,316 (Iraq Body Count Project 2009). These figures were based on a comprehensive survey of reports from major international news agencies. Many other civilian deaths have probably gone unreported. The numbers dead do not include people who died as an indirect result of the war (such as heat exhaustion or heart attacks) which may be as high as 600,000. In addition 1.5 million Iraqi refugees still remain outside the country (Refugees International 2009).

Is the United States a liberating or an occupying power in Iraq? How are children in Iraq processing the U.S. presence? Consider that 10.1 million Iraqis are 14 years of age or younger. Americans must ask, What does it mean to occupy or liberate a country where 40 percent of the population consists of children?

■ VISUAL SUMMARY OF CORE CONCEPTS

■ CORE CONCEPT 1: The ongoing agricultural, industrial, and information revolutions have profoundly shaped the world's economic systems.

A society's economic system is the social institution that coordinates human activity to produce, distribute, and consume goods and services. Three major, ongoing revolutions have shaped the world's economic systems: agricultural, industrial, and information revolutions.

Department of Defense

■ CORE CONCEPT 2: The world's economic systems fall along a continuum whose endpoints are ideal forms of capitalism and socialism.

We can classify the economies of the world as falling somewhere along a continuum that has ideal forms of capitalism and socialism as its extremes. Ideally, capitalism is an economic system in which the means of production are privately owned. It is profit-driven, free of government interference, and governed by the laws of supply and demand. Socialism is an economic system in which the raw materials and the means of producing and distributing goods and services are collectively owned. Socialists believe that government or some worker- or community-oriented organization should play the central role in regulating economic activity on behalf of the people as a whole.

Department of Defense photo by Don S. Montgomery, USN

Department of Defense photo by Cpl. D.W. Yarnall, USMC

■ CORE CONCEPT 3: Capitalists' responses to economic downturns and stagnation have driven a 500-year-plus economic expansion, which has facilitated interconnections between regional and national economies.

World system theory focuses on the forces underlying the development of economic transactions that transcend national boundaries. In particular, world system theorists identify profit-generating strategies that have caused capitalism to dominate and facilitate a global network of economic relationships. As a result, capitalism has spread steadily throughout the globe. In addition, every country of the world has come to play one of three different and unequal roles in the global economy: core, peripheral, or semiperipheral.

© John Coletti/Index Stock Imagery

■ **CORE CONCEPT 4:** Peripheral economies are built upon on a few commodities or a single commodity; thus, they are extremely vulnerable to fluctuations in price and demand.

Peripheral economies operate on the so-called fringes of the world economy. They are highly vulnerable economies, because they build on revenues from a few commodities or even a single commodity. These economies have a dependent relationship with core economies that traces its roots to colonialism.

Department of Defense photo by SPC James P. Johnson, USA

■ **CORE CONCEPT 5:** Core economies are wealthy societies with highly diversified economies and secure, stable governments.

Core economies include the wealthiest, most highly diversified economies with strong, stable governments. When economic activity weakens in core economies, the economies of other countries suffer because exports to core economies decline and price levels fall. A disproportionate share of multinational and global corporations are headquartered in core economies. Such economies have a strong tertiary (service-related) sector, and computer and information technologies play important roles in them.

© VStock LLC/Index Open

■ **CORE CONCEPT 6:** When people believe that power differences are legitimate, those with power possess authority, which may be classified as charismatic, traditional, or legal-rational.

Authority is legitimate power—power that people believe is just and proper. A leader has authority to the extent that people view him or her as being entitled to give orders. Authority can be of three types: (1) traditional, which relies on the sanctity of time-honored norms that govern the selection of someone to a powerful position (such as chief, king, or queen) and that specify responsibilities and conduct appropriate for the individual selected; (2) charismatic, which derives from the exceptional and exemplary qualities of the person who issues the commands; and (3) legal-rational, which derives from a system of impersonal rules that formally specify the qualifications for occupying a powerful position.

U.S. Marine Corps photo by Lance Cpl. Michael J. Ayotte/Released

■ **CORE CONCEPT 7:** Government is an organizational structure that directs and coordinates people acting in the name of a country or some other political entity. That structure may be democratic, authoritarian, totalitarian, or theocratic.

Government is the organizational structure that directs and coordinates people's involvement in the political and economic activities of a country or some other territory, such as a city, county, or state. Forms of government include democracy, totalitarianism, authoritarianism, and theocracy. One measure of a democracy is free and fair elections. In Iraq, voters inked their fingers to identify them as voters and to prevent them from voting more than once.

Department of Defense photo by Spc. Jeffery Sandstrum, U.S. Army

■ **CORE CONCEPT 8:** The power elite consists of people whose positions in the military, the government, and the largest corporations are so high that their decisions affect the lives of millions, even billions, of people.

The power elite are those few people who occupy such lofty positions in the social structure of leading institutions that their decisions affect millions, even billions, of people worldwide. For the most part, the source of this power is legal-rational. In writing about the power elite in the United States, sociologist C. Wright Mills focused on those who occupy the highest positions in the leading institutions: the military, the 200 largest corporations, and the government.

DoD photo by Mass Communication Specialist 1st Class
Chad J. McNeely, U.S. Navy/Released

■ **CORE CONCEPT 9:** Pluralist models view politics as a plurality of special-interest groups competing, compromising, forming alliances, and negotiating with each other for power.

A pluralist model views politics as an arena of compromise, alliances, and negotiation among many competing special-interest groups, and it views power as something dispersed among those groups. Special-interest groups consist of people who share an interest in a particular economic, political, or social issue and who form an organization or join an existing organization to influence public opinion and government policy.

Ben F. Schumin/The Schumin Web

■ **CORE CONCEPT 10:** Empire, imperialism, hegemony, and militarism are concepts that apply to political entities such as governments that can exercise their will over other political entities.

A government is an empire when it controls the political, economic, and cultural development of a group of countries. An imperialistic power controls foreign entities through military force or through political policies and economic pressure. Imperialists justify their control by claiming that they have cultural, political, or economic superiority and that they are using it for the greater good of all humankind. Hegemony is a process by which a power formalizes its dominance over foreign entities—using established institutions, bureaucratic structures, the mass media, and military or police force. A militaristic power believes that military strength, and the willingness to use it, is the source of national and even global security.

Department of Defense photo by Cpl. Michael S. Cifuentes, U.S. Marine Corps

Resources on the Internet

 Sociology: A Global Perspective Book Companion Web Site
www.cengage.com/sociology/ferrante

Visit your book companion Web site, where you will find flash cards, practice quizzes, Internet links, and more to help you study.

CENGAGENOW™

Just what you need to know NOW!

Spend time on what you need to master rather than on information you have already learned. Take a pre-test for this chapter, and CengageNOW will generate a personalized study plan based on your results. The study plan will identify the topics you need to review and direct you to online resources to help you master those topics. You can then take a post-test to help you determine the concepts you have mastered and what you will need to work on. Try it out! Go to www.cengage.com/login to sign in with an access code or to purchase access to this product.

Key Terms

527 group   322
authoritarian government   318
authority   315
capitalism   305
charismatic authority   315
colonization   301
democracy   316
domestication   300
economic system   300
empire   323
goods   300
government   316
gross domestic product (GDP)   313
hegemony   323

imperialistic power   323
insurgents   324
legal-rational authority   316
mechanization   301
militaristic power   323
pluralist model   322
political action committees (PACs)   322
political system   315
post-industrial society   302
power   315
power elite   319
primary sector (of the economy)   313

representative democracy   317
safe debt   310
secondary sector (of the economy)   313
services   300
socialism   306
special-interest groups   322
stupid debt   310
survival debt   310
tertiary sector (of the economy)   313
theocracy   319
totalitarianism   317
traditional authority   315
welfare state   307

12 Family

With Emphasis on Japan

When sociologists study the family, they focus on the many factors that affect its structure and composition.

© Eriko Koga/Taxi Japan/Getty Images

▲ Japan's couples do not reproduce enough to replace the number of people who die. The low total fertility rate is a major national concern, and it has prompted a variety of responses, including delivering urgent appeals to couples to have babies and initiating policies that support women who wish to pursue a career and have a family.

WHY FOCUS ON

Japan?

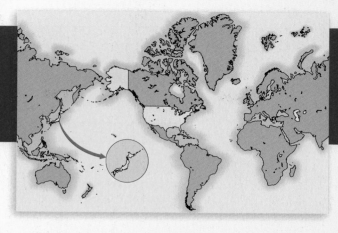

THE FAMILY IS an ever-changing entity. Millions of seemingly personal decisions influence the variety of family arrangements that exist in any country—decisions about (1) whether to have children, and if so, how many to have, when to have them, and how to space them; (2) whether to marry, and if so, when; (3) whether to work for pay; and (4) whether to become a caregiver to dependent relatives. As we will see, these decisions are shaped and constrained by several larger forces, including average life expectancy, employment opportunities, and social norms.

In this chapter, we compare family life in Japan and in the United States. Based on indicators that seem to be associated with family well-being and stability, Japan appears to do better than the United States. The country has lower infant mortality, cohabitation, and divorce rates. Japan has a lower teen birth rate (5.6 live births per 1,000 females age 15–19) and a much smaller percentage of single-parent households. A relatively high percentage of the elderly in Japan live with their children. In addition, Japan has much fewer reported cases of domestic and child abuse than the United States (see Figure 12.1).

On the other hand, people in the United States seem more optimistic about marriage and children than people in Japan. The United States has a higher marriage rate, a smaller percentage of never married women, a higher teen birth rate (41 live births per 1,000 females age 15–19) and a higher **total fertility rate**, the average number of children that a woman bears in her lifetime (see Figure 12.1).

Both the United States and Japan have an **aging population**. That is, the percentage of their people that are age 65

and older is increasing relative to other age groups. Currently, 12.4 percent of the U.S. population and 21 percent of the Japanese population are 65 years old or older. In 2020, almost 30 percent of Japan's population will be 65 and older, compared with 16.6 percent of the U.S. population (U.S. Bureau of the Census 2007). An aging population is one in which family members tend to live longer—and thus have the opportunity to be part of family life for a greater number of years—than ever before.

Japan's low total fertility rate, combined with its long life expectancy and low immigration rate, means that the country has one of the oldest populations in the world. The low total fertility rate is a major national concern, and it has prompted a variety of responses, including condemning young people for being selfish, delivering urgent appeals to couples to have babies, and initiating policies that make it easier than before for women to pursue a career and raise children (Boling 1998; *Japan Times* 2003a).

In the United States the decline in the number of nuclear families—those comprising a husband and wife and their children—relative to other family forms has grabbed headlines. The high teen birth rate and high percentage of single-parent households are considered by many to pose a national problem. In this chapter, we examine some of the major social forces shaping family life in such different ways in the United States and Japan (see Figure 12.1).

FACTS TO CONSIDER

- In Japan the average number of children that a woman bears in her lifetime is 1.2. In the United States that number is 2.1.

- In the United States 28 percent of households with children are headed by a single parent, usually the mother. In Japan 1.4 percent of children live in a single-parent household.

- The teen birth rate in Japan is 5.6 per 1,000 females age 15–19; in the United States the teen birth rate is 41 births per 1,000 females age 15–19.

- In Japan 21 percent of the population is age 65 or older. In the United States 12.4 percent of the population is 65 or older.

- In the United States 13 percent of those 65 and older live with their children or some other relative. In Japan 50 percent of persons 65 and older live with their adult children.*

*References for these statistics can be found in Figure 12.1 and Tables 12.5 and 12.6.

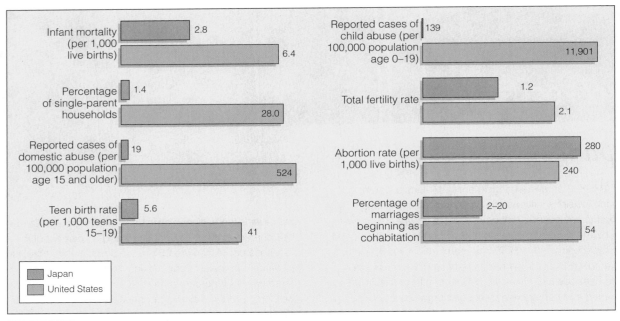

▲ **Figure 12.1**   Indicators of Family Structure, Composition, and Well-Being in Japan and the United States

Sources: U.S. Central Intelligence Agency (2007); Statistics Bureau, Ministry of Internal Affairs and Communications (Japan) (2007); U.S. Department of Health and Human Services (2004, 2006); McCurry (2006); Al-Badr (2006); Scommegna (2002); Iwasawa, Raymo, and Bumpass (2005), United Nations Statistics Division (2004)

Defining Family

■ **CORE CONCEPT 1: An amazing variety of family arrangements exists in the world. This variety makes it difficult to define family.** **Family** is a social institution that binds people together through blood, marriage, law, and/or social norms. Family members are generally expected to care for and support each other. This definition is very general because getting people to agree on a definition is surprisingly difficult. The difficulty is rooted in the fact that, even though every person is a member of a family (if only in the biological sense), "there is no concrete group which can be universally identified as 'the family'" (Zelditch 1964, p. 681). Instead, an amazing variety of family arrangements exists worldwide—a variety reflected in the numerous norms that specify how two or more people can become a family. These norms govern family-related matters such

as who can marry, the number of spouses people can have, the way people connect with spouses, and the ways people trace their ancestors and descendants (see Table 12.1). In light of this variability, we should not be surprised that when people think of family, they often emphasize different dimensions, such as kinship, ideal members, or legal ties.

Kinship

Definitions of *family* that emphasize kinship view the family as comprising members who are linked together by blood, marriage, or adoption. Based on this definition, the size of any given person's family network is beyond comprehension, because one person has an astronomical number of living and deceased kin. Keep in mind that to calculate the number of a person's relatives, one would have to count primary kin (mother, father, sister, brother), secondary kin (mother's mother, mother's father, sister's son, brother's daughter), tertiary kin (brother's daughter's son, mother's sister's son), and beyond (brother's daughter's son's son).

Given that it is virtually impossible to keep track of even one's living relatives, let alone maintain social relationships with them, every society finds ways to exclude some kin from their idea of family. For example, some societies trace family lineage through the maternal or the paternal side only. Selective forgetting and remembering

total fertility rate   The average number of children that a woman bears in her lifetime.

aging population   A population in which the percentage that is age 65 and older is increasing relative to other age groups.

family   A social institution that binds people together through blood, marriage, law, and/or social norms. Family members are generally expected to care for and support each other.

Table 12.1 Norms Governing Family Structure and Composition

Number of Marriage Partners

Monogamy	One husband, one wife
Serial monogamy	Two or more successive spouses
Polygamy	Multiple spouses at one time
Polygyny	One husband, multiple wives at one time
Polyandry	One wife, multiple husbands at one time

Choice of Spouse

Arranged	Parents select their children's marriage partners
Romantic	One selects a marriage partner oneself based on love
Endogamy	Marriage within one's social group
Exogamy	Marriage outside one's social group
Homogamy	Marriage to a person or persons whose social characteristics—such as class, religion, and level of education—are similar to one's own

Authority

Patriarchal	Male dominated
Matriarchal	Female dominated
Egalitarian	Equal authority between sexes

Descent

Patrilineal	Traced through father's lineage
Matrilineal	Traced through mother's lineage
Bilateral	Traced through both mother's and father's lineage

Family Type

Nuclear	Husband, wife, and their immediate children
Extended	Three or more generations
Single-parent	Mother or father living with children
Household	People who share the same residence
Domestic partnership	People committed to each other and sharing a domestic life but not joined in marriage or civil union
Civil union	A legally recognized partnership providing same-sex couples with some of the rights, benefits, and responsibilities associated with marriage

Family Residence

Patrilocal	Wife living with or near husband's family
Matrilocal	Husband living with or near wife's family
Neolocal	Wife and husband live apart from their parents

is another way of excluding some kin. That is, people make conscious or unconscious decisions about which kin they will acknowledge as family and which kin they will "forget" to mention to their children (Waters 1990).

Membership

Definitions of *family* that emphasize membership characteristics view the family in terms of who should be counted as a family member. One of the broadest definitions of *family* is "everyone who lives under one roof and expresses love and solidarity" (Aguilar 1999). Organizations such as the World Congress of Families (2001) argue that the ideal family is the voluntary union of a man and a woman in a lifelong marriage covenant welcoming of children. An **ideal** is a standard against which real cases can be compared. If we simply think about the living and procreation arrangements we observe every day, we quickly realize that many of these arrangements do not match the World Congress of Families' ideal. This fact does not stop people from using such an ideal to evaluate their own and others' living arrangements, and from labeling family arrangements that do not fit the ideal as nontraditional, dysfunctional, immoral or at-risk (Cornell 1990).

Legal Recognition

When legal recognition is the defining criterion of a family, a family is defined as two or more people whose living and/or procreation arrangements are recognized *under the law* as constituting a family. U.S. federal law defines *spouse* as "a person of the opposite sex who is a husband or wife" (Defense of Marriage Act 1996, section 7). Some state, county, and city governments within the United States formally recognize alternatives to legal marriage, such as common-law marriages, domestic partnerships, and civil unions.

Legal recognition of family and marriage arrangements means that the benefits, responsibilities, and rights awarded to those arrangements are enforced by law. In the United States, federal law defines *marriage* as "a legal union between one man and one woman as husband and wife." But some government entities recognize alternatives to that definition, arguing that many people (including gay, lesbian, and heterosexual couples) form lasting, committed, caring, and faithful relationships—and that

ideal A standard against which real cases can be compared.

The people shown in this photo are related to one another by blood or marriage. Yet these members shown represent only a fraction of all the people who share the status of relative.

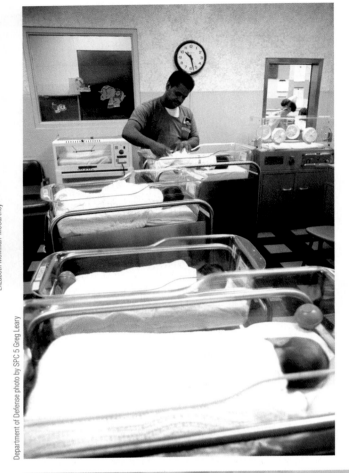

without the legal protections, benefits, and responsibilities associated with civil marriage, such people might suffer numerous obstacles and hardships (Vermont Civil Unions Law 2000).

Functionalist View of Family Life

■ **CORE CONCEPT 2: Family can be defined in terms of the social functions it performs for society.** Because of the debate over what constitutes membership in a family, some sociologists argue that *family* should be defined in terms of social functions. Recall from Chapter 2 that a function is a contribution that a part of society makes to order and stability in the larger society. In this regard, the family performs at least five functions: (1) regulating sexual behavior, (2) replacing the members of society who die, (3) socializing the young, (4) providing care and emotional support, and (5) conferring social status.

Regulating Sexual Behavior

Marriage and family systems specify norms (which can take the form of laws) that regulate sexual behavior. These norms may prohibit sex outside of marriage, or they may specify the characteristics of appropriate sexual and marriage partners. Such norms include laws prohibiting marriage and sexual relationships between specific relatives (such as first cousins), members of different age groups (such as an adult and a minor), and members of different

Families function to replace the members of society who die, by providing a legally sanctioned arrangement that can bring new members into the world.

Ideally, families provide their members with the meaningful social ties that all humans need to thrive both physically and emotionally.

Department of Defense photo by LCPL Keith Underwood, USMC

racial or ethnic groups. Other norms regulate the number of people who may form a marriage (such as monogamy, serial monogamy, and polygamy).

Replacing the Members of Society Who Die

All people eventually die. Thus, for humans to survive as a species, we must replace the members of society who die. Marriage and family systems provide a socially and legally sanctioned environment into which new members can be born and nurtured.

Socializing the Young

Recall that socialization is a learning process that begins immediately after birth. Through this process, "newcomers" develop their human capacities, acquire unique personalities and identities, and internalize the norms, values, beliefs, and language they need to participate in the larger society. The family is the most significant agent of socialization, because it gives society's youngest members their earliest exposure to relationships and the rules of life.

Providing Care and Emotional Support

A family is expected to care for the emotional and physical needs of its members. As we learned in Chapter 4, no matter how old they are, all humans require meaningful social ties to others. Without such ties, people deteriorate both physically and mentally. The human life cycle is such that we all experience at least one stage of extreme dependency (infancy and childhood). Unless we die suddenly, we are also likely to experience in adulthood some level of mental and/or physical deterioration accompanied by varying degrees of dependency.

Conferring Social Status

We cannot choose our family, the quality of the relationship between our parents, or the economic conditions into which we are born (see No Borders, No Boundaries: "Abandoned Offspring of U.S. Servicemen Stationed in Asia"). Among the things we inherit from our parents are a genetic endowment and a social status. For example, the physical characteristics we inherit affect the racial category to which we are assigned. Parents' occupations and incomes are also important predictors of **life chances**—a critical set of potential social advantages, including the chance to survive the first year of life, to live independently in old age, and everything in between.

While everyone might agree that families should fulfill these five functions, we must acknowledge that families often fail to achieve one or more of them, as evidenced by the following facts:

- Family systems do not always regulate sexual activity so that it is confined to a husband and wife. Based on DNA testing, an estimated 2–10 percent of pregnancies produce children who are not the biological offspring of the males believed to have fathered them (Lewin 2001; Seabrook 2001).
- Marriage and family systems do not always succeed in replacing the members of society who die. In at least 96 countries the total fertility rate is below 2.1—the minimum number of live births needed to replace the members who die (U.S. Central Intelligence Agency (2007)). Why a minimum of 2.1 and not 2? Not all babies survive to replace their biological parents. Of course, in societies that have high childhood mortality the total fertility rate must be more than 2.1 (See Global Comparisons: "Countries with Highest and Lowest Total Fertility Rates").

life chances A critical set of potential social advantages, including the chance to live past the first year of life, to live independently in old age, and everything in between.

NO BORDERS, NO BOUNDARIES

Asian Countries (Highlighted in Orange) Where There Are Abandoned Offspring of U.S. Servicemen

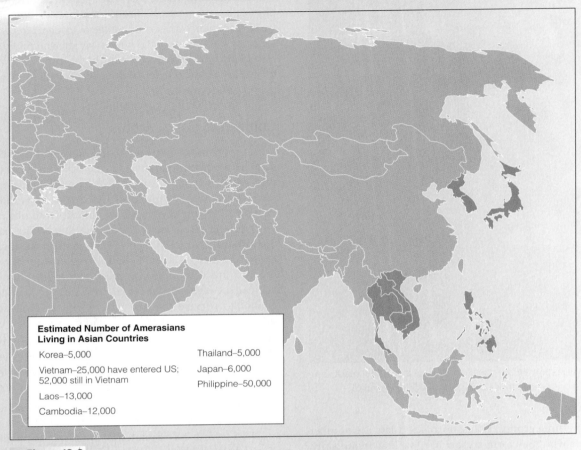

Estimated Number of Amerasians Living in Asian Countries

Korea–5,000

Vietnam–25,000 have entered US; 52,000 still in Vietnam

Laos–13,000

Cambodia–12,000

Thailand–5,000

Japan–6,000

Philippine–50,000

▲ **Figure 12.2**

Source: Amerasian Citizenship Initiative (2007), Karen Downing (2007), Pearl S. Buck International (2007).

- Family members do not always care for one another in positive ways. Domestic and child abuse is a problem in both the United States and Japan. While the number of *reported* cases of such abuse in Japan is very low compared with the number in the United States, many cases go unreported. In one survey taken by the Japanese Cabinet Office's Gender Equality Bureau, 33 percent of women and 17.4 percent of men reported that their partner abused them physically, mentally, or sexually. About 10 percent of women and 2.6 percent of men said they suffered repeated abuse (*Japan Times* 2006).

Conflict View of Family Life

■ **CORE CONCEPT 3:** Family life is not always harmonious. Furthermore, the family passes on social privilege and social disadvantages to its members, thereby perpetuating the system of social inequality. Focusing on the functions of family and marriage systems supports a view that family relationships are largely constructive and harmonious. Conflict theorists argue that although family members often do support one another and have common interests, family

Afghanistan has one of the highest total fertility rates in the world: the average Afghan woman bears 6.4 children during her reproductive life.

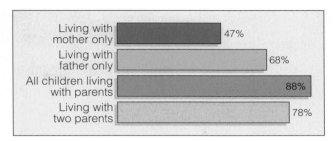

Percentage of U.S. Children with Secure Parental Employment by Living Arrangement, 2006

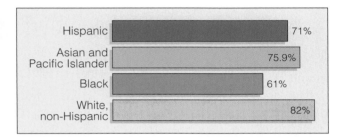

▲ **Figure 12.3**   Percentage of U.S. Children with Secure Parental Employment by Race/Ethnicity Classification, 1998

Source: Childstats.gov (2006).

Conflict theorists emphasize that family relationships are not always constructive and harmonious. Consider the case of U.S. servicemen stationed in Asian countries such as Japan and South Korea who abandon to orphanages children they have fathered with local women. These children are known an Amerasians.

members also have competing interests and some members have the power to exercise their will over other members. Thus choices that some members make do not always benefit everyone in the family or society at large (Cornell 1990). Furthermore, the family perpetuates inequalities by passing on social privilege and social disadvantages to its members. Moreover, marriage and family systems are structured to value productive work and devalue reproductive work, and to maintain and foster racial divisions and boundaries.

Social Inequality

At birth, we inherit a social status that shapes our life chances. Families transfer power, wealth, property, and privilege from one generation to the next. Obviously, parents' income affects the kinds of investments they can make in their children. According to the U.S. Census Bureau, 77 percent of children in the United States live in a household with **secure parental employment**. That is, they reside with at least one parent or guardian who was employed full-time (35 or more hours per week for at least 50 weeks in the past year). The percentage varies by racial classification and living arrangement (see Figure 12.3). Note that it drops to 47 percent for children living with their mothers only and increases to 88 percent for children living with both parents. Children classified as white are more likely than others to live in secure parental employment households, whereas children classified as black are least likely to live in such households. As we might expect, parents or guardians who are securely employed are less likely than others to live in poverty and more likely than others to have access to adequate housing and health care (ChildStats.gov 2006).

secure parental employment A situation in which at least one parent or guardian is employed full-time (35 or more hours per week for at least 50 weeks in the past year).

Countries with Highest and Lowest Total Fertility Rates

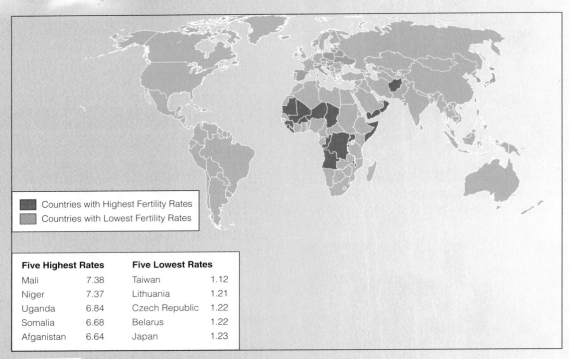

Five Highest Rates		Five Lowest Rates	
Mali	7.38	Taiwan	1.12
Niger	7.37	Lithuania	1.21
Uganda	6.84	Czech Republic	1.22
Somalia	6.68	Belarus	1.22
Afganistan	6.64	Japan	1.23

▲ **Figure 12.4**

This map shows the 20 countries and territories with the highest and the lowest total fertility rates. Taiwan has the lowest total fertility rate, at 1.12; Mali has the highest total fertility rate, at 7.38. That is, the average woman in Taiwan bears 1.12 children in her lifetime, whereas the average woman in Mali bears 7.38 children. The total fertility rate in the United States is 2.09.

Source: U.S. Central Intelligence Agency (2007)

Productive and Reproductive Work

According to Friedrich Engels (1884), the determining factor in history is the "production and reproduction of immediate family life": on one side, **productive work**—"the production of the means of existence, of food, clothing, and shelter and the tools necessary for that production"; on the other side, **reproductive work**—"the production of human beings themselves, the propagation of the species" (pp. 71–72). Reproductive work involves not only bearing children, but caregiving, managing households, and educating children. Both types of activities—production and reproduction—are work: "Renewing life is a form of work, a kind of production, as fundamental to the perpetuation of society as the production of things" (Laslett and Brenner 1989). While we say that reproductive work is valued, it is not usually rewarded highly on an economic level. In addition, reproductive work is disproportionately performed by women, whether or not pay is involved (see Table 12.2). Besides being associated with low wages, reproductive labor often involves poor working

productive work Work that involves "the production of the means of existence, of food, clothing, and shelter and the tools necessary for that production" (Engels 1884, pp. 71-72).

reproductive work Work that involves bearing children, caregiving, managing households, and educating children.

Table 12.2 Total Working in Caregiving Occupations, Percentage of Females in Caregiving Occupations, and Median Weekly Earnings

Occupation	Total Employed (in thousands)	% of Caregivers Who Are Women	Median Weekly Earnings
Registered nurses	2,529	91.3	$971
Elementary and middle school teachers	2,701	82.2	$824
Nursing, psychiatric, and home health aides	1,906	88.9	$395
Child-care workers	1,401	94.2	$345
Maids and housekeepers	1,423	90.3	$348
Teacher assistants	942	92.3	$405
Preschool and kindergarten teachers	690	97.7	$554

Source: U.S. Department of Labor (2006).

Reproductive occupations–whether those of teachers, nurses, or maids–are disproportionately filled by women.

conditions, little or no training, and few benefits, such as health insurance and retirement packages.

Although we have no comparable statistics on reproductive occupations in Japan, we do have data on housework sharing between husbands and wives. On average, Japanese men spend 33 minutes per day doing housekeeping and child-care activities, whereas Japanese women spend 3 hours and 45 minutes per day doing such activities. Compared with 10 years ago, men contribute an average of 10 minutes more per day and women contribute 1 minute less (Statistics Bureau, Ministry of Internal Affairs and Communications [Japan] 2007). Regardless of employment status, Japanese women contribute more hours than their male counterparts to housework and child care (see Table 12.3).

It is difficult to compare the United States and Japan in terms of housework and child-care sharing, if only because homes in the United States tend to be larger than those in Japan and because the governments of the two countries do not use the same method to chart activities. Nevertheless, it does seem that men living in the United States are more willing to perform domestic activities than men living in Japan. On average, American men spend 17.7 hours per week (up from 4.83 hours in 1965) doing housework, whereas American women spend 29.4 hours (down slightly from 30.01 hours in 1965) doing it. Together, American men and women spend an average of 47.1 hours per week doing housework (up from a combined total of 35 hours in 1965) (U.S. Bureau of Labor Statistics 2006, Bianchi et al. 2000).

Several possible explanations for these trends exist. First, the number of hours men living in the United States spend on household tasks has increased as the number of hours women work outside the home has increased. This shift in men's relative contribution has occurred in conjunction with new norms regarding who should do the housework and new expectations that men should increase their share. Second, the number of hours women living in the United States can devote to housework has decreased as more women have become engaged in paid labor. Third, couples have come to rely on service workers to relieve them of some time-consuming tasks, such as cooking (they may purchase take-out food), cleaning (they may hire a housekeeper), and laundry (they may use dry cleaners). Finally, new values and expectations concerning what constitutes necessary housework have emerged in the United States; for example, women are not expected to iron sheets today (Bianchi et al. 2000).

Of course, the amount of time men and women spend on housework and child-care depends on a whole host of factors, including the number of hours of paid employ-

Table 12.3 Average Time Men and Women in Japan Devote to Housework and Child Care (out of 1,440 minutes per day)

Hours of Paid Employment	Men (minutes per day)	Women (minutes per day)	Time Difference between Men and Women (minutes per day)	Time Difference between Men and Women (days per year)
Under 15 hours	97	271	174	44.1
15 to 34 hours	103	324	221	56.0
35 to 39 hours	105	280	175	44.4
40 to 48 hours	115	236	121	30.7
49 to 59 hours	96	197	101	25.6
60 hours and over	81	188	107	27.1

Source: Statistics Bureau, Ministry of Internal Affairs and Communications [Japan] (2007).

Table 12.4 Average Time U.S. Workers with Children under Age Six Spend on Various Activities (out of 1,440 minutes per day)

On what activities do employed women with children under six spend less time (in minutes) than their male counterparts? On what activities do they spend more time (in minutes) than men?

Activity	Men (minutes per day)	Women (minutes per day)	Women's Time Relative to Men's (minutes per day)	Women's Time Relative to Men's (days per year)
Housework	14.4	53.4	+39	+9.9
Food prep/Cleanup	16.2	47.4	+31.2	+7.9
Lawn care	10.8	5.4	−5.4	−1.4
Household management	4.8	7.2	+2.4	+0.6
Consumer goods purchases	40.2	57	+16.8	+4.3
Caring for household member(s)	78.6	130.2	+51.6	+13.1
Caring for nonhousehold member(s)	6.6	6.6	0	0
Paid employment	330.6	243.6	−87	+22.1
Sleeping and personal care	534	562.2	+28.2	+7.1
Eating and drinking	76.8	65.4	−11.4	−2.9
Socializing and communicating	43.8	43.2	−0.6	.2
Watching TV	123	100.2	−22.8	−5.6
Other	161	119	−42	−10.6

Source: U.S. Bureau of Labor Statistics (2006).

ment, marital status, and age. Table 12.4 compares the average amounts of time employed American men and women with children under age six devote to a range of activities, including housework, caregiving, paid employment, and leisure.

In *Having It All* Francine M. Deutsch (1999) concludes that "men and women can decide to share in reproductive work but doing so involves costs for both parties." Deutsch cites the example of a husband, a wife, and their young son Ollie traveling together by plane. The husband is holding Ollie, who is screaming. A male passenger sitting behind

them asks, "What's wrong with the child's mother? Why doesn't the mother take care of the baby?" The husband answers, "Because I'm his father and I am perfectly capable of taking care of him." The husband's decision not to hand his son over to his wife and his wife's decision not to grab Ollie away represent this couple's commitment to share in the reproductive work. Such decisions come at a price, however. Despite informal and subtle (and probably even hostile) pressures from passengers for the wife to intervene and the husband to hand the child over, the couple stood their ground. This scenario illustrates how even well-intentioned couples committed to equality might give in to societal pressures (Gerson 1999).

Racial Divisions and Boundaries

In the United States, we assume that people choose a partner or mate based on love. Upon investigating who marries, however, we find that people's choices are guided by other considerations as well: a potential partner's age, height, weight, income, education, race, sex, social class, and religion, among other things. When the conditions are right, we "allow" ourselves to fall in love. All societies have norms defining who may date or marry whom. These norms may be formal (enforced by law) or informal (enforced by social pressure).

Exogamy refers to norms requiring or encouraging people to choose a partner from a social category other than their own—for example, a partner outside their immediate family and not a partner of the same sex as their own. **Endogamy** refers to norms requiring or encouraging people to choose a partner from the same social category as their own—for example, a partner of the same race, ethnicity, religion, or social class.

In the United States, ethnic, and especially racial, categories have persisted, because most people "choose" partners whom they believe belong to their own ethnic or racial category. At one time, the United States had laws forbidding people to choose otherwise. (See Working For Change: "Dismantling a Big Lie.")

In Japan 95.5 percent of all marriages involve brides and grooms both classified as Japanese. Of the 36,263 marriages between a Japanese and a foreign partner that took place in one year, approximately 28,000—75 percent—involved a Japanese groom (French 2002). These statistics suggest that most Japanese couples practice endogamy. One reason endogamy is so predominant is that Japan's immigration policies have allowed few foreigners into the country. Thus, Japan is a homogenous society, described by its education minister as "one nation, one civilization, one language, one culture and one race" (Burgess 2007). In Japan 99 percent of residents are classified as Japanese. In the past 10 years, only 50,000 work visas per year have been issued to foreigners. This number is low when we consider that demographers have stated that Japan needs to issue 640,000 visas per year just to prevent its population from shrinking (Burgess 2007, Brooke 2004).

Changing Family Structures in Two Cultures

■ **CORE CONCEPT 4: The structure of the family is affected by a society's history and culture.** The struggle over the definition of family systems and the role they play in society

This photo of newly released slaves shows that people of different "races" (in this case, "black" slaves and "white" masters) have been producing offspring together for some time.

Rob Williams

Most Japanese practice endogamy; that is, they choose partners in the same racial category as their own. Still, about 5 percent break from the practice and choose partners in a different racial category.

is not unique to sociologists, but rather something with which educators, policy makers, government leaders, and others struggle. One reason behind these conceptual challenges is that family structures are not static; they adapt to economic, cultural, historical, and social factors. To capture these responses, sociologists track changes in family structure over time and seek to explain them.

exogamy Norms requiring or encouraging people to choose a partner from a social category other than their own.

endogamy Norms requiring or encouraging people to choose a partner from the same social category as their own.

WORKING FOR CHANGE

Dismantling a Big Lie

WHY DO PEOPLE laugh when someone suggests that a black person and a white person are relatives? Recently, a black student of mine named Mike Morgan was invited to the front of the classroom to draw the name of a fellow student out of a hat full of names. Coincidentally, he drew the name of a white female whose last name was also Morgan.

After Mike called out the name, he said without hesitation, "Perhaps we are related." The class broke into laughter, as if it seemed impossible to them that this white student and black student could belong to the same family. The belief that it is impossible for black individuals and white individuals to be brothers, sisters, cousins, parent and child, and so on speaks to a great American tragedy—the 400-year practice of dividing biologically related people (parents and children, grandparents and grandchildren, brothers and sisters) into distinct racial categories based on physical appearance. In the United States, this practice of dividing family members along racial lines began with slavery.

The story of race in America may be a family story, but it is almost never cast in that light. One person who has told the story in that light is Shannon Lanier. When he was a 21-year-old student at Kent State, he published (with photographer Jane Feldman) *Jefferson's Children: The Story of One American Family* (2000). The book is marketed to young readers.

Lanier (who appears black) begins the book by announcing that he is a descendant of Thomas Jefferson—the third president of the United States—and Sally Hemings—an enslaved woman. According to Lanier, Jefferson was his great-great-great-great-great-great-grandfather. Lanier grew up knowing this. His mother told him, and "she had learned [it] from her mama as her mama had learned from hers, and so on, from lips to ears, down through the generations" (Lanier and Feldman 2000, p. 11). Lanier recalls that as a first grader, he tried to share his family history with his classmates on President's Day but his teacher reprimanded him for lying. After all, nowhere in the history books of the time was Jefferson's relationship with Hemings mentioned. While Lanier dreamed of someday

© AP Photo/Richmond Times-Dispatch, Bob Brown

meeting relatives from both sides of his family, he did not believe it could happen.

In 1998 a team of geneticists tested Y-chromosomal DNA evidence and concluded that a male carrying the Jefferson Y chromosome fathered Eston Hemings—the last child born to Sally Hemings, in 1808. Because some 25 Jefferson males were living in Virginia at that time, the DNA evidence alone could not confirm Thomas Jefferson as the father. The Thomas Jefferson Foundation appointed a research committee to look into the matter further. It eventually concluded that the "DNA study, combined with multiple strands of currently available documentary and statistical evidence,[1] indicates a high probability that Thomas Jefferson fathered Eston Hemings, and that he most likely was the father of all six of Sally Hemings's children" (Thomas Jefferson Foundation 2000).

On May 15, 1999, Lanier traveled to the Monticello estate in Virginia to meet descendants of both Martha Jefferson (Thomas Jefferson's wife) and Sally Hemings.[2] He also met photographer Jane Feldman, who was there to document this historic gathering. Lanier and Feldman decided to travel around the country to meet and photograph probable descendants of the Thomas Jefferson–Sally Hemings relationship. Young readers who encounter this family's tale will come to learn that the true story of race in America is a family story. The persistence of Lanier and his family in telling and passing on their family story despite resistance teaches us that "the voice of the intellect is a soft one but it does not rest until it has gained a hearing" (Freud 1928).

---

1. Documentary evidence listed in the Thomas Jefferson Foundation's report includes the following: "Sally Hemings' birth patterns match Thomas Jefferson's Monticello visitation patterns. . . . Several people close to Jefferson or the Monticello community believed he was the father of Sally Hemings' children. . . . Sally Hemings' children had unique access to freedom. Jefferson gave freedom to no other nuclear slave family. . . . Sally Hemings' children bore a striking resemblance to Thomas Jefferson."

2. Sally Hemings was the half-sister of Jefferson's wife, Martha. In fact, Martha Jefferson had five other half-siblings, the offspring of sexual relationships between her father and enslaved women (Morgan 2002).

The Changing Family in the United States

Over the span of 100 years, the structure of the American family has changed quite dramatically (see Table 12.5). In 1900, 80 percent of children lived in two-parent families in which the mother worked on the family farm or in the home. At that time, only 2 percent of children lived in homes where both parents worked outside the home and fewer than 10 percent lived in one-parent homes. Today, about 32 percent of children live in two-parent homes in which the father is the breadwinner and the mother is a full-time homemaker. Sixty percent of those children who live in two-parent homes have parents who both work. Twenty-eight percent of all children live with a single parent, who also is likely to work outside the home. Today, fewer than 2 percent of children live in farm families (down from 43 percent in 1900). Table 12.5 shows that over the past century, average life expectancy has increased by almost 28 years for men and by more than 30 years for women. Infant mortality has dropped from 99.9 deaths to 6.8 deaths per 1,000 live births.

In "Wives and Work: The Sex Role Revolution and Its Consequences," sociologist Kingsley Davis (1984) identifies "as clear and definite a social change as one can find" (p. 401). Between 1900 and 1980 the percentage of women in the labor force rose from less than 20 percent to 60 percent, and the percentage of married women in the labor force rose from 15.4 percent to 53.1 percent. Davis links women's entry (and especially married women's entry) into the paid labor market to a series of social forces including (1) the rise and fall of the breadwinner system, (2) declines in total fertility, (3) increased life expectancy, (4) higher divorce rates, and (5) increased employment opportunities for women.

The Rise of the Breadwinner System

Before industrialization—that is, for most of human history—the workplace comprised the home and the surrounding land. Under these circumstances, the division of labor was based on sex. In nonindustrial societies (which include both hunting-and-gathering and agrarian societies), men provided raw materials through hunting or agriculture, and women processed these materials and took care of the young. This division of labor was not always clear-cut; women also worked in agriculture and provided raw materials by gathering food and hunting small animals.

The Industrial Revolution separated the workplace from the home and altered the division of labor between men and women. It destroyed the household economy by removing economic production from the home and taking it out of women's hands. A man's work, instead of being directly integrated with that of his wife and children in the home or on the surrounding land, was integrated with that of non-kin in factories, shops, and offices. In one sense, the man's economic role became more important to the family, for he was now the link between the family and the wider market economy, but at the same time, his personal participation in the household diminished. His wife, relegated to the home still performed the parental and domestic duties that women had always performed. She bore and reared children, cooked meals, washed clothes, and cared for her husband's personal needs, but to an unprecedented degree her economic role became restricted. She could not produce what the family consumed, because production had been removed from the home (Davis 1984, p. 403.)

Davis calls this new economic arrangement the "breadwinner system." Historically, this system was not typical.

Table 12.5   American Families: 100 Years of Change

Item	1900	1950	2007
Average household size (persons)	4.8	3.4	2.6
Households with 7 or more people (percent)	20.4	4.9	1.2
Living arrangements of children by family status (percent)			
Two-parent farm family	41	17	2
Two-parent nonfarm family	45	79	64
Father breadwinner, mother homemaker	43	56	32
Dual-earner	2	13	60
Single-parent	9	8	28
Not living with parent	5	6	4
Median age at first marriage*			
Men	25.9	22.8	27.1
Women	21.9	20.3	25.3
Life expectancy at birth (years)			
Males	46.3	65.6	74.1
Females	48.3	71.1	79.5
Infant mortality rate (deaths per 1,000 live births)	99.9	29.2	6.4
Life expectancy at age 65			
Men	11.7	12.9	16.3
Women	12.4	15.2	19.2
Percent age 65 and older	4.1	8.1	12.4
Total fertility rate (children born over woman's lifetime)	3.6	3.1	2.1

*These figures are estimates. Since 1990 the United States has not collected data on age at first marriage. In addition, the U.S. government collects almost no information on divorce. We do not know how many divorces occur each year, because five states (Texas, California, Colorado, Indiana, and Louisiana) do not release divorce statistics. Moreover, the federal government does not gather data on circumstances surrounding divorce—the number of children affected, length of marriage, and so on (Hacker 2002).

Source: U.S. Central Intelligence Agency (2007), U.S. Census Bureau (2005, 2007).

Theodor Horydczak Collection, Library of Congress Prints and Photographs Division

This photo illustrates the breadwinner system, an arrangement in which the father worked outside the home and the mother took care of the house and children.

Rather, it was peculiar to the middle and upper classes and was associated with a particular phase of industrialization. In the United States, "the heyday of the breadwinner system was from about 1860 to 1920" (p. 404). This system has been in decline ever since. Davis asks, "Why did the separation of home and workplace lead to the breadwinner system?" The major reason, he believes, is that women had too many children to engage in work outside the home. This answer is supported by the fact that family size, as measured by the number of living members, reached its peak in the period from the mid-1800s to the early 1900s. During this period, infant and childhood mortality declined but the old norms favoring large families persisted.

The Decline of the Breadwinner System

The breadwinner system did not last long because, for one thing, it placed too much strain on husbands and wives. The strains stemmed from several sources. Never before had the roles of husband and wife been so distinct. Never before had women played less than a direct, important role in producing what the family consumed. Never before had men been separated from the family for most of their waking hours. Never before had men been forced to bear the sole responsibility of supporting the entire family. Davis regards these events as structural weaknesses in the breadwinner system. Given the presence of these weaknesses, the system needed strong normative controls to survive: "The husband's obligation to support his family, even after his death, had to be enforced by law and public opinion. Illegitimate sexual relations and reproduction had to

be condemned; divorce had to be punished, and marriage had to be encouraged by making the lot of the 'spinster' a pitiful one" (p. 406).

Davis maintains that the normative controls collapsed not only because of the strains associated with the breadwinner system, but also because of demographic and social changes that accompanied industrialization. These changes included decreased total fertility rates, increased life expectancy, increased divorce rates, and increased opportunities for employment perceived as suitable for women.

Declines in Total Fertility

The decline in total fertility (number of children born over a woman's lifetime) began before married women entered the labor force in large numbers. Davis attributes the decline to two sources: the forces of industrialization, which changed children from economic assets to economic liabilities, and the "desire to retain or advance one's own and one's children's status in a rapidly evolving industrial society" (p. 408). Davis concluded that the decline in total fertility itself changed women's lives so that they had the time to work outside the home, especially after their children entered school. Not only did the number of children born in the average family decrease, but the average age at which women had their last child decreased (the mother's median age at the time of her last child's birth was 40 in 1850; by 1940 it had fallen to 27.3). In addition, the births occurred more closely together than they had before.

Increased Life Expectancy

Given her relatively short life expectancy and the relatively late age at which she had her last child (40), the average woman in 1860 was dead by the time her last child left home. By 1980, given the changes in family size, spacing of children, and age of last pregnancy, the average woman could expect to live 33 years after her last child left home. As a result, the time devoted to child care came to occupy a smaller proportion of a woman's life. In addition, although life expectancy has increased for both men and women, women can expect to live longer than men do. In 1900, women outlived men by about 1.6 years, on average. Today, the average woman outlives the average man by 5.5 years. Yet, because brides tend to be younger on average than their grooms, the average married woman can expect to live approximately 8 to 10 years beyond her husband's death. In addition, the distorted sex ratio caused by males' earlier death decreases the probability of remarriage later in life. Although few women would probably mention their husbands' impending death as a reason for working, the difference in male-female life expectancy may be a background consideration.

Women have demonstrated that they are capable of doing very physical work, but they have often been constrained by the kinds of jobs society perceived them as capable of doing.

Increased Divorce Rate

Davis traces the rise in the divorce rate to the breadwinner system, and specifically to the shift of economic production outside the home:

> With this shift, husband and wife, parents and children, were no longer bound together in a close face-to-face division of labor in a common enterprise. They were bound, rather, by a weaker and less direct mutuality—the husband's ability to draw income from the wider economy and the wife's willingness to make a home and raise children. The husband's work not only took him out of the home but also frequently put him into contact with other people, including young unmarried women [who have always worked] who were strangers to his family. Extramarital relationships inevitably flourished. Freed from rural and small-town social controls, many husbands either sought divorce or, by their behavior, caused their wives to do so (pp. 410–411).

As Davis notes, an increase in the divorce rate preceded married women's entry into the labor market by several decades. He argues that once the divorce rate reached a certain threshold (a 20 percent or greater chance of divorce), more married women seriously considered seeking employment to protect themselves in case of failed marriage. When both husband and wife participate in the labor force, the chances of divorce increase even more. In such a scenario, both partners interact daily with people who are strangers to the family.

Increased Employment Opportunities for Women

Davis believes that married women are motivated to seek work because of changes in the childbearing experience,

the increase in life expectancy, the rising divorce rate, and the inherent weakness of the breadwinner system. However, women can act on these motivations only when their opportunities to work increase. With improvements in machine technology, productivity increased and the physical labor required to produce goods and services decreased. As industrialization matured, it brought a corresponding increase in the kinds of jobs perceived as suitable for women beyond nursing, clerical and secretarial work, and teaching. As women filled these jobs, they increased their economic self-sufficiency and transformed gender roles. There is "nothing like a checking account to decrease someone's willingness to be pushed into marriage or stay in a bad one" (Kipnis 2004).

Davis helps us to understand forces behind the decline of the breadwinner system. He does not say that the present system is free of problems, however. First, it "lacks normative guidelines. It is not clear what husbands and wives should expect of each other. It is not clear what ex-wives and ex-husbands should expect, or children, cohabitants, friends, and neighbors. Each couple has to work out its own arrangement, which means in practice a great deal of experimentation and failure" (Davis 1984, p. 413). Second, although the two-income system gives wives a direct role in economic production, it also requires them to work away from home, a situation that complicates child care. Third, even in the two-income system, the woman tends to remain primarily responsible for domestic matters. Davis maintains that women bear this responsibility because men and women are unequal in the labor force. (At the time Davis wrote his study, women earned 66 cents for every dollar men earned. Today, they earn about 76 cents for every dollar men earn.) Davis believes that as long as women make an unequal contribution to the overall household income, they will do more work around the house.

INTERSECTION OF BIOGRAPHY AND SOCIETY

Conflicts between Career and Family

THE TENSIONS THAT married women face regarding family and employment can be summarized as follows: If they work and have no children, they are considered selfish. If they stay home and raise a family, they are considered underemployed at best. If they work and raise a family, people wonder how they can possibly do either job right. The following comments from students in my introduction to sociology classes speak to this dilemma:

- I have felt pressured since I was little to go to college and have a career, and all I have ever wanted was to be a housewife and mother. Now that I have been in college for three years, I still have no major because my interests are with homemaking—that does not require a degree. It just kills me that people think that if women want to stay at home they are living in the past and that women should be career-driven. I wish society would stop pressuring women to get a career even if they don't want one. If it continues, I feel a lot more women will feel trapped and lost in college like I do.

- After I was born, my mother faced the tensions between being a mother and working outside the home. I was her second child, and she felt pressured to choose between her job and her two children. She quit her job and had two more kids. Later on, when she was ready to go back and get a job, she got a job at Arby's. However, the only reason she worked there was so she could be home to help us get ready for school, and meet us when we got home from school.

- I worked part-time for years, and once my youngest child entered school my husband kept asking when I was going to work full-time. There always was the power issue in our family—who makes the most money gets to make the big decisions. Truthfully, that played

a part in my going to work full-time—so I could have some more power in the decision making.

- My best friend from high school is extremely intelligent and now works in New Hampshire as a district manager. She makes a six-figure salary, and her husband stays home with their two children. He has only a high-school education, but wants to be an author. During the day, he writes (when the kids are asleep) and is working on his first novel. This is his dream. Probably 90 percent of the people who they know disagree with this arrangement. They feel he should be the breadwinner and she should stay home with the children—even though she said being a stay-at-home mom would drive her crazy. My friend and her husband have struggled with this decision. They finally realized that this is their life—she loves her job and he loves to write and care for the kids during the day.

- My mom was pregnant with three kids before the age of 25. Out of six children, only two were actually planned. She, unlike my father, gave up her career to raise us. When she did go back to work, she found that the cost of child care was more than her salary. Today she has lower self-esteem because she doesn't have skills for the work force. She is scared to return to work, but her life as a housewife is boring. She doesn't feel right spending my dad's money on herself.

- I have never wanted children. I've always strived toward success and a career. Everyone tells me that I won't be able to stand being childless and eventually I'll want children. Well, as I think about it now, a career has more appeal than being "Suzy-homemaker." My mother stayed home with my brother and me, and I love her for it. I feel that if I had children I would want to devote myself to them, so for me it is a choice of one over the other—and I choose career.

The problems of the post-breadwinner system are evident when we consider that a large proportion of married women (almost one-third) "choose" to not work outside the home (U.S. Department of Labor 2007). According to Davis, choice reflects the psychosocial costs of employment to married women—costs that include the stress of juggling family and career, finding reliable day care, and anxiety over making those choices. (For more information on how the labor market disadvantages women in the United States, see Chapter 10.)

The Changing Family in Japan

On August 11, 2000, the Tokyo metropolitan government launched a 47-point plan to save the capital's children from "a state of moral and social decay." Among other things, the plan declared every third Saturday to be "family bonding day." On those days, families are asked to eat together and talk to one another. In addition, the plan calls on adults to follow seven essential rules, including the following: "Make children say 'hello,' 'good-bye,' and

About 21 percent of the population in Japan—one in five people—is age 65 or older. That percentage is expected to increase to 30 percent—one in three—by 2020. How do you think an aging population affects family life?

number of children that Japanese women bear over their lifetime dropped from 5.11 to 1.4. The percentage of males age 25–29 who have never married increased from 18.7 to 66.9. For women, that percentage increased from 8 in 1900 to 48 in 2000. Over the past 100 years, average life expectancy increased by 34 years for men and by more than 40 years for women.

In Japan the pressing question is, "Why has total fertility declined to 1.2 births per woman?" Given that a society require 2.1 births per woman to replace themselves, this issue has become increasingly important. The low total fertility rate also helps to explain the high percentage of people age 65 and older (*Japan Times* 2004). The drop in Japanese total fertility is intertwined with a number of social forces, including the fall of the multigenerational household system and the rise of the breadwinner system, the decline in the number of arranged marriages, the rise of the "parasite single" and the "new single concept," and increased employment opportunities, especially for single women.

The Fall of the Multigenerational Family System and the Rise of the Breadwinner System

In 1898 the Japanese government established the Domestic Relations and Inheritance Laws. Under these laws, everyone belonged to a multigenerational household system (known as an *ie*) and was required to register in an official family registry. Even if families of the same ancestor lived apart, they still registered as one *ie* (Takahashi 1999). Legally the household head held authority over other family members and was responsible for the household's management and well-being. Relationships between family members were shaped by the Confucian values of filial piety, faith in the family, and respect for elders. Ideally, women obeyed men, the young obeyed the old, and daughters-in-law obeyed mothers-in-law (Hashizume 2000). First-born sons held privileged status. As future heads of the household, they were the heirs to all of the family's assets. Younger sons were expected to establish their own livelihoods and residences apart from the main household. Even so, the wives of younger sons were expected to help care for their husbands' parents, despite the fact that their parents-in-law lived in a separate household.

Under the *ie* system, a daughter was viewed as a temporary family member until marriage, at which point she moved in with—or more precisely, was absorbed into—her husband's family and assumed his surname (Takahashi 1999). A bride—especially the bride of the oldest son—was known as the "bride of the family." She served and obeyed her husband, his parents, and the household heir. She was responsible for domestic work, including caring for her parents-in-law in their old age. If a household head and his wife produced no children, the couple adopted

'thank you.' Make children tell you about their day. Don't give in when children whine for something—make them endure. Scold other people's children" (*Japan Times* 2000). The Tokyo government announced this plan in response to the many problems it believes are connected to Japan's declining total fertility rate.

According to Tokyo's governor, the low total fertility rate and corresponding changes in family structure have created a generation of overprotected, spoiled, self-centered children who "lack basic principles and an ability to sympathize with others." The governor named classroom collapse, bullying, dating for pay, and ill-mannered children as other disruptive consequences of these trends (*Japan Times* 2000).

The Tokyo initiative is one of many programs and policies being considered throughout Japan in an effort to "rescue the family." At the national level, for example, the government recently reviewed 150 proposals aimed at reversing the declining total fertility rate and encouraging working women to have children. It enacted legislation requiring corporations with 300 or more employees to introduce child-care programs (*Japan Times* 2002). At least 20 local governments have agreed to subsidize fertility treatments (Hoffman 2003; *Japan Times* 2003b). The national government has also expanded health care coverage to include the cost of artificial insemination (*Japan Times* 2007).

Over the span of 100 years, the structure of the Japanese family has changed drastically (see Table 12.6). Infant mortality declined sharply during the last century, from 155 deaths to 3.3 deaths per 1,000 births. The average

Table 12.6 Japanese Families: 100 Years of Change

Item	1900	1950	2007
Average household size (persons)	4.8	3.4	2.6
Percent age 65 and older	5.5	4.9	21
Infant mortality rate (deaths per 1,000 live births)	155	60.1	3.7
Mean age at first marriage			
Women	23	23	28
Men	27	25.9	29.1
Mean age at first birth	—	—	27.9
Total fertility rate (number of children born over average woman's lifetime)	5.11	3.65	1.2
Number of abortions to 100 live births	—	67.6	28.3
Life expectancy at birth (years)			
Males	44.25	59.57	78.3
Females	44.83	62.97	85.2
Life expectancy at age 65 (projected years)			
Males	10.14	11.35	18
Females	11.35	13.36	23
Percentage never married (age 25–29)			
Men	18.7	—	69.3
Women	8.0	—	54
Percentage never married (age 30–34)			
Men	—	—	42.9
Women	—	—	26.6
Percentage of population engaged in agriculture	53.8	41.1	6
Average household size	4.89	4.97	2.6
Percentage of one-person households	6	3.4	20

Source: Statistics Bureau, Ministry of Internal Affairs and Communications (Japan) (2007).

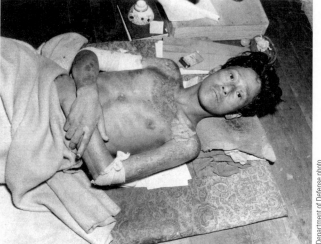

In 1945, during World War II, U.S. forces dropped atomic bombs on the Japanese cities Nagasaki and Hiroshima. The bombs' physical and psychological devastation hastened Japan's surrender that same year. U.S. troops then occupied Japan until 1952, and the occupational authority imposed sweeping changes on Japan's family system.

a child (preferably a male). If the couple had no biological or adopted son, they adopted a son-in-law to marry a daughter, take on the family name, and assume the role of household head. Legally, women could not choose their spouses or own property.

Soon after U.S. forces dropped atomic bombs on the Japanese cities Hiroshima and Nagasaki in 1945, World War II ended; U.S. troops occupied Japan until 1952. Under the direction of General Douglas MacArthur, the occupational authority imposed sweeping changes on Japan's economic, education, legal, and family systems. The *ie* system ended. New laws granting equal rights to women, ending arranged marriage, and abolishing primogeniture were instituted. Women were given the right to vote, to initiate divorce, and to retain custody of children. The Ancestry Registration Law was abolished, and

the nuclear family replaced the *ie* as the legal family unit (Takahashi 1999).

The U.S. government gave economic assistance to its former foe, helped Japan to rebuild, and established a democratic and capitalistic system. Japan literally rose from the ashes of World War II to become one of the top economies in the world and a top trading partner of the United States. The rise of capitalism moved productive functions away from the home. Men assumed the role of breadwinners, working outside the home; women assumed the role of homemakers, responsible for reproductive work, including the care of children and the elderly.

Despite these changes to the family structure, the belief that problems should be managed within the family has persisted. This persistence is evident in the structure of Japan's social welfare system: the Japanese government provides only limited public services to the elderly, relying instead on the caregiving tradition in which daughters-in-law care for their parents-in-law (Hashizume 2000). Caregiving is still a largely private responsibility, with limited government support.

The Japanese government sends mixed messages regarding women's roles. On the one hand, the country needs women in the labor force because the low total fertility rate has not produced enough "new workers." On the other hand, it still counts on women to serve as primary caregivers to the young and old. The government has set ambitious goals to relieve women of caregiving burdens and yet failed to provide adequate funding and other support. For example, it has outlawed discrimination against women and pregnant workers but established no enforcement mechanisms (Boling 1998; Kingston 2001).

The Decline of Arranged Marriages

Between 1955 and 1998 the proportion of all marriages in Japan that were arranged fell from 63 percent to 7 percent. This trend was accompanied by a decline in the proportion of newly married couples who lived with parents from 64 percent to 23 percent (Retherford, Ogawa, and Matsukura 2001). The shift away from arranged marriages has been related in part to an increase in the median age at first marriage. When parents no longer arranged marriages, love became a private matter and young people were left to their own devices. As a result the Japanese marriage market is not as well developed as the U.S. market. Specifically, the country has few voluntary associations and other organized settings in which single men and women can interact. Unlike the United States, Japan does not have a "couples culture." As one young Japanese woman commented, "In the U.S., you are supposed to be with your boyfriend or husband all the time. . . . In Japan, women have their ways of having fun and men have their ways. You're not expected to bring a date everywhere and you don't feel excluded if you're not involved with someone" (Ornstein 2001, p. 34).

Because of the relatively undeveloped marriage market and low societal pressure to date, roughly 45 percent of Japanese women age 16 and older claim to have not even one male friend. Among women who say they hope to marry soon, one in four claims to have no male friends.

Japanese society depends heavily on women to serve as primary caregivers to young and old. As the parents of the man on the left age, it is assumed that his wife will take on the primary responsibility of caring for them.

Among women who have at least one male friend, virtually all report that they have engaged in premarital sex. "The easy availability of premarital sex is another factor that has reduced the urgency of getting married and contributed to increases in the mean age of marriage" (Retherford, Ogawa, and Matsukura 2001, p. 18).

The Rise of the "Parasite Single" and the "New Single Concept"

In Japan a high percentage of working single adults (age 22 and older) live with their parents while contributing little to household expenses. An estimated 90 percent of single women in their twenties and 60 percent of single men in their thirties live with their parents (Retherford, Ogawa, and Matsukura 2001; Yamada 2000; Zielenziger 2002). In the United States, 14 percent of 24- to 34-year-olds live with their parents or grandparents (Lewin 2003). Sociologist Masahiro Yamada (2000) claims that Japanese singles contribute about $239 per month toward household expenses and average between $1,000 and $1,600 in monthly pocket money to spend on name-brand goods, cars, travel, and other desired items (*Japan Echo* 2000b). Yamada coined the phrase "parasite singles" to describe young adults pursuing the "new single concept," enjoying a comfortable life free of the financial and emotional pressures associated with parenting (especially mothering). Among the most stressful pressures that "parasite singles" wish to avoid is that of guiding children through Japan's highly competitive educational system.

This competitiveness can be traced to top corporations' and government ministries' practice of recruiting career employees from a handful of elite universities. Admission to these universities is tied to high scores on very difficult entrance exams. Applicants who have attended elite lower-level schools have a better chance of doing well on those exams. As the percentage of applicants attending universities has increased, so too has the competition to win entry into the elite institutions. The intense competition has given rise to *juku* (cram schools).

Japanese children attend one of 50,000 *juku* after the regular school day, on weekends, and during vacation periods. Approximately 39 percent of all public elementary students (first through sixth grades) and 75 percent of public middle school students attend such schools (Sato 2005). In 1975 the Japanese government capped the number of university enrollments, further increasing the pressure on children to do well academically. The *juku* phenomenon places tremendous financial pressure on the parents who enroll their children into the cram schools, purchase supplementary textbooks and other educational aids, and employ tutors (Brasor 1999). In fact, many mothers work part-time to pay *juku*-related expenses, which can range from $303 to $433 per month (Sato 2005).

Such pressures are reflected in Japanese parents' answers to survey questions about the disadvantages of having children. The top two drawbacks cited are the financial costs and psychological stresses associated with the *juku* system. The *juku* pressure is particularly stressful for mothers, who almost single-handedly guide children through the process. Many working fathers are largely absent from the home, because the workplace is governed by a mentality of "work over family": "Employees feel obligated to work long hours (including weekends), forgo needed time off, and accept sudden job transfers without consideration of family needs" (Newport 2000).

Entrenched Barriers to Employment

In Japan, women make up approximately 40 percent of the paid labor force. Of the women between the ages of 16 and 64, 50 percent work full- or part-time. On average, women in Japan earn 46 cents for every dollar men earn. When we control for occupation, females earn 62 cents for every dollar their male counterparts earn for doing the same or similar work (World Economic Forum 2006). This wage gap varies by age. At one extreme, female employees between the ages of 20 and 24 earn 91 cents for every dollar their male counterparts earn. At the other extreme, women between the ages of 50 and 54 earn 54 cents for every dollar men their age earn. Although women account for 48 percent of Japan's labor force, they compose just 15 percent of all professional and technical workers and 11 percent of all legislators, senior officials, and managers. In contrast, 55 percent of all professional and technical

workers in the United States are women and 46 percent of all legislators, senior officials, and managers are women. Japanese women hold 10 percent of all government positions, whereas American women hold 18 percent of such positions (World Economic Forum 2006).

This inequality is connected to a number of factors. Although Japan's labor laws forbid discrimination against women, it exists at every level, including recruitment, hiring, training, compensation, promotion, and retirement. One of the most important sources of discrimination is the two-track personnel system found in most large companies (Wijers-Hasegawa 2003). One track—managerial—includes employees whose jobs entail decision making and planning and who have been targeted as future executives. The other track—general office—consists of employees who do general office work. At least half of all the corporations with two-track employment systems consider females only for the general office track. In the United States the system is not perfect but the government has initiated support programs for businesswomen and equal opportunity programs to address past discrimination (Hiroko 2003).

One Japanese government report suggests that "Japan is still a developing country in the sense of gender equality" (Bando 2003). The World Economic Forum (2006) ranked Japan 79th out of its 115 member countries in its ability to empower women; it ranked the United States 23rd.

For the most part, Japanese women are expected to quit working when they marry or have children (Boling 1998; *Japan Times* 2003d). When mothers do work, they tend to accept low-paying and insecure employment in exchange for the flexibility needed to meet household and caregiving responsibilities. Although this situation is beginning to change, many corporations allow women only one of two choices: devote their lives to the company or quit upon getting married and having children (Boling 1998). This pressure to leave work or to scale back is further encouraged by a tax system that offers married women incentives to "choose" part-time employment. The government caps the amount of income that married women can earn without paying taxes and being dropped from their husbands' insurance policies.

The Japanese education system also plays a role in perpetuating the male-female wage gap. Of all the students entering four-year colleges, 64 percent are male and 36 percent are female. By comparison, approximately 90 percent of students entering junior colleges are female. Junior colleges "train young women to be skilled, gracious, and responsible homemakers" (Isa 2000). (See Intersection of Biography and Society: "Going Back to Japan After Living in the United States"). Sociologist Kaku Sechiyama argues that the key to establishing a work environment that supports married women, mothers, and women who plan to marry and have children is to create a system that imposes housework, child-rearing, and elder-care duties on men. Then, "employers know that there will be a certain disadvantage regardless of whether they hire men or women" (Takahara 2000).

A Century of Change

■ **CORE CONCEPT 5:** Over the past 100 years, fundamental shifts in the economy, a decline in parental authority, the changing status of children, and dramatic increases in life expectancy have changed the family structure. In reviewing a century of changes to the family structure in Japan and the United States, we can identify a number of common underlying interdependent themes. These themes relate to (1) the fundamental shifts in the economic system, (2) the decline of parental authority, (3) the changing status of children, and (3) dramatic increases in life expectancy.

Fundamental Shifts in the Economic System

In Chapter 8, we learned that social stratification is the system societies use to classify people. When sociologists study stratification, they examine how the categories to which people are assigned affect their perceived social worth and their life chances. They are particularly interested in how individuals who possess one variation of a characteristic (for example, male reproductive organs versus female reproductive organs) are regarded and treated as more valuable than persons who possess another variation of the same characteristic.

Sociologist Randall Collins (1971) offers a theory of sexual stratification to analyze this point. Collins's theory is based on three assumptions. First, people use their economic, political, physical, and other resources to dominate others. Second, any change in the distribution of resources across a society alters the structure of domination. Third, ideology is used to support and justify one group's domination of another.

In the case of males and females, males tend to be physically stronger than females. Collins argues that because of this physical difference, the potential for coercion by males exists in every encounter between males and females. He maintains that the ideology of sexual property, the "relatively permanent claim to exclusive sexual rights over a particular person" (1971, p. 7), lies at the heart of sexual stratification. Collins points out that women have historically been viewed and treated as men's sexual property.

According to Collins, the extent to which women are viewed as sexual property and subordinate to men has historically depended—and still depends—on two important interdependent factors: women's access to agents of violence control (such as the police) and women's position relative to men in the labor market. Based on these factors, Collins identifies four historical economic arrangements: (1) low-technology tribal societies, (2) fortified households, (3) private households, and (4) advanced market economies.

These four arrangements are ideal types; the reality is usually a mixture of two or more types. Each arrangement is broadly characterized by distinct relationships between men and women. We begin our discussion by analyzing the first type of economy, characteristic of **low-technology tribal societies**.

Low-technology tribal societies include hunting-and-gathering societies with technologies that do not permit the creation of **surplus wealth**—that is, wealth beyond what is needed to meet basic human needs, such as food and shelter. In such societies, sex-based division of labor is minimal, because the emphasis is on collective welfare and the belief that all members must contribute to ensure the group's survival. Basic economic tasks for men might include agriculture, carpentry, hunting, and fishing. For women, they might include weaving, cloth making, and oil making (Cote 1997). Although all men and women learn the basic economic tasks deemed appropriate to their sex, they also specialize in certain areas, such as fishing, pigeon hunting, boat building, or wood carving (for men) or producing the most valuable kinds of mats, cloth, medicines, or oils (for women).

Evidence suggests that in hunting-and-gathering societies, women perform more menial tasks and work longer hours than men do. Men hunt large animals, for example, while women gather most of the food and hunt smaller animals. The women might perform routine agricultural work (weeding, transplanting, and gathering crops). Women also do routine fishing or hunting for small prey (Cote 1997).

Because almost no surplus wealth exists, marriage between men and women from different families does little to increase a family's wealth or political power. Consequently, daughters are not treated as "property," in the sense that they are not used as bargaining chips to achieve such aims. Compared with women in fortified households and private household arrangements, women in low-technology tribal societies have more "freedom of self-determination, more social value, and higher-status economic roles" (Cote 1997, p. 8).

Fortified households are preindustrial arrangements characterized by no police force, militia, national guard,

---

low-technology tribal societies Hunting-and-gathering societies with technologies that do not permit the creation of surplus wealth.

surplus wealth Wealth beyond what is needed to meet basic human needs, such as food and shelter.

fortified households Preindustrial arrangements in which a household acts as an armed unit and the head of the household acts as its military commander. The household is characterized by the presence of a nonhouseholder class, consisting of propertyless laborers and servants

INTERSECTION OF BIOGRAPHY AND SOCIETY

Going Back to Japan after Living in the United States

PERHAPS THE BEST way to determine the ways in which biography is shaped by the society in which one lives, is to live in two societies and compare/contrast the corresponding opportunities and constraints. Noriko Ikarashi moved from Japan to the United States with her husband and children for four years and then returned to home. In the essay that follows she reflects on that experience and how living in each place shapes her life.

In 2000, my children and I moved to the United States with my husband. Six months later, one of my friends asked "Why don't you go to Northern Kentucky University to study?" I did not want to miss this chance, because I could never consider being a college student when I returned to Japan. So in the fall of 2000, my life as a college student began. I took an Introduction to Anthropology course, and one of my first homework assignments was to define who I am. I wrote: "I am Japanese. I am a wife. I am a mother. I am a woman. I am 35. I am a student at NKU." Many American students listed personal characteristics: "I am cheerful," "I am optimistic," and "I am tall." However, I did not think to add such characteristics to my list. Why? Two years have gone by since I took the class, and now I think I can answer this question: "Because I am Japanese." The Japanese tend to identify themselves in terms of a social context rather than by personal characteristics. In other words, one's role in relation to others is more important than individual, personal, and physical characteristics. Now, a year is left until my family returns to Japan, and my friends ask me, "Do you want to go back to Japan?" My answer is, "Well, no I don't," because many social obligations and social pressures that I do not feel in the United States await me there.

First of all, "I am the wife" of the oldest son in the Ikarashi family. In Japan, that means my husband and I have to take care of his parents.[1] To meet this obligation, we started to live with them a year before coming to the United States. They expect economic, mental, and physical support from us, especially when they get really old. When I return to Japan, I will have to live with my parents-in-law. In some ways, living with extended family is good, especially for my children. The presence of grandparents at home is beneficial, because grandparents are usually gentle and loving with grandchildren, while it is difficult for parents to be nice and gentle to them all the time. Grandparents and grandchildren play, talk, care for, and influence each other. My parents-in-law are helpful to me, too, because they provide some help with child care and housework. However, the biggest concern revolves around the time my parents-in-law get very old and become disabled. In Japan, nursing

Noriko Ikarashi graduated from Northern Kentucky University at age 37. From an American point of view, this accomplishment is not particularly unusual. In Japan, however, it is the rare woman who attends college at age 37.

homes are not common like in the United States. So I, being the oldest son's wife, have to or am expected to take care of his parents. Deep down, I would rather take care of my own parents, not my in-laws, but that is a social obligation that I am expected to fulfill. If I did not, I would feel very guilty and be accused of being selfish and dishonest.

"I am a mother." In most Japanese families, the mother is almost exclusively responsible for child care. I have heard that women in younger generations tend to share this task with their husbands, but still the Japanese men are not equal partners with their wives. Ideally, mothers should take close care of their children at least until they are three or four years old. If a mother asks somebody to take care of the child from a very young age, and the child's behavior is terribly bad, the mother is often criticized for not taking care of the child by herself.

At the same time, it is difficult to get a husband's help with child care because of men's work schedules. For example, in Japan, we rarely see fathers at parent-teacher conferences. I have never seen men attend a school field trip. While in the United States, my husband, Triesto, comes to our children's activities like many American fathers do, although he does not come often. In Japan, driving back and forth for after-school activities, including juku, a private supplemental school, is completely the mother's work. In addition to these domestic obligations, the mental pressure in everyday life is enormous for Japanese mothers.

Since the Japanese value uniformity, to be too proud of one's own children is a taboo. In reality, mothers are proud of their children, but they avoid expressing it. If someone praises their child—"Your child got 100 percent on the last exam. He is very smart, isn't he?"—the ideal response would be "No, he is not so smart. He was just lucky on the test." Even though you think your child is smart, you should not say it. Although everyone wants to be better than the others, being looked at as such is not helpful in Japan. Being proud of something and expressing that pride to others leads to negative consequences. More often than not, such a person has difficulty maintaining good relationships with others. Therefore, returning to Japan with children who are used to self-expression in the United States is a source of anxiety for me. The Japanese maintain their relationships with others by being the "same as others." My 11-year-old daughter, Anna, told me after visiting Japan, "It was tiring living in Japan because I had to think about how my friends think about my words and actions." Anna explained, "When in Japan I just mentioned to a girl 'You have the best outfit on today.'" However, the girl thought Anna was making fun of her. The word "best" suggests she stood out from the others. My daughter's words hurt the girl's feelings because most Japanese do not value standing out among their friends. Then, the girl told her friends what Anna said, so all of her friends glared at Anna. During socialization, the Japanese learn empathy, so even children try to read how others feel. Anna should have anticipated how the girl would feel before she talked to her. Since the Japanese try to be the same as others, mothers tend to have their children do the same thing as other children do. If her child's friends take English lessons after school, the mother believes that her child should do so as well.

I am 37 now. From the Japanese point of view, I am already too old to have a good job. In the Japanese workplace, age as well as gender discrimination are very common. Therefore, I am in a double-bind situation in trying to get a job when I go back. If you look at job advertisements, they clearly mention an age limit. Usually, well-paying jobs for women limit the age to 35 or younger. For women older than 35, the job market is very cold; they are considered as oba-san, misses or lady. The word has a negative connotation, which adds to the image of "old lady." At 37, I am already in the category of "old woman" in Japan.

There is a tendency in the Japanese job market to value young women. Women who give up good careers for a time because of marriage or pregnancy usually find it hard to reenter the workplace unless they have a specialized skill such as nursing. This is one reason that many Japanese women marry at an older age than their American counterparts and that the birth rate is low. Being a wife and mother is still regarded as an obstacle to a successful career.

Moreover, government policy limits married women's job opportunities. Generally speaking, if a woman earns less than about $10,000 annually, she does not have to pay income tax and can remain as a husband's dependent. In addition, husbands can get a family allowance and dependents' tax credit. If the woman earns $15,000 or more, she has to pay income tax and pay into the social security pension system, and her husband is not entitled to family benefits.

Consequently, her contribution to total family earning will be less than when she earned $10,000 annually. Therefore, many married women "choose" low-paying part-time jobs. If a woman chooses a full-time job, there is no benefit unless she earns more than $20,000 per year.

For women who do not have special skills, and who are also doing domestic work, earning that amount of money is not so easy. Even though I will return to Japan college-educated, I am very anxious about whether I can get a well-paying job given my age and domestic status. To complicate matters, Japanese men tend to dislike well-educated older women. Men regard such women as assertive, a quality not associated with the ideal woman.

"I am a student at NKU." Until I came to the United States, I never thought that I would be a college student. That rarely happens in Japan. Many mothers of my generation are busy bringing up children. I have to save money for my children instead of using it for my education. A 37-year-old married woman who has two children and goes to a university is ridiculous by the Japanese standard, even though some Japanese women might wish to do this. I have almost finished my undergraduate degree, and I want to go to graduate school in Japan. However, going to college, including graduate school, seems simply too difficult for me. I have parents-in-law and children to take care of, and there are many domestic tasks connected with that. Tuition is relatively high. In addition, I have to consider how my female friends, especially neighbors of the same age, see me. If I go to graduate school, they might say, "Wow, that is wonderful," but many of them would really feel uneasy about that. Some of them may think that I am being ridiculous or selfish. Doing something very different from others affects the social relationship that I have with them.

"I am Japanese," so I know I will adapt to the Japanese way again soon. However, I also know that there are many social obstacles in Japan, as I explained. Those obstacles do not necessarily apply to everyone. For instance, right now more than 70 percent of my friends do not live in an extended family. However, many expect to eventually care for parents-in-law. *(Continued on next page)*

INTERSECTION OF BIOGRAPHY AND SOCIETY

Going Back to Japan after Living in the United States (continued)

Epilogue: Since writing this essay, Noriko returned to Japan at the end of December 2003 with her husband and three children. She gave an update on February 27, 2004. Here is what Noriko had to say:

I still have many unopened boxes in my house, but I would like to send a report about what I saw in Japan after I returned. First of all, my 12-year-old daughter is adapting to Japanese school much better than we thought. Yesterday she said, "I often have another opinion to my friend's one. But I don't want to say my opinion, I just keep that in my mind." She knows how to survive. The younger daughter is very flexible, and she learns the Japanese way very quickly.

She walks to school every day with her backpack and a bottle of wheat tea. Recently, a stranger walked around the area where students pass through on their way to come home from school, so now the students return home with a teacher's supervision. This type of trouble happens very often not only in my area, but also all around Japan.

My 41-year-old friend went to a job interview to be a translator and was told, "You are old so we will choose a younger person if you both have the same abilities." When I registered to work at a temporary agency, the first question the interviewer asked was, "How old are you?" I went to a lecture on making society better through gender equality. At the lecture, the speaker noted that many women learned gender roles unknowingly during their socialization. For example, making a cup of tea is the woman's role in Japan. Many women at the seminar answered that they think a cup of tea made by a woman tastes better than one made by a man.

Today, my daughter will take an entrance examination for juku, private after-school supplemental class. We decided to let her go to a public junior high school, so she will eventually need to take high school and college entrance examinations.

Source: Noriko Ikarashi, Northern Kentucky University, Class of 2003.

---

or other peacekeeping organization. Instead, the household acts as an armed unit, and the head of the household acts as its military commander. Fortified households "may vary considerably in size, wealth, and power, from the court of a king or great lord . . . down to households of minor artisans and peasants" (Collins 1971, p. 11). All fortified households, however, share one characteristic: the presence of a nonhouseholder class, consisting of propertyless laborers and servants. In the fortified household, "the honored male is he who is dominant over others, who protects and controls his own property, and who can conquer others' property" (p. 12). Men treat women as sexual property in every sense: Daughters are bargaining chips for establishing economic and political alliances with other households; male heads of household have sexual access to female servants; and women (especially in poorer households) bear many children, who eventually become an important source of labor. In this system, women's power depends on their relationship to the dominant men.

Private households emerge with the establishment of a market economy, a centralized, bureaucratic state, and agencies of social control that alleviate the need for citizens to take the law into their own hands. Under the private household arrangement, men monopolize the most desirable and important economic and political positions.

As you might guess, private households exist when the workplace is separate from the home.

Men are still heads of household but assume the role of breadwinner, and women remain responsible for housekeeping and child rearing. Men, as heads of household, control the property. Until 1848 in the United States, for example, a married woman's legal position was such that any wages she earned and any property she had acquired before or after her marriage belonged to her husband. In the event of divorce, husbands were awarded custody of children. Married women could not sign business contracts, sue, or be sued. Husbands, in turn, were responsible for any debts that their wives assumed (PBS 2000). Similarly, in Japan until the end of World War II, women lacked property rights under the *ie* system.

The decline in the number of fortified households, the separation of the workplace from the home, smaller family size, and the existence of a police force to which women could appeal in cases of domestic violence gave rise to the notion of romantic love as an important ingredient in marriage. In the marriage market associated with private households, men offer women economic security, because they dominate the important, high-paying positions. Women offer men companionship and emotional support and strive to be attractive—that is, to achieve the ideal version of femininity, which might include possessing an

Department of Defense photo by LCPL John P. McGarity, USMC

Sociologist Kaku Sechiyama argues that employers would discriminate against women less if men shared equally in reproductive tasks.

18-inch waist or removing most facial and body hair. At the same time, women try to act as sexually inaccessible as possible, because they offer sexual access to men only in exchange for economic security.

Advanced market economies offer widespread employment opportunities for women. Although women remain far from being men's economic equals, some can enter into relationships with men by offering more than an attractive appearance; they can provide an income and other personal achievements. Because they have more to offer, these women can make demands on men to be sensitive and physically attractive, to meet the standards of masculinity, and to help with reproductive work. In the United States today, almost one-third (31 percent) of all women in dual-earner households where both partners work full-time earn more money than their male partner (AmeriStat 2000a). Moreover, young women in Japan and the United States are the female group whose average income comes closest to equaling that of their male counterparts. This situation may explain why increasing commercial attention has been given in the past decade to women who are evaluating men's appearances—as evidenced by advertisements related to guy watching, bodybuilding, hairstyles, and male skin and cosmetic products.

Decline in Parental Authority

Around the beginning of the 20th century, children learned from their parents and other relatives the skills needed to make a living. As the pace of industrialization increased, jobs moved away from the home and into factories and office buildings. Parents no longer trained their children, because the skills they knew were becoming obsolete.

Children came to expect that they would not make their living in the same way their parents did. In short, as the economic focus shifted from agriculture to manufacturing, the family became less involved in children's lives. Ultimately, the transfer of work away from the home and neighborhood removed opportunities for parents and children to work together.

The separation that characterized the work life of parents away from their children soon spread to other areas of life as parents and their children moved in different social spheres. Parental authority over adult children lost its economic and legal supports. The legal rights of individuals, especially women, were strengthened. Gradually values and norms developed that supported privacy and intergenerational independence (for example, the ideas that elders should not interfere in the lives of their adult children and that a healthy living arrangement is one in which parents and children reside in separate households). Popular opinion pushed governments to establish social security and health insurance programs supporting the elderly. Those policies further reinforced changes in family structure and values. Contributing to these changes was the increased prevalence of employed females, a group that had historically assumed caregiving responsibilities for elderly family members. Ogawa and Retherford (1997) have labeled this new parent–adult child relationship as "intimacy at a distance." The changed relationships between family members—especially the trend away from shared residences—offered fewer "natural" opportunities for intergenerational activities. Education, work, social, and leisure activities became largely age-segregated experiences, to the point that if intergenerational activities were to take place on a societal level, they needed to be planned.

Although some intergenerational programs do exist in the United States, "a rich array of intergenerational initiatives exists in Japanese schools (public and private), community settings, and institutional facilities" (Kaplan and Thang 1997, p. 302). Such programs include Rent-a-Family, which enables seniors with limited or unsatisfactory family ties to "rent the services of actors to play the part of children and grandchildren and share family-like experiences"; other programs integrate senior and youth services (for example, children's libraries and playgrounds are located in or near the open spaces of old-age homes). All such programs create opportunities for sustained and meaningful interaction between the generations.

Intergenerational programs in Japan are structured with one or more goals in mind. Those goals include offering seniors meaningful and productive roles, combating ageism, fostering intergenerational understanding, drawing on the talents and energies of youth and elderly to solve problems, fostering an environment of mutual support and understanding, and establishing a sense of

In primarily agriculture-based economies, children represent an important source of free unskilled labor for the family. In information-service-based economies, the motives for having children have changed from such practical benefits to the intangible, "emotional" services that children can provide.

cultural continuity and respect for tradition (Kaplan and Thang 1997).

Status of Children

As noted earlier, the technological advances associated with the Industrial Revolution and the shift from an agriculture-based economy to a manufacturing-based one not only changed relationships between family members, but also altered the status of children from economic assets to liabilities. Mechanization decreased the amount of physical effort and time needed to produce food and other commodities. Consequently, children lost their economic value. In agriculture, which is a form of extractive economy, children represent an important source of free unskilled labor for the family. This fact may partly explain why the highest total fertility rates in the world occur in places like Afghanistan, Somalia, the Democratic Republic of the Congo, and Niger—places where labor-intensive production remains common (see Intersection of Biography and Society: "The Economic Role of Children in Labor-Intensive Environments").

Demographer S. Ryan Johansson (1987) argues that in high-income economies, couples who choose to have children bring them into the world to provide intangible, "emotional" services—services such as love, companionship, an outlet for nurturing feelings, enhancement of dimensions of adult identity—rather than economic services. The shift away from labor-intensive production has stripped children of whatever economic contribution they might have made to the family (Johansson 1987). In high-income economies, the family's energies have shifted away from production of food and other necessities and toward the consumption of goods and services.

In the United States, largely because of the shift away from labor-intensive production, children have become expensive to rear. Today, depending on income, Americans will spend between $190,000 (average for lowest income) and $381,050 (average for highest income) to house, feed, clothe, transport, and supply medical care to a child until he or she is 18 years old; annual expenses for child rearing will range from $7,500 to $15,700 (U.S. Department of Agriculture 2006). These costs cover only the basics; when we include extras, such as summer camps, private schools,

INTERSECTION OF BIOGRAPHY AND SOCIETY

The Economic Role of Children in Labor-Intensive Environments

CHILDREN ARE OFTEN an important source of cheap labor, especially for labor-intensive, nonmechanized industries. The childhood experiences of Chico Mendes, the son of an Amazon rubber tapper, illustrate the economic role children can take in such an environment.

Childhood for Chico Mendes was mostly heavy work.... If all his siblings had lived to adulthood, Chico would have had seventeen brothers and sisters. As it was, conditions were so difficult that by the time he was grown, he was the oldest of six children—four brothers and two sisters.

When he was five, Chico began to collect firewood and haul water. A principal daytime occupation of young children on the seringal [rubber estate] has always been lugging cooking pots full of water from the nearest river. ... Another chore was pounding freshly harvested rice to remove the hulls. A double-ended wooden club was plunged into a hollowed section of tree trunk filled with rice grains, like an oversize mortar and pestle. Often two children would pound the rice simultaneously, synchronizing their strokes so that one club was rising as the other descended.

By the time he was nine, Chico was following his father into the forest to learn how to tap.... Well before dawn, Chico and his family would rise.... They would grab the tools of their trade—a shotgun, ... a machete, and a pouch to collect any useful fruits or herbs found along the trail. They left before dawn, because that was when the latex was said to run most freely.... As they reached each tree, the elder Mendes grasped the short wooden handle of his rubber knife in two hands, with a grip somewhat like that of a golfer about to putt.... It was important to get the depth [of the cut] just right.... A cut that is too deep strikes the cambium [generative tissue of the tree] and imperils the tree; a cut that is too shallow misses the latex producing layer and is thus a waste....

Chico learned how to position a tin cup, or sometimes an empty Brazilian nut pod, just beneath the low point of the fresh cut on a crutch made out of a small brand. ... The white latex immediately began to dribble down the slash and into the cup.... Chico quickly adopted the distinctive, fast stride of the rubber tapper.... The pace has

evolved from the nature of the tapper's day. The 150 and 200 rubber trees along an estrada [trail] are exasperatingly spread out. A simple, minimal bit of work is required at each tree, but there is often a 100-yard gap between trees. Thus, a tapper's morning circuit can be an 8- to 11-mile hike. And that is just the morning.... Then, they would retrace their steps on the same trail, to gather the latex that had flowed from the trees during the morning. Only rarely would they return home before five o'clock. By then they would be carrying several gallons of raw latex in a metal jar or sometimes in a homemade, rubber-coated sack.

The very best rubber is produced when this pure latex is immediately cured over a smoky fire.... Chico would ladle the latex onto a wooden rod or paddle suspended over a conical oven.... As the layers built up on the rod, the rubber took on the shape of an oversize rugby football. The smoking process would continue into the evening. After a day's labor of fifteen hours or more, only six or eight pounds of rubber were produced.

The tappers often developed chronic lung diseases from exposure to the dense, noxious smoke. Toward December, with the return of the rainy season, the tappers stopped harvesting latex and began collecting Brazil nuts. During the rainy season, the Mendeses often crouched on their haunches around a pile of Brazil nut pods, hacking off the top of each one with a sharp machete blow, then tossing the loose nuts onto a growing pile.... When the Mendeses were not harvesting latex or nuts, they tended small fields of corn, beans, and manioc.

Chico Mendes was an important Brazilian grassroots environmentalist and union activist dedicated to preserving the Amazon rain forest and to improving rubber tappers' economic standard of living as compared with that of cattle ranchers and rubber barons. He was murdered in 1988 by Amazon cattle ranchers. Indians and rubber tappers are now working together against those who misuse or otherwise destroy the resources of the forests.

Source: From *The Burning Season* by Andrew Revkin, pp. 69–76.

Copyright © 1990 by Andrew Revkin. Reprinted by permission of Houghton Mifflin Company.

---

sports, and music lessons, the costs go even higher. Even if children go to work when they reach their teens, they usually spend their earnings on personal goods and services and do not contribute to paying household expenses.

Of course, for many parents, the cost of raising children does not stop when they reach age 18. About 60 per-

cent of U.S. high school graduates enroll in college. The average annual cost of in-state college tuition and fees ranges from $5,132 for public institutions to $20,082 for private institutions (College Board 2005). The high cost of raising children may be one reason that, in a recent national survey of 2200 adults, children were among the

Five generations of family members point to the fact that with increased life expectancy family members live longer and thus have the opportunity to be part of family life for a greater number of years than had previously been possible.

In Japan intergenerational programs that bring together senior citizens and youth are quite popular.

least cited contributors to a successful marriage. Faithfulness, a good sexual relationship, household chore-sharing, income, good housing, shared religious beliefs, and similar tastes and interests were cited more often than children (St. George 2007).

The cost of raising children in Japan is also high. In May 2000, the Japanese government expanded the subsidies it awards to parents with children age 3 and younger to include any child younger than age 6. The subsidies go to parents with annual incomes of less than $43,000 if self-employed and $67,000 if salaried. Qualified parents receive $50 per month if they have one child, $100 per month if they have two children, $200 per month if they have three children, and $100 more per month for each additional child. The subsidies represent one response to the low birth rate in Japan that is tied, at least in part, to the high financial cost of raising children (Sims 2000). In addition to the cost of feeding and clothing children, parents spend an average of $1,100 per year on cram school (*juku*) and fees for private tutors.

Dramatic Increases in Life Expectancy

Since 1900 the average life expectancy at birth has increased by 28 years in the core economies and by 20 years (or more) in the labor-intensive poor countries. In *The Social Consequences of Long Life*, sociologist Holger Stub (1982) describes at least four ways in which gains in life expectancy have altered the composition of the family in the last century. First, the chance that a child will lose one or both parents before he or she reaches 16 years of age has decreased sharply. In 1900 the chance of such an occurrence was 24 percent; today it is less than 1 percent. At the same time, parents can expect their children to survive infancy and early childhood. In 1900, 250 of

every 1,000 children born in the United States died before age 1; 33 percent did not live to age 18. Today 17 of every 1,000 children born die before they reach age 1; less than 5 percent die before reaching age 18.

Second, the potential length of the average marriage has increased. Given the mortality patterns in 1900, newly married couples could expect their marriage to last 23 years before one partner died (assuming they did not divorce). Today, if they do not divorce, newly married couples can expect to be married for 53 years before one partner dies. This structural change may be one factor underlying the currently high divorce rates. At the turn of the last century, death nearly always intervened before a typical marriage had run its natural course. Now, many marriages run out of steam with decades of life remaining for each spouse. When people could expect to live only a few more months or years in an unsatisfying relationship, they would usually resign themselves to their fate. But today the thought of living 20, 30, or even 50 more years in an unsatisfying relationship can provoke decisive action at any age (Dychtwald and Flower 1989, p. 213). According to Stub, divorce dissolves today's marriages at the same rate that death did 100 years ago.

Third, people now have more time to choose and get to know a partner, settle on an occupation, attend school, and decide whether they want children. In fact, nearly all of these areas of experience are occurring later in life than they did in past generations. Some sociologists are labeling the years 18–35 as "transitional adulthood"—a life stage between childhood and adulthood (Lewin 2003). Moreover, an initial decision made in any one of these areas is not final. The amount of additional living time enables individuals to change partners, careers, or educational and family plans—a luxury not available to their counterparts of a century ago. Stub (1982) argues that

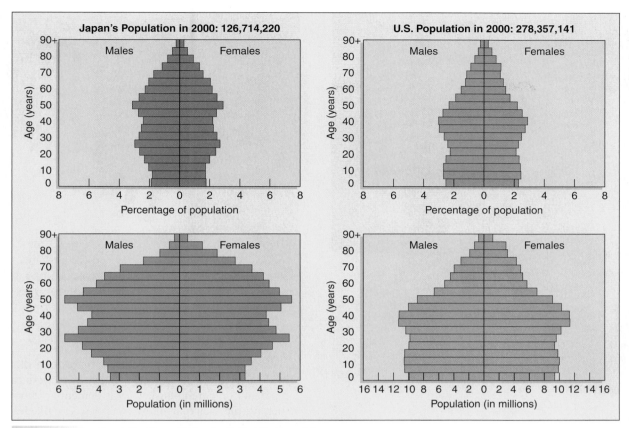

▲ **Figure 12.5**   Population Pyramids for the United States and Japan

Population pyramids can be presented in two ways. One way is to present age groupings as percentages. The other is to present them as raw numbers. Notice that when we use percentages, we can answer questions such as What percentage of the Japanese population is between the ages of 60 and 65? and What percentage of the U.S. population is between the ages of 60 and 65? When we portray age groupings as raw numbers, we can answer questions such as How many people in Japan are between the ages of 60 and 65? and How many in the United States are between those ages?"

Source: U.S. Census Bureau (2004a, 2004b).

the so-called midlife crisis is related to long life, because many people "perceive that there yet may be time to make changes and accordingly plan second careers or other changes in life-style" (p. 12).

Finally, the number of people surviving to old age has increased. In countries such as Japan and the United States, where the total fertility rate is low or declining, the proportion of older people in the population—not merely the number—is increasing. In 1970, approximately 25 percent of all people in their late fifties had at least one surviving parent; in 1980, 40 percent of this group had a surviving parent. Even more astonishing, in 1990, 20 percent of people in their early sixties and 3 percent of people in their early seventies had at least one surviving parent (Lewin 1990).

Much has been written about the growing population of people older than age 65 in the world, especially in terms of the challenges of caring for the disabled and the frail elderly. The emphasis on this segment of the older

population should not obscure the fact that the elderly are a rapidly changing, heterogeneous group. They differ by gender, age (a difference of 30 years or more may separate persons aged 65 from those in their nineties), social class, and health status. While most elderly are not institutionalized, a major issue faced by even the closest of families is how to meet the needs of disabled and frail elderly so as not to constrain investments in children or impose too great a psychological, physical, or time burden on caregivers.

The Caregiver Role in the United States and Japan

■ **CORE CONCEPT 6:** The aging of the population had no historical precedent. The family must find ways to adapt to this situation, especially as regards caregiving. Although there has

Department of Defense photo by JO1 Robert Benson

Department of Defense photo by Chief Master Sgt. Don Sutherland, USAF

The elderly are a heterogeneous population; they vary by sex, age, race, income, and social ties.

always been a small number of people who lived to age 80 or 90, "There is no historical precedent for the aging of our population. We are in the midst of a new phenomenon," and families must adapt to the changing environment (Soldo and Agree 1988, p. 5). Finding ways to care for this population is among the greatest challenges. Here we focus on the elderly and their caregivers. We will detail the caregivers' characteristics, time spent giving care, and the **caregiver burden**, the extent to which caregivers believe that their emotional balance, physical health, social life, and financial status suffer because of their caregiver role (Zarit, Todd, and Zarit 1986). Keep in mind, however, that family caregiving is a complex activity that goes beyond time commitments and perceived burdens.

caregiver burden The extent to which caregivers believe that their emotional balance, physical health, social life, and financial status suffer because of their caregiver role.

When Noriko Yamamoto and Margaret I. Wallhagen (1997) interviewed 26 women in Japan who were caring for parents or parents-in-law with dementia, they were struck by the fact that many interviewees talked about their difficulties "with a big smile. They looked almost happy about having such difficulties of care" (p. 167). The interviews lasted an average of two hours and centered on the following question: Why do you continue to care? The women gave a variety of answers to this question:

- The need to pay back recipients for the sacrifices they made in raising and caring for the caregiver or her husband
- The emotional bond between the caregiver and the recipient
- A desire to live a life free of regret—and the belief that failure to fulfill the caregiver role would lead to a life of regret
- A feeling of accomplishment for a job well done, especially when the recipient expresses appreciation and satisfaction
- Personal growth derived from the caregiving experience

Caregivers in the United States

Contrary to popular belief, most American seniors (96 percent) are not in nursing homes. However, many report a physical, mental, or sensory disability (see Figure 12.6). In the United States 4.3 percent of the population age 65 and older live in nursing homes, and in Japan 1.2 percent does (U.S. Census Bureau 2005; National Center for Health Statistics 2004). Figure 12.7 shows the living

Around the turn of the 20th century, 250 of every 1,000 children born in the United States died before age one.

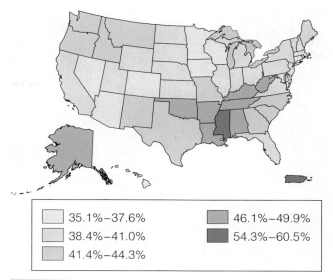

35.1%–37.6%	46.1%–49.9%
38.4%–41.0%	54.3%–60.5%
41.4%–44.3%	

▲ **Figure 12.6** Civilian Noninstitutionalized Persons Age 65 Years and Older Reporting a Physical, Mental, or Sensory Disability, 2005
Which states have the highest percentage of people 65 and older reporting a disability? Which states have the lowest percentage?
Source: U.S. Bureau of the Census (2005).

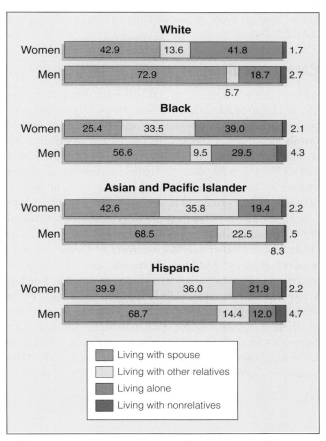

▲ **Figure 12.7** Living Arrangements of People Age 65 and Older, by Racial or Ethnic Classification
Source: U.S. Census Bureau (2005).

arrangements of the U.S. population age 65 and older by race and sex. The percentage of elderly who live with other relatives varies by racial category and gender. Note that between 56 and 72 percent of elderly men live with their spouses, as compared with between 25.5 and 42.9 percent of elderly women. If elderly women are not living with a spouse, they tend to live alone. However, female elderly classified as Asian or Pacific Islander are most likely to live with a relative.

Although most elderly in the United States do not live in nursing homes, one in four does require assistance with daily activities, such as bathing, walking, dressing, and eating. An estimated 17 percent of households (18.5 million) provide some kind of informal care to a person age 50 or older (National Alliance for Caregiving and AARP 2004). That percentage varies by racial classification, with 42 percent of Asian-American households and 19 percent of white households offering such informal care (American Association of Retired Persons 2001). One cannot, however, conclude from these data that nonwhites are more willing to care for elderly members than whites, because whites age 65 and older are the most likely to rate their health as good to excellent. Almost 75 percent of men and women 65 and older who are classified as white rate their health as good to excellent, compared with 60 percent of those classified as black and 65 percent of those classified as Hispanic or Latino (Federal Interagency Forum on Aging-Related Statistics 2000).

Ironically, members of ethnic groups that provide the most care are also the most likely to feel guilty about the amount of care given (American Association of Retired Persons 2001). In the United States approximately 61 percent of all caregivers are women. This percentage holds for all racial groups except Asian Americans; in that group, 54 percent of caregivers are male. About 41 percent of all people caring for seniors are also caring for children younger than age 18. The typical caregiver is a 46-year-

Contrary to a popular notion, most elderly people do not live in nursing homes; rather, they lead active, productive lives.

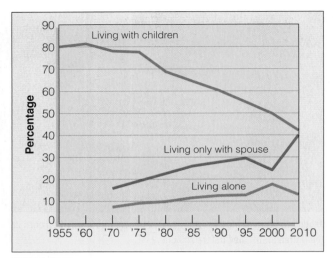

▲ **Figure 12.8**   Living Arrangements of People Age 65 and Older in Japan

Since 1955 the proportion of people 65 and older living with adult children has declined from 80 to 50 percent and is projected to decline to 42 percent by 2010.

Source: Population Reference Bureau (2006).

old employed woman caring for her 76-year-old mother, who lives nearby, for an average of 18 hours per week. The average person who cares for a senior does so for 4.3 years (Jackson 2003; National Alliance for Caregiving and AARP 2004).

Caregivers in Japan

As we have learned, Japan has a longstanding tradition of caring for parents that has been shaped by norms supporting filial piety, women as caregivers, and a patrilineal family structure. Although the percentage of persons age 65 and older living with their adult children has declined since 1955, when 80 percent of elderly individuals in Japan had such a living arrangement, it remains high, at around 50 percent. It is projected to decline to 42 percent by 2010 (see Figure 12.8). In Japan approximately 1.2 percent of all elderly persons reside in nursing homes.

As in the United States, caregivers in Japan are overwhelmingly female. Although specific data on caregiver and recipient characteristics in Japan are not available, we know that 76 percent of elderly Japanese men expect that their wives will take care of them if necessary, followed by their daughters-in-law, and then their daughters. In contrast, elderly Japanese women expect to be taken care of by daughters, followed by their daughters-in-law. "The Japanese government views the persistence of co-resident households [and the tradition of caring] as a unique asset that can be tapped to offset the adverse effects of population aging" (Ogawa and Retherford 1997, p. 76). In fact, the government has built its policies related to caring for the elderly around these traditions. Nevertheless, some signs suggest that these traditions are breaking down.

To date, research on Japan's caregivers has focused more on documenting changes in the caregiving relationship than on the caregiving relationship *per se*, placing particular emphasis on the shift from traditional extended family living to other kinds of housing arrangements. Although a high percentage of elderly still live with their children and expect their daughters-in-law or daughters to care for them, there is a public perception that this trend will not continue. A 1999 survey conducted by the prime minister's office examined 18- to 39-year-old men's and women's views on marriage. About 30 percent of men considered marriage a burden, citing financial factors as the major drawback. About 40 percent of women considered marriage a burden, because of "housework, child rearing, nursing their aged parents-in-law, and coping with the double responsibilities of work and homemaking" (*Japan Echo*, 2000a).

■ Visual Summary of Core Concepts

■ CORE CONCEPT 1: An amazing variety of family arrangements exists in the world. This variety makes it difficult to define family.

The family is a social institution that binds people together through blood, marriage, law, and/or social norms. Family members are generally expected to care for and support each other. This definition is very general because getting people to agree on a definition is surprisingly difficult. Why? Because no single concrete group represents family universally. Instead, an amazing variety of family arrangements exist worldwide—a variety reflected in the numerous norms that specify how two or more people can become a family.

Department of Defense photo by SSGT Michelle Leonard, USAF

■ CORE CONCEPT 2: Family can be defined in terms of the social functions it performs for society.

Some sociologists argue that *family* should be defined in terms of social functions, or its contributions to order and stability in the larger society. Such functions include regulating sexual behavior, replacing the members of society who die, socializing the young, providing care and emotional support, and conferring social status.

Department of Defense photo by SPC 5 Greg Leary

■ CORE CONCEPT 3: Family life is not always harmonious. Furthermore, the family passes on social privilege and social disadvantages to its members, thereby perpetuating the system of social inequality

Conflict theorists argue that the family perpetuates existing inequalities by passing on social privilege and social disadvantages to its members. Moreover, marriage and family systems are structured to value productive work and devalue reproductive work, and to maintain and foster racial divisions and boundaries.

Department of Defense photo by TSGT Curt Eddings

■ CORE CONCEPT 4: The structure of the family is affected by a society's history and culture.

Family structures are not static; they adapt to economic, cultural, historical, and social factors. The American family has been shaped by women's entry into the labor force. From 1900 to 1980, the percentage of women in the labor force rose from less than 20 percent to 60 percent. That change was influenced by a number of factors, including the rise and then subsequent decline of the breadwinner system, declines in total fertility, increased life expectancy, increased divorce rates, and increased employment opportunities for women.

Theodor Horydczak Collection, Library of Congress Prints and Photographs Division

The Japanese family is affected by a total fertility rate below replacement level. This situation explains the high percentage of people in Japan who are age 65 and older. The drop in Japanese total fertility is connected to a number of forces, including the fall of the multigenerational household system and the rise of the breadwinner system, the decline of arranged marriages, the rise of the "parasite single" and the "new single concept," and increased employment opportunities, especially for single women.

■ CORE CONCEPT 5: Over the past 100 years, fundamental shifts in the economy, a decline in parental authority, the changing status of children, and dramatic increases in life expectancy have changed the family structure.

Four economic arrangements affect the family and relationships among its members: low-technology tribal societies, fortified households, private households, and advanced market economies. Because of economic forces that accompanied industrialization and the rise of private households and advanced market economies, children no longer learned from their parents and other relatives the skills needed to make a living. The separation that characterized the work life of parents away from their children soon spread to other areas of life as parents and their children moved in different social spheres. Parental authority over adult children lost its economic and legal supports.

Lori Arviso Alvord, M.D./National Library of Medicine

The technological advances associated with the Industrial Revolution and the shift from an agriculture-based economy to a manufacturing-based one not only changed relationships among family members, but also altered the status of children from economic assets to liabilities.

Increases in life expectancy have also changed the family in fundamental ways. Parents can expect their children to survive infancy and early childhood, children can expect their parents to live a long life, couples can count on a long marriage, and people have more time to settle on a partner, an occupation, and whether to have children.

■ CORE CONCEPT 6: The aging of the population has no historical precedent. The family must find ways to adapt to this situation, especially as regards caregiving.

Finding ways to care for the elderly population is among the greatest challenges. In both the United States and Japan caregivers are overwhelmingly female. The caregiver burden is the extent to which caregivers believe that their emotional balance, physical health, social life, and financial status suffer because of their caregiver role.

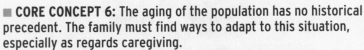

Department of Defense photo by O.J. Sanchez

In the United States most elderly persons do not live in nursing homes, but one in four does require assistance with daily activities, such as bathing, walking, dressing, and eating. Japan has a longstanding tradition of caring for parents that has been shaped by norms supporting filial piety, women as caregivers, and a patrilineal family structure. Although the percentage of persons age 65 and older living with their adult children has declined since 1955, when 80 percent of elderly individuals in Japan had such a living arrangement, it remains high, at around 50 percent. It is projected to decline to 42 percent by 2010.

Resources on the Internet

 Sociology: A Global Perspective Book Companion Web Site

www.cengage.com/sociology/ferrante

Visit your book companion Web site, where you will find flash cards, practice quizzes, Internet links, and more to help you study.

CENGAGENOW™

Just what you need to know NOW!

Spend time on what you need to master rather than on information you have already learned. Take a pre-test for this chapter, and CengageNOW will generate a personalized study plan based on your results. The study plan will identify the topics you need to review and direct you to online resources to help you master those topics. You can then take a post-test to help you determine the concepts you have mastered and what you will need to work on. Try it out! Go to www.cengage.com/login to sign in with an access code or to purchase access to this product.

Key Terms

aging population   332
caregiver burden   360
endogamy   341
exogamy   341
family   332

fortified households   351
ideal   333
life chances   335
low-technology tribal
    societies   351

productive work   338
reproductive work   338
secure parental employment   337
surplus wealth   351
total fertility rate   332

13 Education

With Emphasis on the European Union

When sociologists study education, they focus on the formal and informal social interactions that train, discipline, or shape (or reshape) the mind and body in planned or unplanned ways.

©Age Fotostock

▲ Schooling involves systematic instruction such that it imparts specific skills and ways of thinking.

WHY FOCUS ON

The European Union?

THE EUROPEAN UNION (EU) is an economic and political alliance that began in 1952 with 6 member countries. The alliance has expanded several times to eventually include 27 members in 2007. The EU's goal is to eliminate legal barriers to the free movement of people (including labor), goods, services, and capital across member countries' borders. We focus on the European Union in this chapter for several reasons.

First, the EU is investing heavily in education and research to boost its international competitiveness and to ensure that Europeans have the skills necessary to thrive in the 21st century (Bologna Declaration 1999). The EU is also offering scholarships to attract the world's "super-scholars," and it is working to open its higher education institutions to the rest of the world, thereby challenging the United States' dominance as a host country to international students (Riding 2003; Dillon 2004; Lee 2004).

Second, the U.S. Department of Education routinely compares its students and education system with foreign, especially European, counterparts on a host of attributes, including teachers' salaries, reading scores, scientific literacy, per capita spending on education, and access to educational opportunities. This comparative analysis allows an assessment of U.S. strengths and weaknesses relative to those of other countries.

Third, the United States was the first country in the world to embrace the concept of mass education. In doing so, it broke with the European view that education should be limited to an elite few. Europeans observed the American experiment with mass education, and their early impressions offer important, lasting assessments about the cultural values that the American system of public education promotes. In particular, the U.S. system seems to create students who (1) are preoccupied with knowledge as it applies to income generation and wealth creation, (2) value personal observations over accumulated knowledge and experience with other ways of life, (3) come away with a belief that the ideal person is self-made and able to transcend societal forces, and (4) place high value on educational achievement but not on the dedicated study needed to attain it (Hamilton 1883, Combe 1839).

FACTS TO CONSIDER

- About 80 percent of American 15-year-olds expect to have a high-skilled, white-collar job by age 30. Less than 50 percent of Czech, French, and German 15-year-olds expect to have such a job by that age.

- Less than 10 percent of American high school students are enrolled in vocational programs. Depending on the country, 35.6–80.7 percent of European students are enrolled in such programs.

- The United States is one of the few countries in the world that does not require students to learn at least one other language. In European Union countries mandatory foreign language study begins as early as age five.

- Relative to their European counterparts, U.S. teachers spend more hours in direct contact with students.

Education

■ **CORE CONCEPT 1: Education includes the formal and informal experiences that train, discipline, and shape the mental and physical potentials of the maturing person.** In the broadest sense, **education** includes the experiences that train, discipline, and shape the mental and physical potentials of the maturing person. An experience that educates may be as commonplace as reading a sweater label and noticing that the sweater was made in China or as intentional as performing a scientific experiment to learn how genetic makeup can be altered deliberately through the use of viruses. In view of this definition and the wide range of experiences it encompasses, we can say that education begins when people are born and ends when they die.

Sociologists make a distinction between informal and formal education. **Informal education** occurs in a spontaneous, unplanned way. Experiences that educate informally occur naturally; they are not designed by someone to stimulate specific thoughts or to impart specific skills. Informal education takes place when a child puts her hand inside a puppet and then works to perfect the timing between the words she speaks for the puppet and the movement of the puppet's mouth.

Formal education is a purposeful, planned effort to impart specific skills or information. Formal education, then, is a systematic process (for example, military boot camp, on-the-job training, or smoking cessation classes) in which someone designs the educating experiences. We tend to think of formal education as an enriching, liberating, or positive experience, but it can be an impoverishing and narrowing experience (such as indoctrination or brainwashing). In any case, formal education is considered a success when those being instructed acquire the skills, thoughts, and information that those designing the experience seek to impart.

This chapter is concerned with a specific kind of formal education: schooling. **Schooling** is a program of formal, systematic instruction that takes place primarily in

education In the broadest sense, the experiences that train, discipline, and shape the mental and physical potentials of the maturing person.

informal education Education that occurs in a spontaneous, unplanned way.

formal education A systematic, purposeful, planned effort intended to impart specific skills and modes of thought.

schooling A program of formal, systematic instruction that takes place primarily in classrooms but also includes extracurricular activities and out-of-classroom assignments.

Gaijin Bikers

Department of Defense photo by LCPL Robert D. Fleagle, Jr., USMC

Informal education occurs in spontaneous, unplanned ways—as when this little girl learns to bake by watching her grandmother. Formal education occurs when someone designs the educating experience with an outcome in mind—as when a drill instructor motivates a recruit through an obstacle.

Schooling takes place in classrooms and includes related extracurricular activities and out-of-classroom assignments.

classrooms but also includes extracurricular activities and out-of-classroom assignments. In its ideal sense, "education must make the child cover in a few years the enormous distance traveled by mankind in many centuries" (Durkheim 1961, p. 862). More realistically, schooling represents the means by which instructors pass on the values, knowledge, and skills that they or others have defined as important for success in the world. What constitutes an ideal education—in terms of learning objectives, material, and instructional techniques—varies according to time and place. Much depends on whether schooling is viewed as a mechanism for meeting society's needs or as a mechanism for helping students become independent thinkers, freed of the constraints on thought imposed by family, peers, culture, and nation.

Social Functions of Education

■ **CORE CONCEPT 2:** Schools perform a number of important social functions that, ideally, contribute to the smooth operation of society. Schools perform a number of important functions that serve the needs of society and contribute to its smooth operation. These functions include transmitting skills, facilitating change and progress, contributing basic and applied research, integrating diverse populations, and screening and selecting the most qualified students for what are considered the most socially important careers.

Transmitting Skills. Schools exist to teach children the skills they need to adapt to their environment. To ensure that

this end is achieved, society reminds teachers "constantly of the ideas, the sentiments that must be impressed" on students. Educators must pass on a sufficient "community of ideas and sentiments without which there is no society." These ideas and sentiments may relate to instilling a love of country, training a skilled labor force, or encouraging civic engagement. Without some agreement, "the whole nation would be divided and would break down into an incoherent multitude of little fragments in conflict with one another" (Durkheim 1968, pp. 79, 81).

Facilitating Personal Reflection and Change. Education can be a liberating experience that releases students from the blinders imposed by the accident of birth into a particular family, culture, religion, society, and time in history. Education can broaden students' horizons, making them aware of the conditioning influences around them and encouraging them to think independently of authority. In that sense, schools function as agents of personal change.

Contributing to Basic and Applied Research. Universities employ faculty whose job descriptions require them to do basic and applied research (in addition to teaching and sometimes in lieu of teaching). The following examples of university-based research centers or institutes suggest that universities add to society's knowledge base and influence policy in a variety of fields: the Center for Aging Research at Indiana University, the Artificial Intelligence Center at the University of Georgia, the Institute for the Study of Planet Earth at the University of Arizona, and the Institute for Drug and Alcohol Studies at Virginia Commonwealth University.

Integrating Diverse Populations. Schools function to integrate (for example, to Americanize or Europeanize) people of different ethnic, racial, religious, and family backgrounds. In the United States, schools play a significant role in what is known as the melting-pot process. Recall that the "peopling of America is one of the great dramas in all of human history" (Sowell 1981, p. 3). It involved the conquest of the native peoples, the annexation of Mexican territory along with many of its inhabitants (who lived in what is now New Mexico, Utah, Nevada, Arizona, California, and parts of Colorado and Texas), and an influx of millions of people from practically every country in the world. Early American school reformers—primarily those of Protestant and British backgrounds—saw public education as the vehicle for "Americanizing" a culturally and linguistically diverse population, instilling a sense of national unity and purpose, and training a competent workforce.

The European Union is relying on its schools to facilitate smooth relationships and interactions among 470 million people in 27 member states speaking more than 20

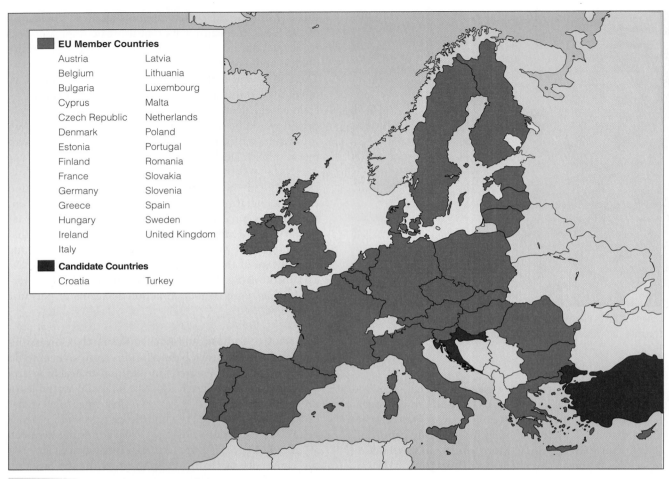

Austin Figure 13.1   Map of European Union Member and Candidate Countries

different languages. Among other things the EU Commission has recommended that schools prepare EU residents to be conversationally proficient in at least two languages beyond the mother tongue (see Figure 13.1).

Screening and Selecting. Schools use tests and grades to evaluate students and reward them accordingly by conferring or withholding degrees, issuing certificates, assigning students to academic tracks, rejecting or admitting students into programs, and giving negative or positive recommendations. Thus the schools channel students toward different career paths. Ideally, they channel the best skilled into the most desirable and important occupations and the least skilled into occupations that are believed to require no special talents.

Solving Social Problems. Societies use education-based programs to address a variety of social problems, includ-

ing parents' absence from the home, racial inequality, drug and alcohol addictions, malnutrition, teenage pregnancy, sexually transmitted diseases, and illiteracy. Although all countries support education-based programs that address social problems, the United States is probably unique in that education is viewed as the *primary* solution to many problems, including childhood obesity, illegal drug use, poverty, hunger, unwanted pregnancy, and so on.

The U.S. belief that education-based programs can be used to solve social problems and address community needs manifests itself in service learning programs and partnerships with community-based organizations to address community, even global, problems and needs. In fact, some leading educators argue that "the primary purpose of the university in the 21st century is to conduct research on pressing problems and to prepare students to address those problems" (Dobelle 2001). (See Working For Change: "Service Learning in Higher Education.")

WORKING FOR CHANGE

Service Learning in Higher Education

THE PRESIDENT'S HIGHER Education Community Service Honor Roll is a response to President George W. Bush's call to service. It builds on and supports the civic engagement mission of our nation's colleges and universities. This recognition program is designed to increase public awareness of the contributions that college students are making within their local communities and across the country through volunteer service. The program also identifies and promotes community service model programs and practices in higher education. Three universities out of 510 applicants won the 2006 Community Service Award for Excellence in General Community Service. They are listed below, along with one of the several projects each listed on its application:

California State University, Monterey Bay (Seaside, CA)
3,773 Students Enrolled

As part of an effort to revitalize the blighted Chinatown area of downtown Salinas, 64 CSUMB students from 13 service-learning courses provided 2,370 hours of service to address the needs of the homeless and other marginalized members of the Soledad Street/Old Chinatown neighborhood. Students' diverse service projects included preparing meals at the local soup kitchen; leading storytelling and drama activities with homeless people and others; conducting a survey of job-training needs and interests of homeless people; establishing a computer lab in the neighborhood; supporting the development of the county's first shelter for single women; and organizing a Soledad Street Beautification Day and a Soupline Forum to build community awareness and support for the revitalization effort.

Elon University (Elon, NC)
5,230 Students Enrolled

Elon became one of the first universities in the nation to accept a challenge from the Eugene Lang Foundation to provide a learning experience that would instill in students an abiding sense of social responsibility and civic concern through Project Pericles. The 35 members of each class who are accepted as Periclean Scholars take a series of courses culminating in a class project of local or global social change. The Periclean Scholars of the class of 2006 focused on raising awareness of issues surrounding the spread of HIV/AIDS locally and in Namibia. The class used the talents of faculty and more than 60 students from various disciplines to produce *Testing Positive*, a short film about abuse toward women and the fear of being tested for HIV. The film is now being shown at freshman orientation and other educational programs at universities in the U.S. and abroad. The issue identified by the class of 2007 is malnutrition, locally and in Honduras.

Indiana University–Purdue University, Indianapolis (Indianapolis, IN)
29,933 Students Enrolled

The School of Dentistry's community oral health program and statewide mobile dental program, Seal Indiana, provides oral health services for qualifying children and disabled adults as well as dental services to homeless people and residents of shelters. In 2005–06, 108 students provided more than 500 hours of dental services.

Source: Corporation for National and Community Service (2007).

Other Functions. Schools perform other, less obvious functions. For one, they function as reliable babysitters, especially for nursery-, preschool-, and grade school–age children. They also function as a dating pool and marriage market, bringing together young people of similar and different backgrounds and ambitions whose paths might otherwise never cross.

The Conflict Perspective

■ **CORE CONCEPT 3:** Any analysis of school systems must focus on unequal power arrangements that allow more-powerful individuals and groups to exert their will over those with less

power. Schools are not perfect: not all minds are liberated; students drop out, refuse to attend, or graduate with skill deficiencies; schools misclassify students; and so on. The point is that the functionalist analysis best describes ideals, which some school systems realize better than others. The conflict perspective draws our attention away from order and stability and to dominant and subordinate group dynamics and to unequal power arrangements. The conflict perspective asks questions like Who writes the curriculum? Who has access to the most up-to-date computer or athletic facilities? Which groups are most likely to drop out of high school, and which to attend college? Who studies abroad? The answers to these questions

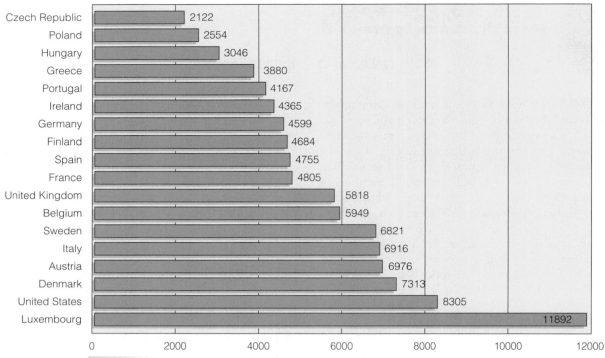

▲ **Figure 13.2** Per-Pupil Spending on Primary and Secondary Education in the United States and Selected EU Countries (in U.S. $)

Source: OECD (2006).

Schools perform many functions, some of which are not obvious, such as bringing together young people of similar and different backgrounds whose paths might otherwise never cross.

suggest that, for the most part, schools simply perpetuate the inequalities of the larger society. This point is obvious when we consider that the poorest schools usually have the highest drop-out rates, lowest high school graduation rates, and lowest college enrollments. In this regard, former Microsoft executive, turned full-time philanthropist, Bill Gates (2006) commented on the failure of the American system of education: "There's an acceptance of a tier-ing approach, where over a third of the students never graduate, and another third are trapped into a situation where they don't have the skills that are going to give them a good lifetime outcome." One factor that contributes to this inequality is, of course, differences in funding.

Funding as a Broad Measure of Inequality

Compared with its EU counterparts, the United States ranks third in per-pupil spending ($7,397 per student). Only Belgium and Denmark spend more. Within the European Union the spending gap between the wealthiest and poorest countries is $6,119 (see Figure 13.2). Note that this gap is somewhat larger than the $5,766 gap separating New York (the highest-spending state, at $11,378 per pupil) and Mississippi (the lowest-spending state, at $5,822 per pupil).

Each country follows distinct funding formulas that draw from federal, state or regional, and/or local sources. In the United States primary and secondary schools receive approximately 7 percent of their funding from the federal government, 49 percent from the state government, and 44 percent from local and other intermediate sources, raised primarily through property taxes (National Center for Education Statistics 2004b). Such a heavy reliance on state revenue is problematic, because less wealthy states generate less tax revenue than do wealthier ones.

Literacy is the ability to understand and use a symbol system, whether it is based on sounds, letters, numbers, pictographs, or some other type of symbol. Until these children learn the symbol system needed to read they are technically illiterate.

Heavy reliance on local revenue is especially problematic, because it creates funding disparities among schools within states. Indeed, the courts in at least 28 states have evaluated (or are evaluating) claims that methods of financing have helped create unequal school systems within the state, or they have ruled that the methods of school financing are unconstitutional (Celis 1992, 1993b).

One way to evaluate areas in which United States school systems fall short or succeed at educating their students is by comparing those systems with EU education systems, using criteria that sociologists have identified as critical to profiling and evaluating education systems. Most of the data for our comparative analysis comes from Organisation for Economic Co-operation and Development (OECD) education-related reports. We begin this comparative analysis by comparing the percentages of the populations in the United States and EU countries that are considered functionally illiterate.

Illiteracy. In the most general and basic sense, **illiteracy** is the inability to understand and use a symbol system, whether it is based on sounds, letters, numbers, pictographs, or some other type of symbol. In the United States (as in all countries) some degree of illiteracy has always existed, but conceptions of what people need to know to be considered literate have varied over time. At one time, Americans were considered literate if they could sign their names and read the Bible. At another time, completing the fourth grade made someone literate. The National Literacy Act of 1991 defines literacy as "an individual's ability to read, write, and speak English and compute and solve problems at levels of proficiency necessary to function

on the job and in society, to achieve one's goals, and to develop one's knowledge and potential" (U.S. Department of Education 1993, p. 3).

■ **CORE CONCEPT 4:** Even in societies where schooling is compulsory, a significant percentage of the population does not possess skills needed to succeed in the society in which they live. People are considered **functionally illiterate** when they do not possess the level of reading, writing, calculating, and other skills needed to succeed (that is, function) in the society in which they live. This point suggests that illiteracy is a product of one's environment. Functional illiteracy includes the inability to use a computer, read a map to find a destination, make change for a customer, read traffic signs, follow instructions to assemble an appliance, fill out a job application, and comprehend the language others are speaking.

Among other things, the OECD (2006) report focuses on three kinds of literacy: reading, mathematical, and scientific. It seeks to determine how successful school systems are at developing these literacies. For example, in the area of mathematical literacy—defined as "an individual's capacity to identify and understand the role that mathematics plays in the world, to make well-founded judgments, and to use and engage with mathematics in ways that meet the needs of that individual's life as a constructive, concerned, and reflective citizen" (p. 72)—the OECD classifies 34.1 percent of U.S. students as illiterate. In comparison with the United States, 5 of the 19 EU countries for which data is available have a higher percentage of 15-year-old students classified as illiterate (see Table 13.1).

Foreign Language Illiteracy. If we focus on languages, of which there may be as many as 9,000, we can see that the potential number of illiteracies is overwhelming, as people cannot possibly be literate in every language. If a person speaks, writes, and reads in only one language, by definition he or she is illiterate in as many as 8,999 languages. For most people such a profound level of illiteracy rarely presents a problem until they find themselves in a setting (such as a war or a business negotiation) that puts them at a disadvantage for not knowing the language others around them are speaking. Both United States and European Union leaders acknowledge the societal benefits that accompany literacy in a language other than the mother

illiteracy The inability to understand and use a symbol system, whether it is based on sounds, letters, numbers, pictographs, or some other type of symbol.

functionally illiterate Lacking the level of reading, writing, and calculating skills needed to function in the society in which one lives.

Table 13.1 Percentage of 15-year-olds in the United States and EU Countries Whom the OECD Classified as Mathematically Illiterate

Students who take the test are classified into one of seven levels: below 1, level 1, level 2, level 3, level 4, level 5, and level 6. Those who are classified as below 1 or at level 1 are considered illiterate. Individual countries have different criteria for mathematical literacy. To be considered literate in the Netherlands, for example, students must be classified at level 4 or better (de Lange 2006).

Country	% Illiterate
Greece	47.5
Spain	39.6
Portugal	37.6
Ireland	35.7
Austria	34.8
United States	34.1
Sweden	33.4
Italy	31.9
Czech Republic	31.7
Hungary	31.6
Luxembourg	28.6
Germany	28.2
France	25.8
Denmark	25.3
Belgium	23.1
Poland	22.0
Netherlands	20.6
Slovakia	19.9
Finland	17.5

Department of Defense photo by Gunnery Sgt. Keith A. Milks, USMC

Look at the writing on the the front of the truck. If you can only read the Arabic, then you are illiterate in the symbol system known as English.

Christopher Brown

This is one of a handful of U.S. students who choose to study abroad in Egypt. Great Britain is the number one choice for U.S. students.

tongue. In fact, the European Commission put forth an action plan calling for every citizen to "have a good command of two foreign languages together with the mother tongue" (Binder 2006). For its part, the U.S. Department of Education (2006) lamented that other nations "have an edge in foreign language instruction, a key to improved national security and global understanding."

In European countries, foreign language instruction is compulsory and can start as early as age five. In the United States, by contrast, 44 percent of high school students study a foreign language (usually for about two years) and only 8 percent of undergraduates take foreign language courses. Study abroad is one way to immerse oneself in a foreign language and culture. Yet, even when Americans study abroad, the most popular destinations are English-speaking countries (see No Borders, No Boundaries: "Study Abroad Destinations").

Because of the European Union's greater emphasis on foreign language instruction, 50 percent of people in EU countries indicate that they can speak at least one language beyond their mother tongue well enough to carry on a conversation (Eurobarometer 2005). While comparable estimates of Americans who possess such conversa-

tional skills do not exist, anecdotal evidence suggests that most high school and college students who have taken foreign language courses cannot claim to have acquired this level of skill.

Education critic Daniel Resnick (1990) argues that the absence of serious foreign language instruction contributes to the parochial nature of American schooling. The almost exclusive attention paid to a single language has deprived students of the opportunity to appreciate the connection between language and culture and to see that language is a tool that enables them to think about the world. According to Resnick the focus on a single language "has cut students off from the pluralism of world culture and denied them a sense of powerfulness in approaching societies very different from their own" (1990, p. 25). It also denies students from non-English-speaking heritages

NO BORDERS, NO BOUNDARIES

Study Abroad Destinations

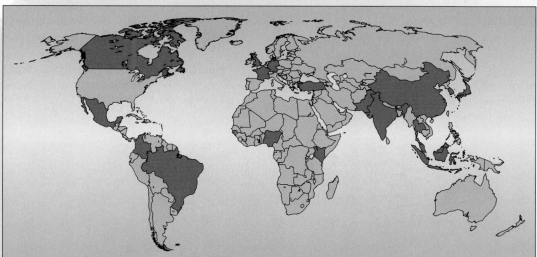

▲ **Figure 13.3** Top 20 Countries (Highlighted in Red) Sending Students to the United States

Over 500,000 foreign students attend college in the United States. The top 20 countries sending students are shown in Figure 13.3. India ranks number 1, sending 76,503 students. Pakistan ranks number 20, sending 5,759 students. Only 2 EU countries are among the top 20 countries: France and Germany.

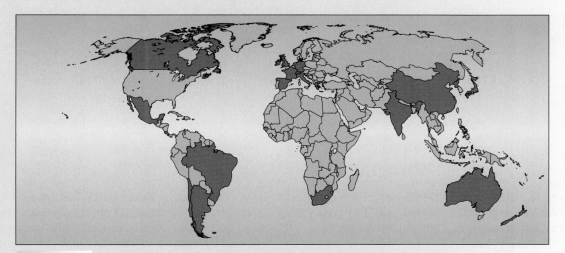

▲ **Figure 13.4** Top 20 Countries (Highlighted in Red) to Which the United States Sent Students

Approximately 205,000 U.S. students study outside the United States each year. The top 20 destinations are highlighted in Figure 13.4. Of the 20 most popular destinations, 8 are EU countries. In fact, 107,231 U.S. students—52 percent of those who study abroad—go to EU countries. Also note that 32,071—15.6 percent—study in the United Kingdom, an English-speaking country.

Source: Institute of International Education (2006).

GLOBAL COMPARISONS

The Legacy of European Colonization on Language Instruction

THE LEGACY OF colonization helps to explain why most people around the world other than Americans speak more than one language. Beginning as early as 1492, Europeans learned of, conquered, or colonized much of North America, South America, Asia, and Africa. In doing so, Europeans imposed their language on the colonized peoples. In many cases, Europeans created countries that pulled together peoples who spoke different languages. Although people learned the new, imposed languages, they also continued to speak their native languages.

Countries where the *official* language was imposed by a European nation that was once a colonial power are highlighted in yellow.

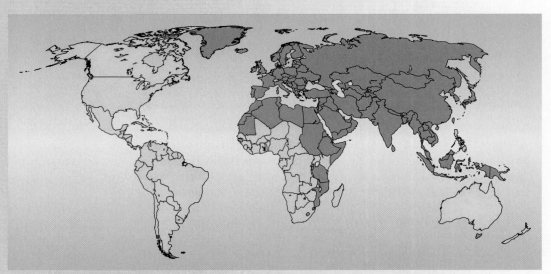

▲ **Figure 13.5**   Non-EU Countries (Highlighted in Yellow) Where a European Language Is the Official Language

Sources: U.S. Central Intelligence Agency (2005); Eurydica (2001).

the means of reflecting on and fully appreciating their ancestors' lives (see Global Comparisons: "The Legacy of European Colonization on Language Instruction.").

The Availability of College

■ **CORE CONCEPT 5:** Only a handful of countries in the world give a significant share of the population the opportunity to attend college. The United States is one of a small group of countries where the college-educated constitute a significant share of the population. Figure 13.6 shows that almost 40 percent of 25- to 34-year-olds in the United States have at least a college education. In some EU countries—Belgium, Finland, and Sweden—the percentage is about the same; in others—notably, Italy (12 percent) and the Czech Republic (11 percent)—the percentage is much smaller.

One distinctive feature of the U.S. education system is that, in theory, anyone can attend a college, even if he or she has not graduated from high school or has received a general equivalency degree (GED). A U.S. Department of Education study found that 400,000 students—2 percent (1 in 50) of all college students—did not have a high school diploma or GED (Arenson 2006). Among those who do graduate from high school, almost 67 percent enroll in college the following year (U.S. Bureau of Labor Statistics 2005). Racial or ethnic classification has little effect on who among high school graduates attends college, but it does affect the chances of dropping out of high school. Keep in mind that 31.9 percent of American students entering ninth grade do not graduate from high school four years later. The dropout rate varies for students classified as Asian (21 percent), white (24 percent), Native American (43 percent),

▲ **Figure 13.6** Postsecondary Education Statistics

The chart shows that 39 percent of Americans between the ages of 25 and 34 have the equivalent of a college degree. France, Finland, Belgium, Sweden, and Spain are countries with similar percentages. The per-student cost to attend college each year is highest in the United States ($20,358).[1,2] Among EU countries, Greece has the lowest per capita cost ($3,402). In the United States only one-third (33.9 percent) of college costs are subsidized with public expenditures. In some EU countries, 90 percent or more of college costs are subsidized with public expenditures. Many European governments are rethinking this policy as greater numbers of students attend college. To see why, consider that in 1985, 14 percent of British high school graduates went on to college; today 40 percent do so.

Country	Per-Student Cost to Attend Postsecondary Education (in U.S. $)[2]	Percentage of 25- to 34-year-old Population Obtaining the Equivalent of at Least a College Degree	Public Expenditure as a Percentage of the Total Cost of Postsecondary Education
Austria	12,854	14	96.7
Belgium	16,771	38	85.2
Czech Republic	5,431	11	85.5
Denmark	11,981	29	97.6
Finland	8,244	38	97.2
France	8,373	34	85.7
Germany	10,989	22	91.8
Greece	3,402	24	99.7
Hungary	7,024	15	76.7
Ireland	11,083	28	79.2
Italy	8,065	12	77.5
Netherlands	11,934	27	77.4
Portugal	4,766	15	92.5
Slovak Republic	4,949	12	91.2
Spain	6,666	36	74.4
Sweden	15,097	37	88.1
United Kingdom	9,657	29	67.7
United States	20,358	39	33.9

1. The cost of college at major public universities can range from $6,054 (University of Louisville) to $15,646 (Penn State University) per year. At a private college, the price can be as high as $41,218 (Sarah Lawrence) per year (New York *Times* 2003).
2. This cost covers housing, food, and other expenses beyond tuition and books.

Source: OECD (2003).

black (45 percent), and Hispanic (47 percent). If we consider dropout rates in terms of the share of *all* 18-year-olds who go to college, the percentage who attend college would fall to about 40 percent (U.S. Department of Education 2007).

For another indicator of the relatively easy access to college found in the United States, consider that many U.S. four-year colleges and universities accept students regardless of deficiencies or poor grades in high school coursework, or low scores received on the ACT or SAT. In fact, ACT scores suggest that only 25 percent of students taking them are prepared for college-level work (Lewin 2005). Almost 80 percent of colleges and universities offer remedial courses for students who lack the skills needed to do college-level work. An estimated 28 percent of entering freshmen take one or more remedial courses in reading, writing, or mathematics (National Center for Education Statistics 2004d).

While the United States makes it easier for people to enroll in college in spite of academic deficiencies, it places more of the funding burden on the private sector. Compared with its EU counterparts, the United States ranks

Department of Defense photo by Photographer's Mate 2nd Class Daniel J. McLain, U.S. Navy

These graduates of the U.S. Naval Academy are among an elite minority of people around the world who have access to a college education.

Stefan Wagner, http://frumpkin.de

Many European governments are reassessing the long-held belief that the cost of a university education should be heavily subsidized. Here 4,000 Germans are demonstrating against increases in tuition fees.

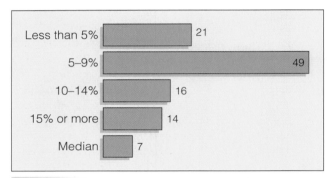

Less than 5%	21
5–9%	49
10–14%	16
15% or more	14
Median	7

▲ **Figure 13.7** Debt Burden Four Years after College as a Percentage of Monthly Income

The chart shows the debt burden, or the monthly student loan payment, as a percentage of monthly income of students who graduated from college in 2001. For that cohort, the average undergraduate loan is $18,900. Note that 14 percent of borrowers use 15 percent or more of their monthly income to pay back this debt (Baum and O'Malley 2003).

last in the percentage of postsecondary education costs paid by public funds. In the 17 EU countries for which we have data, at least 67 percent of postsecondary education costs is paid by public funds. In the United States, that share is 33.9 percent (National Center for Education Statistics 2004c). The U.S. system relies more heavily on private sources—such as school fund-raising, college endowment funds, fees, tuition reimbursement from employers, and bank loans—to offset personal costs (Lyall 2003). As a result, 65 percent of U.S. college students borrow money to pay for college. Among those who take out loans to finance college, 30 percent spend at least 10 percent of their monthly income on paying back the loans (see Figure 13.7). Given this high level of private investment, one should not

be surprised to find that students (and their parents) are preoccupied with the return on this investment.

Despite the debt burden, the U.S. way of funding higher education may have some advantages. Claude Allègre, a former French education minister and reformer, argues that the U.S. university system drives American prosperity and that the French government simply does not invest in higher education. It promises a college education to any high school graduate who passes the baccalaureate exam, but the government does not deliver the facilities, which are considered to be run-down, crowded, and uninspiring (Sciolino 2006). Americans, on the other hand, expect more because they are paying more: "comfortable student residences, gyms with professional exercise equipment, better food of all kinds, more counselors, . . . more high-tech classrooms and campuses that are spectacularly handsome" (Chace 2006).

Curriculum

When we look at school curriculum, we might focus on at least three questions: (1) Is the curriculum centralized or decentralized? (2) How do the schools track students? (3) As students learn the curriculum, what other, "hidden" lessons do they learn?

Centralized versus Decentralized Models

Austria, France, Germany, Greece, Portugal, Spain, and the United Kingdom have a centralized model, or national curriculum, that applies to all types of schools, grades, and subjects. Typically, a national curriculum specifies the subjects, the time that must be allocated to each, and the curriculum content in more or less detail. "The school is responsible for managing and delivering the curriculum" (OECD 2003).

In a decentralized model, the state or region, school districts, or local school systems establish the curriculum and the schools within them are evaluated on whether they meet achievement targets set by some higher authority. A national body may also recommend (as opposed to mandate) minimum requirements regarding subject matter, test scores, and time allocated to the various subjects. The Netherlands, the Czech Republic, Belgium, and Denmark are governed by the decentralized model. The U.S. school system also follows a decentralized model.

No uniform curriculum exists in the United States. Instead, each state sets its own broad curriculum requirements for kindergarten through high school; each school, in turn, interprets and implements these requirements. Consequently, even with state guidelines, the textbooks, assignments, instructional methods, staff qualifications, and material covered may vary across the schools within each state. In addition to curriculum differences across and within states, students within the same school are usually grouped or "tracked" based on test scores or past performance. For example, students enrolled in standard diploma tracks tend to take fewer mathematics courses than students enrolled in college preparatory, honors, or advanced tracks and the kinds of courses they take may differ from the kinds other students take (for example, general math instead of algebra). Similarly, 66 percent of high school students who graduate take English composition rather than creative writing or advanced writing to meet the state English requirement (National Center for Education Statistics 2004a). Keep in mind that the decentralized model does not necessarily allow individual teachers to have greater control over the curriculum. In fact, according to National Education Association (2004) research, 88 percent of U.S. public school teachers want "more influence on curriculum and instruction decisions in their schools."

Tracking

■ **CORE CONCEPT 6:** Most, if not all, education systems sort students into distinct instructional groups according to similarities in past academic performance, performance on standardized tests, or even anticipated performance. Schools arrange students into distinct instructional groups according to similarities in past academic performance, performance on standardized tests, or even anticipated performance. This practice may be known as tracking, ability grouping, or something else. Under such a sorting and allocation system, students may be assigned to separate instructional groups within a single classroom; they may be separated across the entire array of subjects (for example, college preparatory versus general studies tracks); or they may be sorted regarding selected subjects, such as mathematics, science, and English (assigned, for example, to advanced placement, honors, regular, or basic courses in these subjects).

The following rationales underlie ability grouping, streaming, or tracking (hereafter referred to as tracking):

1. Students learn better when they are grouped with those who learn at the same rate. The brighter students are not held back by the slower learners, and the slower learners receive the extra time and special attention needed to correct academic deficiencies.
2. Slow learners develop more-positive attitudes when they do not have to compete with the more academically capable.
3. Groups of students with similar abilities are easier to teach than students of various abilities.

Research suggests that tracking has a positive effect on high-track students, a negative effect on low-track students, and no noticeable effect on middle-track or regular-track students (Ansaloe 2004; Hallinan 1996; Schmidt 2004). In addition, there is little evidence to indicate that placing students in remedial or basic courses contributes to their intellectual growth, corrects their academic deficiencies, prepares them for success in higher tracks, or increases their interest in learning. Instead, special curricula seem to exaggerate and widen differences among students and perpetuate beliefs that intellectual ability varies according to social class and ethnic group (Oakes 1986a, 1986b).

In a classic study, sociologist Jeannie Oakes (1985) investigated how tracking affected the academic experiences of 13,719 middle school and high school students in 297 classrooms and 25 schools across the United States. "The schools themselves were different: some were large, some very small; some in the middle of cities; some in nearly uninhabited farm country; some in the far West, the South, the urban North, and the Midwest. But the differences in what students experienced each day in these schools stemmed not so much from where they happened to live and which school they happened to attend but rather, from differences within each school" (Oakes 1985, p. 2).

Oakes's findings were consistent with the findings of hundreds of other studies of tracking in terms of how students were assigned to groups, how they were treated, how they viewed themselves, and how well they did.

Placement. Poor and minority students were placed disproportionately in the lower tracks.

Treatment. The different tracks were not treated as equally valued instructional groups. Clear differences existed in the quality, content, and quantity of instruction and in classroom climate, as reflected in the teachers' attitude and in student-student and teacher-student relationships. Low-track students were consistently exposed to inferior instruction—watered-down curricula and endless repetition—and to a more rigid, more emotionally strained classroom climate.

Self-image. Low-track students did not develop positive images of themselves, because they were publicly identified and treated as educational discards, damaged merchandise, or unteachable. Overall, among the average and the low-track groups, tracking seemed to foster lower self-esteem and promote misbehavior, higher dropout rates, and lower academic aspirations. In contrast, placement in a college preparatory track had positive effects on academic achievement, grades, standardized test scores, motivation, educational aspirations, and attainment. "And this positive relationship persists even after family background and ability differences are controlled" (Hallinan 1988, p. 260).

Achievement. The brighter students tended to do well regardless of the academic achievements of the students with whom they learned. This finding was reflected in the written answers that teachers and students gave to various questions asked by Oakes and her colleagues. For example, when teachers were asked about the classroom climate, high-track teachers tended to reply in positive terms:

> There is a tremendous rapport between myself and the students. The class is designed to help the students in college freshman English composition. This makes them receptive. It's a very warm atmosphere. I think they have confidence in my ability to teach them well, yet because of the class size—32—there are times they feel they are not getting enough individualized attention. (Oakes 1985, p. 122)

Low-track teachers replied in less positive terms:

> This is my worst class. Kids [are] very slow—underachievers and they don't care. I have no discipline cases because I'm very strict with them and they are scared to cross me. They couldn't be called enthusiastic about math—or anything, for that matter. (p. 123)

Clear differences also appeared in the high-track and low-track students' responses to the question "What is the most important thing you have learned or done so far in this class?" The replies of high-track students centered around themes of critical thinking, self-direction, and independent thought:

- The most important thing I have learned in this [English] class is to loosen up my mind when it comes to writing. I have learned to be more imaginative. (p. 87)
- The most important thing I have learned in this [math] class is the benefit of logical and organized thinking;

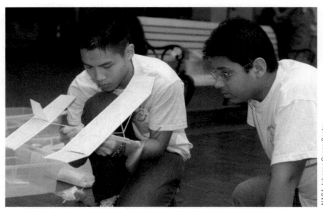

In the United States, students placed in high academic tracks are likely to be exposed to science and math curricula that are different from, and more stimulating than, the curricula taught to lower-track students.

learning is made much easier when the simple processes of organizing thoughts have been grasped. (p. 88)

Low-track students were more likely to give answers that centered on themes of boredom and conformity:

- I think the most important thing is coming into [math] class and getting out folders and going to work. (p. 89)
- To be honest, nothing. (p. 71)
- Nothing I'd use in my later life; it will take a better man than I to comprehend our world. (p. 71)

Although many educators have recognized the problems associated with tracking, efforts to detrack have collided with demands from politically powerful parents of high-achieving or "gifted" students; these parents insist that their children must get something more than the "other" students (Wells and Oakes 1996). As a result, tracking persists, "even though many educators and policymakers acknowledge that students in the low and middle tracks are not held to high enough standards and thus are not adequately prepared for either college or the transition to work" (Wells and Oakes 1996, p. 137).

European Tracking: Vocational versus University Tracks

While EU countries track students according to academic abilities, they do not offer students in lower tracks a watered-down or simpler version of a subject matter. That is, lower-track students do not take basic math while those in higher tracks take algebra. Observations from a German student studying the United States illustrate this:

> In Germany grades 5–13 are considered high school and 1–4 are elementary. After 4th grade student report cards are evaluated and if they are comprised of As and Bs, students are advised to go to what is called the gymnasium. If there are

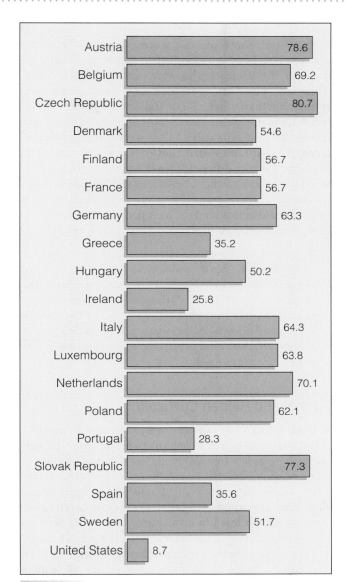

Austria — 78.6
Belgium — 69.2
Czech Republic — 80.7
Denmark — 54.6
Finland — 56.7
France — 56.7
Germany — 63.3
Greece — 35.2
Hungary — 50.2
Ireland — 25.8
Italy — 64.3
Luxembourg — 63.8
Netherlands — 70.1
Poland — 62.1
Portugal — 28.3
Slovak Republic — 77.3
Spain — 35.6
Sweden — 51.7
United States — 8.7

▲ **Figure 13.8** Percentage of Students in Vocational Programs in European Countries and the United States

Source: OCED (2003)

Cs, then the students are advised to go to the 10-year school; if Ds to the 9-year school. Parents can override that advice everywhere except Bavaria, but a gymnasium headmaster can always override a parent. In my class of 21 students, 13 went to the gymnasium (a very high proportion). After the first year, those who had been advised to go to 9- or 10-year schools left because they were failing. While failing students are not forced to leave they must repeat grades in order to stay. Everyone takes the same math in high school and you have to take math every year; there is no option. There is no difference in math classes within schools.

EU schools are also known for vocational tracks. A high percentage—35.6 to 80.7 percent—of secondary-level students in EU countries are enrolled in vocational

When you look at these students classified as Asian, do you think they will be better at math and other academic subjects than students classified as another race? When teachers approach students with such preconceived notions, they can set into motion the dynamics surrounding the self-fulfilling prophecy.

education programs. These programs prepare students for "direct entry, without further training, into a specific occupation" (OECD 2003). Less than 10 percent of American high school students are enrolled in vocational programs (see Figure 13.8). We should not apply our ideas of U.S. vocational education to European vocational education programs. According to Richard Owen (2000), a Sacramento high school principal who visited 10 European countries to learn about their education systems, most vocational programs are equivalent in rigor to U.S. college preparatory programs.

Self-Fulfilling Prophecies

Tracking can create self-fulfilling prophecies by affecting teachers' expectations of the academic potentials and abilities of students placed in each track. The idea of a self-fulfilling prophecy is a deceptively simple yet powerful one that originated from an insight by William I. and Dorothy Swain Thomas: "If [people] define situations as real, they are real in their consequences" ([1928] 1970, p. 572). A **self-fulfilling prophecy** begins with a false definition of a situation.

self-fulfilling prophecy A concept that begins with a false definition of a situation. Despite its falsity, people assume it to be accurate and behave accordingly. The misguided behavior produces responses that confirm the false definition.

The false definition is assumed to be accurate, however, and people behave as if it were true. In the end, the misguided behavior produces responses that confirm the false definition (Merton 1957). A self-fulfilling prophecy in education can occur if teachers and administrators assume that some students are "fast," "average," or "slow" learners and consequently expose them to "fast," "average," and "slow" learning environments. Over time, real differences in quantity, quality, and content of instruction cause many students to actually become (and believe that they are) "slow," "average," or "fast."

> The tragic, often vicious, cycle of self-fulfilling prophecies can be broken. The initial definition of the situation which has set the circle in motion must be abandoned. Only when the original assumption is questioned and a new definition of the situation is introduced, does the consequent flow of events [show the original assumption to be false]. (Merton 1957, p. 424)

In the tradition of symbolic interactionism, Robert Rosenthal and Lenore Jacobson designed an experiment to test the hypothesis that teachers' positive expectations about students' intellectual growth can become a self-fulfilling prophecy. The experiment took place in an elementary school called Oak School, a name given the school to protect its identity. The student body came mostly from lower-income families and was predominantly white (84 percent); 16 percent of the students were Mexican Americans. Oak School sorted students into ability groups based on reading achievement and teachers' judgments.

At the end of a school year, Rosenthal and Jacobson gave a test, purported to be a predictor of academic "blooming," to the students who were expected to return the next year. Just before classes began in the fall, all full-time teachers were given the names of the white and Hispanic students "fast," "average," and "slow" groups who had supposedly scored in the top 20 percent. The teachers were told that these students "will show a more significant inflection or spurt in their learning within the next year or less than will the remaining 80 percent of the children" (Rosenthal and Jacobson 1968, p. 66). Teachers were also told not to discuss the scores with the students or the students' parents. Actually, the names given to teachers were chosen randomly; the differences between the children earmarked for intellectual growth and the other children existed only in the teachers' minds. The students were retested after one semester, at the end of the academic year, and after a second academic year.

formal curriculum The various academic subjects, such as mathematics, science, English, reading, and physical education.

hidden curriculum All the other activities that go on as students learn subject matter, and the "lessons" that those other activities convey about the value and meaning of what the students are learning.

Overall, intellectual gains, as measured by the difference between successive test scores, were greater for the students who had been identified as "bloomers" than they were for the students who were not identified as such. Although "bloomers" benefited in general, some benefited more than others: First- and second-graders, Mexican American children, and children in the middle track showed the largest increases in test scores. The "bloomers" received no special instruction or extra attention from teachers; the only difference between them and the other students was the teacher's belief that the "bloomers" bore watching. Rosenthal and Jacobson speculated that this belief was communicated to "bloomers" in very subtle and complex ways, which they could not readily identify:

> To summarize our speculations, we may say that by what she said, by how and when she said it, by her facial expressions, postures, and perhaps by her touch, the teacher may have communicated to the ["bloomers"] that she expected improved intellectual performance. It is self-evident that further research is needed to narrow down the range of possible mechanisms whereby a teacher's expectations become translated into a pupil's intellectual growth. (p. 180)

Formal and Hidden Curriculum

■ **CORE CONCEPT 7:** Teachers everywhere teach two curricula simultaneously: a formal one and a hidden one. Teachers everywhere teach two curricula simultaneously: a formal one and a hidden one. The various academic subjects—mathematics, science, English, reading, physical education, and so on—make up the **formal curriculum**. Students do not learn in a vacuum, however. As teachers instruct students and as students learn the subject matter and complete their assignments other activities are occurring around them. These other activities are the **hidden curriculum** and include the teaching method, the types of assignments and tests, the tone of the teacher's voice, the attitudes of classmates, the number of students absent, the frequency of the teacher's absences, the number of interruptions during a lesson, and the criteria teachers use to assign grades (see Intersection of Biography and Society: "European Students Studying in the United States Comment on American Teachers, Tests, and Study Habits"). These so-called extraneous factors function to convey messages to students not only about the value of the subject, but also about the values of society, the place of learning in their lives, and their role in society—as the case of spelling baseball illustrates.

Spelling Baseball. Jules Henry (1965) uses typical classroom scenes (acquired from thousands of hours of participant observation), such as a session of "spelling baseball," to demonstrate the seemingly ordinary way hidden curriculum is transmitted and to show how students are

INTERSECTION OF BIOGRAPHY AND SOCIETY

European Students Studying in the United States Comment on American Teachers, Tests and Study Habits

THREE EUROPEAN STUDENTS attending college in the United States comment on how the U.S. system is different from that in their home countries. Their comments reinforce the idea that the society shapes biography (personal experiences and interactions) in dramatic ways.

College Professors

Germany: American teachers do not seem to ask their students to do as much work as German teachers do. In the United States the homework is more like busy work such as a scavenger hunt on a Web site or looking at a catalog to find out what the requirements are for a particular major. These are things that I would do on my own but the students here need to be told to do them.

In Germany we are taught to be independent and in the United States it is the opposite. It was nice in the beginning because I got straight "As" and it helped with my confidence but now I want to be more challenged. In Germany I was an OK student but in the United States I am a straight A student. American teachers are very supportive and helpful with keeping students up with the lessons. I feel like I can ask my American teachers anything and they will do whatever they can to help. It's not so much that I would not ask a German teacher something but here I never feel that I ask a stupid question.

Poland: American teachers are more helpful, more approachable, and they seem more equal to students. You would never ask a professor in Poland how they are or call them by their first name. Here you can walk in and talk to a professor about personal things such as what you did over the weekend; that would never happen in Poland. I feel I can ask American teachers for help and I will get it.

Spain: American teachers are more friendly and accessible. They are always asking about personal things like what happened over the weekend. American teachers talk to students like they are kids, not adults. In the United States professors seem to want to be friends with the students; in Spain they are just professors, not friends too.

Tests

Germany: In Germany it was essay only. Here the tests are too easy with multiple choice and true/false. At first it was a good feeling that I was always doing well but now I just want to be more challenged.

Poland: Tests are a different format here. In Poland they were all essay questions, no multiple choice. We also would have one test at the end of the semester that covered everything. Here we have three or four tests.

Spain: In my entire education in Spain, I have only taken one multiple-choice test. They are not very common. Even essay tests are very easy in the United States. If American students write something even remotely related to the answer, teachers give them some points just for their effort.

Study Habits

Germany: American students are all about memorizing. My roommate memorizes her flash cards the night before the test. At home it is more about understanding the subject, not just a specific date that something happened. We read other books beside the textbook so we can understand the material but American students just study their notes.

Poland: American students spend less time studying than students in Poland. Polish students discuss serious topics for fun. I guess fun is the wrong word. We talk about serious topics because they are serious and need attention. When American students get together to talk, it is more about what was on the headline news or what was going on in class. We don't get into groups and do assignments or have discussions. We are more independent in Poland.

Spain: It's easier in the United States. In Spain, students cannot study two days and expect to do well on a test. . . . In Spain we study almost a month before taking exams and have a week of tests every few months. Even in September when school starts we have a week of exams so we have to study over the summer. In high school, we take 10 subjects over the course of a school year and if someone fails two or three exams, he or she has to take the whole year again. American students eat, sleep, and listen to music while they study. I might listen to some music when I study but it would be relaxing music like classical, but not rock.

Sources: Missy Gish, Northern Kentucky University, Class of 2005 (interviewer); Isabell Haage, Class of 2007 (Germany); Anna Nowak, Class of 2005 (Poland); and Cristina Gonzalez, Class of 2004 (Spain).

Department of Defense photo by A1C Joshua E. Coleman, USAF

Jules Henry maintains that when American children are sent to the chalkboard, especially to do math problems, they learn to "fear failure" and "envy success." If they get the problem wrong, a classmate will be asked to correct the mistake; hence, one student's success is achieved at the expense of another's failure.

exposed simultaneously to the two curricula. Although Henry observed this scene in 1963, his observations hold even today:

> The children form a line along the back of the room. They are to play "spelling baseball," and they have lined up to be chosen for the two teams. There is much noise, but the teacher quiets it. She has selected a boy and a girl and sent them to the front of the room as team captains to choose their teams. As the boy and girl pick the children to form their teams, each child chosen takes a seat in orderly succession around the room. Apparently they know the game well. Now Tom, who has not yet been chosen, tries to call attention to himself in order to be chosen. Dick shifts his position to be more in the direct line of vision of the choosers, so that he may not be overlooked. He seems quite anxious. Jane, Tom, Dick, and one girl whose name the observer does not know, are the last to be chosen. The teacher even has to remind the choosers that Dick and Jane have not been chosen.
>
> The teacher now gives out words for the children to spell, and they write them on the board. Each word is a pitched ball, and each correctly spelled word is a base hit. The children move around the room from base to base as their teammates spell the words correctly. The outs seem to increase in frequency as each side gets near the children chosen last. The children have great difficulty spelling "August." As they make mistakes, those in the seats say, "No!" The teacher says, "Man on third." As a child at the board stops and thinks, the teacher says, "There's a time limit; you can't take too long, honey." At last, after many children fail on "August," one child gets it right and returns, grinning with pleasure, to her seat. . . . The motivation level in this game seems terrific. All the children seem to watch the board, to know what's right and wrong, and seem quite keyed

up. There is no lagging in moving from base to base. The child who is now writing "Thursday" stops to think after the first letter and the children snicker. He stops after another letter. More snickers. He gets the word wrong. There are frequent signs of joy from the children when their side is right. (Henry 1965, pp. 297–298)

According to Henry, learning to spell is not the most important lesson that students learn from this exercise. They are also learning important cultural values from the way in which spelling is being taught; that is, they are learning to fear failure and to envy success. In exercises such as spelling baseball, "failure is paraded before the class minute upon minute" (p. 300), and success is achieved after others fail. And, "since all but the brightest children have the constant experience that others succeed at their expense they cannot but develop an inherent tendency to hate—to hate the success of others" (p. 296).

In an exercise such as spelling baseball, students also learn to be "absurd." To be absurd, as Henry defines it, means to make connections between unrelated things or events, to do assignments without understanding the purpose, to do assignments that have no purpose. From Henry's point of view, spelling baseball teaches students to be absurd, because no logical connection exists between learning to spell and baseball.

If we reflect that we do not settle a baseball game by converting it into a spelling contest, we see that "baseball is bizarrely irrelevant to spelling" (p. 300). Yet, most students participate in classroom exercises such as spelling baseball without questioning their purpose. Although some children may ask, "Why are we doing this? What is the point?" and may be told, "So you can learn to spell" or "To make spelling fun," few children challenge the purpose of this activity further. Students go along with the teacher's request and play the game as if spelling is related to baseball because, according to Henry, they are terrified of failure and want so badly to succeed.

The Lessons of Spelling Baseball. Henry argues that classroom activities such as spelling baseball prepare students to fit into a competitive and consumption-oriented culture. Because the U.S. economy depends on consumption, the country benefits if its citizens purchase nonessential goods and services. The assignments that children undertake in school do not prepare them to question false or ambiguous statements made by advertisers; schools do not properly prepare demanding individuals to "insist that the world stand up and prove that it is real" (p. 49). Henry argues that this sort of training—this hidden curriculum—makes possible an enormous amount of selling that otherwise could not take place:

> In order for our economy to continue in its present form people must learn to be fuzzy-minded and impulsive, for if they were clear-headed and deliberate, they would rarely put their hands

Jules Henry argues that classroom activities such as spelling baseball prepare Americans to fit into a competitive and consumption-oriented culture. Henry argues that this sort of learning makes possible an enormous amount of selling that otherwise could not take place.

Teachers everywhere teach two curricula simultaneously: the subject matter and the hidden lessons conveyed through such things as tone of voice and teaching method.

in their pockets. . . . If we were all logicians the economy [as we know it] could not survive, and herein lies a terrifying paradox, for in order to exist economically as we are we must . . . remain stupid. (p. 48)

According to Henry, students who have the intellectual strength to see through absurd assignments such as spelling baseball and who find it impossible to accept such assignments as important may rebel against the system, refuse to do the work, drop out, or eventually think of themselves as stupid. We cannot know how many students perform poorly in school because they cannot accept the manner in which subjects are taught.

The Third International Mathematics and Science Study, which reviewed math instruction in nationally representative samples of German, Japanese, and American classrooms, lends support to Henry's idea that U.S. students do assignments with no purpose.

Hidden Curriculum in Math Lessons. For this study, researchers filmed 100 German, 50 Japanese, and 81 U.S. math teachers teaching the subject matter. The findings reveal the "hidden lessons" about the importance of math that teachers convey by the way they teach the subject. The

researchers rated almost 90 percent of math lessons taught by U.S. teachers as low in quality compared with 40 percent of German teachers' lessons and 13 percent of Japanese teachers' lessons. No math lesson taught by American teachers was rated as high quality; 23 percent of German lessons were rated as high quality and 30 percent of Japanese lessons.

Researchers identified a number of areas where American teachers fell short: (1) they spent too little time on individual math topics and covered too many topics; (2) they taught lessons in a very dry way, essentially showing students how to do a problem and then assigning a long list of similar problems to complete in class or for homework; and 3) they presented problems that were of little interest to students. In contrast, German and Japanese math teachers spent longer amounts of time on each topic and systematically built upon it. They also presented and assigned more-interesting and more-realistic problems:

They give kids problems in a more realistic way. You have this piece of property. You have to divide it in half. How would you do it? And then they have the kids sit there and really struggle. It's not an easy answer. It's something that they may have to work out for 20 minutes, not sure they have the right answer. Then they have a class discussion where kids will get up and say, "I think this is the way to do it." That kind of discussion gets kids emotionally involved in the answer. Brain research says the more emotional you are about something that you're learning, the better you're going to remember it. So those kids are getting excited. They're fighting over their answer. Then the teacher finally gets up and says, "Well, I think this is the best way. Now watch while I do this." And all the kids that have been sitting there struggling suddenly say, "He's right! That is the right way." And they remember it as a result (Online Newshour 1996).

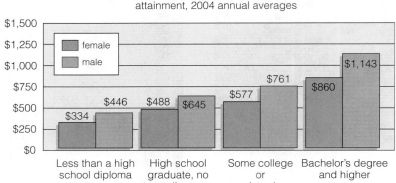

Figure 13.9 Average Weekly Earnings of Full-Time Workers Age 25 and Older by Sex and Educational Attainment, 2004

Textbooks. Perhaps the dryness of math lessons and other assignments can be traced to an American invention—the textbook. Early American education reformers, such as Benjamin Rush, Thomas Jefferson, and Noah Webster, insisted that schools represented an important mechanism by which a diverse population could acquire a common culture. They believed that the new, "perfectly homogeneous" American was one who studied at home (not abroad) and who used American textbooks. To use Old World books "would be to stamp the wrinkles of decrepit age upon the bloom of youth" (Webster, quoted in Tyack 1996, p. 32).

The first textbooks in the United States were modeled after catechisms, short books covering religious principles written in question-and-answer format. Each question had one answer only; in repeating this answer, the student strictly adhered to the question's wording (Potter and Sheared 1918). This format discouraged readers from behaving as active learners "who could frame questions, interpret materials, and reflect on the significance of what was presented. . . . No premium was placed on generating and inventing ideas or arguing about the truth or value of what others had written" (Resnick 1990, p. 18). The reader's job was to memorize the "right" answers for the questions.

Under this model, not surprisingly, textbooks tended to be written so that the primary reason to read them was to find the "right" answers to the accompanying questions. The influence of catechisms on learning today becomes evident whenever students are assigned to read a chapter and answer a list of questions or work a group of problems at the end. Many students discover that they do not need to read the material to answer the questions or work the problems at the end; instead, they can simply skim the text until they find key words or formulas that correspond to the question wording, and then copy the surrounding sentences or apply the formulas. In the case of math, most American students come to learn through the hidden curriculum that to succeed they simply have to apply the formulas correctly; since the problems are not meaningful, then math must not be.

The Promise of Education

■ **CORE CONCEPT 8:** All education systems promise that increased opportunities come with education. Sometimes these opportunities do not materialize. Most Americans (as early as first grade) are taught to equate education with increased job opportunities and higher salaries. This belief has a basis in reality. Figure 13.9 shows that income rises with the level of education obtained. On average, both males and females with four-year college degrees earn more than their same-sex counterparts who have less education. On the other hand, earnings are still affected by one's gender and other characteristics, such as race (see Chapters 9 and 10).

In the United States the connection between a college education, job opportunities, and higher salaries is not always realized on a personal level. As many as 38 percent of college graduates are underemployed; that is, they do not believe a college degree is required for the job they are in four years after graduation. To complicate matters, many workers without a four-year college degree earn more than the average college graduate. For example, 62 percent of workers without a bachelor's degree earn less than $572 per week—but so do 23 percent of college graduates (Mariani 1999). Sociologist John Reynolds points out that American parents and high school counselors send the message that a college degree is the only option for obtaining a well-paying job, when in fact skilled electricians and plumbers earn more than many college-educated professionals (Reuters 2006).

THE PROMISE OF EDUCATION

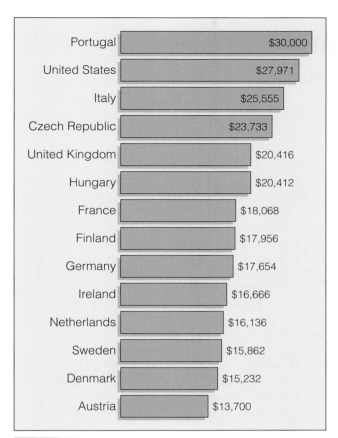

▲ **Figure 13.10** Average Earnings of College-Educated Workers for Every $10,000 Earned by Workers with Less Than a High School Education (U.S. $)

Source: OECD (2003).

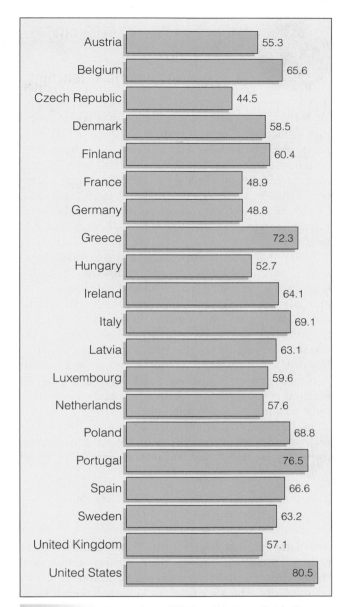

▲ **Figure 13.11** Percentage of 15-Year-Olds Expecting to Have a High-Skill, White-Collar Job by Age 30, United States and Selected EU Countries

Source: OECD (2003).

Figure 13.10 shows that the financial rewards associated with a college education relative to the rewards gained by workers who do not complete high school vary by country. In Portugal, for every $10,000 workers with less than a high school education earn, college-educated workers earn $30,000 (a difference of $20,000). In Austria college-educated workers earn $13,700 for every $10,000 earned by workers with less than a high school education (a difference of $3,700). In the United States the difference is $17,971.

While education does not always deliver promised opportunities in EU countries, it does seem that U.S. students have less realistic expectations than their European counterparts regarding career opportunities that come with education. Figure 13.11 shows that 80 percent of American 15-year-olds expect to occupy a high-skilled, white-collar job by age 30. Between 44.5 and 76.5 percent of EU 15-year-olds hold such expectations. Of course, it is impossible for 80 percent of all 30-year-olds to hold high-skilled, white-collar jobs—if only because not enough of those positions exist to meet the expectations. Therefore,

we might speculate that as those American 15-year-olds grow older, many learn that, regardless of their educational achievement, a high-status, well-paying, white-collar job is not guaranteed. Consequently, many feel ambivalence toward the system of education, which they may perceive as letting them down.

One might hypothesize that feelings of ambivalence are lower among EU students because their education systems seem to promote more-realistic expectations. Recall that 60 percent of French students attend vocational schools.

INTERSECTION OF BIOGRAPHY AND SOCIETY

An American Teaching in French Vocational Schools

WHILE IN FRANCE I taught English at two technical high schools or, as we call them in the good old US of A, vocational. My students were learning to be electricians, plumbers, mechanics, and so forth, having been engaged in these pursuits for years. In such circumstances, learning English is simply not a priority for them. (On the other hand, anyone wishing to become involved in business, finance, politics, etc., necessarily has to become fluent in English.)

Vocational schools are common in the socialist-oriented French education system. Although technical schools provide a solid education and valuable training and guarantee a respectable occupation upon completion, there is something about them that is unsettling to the American psyche. Socialized from birth to realize the unequivocal ambition of the American dream, every red-blooded American child strives to become a self-made person, to go from rags to riches and ascend to the highest echelons of society, borne by the winds of their own ingenuity, vitality, and determination.

I learned that such fantastical dreams are not, in general, shared by the French by way of one of the first assignments I presented to my students. They were supposed to write a short paragraph on how they see themselves in ten years. They wrote things like, "in ten years I will be a plumber living in Nice with my wife. Hopefully we'll have enough money to buy the latest model Peugeot or Fiat." Or, "I will be fixing air conditioners with my brother and cousins in Antibes." They were supremely practical, without a trace of far-flung expectations. I am quite certain that if the same question were put to American teenagers, one would have a class full of future athletes, film stars, and millionaire businesspeople. An even safer bet, I think, is that there would be no mention of Peugeots or Fiats. Perhaps this distinction implies a greater maturity on the part of the French pupils but, at the risk of revealing my hopeless Americanism, I could not help but feel a little sorry for them.

On the whole I was consistently impressed with the French students' enthusiasm, certainly not for learning English, but for life in general. They were always upbeat, and exhibited, almost to a student, a lively and good-natured if extremely mischievous sense of humor.

Source: Christopher Brown, Northwestern University, Class of 2004.

Anecdotal evidence suggests that French students make realistic projections about their future occupational status. Still, unlike their American vocational school counterparts, who are labeled as "failures," French vocational students study English as a second language and are encouraged to be engaged citizens (see "Intersection of Biography and Society: An American Teaching in French Vocational Schools").

Demands on Schools and Teachers

■ **CORE CONCEPT 9: Schools are a stage on which society's crises play out.** Every government seems to think its education system is failing in major ways. Singapore, which boasts the best math students in the world, believes it needs to improve before others catch up; Japan worries that its exam system puts too much pressure on students; India and China worry that their growing armies of engi-neers and customer sales forces have not had enough humanities courses (Friedman 2006). The United States is no exception. Throughout U.S. history, it seems as if public education has always been in a state of crisis and under reform. Between 1880 and 1920, for example, the public was concerned about whether schools were producing qualified workers for the growing number of factories and businesses and whether the schools were instilling a sense of national identity (that is, patriotism) in immigrant and ethnically diverse populations. When the United States became involved in major wars—World War I, World War II, the Korean conflict, and the Vietnam War—the public voiced concerns about whether the schools were turning out recruits physically and mentally capable of defending the American way of life.

When the Soviet Union launched its Sputnik satellites in the mid-1950s, Americans were forced to consider the possibility that their public schools were not educating students in mathematics and sciences as well as the Soviet schools were. In the 1960s, civil rights events forced Americans to question whether they were offering chil-

On average, American teachers spend more than 1,000 hours each school year interacting with students. Their EU counterparts spend 553–750 hours.

dren from less advantaged ethnic groups and social classes the knowledge and skills that would allow them to compete economically with children from more advantaged groups.

Beginning in the late 1970s and continuing throughout the 1980s and 1990s, many critics both at home and abroad maintained that the U.S. system of education was inadequate for meeting the challenges of competing in a global economy. The system was deemed so inadequate that this warning was issued: "If an unfriendly foreign power had attempted to impose on America the mediocre education performance that exists today, we might well have viewed it as an act of war (National Commission on Excellence in Education 1983, p. 5). Many employers claimed that they were unable to find enough workers with the level of reading, writing, mathematical, and critical thinking skills needed to function adequately in the workplace. Reform efforts focused on increasing "seat time," the amount of time students devoted to academic activities, either in school or at home. The rationale came from findings that U.S. students devoted considerably less time to their academic activities than did their foreign counterparts.

In the 21st century the focus is on leaving no child behind and on educating workers for a global economy. There is also an emphasis on accountability or measuring effectiveness. Testing students and teachers is presented as the most cost-effective method of holding students, schools, and parents accountable to a specified standard.

The ongoing nature of the education "crisis" in the United States suggests that schools are very visible and highly vulnerable, and that they are a "stage on which

a lot of cultural crises get played out" (Lightfoot 1988, p. 3). Inequality, racial segregation, poverty, chronic boredom, family breakdown, unemployment, and illiteracy are "crises" that transcend the school environment. Yet, we confront them whenever we go into the schools (Lightfoot 1988). Consequently, the schools seem to be both a source of and a solution for our problems. Of course, this relentless focus on schools invites a corresponding focus on teachers (see Figure 13.12).

Teachers' jobs are complex; teachers are expected to undo learning disadvantages generated by inequalities in the larger society and to handle an array of discipline problems. A recent EU report on teacher workload pointed out that "stress seems to be intrinsic to teaching." The report named 45 sources of stress, including stress generated by mandated changes in curriculum, inadequate salaries, increased class size, lack of parental involvement, excessive paperwork, and lack of student motivation, attention, or interest. High stress is connected to absenteeism, turnover, and poor teaching (Education International and European Trade Union Committee for Education 2007). Likewise, U.S. teachers name uncompleted homework assignments, cheating, stealing, drugs and alcohol, truancy, and absenteeism as very serious or fairly serious problems in their school.

Compared with their counterparts in Germany, U.S. teachers more often name "uninterested students," "uninterested parents," "low student morale," "tardiness," and "intimidation or verbal abuse of teachers/staff" as problems that limit teaching effectiveness and disrupt the learning environment. Unlike their EU counterparts, U.S. teachers work in environments that place high value on school sports, and thus on athletic achievement—which, according to sociologist James Coleman, is central to the values underlying the formation of adolescent subcultures.

The Adolescent Subculture

■ **CORE CONCEPT 10:** Because of the high value placed on school-sponsored sports teams, the adolescent society in the United States penalizes academic achievements while rewarding boys for athletic achievements and girls for achieving popularity with boys, especially with boys who are athletes. Around the turn of the 20th century—the early decades of late industrialization—fewer than 10 percent of teenagers 14 to 18 years old attended high school in the United States. Young people attended elementary school to learn the three Rs; they then learned the skills needed to make a living from their parents or from their neighbors. As the pace of industrialization increased, jobs began to move away from the home and the neighborhood and into factories and office buildings. Parents no longer trained their children, because the skills they knew were becoming obsolete. Children therefore came to expect that they would not make a living in the same way as their parents.

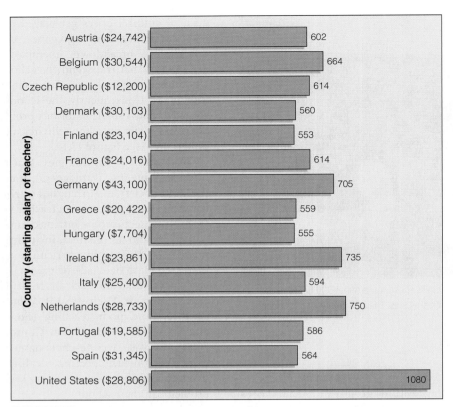

▲ **Figure 13.12** Contact Time in Hours per Academic Year for Upper Secondary School Teachers in EU Countries and the United States

The number of hours teachers spend in direct contact with students is one rough indicator of the relative demands and workloads put on teachers. We see from the bar graph that U.S. teachers spend 330–527 more hours per year in direct contact with high school students than their EU counterparts.

The bars above show the annual number of hours involving direct contact between high school students and teachers in the United States and EU countries. Note that the United States is the only country in which teachers average more than 1,000 contact hours with students. To the left of the bars is a list of the average starting teachers' salaries by country. Note that the average starting salary of U.S. teachers ranks fifth among the 16 salaries listed.

Source: OECD (2006).

In short, as the economic focus in the United States shifted from predominantly farm and small-town work environments to the factory and the office, the family became less involved in the training of its children and, by extension, less involved in children's lives. The transfer of work away from the home and neighborhood removed opportunities for parents and children to work together. Under this new arrangement, family occasions became events that were consciously arranged to fit everyone's work schedule.

According to Coleman, this shift in training from the family to the school cut adolescents off from the rest of society and forced them to spend most of the day with members of their own age group. Adolescents came "to constitute a small society, one that has most of its impor-

tant interactions within itself, and maintains only a few threads of connection with the outside adult society" (Coleman, Johnstone, and Jonassohn 1961, p. 3).

Coleman surveyed students from 10 high schools in the Midwest to learn about adolescent society. He selected schools representative of a wide range of environments: Five schools were in small towns, one in a working-class suburb, one in a well-to-do suburb, and three in cities of varying sizes. Also, one of the schools was an all-male Roman Catholic school. Coleman was interested in the adolescent status system, a classification of achievements resulting in popularity, respect, acceptance into the crowd, praise, awe, and support, as opposed to isolation, ridicule, exclusion from the crowd, disdain, discouragement, and disrespect.

U.S. Marines in Japan photo by Lance Cpl. Cathryn D. Lindsay

Tim Lindenbaum

Coleman's research on adolescent subcultures, published in 1961, is considered a classic study. Do his conclusions about the importance for boys of sports and the importance for girls of success with boys (accomplished through being a cheerleader or being good-looking) still hold true in U.S. high schools?

To learn about this system, he asked students questions similar to the following:

- How would you like to be remembered—as an athlete, as a brilliant student, as a leader in extracurricular activities, or as the most popular student?
- Who is the best athlete? The best student? The most popular? The boy the girls go for most? The girl the boys go for most?
- Which person in the school would you like most to date? To have as a friend? What does it take to get in with the leading crowd in this school?

Based on the answers to these and other questions, Coleman identified a clear pattern common to all 10 schools. "Athletics was extremely important for the boys, and social success with boys [accomplished through being a cheerleader or being good-looking] was extremely important for girls" (1960, p. 314).

Coleman found that girls in particular did not want to be considered as good students, "for the girl in each grade in each of the schools who was most often named as best student has fewer friends and is less often in the leading crowd than is the boy most often named as best student"

(Coleman 1960, p. 338). A popular boy could often be a good student or dress well or have enough money to meet social expenses, but to be truly admired he also had to be a good athlete. Coleman also found that the peer group had more influence over and exerted more pressure on adolescents than did their teachers, and a significant number of adolescents were influenced more by the peer group than by their parents.

Why does the adolescent society penalize academic achievement in favor of athletic and other achievements? Coleman maintained that the manner in which students are taught contributes to their lack of academic interest: "They are prescribed 'exercises,' 'assignments,' 'tests,' to be done and handed in at a teacher's command" (1960, p. 315). This type of academic work does not require creativity but rather conformity.

Students show their discontent by choosing to become involved in and to acquire things they can call their own: athletics, dating, clothes, cars, and extracurricular activities. This reaction is inevitable given the passive roles that students play in the classroom:

> One consequence of the passive, reactive role into which adolescents are cast is its encouragement of irresponsibility. If a group is given no authority to make decisions and take action on its own, the leaders need show no responsibility to the larger institution.
>
> Lack of authority carries with it lack of responsibility; demands for obedience generate disobedience as well. But when a person or group carries the authority for his own action, he

carries responsibility for it. In politics, splinter parties which are never in power often show little responsibility to the political system; a party in power cannot show such irresponsibility. . . . An adolescent society is no different from these. (Coleman 1961, p. 316)

Athletics is one of the major avenues open to adolescents, especially males, in which they can act "as a representative of others who surround [them]" (1961, p. 319). Others support this effort, identify with the athletes' successes, and console athletes when they fail. Athletic competition between schools generates an internal cohesion among students as no other event can. "It is as a consequence of this that the athlete gains so much status: he is doing something for the school and the community" (p. 260). Coleman argues that because athletic achievement is so widely admired, everyone with some athletic ability will try to develop it. In contrast, because academic pursuits go relatively unrewarded, "those who have the most ability may not be motivated to compete" (p. 260). This reward structure suggests that the United States does not draw into the competition everyone who has academic potential.

Coleman's findings should deliver the message that the peer group represents a powerful influence on learning, but they should not leave the impression that the peer group's world does not overlap with the family or the classroom. In fact, it seems more appropriate to consider the interrelationships of the multiple contexts of students' lives. From data collected during interviews and observations of 54 ethnically and academically diverse youth in four urban desegregated high schools in California, educators Patricia Phelan, Ann Locke Davidson, and Hanh Cao Yu (1991, 1993; Phelan and Davidson 1994) generated a model of the interrelationships between students' family, peer, and school worlds. The Students' Multiple Worlds Model describes students' perceptions of boundaries and borders between these three worlds the adaptation strategies students employ as they move from one context to another. In this regard researchers uncovered four patterns that students follow as they move between school, family, and peers:

1. *Congruent Worlds/Smooth Transitions*—peer groups, classrooms, and family are harmonious and uncomplicated, as shared values, beliefs, and expectations override differences among the three groups.
2. *Different Worlds/Border Crossings Managed*—the three groups do not share similar beliefs, values, and expectations but adolescents have acquired successful strategies for adjusting and reorienting when moving between peer groups, school, and family.
3. *Different Worlds/Border Crossings Resisted*—crossing boundaries between family, peers, and school is painful for these students, who only feel comfortable in one of the three contexts and dismiss the other two as irrel-

evant. (Phelan, Davidson, and Yu 1991, 1993; Phelan and Davidson 1994)
4. *Different Worlds/Border Crossing Difficult*—family, peer, and school worlds are viewed as totally distinct from another and adolescents have no strategies for managing border crossing.

Family Background

■ **CORE CONCEPT 11: Family background is the most important factor in explaining academic success.** Coleman (1966) was the principal investigator behind the congressionally-mandated report *Equality of Educational Opportunity*, popularly known as the Coleman Report. Coleman's research focused on the degree to which public education was segregated and on inequalities of educational opportunity in the United States. Coleman and six colleagues surveyed 570,000 students and 60,000 teachers, principals, and school superintendents in 4,000 schools across the United States.

Students filled out questionnaires about their home background and educational aspirations and took standardized achievement tests intended to measure verbal ability, nonverbal ability, reading comprehension, mathematical ability, and general knowledge. Teachers, principals, and superintendents answered questionnaires about their backgrounds, training, attitudes, school facilities, and curricula.

Coleman found that a decade after the Supreme Court's famous desegregation decision in 1954 (*Brown v. Board of Education*), U.S. schools were still largely segregated: 80 percent of white children attended schools that were 90 to 100 percent white, and 65 percent of black students attended schools that were more than 90 percent black. Almost all students in the South and the Southwest attended schools that were 100 percent segregated. Although Mexican Americans, Native Americans, Puerto Ricans, and Asian Americans also attended primarily segregated schools, they were not segregated from whites to the degree that blacks were. The Coleman Report also found that white teachers taught black children, but that black teachers did not teach whites: approximately 60 percent of the teachers who taught black students were black, whereas 97 percent of the teachers who taught white students were white. When the characteristics of teachers who taught white student were compared with those of teachers who taught black students, the study found no significant differences in professional qualifications (as measured by degree, major, and teaching experience).

Coleman did find sharp differences across racial groups with regard to verbal ability, nonverbal ability, reading comprehension, mathematical achievement, and general information as measured by the standardized tests. The white students scored highest, followed by Asian Americans, Native Americans, Mexican Americans, Puerto Ricans, and African Americans.

Contrary to Coleman's expectations, on average no significant differences in quality existed between schools attended predominantly by the various racial groups and schools attended by whites. (Quality was measured in terms of age of buildings, library facilities, laboratory facilities, number of books, class size, expenditure per pupil, extracurricular programs, and the characteristics of teachers, principals, and superintendents.) Surprisingly, any variations in school quality that Coleman found did not have much effect on the students' test scores.

Test scores were affected, however, by family background and related attributes. The average minority student was likely to come from an economically and educationally disadvantaged household and was likely to attend school with similarly disadvantaged students. Fewer of his or her classmates would complete high school, maintain high grade-point averages, enroll in college preparatory curricula, or be optimistic about their future.

The Coleman Report examined the progress of disadvantaged blacks who had participated in school integration programs and found that their test scores were higher than those of their disadvantaged counterparts who did not participate. Clearly the important variable is disadvantaged environment and not student's race.

Based on his research Coleman concluded that schools have little influence on academic achievement that is independent of students' background and the general social context, and this very lack of an independent effect means that the inequalities imposed on children by their home, neighborhood, and peer environment are carried along to become the inequalities with which they confront adult life (1966, p. 325).

Coleman's findings were used to support busing as a means of achieving educational equality. Although Coleman initially supported this policy, he later retracted his endorsement, after busing hastened "white flight," or the migration of middle-class white Americans from cities to suburbs. This migration merely worsened the racial segregation in city and suburban schools.

As the number of white students dropped sharply in relation to that of black students, the positive effects of desegregation proved to be short-lived. Consequently, economically and educationally disadvantaged blacks were sent from their deficient schools into equally deficient lower-class and lower-middle-class white neighborhoods. Coleman adamantly maintained that court-ordered busing alone could not achieve integration: "With families sorting themselves out residentially along economic and racial lines, and with schools tied to residence, the end result is the demise of the common school attended by children from all economic levels. In its place is the elite suburban school, . . . the middle-income suburban school, the low-income suburban school, and the central-city schools of

Tina Collins, Fort Lee School Liaison Officer/U.S. Army

School segregation appears to have changed very little over the past four decades. In 1968 the federal government reported that 76 percent of black students and 55 percent of Hispanic or Latino students attended predominantly minority schools. More than thirty years later, the Harvard Project on School Desegregation found those figures to be 70 percent for blacks and 76 percent for Hispanics or Latinos. At the Walnut Hill Elementary School in Petersburg, Virginia, 92 percent of the students are classified as black (greatschools.net 2009).

several types—low-income white schools, middle-income white schools, and low- or middle-income black schools" (1977, pp. 3–4).

The findings of this study do not imply that students are trapped by their family background. Indeed, Coleman never claimed that family background explains all of the variations in test scores. He did note, however, that it was the *most important factor* in his study. Coleman's findings are very relevant today; four decades after they were published, school-based segregation is still a problem. In 1968 the federal government reported that 76 percent of black students and 55 percent of Hispanic or Latino students attended predominantly minority schools (schools in which 50 percent or more of the students are black, Asian, Native American, and/or Hispanic or Latino). Today those figures are 70 percent for blacks and 76 percent for Hispanics or Latinos. Illinois, Michigan, New York, and California have been identified as the most segregated states; more than 50 percent of their schools are composed of 90 to 100 percent minority students. As in the 1960s, black and other minority students are significantly more likely to find themselves in schools in which overall academic achievement is undervalued and low (Celis 1993; Civil Rights Project 2001; Schemo 2001).

Jason Clarke

A number of studies indicate that the home environment is an important variable—though not the only one—in explaining academic success.

Segregating minority student populations is not unique to the United States. In the European Union, first- and second-generation immigrants—who are likely to be Muslims from Africa, the Middle East, or Asia—"often attend schools with economically, socially, and culturally disadvantaged student populations." In fact, 25 percent of second-generation immigrant students are enrolled in schools in which at least 50 percent of the student body is second-generation (OECD 2006).

Likewise, many studies confirm the importance of family background in educational achievement (Hallinan 1988). For example, the International Association for the Evaluation of Educational Achievement (2003) tested students in 22 countries (13 of which belong to the EU) on six subjects. It found that the "home environment is a most powerful factor in determining the level of school achievement of students, student interest in school learning, and the number of years of schooling the children will receive" (Bloom 1981, p. 89; Ramirez and Meyer 1980). Nevertheless, in this international study, in the Coleman study, and in other studies, home background (defined in terms of parents' race, income, education, and occupation) explains

▲ **Figure 13.13** Literacy Scores by Highest Educational Level of Either Parent

The data below show that students with at least one university-educated parent do better on reading literacy tests than students with parents who completed lower secondary education. In the United Kingdom, the test score gap between the two groups is only 15 points; in the Netherlands, the gap is 137 points. Among the students with at least one university-educated parent, the test scores range from a low of 518 to a high of 603 (an 85-point difference). Among students with two relatively uneducated parents, the scores range from a low of 455 to a high of 588 (a 133-point difference). British students with parents who did not complete secondary education score higher than students with at least one university-educated parent in Cyprus, the Czech Republic, Germany, Greece, Hungary, Italy, the Slovak Republic, and Sweden.

Country	At Least One Parent Completed University-Level Education		Parents Completed Lower Secondary Education Only		Gap Between Two Groups
	Percentage of Students	Average Score	Percentage of Students	Average Score	
Cyprus	38	518	12	458	60
Czech Republic	22	569	4	499	70
Germany	27	575	12	501	74
Greece	23	572	12	501	71
Hungary	30	582	8	492	90
Italy	18	569	30	520	49
Netherlands	12	592	45	455	137
Slovak Republic	19	556	4	465	91
Sweden	35	583	6	525	58
United Kingdom	36	603	35	588	15

Source: OECD (2003).

only about 30 percent of the variation in students' achievement (see Figure 13.13). Consequently, factors other than socioeconomic status must affect academic performance:

In most, if not all, societies, children and youth learn more of the behavior important for constructive participation in the society outside of school than within. This fact does not diminish the importance of school but underlines the nation's dependence on the home, the workplace, community institutions, the peer group, and other informal settings to furnish a major part of the education required for a child to be successfully inducted into society. Only by clear recognition of the school's special responsibilities

can it be highly effective in educating its students (Tyler, quoted in Purves 1974, p. 74c). Nevertheless, the 2002 No Child Left Behind Act mandated that schools take responsibility for student failures regardless of the social, economic, physical, or intellectual disadvantages the students bring with them to school (Schemo 2006).

In view of these findings, schools have a special responsibility not to duplicate the inequalities outside the school. Over the past three decades, researchers have found that "schools exert some influence on an individual's chances of success, depending on the extent to which they provide equal access to learning" (Hallinan 1988, pp. 257–258).

■ VISUAL SUMMARY OF CORE CONCEPTS

■ CORE CONCEPT 1: Education includes the formal and informal experiences that train, discipline, and shape the mental and physical potentials of the maturing person.

Education includes the experiences that train, discipline, and shape the mental and physical potentials of the maturing person. Sociologists make a distinction between formal and informal education. Informal education occurs in a spontaneous, unplanned way; it is not designed by someone to stimulate specific thoughts or to impart specific skills. Formal education encompasses a purposeful, planned effort aimed at imparting specific skills or information. Formal education is considered a success when those being instructed acquire the skills, thoughts, and information that those designing the experience seek to impart.

Gaijin Bikers

■ CORE CONCEPT 2: Schools perform a number of important social functions that, ideally, contribute to the smooth operation of society.

Schools perform a number of important functions that serve the needs of society and contribute to its smooth operation. These functions include transmitting skills, liberating minds, facilitating personal change and reflection, performing basic and applied research, integrating diverse populations, solving social problems, and screening and selecting the most qualified students for the most socially important careers. Schools perform other, not so obvious functions; they function as reliable babysitters and as dating pools and marriage markets.

Joe Mabel

■ **CORE CONCEPT 3:** Any analysis of school systems must focus on unequal power arrangements that allow more-powerful individuals and groups to exert their will over those with less power.

Schools are not perfect: not all minds are liberated; students drop out, refuse to attend, or graduate with skill deficiencies; schools misclassify students; and so on. The point is that the functionalist analysis best describes ideals, which some school systems realize better than others. The conflict perspective draws our attention to the dynamics among dominant and subordinate groups and to unequal power arrangements in which those with more wealth and power can exercise their will over others with less status.

Department of Defense

■ **CORE CONCEPT 4:** Even in societies where schooling is compulsory, a significant percentage of the population does not possess skills needed to succeed in the society in which they live.

People are considered functionally illiterate when they do not possess the level of reading, writing, calculating, and other skills needed to succeed (that is, function) in the society in which they live. This point suggests that illiteracy is a product of one's environment. Functional illiteracy includes the inability to use a computer, read a map to find a destination, make change for a customer, read traffic signs, follow instructions to assemble an appliance, fill out a job application, and comprehend the language others are speaking. If we focus on languages, of which there may be as many as 9,000, we can see that the potential number of illiteracies is overwhelming, as people cannot possibly be literate in every language.

Department of Defense

■ **CORE CONCEPT 5:** Only a handful of countries in the world give a significant share of the population the opportunity to attend college.

The United States is one of a small group of countries where the college-educated constitute a significant share of the population. Almost 40 percent of 25- to 34-year-olds in the United States have at least a college education. In some EU countries the percentage is about the same, but in others—notably, Italy (12 percent) and the Czech Republic (11 percent)—the percentage is much smaller. One distinctive feature of the U.S. education system is that, in theory, anyone can attend a college if he or she has graduated from high school or has received a general equivalency degree (GED). For another indicator of the relatively easy access to college in the United States, consider that many four-year colleges and universities accept students regardless of what courses they took, what grades they earned in high school, or what scores they received on the ACT or SAT.

White House photo by Paul Morse

■ **CORE CONCEPT 6:** Most, if not all, educational systems sort students into distinct instructional groups according to similarities in past academic performance, performance on standardized tests, or even anticipated performance.

Schools arrange students into distinct instructional groups according to similarities in past academic performance, performance on standardized tests, or even beliefs about how some in a particular sex, race, or other category are projected to perform. This practice may be known as tracking, ability grouping, streaming or some other name. Advocates of tracking argue that students learn better when they are grouped with those who learn at the same rate; slow learners develop more-positive attitudes when they do not have to compete with the more academically capable; and groups of students with similar abilities are easier to teach than students of various abilities. Research suggests that tracking has a positive effect on high-track students, a negative effect on low-track students, and no noticeable effects on middle-track or regular-track students. While EU countries track students according to academic abilities, they do not offer students in lower tracks a watered-down or simpler version of the subject matter.

Christopher Lim Mu Yao

■ **CORE CONCEPT 7:** Teachers everywhere teach two curricula simultaneously: a formal one and a hidden one.

The various academic subjects—mathematics, science, English, reading, physical education, and so on—make up the formal curriculum. The hidden curriculum consists of the other activities that go on as students learn the subject matter and of the "lessons" that those other activities convey about the value and meaning of what the students are learning. The teaching method, the types of assignments and tests, the tone of the teacher's voice, the attitudes of classmates, the number of students absent, the frequency of the teacher's absences, and the number of interruptions

Department of Defense photo by Lisa Derek M. Poole, USN

during a lesson are all factors in the classroom as students learn the formal curriculum. These so-called extraneous factors function to convey messages to students not only about the value of the subject, but also about the values of society, the place of learning in their lives, and their role in society.

■ **CORE CONCEPT 8:** All education systems promise that increased opportunities come with education. Sometimes these opportunities do not materialize.

Most Americans (as early as first grade) are taught to equate education with increased job opportunities and higher salaries. This belief has a basis in reality. On average, both male and female college graduates earn more than their same-sex counterparts who have less education. On the other hand, earnings are still affected by one's gender and other characteristics, such as race. In the United States the connection between a college education, job opportunities, and higher salaries is not always realized on a personal level. As many as 38 percent of college graduates are underemployed. To complicate matters, many workers without a four-year college degree earn more than the average college graduate. The financial rewards associated with a college education relative to the rewards gained by workers who do not complete high school vary by country.

Department of Defense

■ **CORE CONCEPT 9: Schools are a stage on which society's crises play out.**

The ongoing nature of the education "crisis" in the United States suggests that schools are very visible and highly vulnerable, and that they are a "stage on which cultural crises get played out" (Lightfoot 1988, p. 3). Inequality, racial segregation, poverty, chronic boredom, family breakdown, unemployment, and illiteracy are "crises" that transcend the school environment. Yet, we confront them whenever we go into the schools. Consequently, the schools seem to be both a source of and a solution for our problems. In the case of racial inequality, segregated schools contributed to that inequality; integrating the schools became a solution, but not without controversy, as the 1959 protest against the integration of Central High School (Little Rock, Arkansas) showed. Of course, this relentless focus on schools increases demands made of teachers, who are expected to undo learning disadvantages generated by inequalities in the larger society and to handle an array of discipline problems.

■ **CORE CONCEPT 10: Because of the high value placed on school-sponsored sports teams, adolescent society in the United States penalizes academic achievements while rewarding boys for achieving athletic success and girls for achieving popularity with boys, especially with boys who are athletes.**

Sociologist James Coleman defined the adolescent subculture as a "small society, one that has most of its important interactions within itself, and maintains only a few threads of connection with the outside adult society." In this society "athletics was extremely important for boys, and social success with boys [accomplished through being a cheerleader or being good-looking] was extremely important for girls." Coleman also found that the peer group had more influence over and exerted more pressure on adolescents than did their teachers, and a significant number of adolescents were influenced more by the peer group than by their parents. Students show their discontent with school by choosing to become involved in and to acquire things they can call their own: athletics, dating, clothes, cars, and extracurricular activities.

■ **CORE CONCEPT 11: Family background is the most important factor in explaining academic success.**

Family background is the *most important factor* in Coleman's study. Many studies confirm the importance of family background in educational achievement. For example, the International Association for the Evaluation of Educational Achievement (2003) tested students in 22 countries (13 of which belong to the EU) on six subjects. It found that the "home environment is a most powerful factor in determining the level of school achievement of students, student interest in school learning, and the number of years of schooling the children will receive" (Bloom 1981, p. 89; Ramirez and Meyer 1980). In most, if not all, societies, children and youth learn more of the behavior important for constructive participation in the society outside of school than within. This fact does not diminish the importance of school but underlines the nation's dependence on the home, the workplace, community institutions, the peer group, and other informal settings to furnish a major part of the education required for a child to be successfully inducted into society. Only by clear recognition of the school's special responsibilities can it be highly effective in educating its students.

Resources on the Internet

 Sociology: A Global Perspective Book Companion Web Site

www.cengage.com/sociology/ferrante

Visit your book companion Web site, where you will find flash cards, practice quizzes, Internet links, and more to help you study.

CENGAGENOW™

Just what you need to know NOW!

Spend time on what you need to master rather than on information you have already learned. Take a pre-test for this chapter, and CengageNOW will generate a personalized study plan based on your results. The study plan will identify the topics you need to review and direct you to online resources to help you master those topics. You can then take a post-test to help you determine the concepts you have mastered and what you will need to work on. Try it out! Go to www.cengage.com/login to sign in with an access code or to purchase access to this product.

Key Terms

education   368
formal curriculum   382
formal education   368

functionally illiterate   373
hidden curriculum   382
illiteracy   373

informal education   368
schooling   368
self-fulfilling prophecy   381

14 Religion

With Emphasis on the Islamic Republic of Afghanistan

When sociologists study religion, they do not study whether God or some other supernatural force exists, whether certain religious beliefs are valid, or whether one religion is better than another. Instead, they focus on the social aspects of religion, such as the characteristics common to all religions and the ways in which people use religion to justify almost any kind of action.

Department of Defense photo by Tech. Sgt. Cecilio M. Ricardo Jr., U.S. Air Force

▲ Because of its mountainous and inhospitable terrain, military analysts call Afghanistan one of the toughest places in the world to wage a war.

WHY FOCUS ON

Afghanistan?

ON SEPTEMBER 20, 2001, nine days after the September 11 terrorist attacks on the United States, President George W. Bush answered a question many Americans were asking: Who attacked the United States? He described those who hijacked the commercial aircraft and turned them into "cruise missiles" as belonging to a collection of loosely affiliated terrorist organizations led by Osama bin Laden and known as al-Qaida. Al-Qaida members, who are believed to be scattered across more than 60 countries, learned the tactics of terrorism in Afghanistan and other places. That night, Bush demanded that the Taliban (an Islamic fundamentalist group that took control of Afghanistan in 1996) close all terrorist training camps, turn over all al-Qaida leaders living in Afghanistan, and grant the U.S. government full access to the camps (Bush 2001a). The demands were not met, so on October 7, 2001, the United States began air strikes on Taliban military installations and al-Qaida training camps.

For many Americans, the September 11 attacks represented the most devastating in a long line of attacks in which some militant Islamist group—such as the Islamic Jihad, Hezbollah (Party of God), or al-Qaida—was responsible. In other words, each attack was reduced to the actions of religious fanatics who were acting solely from "primitive and irrational" religious conviction. This perspective masks more-important political, geographic, and economic factors, and it causes

The September 11 attacks represented the most devastating in a long line of attacks on Americans, including the 1995 fuel truck explosion that killed 19 and injured more than 260 members of the U.S. military stationed at the King Abdul Aziz Air Base in Saudi Arabia.

Department of Defense photo by SRA Sean Worrell

observers to lose sight of larger questions: Is the Taliban's version of Islamic law consistent with "Islamic principles"? How did the Taliban rise to power in Afghanistan? How did the al-Qaida network come to be connected with Afghanistan? Why do the religious affiliations of the Taliban and of al-Qaida members receive more attention than the social, economic, and political factors driving their policies and actions? Such questions are largely absent from public discussion and debate. Yet, "lurking behind every terrorist act is a specific political antecedent. That does not justify either the perpetrator or his political cause. Nevertheless, the fact is that almost all terrorist activity originates from some political conflict and is sustained by it as well" (Brzezinski 2002).

In this chapter, we consider the answers to the questions mentioned above. In doing so, we will come to realize that religious affiliation explains little about the causes behind terrorist acts or the wars against terrorism. Rather, it is the social, economic, and political circumstances that cause people to draw upon religion to justify responses defined as terrorism. The sociological perspective is useful because it allows us to step back and view in a detached way an often emotionally charged subject. Detachment and objectivity are necessary if we wish to avoid making sweeping generalizations about the nature of religions, such as Islam, that are unfamiliar to many of us.

- When the attack on the World Trade Center towers began at 8:40 a.m. on September 11, 2001, Saif Rahman, a Muslim American, was just arriving to begin work at a law office in northern Virginia. According to Rahman (2001), "One of the attorneys grabbed me, and he said, 'Have you seen what's happened? Have you heard?' I said, 'What?'

And he said, 'A plane's hit the World Trade Center.' I ran to the TV set in our conference room and I saw the second plane hit, and immediately my heart dropped. I said, 'Oh, no, I just hope this is not someone who claims to be a Muslim.'"

- On October 7, 2001, President George W. Bush announced that U.S. military forces had begun strikes against targets in Afghanistan. Bush noted that the strikes were merely the beginning of a larger, even global campaign against terrorism. "Today we focus on Afghanistan, but the battle is broader. The U.S. did not ask for this fight, but we will win it" (*Online Newshour* 2001).

What Is Religion?

■ **CORE CONCEPT 1: When sociologists study religion, they are guided by the scientific method and by the assumption that no religions are false.** When sociologists study religion, they do not investigate whether God or some other supernatural force exists, whether certain religious beliefs are valid, or whether one religion is better than another. Sociologists cannot study such questions, because they adhere to the scientific method, which requires them to study only observable and verifiable phenomena. Instead, they investigate the social aspects of religion, focusing on the characteristics common to all religions, the functions and dysfunctions of religion, the conflicts within and between religious groups, the way religion shapes people's behavior and their understanding of the world, and the way religion is intertwined with social, economic, and political issues. We begin with a definition of religion. Defining religion is a surprisingly difficult task, with which sociologists have been greatly preoccupied.

What makes something a religion? In the opening sentences of *The Sociology of Religion* Max Weber (1922) states, "To define 'religion,' to say what it is, is not possible at the start of a presentation such as this. Definition can be attempted, if at all, only at the conclusion of the study" (p. 1). Despite Weber's keen interest in, and extensive writings about, religious activity, he could offer only the broadest of definitions: religion encompasses those human responses that give meaning to the ultimate and inescapable problems of existence—birth, death, illness, aging, injustice, tragedy, and suffering (Abercrombie and Turner 1978). To Weber, the hundreds of thousands of religions, past and present, represented a rich and seemingly endless variety of responses to these problems. In view of this variety, he believed that no single definition could hope to capture the essence of religion.

Department of Defense photo by PH3 William J. Davis, USN

Religion offers a life-affirming response to important human events, such as births, deaths, and illnesses. Christians celebrate baptism as a process for purifying and welcoming "new" members into the church.

Like Max Weber, Émile Durkheim believed that *religion* is difficult to define. In the first chapter of his book *The Elementary Forms of the Religious Life*, Durkheim ([1915] 1964) cautions that when studying religions, sociologists must assume that "there are no religions which are false" (p. 3). Like Weber, Durkheim believed that all religions are true in their own fashion; all address the problems of human existence, albeit in different ways. Consequently, he said, those who study religion must first rid themselves of all preconceived notions of what religion should be. We cannot study religion using standards that reflect our own personal experiences and preferences.

Consider that many critics view the *hijab*, the traditional head covering of Muslim women, as the primary evidence

It is a challenge to see through one's own preconceptions about what is "right" in everyday life. A Western woman may look on the traditional Muslim women's head covering, the *hijab*, as a sign of sexual oppression. A Muslim woman may look on a Western woman's garb as oppression of women imposed by media pressures to display themselves as sex objects.

that these women are severely oppressed. Although women in Islamic countries certainly do not have the same rights as men, critics should not be so quick to assume that the *hijab* is the *source* of oppression (Kristof 2002), especially when we consider the view that some Muslim women hold toward American dress customs:

> If women living in Western societies took an honest look at themselves, such a question [as why Muslim women are covered] would not arise. They are the slaves of appearance and the puppets of male chauvinistic society. Every magazine and news medium (such as television and radio) tells them how they should look and behave. They should wear glamorous clothes and make themselves beautiful for strange men to gaze and gloat over them. So the question is not why Muslim women wear hijab, but why the women in the West, who think they are so liberated, do not wear hijab. (Mahjubah 1984)

The discussion of the *hijab* shows that preconceived notions of what constitutes religion and uninformed opinions about the meaning of religious symbols and practices can close people off to a wide range of religious beliefs and experiences.

Essential Features of Religion

■ **CORE CONCEPT 2:** All religions have at least three essential characteristics: beliefs about the sacred and the profane, rituals, and a community of worshipers. In formulating his ideas about religion, Durkheim remained open to the many varieties of religious experiences throughout the world

(see Global Comparisons: "The World's Predominant Religions"). He identified three essential features that he believed were common to all religions, past and present: (1) beliefs about the sacred and the profane, (2) rituals, and (3) a community of worshipers. Thus Durkheim defined religion as a system of shared rituals and beliefs about the sacred that bind together a community of worshipers.

Beliefs about the Sacred

At the heart of all religious belief and activity stands a distinction between two domains: the sacred and the profane. The **sacred** includes everything that is regarded as extraordinary and that inspires in believers deep and absorbing sentiments of awe, respect, mystery, and reverence. These sentiments motivate people to safeguard what is sacred from contamination or defilement. To find, preserve, or guard what they consider sacred, people have gone to war, sacrificed their lives, and performed other life-endangering acts (Turner 1978).

Definitions of what is sacred vary according to time and place. Sacred things may include objects (such as chalices, scriptures, and statues), living creatures (such as cows, ants, and birds), elements of nature (such as rocks, mountains, trees, the sea, the sun, the moon, and the sky), places

sacred A domain of experience that includes everything regarded as extraordinary and that inspires in believers deep and absorbing sentiments of awe, respect, mystery, and reverence.

GLOBAL COMPARISONS

The World's Predominant Religions

THE MAP SHOWS which of the world's major religions predominates in each country. Keep in mind that other religions are also practiced in each country, some by many people. In the United States, for example, the population is believed to be 52 percent Protestant, 24 percent Roman Catholic, 2 percent Mormon, 1 percent Jewish, 1 percent Muslim, 10 percent some other religion, and 10 percent no religion. Likewise, Latin American countries have many Protestants, although the predominant religion is Roman Catholicism. The term *syncretism* refers to compatible combinations of belief systems, such as Confucianism, Buddhism, Taoism, and Shinto in Japan.

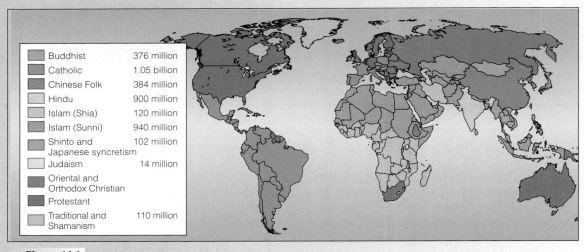

Buddhist	376 million
Catholic	1.05 billion
Chinese Folk	384 million
Hindu	900 million
Islam (Shia)	120 million
Islam (Sunni)	940 million
Shinto and Japanese syncretism	102 million
Judaism	14 million
Oriental and Orthodox Christian	
Protestant	
Traditional and Shamanism	110 million

▲ **Figure 14.1**

Source: U.S. Central Intelligence Agency (2007).

(such as churches, mosques, synagogues, and birthplaces of religions or their founders), days that commemorate holy events, abstract forces (such as spirits), persons (Jesus Christ, the Buddha, Moses, Muhammad, Zarathustra, and Nanak), states of consciousness (such as wisdom and oneness with nature), past events (such as the crucifixion and resurrection of Jesus, the escape of Hebrews from Egypt, and the birth of the Buddha), ceremonies (such as baptism, marriage, and burial), and other activities (holy wars, just wars, confession, fasting, and pilgrimages).

Durkheim ([1915] 1964) maintains that sacredness springs not from the item, ritual, or event itself, but rather from its symbolic power and from the emotions that people experience when they think about the sacred

thing or when they are in its presence. These emotions can be so strong that believers feel part of something larger than themselves and become outraged when other people behave inappropriately in the presence of the sacred thing. Ideas about what is sacred are such an important element of religious activity that many researchers classify religions according to the type of phenomenon that their followers consider sacred. One such typology of religion includes three categories: sacramental, prophetic, and mystical (Alston 1972).

Sacramental, Prophetic, and Mystical Religions

In **sacramental religions**, followers seek the sacred in places, objects, and actions believed to house a god or a spirit. These locations may include inanimate objects (such as relics, statues, and crosses), animals, trees or other plants, foods, drink (such as wine and water), places,

sacramental religions Religions in which the sacred is sought in places, objects, and actions believed to house a god or spirit.

The Hindu in the photo is bathing in the river Ganga (also known in the West as the Ganges). Hindus believe that bathing in the river assists in their attainment of salvation.

One of the Five Pillars of Islam is a pilgrimage to Mecca, Saudi Arabia, if a Muslim is physically and financially able.

and certain processes (such as the way people prepare for a hunt or perform a dance). Sacramental religions include various forms of Native American spirituality. An excerpt from the "Statement of Walter Echo-Hawk before the United States Commission on Civil Rights" describes such sacramental qualities:

> (1) It is important to note that there are probably as many native religions as there are Indian tribes in this country. (2) None of these religions or religious tenets [has] been reduced to writing in a holy document such as the Bible or Koran. (3) None of these religions [has] man-made churches in the Judeo-Christian sense; rather, the native religions are practiced in nature, at sacred sites, or in temporary religious structures—such as a tepee or sweat lodge. (4) The religious beliefs are tied to nature, the spiritual forces of nature, the natural elements, and the plants and creatures which make up the environment. . . . Natives are dependent upon all these things in order to practice their many religious ceremonies, rituals and religious observances. (Echo-Hawk 1979, p. 280)

In **prophetic religions**, the sacred revolves around items that symbolize historic events or around the lives, teachings, and writings of great people. Sacred books, such as the Christian Bible, the Muslim Qur'an, and the Jewish Tanakh, hold the records of these events and revelations. In the case of historic events, God or some other higher being is believed to be directly involved in the course and outcome of the events (such as a flood, the parting of the Sea of Reeds, or the rise and fall of an empire). In the case of great people, the lives and inspired words of prophets or messengers reveal a higher state of being, "the way," a set of ethical principles, or a code of conduct. Followers then seek to live accordingly. Some of the best-known prophetic religions include Judaism, as revealed to Abraham in Canaan and to Moses at Mount Sinai, Confucianism (founded by Confucius), Christianity (founded by the earliest followers of Jesus Christ), and Islam (founded by Muhammad). The set of ethical principles may include the Ten Commandments of Judaism and Christianity or the Five Pillars of Islam. Muslim tenets include the following:

- The declaration of faith known as the *shahadah* ("There is no god but Allah, and Muhammad is his messenger.")
- Obligatory prayer known as *salah* (performed five times per day)

prophetic religions Religions in which the sacred revolves around items that symbolize significant historical events or around the lives, teachings, and writings of great people.

- Almsgiving (Each year, devout Muslims set aside a percentage of their accumulated wealth to assist the poor and sick.)
- A pilgrimage to the city of Mecca, Saudi Arabia, known as *hajj* (if one is physically and financially able)

In **mystical religions**, followers seek the sacred in states of being that can exclude all awareness of their existence, sensations, thoughts, and surroundings. In such states, mystics become caught up so fully in the transcendental experience that earthly concerns seem to vanish. Direct union with the divine forces of the universe assumes the utmost importance. Not surprisingly, mystics tend to become involved in practices such as fasting or celibacy to separate themselves from worldly attachments.

In addition, they meditate to clear their minds of worldly concern, "leaving the soul empty and receptive to influences from the divine" (Alston 1972, p. 144). Buddhism and philosophical Hinduism are two religions that emphasize physical and spiritual discipline as a means of transcending the self and earthly concerns.

Keep in mind that the distinctions between sacramental, prophetic, and mystical religions are not clear-cut. In fact, most religions in each of these categories incorporate or combine elements of the other categories. Consequently, most religions cannot be assigned to a single category.

Founded in the sixth and fifth centuries BC by the Buddha, Siddhartha Gautama, Buddhism has an estimated 376 million followers. Buddhism teaches that suffering is an inevitable part of human existence; desires and feelings of self-importance cause suffering; nirvana is achieved through meditation, karma, and righteous actions, thoughts, and attitudes (New York Public Library 1993).

Beliefs about the Profane

According to Durkheim ([1915] 1964), the sacred encompasses more than the forces of good: "There are gods of theft and trickery, of lust and war, of sickness and of death" (p. 420). Evil and its various representations are, however, generally portrayed as inferior and subordinate to the forces of good: "In the majority of cases we see the good victorious over evil, life over death, the powers of light over the powers of darkness" (p. 421). Even so, Durkheim considers superordinary evil phenomena to fall within the category of the sacred, because they are endowed with special powers and serve as the objects of rituals (such as confession to rid one of sin, baptism to purify the soul, penance for sins, and exorcism to rid one of evil) designed to overcome or resist the negative influences of such phenomena.

Religious beliefs, doctrines, legends, and myths detail the origins, virtues, and powers of sacred things and describe the consequences of mixing the sacred with the profane. The **profane** encompasses everything that is not considered sacred, including things opposed to the sacred (such as the unholy, the irreverent, and the blasphemous) and things that stand apart from, though not necessarily in opposition to, the sacred (such as the ordinary, the commonplace, the unconsecrated, the temporal, and the bodily) (Ebersole 1967).

Believers often view contact between the sacred and the profane as being dangerous and sacrilegious, as threatening the very existence of the sacred, and as endangering the fate of the person who makes or allows such contact. Consequently, people take action to safeguard sacred things by separating them from the profaner. For example, some people refrain from speaking the name of God when they feel frustrated; others believe that a woman must cover her hair or her face during worship and that a man must remove his hat during worship.

The distinctions drawn between the sacred and the profane do not mean that a person, an object, or an idea cannot pass from one domain to another or that something profane cannot ever come into contact with the sacred. Indeed, such transformations and contacts are authorized by religions through rituals—the active and most observable side of religion.

mystical religions Religions in which the sacred is sought in states of being that, at their peak, can exclude all awareness of one's existence, sensations, thoughts, and surroundings.

profane A term describing everything that is not sacred, including things opposed to the sacred and things that stand apart from the sacred, albeit not in opposition to it.

Under the Taliban version of Islamic law, women who appear in public must be covered from head to toe. Many Afghan women still cover themselves out of fear, by convention, or because they live in areas that still restrict their behavior. An American contractor working in Afghanistan asked this woman to hold up a sign saying hi to a friend of his in the United States.

Rituals include kneeling in prayer facing Mecca five times each day and submerging the entire body in the Arabian Sea to experience baptism.

Rituals

In the religious sense, **rituals** are rules that govern how people behave in the presence of the sacred. These rules may take the form of instructions detailing the appropriate context for worship, the roles of various participants, acceptable dress, and the precise wording of chants, songs, and prayers. Participants must follow the instructions closely to achieve a specific goal, whether it be to purify the participant's body or soul (as through confession, immersion, fasting, or seclusion), to commemorate an important person or event (as by making a pilgrimage to Mecca or celebrating Passover or the Eucharist), or to transform profane items into sacred items (for example, changing water to holy water and bones to sacred relics (Smart 1976, p. 6).

Rituals can be as simple as closing one's eyes to pray or having one's forehead marked with ashes. Alternatively, they can be as elaborate as fasting for three days before entering a sacred place to chant, with head bowed, a particular prayer for forgiveness. Although rituals are often enacted in sacred places, some are codes of conduct aimed at governing the performance of everyday activities, such as sleeping, walking, eating, defecating, washing, and dealing

with members of the opposite sex. The Qur'an lays down a code of modesty for women:

> And say to the believing women that they should lower their gaze and guard their modesty; and that they should not display their beauty and ornaments except what (must ordinarily) appear thereof; that they should draw their veils over their bosoms and not display their beauty save to their husbands, or their fathers or their husbands' fathers, or their sons or their husbands' sons, or their brothers or their brothers' sons, or their sisters' sons, or their women, or the slaves whom their right hands possess, or male servants free of physical desire, or small children who have no sense of sex; and that they should not stamp their feet in order to draw attention to their hidden ornaments. And O believers! Turn all together towards Allah, that you may attain bliss. (24:31)

According to Durkheim, the nature of the ritual is relatively insignificant. Rather, the important element is that the ritual is shared by a community of worshipers

rituals Rules that govern how people must behave in the presence of the sacred to achieve an acceptable state of being.

Roman Catholic missionaries traveled to Brazil to convert native people to Christianity as early as the 16th century. In the process, they established churches where the converted could gather. The churches pictured were established by such missionaries working in Brazil.

and evokes certain ideas and sentiments that help individuals feel themselves to be part of something larger than themselves.

Community of Worshipers

Durkheim uses the word **church** to designate a group whose members hold the same beliefs regarding the sacred and the profane, who behave in the same way in the presence of the sacred, and who gather in body or spirit at agreed-on times to reaffirm their commitment to those beliefs and practices. Obviously, religious beliefs and practices cannot be unique to an individual; they must be shared by a group of people. If not, then the beliefs and practices would cease to exist when the individual who held them died or if he or she chose to abandon them. In the social sense, then, religion is inseparable from the idea of church. The gathering and the sharing create a moral community and allow worshipers to share a common identity. The gathering need not take place in a common setting, however. When people perform a ritual on a given day or at given times of day, the gathering may be spiritual rather than physical.

church A group whose members hold the same beliefs about the sacred and the profane, who behave in the same way in the presence of the sacred, and who gather in body or spirit at agreed-on times to reaffirm their commitment to those beliefs and practices.

ecclesia A professionally trained religious organization, governed by a hierarchy of leaders, that claims everyone in a society as a member.

Durkheim ([1915] 1964) uses the term *church* loosely, acknowledging that it can assume many forms: "Sometimes it embraces an entire people . . . sometimes it embraces only a part of them . . . sometimes it is directed by a corps of priests, sometimes it is almost completely devoid of any official directing body" (p. 44). Sociologists have identified at least five broad types of religious organizations or communities of worshipers: ecclesiae, denominations, sects, established sects, and cults. As with most classification schemes, these categories overlap on some characteristics because the classification criteria for religions are not always clear.

Ecclesiae

An **ecclesia** is a professionally trained religious organization, governed by a hierarchy of leaders, that claims everyone in a society as a member. Membership is not voluntary; it is the law. Consequently, considerable political alignment exists between church and state officials, so that the ecclesia represents the official church of the state. Ecclesiae formerly existed in England (the Church of England [Anglican], which remains the official state church), France (the Roman Catholic Church), and Sweden (the Church of Sweden [Lutheran]). The Afghan constitution signed in 2004 declares the country to be an Islamic republic, makes Islam the official religion, and announces that "no law can be contrary to the sacred religion of Islam." The Afghan government, however, guarantees non-Muslims the right to "perform their religious ceremonies within the limits of the provisions of law" (Feldman 2003).

Individuals are born into ecclesiae, newcomers to a society are converted, and dissenters are often persecuted. Those who do not accept the official religious view tend

Tom Kaelin

This child is learning the important Catholic ritual of dipping one's fingers into holy water and then using those fingers to make the sign of the cross.

Department of Defense photo by LCPL Kevin C. Quihuis, Jr., USMC

This monument displays a passage from the Qur'an. Islam (Arabic for "submission to God") was founded around AD 610 by the prophet Muhammad, who claimed to have received the Qur'an from Allah. The two main divisions of Islam are the Sunni and the Shia. In addition, sects exist within the Sunni branch (the Wahabis) and Shiite branch (the Assassins, the Druze, and the Fatimids) (New York Public Library 1993).

to emigrate or to occupy a marginal status. The ecclesia claims to be the one true faith and often does not recognize other religions as valid. In its most extreme form, it directly controls all facets of life.

Denominations

A **denomination** is a hierarchical religious organization in a society in which church and state usually remain separate; it is led by a professionally trained clergy. In contrast to an ecclesia, a denomination is one of many religious organizations in society. For the most part, denominations tolerate other religious organizations; they may even collaborate with other such organizations to solve problems in society. Although membership is considered to be voluntary, most people who belong to denominations did not choose to join them. Rather, they were born to parents who were members. Denominational leaders generally make few demands on the laity (church members who are not clergy), and most members participate in limited and specialized ways.

For example, members may choose to send their children to church-operated schools, attend church on Sundays and religious holidays, donate money to the church, or attend church-sponsored functions. But the leaders of a denomination do not oversee all aspects of members' lives. Although laypeople vary widely in lifestyle, denominations frequently attract people of particular races, ethnicities, and social classes.

Major denominations in the world include Buddhism, Christianity, Confucianism, Hinduism, Islam, Judaism, Shinto, and Taoism. Each is predominant in different areas of the globe. For example, Christianity predominates in Europe, the Americas, and Australia and Oceania; Islam predominates in the Middle East and North Africa; and Hinduism predominates in India.

Sects and Established Sects

A **sect** is a small community of believers led by a lay ministry; it has no formal hierarchy, or official governing body, to oversee its various religious gatherings and activities. Sects are typically composed of people who broke away from a denomination because they came to view it as corrupt. They then created the offshoot in an effort to reform the religion from which they separated.

denomination A hierarchical religious organization, led by a professionally trained clergy, in a society in which church and state are usually separate.

sect A small community of believers led by a lay ministry, with no formal hierarchy or official governing body to oversee its various religious gatherings and activities. Sects are typically composed of people who broke away from a denomination because they came to view it as corrupt.

All of the major religions encompass splinter groups that have sought at one time or another to preserve the integrity of their religion. In Islam, for example, the most pronounced split occurred about 1,300 years ago, approximately 30 years after the death of Muhammad. The split related to Muhammad's successor. The Shia maintained that the successor should be a blood relative of Muhammad; the Sunni believed that the successor should be selected by the community of believers and need not be related to Muhammad by blood. After Muhammad's death, the Sunni (encompassing the great majority of Muslims) accepted Abu-Bakr as the caliph (successor). The Shia supported Ali, Mohammad's first cousin and son-in-law, and they called for the overthrow of the existing order and a return to the pure form of Islam. Today Shiite Islam predominates in the Islamic Republic of Iran (95 percent), whereas Sunni Islam predominates in the Islamic Republics of Pakistan (77 percent) and Afghanistan (84 percent).

The divisions within Islam have existed for so long that Sunni and Shia have become recognized as **established sects**—groups that have left denominations or ecclesiae and have existed long enough to acquire a large following and widespread respectability. In some ways, established sects resemble both denominations and sects. As you might expect, several divisions have formed within each established sect as new splinter groups have attempted to reform some policy, practice, or position held by the religious organization from which they have separated.

Similarly, several splits have occurred within Christianity. During a period from about the 11th century to the early 13th century, for example, the Greek-language Eastern churches (then centering on Constantinople) and the Latin-language Western church (centering on Rome) gradually drifted apart over issues such as the papal claim of supreme authority over all Christian churches in the world. The Protestant churches owe their origins largely to Martin Luther (1483–1546), who also challenged papal authority and protested against many practices of the medieval Roman Catholic Church. In addition, divisions exist within various Protestant sects and among Catholics. Offshoots of the Roman Catholic Church, for example, include the various Old Catholic churches. Theoretically, people are not born into sects, they convert; they choose

Department of Defense photo by T3C Irving Katz

Hundreds of stolen Jewish and Hebrew books and "Saphor Torahs" (Sacred Scrolls) were discovered in the cellar of the Race Institute in Germany.

membership later in life, when they are considered able to decide for themselves. Sects vary on many levels, including the degree to which they view society as religiously bankrupt or corrupt and the extent to which they take action to change people in society.

Cults

Generally, **cults** are very small, loosely organized religious groups, usually founded by a charismatic leader who attracts people by virtue of his or her personal qualities. Because the charismatic leader plays such a central role in attracting members, cults often dissolve after the leader dies. Consequently, few cults last long enough to become established religions. Even so, a few manage to survive, as evidenced by the fact that the major world religions began as cults. Because cults form around new and unconventional religious practices, outsiders tend to view them with considerable suspicion.

Cults vary in terms of their purpose and the level of commitment that their leaders demand of converts. They may draw members by focusing on highly specific but eccentric interests, such as astrology, UFOs, or transcendental meditation. Members may be attracted by the promise of companionship, a cure for illness, relief from suffering, or enlightenment.

A cult may meet infrequently and strictly voluntarily (as at conventions or monthly meetings). In some cases,

established sects Religious organizations, resembling both denominations and sects, that have left denominations or ecclesiae and have existed long enough to acquire a large following and widespread respectability.

cults Very small, loosely organized groups, usually founded by a charismatic leader who attracts people by virtue of his or her personal qualities.

however, the cult leader may require members to break all ties with family, friends, and jobs and thus to rely exclusively on the cult to meet all of their needs.

Civil Religion

■ **CORE CONCEPT 3:** Civil religion is an institutionalized set of beliefs about a nation's past, present, and future and a corresponding set of rituals that take on a sacred quality and elicit feelings of patriotism. The dynamics of civil religion are most notable during times of crisis and war and on national holidays. Durkheim's definition of *religion* highlights three essential characteristics: beliefs about the sacred and the profane, rituals, and a community of worshipers. Critics argue that these characteristics are not unique to religious activity. This combination of characteristics, they say, applies to many gatherings (for example, sporting events, graduation ceremonies, reunions, and political rallies) and to many political systems (for example, Marxism, Maoism, and fascism). On the basis of these characteristics alone, it is difficult to distinguish between an assembly of Christians celebrating Christmas, a patriotic group supporting the initiation of a war against another country, and a group of fans eulogizing James Dean or Elvis Presley. In other words, religion is not the only unifying force in society that incorporates the three elements defined by Durkheim as characteristic of religion. Civil religion represents another such force that resembles religion as Durkheim defined it.

Civil religion is an institutionalized set of beliefs about a nation's past, present, and future and a corresponding set of rituals. Both the beliefs and the rituals take on a sacred quality and elicit feelings of patriotism. Civil religion forges ties between religion and a nation's needs and political interests (Bellah 1992, Hammond 1976, Davis 2002). A nation's values (such as individual freedom and equal opportunity) and rituals (such as parades, fireworks, singing the national anthem, and 21-gun salutes) often assume a sacred quality. Even in the face of internal divisions based on race, ethnicity, region, or gender, national beliefs and rituals can inspire awe, respect, and reverence for the country. These sentiments are most evident during times of crisis and war and on national holidays that celebrate important events or people (such as Thanksgiving, Presidents' Day, Martin Luther King Jr. Day, and Independence Day), in the presence of national monuments or symbols (the flag, the Capitol, the Lincoln Memorial, the Vietnam Memorial).

In times of war, presidents offer a historical and mythological framework that gives the country's involvement in the war moral justification and offers the public a vision and an identity for the country. Sociologist Roberta Cole (2002) argues that America's civil religion found its voice in a 19th-century political doctrine known as *manifest*

Department of Defense photo by MCSN Christopher A. Lussier, USN

The faces pictured are part of the Navy Memorial in Washington, D.C. The American flag and uniform dress symbolize that those pictured died for a higher cause than self-interest. We may not question what kind of people they were, apart from their military service, as death in the name of their country allows them to escape such scrutiny.

destiny. While the term was first used in 1845, it expressed a longstanding ideology that the United States, by virtue of its moral superiority, was destined to expand across the North American continent to the Pacific Ocean and beyond (Chance 2002). Manifest destiny included the beliefs that the United States had a divine mission to serve as a democratic model to the rest of the world, that the country was a redeemer exerting its good influence upon other nations, and that it represented hope to the rest of the world—not just the hope that others would want to and could become like the United States, but the hope that the United States would come to the rescue whenever a problem arose (Cole 2002). In 1835 Alexis de Tocqueville observed this longstanding belief among Americans that their country was unique:

> For 50 years, it has been constantly repeated to the inhabitants of the United States that they form the only religious, enlightened, and free people. They see that up to now, democratic institutions have prospered among them; they therefore have an immense opinion of themselves, and they are not far from believing that they form a species apart in the human race.

civil religion An institutionalized set of beliefs about a nation's past, present, and future and a corresponding set of rituals. Both the beliefs and the rituals take on a sacred quality and elicit feelings of patriotism. Civil religion forges ties between religion and a nation's needs and political interests.

The painting portrays those who participated in the United States' westward expansion as fulfilling an almost divine mission, represented by the guardian angel-like figure watching over them. Of course, westward expansion was not the peaceful process depicted here.

Library of Congress Prints and Photographs Division

Civil Religion and the Cold War

The cold war (1945–1989) included an arms race, in which the Soviet Union and the United States competed to match and then surpass any advances made by each other in the number and technological quality of nuclear weapons. While the United States and the Soviet Union fell short of direct, full-scale military engagement, they took part in as many as 120 proxy wars fought in developing countries. In many of these conflicts, the United States and the Soviet Union supported opposing factions by providing weapons, military equipment, combat training, medical supplies, economic aid, and food. Three of the best-known proxy wars were fought in Korea, Vietnam, and Afghanistan. Soviet and American leaders justified their direct or indirect intervention on the grounds that it was necessary to contain the spread of the other side's economic and political system, to protect national and global security, and to prevent the other side from shifting the balance of power in favor of its system.

From 1945 through 1989, the foreign and domestic policies of the United States were largely shaped by cold war dynamics—specifically, a professed desire to save the world from Soviet influence and the spread of communism. Robert S. McNamara (1989), U.S. Secretary of Defense under Presidents Kennedy and Johnson, remarked that "on occasion after occasion, when confronted with a choice between support of democratic governments and support of anti-Soviet dictatorships, we have turned our backs on our traditional values and have supported the antidemocratic," brutally repressive, and totalitarian regimes (p. 96). President George W. Bush (2003b) agreed with McNamara's assessment when he acknowledged in a speech to the British people that

> we must shake off decades of failed policy on the Middle East. Your nation and mine in the past have been willing to make a bargain to tolerate oppression for the sake of stability. Long-standing ties often led us to overlook the faults of local elites. Yet this bargain did not bring stability or make us safe. It merely bought time while problems festered and ideologies of violence took hold.

The United States and Muslims as Cold War Partners

The cold war between the United States and the Soviet Union made Afghanistan a focus of those two countries' conflict (see Figure 14.2). When the Soviet Union invaded Afghanistan in 1979 and put its Afghan supporters in charge, the United States supported Islamic guerrillas, known as the *mujahideen*, by funneling money through Pakistan. At that time, Pakistani president Zia's goal was to turn Pakistan into the leader of the Islamic world and then use that leverage to cultivate an Islamic opposition to Soviet expansion into central Asia. Zia's aims fit well with the United States' cold war goals of containing the Soviet Union. If the United States could show the Soviet Union that the entire Muslim world was its partner, then the United States would indeed be a force to fear (Rashid 2001).

The U.S. Central Intelligence Agency (CIA) worked with its Pakistani equivalent, the Inter-Services Intelligence Agency, on a plan to recruit radical Muslims from all over the world to fight with their Afghan brothers against the Soviet Union. An estimated 35,000 Muslims from 43 countries—primarily in central Asia, North and East Africa, and the Middle East—heeded the call. Thousands more came to Pakistan to study in *madrassas* (Muslim schools) in Pakistan and along the Afghan border (Rashid 2001).

Military training camps staffed with U.S. advisors helped train the guerrillas, and the *madrassas* offered a place for the most radical Muslims in the world to meet, exchange ideas, and learn about Islamic movements in one another's countries. Among those who came to Afghanistan was Osama bin Laden. At that time the pressing question for the United States, as asked by national security advisor Zbigniew Brzezinski (2002), was, "What was more important in the world view of history? The possible creation of an armed, radical Islamic movement or the fall of the Soviet Empire? A few fired-up Muslims or the liberation of Central Europe and the end of the Cold War?" These Pakistani- and U.S.-supported recruiting and military centers would eventually evolve into al-Qaida ("the base").

In 1989, the year the term al-Qaida was first used, Osama bin Laden had taken over as the centers' leader.

The Historical Context of Afghanistan

BEFORE THE 19TH century, mountainous Afghanistan lay in the path of invaders from China, Persia (ancient Iran), and the Indian subcontinent. In the 19th and 20th centuries, the country became a battleground for the British and Russian empires and, after World War II, for the United States and the Soviet Union. In 1979 the Soviet Union invaded Afghanistan to support a secular government.

Supported materially by the United States, Afghanistan's military resistance to the Soviets was mobilized in large part through religious institutions that proclaimed a "holy war." The Soviet exit in 1989 left a ravaged country full of rival factions fighting for control. When the Taliban government took power in 1996, it justified many of its new policies on Islamic grounds. Westerners tend to see such policies as simply fanatical and irrational and to overlook the history that led up to them.

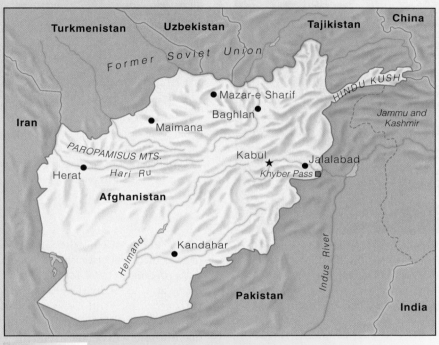

▲ Figure 14.2

That same year, Soviet troops withdrew from Afghanistan, leaving behind

> an uneasy coalition of Islamist organizations intent on promoting Islam among all non-Muslim forces. [They also] left behind a legacy of expert and experienced fighters, training camps and logistical facilities, elaborate trans-Islam networks of personal and organizational relationships, a substantial amount of military equipment, . . . and most importantly, a heavy sense of power and self-confidence based on what [they] had achieved, and a driving desire to move on to other victories (Huntington 2001, p. A12).

To help measure the legacy of U.S.-supported training camps, consider that "key leaders of every major terrorist attack, from New York to France to Saudi Arabia, inevitably turned out to be veterans of the Afghan War" (Mamdani 2004).

Civil Religion and Osama bin Laden

In 1990, at the request of the Saudi government, the United States government sent 540,000 troops to the Persian Gulf region after Iraqi troops invaded Kuwait. In a presidential address, George H. W. Bush (1991) described the United States in sacred terms:

> I come to this House of the people to speak to you and all Americans, certain that we stand at a defining hour. Halfway

Department of Defense photo by SSGT. Nolan

The United States offered medical treatment and other assistance to mujahideen who were wounded fighting the Soviet Union. Here a U.S. medic plays cards with two wounded "freedom fighters," the term the American government used to characterize the anti-communist guerrillas, en route to a U.S. airbase in Germany for specialized medical care.

Farm Security Administration and Office of War Information Collection, Library of Congress Prints and Photographs Division

In 1941, the year when these children were reciting with hand over heart the Pledge of Allegiance, they said, "one Nation indivisible, with Liberty and Justice for all." The words "under God" were added in the 1950s during the cold war. At that time, the U.S. government started stamping "In God We Trust" on its coins.

around the world, we are engaged in a great struggle in the skies and on the seas and sands. We know why we're there: We are Americans, part of something larger than ourselves. For two centuries, we've done the hard work of freedom. And tonight, we lead the world in facing down a threat to decency and humanity. . . . Yes, the United States bears a major share of leadership in this effort. Among the nations of the world, only the United States of America has both the moral standing and the means to back it up. We're the only nation on this Earth that could assemble the forces of peace. This is the burden of leadership and the strength that has made America the beacon of freedom in a searching world.

Osama bin Laden, who had hoped to raise a force composed of Afghan War veterans to fight Iraq, was infuriated with the Saudi royal family for calling upon the United States for help. He appealed to Muslim clerics to issue a *fatwa* (ruling or decree) condemning the stationing of non-Muslim troops in Saudi Arabia. His request was

denied, and eventually the Saudi royal family, tired of bin Laden's incessant criticism, revoked his citizenship.

According to bin Laden (2001a), the "U.S. knows that I have attacked it, by the grace of God, for more than ten years now. . . . Hostility toward America is a religious duty and we hope to be rewarded for it by God. I am confident that Muslims will be able to end the legend of the so-called superpower that is America." Osama bin Laden claimed credit for the 1993 World Trade Center bombings and attacks on U.S. soldiers in Somalia, the 1998 attacks on U.S. embassies in East Africa, and the 1998 attack on the USS *Cole*. While not formally claiming credit for the September 11, 2001, attacks on the United States, bin Laden (2001b) condoned them with the following religiously charged words: "Here is America struck by God Almighty in one of its vital organs, so that its greatest buildings are destroyed. Grace and gratitude to God. America has been filled with horror from north to south and east to west, and thanks be to God."

Civil Religion and the War on Terror

On September 20, 2001, President George W. Bush indicated that a global war on terror would begin with air strikes against al-Qaida and Taliban strongholds in Afghanistan. The enemy was larger than Afghanistan,

Consider this legacy of the U.S.-supported training camps in Afghanistan during the cold war: key leaders behind every major terrorist attack on U.S. interests since then were veterans of the Afghan War.

Five members of al-Qaida hijacked American Airlines flight 77 and then crashed the aircraft into the Pentagon, killing the 64 passengers onboard and 125 people on the ground. The Pentagon was one of three targets on September 11, 2001; the others being the twin towers of the World Trade Center and presumably the White House (which was spared when passengers brought their aircraft down in a Pennsylvania field).

After the U.S. military invaded and occupied Iraq, it destroyed this compound, which was suspected of housing weapons of mass destruction. After the fact, U.S. troops inspected the site and declared it clear.

however; it was a "radical network of terrorists" and the governments (as many as 60) that supported them. Bush (2001a) indicated that the war would not end "until every terrorist group has been found, stopped, and defeated."

On March 22, 2003, Bush announced the beginning of Operation Iraqi Freedom. He described the mission as clear: "to disarm Iraq of weapons of mass destruction, to end Saddam Hussein's support of terrorism (it has aided, trained, and harbored terrorists, including operatives of al-Qaida), and to free the Iraqi people." We know now that Iraq had no weapons of mass destruction and no substantiated links to al-Qaida. Our purpose here is *not* to address the question of whether the wars in Iraq and Afghanistan have been just or whether the war effort has succeeded. Rather, we focus on the language that presidents use to justify war and to articulate a national identity in time of war. Usually during such times, the nation assumes a sacred quality and the president projects a moral certitude, as evidenced in the following statements. Some critics liken this moral certitude to "a kind of fundamentalism" and a "dangerous messianic brand of religion, one where self-doubt is minimal" (Hedges 2002).

- Every nation in every region now has a decision to make. Either you are with us or you are with the terrorists. (Bush 2001a)
- We'll meet violence with patient justice—assured of the right men, of our cause, and confident of the victories to come. In all that lies before us, may God grant us wisdom and may He watch over the United States of America. (Bush 2001b)
- We did not ask for this mission, but we will fulfill it. The name of today's military operation is Enduring

NO BORDERS, NO BOUNDARIES

Countries with Significant Percentages of Muslims

THE PROPHET MUHAMMAD'S first recitations of the Qur'an occurred in Arabia around AD 610. Islam's spread has made it one of the world's major religions. The map shows countries where a significant percent-age of people practice Islam. Here *significant percentage* is defined as "at least 2 percent of the population (or 1 in every 50 persons)."

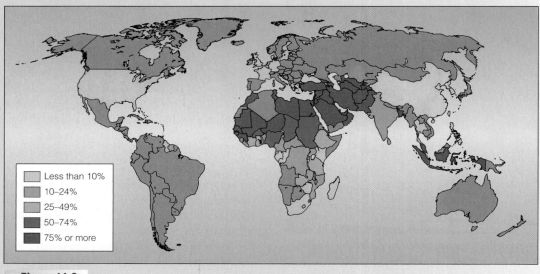

Less than 10%
10–24%
25–49%
50–74%
75% or more

▲ **Figure 14.3**

Source: U.S. Central Intelligence Agency (2007).

Freedom. We defend not only our precious freedoms, but also the freedoms of people everywhere. (Bush 2001b)

● Operation Iraqi Freedom was carried out with a combi-nation of precision and speed and boldness the enemy did not expect, and the world had not seen before. (Bush 2003a)

● Those we lost were last seen on duty. Their final act on this earth was to fight a great evil and bring liberty to others. All of you . . . have taken up the highest calling of history, . . . and wherever you go, you carry a message of hope—a message that is ancient and ever new. In the words of the prophet Isaiah, "to the captives 'come out,' and to those in darkness, 'be free.'" (Bush 2003a)

The larger point of these examples is that the traits Durkheim cites as characteristic of religion apply to other events, relationships, and forces within society that many

people might not define as religious. One might argue that perhaps sociologists should develop a narrower, less inclu-sive definition of *religion* than that proposed by Durkheim. Narrow definitions are problematic as well, however. Sup-pose that we narrow the definition of *religion* to "the belief in an ever-living God." This definition would exclude polytheistic religions, such as Hinduism, which has more than 640 million adherents. It would also exclude reli-gions in which a deity plays little or no role, such as Bud-dhism, which has more than 300 million followers. As you can see, narrow definitions of *religion* do not necessarily improve on broad ones.

Despite its shortcomings, Durkheim's definition of *reli-gion* remains one of the best and most widely used. No sociologist with any standing in the discipline can study religion without encountering and addressing Durkheim's definition. Besides proposing a definition of *religion*, Durkheim wrote extensively about the functions of religion. His work laid the foundation for the functionalist perspective on religion.

The Functionalist Perspective

■ **CORE CONCEPT 4:** The functionalist perspective maintains that religion serves vital social functions for the individual and for the group. Some form of religion appears to have existed for as long as humans have lived (at least 2 million years). In view of this fact, functionalists maintain that religion must serve some vital social functions for the individual and for the group. On the individual level, people embrace religion in the face of uncertainty; they draw on religious doctrine and ritual to comprehend the meaning of life and death and to cope with misfortunes and injustices (such as war, drought, and illness).

Life would be intolerable without reasons for existing or without a higher purpose to justify the trials of existence (Durkheim 1951). Try to imagine, for example, how people might cope with the immense devastation and destruction resulting from decades of war. When Soviet troops invaded Afghanistan in 1979, they attacked civilian populations, burned village crops, killed livestock, used lethal and non-lethal chemical weapons, planted an estimated 10 million mines, and engaged in large-scale high-altitude carpet bombing. "In the countryside it was standard Soviet practice to bombard or even level whole villages suspected of harboring resistance fighters. Sometimes women, children, and old men were rounded up and shot. This devastation of towns and villages forced many civilians to seek refuge in Kabul, whose prewar population of less than 1 million swelled to nearly 2 million" (Kurian 1992, p. 5).

Even after the Soviets withdrew from Afghanistan in 1989, the civil war continued, as various political parties competed to fill the power vacuum. Table 14.1 summarizes the tragic results of more than 20 years of war in this country. In light of this situation, is it any wonder that Afghan people might turn to religion to cope with the devastation and restore a sense of order out of chaos?

Besides turning to religion in the face of intolerable circumstances, people rely on religious beliefs and rituals to help them achieve a successful outcome (such as the birth of a healthy child or a job promotion) and to gain answers to questions of meaning: How did we get here? Why are we here? What happens to us when we die? According to Durkheim, people who have communicated with their God or with other supernatural forces (however conceived) report that they gain the inner strength and the physical strength to endure and to conquer the trials of existence:

> It is as though [they] were raised above the miseries of the world. . . . Whoever has really practiced a religion knows very well . . . these impressions of joy, of interior peace, of serenity, of enthusiasm, which are, for the believer, an experimental proof of his beliefs. (Durkheim 1915 pp. 416–417)

Table 14.1   Profile of the Islamic Republic of Afghanistan

Department of Defense photo by SPC Christopher Barnhardt, USA

Almost half of all Afghan children are malnourished. These four-year-old twins have not grown since infancy.

Before the United States began its military attacks on Afghanistan in 2001, the country was already devastated. Barnett Rubin (1996) argues that no one paid more for the U.S. cold war victory than did Afghanistan and its people: "Millions of unknown people sacrificed their homes, their land, their cattle, their health, their families, with barely hope of success or reward, at least in this world" (p. 21). Following the Soviet withdrawal, the country experienced a decade of civil war, Taliban rule, and severe drought (1999–2001). One has to go back to the early 1800s to find a time when Afghanistan could be considered a unified country (Halloran 2004).

Population	31.9 million
Persons dependent on food aid	6.0 million
Refugees	2.2 million
Deaths as a result of war	500,000 military, 1.5 million civilian (since 1979)
Life expectancy at birth	Males: 43.6 years; females: 43.9 years
Malnutrition	49.3% of children younger than age 5 underweight for age
Access to drinking water	60% of households with no safe drinking water
Literacy rate	Males: 43.2%; females: 14.1%; overall: 28.7%
Maternal mortality	Catastrophic levels[1]
Infant mortality	257 babies per 1,000 live births before age 5
Birth rates	47.27 births per 1,000 people[2]
Illicit drugs	World's largest producer of opium[3]

[1]Catastrophic means that Afghanistan has one of the highest maternal mortality rates in the world.
[2]The highest birth rate possible is around 50 per 1,000 people. This means that Afghan women are having babies as fast as biologically possible. To put it another way, millions of women are in a "constant cycle of pregnancy and birth" (Gall 2003).
[3]Opium production may account for one-third of Afghanistan's GDP; 80–90% of heroin consumed in Europe comes from Afghan opium.

Sources: U.S. Central Intelligence Agency (2007); United Nations High Commissioner for Refugees (2005).

Religion functions in several ways to promote group unity and solidarity. First, shared doctrines and rituals create emotional bonds among believers. Second, theoretically, all religions strive to raise individuals above themselves—to help them achieve a life better than they would lead if left to their own impulses. In this sense, religion offers ideas of proper conduct that carry over into everyday life. When believers violate this code of conduct, they feel guilt and remorse. Such feelings, in turn, motivate them to make amends. Third, although observance of many religious rituals functions to alleviate individual anxieties, uncertainties, and fears, it also establishes, reinforces, or renews social relationships—thereby binding individuals to a group. Finally, religion functions as a stabilizing force in times of severe social disturbance and abrupt change. During such times, many regulative forces in society may break down. In the absence of such forces, people are more likely to turn to religion in search of a force that will bind them to a group. This tie helps people think less about themselves and more about some common goal (Durkheim 1951)—whether that goal is to work for peace or to participate more fervently in armed conflict. From a functionalist point of view, one could argue that the Taliban enforced their version of strict Islamic law in Afghanistan because they believed that drastic policies would restore order after decades of chaos wrought by war.

That religion functions to meet individual and societal needs, and that people create sacred objects and rituals, led Durkheim to reach a controversial but thought-provoking conclusion: the "something out there" that people worship is actually society.

Society as the Object of Worship

■ **CORE CONCEPT 5:** The variety of religious responses is endless, because people play a fundamental role in determining what is sacred and how they should act in its presence. If we operate under the assumptions that all religions are true in their own fashion and that the variety of religious responses is virtually endless, we find support for Durkheim's conclusion that people create everything encompassed by religion—gods, rites, sacred objects. That is, people play a fundamental role in determining what is sacred and how to act in the presence of the sacred. Consequently, at some level, people worship what they (or their ancestors) have created. This point led Durkheim to conclude that the real object of worship is society itself—a conclusion that many critics cannot accept (Nottingham 1971).

Let us give Durkheim the benefit of the doubt, however, and ask, Is there anything about the nature of society that makes it deserving of such worship? In reply to this question, Durkheim gave what sociologist W. S. F. Pickering (1984) calls a "virtual hymn to society, a social Gloria in

Excelsis" (p. 252). Durkheim maintained that society transcends the individual life, because it frees people from the bondage of nature (as in "nature versus nurture"). How does it accomplish this task? Chapter 4 presented cases showing the consequences of extreme isolation, neglect, and limited social contact. Such cases make it clear that "it is impossible for a person to develop without social interaction" (Mead 1940, p. 135). In addition, studies of mature and even otherwise psychologically and socially sound persons who experience profound isolation—astronauts orbiting alone in space, prisoners of war placed in solitary confinement, individuals who volunteer to participate in scientific experiments in which they are placed in deprivation tanks—show that when people are deprived of contact with others, they lose a sense of reality and personal identity (Zangwill 1987). The fact that we depend so strongly on society supports Durkheim's view that for the individual "it is a reality from which everything that matters to us flows" (Durkheim, cited in Pickering 1984, p. 252).

Durkheim does not, however, claim that society provides us with perfect social experiences: "Society has its pettiness and it has its grandeur. In order for us to love and respect it, it is not necessary to present it other than it is. If we were only able to love and respect that which is ideally perfect, . . . God Himself could not be the object of such a feeling, since the world derives from Him and the world is full of imperfection and ugliness" (quoted in Pickering 1984, p. 253). Durkheim observed that whenever any group of people has strong conviction, that conviction almost always takes on a religious character. Religious gatherings and affiliations become ways of affirming convictions and mobilizing the group to uphold them, especially when the group is threatened.

A Critique of the Functionalist Perspective of Religion

■ **CORE CONCEPT 6:** The functionalist perspective tends to underestimate the negative ways in which people use religion. To claim that religion functions as a strictly integrative force is to ignore the long history of wars between different religious groups and the many internal struggles between factions within the same religious group. For example, although the Afghan *mujahideen* united to oppose the Soviet occupation and its secular government, many competing factions existed within the *mujahideen*. After the Soviets withdrew from Afghanistan, the former *mujahideen* commanders became the major power brokers, and each took control of different cities outside Kabul. At this point, as had happened in the past, the same rugged terrain that made it impossible for the Soviets to gain control over the entire country likewise made it difficult for any one internal group to consolidate its power. Tribal elders and religious students, in turn, tried to wrestle control

INTERSECTION OF BIOGRAPHY AND SOCIETY

Personal Images of Jesus

THE RELIGIOUS PAINTER and illustrator Warner Salman (pictured on the right) created the image of Jesus Christ (in the background) in 1940. The number of times this image has been reproduced on "church bulletins, calendars, posters, book marks, prayer cards, tracts, buttons, stickers and stationery" is more than 500 million (Grimes 1994). As a result, many people in the United States have come to think of Jesus' physical appearance as such. Given that Jesus was born in Bethlehem (according to the Christian Bible), a town in the Middle East, is the Salman image the most accurate representation of how Jesus might have looked? Archaeological evidence suggests that the average man at the time of Jesus was 5 feet, 3 inches tall and weighed approximately 110 pounds (Gibson 2004). Student comments suggest that many accept Salman's image but others have come to question it:

© AP Photo

- Whenever I think about what Jesus looks like, I always see Salman's image. However, I believe that Jesus cannot look like that because of the geographic region in which he was born. I like to think that maybe God is female.
- I really can't believe that my image of Jesus is a man-made one. Warner Salman's image has been in my head so long I cannot even comprehend another image.
- It is shocking to me to learn that Jesus probably had dark skin, hair, and facial features, because I have always imagined Jesus to look like Salman's paintings.

- My image of Jesus used to be that of a Caucasian male, but I remember reading in the Bible that he had "hair like sheep's wool and dark skin."
- The Bible does not give an exact physical description of Jesus, but it does say that he was "unattractive to the eye." People think that he was beautifully pale with long wavy brown hair, when really he was unattractive. Now, no one really knows what he looks like. That's why I belong to a church that's Christian nondenominational and that doesn't display or worship pictures of Jesus.

from rebel commanders (U.S. Central Intelligence Agency 2001).

Eventually, the Taliban, with the help of the Pakistani government, rose to power and came to control 90 percent of the country. Its ultimate aim was to establish a pure Islamic state. At first, the Taliban seemed a welcome relief to the chaos of decades of war. Later, its strict interpretation of Islamic laws, which were enforced by amputations and public executions, created widespread resentment among the Afghan people.

This example suggests that religion is not entirely an integrative force. If it were, then it could not be used as the justification for destroying persons who did not support a religion or version of a religion. An Amnesty International

(1996c) document, *Afghanistan: Grave Abuses in the Name of Religion*, outlines numerous human rights violations committed by the Taliban in the name of religion. Those abuses included "indiscriminate killings, arbitrary and unacknowledged detention of civilians, physical restrictions on women for reasons of their gender, the beating and ill-treatment of women, children and detainees, deliberate and arbitrary killings, amputations, stoning and executions."

The functionalist perspective tends to overemphasize the constructive consequences associated with religions' unifying, bonding, and comforting functions. Strict functionalists, who focus only on the consequences that lead to order and stability, tend to overlook the fact that religion can also unify, bond, and comfort believers in such a way that it supports war and other forms of conflict. The conflict perspective acknowledges the unifying and comforting functions of religion, but views such functions as ultimately problematic.

The Conflict Perspective

■ **CORE CONCEPT 7:** Conflict theorists focus on ways in which people use religion to repress, constrain, and exploit others. Scholars who view religion from the conflict perspective focus on how religion turns people's attention away from social and economic inequality. This perspective stems from the work of Karl Marx, who believed that religion was the most humane feature of an inhumane world and that it arose from the tragedies and injustices of human experience. He described religion as the "sigh of the oppressed creature, the sentiment of a heartless world, and the soul of soulless conditions. It is the opium of the people" (Pelikan and Fadiman 1990, p. 80). According to Marx, people need the comfort of religion to make the world bearable and to justify their existence. In this sense, he said, religion is analogous to a sedative.

Even though Marx acknowledged the comforting role of religion, he focused on its repressive, constraining, and exploitative qualities. In particular, he conceptualized religion as an ideology that justifies the status quo. That is, religion is used to rationalize existing inequities or downplay their importance. This aspect of religion is especially relevant with regard to the politically and economically disadvantaged. For them, Marx said, religion serves as a source of false consciousness. That is, religious teachings encourage the oppressed to accept the economic, political, and social arrangements that constrain their chances in this life because they are promised compensation for their suffering in the next world.

This kind of ideology led Marx to conclude that religion justifies social and economic inequities and that religious teaching inhibits protest and revolutionary change. He went so far as to claim that religion would be unnecessary in a truly classless society—that is, a propertyless society providing equal access to the means of production. In the absence of material inequality, exploitation and injustice—the experiences, he claimed, that cause people to turn to religion—would not occur. In sum, Marx believed that religious doctrines shift people's attention away from unjust political and economic arrangements, and these doctrines rationalize and defend the political and economic interests of the dominant social classes. For some contemporary scholars, this legitimating function explains why most religions allow only a specific category of people—men—to be leaders and to handle sacred items.

Consider a more extreme example of religion-inspired inequity: after the Taliban took control of Afghanistan, they placed, in the name of Islam, severe restrictions

Ironically, in its quest to support groups that would fight against the Soviet occupation of Afghanistan, the United States supported producers of opium poppies. As the mujahideen insurgents pushed the Soviets out, they ordered peasants to plant opium. U.S. support helped to turn Afghanistan into the world's largest producer of opium and processed heroin. Today the Afghan economy depends on opium, and profits from its production fund Taliban operations against the United States (Judah 2001).

on women and the population in general. Women were required to appear covered from head to toe; they had to stay home unless accompanied by a close male relative and could not work outside the home or go to school. Nonreligious schools were closed; music, television, and other forms of entertainment were banned; homosexuals were killed. The Taliban's funding came from two major sources: Osama bin Laden and revenue from the production of opium for heroin (Judah 2001).

Sometimes believers twist religion in ways that serve the interests of dominant groups or of groups seeking dominance. During the days of slavery, for example, some Christians prepared special catechisms for slaves to study. Such catechisms included questions and answers like these:

Q: What did God make you for?
A: To make a crop.
Q: What is the meaning of Thou shalt not commit adultery?
A: To serve our heavenly Father, and our earthly Master, obey our overseer, and not steal anything. (Wilmore 1972, p. 34)

A Critique of the Conflict Perspective of Religion

■ **CORE CONCEPT 8:** Conflict theory underestimates the way people are inspired by religion to confront social and economic inequalities. The major criticism leveled at Marx

WORKING FOR CHANGE

Faith-Based Organizations in the United States

BASIC STATISTICS SUGGEST that faith-based organizations operate as significant agents of change in American life:

- Faith ministers to the less fortunate. Eighty percent of the more than 300,000 religious congregations in America provide services to those in need.
- Faith shapes lives. Over 90 percent of urban congregations provide social services, ranging from preschool to literacy programs to health clinics.
- Faith shepherds communities. Polls estimate that between 60 and 90 percent of America's congregations provide at least one social service, and about 75 percent of local congregations provide volunteers for social service programs.
- Faith nurtures children. One out of every six child care centers in America is housed in a religious facility. The nation's largest providers of child care services are the Roman Catholic Church and the Southern Baptist Convention.

- And perhaps most significantly, faith inspires the faithful to love their neighbors as they'd love themselves. Nearly one-quarter of all Americans volunteer their time and effort through faith-based organizations. Many of America's best ideas—and best results—for helping those in need have come not from the federal government but from grassroots communities, private and faith-based organizations of people who know and care about their neighbors. For years, America's churches and charities have led the way in helping the poor achieve dignity instead of despair, self-sufficiency instead of shame.

Sources: Excerpted from "President Bush's Faith-Based and Community Initiative," www.whitehouse.gov/fbci (August 1, 2004); prepared remarks of Attorney General John Ashcroft, White House Faith-Based and Community Initiatives Conference (June 13, 2004).

---

and the conflict perspective of religion is that, contrary to that perspective, religion is not always a sign or tool of oppression. Sometimes religion has been used as a vehicle for protesting or working to change social and economic inequities (see Working For Change: "Faith-Based Organizations in the United States"). Consider that 95 Christian leaders sent President George W. Bush a letter emphasizing the fundamental moral responsibility the Bible states that humans have for protecting the environment; they pointed out that the book of Genesis "records that God beholds creation as 'very good' and commands us to 'till and tend the garden'" (New York *Times* 2004).

Liberation theology represents one such approach to religion. Liberation theologians maintain that Christians have a responsibility to demand social justice for the marginalized peoples of the world, especially landless peasants and the urban poor, and to take an active role at the grassroots level to bring about political and economic justice. Ironically, this interpretation of Christian faith and practice is partly inspired by Marxist thought, in that it advocates raising the consciousness of the poor and teaching them to work together to obtain land and employment and to preserve their cultural identity.

Sociologist J. Milton Yinger (1971) identifies at least two interrelated conditions under which religion can become a vehicle of protest or change. In the first condition, a government or other organization fails to advance its ideals (such as equal opportunity, justice for all, or the right to bear arms). In the second condition, a society becomes polarized along class, ethnic, or sectarian lines. In such cases, disenfranchised or disadvantaged groups may form sects or cults and "use seemingly eccentric features of the new religion to symbolize their sense of separation" and to rally their followers to fight against the establishment, or the dominant group (p. 111). In the United States, one religion that emerged in reaction to society's failure to ensure equal opportunity was the Nation of Islam.

liberation theology A religious movement based on the idea that organized religions have a responsibility to demand social justice for the marginalized peoples of the world, especially landless peasants and the urban poor, and to take an active role at the grassroots level to bring about political and economic justice.

Conflict theorists underestimate the ways in which religion inspires people to confront social and economic inequalities. This nun helps run the Sisters of Charity Baby Clinic in Madagascar.

Elijah Muhammad, the man who succeeded the founder of the Nation of Islam, is shown here addressing his followers in 1964, at the height of the civil rights movement. One of his best-known followers is in attendance: Cassius Clay, who later changed his name to Muhammad Ali.

In the 1930s, black-nationalist leader W. D. Fard (who went by a variety of names, including Wallace Fard Muhammad) founded the Nation of Islam and began preaching in the Temple of Islam in Detroit. (When Fard disappeared in 1934, he was replaced by his chosen successor, Elijah Mohammed.) According to Fard, the white man was the personification of evil, and black people, whose religion had been stripped from them upon enslavement, were Muslim. In addition, he taught that the way out was not through gaining the "devil's" (that is, the white man's) approval but through self-help, discipline, and education. Members received an X to replace their "slave name" (hence, Malcolm X). In the social context of the 1930s, this message was very attractive:

> You're talking about Negroes. You're talking about niggers, who are the rejected and the despised, meeting in some little, filthy, dingy little [room] upstairs over some beer hall or something, some joint that nobody cares about. Nobody cares about these people. . . . You can pass them on the street and in 1930, if they don't get off the sidewalk, you could have them arrested. That's the level of what was going on. (National Public Radio 1984a)

The Nation of Islam is merely one example of a religious organization working to improve life for African Americans. Historically, African American churches

have reached out to millions of black people who have felt excluded from the U.S. political and economic system (Lincoln and Mamiya 1990). For example, African American churches did much to achieve the overall successes of the civil rights movement. Indeed, some observers argue that the movement would have been impossible if the churches had not become involved (Lincoln and Mamiya 1990).

The Interplay between Economics and Religion

■ **CORE CONCEPT 9:** Modern capitalism emerged and flourished in Europe and the United States because Calvinism supplied an ideologically supportive spirit or ethic. Max Weber wanted to understand the role of religious beliefs in the origins and development of **modern capitalism**—an economic system that involves careful calculation of costs of production relative to profits, borrowing and lending money, accumulating all forms of capital, and drawing workers from an unrestricted global labor pool (Robertson 1987).

In his book *The Protestant Ethic and the Spirit of Capitalism*, Weber (1958) asked why modern capitalism emerged and flourished in Europe rather than in China or India (the two dominant world civilizations at the end of the 16th century). He also asked why business leaders and capitalists in Europe and the United States were overwhelmingly Protestant.

To answer these questions, Weber studied the major world religions and some of the societies in which these religions were practiced. He focused on how norms generated by different religious traditions influenced the adherents' economic orientations and motivations. Based

modern capitalism An economic system that involves careful calculation of costs of production relative to profits, borrowing and lending money, accumulating all forms of capital, and drawing labor from an unrestricted global labor pool.

on his comparisons, Weber concluded that a branch of Protestant tradition—Calvinism—supplied a "spirit" or an ethic that supported the motivations and orientations required by capitalism.

Unlike other religions that Weber studied, Calvinism emphasized **this-worldly asceticism**—a belief that people are instruments of divine will and that God determines and directs their activities. Consequently, Calvinists glorified God when they accepted a task assigned to them, carried it out in an exemplary and disciplined fashion, and did not indulge in the fruits of their labor (that is, when they did not use money to eat, drink, or otherwise relax to excess). In contrast, Buddhism, a religion that Weber defined as the Eastern parallel and opposite of Calvinism, "emphasized the basically illusory character of worldly life and regarded release from the contingencies of the everyday world as the highest religious aspiration" (Robertson 1987, p. 7). Calvinists conceptualized God as all-powerful and all-knowing; they also emphasized **predestination**—the belief that God has foreordained all things, including the salvation or damnation of individual souls. According to this doctrine, people could do nothing to change their fate. To compound matters, only relatively few people were destined to attain salvation.

Weber maintained that such beliefs created a crisis of meaning among adherents as they tried to determine how they should behave in the face of their predetermined fate. Such pressures led them to look for concrete signs that they were among God's chosen people, destined for salvation. Consequently, accumulated wealth became an important indicator of whether one was among the chosen. At the same time, this-worldly asceticism "acted powerfully against the spontaneous enjoyment of possessions; it restricted consumption, especially of luxuries" (Weber 1958, p. 171). Frugal behavior encouraged people to accumulate wealth and make investments—important actions for the success of capitalism.

This calculating orientation was not an official part of Calvinist doctrine *per se*. Rather, it grew out of and was supported by Calvinist asceticism and predestination. Given this distinction, we must not misread the role that Weber attributed to the Protestant ethic in supporting the rise of a capitalist economy. According to Weber, the ethic was a significant ideological force; it was not the sole cause of capitalism but *"one of the causes of certain aspects of capitalism"* (Aron 1969, p. 204). Unfortunately, many people who encounter Weber's ideas overestimate the importance that he assigned to the Protestant ethic for achieving economic success, drawing a conclusion that Weber himself never reached: The reason that some groups and societies are disadvantaged is simply that they lack this ethic.

Finally, note that Weber was writing about the origins of industrial capitalism, not about the form of capitalism that exists today, which places a heavy emphasis on consumption and self-indulgence. He maintained that once

This print, whose setting is New York City's Trinity Church, shows men who were considered empire builders in U.S. history: James J. Hill, Andrew Carnegie, Cornelius Vanderbilt, John D. Rockefeller, J. Pierpont Morgan, Jay Cooke or Edward H. Harriman, and Jay Gould. Notice the caption below: "Those Christian men to whom God in his infinite wisdom has given control of the property interests of the country."

established, capitalism would generate its own norms and become a self-sustaining force. In fact, Weber argued, "Capitalism produces a society run along machine-like, rational procedures without inner meaning or value and in which men operate almost as mindless cogs" (Turner 1974, p. 155). In such circumstances, religion becomes an increasingly insignificant factor in maintaining the capitalist system.

Some sociologists argue that industrialization and scientific advances—both driving forces of capitalism—cause society to undergo unrelenting secularization—a process in which religious influences become increasingly irrelevant not only to economic life, but also to most aspects of social life. Others argue that as religion becomes less relevant to economic and social life in general, a significant number of people take on a fundamentalist view; that is, they reexamine their religious principles in an effort to identify and return to the most basic principles (from which believers have departed) and to hold those principles up as the definitive and guiding blueprint for life.

this-worldly asceticism A belief that people are instruments of divine will and that God determines and directs their activities.

predestination The belief that God has foreordained all things, including the salvation or damnation of individual souls.

Secularization and Fundamentalism

■ **CORE CONCEPT 10:** Secularization and fundamentalism fuel each other's growth. Secularization, a process by which religious influences on thought and behavior are reduced, invites a fundamentalist response—a belief in the timelessness of sacred writings and a belief that such writings apply to all kinds of environments. Secularization and fundamentalism are processes that have become increasingly popular. Each has expanded in spite of the other's growth, or possibly in opposition to it. This section examines these two opposing trends.

Secularization

In the most general sense, **secularization** is a process by which religious influences on thought and behavior are reduced. It is difficult to generalize about the causes and consequences of secularization because they vary across contexts. Americans and Europeans tend to associate secularization with an increase in scientific understanding and in technological solutions to everyday problems of living. In effect, science and technology assume roles that were once filled by religious belief and practice. Muslims, in contrast, tend not to attribute secularization to science or to modernization; indeed, many devout Muslims are physical scientists.

From a Muslim perspective, secularization is a Western-imposed phenomenon—specifically, a result of exposure to what many people in the Middle East consider the most negative of Western values. This point is illustrated by the following two observations—one by a Muslim student attending college in Great Britain and another by a Muslim artist:

- If I did not watch out [while I was in college], I knew that I would be washed away in that culture. In one particular area, of course, was exposure to a society where free sexual relations prevailed. There you are not subject to any control, and you are faced with a very serious challenge, and you have to rely upon your own strength, spiritual strength to stabilize your character and hold fast to your beliefs. (National Public Radio 1984b)
- You can't be against foreign influence, because art and culture and science are things that you learn from peo-

secularization A process by which religious influences on thought and behavior are reduced.

subjective secularization A decrease in the number of people who view the world and their place in it from a religious perspective.

U.S.-led troops distribute dolls and other toys to orphaned Afghan girls. What does it mean to receive such toys, which challenge Islamic beliefs about modest dress and other aspects of life?

ple and [that] we exchange. It depends on what kind of foreign influence is being exerted. For example, if we speak of the United States, the U.S. has a tradition of science and of art and culture . . . which can be extremely useful and can enrich our own culture if there is an exchange. But the type of culture that we are getting from the U.S. at the moment is the *Dallas [TV]* series, the cowboy serials, crimes, commodity values, big cars, luxury, things like that. This kind of culture can't help. (National Public Radio 1984b)

While *secularization* is a broad term used to describe the decline of religious influences over everyday life, **subjective secularization** describes a decrease in the number of people who view the world and their place in it from a religious perspective. In other words, paradigms shift from an understanding of the world grounded in religious faith to an understanding grounded in observable evidence and the scientific method. In the face of uncertainty, secular thinkers do not turn to religion or to a supernatural power to intervene; rather, they rely on human intervention or scientific explanation.

Consider the case of the lightning rod. For centuries, the Christian Church maintained that lightning was a visible sign of divine wrath. The most common targets of lighting were the bell towers of churches and cathedrals, as they were the tallest structures. When bell towers were damaged or destroyed, clergy launched campaigns to address local wickedness and to raise money to repair the towers. When Ben Franklin invented the lightning rod, the clergy was forced to admit either that Franklin had the power to thwart divine will or that lightning was simply a natural phenomenon (Stark and Bainbridge 1985).

Considerable debate exists over the extent to which secularization is taking place. Data collected by the Gallup Organization over the past 25 years show little change in the degree of importance that Americans assign to religion. Polls show that 88 percent say they have a religious preference, almost 63 percent are members of one of the more than 250,000 places of worship within the United States (Lyons 2005; White House press release 1995), and 40 percent indicate that they have attended church in the past seven days. According to polls, 57 percent of Americans state that religion is very important in their lives (Gallup 2006), 82 percent "believe in God" (CBS News Poll 2006), and 43 percent claim to be "born again" or "evangelical" (Gallup 2006). Americans classified as black are more likely than those classified as white to claim "born-again" status (63 percent versus 39 percent) (Winseman 2005).

Fundamentalism

In "Popular Conceptions of Fundamentalism," anthropologist Lionel Caplan (1987) offers his readers an extremely clear overview of a complex religious phenomenon: **fundamentalism**—a belief in the timelessness of sacred writings and a belief that such writings apply to all kinds of environments. In its popular usage, the term *fundamentalism* is applied to a wide array of religious groups around the world, including the Moral Majority in the United States, Orthodox Jews in Israel, and various Islamic groups in the Middle East.

Religious groups labeled as fundamentalist are usually portrayed as "fossilized relics . . . living perpetually in a bygone age" (Caplan 1987, p. 5). Americans frequently employ this simplistic analysis to explain events in the Middle East, especially the causes of political turmoil that threatens the interests of the United States (including its demand for oil). Such oversimplification misrepresents fundamentalism, however, and it cannot explain the widespread appeal of contemporary fundamentalist movements within several of the world's religions.

The Complexity of Fundamentalism

Fundamentalism is a more complex phenomenon than popular conceptions would lead us to believe. It is impossible to define a fundamentalist in terms of age, ethnicity, social class, political ideology, or sexual orientation, because this kind of belief appeals to a wide range of people. Moreover, fundamentalist groups do not always position themselves against those in power; in fact, they are equally likely to be neutral or to support existing regimes fervently. Perhaps the most important characteristic of fundamentalists is their belief that a relationship with God, Allah, or some other supernatural force provides answers to personal and social problems. In addition, fun-

damentalists often wish to "bring the wider culture back to its religious roots" (Lechner 1989, p. 51).

Caplan (1987) identifies a number of other traits that seem to characterize fundamentalists. First, fundamentalists emphasize the authority, infallibility, and timeless truth of sacred writings as a "definitive blueprint" for life (p. 19). This characteristic does not mean that a definitive interpretation of sacred writings actually exists. Indeed, any sacred text has as many interpretations as there are groups that claim it as their blueprint. Even members of the same fundamentalist organization may disagree about the true meaning of the texts they follow.

Second, fundamentalists usually conceive of history as a "process of decline from an original ideal state, [and] hardly more than a catalog of the betrayal of fundamental principles" (p. 18). They conceptualize human history as a "cosmic struggle between good and evil": the good results from one's dedication to principles outlined in sacred scriptures, and the evil is an outcome of countless digressions from sacred principles. To fundamentalists, truth is not a relative phenomenon; it does not vary across time and place. Instead, truth is unchanging and knowable through the sacred texts.

Third, fundamentalists do not distinguish between the sacred and the profane in their day-to-day lives. Religious principles govern all areas of life, including family, business, and leisure. Religious behavior, in their view, does not take place only in a church, a mosque, or a temple.

Fourth, fundamentalist religious groups emerge for a reason, usually in reaction to a perceived threat or crisis, whether real or imagined. Consequently, any discussion of a particular fundamentalist group must include some reference to an adversary.

Fifth, one obvious concern for fundamentalists is the need to reverse the trend toward gender equality, which they believe is symptomatic of a declining moral order. In fundamentalist religions, women's rights often become subordinated to ideals that the group considers more important to the well-being of the society, such as the traditional family or the "right to life." Such a priority of ideals is regarded as the correct order of things.

Islamic Fundamentalism

In *The Islamic Threat: Myth or Reality?* (1992), professor of religious studies John L. Esposito maintains that most Americans' understanding of fundamentalism does not apply very well to contemporary Islam. The term *fundamentalism* has its roots in American Protestantism and the

---

fundamentalism A belief in the timelessness of sacred writings and a belief that such writings apply to all kinds of environments.

20th-century movement that emphasized the literal interpretation of the Bible.

Fundamentalists are portrayed as static, literalist, retrogressive, and extremist. Just as we cannot apply the term *fundamentalism* to all Protestants in the United States, we cannot apply it to the entire Muslim world, especially when we consider that Muslims make up the majority of the population in at least 45 countries. Esposito believes that a more appropriate term is **Islamic revitalism** or *Islamic activism*. The form of Islamic revitalism may vary from one country to another, but it seems to be characterized by the following responses to the belief that existing political, economic, and social systems have failed: disenchantment with, and even rejection of, the West; soul-searching; a quest for greater authenticity; and a conviction that Islam offers a viable alternative to secular nationalism, socialism, and capitalism (Esposito 1986).

Esposito (1986) asks, "Why has religion [specifically Islam] become such a visible force in Middle East politics?" He believes that Islamic revitalism represents a "response to the failures and crises of authority and legitimacy that have plagued most modern Muslim states" (p. 53). Recall that after World War I, France and Great Britain carved up the Middle East into nation-states, drawing the boundaries to meet the economic and political needs of Western powers. Lebanon, for example, was created in part to establish a Christian tie to the West; Israel was envisioned as a refuge for persecuted Jews when no country seemed to want them; the Kurds received no state; Iraq became virtually landlocked; and resource-rich territories were incorporated into states with very sparse populations (for example, Kuwait, Saudi Arabia, the United Arab Emirates). Their citizens viewed many of the leaders who took control of these foreign creations "as autocratic heads of corrupt, authoritarian regimes that [were] propped up by Western governments and multinational corporations" (p. 54).

When Arab armies from six states lost "so quickly, completely, and publicly" in a war with Israel in 1967, Arabs were forced to question the political and moral structure of their societies (Hourani 1991, p. 442). Had the leaders and the people abandoned Islamic principles or deviated too far from them? Could a return to a stricter Islamic way of life restore confidence to the Middle East and give it an identity independent of the West? Questions of social justice also arose. Oil wealth and modernization poli-

An Afghan child holds up a leaflet warning against picking up unexploded ordinance. The warning extends to unexploded ordinance that has accumulated over the last 25 years of war. In light of history, is it any wonder that Afghans have rejected nationalism, socialism, and capitalism in favor of Islam?

Department of Defense photo by SSGT Cecilio Ricardo, USAF

Islamic revitalism Responses to the belief that existing political, economic, and social systems have failed—responses that include a disenchantment with, and even a rejection of, the West; soul-searching; a quest for greater authenticity; and a conviction that Islam offers a viable alternative to secular nationalism, socialism, and capitalism.

cies had led to rapid increases in population and urbanization and opened up a vast chasm between the oil-rich countries, such as Kuwait and Saudi Arabia, and the poor, densely populated countries, such as Egypt, Pakistan, and Bangladesh. Western capitalism, which was seen as one of the primary forces behind these trends, seemed blind to social justice, instead promoting unbridled consumption and widespread poverty. Likewise, Marxist socialism (a godless alternative) had failed to produce social justice. It is no wonder that the Taliban and other Muslim groups in Afghanistan rejected secular nationalism, Western capitalism and Marxist socialism. After all, the disintegration of Afghanistan was a direct product of the cold war between the United States and the Soviet Union.

For many people, Islam offers an alternative vision for society. According to Esposito (1986), five beliefs guide Islamic activists (who follow many political persuasions, ranging from conservative to militant):

1. Islam is a comprehensive way of life relevant to politics, law, and society.
2. Muslim societies fail when they depart from Islamic ways and follow the secular and materialistic ways of the West.
3. An Islamic social and political revolution is necessary for renewal.
4. Islamic law must replace laws inspired or imposed by the West.
5. Science and technology must be used in ways that reflect Islamic values, to guard against the infiltration of Western values.

Muslim groups differ dramatically in their beliefs about how quickly and by what methods these principles should be implemented. Most Muslims, however, are willing to work within existing political arrangements; they condemn violence as a method of bringing about political and social change.

The information presented in this section points out the complex interplay of religion with political, economic, historical, and other social forces. Fundamentalism cannot be viewed in simple terms; rather, any analysis must consider the context. Focusing on context allows us to see that fundamentalism can represent a reaction to many events and processes, including secularization, foreign influence, failure or crisis in authority, the loss of a homeland, and rapid change.

Jihad and Militant Islam

In thinking about the meaning of *jihad*, it is important to distinguish between religious and political *jihad*. Many Islamic scholars have pointed out that in the religious sense of the word, true *jihad* is the "constant struggle of Muslims to conquer their inner base instincts, to follow the path to God, and to do good in society" (Milten 2002). But as Daniel Pipes (2003) points out in *Militant Islam Reaches America*, *jihad* as used by those who lead political organizations such as Egyptian Islamic *Jihad*, Yemeni Islamic *Jihad*, and International Islamic Front for the *Jihad* against Jews and Christians means "armed struggle against non-Muslims" and against "Muslims who fail to live up to the requirements of their faith" (p. 264). Militant Islam is an "aggressive totalitarian ideology that ultimately discriminates barely, if at all, among those who stand in its path" (p. 249). In other words, non-Muslims as well as Muslims (who do not share the militants' outlook or who happen to be in the wrong place at the wrong time) can be targets of attack.

How many militant Islamist political *jihadists* exist in the world today? Some estimates follow:

- 15,000—based on the number believed to have been trained in al-Qaida training camps
- 5,000 living in the United States—based on FBI figures created in response to pressure from Congress to identify a number (Scheiber 2003)
- 100,000 or more—based on the U.S. State Department's terrorist watch list or "no-fly list" (Lichtblau 2003)
- Several thousand—the number of people believed to make up the *inner core of* militant Islamist organizations (Pipes 2003).

Daniel Pipes suggests that it is also important to go beyond the core believers to include second and third rings of adherents. Based on "election data, survey research, and anecdotal evidence, and the opinions of informed observers," Pipes estimates that the second ring consists of 100 to 150 million persons worldwide who support militant Islam but are not part of the core. A third ring consists of an estimated 500 million persons "who do not accept all the particulars" of militant Islam but are sympathetic and supportive of the anti-American stance. "That such a multitude hates the United States is sobering indeed" (Pipes 2003, p. 248).

■ VISUAL SUMMARY OF CORE CONCEPTS

■ **CORE CONCEPT 1: When sociologists study religion, they are guided by the scientific method and by the assumption that no religions are false.**

When sociologists study religion, they adhere to the scientific method, which requires them to study only observable and verifiable phenomena. When studying religions, sociologists must assume that no religions are false, and they must rid themselves of all preconceived notions of what religion should be.

Steve Evans

■ **CORE CONCEPT 2: All religions have at least three essential characteristics: beliefs about the sacred and the profane, rituals, and a community of worshipers.**

Durkheim believed that all religions (past and present) have three essential features: (1) beliefs about the sacred and the profane, (2) rituals, and (3) a community of worshipers. Ideas about what is sacred are such an important element of religious activity that many researchers classify religions as sacramental, prophetic, or mystical, according to the type of phenomenon their followers consider sacred. Sociologists have identified at least five broad types of religious organizations or communities of worshipers: ecclesiae, denominations, sects, established sects, and cults.

DoD photo by SPC Preston E. Cheeks, USA

■ **CORE CONCEPT 3: Civil religion is an institutionalized set of beliefs about a nation's past, present, and future and a corresponding set of rituals that take on a sacred quality and elicit feelings of patriotism. The dynamics of civil religion are most notable during times of crisis and war and on national holidays.**

Civil religion forges ties between religion and a nation's needs and political interests. Even in the face of internal divisions, national beliefs and rituals can inspire awe, respect, and reverence for country. In times of war, presidents have woven the situation into a historical and mythological framework that gives the country's involvement in the war moral justification and offers the public a vision and an identity for the country. America's civil religion can be traced to a 19th-century political doctrine known as *manifest destiny*—the belief that the United States had a divine mission to serve as a democratic model to the rest of the world, that the country was a redeemer exerting its good influence upon other nations, and that it represented hope to the rest of the world.

Department of Defense photo by MCSN Christopher A. Lussier, USN

■ **CORE CONCEPT 4:** The functionalist perspective maintains that religion serves vital social functions for the individual and for the group.

That some form of religion appears to have existed for as long as humans have lived encourages functionalists to maintain that religion must serve some vital social functions for the individual and for the group. On the individual level, people embrace religion in the face of uncertainty, in intolerable circumstances, and to achieve a successful outcome. Religion functions in several ways to promote group unity and solidarity: (1) shared doctrines and rituals create emotional bonds among believers; (2) all religions strive to raise individuals above themselves; (3) religious rituals function to alleviate individual anxieties, uncertainties, and fears; (4) religion functions as a stabilizing force in times of severe social disturbance and abrupt change.

■ **CORE CONCEPT 5:** The variety of religious responses is endless, because people play a fundamental role in determining what is sacred and how they should act in its presence.

If we operate under the assumptions that all religions are true in their own fashion and that the variety of religious responses is virtually endless, we must realize the role people play in creating religion and in determining what is sacred and how to act in the presence of the sacred. Consequently, at some level, people worship what they (or their ancestors) have created.

■ **CORE CONCEPT 6:** The functionalist perspective tends to underestimate the negative ways in which people use religion.

To claim that religion functions as a strictly integrative force is to ignore the long history of wars between different religious groups and the many internal struggles between factions within the same religious group.

■ **CORE CONCEPT 7:** Conflict theorists focus on ways in which people use religion to repress, constrain, and exploit others.

Conflict theorists focus on how religion turns people's attention away from social and economic inequality and on religion's repressive, constraining, and exploitative qualities. From this point of view, religion is used to rationalize existing inequities or to downplay their importance.

■ **CORE CONCEPT 8:** Conflict theory underestimates the way people are inspired by religion to change social and economic inequalities.

Religion is not always a sign or tool of oppression; it has been used as a vehicle for protesting or working to change social and economic inequities. In particular, liberation theologians maintain that they have a responsibility to demand social justice for the marginalized peoples of the world, especially landless peasants and the urban poor, and to take an active role at the grassroots level to bring about political and economic justice.

■ **CORE CONCEPT 9:** Modern capitalism emerged and flourished in Europe and the United States because Calvinism supplied an ideologically supportive spirit or ethic.

Max Weber focused on understanding how norms generated by different religious traditions influenced adherents' economic orientations and motivations. Based on his comparisons, Weber concluded that a branch of Protestant tradition—Calvinism—supplied a "spirit" or an ethic that supported the motivations and orientations required by capitalism. In particular, Calvinism emphasized this-worldly asceticism—a belief that people are instruments of divine will and that God determines and directs their activities. This belief led followers to look for concrete signs that they were among God's chosen people, destined for salvation. Consequently, accumulated wealth became an important indicator of whether one was among the chosen, and frugal behavior encouraged people not only to accumulate wealth but also to make investments—important actions for the success of capitalism.

THE EMPIRE BUILDERS

■ **CORE CONCEPT 10:** Secularization and fundamentalism fuel each other's growth. Secularization, a process by which religious influences on thought and behavior are reduced, invites a fundamentalist response—a belief in the timelessness of sacred writings and a belief that such writings apply to all kinds environments.

Secularization is a broad term used to describe the decline of religious influences over everyday life, as indicated by a decrease in the number of people who view the world and their place in it from a religious perspective. In the face of uncertainty, secular thinkers do not turn to religion or to a supernatural power to intervene; rather, they rely on human intervention or scientific explanation.

Department of Defense photo by SPC Leslie Angulo, USA

Fundamentalism is a belief in the timelessness of sacred writings and a belief that such writings apply to all kinds of environments. They conceive of history as a process of decline and betrayal of fundamentalist principles, and they often wish to restore the wider culture to its religious roots. Fundamentalists do not distinguish between the sacred and the profane in their day-to-day life, as religious principles govern all areas of life.

Resources on the Internet

Sociology: A Global Perspective Book Companion Web Site

www.cengage.com/sociology/ferrante

Visit your book companion Web site, where you will find flash cards, practice quizzes, Internet links, and more to help you study.

CENGAGENOW™

Just what you need to know NOW!

Spend time on what you need to master rather than on information you have already learned. Take a pre-test for this chapter, and CengageNOW will generate a personalized study plan based on your results. The study plan will identify the topics you need to review and direct you to online resources to help you master those topics. You can then take a post-test to help you determine the concepts you have mastered and what you will need to work on. Try it out! Go to www.cengage.com/login to sign in with an access code or to purchase access to this product.

Key Terms

church   408
civil religion   411
cults   410
denomination   409
ecclesia   408
established sects   410
fundamentalism   425

Islamic revitalism   426
liberation theology   421
modern capitalism   422
mystical religions   406
predestination   423
profane   406
prophetic religions   405

rituals   407
sacramental religions   404
sacred   403
sect   409
secularization   424
subjective secularization   424
this-worldly asceticism   423

15 Population and Urbanization

With Emphasis on India

When sociologists study populations, they consider the rates at which people are born, die, and move into or out of a country or other designated geographical area. They also consider the social factors that influence these rates.

Center for Disease Control/Chris Zahniser, B.S.N., R.N., M.P.H.

▲ India is one of only two countries in the world with a population of over one billion; 41 percent of its population is age 19 or under.

WHY FOCUS ON

India?

INDIA HAS THE second largest population in the world, after the People's Republic of China. In the next 50 years, the country is projected to add 500 million more people, becoming the world's most populous country in 2050. At the time of this writing, India's population stands at 1.13 billion and China's population totals 1.32 billion. The country with the third largest population in the world—the United States, with approximately 302 million people—has 828 million fewer people than does India (see Table 15.1). Although India accounts for 17 percent of the world's population, it covers only 2.4 percent of the world's land mass. India is about one-third the physical size of the United States (U.S. Department of State 2007), and yet it has a population larger than the entire continent of Africa. Given India's large population and relatively small physical size, you may be surprised that only 28 percent of its population lives in urban areas.

Indian government officials pronounced the birth of India's 1 billionth living member as "a moment of celebration, a moment to ponder" (Misra 2000). On the one hand, it was "a moment of celebration," because this milestone could have been reached much sooner had the Indian government not been the first in the world to adopt a national family planning program in the early 1950s (Kapoor 2000). At that time, India's population was 370 million and was increasing by a rate of 1.9 percent per year. If the country had maintained that rate of increase, its population would have reached 1 billion in 1989. Instead, it reached that mark in 2000, 11 years later.

On the other hand, the birth of its billionth member was "a moment to ponder," because at its current rate of growth, India's population will double to 2 billion in 43.8 years. In addition, almost one-third of India's current population—343 million people—is classified as *hungry.* That is, these people consume 80 percent or less of the minimum energy requirements (Kumar 1999). About 25 percent of India's population has an income below the poverty level. The country faces severe strains on its natural resources and environment, including deforestation, soil erosion, overgrazing, desertification, air pollution, and water pollution (U.S. Central Intelligence Agency 2007).

How did India's population grow from 370 million in 1950 to 1 billion 50 years later? How do we know that India's population will surpass that of China in another 50 years? How can

we be sure that a third country will probably never reach 1 billion in population in this century? Why does India have a relatively low level of urbanization? These questions can be answered rather easily, provided that one has a basic understanding of concepts and principles that guide sociological analysis of population.

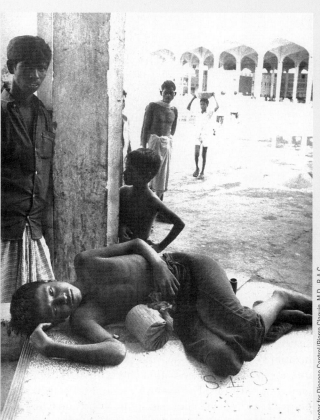

This young man in India is suffering so greatly from malnutrition that he is too weak to sit up to beg for food. Improper nutrition weakens the human immune system, making the malnourished susceptible to all forms of disease.

- On May 11, 2000, India's one billionth resident was born at 5:05 a.m. The country has added more than 130 million people since passing that population milestone.

- India is only the second country in the world, after the People's Republic of China, to pass the 1 billion population mark (Census of India 2001).

- The United States has the third largest population in the world, with 302 million people.

- Every hour, 467 babies are born in the United States while 2,788 are born in India.

- To appreciate the difference between the population sizes of India and the United States, consider that counting at the rate of 1 unit per second to 302 million would take almost 9 years; counting to 1.13 billion would take 35 years.

Table 15.1 The World's Most Populous Countries, 2007

Rank	Country	Population	% of World's Population
1	People's Republic of China	1,321,000,000	20.1
2	India	1,129,000,000	17.1
3	United States	301,819,700	4.6
4	Indonesia	234,950,000	3.6
5	Brazil	186,500,000	2.8
6	Pakistan	163,630,000	2.5
7	Bangladesh	150,500,000	2.3
8	Russia	141,400,000	2.2
9	Nigeria	134,700,000	2
10	Japan	127,720,000	1.9

Source: U.S. Central Intelligence Agency 2007

The Study of Population

■ **CORE CONCEPT 1:** Demography, a subspecialty within sociology, focuses on births, deaths, and migration—major factors that determine population size and rate of growth. **Demography** is a subspecialty within sociology that focuses on the study of human populations, particularly on their size and rate of growth. Population size is determined by births, deaths, age-sex composition, and migration. Most organizations—private, public, and governmental—have an interest in knowing population characteristics, if only for planning purposes. For example, school officials need to know the size of the school-age population and whether it is projected to decline or increase, as this will affect decisions to expand or consolidate the number of schools. Health care planners need to know the size of the population age 65 and older and whether it is projected to decline or increase, as this age group has some of the greatest health care needs. As a final example, city planners need to know the size of the population and whether that population is expected to decline or increase as the result of in- or out-migration, if only to determine the demand for city services like garbage collection (see Working For Change: "The U.S. Census Bureau"). Although we are using basic demographic concepts and principles to study India's population, they can be applied to any population—from that of a neighborhood block to that of the world.

Births

Births add new people to a population. Each year, India adds through births approximately 27 million people and the United States adds about 4.2 million. For comparison, demographers often convert the number of births into a

demography A subspecialty within sociology that focuses on the study of human populations, particularly on their size and rate of growth.

WORKING FOR CHANGE

The U.S. Census Bureau

EVERY TEN YEARS, the U.S. Census Bureau surveys the population of the United States. Among other things, the information gathered allows us to know how many people were born since the last census, moved from one location to another within the United States, and moved into the United States from a foreign country, as well as the age-sex composition of the population. The Census Bureau normally employs nearly 12,000 people, but it expands its workforce dramatically when the census is taken every 10 years. About 860,000 temporary workers were hired for Census 2000.

Local and state governments use statistical information the Census Bureau provides to distribute money to cities and towns. The census data help officials plan and fund items such as these:

- School construction
- Transportation systems
- Police and fire services
- Public housing construction
- Public utilities

Data analysts use census data to make determinations such as these:

- How cities are growing and changing
- Sites for new businesses.
- Whether a community has the workforce a company needs
- Future demand for products
- Sites for day care and nursing care

Source: Adapted from U.S. Census Bureau (2007)

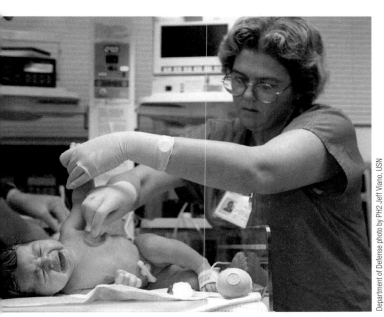

Birth is one of two ways that societies add people; the other is in-migration.

Department of Defense photo by PH2 Jeff Viano, USN

rate. The **crude birth rate** is the annual number of births per 1,000 people in a designated area. That designated area may be the world, a region of the world, a country, or a neighborhood. To calculate the birth rate, we divide the number of births in a year by the size of the area's popula-

tion at the onset of that year and then multiply that figure by 1,000. In 2007 India's crude birth rate was 24 per 1,000 population; the United States' rate was 14 per 1,000.

Sometimes demographers want to know the birth rate for a specific age group within the population. Of particular interest is the birth rate among women of childbearing age (15–54 years old). This rate is called the **age-specific birth rate**. In 2007, the 300.5 million women of childbearing age in India gave birth to 27,116,788 babies. The age-specific birth rate for this group was 90.2 per 1,000 women. In the United States, the age-specific birth rate was 43.5 per 1,000.

In addition to the birth rate, demographers are interested in the **total fertility rate**, which states the average number of children that women in a specific population bear over their lifetime. The average woman in India bears 2.6 children over her lifetime, whereas the average woman in the United States bears 2.1 children. When we consider

crude birth rate The annual number of births per 1,000 people in a designated geographic area.

age-specific birth rate The annual number of births per 1,000 women of a specific age group.

total fertility rate The average number of children that women in a specific population bear over their lifetime.

Death reduces a society's population size. When deaths outnumber births and in-migration, societies decline in size.

Joe Zachariah

that a woman has the potential to bear 20 to 25 children over her lifetime, we are led to ask, "What factors help to explain why most women do not have this many children?" We will address this question later in the chapter.

Deaths

Deaths reduce the size of a population. Each year, India loses about 7.9 million people to death; the United States loses 2.4 million people. This loss is often expressed as a rate. The **crude death rate** is the annual number of deaths per 1,000 people in a designated area. Like the crude birth

---

crude death rate The annual number of deaths per 1,000 people in a designated geographic area.

infant mortality rate The annual number of deaths of infants one year old or younger for every 1,000 such infants born alive.

migration The movement of people from one residence to another.

migration rate A rate based on the difference between the number of people entering and the number of people leaving a designated geographic area in a year. We divide that difference by the size of the relevant population and then multiply the result by 1,000.

push factors The conditions that encourage people to move out of a geographic area.

pull factors The conditions that encourage people to move into a geographic area.

---

rate, it is calculated by dividing the number of deaths in a year by the area's population size at the onset of that year and then multiplying that number by 1,000 (see Table 15.2). The crude death rate for India is 7 deaths per 1,000 population, while the rate for the United States is 8 deaths per 1,000 population. One reason India has a lower death rate is that its population has a greater proportion of young people than the U.S. population has.

As with birth rates, we can calculate the death rates for specific segments of the population, such as for men, for women, or for specific age categories such as one year old or younger.

The death rate among children one year old or younger is called the **infant mortality rate**. Infant mortality is calculated by dividing the number of deaths among those one year old or younger by the total number of births in that year and then multiplying that result by 1,000. In the United States, the infant mortality rate is approximately 6 per 1,000; in India the rate is 40 per 1,000. Thus, for every 1,000 babies born in India, 40 die before they reach the age of one year.

Migration

Migration is the movement of people from one residence to another. That movement increases a population if the people are moving in, reduces the population if they are moving out, or makes no difference if they are simply moving within the geographic area of interest. Each year about 56,500 more people leave India than move into the country. By comparison, almost 1 million more people enter the United States than leave it each year. Often migration is expressed as a rate. To calculate the **migration rate**, we first determine the difference between the number of people entering and the number of people leaving a designated geographic area in a year. Next we divide that difference by the size of the relevant population, and then we multiply the result by 1,000. We can calculate the migration rate for towns, cities, counties, states, countries, or any other region of the world.

Migration results from two factors. **Push factors** are the conditions that encourage people to move out of an area. Common push factors include religious or political persecution, discrimination, depletion of natural resources, lack of employment opportunities, and natural disasters (droughts, floods, earthquakes, and so on). A dramatic example of a push factor was the 2005 Hurricane Katrina, which pushed 60 percent of New Orleans' population out of the city, changing overnight the city's size from 454,863 to 187,525 (U.S. Bureau of the Census 2006). If we consider the entire Gulf Coast population, the number of people pushed out of the area exceeds one million (Nossiter 2006)

Pull factors are the conditions that encourage people to move into an area. Common pull factors include

Department of Defense photo by SPC Eric D. Moore, Wyarng
Department of Defense photo by CMSGT Gonda Moncada, USAF

In 2005 Hurricane Katrina pushed more than one million people out of the Gulf Coast region of the United States. Some with cars were able to leave; many without cars were stranded in their homes or made their way to shelters such as the Super Dome.

employment opportunities, favorable climate, and tolerance. Migration can be placed into two broad categories: international and internal.

International Migration. International migration involves the movement of people between countries. Demographers use the term **emigration** to denote the departure of individuals from one country or other geographic area to take up residence elsewhere, and the term **immigration** to denote the entrance of individuals into a country or other geographic area of which they are not natives to take up residence there. Most governments restrict the numbers of people who can immigrate. Sometimes governments encourage the immigration of certain categories of people, such as nurses, to fill occupations characterized by a shortage of workers. An estimated 110,000 nursing positions go unfilled each year in the United States. To relieve the shortage, the United States has recruited nurses from Canada, Ireland, the Philippines, and, most recently, India (Rai 2003).

Worldwide, three major flows of intercontinental (and thus, international) migration occurred from 1600 and through the early 20th century:

- The massive exodus of European peoples to North America, South America, Asia, and Africa to establish colonies and commercial ventures. In some cases, the colonizers eventually displaced native peoples and established independent countries (as occurred in what are now the United States, Brazil, Argentina, Canada, New Zealand, Australia, and South Africa).
- The smaller flow of Asian migrants to East Africa, the United States (including Hawaii, which did not become

a state until 1959), and Brazil, where they provided cheap labor for major transportation and agricultural projects.
- The forced migration of some 11 million Africans by Spanish, Portuguese, French, Dutch, and British slave traders to the United States, South America, the Caribbean, and the West Indies.

India became entangled in these international migration flows after the British Empire abolished slavery in 1833 and the French abolished it in 1844. After abolition, the British established an indentured labor system that encouraged Indians to migrate from India to other colonies to replace the lost slave labor. From 1883 to 1914, hundreds of thousands of Indians (mostly males) left for Mauritius, Jamaica, Natal, British Guiana, French Guiana, Burma, South Africa, and other destinations. Emigration rates for Indians were especially high during major famines and epidemics in India, such as occurred in 1837 (when 8.5 million Indians died), 1861 (when 13.3 million died), 1866 (when 16.2 million died), 1874 (when 17.7 million died), and 1877 (when 26.9 million died) (Chailiand and Rageau 1995).

Today the government of India estimates that 20 million Indians are living abroad and that their combined income represents 35 percent of India's gross domestic

emigration The departure of individuals from one country or other geographic area to take up residence elsewhere.

immigration The entry of individuals into a country or other geographic area of which they are not natives to take up residence there.

INTERSECTION OF BIOGRAPHY AND SOCIETY

Moving to the United States from Liberia

THERE ARE MORE than 33,000 non-U.S. citizens serving in the U.S. military, which means that at some point in their life each immigrated to the United States. One among the 33,000 is a Liberian native named Nimley Tabue. Tabue's parents came from different tribes. He said his parents' tribal differences did not affect his family until a war between the tribes erupted in 1989. "My father refused to kill, so [rebels] tried to kill him," Tabue said.

Tabue remembers fleeing through the country for three days as a child. "We stopped by a river once to get some water," said Tabue, who was with his mother and siblings at the time. "I held my 4-month-old brother in my arms as he died." According to Tabue, his father, Aloysius Tabue, traveled to America searching for ways to improve his family's life, and he called home often. "I learned about the Marines from my father," Tabue said. "He would say, 'If you guys come over here, make sure you do something with your life. The Marines will give you something no other service can.'"

Because of the ongoing war around him, school became less of a priority, and Tabue was taken out of school following the second grade. He, along with his mother and sister, came to Chicago to live with his father. At 12 years old, Tabue jumped back into the school swing. But after four years without touching a book, school presented a new challenge. "I forgot how to do math, and my English was bad," Tabue said. "I had to go to school over the summer and take extra classes."

After years of extra classes, Tabue's name was added to the high school honor roll. Tabue had not planned on leaving Chicago, but he remembered what his father had always told him about the Corps. "He told me, 'This is where they separate the men from the boys,'" Tabue said. Adjusting to boot camp was harder than any English class. . . . "The first day was horrible. I almost lost my tem-

People move for reasons. Sociologists use push and pull factors to describe those reasons. Those push and pull factors are often the remote and impersonal forces of history. In the case of the Liberian man profiled above, the push factor was civil war within his home country and the pull factor was opportunity in the United States, especially the chance to join the Marine Corps.

per when the drill instructor got in my face. . . . But I told myself it was just a mind game. I had trouble speaking in third person (as required in boot camp). Instead of saying 'This recruit requests permission to use the head,' I would say, 'I would like to use the head.' Drill instructors didn't really like that." When the Crucible—the grueling 54-hour field exercise that is the culmination of boot camp—came, Tabue found his role in the platoon. "He stepped up," drill instructor Marine Staff Sgt. Nathan Nofziger said. "He wasn't a squad leader, but he acted as one." After Marine Corps recruit training, Tabue will become a mortar man in the Marine Corps Reserve.

Source: "West African Immigrant Heeds Father's Words, Joins U.S. Marines" by Lance Cpl. Dorian Gardner, USMC Special to American Forces Press Service.

---

product. For example, an estimated 3–4 million Indians work in the Persian Gulf region and send home $7 billion in remittances each year (Waldman 2003a, 2003b). In contrast to these numbers, an estimated 6.6 million Americans (excluding military and government personnel) live abroad in more than 160 countries; it is not known how much remittance income they send home (see No Borders, No Boundaries: "Countries That Attract the Most Americans Living Abroad").

internal migration The movement of people within the boundaries of a single country—from one state, region, or city to another.

Internal Migration. In contrast to international migration, **internal migration** involves movement of people

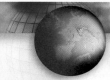

NO BORDERS, NO BOUNDARIES

Countries That Attract the Most Americans Living Abroad

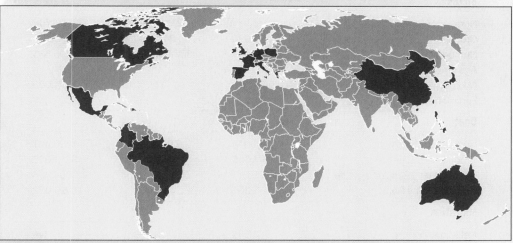

▲ **Figure 15.1**

The map shows the top 25 countries in which Americans living abroad reside. The table gives the estimated numbers of Americans living in each of the top 10 destination countries.

Country	Number of Americans	Country	Number of Americans
Mexico	1,036,300	Italy	166,967
Canada	687,700	Philippines	105,000
United Kingdom	224,000	Australia	102,800
Germany	210,880	France	101,750
Israel	184,195	Spain	94,513

Note: The latest breakdown on Americans living abroad by country is from 1999. Since that time the number is believed to have increased from 4.1 million to 6.6 million (a 61 percent increase).

Sources: U.S. Department of State (2002), Association of Americans Resident Overseas (2007).

within the boundaries of a single country—from one state, region, or city to another. Demographers use the term **in-migration** to denote the movement of people into a designated area and the term **out-migration** to denote the movement of people out of a designated area. One major type of internal migration is the rural-to-urban movement (urbanization) that accompanies industrialization. This type of internal migration will be discussed in the second half of this chapter.

The United States is a country characterized by high rates of internal migration. Consider that each year 43.4 million Americans move (change residences). More than 56 percent of that number move from one residence to another within the same county. Approximately 20 percent move from one county to another within the same state. Another

in-migration The movement of people into a designated geographic area, such as a country, region, or city.

out-migration The movement of people out of a designated geographic area, such as a country, region, or city.

Table 15.2   Key Demographic Indicators for the United States, India, and the World

	United States	India	World
Total Population (July 2007)	301,139,947	1,129,866,154	6,602,224,175
Births			
Number of Births	4,218,971	27,116,788	132,704,706
Crude Birth Rate	14.01/1,000	24/1,000	20.1/1,000
Number of Women Age 15–54	97,061,559	300,527,000	1,880,405,535
Age-Specific Birth Rate (Women Age 15–54)	43.5/1,000	90.2/1,000	70.6/1,000
Fertility Rate	2.1	2.9	2.6
Deaths			
Number of Deaths	2,409,119	7,909,063	54,798,461
Crude Death Rate	8/1,000	7/1,000	8.3/1,000
Deaths Within First Year of Life	25,313	316,362	2,384,808
Infant Mortality Rate	6/1,000	40/1,000	43.5/1,000
Migration			
Net Migration	933,534	-56,493	NA
Migration Rate	3.1/1,000	-0.00005	NA
Population Growth			
Rate of Natural Increase	0.6	1.7	1.16
Population Change (July 2007–June 2008)*	2,743,384	19,151,232	72,906,245
Growth Rate	0.9	1.7	1.2
Total Population (June 31, 2008)	303,883,333	1,149,017,386	6,680,130,420

* Population change = births − deaths + net migration.

Sources: U.S. Bureau of the Census 2007; U.S. Central Intelligence Agency 2007.

20 percent (8.7 million people) move from one state to another (U.S. Census Bureau 2000). Yet, despite this high interstate mobility, 70 percent or more of residents in some states—Ohio, Pennsylvania, Kentucky—have lived there all their life (U.S. Census Bureau 2000).

An estimated 30 percent of India's total population has been involved in some sort of migration in the past 10 years. This migration has been dominated, however, by short-distance rural-to-rural movements among females upon marriage. Usually the woman moves from her village to a nearby village where her husband lives (Stephenson et al. 2003).

natural increase The number of births minus the number of deaths occurring in a population in a year.

rate of natural increase The number of births minus the number of deaths occurring in a population in a year, divided by the size of the population at the beginning of the year.

Population Growth

The population size of a geographic area constantly changes, depending on births, deaths, and migration flows. Demographers calculate annual growth in population size according to the following formula: (number of births − number of deaths) + (in-migration − out-migration). Each year, India's population grows by about 19 million and the U.S. population grows by about 2.7 million (see Table 15.2). To determine the rate of population growth in India and the United States, divide each of these numbers by the country's population size at the beginning of the year. The annual growth rate for India is 1.5 percent, whereas the rate for the United States is 0.9 percent.

Sometimes sociologists choose to focus on a population's **natural increase**, the number of births minus the number of deaths occurring in a year. When we divide this number by the size of a population at the beginning of the year, we establish the **rate of natural increase**. The natural increase for the planet is about 0.6 percent per year. At this annual rate of natural increase, the world's population will double from its current size of 6.6 billion in 60 years.

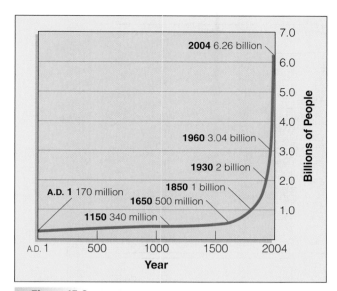

▲ Figure 15.2   World Population Growth, AD 1 to 2007
The graph shows that the world's population reached 1 billion around 1850. Eighty years later, the population doubled to 2 billion. Today the world's population stands at 6.6 billion.

Doubling time is the estimated number of years required for a country's population to double in size. India, with a natural growth rate of 1.7 percent, will double its population of 1.13 billion in about 49 years. The United States, with a natural growth rate of 0.6 percent, will double its population in about 126 years. However, if we factor in population increase due to immigration, the United States' growth rate is 0.9 percent and its doubling time is estimated at 80 years.

Figure 15.2 shows world population growth since AD 1. Note that the population has doubled five times in the last 2,000 years and that the time between the doublings has decreased dramatically, even alarmingly. The graph shows that it took approximately 1,150 years for the world's population to double from 170 million in AD 1 to 340 million in 1150. In 1960 the world's population reached 3.04 billion, and it took just 30 years to double to 6.26 billion.

From 1920 to 1930 the world's population reached 2 billion people, taking less than 100 years to double from 1 billion in 1850. Because of this dramatic increase, demographers sought to explain and predict population growth by putting forth the theory of the demographic transition.

Age-Sex Composition

■ **CORE CONCEPT 2: The age-sex composition of a population helps demographers predict birth, death, and migration rates.** A population's age and sex composition is commonly depicted as a **population pyramid**, a series of horizontal bar graphs, each representing a different five-year age cohort. A **cohort** is a group of people born around the same time—in this case, within a five-year time frame—who share common experiences and perspectives by virtue of the time they were born. To create a population pyramid, we construct two bar graphs for each cohort—one for males and the other for females. We place the bars end to end, separating them by a line representing zero. Typically, the left side of the pyramid depicts the number or percentage of males that make up each cohort, and the right side depicts the number or percentage of females. We stack the bar graphs according to age—the age 0–4 cohort forming the base of the pyramid and the age 70–90+ cohort forming the apex. The population pyramid allows us to compare the sizes of the cohorts and to compare the numbers or percentages of males and females in each cohort.

The population pyramid offers a snapshot of the number of males and females in the various cohorts at a particular time. Generally, a country's population pyramid approximates one of three shapes: expansive, constrictive, or stationary. An **expansive pyramid** is triangular; it is broadest at the base, and each successive bar is smaller than the one below it. The relative sizes of the cohorts in expansive pyramids show that the population is increasing and consists disproportionately of young people. A **constrictive pyramid** is narrower at the base than in the middle. This shape shows that the population consists disproportionately of middle-aged and older people. A **stationary pyramid** is similar to a constrictive pyramid, except that all cohorts other than the oldest are roughly the same size (see Figure 15.3).

Knowing age-sex composition helps demographers predict birth, death, and migration rates. For example, almost 60 percent of India's population consists of men

doubling time The estimated number of years required for a country's population to double in size.

population pyramid A series of horizontal bar graphs, each representing a different five-year age cohort, that allows us to compare the sizes of the cohorts.

cohort A group of people born around the same time (such as a specified five-year period) who share common experiences and perspectives by virtue of the time they were born.

expansive pyramid A triangular population pyramid that is broadest at the base, with each successive cohort smaller than the one below it. This pyramid shows that the population consists disproportionately of young people.

constrictive pyramid A population pyramid that is narrower at the base than in the middle. It shows that the population consists disproportionately of middle-aged and older people.

stationary pyramid A population pyramid in which all cohorts (except the oldest) are roughly the same size.

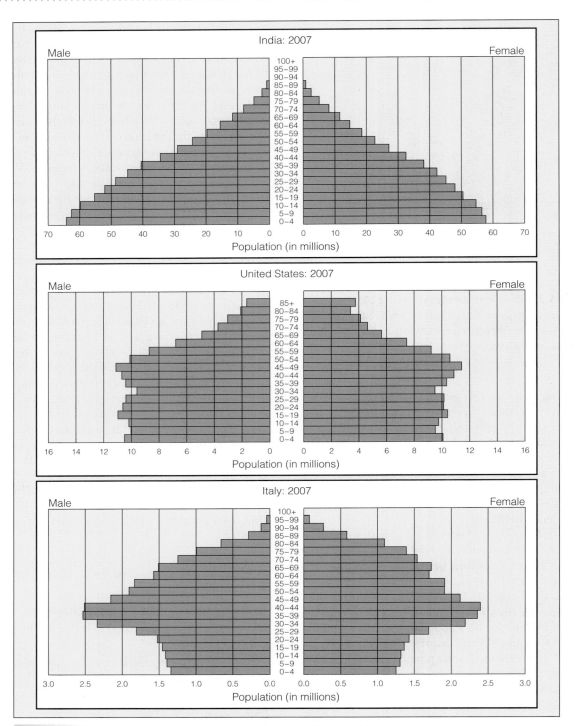

▲ **Figure 15.3**   **Expansive, Constrictive, and Stationary Population Pyramids**

(a) India's population pyramid is classified as expansive, because it is broadest at the base, and each succes-sive bar is smaller than the one below it. The relative sizes of the cohorts indicate that India's population is increasing and consists disproportionately of young people.

(b) The United States age-sex distribution yields a nearly stationary pyramid, because, except for the older categories, each cohort is roughly the same size.

(c) The age-sex distribution for Italy can be labeled constrictive, because it is narrower at the base than in the middle—showing that the population consists disproportionately of middle-aged and older people.

Source: U.S. Census Bureau (2007).

The sex ratio can be used to describe any group of people. The Epidemic Intelligence Service (EIS) graduating class of 2004 consists of medical doctors, researchers, and scientists who belong to epidemic surveillance and response units that travel all over the world. Pictured are 30 males and 52 females—a sex ratio of 17 females for every 10 males.

and women age 15–54—the childbearing age for women. This fact helps explain why India's birth rate is higher than that of the United States and how India's population is growing by approximately 15.4 million people each year. Understanding age-sex composition also helps explain why the death rate in India is lower than the U.S. death rate. India has an official death rate of 8, while the United States has one of 7. Approximately 12.4 percent of the U.S. population is older than age 65, compared with only 5 percent of India's population.

If we know age-sex composition, we can calculate the **sex ratio**—the number of females for every thousand males (or another preferred constant, such as 10, 100, or 10,000). India's population pyramid shows about 65 million males and 59 million females age 0–4. That translates into 900 females for every 1,000 males. Sex ratios that heavily favor males are affected by the practice of female infanticide, the general neglect of females, maternal mortality, and migration patterns. Cultural beliefs and values about the worth of women relative to men also play an important role. In India (as in China) sons are valued more than daughters. Why? By tradition, a male and his bride live with his parents and support and care for them in their old age. When a son marries, the bride's family pays his family a dowry (Rhode 2003). This practice may explain why mobile ultrasound companies advertise with the slogan "Pay 500 rupees now and save 50,000 later" (*Economist* 2003).

The Theory of Demographic Transition

■ **CORE CONCEPT 3:** The demographic transition links the birth and death rates in Western Europe and North America to the level of industrialization and economic development. In the 1920s and early 1930s, demographers observed birth and death rates in various countries. They soon noticed that both birth and death rates were high in Africa, Asia, and South America. In eastern and southern Europe, death rates were declining and birth rates remained high. In Western Europe and North America, birth rates were declining and death rates were low. At that time, demographers observed that Western Europe and North America had the following sequence of birth and death rates:

1. Birth and death rates remained high until the mid-18th century, when death rates began to decline.
2. As the death rates decreased, the population grew rapidly, because more births than deaths occurred. The birth rates began to decline around 1800.
3. By 1920, both birth and death rates had dropped below 20 per 1,000 (see Figure 15.4).

Based on these observations, demographers put forth the theory of the demographic transition. They proclaimed that the characteristics of a country's birth and death rates are linked to its level of industrial or economic development, and they hypothesized that the less economically and industrially developed countries would follow this pattern.

Note that this three-stage model documents the general situation; it should not be construed as a detailed description of the experiences of any single country. Even so, we can say that all countries have followed the essential pattern of the demographic transition, although they have differed in the timing of the declines and the rates at which their populations have increased since death rates began to fall. The theory of the demographic transition also sought to explain the events that caused birth and death rates to drop in Western Europe and North America, and to predict when these declines would occur in the rest of the world.

Stage 1: High Birth and Death Rates

For most of human history—the first 2 to 5 million years—populations grew very slowly, if at all. The world population remained at less than 1 billion until around AD 1850, when it began to grow explosively. Demographers speculate that growth until that time was slow because **mortality crises**—violent fluctuations in the death rate, caused

sex ratio The number of females for every thousand males (or another preferred constant, such as 10, 100, or 10,000).

mortality crises Violent fluctuations in the death rate, caused by war, famine, or epidemics.

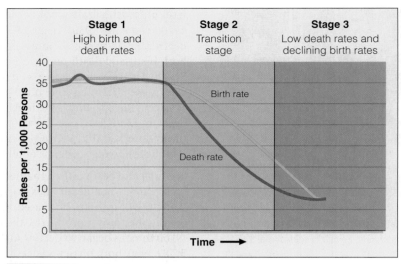

Stage 1
High birth and
death rates

Stage 2
Transition
stage

Stage 3
Low death rates and
declining birth rates

Birth rate

Death rate

Rates per 1,000 Persons

Time →

▲ **Figure 15.4**   **The Demographic Transition**
The theory of the demographic transition can be represented by a graph of the historical changes in birth and death rates in Western Europe and the United States.

Center for Disease Control

The plague known as the Black Death hit Europe, the Middle East, and Asia in the mid-14th century, recurring periodically for approximately 300 years. The plague's name came from one of its symptoms: gangrene.

by war, famine, or epidemics—were a regular feature of life. Stage 1 of the demographic transition is often called the stage of high potential growth: if something happened to cause the death rate to decline—for example, improvements in agriculture, sanitation, or medical care—the population would increase dramatically. In this stage, life is short and brutal; the death rate almost always exceeds 50 per 1,000. When mortality crises occur, the death rate seems to have no limit. Sometimes half the population is affected, as when the Black Death struck Europe, the Middle East, and Asia in the mid-14th century (the plague recurred periodically for approximately 300 years). Within 20 years of its onset, the plague killed an estimated three-fourths of all people in the affected populations.

Another mortality crisis—but one that has not received as much attention as the Black Death—affected the indigenous populations of North America when Europeans arrived in the 15th century. A large proportion of the native population died because they had no resistance to diseases such as smallpox, measles, tuberculosis, and influenza, which the colonists brought with them. Historians continue to debate what proportion of the native population died because of this contact; estimates range from 50 to 90 percent.

In stage 1, then, average life expectancy at birth remained short—perhaps 20 to 35 years— with the most vulnerable groups being women of reproductive age, infants, and children younger than age 5. It is believed that women gave birth to large numbers of children and that the crude birth rate was about 50 per 1,000—the highest rate recorded, and thus believed possible, for humans. Families remained small, however, because one of every three infants died before reaching age 1, and another died before reaching adulthood. If the birth rate had not remained high, the society would have become extinct. Demographer Abdel R. Omran (1971) estimates that in societies where life expectancy at birth is 30 years, each woman must have an average of seven live births to ensure that two children survive into adulthood. She must bear six sons to ensure that at least one son survives into adulthood. In Western Europe before 1650, high mortality rates were associated closely with food shortages and famines. Even when people did not die directly from starvation, they died from diseases that preyed on their weakened physical state.

Thomas Malthus ([1798] 1965), a British economist and an ordained Anglican minister, concluded that "the

Malthus used the misleading phrase "positive checks" for events that increase death, such as a massive landslide that destroyed an entire Filipino village. Here U.S. Marines search for 1,000 missing persons.

Preserving foods in air-tight jars, cans, or pouches and then heating to destroy contaminating microorganisms improved the nutritional status of the population in industrialized societies, leading to lower death rates.

power of population is so superior to the power in the earth to produce subsistence for man, that premature death must in some shape or other visit the human race" (p. 140). According to Malthus, **positive checks** served to keep population size in line with the food supply. He defined *positive checks* as events that increase deaths, including epidemics of infectious and parasitic diseases, war, famine, and natural disasters. An estimated 20.1 million people died over the past century due to drought (10 million), floods (6.9 million), windstorms (1.2 million), earthquakes (1.9 million), volcanoes (96,000), and landslides (54,000) (Marsh 2005). Malthus believed that the only moral ways to prevent populations from growing beyond what the food supply could support were delayed marriage and celibacy.

Stage 2: Transition

Around 1650, mortality crises became less frequent in Western Europe; by 1750, the death rate there had begun to decline slowly. This decline was triggered by a complex array of factors associated with the onset of the Industrial Revolution. The two most important factors were (1) increases in the food supply, which improved the nutritional status of the population and increased its ability to resist diseases, and (2) public health and sanitation measures, including the use of cotton to make clothing and new ways of preparing food. The following excerpt elaborates on these trends:

> The development of winter fodder for cattle was important; fodder allowed the farmer to keep his cattle alive during the winter, thereby reducing the necessity of living on salted meats during half of the year. . . . [C]anning was discovered in the

positive checks Events that increase deaths—including epidemics of infectious and parasitic diseases, war, famine, and natural disasters—and thus keep population size in line with the food supply.

It is not uncommon for people born from 1945 to 1963 in the United States to have five to nine siblings (*left*). People born after 1980 are more likely to have just one sibling (*right*).

early nineteenth century. This method of food preservation laid the basis for new and improved diets throughout the industrialized world. Finally, the manufacture of cheap cotton cloth became a reality after mid-century. Before then, much of the clothes were seldom if ever washed, especially among the poor. A journeyman's or tradesman's wife might wear leather stays and a quilted petticoat until they virtually rotted away. The new cheap cotton garments could easily be washed, which increased cleanliness and fostered better health. (Stub 1982, p. 33)

Contrary to popular belief, advances in medical technology had little influence on death rates until the turn of the 20th century—well after improvements in nutrition and sanitation had caused dramatic decreases in deaths due to infectious diseases. Over a 100-year period, the death rate fell from 50 per 1,000 to less than 20 per 1,000, and life expectancy at birth increased to approximately 50 years of age. As the death rate declined, fertility remained high. Fertility may even have increased temporarily, because improvements in sanitation and nutrition enabled women to carry more babies to term. With the decrease in the death rate, the **demographic gap**—the difference

demographic gap The difference between a population's birth rate and death rate.

urbanization An increase in the number of cities in a designated geographic area and growth in the proportion of the area's population living in cities.

between the birth rate and the death rate—widened, and the population grew substantially.

Accompanying the unprecedented growth in population was **urbanization**, an increase in the number of cities and growth in the proportion of the population living in cities. (As recently as 1850, only 2 percent of the world's people lived in cities with populations of 100,000 or more.) Around 1880, fertility began to decline. The factors that caused birth rates to drop are unclear and continue to inspire debate among demographers. But one thing is clear: the decline was not caused by innovations in contraceptive technology, because the methods available in 1880 had been available throughout history. Instead, the decline in fertility seems to have been associated with several other factors.

First, the economic value of children declined in industrial and urban settings, as children no longer represented a source of cheap labor but rather became an economic liability to their parents. Second, with the decline in infant and childhood mortality, women no longer had to bear a large number of children to ensure that a few survived. Third, a change in the status of women gave them greater control over their reproductive life and made childbearing less central to their life.

Stage 3: Low Death Rates and Declining Birth Rates

Around 1930, both birth and death rates fell to less than 20 per 1,000, and the rate of population growth slowed considerably. Life expectancy at birth surpassed 70 years—an

The British Empire used its colonies to supply the world with goods: "the gold and diamonds of South Africa; the wood, wheat, butter, and meat of Australia; the wheat, fish, and timber of Canada; the sugar of the West Indies; the rubber and tin of Malaysia; the wheat, cotton, jute, rice, and tea of India" (Demangeon 1925, p. 14). Here we see female workers in a Bombay textile mill and **tea** pickers in the Himalayas of India.

Library of Congress Prints and Photographs Division

Library of Congress Prints and Photographs Division

unprecedented statistic. The remarkable successes in reducing infant, childhood, and maternal mortality rates permitted accidents, homicides, and suicide to become the leading causes of death among young people. The reduction of the risk of dying from infectious diseases ensures that people who would have died of infectious diseases in an earlier era can survive into middle age and beyond, when they face an elevated risk of dying from degenerative and environmental diseases (such as heart disease, cancer, and strokes). For the first time in history, people age 50 and older account for more than 70 percent of annual deaths. Before stage 3, infants, children, and young women accounted for the largest share of deaths (Olshansky and Ault 1986).

As death rates decline, disease prevention becomes an important issue. The goal is to live not merely a long life but a "quality life" (Olshansky and Ault 1986; Omran 1971). As a result, people become conscious of the link between health and lifestyle (sleep, nutrition, exercise, and drinking and smoking habits). In addition to low birth and death rates, stage 3 is distinguished by an unprecedented emphasis on consumption (made possible by advances in manufacturing and food production technologies).

Theoretically, Japan and all of the countries of Western Europe and North America now occupy stage 3 of the demographic transition. At one time, some sociologists and demographers maintained that the so-called developing countries would follow this model of development. But industrialization in these countries has differed so fundamentally from that in Western Europe and the United States that they are unlikely to follow the exact same path.

Industrialization: An Uneven Experience

■ **CORE CONCEPT 4:** Industrialization was not confined to Western Europe and North America. It pulled people from across the planet into a worldwide division of labor and created long-lasting, uneven economic relationships between countries. As we learned elsewhere in this textbook (see Chapters 1, 5, and 6), the Industrial Revolution was not unique to Western Europe and the United States. In fact, during this revolution, people from even the most seemingly isolated and remote regions of the planet became part of a worldwide division of labor. Industrialization's effects were not uniform; they varied according to country and region of the world.

With regard to industrialization the countries of the world are commonly placed into two broad categories, such as "developed" and "developing." Comparable but equally misleading terms for this dichotomy include *industrialized/industrializing* and *first world/third world*. These terms are misleading because they suggest that a country is either industrialized or not industrialized. The dichotomy implies that a failure to industrialize is what makes a country poor, and it camouflages the fact that as Europe and North America plunged into industrialization, they took possession of Asia, Africa, and South America—establishing economies there that served the industrial needs of the colonizers, not the needs of the colonized. The point is that countries we label as "developing" or "industrializing" were actually part of the Industrial Revolution from the beginning.

India, for example, was considered the "crown jewel" of the British Empire, because "she has been for her masters the richest and most valuable of their colonies of exploitation" (Demangeon 1925, p. 238) (see Global

World Map of the British Empire and Commonwealth

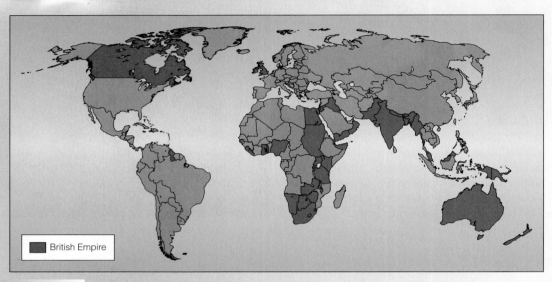

British Empire

▲ **Figure 15.5**

The territories the map shown in red were part of the British Empire and Commonwealth (including all the dominions, colonies, protectorates, and mandates) when it was at its peak in the 1930s. The British Empire lasted 350 years; at its peak it controlled (on some level) 25 to 30 percent of the world's land and 25 percent of its population.

Comparisons: "World Map of the British Empire and Commonwealth").

The World Bank, the United Nations, and other international organizations use a number of indicators to distinguish between so-called developed and developing countries, including the following: doubling time, infant mortality, total fertility, per capita income, percentage of the population engaging in agriculture, and per capita energy consumption. Instead of the term *developing*, *industrializing*, or *third-world* we will use the term **labor-**

intensive poor economies. Instead of the term *developed*, *industrialized*, or *first-world* we will use the term **core economies**. Table 15.3 shows how labor-intensive poor economies differ from core economies on a number of important indicators, such as per capita electricity consumption and doubling time.

The Demographic Transition in Labor-Intensive Poor Economies

■ **CORE CONCEPT 5:** The theory of the demographic transition does not apply to India and most other former colonies. Birth rates have remained high relative to death rates and have taken longer to decline; the level of rural-to-urban migration has been unprecedented. Colonization and its legacy help explain why the model of the demographic transition does not exactly apply to India and most other former colonies. Labor-intensive poor economies differ from core economies in several characteristics: in particular, they have experienced relatively high birth rates despite declines

labor-intensive poor economies Economies that have a lower level of industrial production and a lower standard of living than core economies. They differ markedly from core economies on indicators such as doubling time, infant mortality, total fertility, per capita income, and per capita energy consumption.

core economies Economies that have a higher level of industrial production and a higher standard of living than labor-intensive poor economies. They include the wealthiest, most highly diversified economies in the world.

Table 15.3   Demographic Differences between Labor-Intensive Poor Economies and Core Economies

Labor-intensive poor economies differ markedly from core economies on a number of important indicators, including doubling time, infant mortality, total fertility, and per capita income.

	Population Doubling Time (years)	Infant Mortality (per 1,000)	Total Fertility	Per Capita Income ($U.S.)
Labor-Intensive Poor Economies				
Afghanistan	20.7	157.4	6.6	800
Haiti	41.9	63.9	4.9	1,800
India	46	40	2.9	3,700
Core Economies				
United States	76	6	2.1	43,500
Japan	636	2.8	1.2	33,100
Germany	1,750	4.1	1.4	31,400

Source: U.S. Central Intelligence Agency (2007).

in their death rates, resulting in more-rapid population growth and unprecedented levels of rural-to-urban migration (urbanization).

Birth and Death Rates

Birth rates, although still relatively high, are declining in most labor-intensive poor economies. The demographic gap remains wide, however. Figure 15.6 shows high birth rates in India until the 1920s, when the rates began a gradual decline to the present level of 23.3 per 1,000.

Sociologists Bernard Berelson (1978) and John Samuel (1997) have identified some important "thresholds" associated with declines in fertility:

1. Less than 50 percent of the labor force is employed in agriculture. (The economic value of children decreases in industrial and urban settings.)
2. At least 50 percent of persons between the ages of 5 and 19 are enrolled in school. (Especially for women, education "widens horizons, sparks hope, changes status concepts, loosens tradition, and reduces infant mortality" (Samuel 1997).
3. Life expectancy is at least 60 years. (With increased life expectancy, parents can expect their children to survive infancy and early childhood.)
4. Infant mortality is less than 65 per 1,000 live births. (When parents have confidence that their babies and children will survive, they limit the size of their families.)
5. Eighty percent of the females between the ages of 15 and 19 are unmarried. (Delayed marriage is important when it is accompanied by delayed sexual activity or protected premarital sex.)

Center for Disease Control/Chris Zahniser, B.S.N., R.N., M.P.H.

Almost 60 percent of India's labor force makes a living in the agricultural sector of the economy.

In India total fertility has declined during the past four decades from 5.9 children per woman to 2.9. Two of the thresholds Berelson names (life expectancy and infant mortality) have been met in India. Note the current data for India:

1. Almost 60 percent of India's labor force is employed in agriculture.
2. About 47 percent of Indian women and 59 percent of Indian men have at least a secondary education.
3. Life expectancy in India is 69 years.

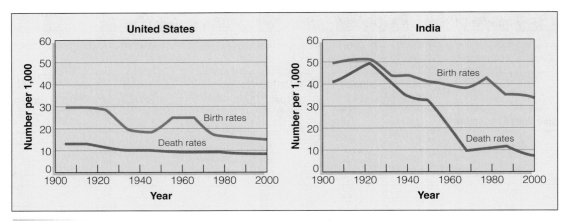

▲ **Figure 15.6** Crude Birth and Death Rates: United States and India

These graphs show that as the death rate declined, the demographic gap was much wider and more persistent in India than in the United States. The U.S. graph shows dramatically how the 1950s earned the name "baby boom." Notice that there is a wider gap between India's birth and death rates which suggests that India's population growth increased more dramatically than did U.S. population growth.

4. In India infant mortality is 40 per 1,000 live births.
5. Almost all women in India are married by age 19.

Another factor responsible for declining birth and fertility rates is a country's family planning policies. At times, India's population policies and programs have been taken to disastrous extremes. The most well-known example occurred during the Emergency Period of 1975–1977, when the government, under Prime Minister Indira Gandhi, promoted forced sterilizations and vasectomies and offered people "incentives" to "volunteer" for these procedures. Although forced and targeted sterilizations are no longer official government policy, female sterilization appears to be a major form of contraception. In 2002–2003, an estimated 4.6 million tubal ligations were performed in India, compared with only 114,426 vasectomies (New York Times 2003). Of course, India's government is not the only one in the world to establish contraceptive policies. Until 1999 the Japanese government banned low-dose oral contraception, arguing that its use promoted promiscuity and reduced reliance on condoms (both promiscuity and failure to use condoms increase the risk of contracting HIV and other sexually transmitted diseases) (Queenan 1999).

As a second example, the U.S. government designates $80 million per year for sex education programs that must teach an abstinence-only (no premarital sex) curriculum and must emphasize "that sexual activity outside of marriage is likely to have harmful psychological and physical effects" (Lancet 2002, p. 97).

demographic trap The point at which population growth overwhelms the environment's carrying capacity.

Death Rates

Death rates in the labor-intensive poor economies have declined much more rapidly than they did in the core economies. Demographers attribute the relatively rapid decline to cultural diffusion. That is, the labor-intensive poor economies imported Western technology—such as pesticides, fertilizers, immunizations, antibiotics, sanitation practices, and higher-yield crops—which caused an almost immediate decline in the death rates. In India the death rate was so high at the beginning of the 20th century that the gap between the birth rate and the death rate was relatively small. Around 1920 the death rate began to steadily decline because of medical advancements, especially mass inoculations.

The swift decline in death rates and relatively slower decline in birth rates has caused the populations in India and other labor-intensive poor economies to grow very rapidly. Some demographers believe that such countries may be caught in a **demographic trap**—the point at which population growth overwhelms the environment's carrying capacity:

Once populations expand to the point where their demands begin to exceed the sustainable yield of local forests, grasslands, croplands, or aquifers, they begin directly or indirectly to consume the resource base itself. Forests and grasslands disappear, soils erode, land productivity declines, water tables fall, or wells go dry. This in turn reduces food production and incomes, triggering a downward spiral. (Brown 1987, p. 28)

A United Nations World Food Programme (WFP) report takes issue with the idea that rapid population growth, by itself, overwhelms the environment's carrying capacity. In fact, the WFP argues that enough food exists to nourish the estimated 800 million people in the world who go to bed hungry each night. The WFP maintains that "many poor people simply do not have the money to buy enough food" (UN World Food Programme 2001).

The world economy is structured so that access to food, and to the economic resources needed to acquire adequate amounts of food, remains highly uneven across the globe. Consider that an estimated 350 million Indians are classified as malnourished. Yet, over the years, the Indian government has stored 58.1 metric tons of surplus wheat and rice, much of which has rotted. The root of the crisis is a government policy of buying grain from farmers at high prices while reducing food subsidies to consumers. This policy has resulted from pressure exerted by farm lobbyists, who seek higher prices for farmers, and international lenders, who advise the Indian government to reduce subsidies to consumers (Sengupta 2006; Waldman 2002a). This food surplus situation undermines the belief that famines are caused by shortages of food or drastic declines in food production. Upon closer examination, we find that the cause is soaring food prices and low wages, which combine to make it hard for low-income people to buy the food (Massing 2003).

Regarding food shortages, we could also point to problems in the ways that chemical technologies and genetically modified seeds have been used. During the 1960s and 1970s the Indian government encouraged chemical companies such as Union Carbide to locate in the country. These firms were supposed to be part of the Green Revolution, a plan to relieve chronic food shortages and help the country become self-sufficient in food production through agricultural technologies, including treated seeds, pesticides, and fertilizers (Derdak 1988). In addition, the chemical companies used local labor and regional raw materials, thereby creating employment opportunities. The manufactured chemicals were used to prevent malaria and other insect-borne diseases and to protect crops and harvests from insects, rodents, and diseases.

Agricultural yields have indeed been impressive; India is now the second largest grain producer in the world (Indo-Asian News Service 2005). Nevertheless, the means chosen to achieve self-sufficiency in food production have had negative long-term consequences. For example, only the wealthiest Indian farmers have been able to purchase the chemicals. In conjunction with mechanization, the chemicals have allowed these farmers to farm more efficiently. As a result, they have pushed the poor, small farmers, who have been unable to compete, off the land and out of business:

> Ironically, as food production rises, so do the numbers of hungry and poor in that the poor are pushed off the land and are unable to buy the food produced. In India, Monsanto has promoted their herbicide Roundup—a big part of their GM package—partly on the grounds that it will reduce costs by

reducing the need for labor. That is, instead of income going to local people for weeding the fields, wealth will instead flow to the corporation. (Christian Aid 2005)

In the end, millions of poor farmers have migrated to the cities in search of work, but the cities have been unable to absorb them. Moreover, agricultural chemicals and genetically modified seeds have effectively destroyed India's biodiversity. Before the Green Revolution, India produced some 200,000 varieties of rice (*Times* of India 2005). Now, those varieties may have been reduced by as much as 75 percent.

The speed at which death rates have declined in relation to birth rates in labor-intensive poor economies constitutes only one of several important differences that distinguish these countries from the core economies. Another important difference concerns urbanization.

Urbanization

■ **CORE CONCEPT 6:** Urbanization is a transformative process by which people migrate from rural to urban areas and change the way they use land, interact, and make a living. Urbanization encompasses (1) the process by which a population becomes concentrated in urban areas and (2) the corresponding changes in land use, social interaction, economic activity, and landscape. What constitutes an urban area varies by country. Some countries, such as the United States, define a city or town as urban if its population is at least 2,500. Other countries set the population at 20,000. In addition to population size, urban areas may be defined on the basis of population density or according to the percentage of the population working in the manufacturing or service sectors (Population Reference Bureau 2001).

The world has 337 **agglomerations**, urban areas with populations of 1 million or more (Mongabay.com 2007). Of these, 24 lie within India and nine within the United States. Within the agglomeration category is the **mega city**. This term was first introduced by the United Nations in the 1970s to designate agglomerations of 8 million or more people. In the 1990s the U.S. government revised the definition of a mega city to include populations of at least 10 million people. According to this new definition, 23 mega cities exist in the world. Three lie in the United States: New York (29.9 million in urban area, 8.1 million in city), Los Angeles (16.5 million in urban area, 3.8 million in city),

---

agglomerations Urban areas with populations of 1 million or more.

mega city An agglomeration of at least 8 million (UN definition) or 10 million (U.S. definition) people.

One of the most recognizable symbols of urbanization is traffic jams, such as this one on the road between Nasik and Bombay, India. One notable difference between Indian and U.S. traffic patterns relates to the number of passengers per vehicle. Notice that most cars in this photo are full.

and Chicago (10.8 million in urban area, 2.9 million in city). Three lie in India: Bombay (12.7 million in urban area), Calcutta (11.3 million in urban area), and Delhi (16.7 million in urban area, 10 million in city). Approximately 4.2 percent (26.6 million) of all people in the world live in mega cities (Brockerhoff and Brennan 1998). There are 2,747 smaller urbanized areas with populations from 100,000 to 999,999 residents. Approximately one-third of the world's population lives in an urban area with a population of 100,000 or more (Mongabay.com 2004).

Urbanization in Labor-Intensive Poor Economies versus Core Economies

Urbanization in labor-intensive poor economies differs in several major ways from urbanization in core economies. At comparable points in the demographic transition, the rate of urbanization in labor-intensive poor economies far exceeds that of the core economies. Consider that during the 25 years of its most rapid growth, New York City

over-urbanization A situation in which urban misery—poverty, unemployment, housing shortages, insufficient infrastructure—is exacerbated by an influx of unskilled, illiterate, and poverty-stricken rural migrants, who have been pushed into cities out of desperation.

increased its population by 2.3 million. During the 25 years of its most rapid growth, Bombay, India, added 11.2 million people (Brinkhoff 2001).

Why such a difference? For one thing, "new worlds" existed to siphon off the population growth of Europe (Light 1983). Millions of Europeans who were pushed off the land were able to migrate to sparsely populated places, such as North America, South America, South Africa, New Zealand, and Australia. If the people who fled Europe for other lands in the 18th and 19th centuries had been forced to make their living in the European cities, the conditions there would have been much worse than they actually were:

> Ireland provides the most extreme example. The potato famine of 1846–1849 deprived millions of peasants of their staple crop. Ireland's population was reduced by 30 percent in the period 1845–1851 as a joint result of starvation and emigration. The immigrants fled to industrial cities of Britain, but Britain did not absorb all the hungry Irish. North America and Australia also received Irish immigrants. Harsh as life was for these impoverished immigrants, the new continents nonetheless offered them a subsistence that Britain was unable to provide. (Light 1983, pp. 130–131)

In India, the problem of urbanization is compounded by the fact that many people who migrate to the cities come from some of the most economically precarious sections of India. In fact, most rural-to-urban migrants are not pulled into the cities by employment opportunities; rather, they are forced to move there because they have no alternatives. When these migrants come to the cities, they face not only unemployment, but also a shortage of housing and a lack of services (electricity, running water, waste disposal). One distinguishing characteristic of cities in labor-intensive poor economies is the prevalence of slums and squatter settlements, which are much poorer and larger than even the worst slums in the core economies.

> It is a familiar sight in so-called underdeveloped countries to find somewhere, in the midst of great poverty, . . . a gleaming, streamlined new factory, created by foreign enterprise. . . . Immediately outside the gates you might find a shanty town of the most miserable kind teeming with thousands of people, most of whom are unemployed and do not seem to have a chance of ever finding regular employment of any kind. (Schumacher 1985, p. 490)

Sociologist Kingsley Davis uses the term **over-urbanization** to describe a situation in which urban misery—poverty, unemployment, housing shortages, and insufficient infrastructure—is exacerbated by an influx of unskilled, illiterate, and poverty-stricken rural migrants, who have been pushed into cities out of desperation. In this regard, the United Nations estimates that one billion people worldwide live in slums lacking essential services such as water and sanitation (Dugger 2007).

One vivid and dramatic example is the city of Bhopal, India. This was the site of one of the worst industrial accidents in human history in 1984, when approximately 40 tons of a highly toxic, volatile, inflammable gas used in making pesticides escaped from a Union Carbide storage tank and blanketed the densely populated city. Some 800,000 residents awoke coughing and vomiting, with eyes burning and watering. The accident is known to have killed 22,149 people and to have injured an estimated 578,000. More than 20 years later, 10 to 15 people die each month from injuries related to the accident, and most of the injured live with chronic side effects of their exposure, including lung and kidney damage, visual impairment, and skin diseases and eruptions (Everest 1986; Sharma 2000). More than 20 years after the accident, Union Carbide has still not cleaned up the Bhopal site. Moreover, the company has not revealed the composition of the released gas, claiming trade secrets as the reason (Bhopal.net 2004; Waldman 2002b).

The population of Bhopal stood at 102,000 in 1966. After Union Carbide and other industries settled there in the 1960s, the population grew to 385,000 in 1971, 670,000 in 1981, and 800,000 in 1984. At the time of the accident, approximately 20 percent of Bhopal's 800,000 residents lived in squatter settlements. The location of two of these settlements—directly across from the Union Carbide plant—explains why the deaths occurred disproportionately among the poorer residents. The people who lived in these settlements were paid poverty-level wages, which prevented them from acquiring decent living quarters.

In 2004 a tsunami hit six countries around the Indian Ocean, including coastal areas of India, leaving more than 8,000 people there dead.

Urban versus Nonmetropolitan

According to the latest American Housing Survey (2005), the United States had 124.4 million year-round housing units. Of these units, 94.5 million (76 percent) lie in a **metropolitan statistical area (MSA)**. An MSA includes one or more cities with at least 50,000 residents, surrounded by densely populated counties. In the United States, 261 geographical areas are classified as MSAs. The 94.5 million housing units can be further classified as being either part of a central city or a suburb. A **central city** is the largest city within an MSA. In some MSAs two or more cities are designated as central cities (U.S. Census Bureau 2000b). Some 35.8 million housing units lie in central cities in the United States.

Almost 60 million U.S. housing units lie in a **suburb**, an urban area outside the political boundaries of a city. Almost 30 million housing units lie in geographical areas classified as **nonmetropolitan**—geographical areas beyond the political boundaries of a central city and its suburbs.

As you might imagine, it is difficult to generalize about central cities, because they can range in size from 50,000 to almost 8 million residents. Likewise, suburbs can be part of cities or towns that range in size from 2,500 to 250,000 residents. To complicate matters further, the relationship between suburbs and central cities is so interdependent that it is impossible to think of them as distinct geographic units. The U.S. Conference of Mayors issued a report encouraging community leaders and the public to view cities and their suburbs as socially and economically interdependent units. In fact, the seven largest metropolitan areas in the United States are among the world's

metropolitan statistical area (MSA) One or more cities with at least 50,000 residents, surrounded by densely populated counties.

central city The largest city within a metropolitan statistical area (MSA).

suburb An urban area outside the political boundaries of a city.

nonmetropolitan Characteristic of a geographical area beyond the political boundaries of a central city and its suburbs.

30 largest economies. For example, the economy of the New York metropolitan area ranks 14th in the world, just behind the economy of South Korea, and it ranks higher than the economies of Australia, the Netherlands, Taiwan, and Russia (Schmitt 2001). The other six U.S. metropolitan areas ranking among the world's top 30 economies are Los Angeles (16th), Chicago (18th), Boston (23rd), Washington, D.C. (27th), Philadelphia (29th), and Houston (30th).

Regions classified as nonmetropolitan are equally diverse. They include rural communities, for example. Our views of rural communities may be influenced by popular conceptions (or misconceptions) of these places as being insular and having low crime rates, strong social connections among members, few employment opportunities, distrust of outsiders, high poverty rates, low literacy rates, and few social services and amenities. In reality, communities classified as rural may differ greatly from one another. One rural area may be economically dependent on a single factory there; another rural community may be the site of a regional or state university.

India's equivalent of a nonmetropolitan area is the village, a settlement containing fewer than 5,000 people. An estimated 500,000 villages lie across India, and 80 percent of them have less than 1,000 residents. Village farmers grow wheat, rice, lentils, vegetables, fruits, and many other crops to feed themselves and the nation. While the word *village* may elicit images of self-sufficient communities, an Indian village is usually connected to other villages and to

Robert K. Wallace

An interdependent relationship exists between a city and its suburbs. The bridges connecting northern Kentucky to downtown Cincinnati illustrate this interdependence.

urban areas. Village landowners may hire tenant farmers and laborers to till the soil; villagers who live near urban areas may commute to cities to work. Some migrate to the cities to work for months at a time (Library of Congress (2001a, 2001b).

■ VISUAL SUMMARY OF CORE CONCEPTS

■ **CORE CONCEPT 1:** Demography, a subspecialty within sociology, focuses on births, deaths, and migration—major factors that determine population size and rate of growth.

Population size is determined by births, deaths, and migration. Births add new people to a population and can be expressed as a crude birth rate or an age-specific birth rate. Deaths reduce the size of a population. As with the birth rate, we can calculate the death rate for entire populations or for specific segments. Migration is the movement of people from one residence to another. That movement adds new people to a population if they are moving in, reduces the population size if people are moving out, or makes no difference. Migration, which can be international or internal, results from push and pull factors.

Center for Disease Control/Chris Zahniser, B.S.N., R.N., M.P.H.

■ **CORE CONCEPT 2:** The age-sex composition of a population helps demographers predict birth, death, and migration rates.

A population's age and sex composition is commonly depicted as a population pyramid, which offers a snapshot of the number of males and females in the various age cohorts at a particular time. Generally, a country's population pyramid approximates one of three shapes: expansive, constrictive, or stationary. Knowing age-sex composition helps demographers predict birth, death, and migration rates and calculate the sex ratio.

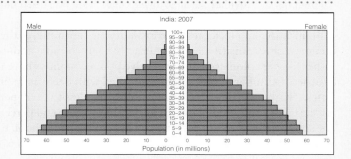

■ **CORE CONCEPT 3:** The demographic transition links the birth and death rates in Western Europe and North America to the level of industrialization and economic development.

The theory of the demographic transition connects the fall in a population's birth and death rates to level of industrialization and economic development. Birth and death rates in Western Europe and North America changed in the following sequence: (1) Birth and death rates remained high until the mid-18th century, when death rates began to decline. (2) As the death rates decreased, the population grew rapidly, because more births than deaths occurred. The birth rates began to decline around 1800. (3) By 1920 both birth and death rates had dropped below 20 per 1,000. Based on these observations, demographers put forth the theory of the demographic transition.

Department of Defense photo by CMSGT Don Sutherland

■ **CORE CONCEPT 4:** Industrialization was not confined to Western Europe and North America. It pulled people from across the planet into a worldwide division of labor and created long-lasting, uneven economic relationships between countries.

The Industrial Revolution was not unique to Western Europe and North America. In fact, during this revolution people from even the most seemingly isolated and remote regions of the planet became part of a worldwide division of labor. Industrialization's effects were not uniform; rather, they varied according to country and region of the world. Consequently, with regard to industrialization, we can place the countries of the world into two broad categories: core economies and labor-intensive poor economies.

Library of Congress Prints and Photographs Division

■ **CORE CONCEPT 5:** The theory of the demographic transition does not apply to India and most other former colonies. Birth rates have remained high relative to death rates and have taken longer to decline; the level of rural-to-urban migration has been unprecedented.

Colonization and its legacy help explain why the model of the demographic transition does not exactly apply to India and most other former colonies. Labor-intensive poor economies have not reached a number of milestones associated with declines in birth rates. These milestones include the following: less than 50 percent of the labor force is employed in agriculture, at least 50 percent of persons between the ages of 5 and 19 are enrolled in school, life expectancy is at least 60 years, infant mortality is less than 65 per 1,000 live births, and 80 percent of the females between the ages of 15 and 19 are unmarried.

Center for Disease Control/Chris Zahniser, B.S.N., R.N., M.P.H.

■ **CORE CONCEPT 6:** Urbanization is a transformative process by which people migrate from rural to urban areas and change the way they use land, interact, and make a living.

The world has 337 agglomerations, urban areas with populations of 1 million or more. Urbanization in labor-intensive poor economies far exceeds that in the core economies, if only because no "new worlds" exist to siphon off such countries' population growth as existed for Europe. In addition, rural migrants tend to come from the most economically precarious segments of the population. The result is over-urbanization, a situation in which urban misery—poverty, unemployment, housing shortages, and insufficient infrastructure—is exacerbated by an influx of unskilled, illiterate, and poverty-stricken rural migrants, who have been pushed into cities out of desperation.

Yann Forget

Resources on the Internet

 ***Sociology: A Global Perspective* Book Companion Website**
www.cengage.com/sociology/ferrante

Visit your book companion Web site, where you will find flash cards, practice quizzes, internet links, and more to help you study.

CENGAGENOW™

Just what you need to know NOW!

Spend time on what you need to master rather than on information you have already learned. Take a pre-test for this chapter, and CengageNOW will generate a personalized study plan based on your results. The study plan will identify the topics you need to review and direct you to online resources to help you master those topics. You can then take a post-test to help you determine the concepts you have mastered and what you will need to work on. Try it out! Go to www.cengage.com/login to sign in with an access code or to purchase access to this product.

Key Terms

16 Social Change

With Emphasis on Greenland

When sociologists study social change, they ask at least three key questions: What has changed? What factors triggered that change? What are the consequences of that change?

© Design Pics Inc./Alamy

▲ Polar bears standing on melting ice floes is one image popularly associated with climate change.

Why Focus on

Greenland?

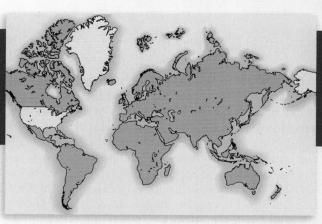

IN 2007 THE Intergovernmental Panel on Climate Change (a joint venture between two United Nations agencies, the World Meteorological Organization and the Environment Programme) issued the report *Climate Change 2007*. The report took six years to write and involved 1,250 authors and 2,500 scientific expert reviewers from 130 countries. The report's message is summarized in this Reuters headline: "U.N. Climate Panel Says Warming Is Man-Made." That same year, a bipartisan U.S. congressional delegation visited Greenland to see "first-hand evidence that climate change is a reality." After the trip, one U.S. representative remarked, "There is just no denying it. We saw the impact on the local people: on their hunting, their fishing, their economic survival" (Kirschbaum 2007).

While climate change is a complex phenomenon, some of the most publicized and vivid images center on one element of the process—melting ice sheets—which the UN report names as "very likely" (90–99 percent probability) contributing to sea-level rise. Images of ice sheets collapsing and polar bears seemingly stranded on ice floes turn our attention to places like Greenland, which features prominently in the former vice president Al Gore's documentary film *An Inconvenient Truth*.

Greenland, a territory of Denmark near the North Pole, possesses the largest reservoir of freshwater on the planet after Antarctica. If Greenland's ice sheet, which covers about 85 percent of the territory, melted entirely, it would release enough water to raise the global sea level by almost 23 feet. In reality,

the sea level has risen by .07 inches per year since 1961, for a total of 3.96 inches. Melting ice from Greenland and Antarctica have "very likely" contributed between .06 inches to .33 inches of that rise (representing 1.5 to 8.3 percent of total sea rise in the past 55 years) (United Nations 2007). Given Greenland's freshwater resources, we should not be surprised to learn that a senior UN water advisor pointed to Greenland as a source of freshwater to meet the world's increasing water shortage. In fact, the Canadian-based Iceberg Corporation of America and a Greenland-based international shipping company have been harvesting Greenland's ice and producing spring water from it for at least 15 years (Business Wire 2000).

We use the concepts and theories in this chapter to answer three important questions: (1) What about social activity has changed since 1750, making it heavily dependent upon fossil fuels? (2) What factors trigger changes in social activity? (3) What are the consequences of those changes? It is not until we answer the third question that we will consider Greenland's situation—specifically how fossil-fuel–dependent social activities and climate change have affected that country. Note that from here on out, we use the term *climate change*, in keeping with the UN report, rather than *global warming*.

FACTS TO CONSIDER

- Since 1750, humans have burned fossil fuels to transport people and goods, first in trains and then also in cars, buses, and planes; to run appliances and light the night; and to regulate temperatures in houses, office buildings, and other indoor environments.

- In 2007 the United Nations issued a report that involved 1,250

  authors and 2,500 scientific experts from 130 countries. That report announced that climate change can no longer be denied or doubted and that human or social activity since 1750 has "very likely" caused the rise in the planet's temperatures.

- The Petition Project (2007) challenges the idea that climate change has been man-made and fossil-fuel

  driven. The project claims support from 17,100 basic and applied scientists—all from one country, the United States.

- The United States accounts for 4.6 percent of the world's population and contributes 25 percent of all greenhouse gas emissions.

Social Change

■ **CORE CONCEPT 1:** When sociologists study any social change, they take particular interest in identifying tipping points—situations in which previously rare events snowball into commonplace ones. Sociologists define **social change** as any significant alteration, modification, or transformation in the organization and operation of social activity. When sociologists study change, they must first identify the social activity that has changed or is changing. The list of possible topics is virtually endless. It includes changes in the division of labor; in how people communicate; in the amounts of goods and services people produce, sell, or buy from others; in the size of the world population; and in the average life span (Martel 1986).

Sociologists are particularly interested in **tipping points**, situations in which a previously rare (or seemingly rare) event, response, or opinion becomes dramatically more common. The process by which the rare becomes commonplace is gradual; at first the change seems so small that few if any people notice it, and if they notice it, they dismiss it as an aberration or simply reject it out of hand. But at some point a critical mass is reached so that the next small increment of change "tips" the system in a dramatic way. The concept "tipping point" is borrowed from physics, which applies it to a situation in which very small weights are continually added to a balanced object. At some point when adding the weights, the object suddenly and completely topples. The moment before this occurs is the tipping point.

In sociology the concept has been applied to white flight—a situation in which residents classified as white decide en masse to move out of a neighborhood because there is "one too many" new residents classified as black. Researchers who studied white flight in the late 1950s and early 1960s found that white families did not move out when the first black family moved in; they stayed as long as the number of black families remained very small; whites moved out en masse at the point when they judged there to be "one too many" black households (Grodzins 1958).

Scientists predict that if climate change continues at the current pace, the ice covering about 85 percent of Greenland will melt in the next 1,000 years (Lovgren 2004). While this seems like a long time, scientists warn that we may be approaching a tipping point. In the 12-year period 1995–2006, 11 years rank among the 12 warmest recorded since 1850. The total rise in global surface temperature in

social change Any significant alteration, modification, or transformation in the organization and operation of social life.

tipping points Situations in which a previously rare (or seemingly rare) event, response, or opinion becomes dramatically more common.

Until 1971 the United States produced more oil than it consumed. This 1944 image shows workers managing large valves that regulated the flow of oil into tanker ships. The oil was to be shipped for use by the U.S. armed forces and allies. At that time the United States was producing 1.6 billion barrels of oil per year.

the 100-year period 1906–2006 is equivalent to 1.3 degrees Fahrenheit; the 50-year period 1956–2006 has contributed 1.17 degrees Fahrenheit to that total, accounting for 90 percent of the increase. Scientists seek to identify the point at which the next slight rise in global surface temperature will trigger a far greater rise. For example, if Siberia's frozen peat bog was to thaw rapidly, it could release billions of tons of the greenhouse gas methane. That thawing could trigger a far greater increase in global surface temperature (Sample 2005).

Sociologists who study changes in human social activity since 1750 are likewise interested in tipping points. Possible tipping points with regard to U.S. dependence on oil are the year when the United States became dependent on foreign countries to meet its domestic oil consumption needs, the year in which its dependence on foreign sources reached 50 percent or more, and the year that U.S. oil production peaked.

Between 1957 and 1963 the United States produced 4 million more barrels of oil per day than it consumed. By

Table 16.1	Fossil Fuel Consumption since 1850 and U.S. Share of World Consumption			
Energy Source	Year First Used for Industrial Purposes	Amount Used in 1850	Amount Used in 2006	% Consumed by U.S.
Coal	1748	994 million tons	5 billion tons	19
Oil	1859	5,475 barrels	30.2 billion barrels	25
Natural gas	1821		2,819 trillion cubic meters	22

Sources: World Coal Institute 2007; U.S. Central Intelligence Agency 2007.

1970 the amount of surplus oil had dropped to 1 million barrels per day. In 1971 the United States turned to foreign sources for the first time in its history to meet domestic demand. In that year, the United States produced 11.3 million barrels per day, an amount it would never exceed (Yergin 1991). In 1993 U.S. dependence on foreign oil exceeded 50 percent of demand; today that dependence hovers around 60 percent. Currently the United States produces 40 percent of the oil it needs—8 million barrels per day. Total U.S. demand is expected to rise from 20 million barrels per day to 28.3 million by 2025 (U.S. Department of Energy 2007). Sociologists are interested in how these landmark events are connected to such topics as U.S. military involvement around the world (especially in the Middle East) and U.S. relations with China and India (countries with increasing oil demand).

Upon identifying a topic, sociologists ask at least three key questions:

- What has changed?
- What factors triggered that change?
- What are the consequences of the change?

Library of Congress Prints and Photographs Division

This 1910 photograph shows a West Virginia coal miner operating the brakes on a motor train carrying coal. The United States has such abundant supplies of coal (unlike oil and natural gas) that it produces more coal each year than it consumes.

Changes in Social Activity

■ **CORE CONCEPT 2:** A number of key sociological concepts help sociologists answer the question "What about social activity has changed since 1750?" Those concepts describe ongoing changes: industrialization and mechanization, globalization, rationalization, McDonaldization, urbanization, and the information explosion. Social change is an important sociological topic. In fact, sociology first emerged as a discipline attempting to understand social change. Recall that the early sociologists were obsessed with understanding the nature and consequences of the Industrial Revolution—an event that triggered dramatic and seemingly endless changes in every area of human social activity. The discipline of sociology offers a number of broad concepts to help us identify and describe the way social activity has changed since 1750. These concepts include industrialization and mechanization, globalization, rationalization, McDonaldization, urbanization, and the information explosion.

Industrialization and Mechanization

We learned in Chapter 1 that the Industrial Revolution is an ongoing process that started as early as 1300 but gained dramatic momentum between 1750 and 1850. The most critical factor driving the momentum was mechanization—the addition of external sources of power, such as coal, oil, and natural gas, to hand tools and modes of transportation. Among other things, these new energy sources replaced wind- and human-powered sailboats with steamships and then freighters, and they replaced horse-drawn carriages with trains, cars, and trucks. The Industrial Revolution turned us into a "hydrocarbon society"—one in which the use of fossil fuels shapes virtually every aspect of our personal and social lives (see Table 16.1). The energy source that lights the night, cools our houses and offices, and powers appliances and tools is likely coal; the energy source that heats our houses,

About 168 million barrels of oil per day (8 percent of global production) are needed to make plastic products (such as containers and toys) and plastic-encased products (such as computers, televisions, and radios) (Fiske 2007).

That 20 million shipments of fresh, frozen, and processed food enter the United States each year is one measure of global interdependence.

workplaces, and water is likely natural gas; and the energy source that enables trains, planes, cars, and buses to move people and goods short and long distances is most certainly oil. When we burn fossil fuels to make the energy to do all these things, we emit greenhouse gases (such as carbon dioxide and methane) into the air.

global interdependence A situation in which the social, political, financial, and cultural lives of people around the world are so intertwined that one country's problems—such as unemployment, drug abuse, environmental pollution, and the search for national security in the face of terrorism—are part of a larger global situation.

globalization The ever-increasing flow of goods, services, money, people, information, and culture across political borders.

Globalization

In the most general sense, **global interdependence** is a situation in which social activity transcends national borders and in which one country's problems—such as unemployment, drug abuse, water shortages, natural disasters, and the search for national security in the face of terrorism—are part of a larger global situation. Because the level of global interdependence is constantly changing, it is part of a dynamic process known as **globalization**—the ever-increasing flow of goods, services, money, people, information, and culture across political borders (Held, McGraw, Goldblatt, and Perraton 1999). Sociologists debate the events that triggered globalization. Theoretically, one could trace its origins back 5 million years to East Africa (believed to be the cradle of human life) and to a time when humans began to spread out and eventually populate and dominate the planet. Other potential dates that mark the start of globalization include the invention of the printing press (1436) and the steam engine (1769). Regardless of the date, there is no question that fossil fuels facilitated globalization. Obviously, humans use trains, cars, buses, boats, planes, phones, and the Internet to deliver people, products, services, and information across national borders (see No Borders, No Boundaries: "Facts about UPS, a Global Package Delivery System"). And as globalization has increased, so has the demand for fossil fuels. One of the most profound measures of global interdependence is that 25,000 shipments of imported food products enter the United States each day. That translates into 20 million shipments per year. For example, "92 percent of all fresh and frozen seafood consumed is imported; 52 percent of

No Borders, No Boundaries

Facts about UPS, a Global Package Delivery System

ONE CORPORATION WHOSE 427,000 employees facilitate indirect interaction between package senders and receivers is United Parcel Service (UPS), founded in 1907. That corporation depends on fossil-fueled vehicles to deliver 1.8 million packages per day to more than 200 countries.

	Employees	U.S.-based	360,000
		Foreign-based	67,000
	Daily Flights	International	796
		Domestic	1,130
	Delivery Fleet	Cars, vans, tractors, motorcycles	94,542
		UPS jet aircraft	284
		Chartered aircraft	323
	Countries served		200+
	Areas served in U.S.		Every address
	Packages Delivered per Day	International	1.8 million
		Domestic	13.8 million
	Packages Picked Up per Day		1.8 million
	Hits on UPS.com per day		15 million

the grapes; 75 percent of the apple juice; and 72 percent of the mushrooms" (Online Newshour 2007).

Rationalization

We learned in previous chapters that **rationalization** is a process whereby thought and action rooted in emotion (such as love, hatred, revenge, or joy), superstition, respect for mysterious forces, and tradition are replaced by value-rational thought and action. Value-rational thought and

action involve striving to find the most efficient way to achieve a valued goal or result (Freund 1968). In the context of industrialization, "the most efficient way" means

rationalization A process whereby thought and action rooted in emotion, superstition, respect for mysterious forces, and tradition are replaced by thought and action grounded in the logical assessment of cause and effect or the means to achieve a particular end.

the fastest and most cost-effective way to achieve a profit (the valued goal).

One important example of rationalization is the profit-making strategy known as **planned obsolescence**, which involves producing goods that are disposable after a single use, have a shorter life cycle than the industry is capable of producing, or go out of style quickly even though the goods can still serve their purpose (Gregory 1947). The market offers an endless number of disposable (single-use) products, including paper cups, paper towels, diapers, cameras, razors, plastic utensils, and paper table cloths. And the market offers many other products that do not seem to last as long as they once did. For example, refrigerators, ovens, washers, and dryers built since 2000 are expected to last 8–12 years, while those built in the 1970s and 1980s lasted 20 years or more (Repair2000.com). Many people buy a new car even though their old car is still in excellent-to-good condition. Similarly, people tend to buy new clothes before they wear out the clothes they already have. Simply consider that we use fossil fuels to manufacture all such products, to deliver them to retailers or to buyers' houses, to make them operate, and to haul them away when owners decide to discard them.

Sociologist Max Weber used the term *rationalization* to refer to the way in which daily life is organized socially to accommodate large numbers of people, and not necessarily to accommodate the way that individuals actually think (Freund 1968). On an individual level, people may be aware that they do not "need" an automobile before the one they drive wears out, but they may find the urge to buy a new car irresistible. Likewise, people may know that as the automobile burns fuel it emits greenhouse gases and other pollutants; they may know they should carpool, walk, or drive a more fuel-efficient car. Still, they may find the efficiency and convenience the automobile offers irresistible. To compound matters, American society is organized around the automobile, so making alternative forms of transportation convenient and accessible is not something in which the American taxpayer has invested. As Weber argued, once people identify a valued goal and decide on the means (actions) to achieve it, they seldom consider less profitable or slower ways to reach the goal.

planned obsolescence   A profit-making strategy that involves producing goods that are disposable after a single use, have a shorter life cycle than the industry is capable of producing, or go out of style quickly even though the goods can still serve their purpose.

McDonaldization   A process whereby the principles governing the fast-food industry come to dominate other sectors of the American economy, society, and the world.

Missy Gish

Pharmacies have borrowed the concept of "drive-thru" service from the fast-food industry. This phenomenon is known as McDonaldization.

The McDonaldization of Society

In Chapter 6 we studied the organizational trend **McDonaldization**, a process whereby the principles governing the fast-food industry come to dominate other sectors of the American economy, society, and the world (Ritzer 1993). These principles are (1) efficiency, (2) quantification and calculation, (3) predictability, and (4) control. Efficiency means offering a product or service that can move consumers quickly from one state of being to another (say, from hungry to full, from fat to thin, from uneducated to educated, or from sleeplessness to sleep). Quantification and calculation mean providing numerical indicators so customers can easily evaluate a product or service (for example: We deliver within 30 minutes! Lose 10 pounds in 10 days! Earn a college degree in 24 months! Limit menstrual periods to four times—or to no times—a year! Obtain eyeglasses in one hour!). Predictability means ensuring that a service or product will be the same no matter where or when it is purchased. Control means planning out in detail the process of producing and acquiring a service or product (for example, by assigning a limited task to each worker, by filling soft drinks from dispensers that automatically shut off, or by having customers stand in line).

There is no question that McDonaldization is a fossil fuel–driven phenomenon. For example, pharmacies, banks, and car wash businesses have adopted "drive-thru" services to facilitate their goal of moving customers from one state of being to another quickly. A bank customer drives up to the window or ATM with the goal of adding money to a checking account, deposits a check, and leaves with a receipt showing the higher balance to be recorded the next business day. A pharmacy customer pulls up to the window with the goal of feeling better, picks up a prescription, and drives away with medicine to help manage or treat a health condition. A driver pulls up to a self-service car wash with the goal of cleaning the outside of the car, inserts money into a machine, moves the car through a wash cycle, and drives away with a clean exterior. In all cases, services are rendered while automobile motors run. Of course, "drive-thru" service is not all there is to McDonaldization. Whatever the service offered—a college degree in 18 months, a medical checkup integrated into a one-stop shopping establishment, matchmaking with success guaranteed in six weeks, a prepaid funeral, or the cheapest air flight—we can find one or more of the McDonaldization principles operating.

Urbanization

Another fossil fuel–driven phenomenon is **urbanization**—a transformative process in which people migrate from rural to urban areas and change the way they use land, interact, and make a living. In 1900, 13 percent of the world's population was considered urban; that percentage increased to 29 percent in 1950 and reached 40 percent in 2005 (United Nations 2005). From a global perspective, urban populations include not only city dwellers but also suburbanites and even residents of small towns. No uniform definition states what constitutes an urban environment. Some countries define *urban* in terms of the number of inhabitants; others define it in terms of a specific number of people per square mile; still others may define it in terms of the percentage of people engaged in nonagricultural activities. In spite of these definitional problems, it is clear that industrialization has shifted a significant percentage of the population away from labor-intensive agricultural occupations into manufacturing, information, and service occupations—all of which depend heavily on fossil fuels not only to make, distribute, and deliver goods and services but to gather employees together at a workplace (World Resource Institute 1996).

Spatial sociologists argue that highways and automobiles have created urban sprawl and have made it difficult to distinguish between city, suburbs, and nonurban environments (see Table 16.2). Urban sprawl spreads development beyond cities by as much as 40 or 50 miles; puts considerable distance between homes, stores, churches, schools, and workplaces; and makes people automobile

Table 16.2	Work-Related Transportation Statistics

The table below shows that the United States has 133.1 million driving-age workers. Of these, 77 percent drive alone to work and take an average of 25 minutes to get there. About one-third of workers indicate that they live in a three-vehicle household.

Workers Age 16 and Older	133,091,043
Means of Transportation to Work	**% of Workers**
Car, truck, or van	87.7
Driving alone	77
Carpooling	10.7
Public transportation (excluding taxicab)	4.7
Walking	2.5
Bicycling	0.4
Taxicab, motorcycle, or other means	1.2
Working at home	3.6
Travel Time to Work (minutes)	
Less than 10	14.7
10 to 14	14.3
15 to 19	15.5
20 to 24	14.5
25 to 29	6.1
30 to 34	13.2
35 to 44	6.4
45 to 59	7.5
60 or more	7.9
Mean travel time to work: 25.1 minutes	
Vehicles Available	
None	4.1
1	20.9
2	43.2
3 or more	31.8

Source: U.S. Census Bureau 2005.

dependent (Sierra Club 2007). In addition, the automobile and highway have allowed people to live in more space than they need. In the past 30 years, the average house size has increased from 1,400 to 2,330 square feet (National Association of Homebuilders 2007).

The Information Explosion

Sociologist Orrin Klapp (1986) wrote about the **information explosion**, an unprecedented increase in the amount

urbanization A transformative process in which people migrate from rural to urban areas and change the way they use land, interact, and make a living.

information explosion An unprecedented increase in the amount of stored and transmitted data and messages in all media (including electronic, print, radio, and television).

Spatial sociologists argue that highways and automobiles have created urban sprawl and have made it difficult to distinguish between city, suburbs, and nonurban environments.

© AbleStock/Index Open

The information explosion gives people access to an unprecedented amount of words, images, and sounds. This information is generated and delivered much faster than the brain can process it.

Department of Defense photo by SGT Jim Greenhill, USA

of stored and transmitted data and messages in all media (including electronic, print, radio, and television). One can argue that the information explosion began with the invention of the printing press. Today the information explosion is driven by the Internet, a vast fossil fuel–powered computer network linking billions of computers around the world. The Internet has the potential to give users access to every word, image, and sound that has ever been recorded (Berners-Lee 1996).

Although it is virtually impossible to catalog all the changes associated with the Internet, we can say that it has (1) speeded up old ways of doing things, (2) given individuals access to the equivalent of a printing press, (3) allowed users to bypass the formalized hierarchy devoted to controlling the flow of information, (4) changed how students learn, and (5) allowed people around the world to exchange information on and communicate about any topic of interest.

Klapp used a metaphor to describe the challenge of sorting through and keeping up with the massive amounts of information generated. He envisioned a researcher "seated at a table fitting pieces of a gigantic jigsaw puzzle. From a funnel overhead pieces are pouring onto the table faster than one can fit them. Most of the pieces do not match up. Indeed, they do not belong to the same puzzle" (p. 110). The pieces falling from overhead represent information accumulating at a pace that overwhelms the brain's capacity to organize and evaluate it. While computer software and telecommunications technologies have increased the speed of data generation, flow, and exchange, people must still read, discuss, and contemplate the information to give it meaning. These activities are very slow compared with the speed at which the information is generated.

To complicate matters, the information is often distorted or exaggerated. Klapp offers two explanations for this fact. First, new technologies permit the existence of large numbers of magazines, newspapers, journals, radio stations, television channels, and Web sites. Message senders must therefore compete for our attention by enticing us to read and listen to their message with eye-catching images, misleading titles, and shocking stories. The titles might catch our attention, but they usually mask rather than reveal a more complex reality.

A second reason the information is often distorted or exaggerated is **dearth of feedback**. Much of the information generated is not subjected to honest, constructive criticism, because there are too many messages and not enough critical readers and listeners to evaluate it before the material is released or picked up by the popular media. Without feedback, the information producers cannot correct their mistakes; as a result, the information has questionable value.

Triggers of Social Change

When we think about what triggers a specific social change, we usually cannot pinpoint a single factor. Change tends to result from a sequence of events. An analogy may help clarify this point. Suppose that a wide receiver, after catching the football and running 50 yards, is tackled at the 5-yard line by a cornerback. One could argue that the cornerback *caused* the receiver to fall to the ground. Such an account, however, does not fully explain the cause. For one thing, the tackle was not the act of one person seizing and throwing his weight onto the person with the ball; it was more complex than that. The teammates of both the wide receiver and the tackler determined how that play developed and ended. Furthermore, the wide receiver was doing everything in his power to elude the tackler's grasp (Mandelbaum 2007).

Similarly, social change—whether it be globalization, industrialization, urbanization or something else—is part of an endless sequence of interrelated events. Even though it is difficult to separate the causes of social change from this sequence, we can identify key factors that trigger change in general and have triggered increased dependence on fossil fuels in particular. The triggers include innovations, revolutionary ideas, conflict, the pursuit of profit, and social movements.

Innovations

■ **CORE CONCEPT 3:** An invention can trigger changes in social activity. For an invention to emerge, however, the cultural base must be large enough to support it. **Innovation** is the invention or discovery of something, such as a new idea, process, practice, device, or tool. Innovations can be placed into two broad categories: basic or improving. The distinction between the two is not always clear-cut, however. **Basic innovations** are revolutionary, unprecedented, or groundbreaking inventions or discoveries that form the basis for a wide range of applications. Basic innovations include the cotton gin, steam engine, and first-generation PC (personal computer). The discoveries of the

Department of Defense photo by SPC Preston E. Cheeks, USA

Each year, fossil fuel-powered airplanes carry an estimated 4.4 billion passengers to destinations around the world.

industrial uses for coal (1748), natural gas (1821), and oil (1853) certainly qualify as basic innovations. **Improving innovations**, by comparison, are modifications of basic inventions that improve upon the originals—for example, making them smaller, faster, more user-friendly, more efficient, or more attractive.

Each "upgrade" of the 1903 Wright Flyer (the first successful airplane, which Orville and Wilbur Wright designed, built, and flew) increased the airplane's capacity to fly farther, higher, faster, and with more passengers. Thirty years and many innovations after the 1903 Wright Flyer, Boeing unveiled the first modern passenger airliner, which could carry 10 passengers at the speed of 155 miles per hour. In 1958 the jet age arrived when Boeing unveiled the first U.S. passenger jet, capable of carrying 181 passengers at the speed of 550 miles per hour. In 1969, wide-body

dearth of feedback A situation in which much of the information released or picked up by the popular media is not subjected to honest, constructive criticism, because the critical audience that exists is too small to evaluate the information before it is used.

innovation The invention or discovery of something, such as a new idea, process, practice, device, or tool.

basic innovations Revolutionary, unprecedented, or groundbreaking inventions or discoveries that form the basis for a wide range of applications.

improving innovations Modifications of basic inventions that improve upon the originals—for example, making them smaller, faster, less complicated, more efficient, more attractive, or more profitable.

NO BORDERS, NO BOUNDARIES

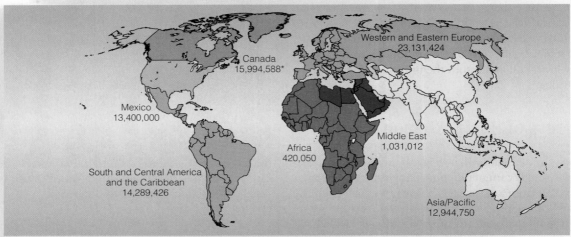

Number of Airline Passengers Traveling between U.S. and Foreign Airports Annually, by Region

Canada
15,994,588*

Western and Eastern Europe
23,131,424

Mexico
13,400,000

Middle East
1,031,012

Africa
420,050

South and Central America
and the Caribbean
14,289,426

Asia/Pacific
12,944,750

*Does not include border crossings by car or foot and excludes departures.

▲ **Figure 16.1**   The map shows by region the number of airline passengers traveling between U.S. and foreign airports each year.

Source: International Trade Administration (2007).

jets, capable of seating 450 passengers, made their debut (Airport Transport Association 2001). The significance of this series of improving innovations is evident when we consider that in 2006 a total of 4.4 billion passengers traveled by airplane (Airports Council International 2007). The number of international passengers reached 842 million, up from an estimated 50,000 passengers in 1950 (World Tourism Organization 2007). This number is expected to increase to 1.6 billion by 2020.

The Cultural Base and the Rate of Change. Anthropologist Leslie White (1949) maintained that once a basic or an improving innovation has emerged, it becomes part of the **cultural base**, which he defined as the number of existing innovations. The size of the cultural base determines the rate of change. White defined an **invention** as a synthesis of existing innovations. For example, the first successful

cultural base The number of existing innovations, which forms the basis for further inventions.

invention A synthesis of existing innovations.

airplane was a synthesis of many preceding innovations, including the gasoline engine, the rudder, the glider, and the wheel.

White suggested that the number of innovations in the cultural base increases geometrically—1, 2, 4, 8, 16, 32, 64, and so on. (Geometric growth is equivalent to a state of runaway expansion.) He argued that for an invention to emerge, the cultural base must be large enough to support it. If the Wright brothers had lived in the 14th century, for example, they could never have invented the airplane, because the cultural base did not contain the ideas, materials, and innovations to support its invention. The seemingly runaway expansion, or increase in the volume, of inventions prompted White to ask a question: Are people in control of their inventions, or do inventions control people? For all practical purposes, he believed that inventions control us. White cited two arguments to support this conclusion. First, he suggested that the old adage "Necessity is the mother of invention" is naive. In too many cases, the opposite idea—that invention is the mother of necessity—is true. That is, an invention becomes a necessity because we find uses for the invention after it comes into being: "We invent the automobile to get us between two points faster, and suddenly we find we have to build new roads. And that means we have to invent traffic regulations and put in stop lights [and build garages]. And then we have to create a whole new organization called the Highway Patrol—and all we thought we were doing was inventing cars" (Norman 1988, p. 483).

Second, White (1949) argued that when the cultural base is capable of supporting an invention, then the invention will come into being whether people want it or not. White supported this conclusion by pointing to **simultaneous-independent inventions**—situations in which more or less the same invention is produced by two or more persons working independently of one another at about the same time (sometimes within a few days or months). He cited some 148 such inventions—including the telegraph, the electric motor, the steamboat, the car, and the airplane—as proof that someone will make the necessary synthesis if the cultural base is ready to support a particular invention. In other words, the light bulb and the airplane would have been developed regardless of whether Thomas Edison and the Wright brothers (the people we traditionally associate with these inventions) had ever been born. According to White's conception, inventors may be geniuses, but they must also be born in the right place and the right time— that is, in a society with a cultural base sufficiently developed to support their inventions. None of the inventors associated with the inventions just mentioned could have delivered their products if fossil fuels had not already been adapted for industrial use.

According to White's theory, if the parts are present, someone will eventually put them together. The implications are that people have little control over whether an invention comes into being and that they adapt to inventions after the fact. Sociologist William F. Ogburn (1968) calls the failure to adapt to a new invention *cultural lag*.

Cultural Lag. In his theory of cultural lag, Ogburn (1968) distinguishes between material culture and nonmaterial culture. Recall from Chapter 3 that material culture includes tangible things—including resources (such as oil, coal, natural gas, trees, and land), inventions (such as paper and the automobile), and systems (such as factories and package delivery)—that people have produced or, in the case of resources such as oil, have identified as having the properties to serve a particular purpose. Nonmaterial culture, by contrast, includes intangible things, such as beliefs, norms, values, roles, and language. Although Ogburn maintains that both material and nonmaterial culture are important agents of social change, his theory of cultural lag emphasizes the material component, which he suggests is the more important of the two. The case of the automobile illustrates how this piece of material culture changed the United States.

The availability of cheap energy and an inexpensive, mass-produced car soon transformed the American land-scape. Suddenly there were roads everywhere—paid for, naturally enough, by a gas tax. Towns that had been too small for the railroads were now reachable by roads, and farmers could get to once unattainable markets. Country stores that stood at old rural crossroads and sold every conceivable kind of merchandise were soon replaced by specialized stores, catering to people who could drive off and shop where they wanted. Families that had necessarily been close and inwardly focused, in part because there was nowhere but home to go at night, weakened somewhat when family members could get in their cars and take off to do whatever they wanted to do (Halberstam 1986, pp. 78–79).

Ogburn believes that one of the most urgent challenges facing people today is the need to adapt to material innovations in thoughtful and constructive ways. He uses the term **adaptive culture** to describe the nonmaterial component's role in adjusting to material innovations. One can argue that Americans adapted easily to the automobile because it supported deeply-rooted norms—values and beliefs that applied to a nation composed mostly of immigrants, who by definition had separated from their native lands and traditions. The automobile simply extended a tradition of forsaking, voluntarily or involuntarily, "the region and habits of their parents" and striking out on their own (Halberstam 1986, p. 79). On the other hand, calls to cut back driving to lessen the nation's dependence on foreign sources of oil have made little to no impact on most Americans. They resist changing the norms that govern their driving habits, a value system that defines the car as the measure of personal freedom and independence, and a belief system that holds the car to be the most efficient method of transportation.

The case of the automobile suggests that adjustments are not always immediate. Sometimes they take decades; sometimes they never occur. Ogburn uses the term **cultural lag** to refer to a situation in which adaptive culture fails to adjust in necessary ways to a material innovation. Despite Ogburn's emphasis on material culture as the key force driving change, he is not a **technological determinist**—someone who believes that humans have

---

simultaneous-independent inventions Situations in which more or less the same invention is produced by two or more persons working independently of one another at about the same time.

adaptive culture The portion of nonmaterial culture (norms, values, and beliefs) that adjusts to material innovations.

cultural lag A situation in which adaptive culture fails to adjust in necessary ways to material innovation.

technological determinist Someone who believes that human beings have no free will and are controlled entirely by their material innovations.

no free will and are controlled entirely by their material innovations. Instead, he notes that people do not adjust to new material innovations in predictable and unthinking ways; rather, they choose to create them and only then choose how to use them. Ogburn argues that if people have the power to create material innovations, then they also have the power to destroy, ban, regulate, or modify those innovations. The challenge lies in convincing people that they need to address an innovation's potentially disruptive consequences before those consequences materialize.

In our discussion about innovations as triggers of social change, we have emphasized material inventions (such as devices and tools). But innovations can also consist of nonmaterial inventions, such as a revolutionary idea.

Revolutionary Ideas

■ **CORE CONCEPT 4:** Social change occurs when someone breaks away from or challenges a paradigm. A scientific revolution occurs when enough people in the community break with the old paradigm and orient their research or thinking according to a new paradigm. In *The Structure of Scientific Revolutions*, Thomas Kuhn (1975) maintains that people tend to perceive science as an evolutionary enterprise. That is, they imagine scientists as problem solvers building on their predecessors' achievements. Kuhn takes issue with this evolutionary view, arguing that some of the most significant scientific advances have been made when someone has broken away from or challenged a paradigm. He defines **paradigms** as the dominant and widely accepted theories and concepts in a particular field of study. Paradigms gain their status not because they explain everything, but rather because they offer the "best way" of looking at the world at that time. On the one hand, paradigms are important thinking tools; they bind a group of people with common interests into a scientific or national community. Such a community cannot exist without agreed-on paradigms. On the other hand, paradigms can act as blinders, limiting the kinds of questions that people ask and the observations that they make.

The explanatory value—and hence the status—of a paradigm is threatened by any **anomaly**, an observation that the paradigm cannot explain. The existence of an anomaly by itself usually does not persuade people to abandon a particular paradigm. According to Kuhn, before people discard an old paradigm, someone must articulate an alternative paradigm that accounts convincingly for the anomaly. He hypothesizes that the people most likely

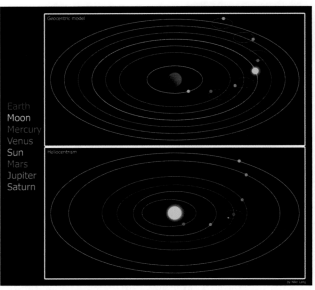

New paradigms, such as the one advanced by Copernicus that upset the paradigm of humankind as "the center of the universe," caused converts to see the world in an entirely new light and to wonder how they could possibly have taken the old paradigm seriously.

to put forth new paradigms are those who are least committed to the old paradigms—the young and those new to a field of study.

A scientific revolution occurs when enough people in the community break with the old paradigm and change their research or thinking to favor the new paradigm. Kuhn considers a new paradigm to be incompatible with the one it replaces, because it "changes some of the field's most elementary theoretical generalizations" (1975, p. 85). The new paradigm causes converts to see the world in an entirely new light and to wonder how they could possibly have taken the old paradigm seriously. "When paradigms change, the world itself changes with them. Led by a new paradigm, scientists adopt new instruments and look in new places" (p. 111).

Perhaps the best example of a scientific revolution can be found in the work of Nicolaus Copernicus, author of *On the Revolutions of the Heavenly Spheres* (1543), which challenged a long-held belief that Earth (and by extension, humankind) was the stationary center of the solar system, with the sun, moon, and planets revolving around it. Copernicus maintained that the stationary center of the solar system was the sun and that Earth and the other planets revolved around that center. Copernicus's ideas did not take hold immediately. In 1633 (90 years later), powerful church inquisitors threatened to torture and kill Galileo, who had embraced Copernicus's theory, if he did not renounce it; upon renouncing it, Galileo was imprisoned for life. "Of all discoveries and opinions, none may

paradigms The dominant and widely accepted theories and concepts in a particular field of study.

anomaly An observation that a paradigm cannot explain.

have exerted a greater effect on the human spirit than the doctrine of Copernicus. The world had scarcely become known as round and complete in itself when it was asked to waive the tremendous privilege of being the center of the universe. Never, perhaps, was a greater demand made on mankind—for by this admission so many things vanished in mist and smoke!" (Goethe 2004).

We can apply Kuhn's theory to the paradigm governing capitalism—a paradigm that defines profit as the measure of success. Thus, capitalism succeeds when it creates new markets to sell products and when it lowers production costs by finding the cheapest sources of labor and raw materials. The anomaly in this paradigm is this: if you pay workers too little, they cannot afford to purchase the products of their labor, thus denying capitalists a market. Henry Ford was one of the first people to recognize this anomaly. In 1913—a year of great advances in mechanization and of high labor turnover—Ford doubled the pay of his autoworkers, from $2.50 to $5.00 per day. Ford was attacked for this move as a traitor to his class. The *Wall Street Journal* labeled it an "economic crime" and the *New York Times* called it "distinctly utopian" (Halberstam 1986). Eventually, other capitalists were forced to match Ford's wages. Higher wages, of course, allowed workers to purchase cars and other mass-produced products. The Ford "revolution" was one factor that fueled production of automobiles. In 1913 about 600,000 cars and trucks were produced worldwide; today 70 million are produced each year (International Organization of Motor Vehicle Manufacturers, 2007).

Conflict

■ **CORE CONCEPT 5:** Conflict—especially when it involves efforts to resolve a dispute or gain the upper hand—can create new norms, relationships, ways of thinking, and innovations. Sociologist Lewis Coser (1973) argues that conflict occurs whenever a group takes action to increase its share of wealth, power, prestige, or some other valued resource and when those actions are resisted by people who benefit from the current distribution system. In other words, those who control valued resources strive to protect their interests from those who hope to gain a share.

Conflict, whether it involves violent clashes or public debate, is both a consequence and a cause of change. In general, any kind of change has the potential to trigger conflict between those who stand to benefit from the change and those who stand to lose because of it. When the bicycle was invented in the 1840s, horse dealers organized against it because it threatened their livelihood. Physicians declared that people who rode bicycles risked getting "cyclist sore throat" and "bicycle stoop." Church groups protested that bicycles would swell the ranks of "reckless" women (because bicycles could not be ridden sidesaddle).

War—a most extreme form of conflict—is responsible for the invention of lifesaving technologies. In fact, the kinds of injuries sustained by soldiers in World War II pushed doctors to develop a system for collecting and preserving blood plasma. Of course, the plasma arrived on the battlefield via fossil fuel-powered vehicles.

Conflict can lead to change as well. It can be a constructive and invigorating force that prevents a social system from becoming stagnant, unresponsive, or inefficient. Conflict—especially when it involves efforts to resolve a dispute or gain the upper hand—can create new norms, relationships, and ways of thinking. It can also generate new, more-efficient communication technologies. The Internet, for example, emerged as an outgrowth of the cold war, when U.S. government leaders pulled together scientists from three sectors—military, industrial, and academic—to coordinate their research and expertise and thereby serve the war effort. Because these scientists worked in offices and laboratories all around the United States, Department of Defense officials worried about the consequences of an attack on a military laboratory, defense contractor, or university site. The officials realized that they needed a computer network that would allow information stored at one site to be transferred to another site in the event of an attack, especially a nuclear attack. At the same time, the computer network had to be designed so that if one or more parts of the network failed or were knocked out by a bomb, the other parts could continue to operate.

Such a design meant that no central control could exist for the network. After all, if central control was destroyed, the entire network would crash. The Internet began in the late 1960s as ARPANET (Advanced Research

Projects Agency), linking four universities: the University of California–Los Angeles (UCLA), the University of California–Santa Barbara (UCSB), the University of Utah, and Stanford University. Thus, the Internet was originally designed to (1) transfer information from one site to another quickly and efficiently in the event of war and (2) create an information-sharing system without central control. Today an estimated 1.1 billion people—16 percent of the world's population—are connected to the fossil fuel–driven Internet (worldinternetstats.com 2007).

The Pursuit of Profit

■ **CORE CONCEPT 6:** The profit-driven capitalist system drives change as it seeks to revolutionize production, create new products, and expand markets. Karl Marx believed that an economic system, capitalism, ultimately caused the explosion of technological innovation and the enormous and unprecedented increase in the amount of goods and services produced during the Industrial Revolution. In a capitalist system, profit is the most important measure of success. To maximize profits, the successful entrepreneur reinvests them to expand consumer markets and to obtain technologies that allow products and services to be produced in the most cost-effective manner. The capitalist system acts as a vehicle of change, because the instruments of production must be revolutionized constantly. Marx believed that the capitalist thirst for profit "chases the bourgeoisie over the whole surface of the globe" (Marx [1881] 1965, p. 531). Marx's theories influenced a group of contemporary sociologists—world system theorists—to write about capitalism as the key agent of change underlying global interdependence.

Immanuel Wallerstein (1984) is the sociologist most frequently associated with world system theory, a modern theory about capitalism. According to Wallerstein, by the late 19th century, the capitalist world economy included virtually the whole inhabited earth, and it is now striving to overcome the technological limits to cultivating the remaining corners—the deserts, the jungles, the seas, and indeed the other planets of the solar system (p. 165).

How has capitalism helped create a fossil fuel–dependent global network of economic relationships? One answer lies in the profit-generating strategies capitalists use to respond to economic stagnation and downturns. These strategies include (1) finding new markets, (2) cre-

ating new products that consumers will feel the "need" to buy (such as computers, MP3 players, and cell phones), (3) improving products and thus making old versions obsolete, and (4) instituting new ways to put "cash" into consumers' hands so they can buy more goods and services (such as no payments for 1 year and 40-year loans). Consider that fossil fuels are essential to manufacturing, powering, and delivering the new products, and in discarding the old.

Social Movements

■ **CORE CONCEPT 7:** Social movements occur when enough people organize to resist a change or to make a change. A **social movement** is formed when a substantial number of people organize to make a change, resist a change, or undo a change in some area of society. It depends on (1) an actual or imagined condition that enough people find objectionable; (2) a shared belief that something needs to be done about the condition; and (3) an organized effort to attract supporters, articulate the condition, and define a strategy for addressing the condition. Usually those involved in social movements work outside the system to advance their cause, because the system has failed to respond. To draw attention to their cause and accomplish their objectives, supporters may strike, demonstrate, walk out, sit in, boycott, go on hunger strikes, riot, or terrorize. Note that "oil executives never hold demonstrations in the street, stage boycotts, or engage in collective violence to advance their interests. They don't have to; their resources are so massive and their links with government representatives so direct that they can get things done without these less-than-fully institutionalized means" (Goode 1992, p. 408).

Social movements include the environmental movement, the civil rights movement, the women's movement, Amnesty International, the abortion rights movement, and the antiabortion movement. Each movement encompasses dozens to hundreds of specific groups that have organized to address the conditions they find objectionable. The environmentalist movement, for example, consists of thousands of specific organizations devoted to reducing air and water pollution, preserving wilderness, protecting endangered species, and limiting corporate activities that harm the environment (see Working For Change: "Protecting the Environment").

Social movements can be placed into four broad categories, depending on the scope and direction of change being sought: regressive, reformist, revolutionary, and counterrevolutionary. Definitions and global examples of each category follow. Keep in mind that the distinctions between the four categories are not always clear-cut. As a result, you might find that some examples fit into more than one category.

Regressive or reactionary movements seek to turn back the hands of time to an earlier condition or state

social movement A situation in which a substantial number of people organize to make a change, resist a change, or undo a change in some area of society.

regressive or reactionary movements Social movements that seek to turn back the hands of time to an earlier condition or state of being, one sometimes considered a "golden era."

WORKING FOR CHANGE

Protecting the Environment

THOUSANDS OF ORGANIZATIONS around the world are dedicated to protecting, restoring, and nurturing the environment. Three such organizations with their corresponding mission statements follow.

Global Water

http://www.globalwater.org/index.htm

Global Water is based upon the belief that the lack of access to safe drinking water is the primary cause of hunger, disease and poverty throughout the world. Founded in 1982, Global Water is an international, non-profit humanitarian organization focused on creating safe water supplies, sanitation facilities and related health programs for rural villagers in developing countries. Our program is designed to provide safe water supplies in rural villages to enable the rural poor to help themselves. To achieve this goal, Global Water's strategy is to provide permanent solutions to a region's water needs by providing appropriate equipment (to include state-of-the-art technology) to:

- Secure, purify, store and distribute new sources of water for domestic uses and agricultural purposes;
- Drill new water wells to allow access to groundwater.

Arctic Council

http://arctic-council.org/

The Arctic Council is an intergovernmental forum for addressing many of the common concerns and challenges faced by the Arctic states: Canada, Denmark (including Greenland and the Faroe Islands), Finland, Iceland, Norway, the Russian Federation, Sweden and the United States.

From the beginning, Arctic governments and indigenous peoples joined together to make environmental monitoring and assessment a key element of the Arctic Council's agenda. Groundbreaking reports have been prepared and have attracted global attention to the state of the Arctic environment. The approach of the Council encourages continuous dialogue among scientists, policy planners, Arctic residents and political level decision-makers. The decision-making of the Council is heavily based on the scientific work done under the umbrella of the Council and also influenced by the traditional knowledge of indigenous peoples.

Pew Center on Global Climate Change

http://www.pewclimate.org/about/

The Pew Center on Global Climate Change brings together business leaders, policy makers, scientists, and other experts to bring a new approach to a complex and often controversial issue. Our approach is based on sound science, straight talk, and a belief that we can work together to protect the climate while sustaining economic growth.

Sources: Global Water (2007), Arctic Council (2007) and PEW Center for Global Climate Change (2007).

---

of being, one sometimes considered a "golden era." The International Forum on Globalization (2007) represents such an effort. A global alliance of 60 organizations in 25 countries, its aims to seriously question the widely held belief that a globalized economy "lifts all boats" and to "reverse the globalization process by encouraging ideas and activities that revitalize local economies and communities, and ensure long-term ecological stability." Among other things, it advocates "local self-reliance in food production" and critical examination of trade policies that "undermine local food security by emphasizing an import-export, [fossil fuel–driven] model, making people dependent on food sources thousands of miles away."

Reformist movements target a specific feature of society as needing change. The nonprofit organization Polar Bears International (2007) focuses on saving the polar bear population and its habitat from extinction. Its goals are to use education and research to conserve the world's polar bears, offer educational resources to the public, encourage constructive dialogue, and build an international organization dedicated to saving this population.

Revolutionary movements seek broad, sweeping, and radical structural changes to a society's basic social institutions or to the world order (Benford 1992). The Earth

reformist movements Social movements that target a specific feature of society as needing change.

revolutionary movements Social movements that seek broad, sweeping, and radical structural changes to a society's basic social institutions or to the world order.

Liberation Front (ELF) is an underground eco-defense movement with no formal leadership or membership. Its members (who sometimes form cells) anonymously and autonomously engage in economic sabotage, including property destruction and guerrilla warfare, against those seen as exploiting and destroying the natural environment. ELF members have made news for setting fire to SUVs on dealership parking lots, a horse slaughterhouse, a scientific research center, a logging company, and a ski resort. Radical environmentalists claim to have committed 1,100 acts of arson and vandalism without killing a single person (Goldman 2007).

Counterrevolutionary movements seek to maintain a social order that reform and revolutionary movements are seeking to change. The Petition Project (2007) qualifies as such a movement, as it seeks to challenge reformist and revolutionary movements demanding reductions in greenhouse gas emissions. It recruits basic and applied scientists to sign a Global Warming Petition urging the U.S. government to reject the Kyoto Protocol, an international agreement to limit greenhouse gas emissions. The signers of the petition believe that *limiting* greenhouse gases will actually harm the environment: "There is no convincing scientific evidence that human release of carbon dioxide, methane, or other greenhouse gasses is causing or will, in the foreseeable future, cause catastrophic heating of the Earth's atmosphere and disruption of the Earth's climate. Moreover, there is substantial scientific evidence that increases in atmospheric carbon dioxide produce many beneficial effects upon the natural plant and animal environments of the Earth." The movement's Web site lists the names of 17,100 basic and applied American scientists who have signed the petition.

On the surface, it would appear that social movements form when enough people feel deprived in an objective or a relative sense. **Objective deprivation** is the condition of those who are the worst off or most disadvantaged—the people with the lowest incomes, the least education, the lowest social status, the fewest job opportunities, and so on. It would seem only logical that the objectively deprived would benefit from cars that are more fuel efficient or from public transportation. **Relative deprivation** is a social condition that is measured not by objective standards, but rather by comparing one group's situation with the situations of groups who are more advantaged. Someone earning an annual income of $100,000 is not deprived in any objective sense. He or she may, however, feel deprived relative to someone making $300,000 or more per year (Theodorson and Theodorson 1969).

The research on social movements shows that the objectively deprived are less likely than the relatively deprived to form or join social movements to address their condition. We can also make the case that every society includes some people who are or feel deprived; most of them have not taken part in a social movement. In some instances, people form or join social movements not to address real or imagined personal deprivations, but rather to address larger moral issues, such as abortion and the treatment of animals. The point is that deprivation alone cannot explain why people form or join social movements or how social movements take off. Sociologist Ralf Dahrendorf (1973) offers one description that attempts to capture the life of a social movement.

In trying to understand how people come to take action, Dahrendorf asks two questions: What is the structural source of conflict? and What forms can conflict take? Dahrendorf's answers rest on the following assumptions. First, every society possesses formal authority structures (such as a state, a corporation, the military, the judicial system, and a school system). Usually, clear dichotomies exist between those who control the formal system of rewards and punishments—and thus have the authority to issue commands—and those who must obey the commands or face the consequences (loss of job, jail, low grades, and so on). Second, a distinction between "us" and "them" arises naturally from the unequal distribution of power.

Using these assumptions, we can trace the structural origins of conflict to authority relations. Conflict can assume many forms. It can be mild or severe; "it can even disappear for limited periods from the field of vision of a superficial observer" (p. 111). As long as an authority structure exists, however, conflict cannot be abolished.

Dahrendorf outlines a three-stage model of conflict. Progression from one stage to another depends on many things. In the first stage, every authority structure contains at least two groups with different interests. Those with power have an interest in preserving the system; those without power have an interest in changing the system. These different interests remain below the surface, however, until those without power decide to organize. "It is immeasurably difficult to trace the path on which a person . . . encounters other people just like himself, and at a certain point . . . [says] 'Let us join hands, friends, so that they will not pick us off one by one'" (Dahrendorf 1973, p. 240).

Often, a significant event makes seemingly powerless people aware that they share an interest in changing the

---

counterrevolutionary movements Social movements that seek to maintain a social order that reformist and revolutionary movements are seeking to change.

objective deprivation The condition of the people who are the worst off or most disadvantaged—those with the lowest incomes, the least education, the lowest social status, the fewest opportunities, and so on.

relative deprivation A social condition that is measured not by objective standards, but rather by comparing one group's situation with the situations of groups who are more advantaged.

system. Václav Havel, a playwright and former president of the Czech Republic, believes that the Chernobyl nuclear accident may have played an important role in bringing about the anti-Communist revolutions in central Europe and the end of the cold war. In 1986 Chernobyl, a town in Ukraine, was the site of a nuclear power plant meltdown, the most serious kind of accident that can occur at a nuclear power plant. Although Ukraine, Russia, and Belarus suffered the most radioactive contamination, areas as far away as Sweden were affected as well. Havel maintains that after the Chernobyl meltdown, people in what was then Communist-controlled Czechoslovakia dared to complain openly and loudly to one another (Ash 1989).

At other times, people organize because they have nothing left to lose. "You don't need courage to speak out against a regime. You just need not to care anymore—not to care about being punished or beaten. A point is reached where enough people don't care anymore about what would happen to them if they speak out" (Reich 1989, p. 20).

In the second stage of conflict, if those without authority have opportunities to communicate with one another, some freedom to meet together, the necessary resources, and a leader, then they organize. At the same time, those in positions of authority often use the power of their positions to censor information, restrict resources, and undermine leaders' attempts to organize.

Resource mobilization theorists maintain that having a core group of sophisticated strategists is key to getting a social movement off the ground. Effective strategists can harness the disaffected's energies, attract money and supporters, capture the news media's attention, forge alliances with those in power, and develop an organizational structure. Cell phones, text messaging, and the Internet have made organizing easier by allowing interested parties to connect in ways that "defy gravity and time" (Lee 2003).

In the third stage of conflict, those seeking change enter into direct conflict with those in power. The capacity of the ruling group to stay in power and the amount and kind of pressure exerted from below then affect the speed and the depth of change. The intensity of the conflict can range from heated debate to violent civil war. It depends on many factors, including the belief that change is possible and the ability of those in power to control the conflict.

If protestors believe that their voices will eventually be heard, the conflict is unlikely to become violent or revolutionary. If those in power decide that they cannot compromise and they proceed to mobilize all of their resources to thwart protests, two results are possible. First, the protesters may believe that the sacrifices they will have to make to continue protests are too great, so they will withdraw from the fray. Alternatively, the protesters may decide to meet the "enemy" head-on, in which case the conflict may become bloody. If the power differential too greatly favors one side, the protestors or their opponents may resort to

terrorism—the systematic use of anxiety-inspiring violent acts by clandestine or semi-clandestine individuals, groups, or state-supported actors.

Consequences of Change

■ **CORE CONCEPT 8:** Sociological concepts and theories can be applied to evaluating the consequences of any social change. In the 16 chapters of this textbook, we have covered sociological concepts and theories that can be applied to evaluating the consequences of any social change. We close the book by selecting one key idea from each chapter to assess how climate change is affecting Greenland and how outsiders have pushed or pulled Greenland into the global arena. The chapter-specific ideas suggest questions that can help guide analysis of social change.

What kinds of social interactions give insights into climate change's effect on Greenland? (Chapter 1)

Sociologists are compelled to study social interactions, whether these interactions occur on a local scale or a global scale. With regard to Greenland, sociologists would be interested in almost any kind of social interaction. The following three examples highlight interactions between Greenlanders (especially the island's indigenous population, the Inuit) and "outsiders" brought about by climate change:

- Two 19-year-old men hiking through Greenland with Inuit hunter guides as part of a 22,000-mile North Pole–South Pole trek using human muscle- and wind-powered modes of transportation (foot, skis, bicycles, and sailboats) to draw attention to climate change (BBC News 2007).
- An Inuit Greenlander (who happens to be president of the Inuit Circumpolar Conference representing 155,000 Inuit living between Greenland and the Bering Strait) speaking to the British Airport Authority to protest its proposal to increase the number of flights in and out of Stansted Airport by as many as 264,000 per year, with a goal of serving 34 million passengers annually. He

resource mobilization A situation in which a core group of sophisticated strategists works to harness a disaffected group's energies, attract money and supporters, capture the news media's attention, forge alliances with those in power, and develop an organizational structure.

terrorism The systematic use of anxiety-inspiring violent acts by clandestine or semi-clandestine individuals, groups, or state-supported actors for idiosyncratic, criminal, or political reasons.

Ecotourists visit Greenland to observe icebergs and to witness the effects of climate change.

No roads connect towns and settlements in Greenland; boats and planes are the modes of transportation. In the winter, people travel using dogsleds and snowmobiles.

remarks, "Our side of the story has to be told to the people of England. It is important because the Arctic is also your backyard. Everything we do, particularly industrial plans, should be looked at in terms of how it impacts on climate change" (*Evening Standard* 2007).

- The 30,000 annual ecotourists visiting Ilulissat, Greenland, a town of 5,000 people, to observe the small island-size icebergs floating outside its harbor and to observe the site where 7 percent of the meltwater from Greenland's ice sheet empties into open seas (*Economist* 2007)

How do sociologists frame a discussion about Greenland and climate change? (Chapter 2)

Each of the three major sociological theories—functionalist, conflict, and symbolic interaction—offers a central question to help guide thinking and a vocabulary for answering that question.

A functionalist asks, What are the anticipated (manifest) and unintended (latent) functions and dysfunctions of climate change on Greenland? One manifest function is an economic boom associated with a lengthened shipping season (once four months long and now eight months long), which allows goods to move into and out of Greenland. A latent function is the emergence of working alliances between Inuit Greenlanders and tropical island peoples, who face cultural extinction from rising sea levels. Another latent function is a growing interest in Greenland, the Arctic, and Antarctica, so that popular films are set in or give prominent attention to these locations. Such films include *The Last Winter* (a horror film), *The Golden Compass*, *Pirates of the Caribbean*, *Arctic Tale*, *An Inconvenient Truth*, and *The March of the Penguins*. One manifest dysfunction is a growing tourism industry,

in which the number of tourists visiting Greenland each year overwhelms the resident population of towns visited. Finally, a latent dysfunction connected to climate change is the loss of status among Inuit elders, who can no longer predict the weather (see the discussion of symbolic interaction following the next paragraph).

A conflict theorist asks, Who benefits from climate change, and at whose expense? Conflict theorists key in on the many industries that have expanded in or moved operations to Greenland because of the warming climate. These commercial interests include zinc, lead, and uranium mining companies; oil drilling and exploration companies; and water companies. From the conflict perspective, such companies and their customers will no doubt benefit at the expense of Greenland and its culture.

Finally, a symbolic interactionist asks, How do the involved parties experience, interpret, influence, and respond to what they and others are doing as interaction occurs? Symbolic interactionists are particularly interested in ways interaction among Greenlanders is changing because of climate change. For example, Inuit Greenlanders no longer turn to their elders for weather forecasts; because of climate change, the signs that once allowed the elders to accurately predict the weather no longer apply.

How is the culture of Greenland's Inuit and of other Arctic peoples changing because of climate change? (Chapter 3)

Sociologists define *culture* as the way of life of a people; more specifically, culture includes the human-created

INTERSECTION OF BIOGRAPHY AND SOCIETY

Cultural Change in the Arctic

THE *ARCTIC CLIMATE IMPACT ASSESSMENT* (Huntington and Fox 2004), a scientific report commissioned by the Arctic Council (an intergovernmental forum of eight Arctic nations), includes a chapter about the changing Arctic from the perspective of indigenous peoples. The following observations illustrate this perspective:

- On the tundra the reindeers used to run towards people, but now they run away. The reindeers are our children. In the olden times when we used to have just the reindeers the air was clean. How should I explain? Now they drive around in skidoos and you can smell the gasoline, yuck! What did they herd with? The reindeer! Now they have started to herd with skidoos.
- The weather has changed. For instance, elders will predict that it might be windy, but then it doesn't become windy. And then it often seems like its going to be very calm and then it suddenly becomes windy. So their predictions are never correct any more—the predictions according to what they see haven't been true.

- The river ice breaks up so much earlier; it used to be in mid-June, and now it has been as early as mid-May. There is not so much snow, and the snow we get melts much earlier. In terms of summertime, in the course of my own life, it has been rare that we ever used to wear shorts and T shirts because it never got warm enough. But today because there are such long heat waves where it was 86 degrees Fahrenheit for an entire month, the whole community goes to the beach and swims. It gets so hot that bugs do not even come around anymore (Watt-Cloutier 2004).
- The direct heat from the sun is warmer; it is not the same anymore and you can't help but notice that. It is probably not warmer overall, but the heat of the sun is stronger. The reason why I mention the fact the sun seems warmer is because another piece of evidence to that is that we get some skin diseases or some skin problems. Because I think in the past when Peter (an elder) was a young boy, we never seemed to have these skin problems and I see them more and more these days.

strategies for adjusting to the environment and to the creatures (including humans) that are part of that environment. Sociologists are interested in how climate change is affecting Greenland's Inuit and the other Arctic peoples who have adapted to an extreme weather environment (see Intersection of Biography and Society: "Cultural Change in the Arctic"). One change affecting the Inuit is declines in, and even the gradual extinction of, marine species. This loss disrupts or destroys their hunting—and, by extension, their eating—habits. It is difficult for people to understand that changing or eliminating hunting means changing or eliminating a way of life: "People think, 'Oh they are just killing animals.' . . . When we go out on the land and teach [our children] to hunt, it's not just about aiming the gun and skinning the seal. It is also teaching courage and patience and how not to be impulsive and to use sound judgment. It is character skills that are transferable to the modern world" (Watt-Cloutier 2004). In response to shrinking food resources, Inuit are relying more on expensive, processed foods, which are changing their diet and overall physical health (Associated Press 2005).

How do ingroup and outgroup memberships related to climate change shape identity? (Chapter 4)

Sociologists use the term *ingroup* to describe a group with which people identify and to which they feel closely attached—particularly when that attachment is founded on opposition to another group known as an outgroup. An outgroup is a group toward which ingroup members feel a sense of separateness, opposition, or even hatred. By definition, ingroups cannot exist without outgroups and one person's ingroup is another person's outgroup. With regard to climate change, sociologists would note the following ingroup-outgroup formation. The Inuit are teaming up with small tropical island peoples (in Fiji, French Polynesia, and Caribbean countries) to speed up international action on climate change. Although the temperatures in the Arctic and in small tropical islands are quite different, both Arctic peoples and tropical islanders live in coastal communities threatened by rising sea levels (Weber 2007). This imminent threat gives seemingly diverse peoples a common identity. For them the outgroup consists of

Peoples who live on small tropical islands are teaming up with Arctic peoples to speed up international action on climate change. This Solomon Islander is looking for his belongings around his house, which was destroyed by an 8.1-magnitude earthquake and a tsunami on April 2, 2007.

Greenland is the largest island in the world. Vikings from Iceland settled in Greenland around the 10th century.

people who live in countries with the highest greenhouse gas emissions, especially the United States and China. The Inuit have filed a complaint with the Inter-American Commission on Human Rights that excessive U.S. greenhouse gas emissions have endangered their basic human rights (Brownwell 2004). The coalition between Arctic peoples and tropical islanders foreshadows the worldwide emergence of two groups with opposing interests. This Arctic-small island coalition foreshadows the worldwide emergence of two groups with opposing interests—one group seeking to protect their natural environment and the other group resisting any changes to their fossil-fuel-dependent lifestyles.

What social forces bring Greenlanders into interaction with outsiders and shape the relationships between the two groups? (Chapter 5)

The division of labor is an important force that pulls people into interaction with one another. It breaks down the making of a product or rendering of a service into special-ized tasks; each task is performed by workers trained to do just that task. The workers do not have to live near each other; they often live in different parts of the country and different parts of the world. Not only are the tasks geographically dispersed, but the parts and materials needed to manufacture products come from many locations around the world. Of course, sociologists look to identify the resources of Greenland that pulled its people into the global division of labor. From the 16th through the late 19th centuries, whalers from many European nations were drawn to Greenland's waters to hunt for bowhead whales. They could extract 44,000–66,000 pounds of oil from every whale. The oil was used to make lamp fuel, lubricants, soap products, ship tar, varnish, paint, and cosmetics. European interest in Greenland's whales ended in the late 19th century, after a new kind of oil (petroleum) was discovered and kerosene was introduced as a cheaper fuel for lighting lamps (Greenland Tourism and Business Council 2007).

Today outsiders are pulled to Greenland to capitalize on easier access to the country's natural resources. A case in point is the mining company Angus and Ross; it recently announced that Greenland's melting ice will allow it to reopen the Black Angel Mine, which closed in 1990. The once 4-month-long shipping season is now 8 months long and is expected to increase to 12 months long (Haines 2007). The company plans to extract four million tons of high-grade lead and zinc. Zinc is used to protect iron structures and to make dry batteries, lightweight coins, paints, rubber products, cosmetics, pharmaceuticals, plastics, soaps, textiles, X-ray and TV screens, and fluorescent lights. Lead is used to make such products as storage batteries, cable covering, plumbing products,

Department of Defense photo by Tech. Sgt. Jose Hernandez

Due to climate change, the once 4-month-long shipping season in Greenland is now 8 months long and is expected to increase to 12 months long.

Library of Congress Prints and Photographs Division

When the Europeans and Americans depended on whaling, it was not considered deviant to kill whales. In fact, whaling was celebrated. The discovery of kerosene and other fossil fuels allowed Westerners to care about whales and condemn those who continued to hunt them.

ammunition, and the antiknock compound in some gasolines (Winter 2007).

Because of climate change, what new formal organizations have emerged in Greenland? (Chapter 6)

Formal organizations are viewed as coordinating mechanisms, because they bring together people, resources, and technology and then channel social activity toward achieving a specific outcome. In 2004–2005 the government of Greenland joined with four large corporations—Air Greenland, KNI, Royal Greenland, and Greenland Tourism and Business—to create an export promotion strategy and an international branding strategy for Greenland. The purpose of the international branding is to market "positive associations and expectations" about Greenland abroad so that whenever the word *Greenland* comes up, people think of the island and its products as very special" (Greenland Home Rule Government 2007). Among other things, the government-corporate coalition has issued the brochure *Invest in Greenland*. The brochure opens with these words: "Greenland is currently experiencing a development that is presenting the Greenlandic and international business community with plenty of opportunities for investment in international growth sectors. Whereas fishing and ancillary industries have been the overwhelmingly dominant business for many years, a significant growth is now taking place in . . . tourism, transport, mineral mining, energy-intensive industry such as the aluminum industry and exclusive food production" (p 4).

An agreement has been signed with Alcoa—the world's leading producer of aluminum, with 122,000 employees in 44 countries—to construct an aluminum smelter. The company serves the aerospace, automotive, construction,

and commercial transportation industries, among others (CNW Telbec 2005). A Greenland Home Rule Government (2007) spokesperson commented, "The Greenlandic People has always felt close to nature. The animals of the sea and land are precious to us. We are confident that the project's environmental issues can be solved. We are convinced this is just as important to Alcoa as it is to Greenland."

The increased interest in Greenland, of course, extends to the entire Arctic region, which is considered to be one of the last energy frontiers in the world. As the ice recedes and temperatures warm, the Arctic's mineral riches (especially oil and natural gas) are becoming more accessible. Russia, the United States, Canada, Norway, and Denmark already have control over some territory in the region and are seeking to claim more territory. Most recently Russia planted a titanium capsule containing the Russian flag as a first step in claiming as much as one-half of the Arctic Ocean floor as its natural territory (Associated Press 2007; Chivers 2007).

How do ideas about what constitutes deviance relate to outsiders' interest or lack of interest in Greenland? (Chapter 7)

Sociologists maintain that the only characteristic common to all forms of deviance is that some social audience challenges or condemns a behavior or appearance because it departs from established norms. In light of this fact, it is difficult to generate a precise list of deviant behaviors, because a behavior considered deviant by some people may not be considered deviant by others. Likewise, a behavior

Department of Defense photo by Capt. Mark Harper, USAF

Imagine living in a society where water is not available with the turn of a faucet handle. This contractor in Tanzania must "carry" water to the site where he is mixing concrete.

considered deviant at one time and place may not be considered deviant at another. Such is the case with whaling. During the 17th, 18th, and 19th centuries, European and American whalers killed tens of thousands of whales in Arctic waters for commercial purposes, with little effective resistance from environmental or animal rights groups. In 1946 the International Whaling Commission (IWC) was established with the charge of governing the conduct of whaling throughout the world. Since then, it has taken such measures as banning the whaling of certain species, designating whale sanctuaries, setting limits on the number and size of whales hunted, and specifying the opening and closing of whaling seasons (International Whaling Commission 2007). Indigenous peoples such as Greenland's Inuit are permitted to catch a limited number of whales on a not-for-profit basis for cultural reasons. The IWC sets quotas and reviews them every five years.

In 2007 the IWC honored West Greenland's request to increase the number of minke whales that could be killed from 175 to 200 (Kazinform 2007).

How is climate change shaping life chances in Greenland and elsewhere? (Chapter 8)

Sociologists define *life chances* as a critical set of potential social advantages, including everything from the chances that a person will survive through the first year of life to the chances that a person will live a long life. Obviously, we had no control over which of the world's 243 countries we were born in, but that country has had a profound effect on our life chances (see Global Comparisons: "Global Access to Sustainable Water"). Climate change is expected to affect access to water as regions susceptible to drought become even drier and regions susceptible to flooding get even more rain. People who live in Greenland have the greatest access to sustainable water— the equivalent of 2.8 billion gallons available to each person each year. People who live in Kuwait have the lowest access, with the equivalent of 2,640 gallons available to each person each year. These figures translate into 7.2 gallons per day per person in Kuwait versus 7.6 million gallons per day per person in Greenland. For countries such as Kuwait to survive, they must import water or share water sources with other countries (UNESCO 2003).

How is racial stratification in Greenland affected by the conditions under which outside racial groups make contact with the Inuit? (Chapter 9)

Sociologists define *race* as a vast collectivity of people more or less bound together by shared and selected history, ancestors, and physical features; socialized to think of themselves as a distinct group; and regarded by others as such. The definition suggests that race is a product of emphasizing or feeling connected to a history shared by a certain broad category of ancestors who were usually forced by laws and other social practices to become socially distinct and separate from other broad categories of ancestors. The story of how Inuit, Danes, and Americans came to interact with one another in Greenland shapes race relations there today.

Greenland's population consists of Inuit and Greenland-born whites (80 percent) and immigrant Danes and Americans of all racial classifications stationed at Thule Air Base. Once a colony of Denmark, Greenland is now a self-governed Danish territory. The United States has maintained a military presence in Greenland since World War II. The Inuit and Europeans have mixed to the point that it is difficult to classify the Inuit—or Europeans, for that matter—as "pure."

GLOBAL COMPARISONS

Global Access to Sustainable Water

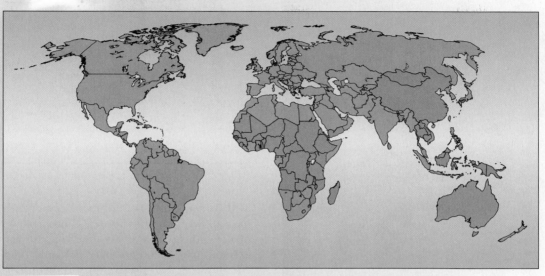

▲ **Figure 16.2**   Countries (Highlighted in Orange) Where at Least 20 Percent of the Population Do Not Have Access to Sustainable Water

The Inuit people are considered the indigenous population of Greenland. This Inuit works at Thule Air Base.

Department of Defense photo by TSGT Jose Hernandez

Sociologists are interested in how racial groups are implicitly or explicitly ranked on a scale of social worth. One snapshot of this ranking can be found in the way Thule Air Base is staffed. Today about 120 American military personnel are stationed at the base. Additional base personnel include 500–600 Danish civilians and only 100 Inuit Greenlander civilians. The Danes, who trained in universities and trade schools in their home country, work in the higher-skilled occupations; the Inuit Greenlanders, having little access to higher education or vocational training, work as truck drivers, taxi drivers, cooks, cashiers, and janitors (Mahr 2004).

What is the sex composition of Greenland? How might it be affected by climate change? (Chapter 10)

The sex composition of a society offers some clues about how it is organized and about inequalities that may exist between males and females. Table 16.3 shows the number of females per 100 males by age category in Greenland. Notice that, irrespective of age, 89 females live there for every 100 males. Males age 30–49 and 50–64 so outnumber their female counterparts that there are 84 females for every 100 males in the former category and only 72 females for every

Department of Defense photo, Thule Air Base, USAF

U.S. military presence in Greenland has set the conditions under which Americans interact with the local population.

Table 16.3	Number of Males and Females in Greenland by Age Category		
Age Category	Males	Females	Females per 100 Males
0–6	3,117	3,027	97
7–14	4,138	4,038	98
15–17	1,380	1,347	98
18–24	2,930	2,769	95
25–59	15,965	12,862	80
60–66	1,591	1,205	76
67+	1,198	1,402	117
Total	30,319	26,650	88

Source: U.S. Census Bureau (2007)

100 males in the latter. The only category in which females outnumber males is age 65-plus, which has 112 females for every 100 males. We might hypothesize that the greater number of males age 30–64 than females that age is related to immigration patterns: Greenland draws males to work in the mining and fishing industries. The larger number of females age 65 and older than males that age reflects differences in life expectancy that favor women. Sociologists hypothesize that the imbalance between the numbers of males and females age 30–64 will increase as more males are "pulled" to new industries locating in Greenland to take advantage of the longer shipping season accompanying climate change.

How did the U.S. military-industrial complex pull Greenland into the international arena? (Chapter 11)

Sociologist C. Wright Mills wrote about the connection between government, industry, and the military in *The Power Elite* (1956). The power elite comprises the few people who occupy such lofty positions in the social structure of leading institutions that their decisions have affected millions, even billions, of people worldwide. According to

Mills, since World War II, rapid advances in technology have allowed power to become concentrated in the hands of a few; those with access to such power can exercise an extraordinary influence over not only their immediate environment, but also millions of people, tens of thousands of communities, entire countries, and the globe.

U.S. military presence in Greenland can be traced to World War II, after German forces occupied Denmark in 1940. But its military presence did not end after that war. In 1951 Denmark and the United States signed a defense treaty establishing the 339,000-acre Thule Air Base to be used in the event of a Soviet attack on the United States. Native Greenlanders had no say in the matter, as Greenland was a colony of Denmark. Operation Blue Jay, the code name for the massive secret effort to build the base, attracted 12,000 workers to Greenland. In 1953 the Danish government displaced an entire Inuit village to expand the base. The U.S. military stored nuclear weapons at Thule, but it did not officially acknowledge this fact until 1996, even though a B-52 bomber carrying four nuclear bombs crashed near the base in 1968, spewing plutonium over the ice (Mahr 2004). Thule Air Base is home to the U.S. Ballistic Missile Early Warning System, and Greenland is part of the Distance Early Warning line, stretching from Alaska and northern Canada to Greenland (Archer and Scrivener 1983). Radar and satellite technologies are in place to warn of an impending ballistic missile attack against North America.

How might climate change affect Greenland's fertility rate? (Chapter 12)

Sociologists define *total fertility* as the average number of live children that women bear in their lifetime. Since 1996 total fertility in Greenland has declined from 2.51 to 2.07 (U.S. Census Bureau 2007). In the rural settlements the rate is 3.3, suggesting that the shift from rural to urban-

In its effort to win the cold war against the Soviet Union, the U.S. military established installations in the Arctic region, including the U.S. Ballistic Missile Early Warning System in Greenland.

Less than 60,000 people live in Greenland, the world's largest island, which is 80 percent ice-capped. Most Greenlanders live in small settlements along the coast.

rural environments has altered the status of children from economic assets to liabilities. Most of Greenland's population lives in towns, and a small percentage lives in rural settlements. Settlements are isolated, with few employment opportunities and very limited access to goods and services. By contrast, towns offer more job opportunities and access to a variety of goods and services. Sociologists would predict that as Greenland opens its borders to various foreign corporations and associated employment opportunities, total fertility will decline further.

What are formal and informal ways outsiders are coming to learn about Greenland, other Arctic cultures, and climate change? (Chapter 13)

In the broadest sense, education includes experiences that train, discipline, and shape the mental and physical potentials of the maturing person. Sociologists make a distinction between informal and formal education. Informal education occurs in a spontaneous, unplanned way, so that learning occurs naturally. In other words, someone does not deliberately design the learning experiences to stimulate specific thoughts or to impart a specific skill. Informal education about Greenland occurs when news audiences who do not consciously seek out stories on Greenland are exposed to such stories. When audiences hear on the nightly news that Santa Claus has left his North Pole home and is crossing Greenland and heading to the United States to deliver presents, audiences come to associate Greenland with the North Pole.

Formal education encompasses a purposeful, planned effort to impart specific knowledge or skills. Certain muse-

Through a process known as informal education many Americans have come to associate Greenland with Santa Claus, the Arctic, and the North Pole.

ums qualify as formal educators, because they are dedicated to preserving and displaying some aspect of human culture. Each shares its collections with the public so people can learn about the culture to which the museum is devoted. One such museum in the United States—the Jensen Arctic Museum in Monmouth, Oregon—was founded in 1985 by the adventurer, educator, collector, and philanthropist Paul H. Jensen. The purpose of the museum is to convey "his love of the arctic people, their art, lifestyle and environment" and to help visitors appreciate "the culture and ingenuity of a People who blended into a difficult and often unforgiving arctic environment" (Jensen Arctic Museum 2007).

What religions did outsiders bring to Greenland? (Chapter 14)

The major religions of Greenland are Lutheran Christianity and shamanism. Close ties exist between church and state in Denmark. Paragraph 4 of the Danish constitution identifies the Evangelical Lutheran Church as the national church of Denmark. Given that Greenland is a former colony of Denmark and that a significant share of Greenland's population is Danish, we should not be surprised to learn that the Lutheran Church has considerable influence in Greenland. For almost 300 years, the Danish Church has established missions in other lands, including Greenland. Today 500 missionaries are involved with 35 Danish government- and donor-subsidized missionary societies in Africa, India, Asia, South America, and Europe (Council on Interchurch Relations 2007).

Shamanism is the traditional religion of the Inuit, who do build sacred buildings known as churches. They consider nature sacred and themselves as children of nature. For the Inuit everything has a soul and is spiritually connected. The universe is in harmony, and the powers of nature are neutral toward humans. When evil (which can take such forms as bad hunting, bad weather, or illness) occurs, the source is almost always people's bad behavior (Mikaelsen 2007).

What is the population size of Greenland, and is the population increasing or decreasing because of climate change? (Chapter 15)

Population size is determined by births, deaths, and migration. Births add new people to a population; deaths reduce its size. Migration is the movement of people from one residence to another. Such movement adds new people to a population if they are moving in from another geographic area; it reduces the population if people are moving out; or it makes no difference if they are moving within the same area. In 2005, the last year for which data was available, the population of Greenland was 56,969. The annual birth

Table 16.4	Number of People Moving Into and Out of Greenland, 2005
Migration Category	**People**
Total in-migration	2,388
Born in Greenland	826
Foreign-born	1,562
Total out-migration	2,733
Born in Greenland	1,141
Foreign-born	1,592
Net Immigration	-345

Source: Greenland Home Rule Government (2007).

rate was 15 births per 1,000 people, for a total of 854 births; the annual death rate was 10 deaths per 1,000 people, for a total of 569 deaths. These figures mean that 285 more people were born in 2005 than died (U.S. Census Bureau 2007; Greenland Home Rule Government 2007). However, Greenland's net migration (the difference between the number of people who moved into the territory versus the number who moved out) was -345, which means that more people moved out than moved in. In 2005 Greenland's population decreased by 60 people. One might predict that in- and out-migrations will increase as more industries expand or establish operations in Greenland.

In light of the information explosion, how does one identify credible sources about climate change? (Chapter 16)

Consider that entering the term "global warming" into the search engine Google pulls up 7.6 million sites, with Wikipedia listed first. It would take a reader more than two years to review just the titles (assuming the reader could process a title every second). Given the level of competition among message senders to attract consumers, what strategies do they use to increase the chances that someone will pay attention to a particular message? Two of the most common strategies are (1) keeping the message short and simple, and (2) using eye-catching headlines and images. One of the most vivid images used to represent the potentially devastating consequences of climate change are polar bears who appear to be stranded on ice floes. Such images give the impression that the bears are stuck with nowhere to go. The images do not disclose the facts that polar bears use floating ice as a platform for catching seals and that the bears can swim long distances (perhaps 150 miles or more). These facts do not mean that the polar bears face no threats; their numbers have declined by 25 percent in the past 20 years (Polar Bears International 2007; Mouland 2007). The images do, however, oversimplify climate change's effect on these animals. Specifically, as warming melts more and more sea ice, polar bears must make riskier and longer swims to reach a solid platform; consequently, they are much thinner than they would otherwise be (Nicklen 2007).

A complete understanding of climate change involves knowing about Greenland's melting ice sheets, but it also involves knowing about the world's glaciers, the massive deforestation of the Amazon, changes to the North Atlantic Current, the ozone hole and ozone thinning, Asian monsoons, and much more (Whitty 2006).

The challenge of sorting through massive amounts of material and weighing opposing viewpoints is exacerbated when message senders engage in name calling and outright dismissal of another side's viewpoints. With regard to climate change, there appear to be two opposing camps. Both agree that the planet is warming, that ice sheets are melting, and that greenhouse gas emissions have risen. The debate is over whether the climate change is man-made or part of a natural cycle, and over whether greenhouse gas emissions are potentially dangerous or beneficial.

The 1,250 authors and 2,500 scientific expert reviewers from 130 countries involved with the UN report *Climate Change 2007* maintain that pre-industrial global atmospheric concentrations of carbon dioxide ranged from 180 to 300 parts per million (ppm). In 2005, carbon dioxide concentrations were 379 ppm. Thus, the man-made contribution could be as small as 79 ppm or as large as 199 ppm. The UN report argues that the current levels of greenhouse gases depart so much from any natural variability (180 to 300 ppm) that they must be man-made (Intergovernmental Panel on Climate Change 2007a). Petition Project signers and others take issue with this assessment, arguing that the planet has experienced ice ages, sea level rises, glacial melting, and higher greenhouse gas emissions before. Therefore, what we call "global warming" or "climate change" may be part of a natural cycle.

Since most people do not have the scientific background to evaluate the two arguments, proponents of each side often use simplistic images to sway people. In assessing the two views, keep in mind that the debate is over who or what is responsible for climate change (humans or nature?). It is also a high-stakes debate about whether humans should change the way they have organized almost every social activity. If the phenomenon is natural, then no amount of effort is likely to change the outcome, so why make changes?

A sociologist interested in objectively evaluating the two viewpoints would probably find it useful to check authors', reviewers', and petition signers' credentials. The Petition Project Web site gives an alphabetized list of signers who reject the idea of human contribution to climate change and who question whether greenhouse gas emissions are problematic for the planet. Along with each signer's name, only the most basic academic credentials (such as MS and PhD) are provided. The list provides no corporate, university, or other affiliations—making it difficult to learn about signers' qualifications. I selected 10 signers at random and used an Internet search engine to help me identify their affiliations and their other qualifications. Four of the 10 names yielded results: the first signer appears to have died in 2004; the second works for Conoco Phillips (an integrated petroleum company); the third specializes in airlines, hotels, rental cars, and cruise line forecasting; and the fourth is a city attorney.

The names of authors and reviewers associated with the *Climate Change 2007* report are listed, along with their academic credentials and their affiliations. I selected 10 names at random and used an Internet search engine to verify their affiliations and credentials. All 10 names yielded results. The affiliations included National Center for Atmospheric Research (Boulder, Colorado), Netherlands Environmental Assessment Agency, and Canadian Centre for Climate Modeling and Analysis. While a sample of 10 names cannot yield definitive results about potential biases driving people affiliated with either camp, it does suggest that further investigation is warranted before judging which source is more credible.

■ VISUAL SUMMARY OF CORE CONCEPTS

■ **CORE CONCEPT 1:** When sociologists study any social change, they take particular interest in identifying tipping points—situations in which previously rare events snowball into commonplace ones.

Social change is any significant alteration, modification, or transformation in the organization and operation of social activity. When sociologists study change, they must first identify the social activity that has changed or is changing. Sociologists are particularly interested in tipping points, a situation in which a previously rare (or seemingly rare) event, response, or opinion becomes dramatically more common.

© Design Pics Inc./Alamy

■ **CORE CONCEPT 2:** A number of key sociological concepts help sociologists answer the question "What about social activity has changed since 1750?" Those concepts describe ongoing changes: industrialization and mechanization, globalization, rationalization, McDonaldization, urbanization, and the information explosion.

The greatest change in social activity since 1750 is the extent to which that activity depends on fossil fuels. The most critical factor driving the Industrial Revolution was mechanization—the addition of external sources of power, such as coal, oil, and natural gas, to hand tools and modes of transportation. There is no question that fossil fuels facilitated globalization. Obviously, humans use trains, cars, buses, boats, planes, phones, and the Internet to deliver people, products, services, and information across national borders. The rationalization driving the profit-making strategy known as planned obsolescence depends on fossil fuels to manufacture products, to deliver them to retailers or to buyers' houses, to make them operate, and to haul them away when owners decide to discard them. Other processes that could not occur without fossil fuels are McDonaldization, urbanization, and the information explosion.

© AbleStock/Index Open

■ **CORE CONCEPT 3:** An invention can trigger changes in social activity. For an invention to emerge, however, the cultural base must be large enough to support it.

Innovation is the invention or discovery of something—a new idea, process, practice, device, or tool. Anthropologist Leslie White argued that for an invention to emerge, the cultural base must be large enough to support it. If the Wright brothers had lived in the 14th century, for example, they could never have invented the airplane, because the cultural base did not contain the ideas, materials, and innovations to support its creation. When the cultural base is capable of supporting an invention, then the invention will come into being whether people want it or not. When this happens, society experiences what sociologist William F. Ogburn calls cultural lag; that is, the adaptive culture (norms, values, and beliefs) fails to adjust in necessary ways to a material innovation.

Library of Congress Prints and Photographs Division

■ **CORE CONCEPT 4:** Social change occurs when someone breaks away from or challenges a paradigm. A scientific revolution occurs when enough people in the community break with the old paradigm and orient their research or thinking according to a new paradigm.

According to Thomas Kuhn, some of the most significant scientific advances have been made when someone has broken away from or challenged a paradigm—a dominant and widely accepted theory or concept in a particular field of study. Before people discard an old paradigm, someone must articulate an alternative paradigm that accounts convincingly for all anomalies (observations the old paradigm cannot explain). The people most likely to put forth new paradigms are those who are least committed to the old paradigms.

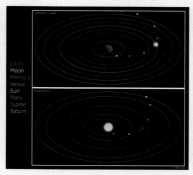
Niko Lang

■ **CORE CONCEPT 5:** Conflict—especially when it involves efforts to resolve a dispute or gain the upper hand—can create new norms, relationships, ways of thinking, and innovations.

Conflict is both a consequence and a cause of change. Change can trigger conflict, and conflict can lead to change. Conflict can be a constructive and invigorating force that prevents a social system from becoming stagnant, unresponsive, or inefficient. Conflict can create new norms, relationships, and ways of thinking. It can also generate new, more-efficient technologies.

Department of Defense photo by Wever

■ **CORE CONCEPT 6:** The profit-driven capitalist system forces change as it seeks to revolutionize production, create new products, and expand markets.

In a capitalist system, profit is the most important measure of success. To maximize profit, the successful entrepreneur must respond to economic stagnation and downturns with profit-generating strategies. Historically, such strategies have included lowering wages, introducing labor-saving technologies, finding new markets, and creating new products that consumers will feel the "need" to buy.

Library of Congress Prints and Photographs Division

■ **CORE CONCEPT 7:** Social movements occur when enough people organize to resist a change or to make a change.

A social movement depends on (1) an actual or imagined condition that enough people find objectionable; (2) a shared belief that something needs to be done about the condition; and (3) an organized effort to attract supporters, articulate the condition, and define a strategy for addressing the condition. Social movements can be classified as regressive, reformist, revolutionary, and counterrevolutionary.

Department of Defense photo by SSGT
Staci L. Rosenberger, USAF

■ **CORE CONCEPT 8: Sociological concepts and theories can be applied to evaluating the consequences of any social change.**
The sociological concepts and theories covered in this textbook can be applied to evaluating the consequences of any social change. Key concepts are important thinking tools, because they suggest questions that can guide analysis of any social change.

Library of Congress Prints and Photographs Division

Resources on the Internet

***Sociology: A Global Perspective* Book Companion Website**
www.cengage.com/sociology/ferrante

Visit your book companion Web site, where you will find flash cards, practice quizzes, internet links, and more to help you study.

CENGAGE**NOW**™

Just what you need to know NOW!

Spend time on what you need to master rather than on information you have already learned. Take a pre-test for this chapter, and CengageNOW will generate a personalized study plan based on your results. The study plan will identify the topics you need to review and direct you to online resources to help you master those topics. You can then take a post-test to help you determine the concepts you have mastered and what you will need to work on. Try it out! Go to www.cengage.com/login to sign in with an access code or to purchase access to this product.

Key Terms

Afghanistan

Source: U.S. Central Intelligence Agency, *World Factbook*, 2007

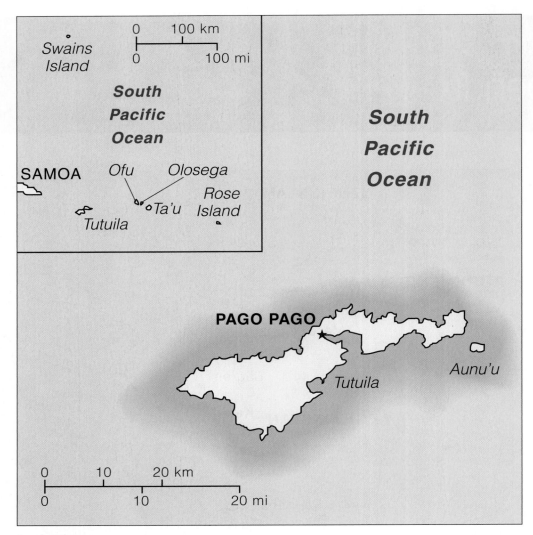

American Samoa

Source: U.S. Central Intelligence Agency, *World Factbook*, 2007

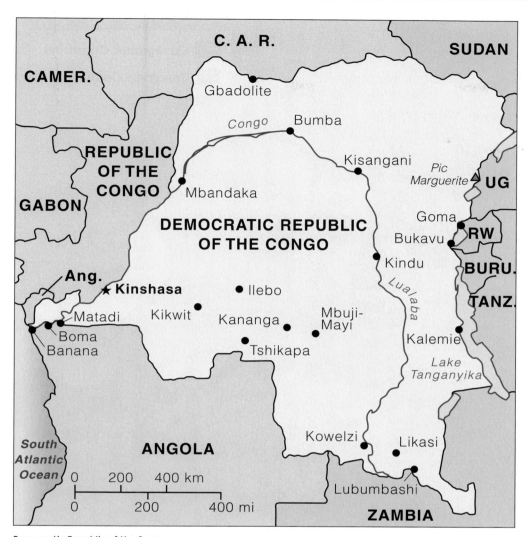

Democratic Republic of the Congo

Source: U.S. Central Intelligence Agency, *World Factbook*, 2007

European Union
Source: U.S. Central Intelligence Agency, *World Factbook*, 2007

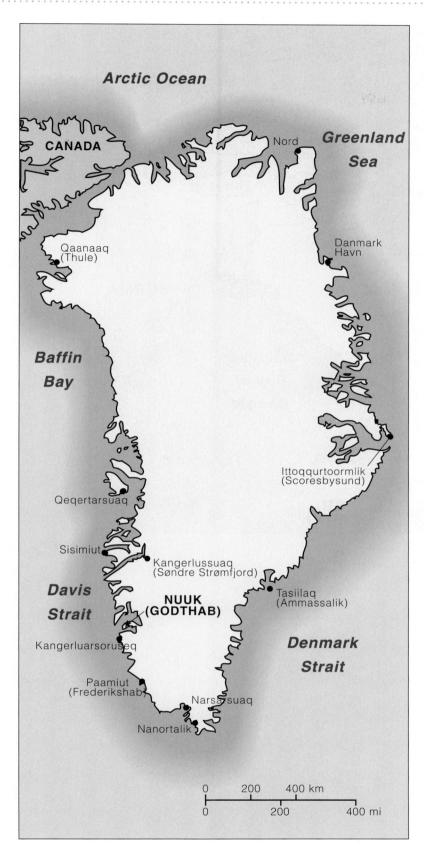

Greenland

Source: U.S. Central Intelligence Agency, *World Factbook*, 2007

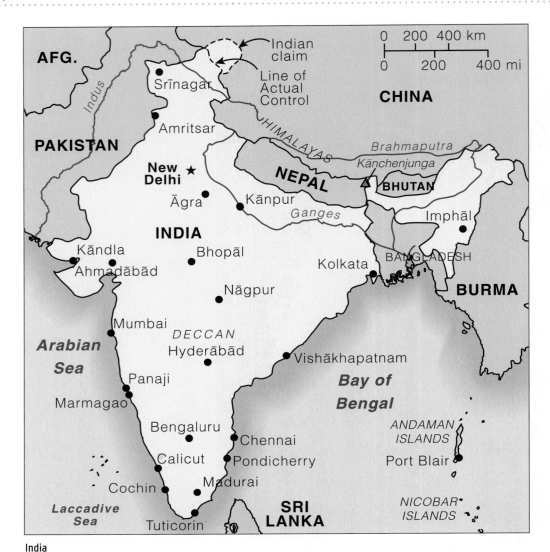

India

Source: U.S. Central Intelligence Agency, *World Factbook*, 2007

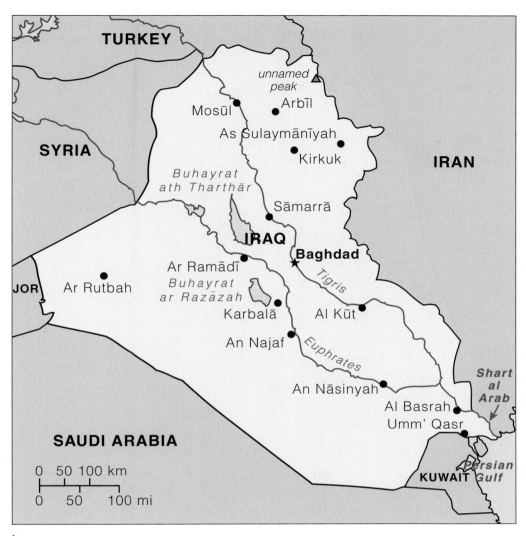

Iraq

Source: U.S. Central Intelligence Agency, *World Factbook*, 2007

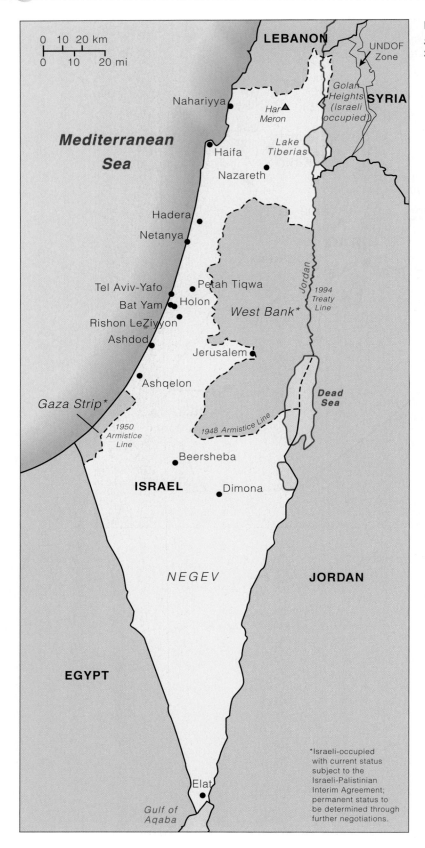

Israel, West Bank, Gaza

Source: U.S. Central Intelligence Agency, *World Factbook,*
2007

Japan

Source: U.S. Central Intelligence Agency, *World Factbook*, 2007

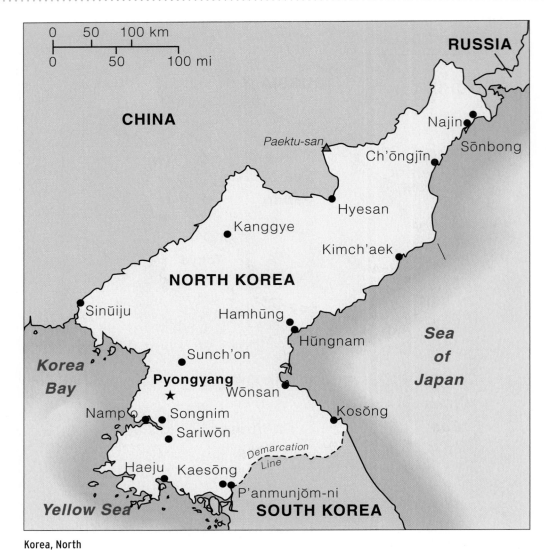

Korea, North

Source: U.S. Central Intelligence Agency, *World Factbook*, 2007

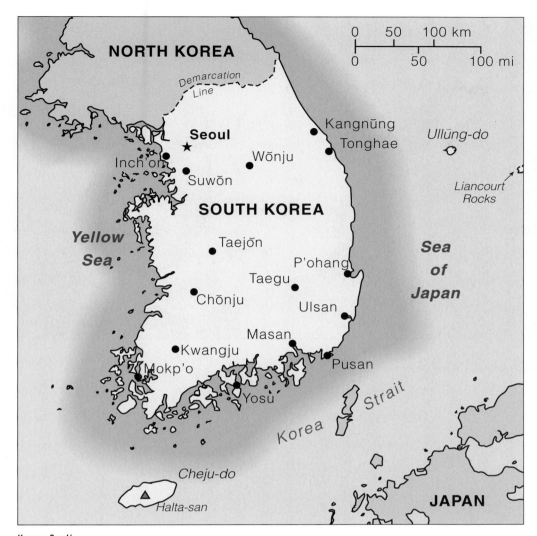

Korea, South

Source: U.S. Central Intelligence Agency, *World Factbook*, 2007

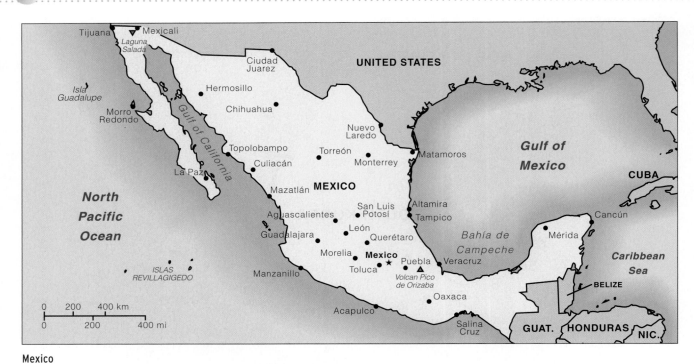

Mexico

Source: U.S. Central Intelligence Agency, *World Factbook*, 2007

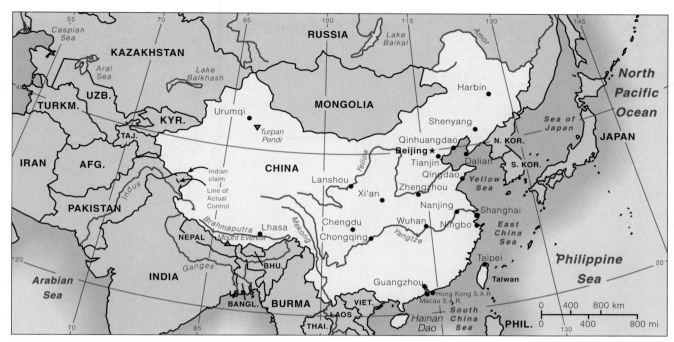

Peoples Republic of China

Source: U.S. Central Intelligence Agency, *World Factbook*, 2007

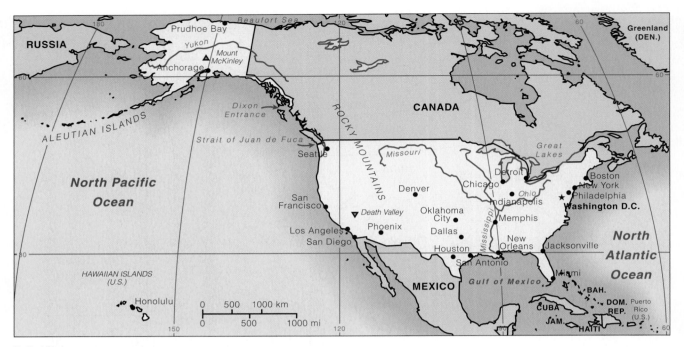

United States

Source: U.S. Central Intelligence Agency, *World Factbook*, 2007

Key Concepts

527 group A tax-exempt advocacy organization that seeks to influence federal elections by running issue related advertisements criticizing the record of a candidate or by mobilizing voters to register and vote.

absolute poverty A situation in which people lack the resources to satisfy the basic needs no person should be without.

absorption assimilation A process by which members of a minority group adapt to the ways of the dominant culture.

achieved characteristic A status acquired through some combination of personal choice, effort, and ability.

achieved statuses Social statuses acquired through some combination of personal choice, effort, and ability. A person's marital status, occupation, and educational attainment are considered examples of achieved statuses.

adaptive culture The portion of nonmaterial culture (norms, values, and beliefs) that adjusts to material innovations.

affectional action Social action that pursues a goal in response to an emotion such as revenge, love, or loyalty.

agents of socialization Significant others, primary groups, ingroups and outgroups, and institutions that (1) shape our sense of self or social identity, (2) teach us about the groups to which we do and do not belong, (3) help us to realize our human capacities, and (4) help us negotiate the social and physical environment we have inherited.

age-specific birth rate The annual number of births per 1,000 women of a specific age group.

aging population A population in which the percentage that is age 65 and older is increasing relative to other age groups.

alienation A state of being in which human life is dominated by the forces of its inventions.

altruistic A state in which the ties attaching the individual to the group are such that he or she has no life beyond the group and strives to blend in with the group to have a sense of being.

altruistic suicide Suicide resulting from social ties so strong that the self has no life apart from the group.

anomaly An observation that a paradigm cannot explain.

anomic A state in which the ties attaching the individual to the group are disrupted due to dramatic changes in economic circumstances.

anomic suicide Suicide resulting from social ties disrupted by dramatic change in economic circumstances.

ascribed characteristic Any physical trait that is biological in origin and/or cannot be changed, to which people assign overwhelming significance. (ch. 2)

ascribed characteristics Attributes people have at birth, develop over time, or possess through no effort or fault of their own. (ch. 8)

ascribed statuses Social positions assigned on the basis of attributes people possess through no fault of their own—those attributes are acquired at birth (such as skin shade, sex, or hair color), develop over time (such as height, weight, baldness, wrinkles, or reproductive capacity), or possess through no effort or fault of their own (such as the country into which one is born and religious affiliation "inherited" from parents).

assimilation A process by which ethnic or racial distinctions between groups disappear because one group is absorbed into another group's culture or because two cultures blend to form a new cultural system.

authoritarian government A system of government in which there is no separation of power and a single person (dictator), group (family, military, single party), or social class holds all power.

authority Legitimate power in which people believe that the differences in power are just and proper—that is, people view a leader as being entitled to give orders.

back stage The area of everyday life out of an audience's sight, where individuals can do things that would be inappropriate or unexpected on the front stage.

basic innovations Revolutionary, unprecedented, or groundbreaking inventions or discoveries that form the basis for a wide range of applications.

beliefs Conceptions that people accept as true, concerning how the world operates and where the individual fits in relationship to others.

biography All the day-to-day activities from birth to death that make up a person's life.

bourgeoisie In Marxist theory, owners of the means of production who exploit the labor of the proletariat. (ch. 1)

bourgeoisie The owners of the means of production (such as land, machinery, buildings, and tools), who purchase labor. (ch. 2)

brain drain The emigration from a country of the most educated and most talented people.

bureaucracy An organization that strives to use the most efficient means to achieve a valued goal.

capitalism An economic system in which the raw materials and the means of producing and distributing goods and services remain privately owned.

caregiver burden The extent to which caregivers believe that their emotional balance, physical health, social life, and financial status suffer because of their caregiver role.

caste system A system of social stratification in which people are ranked on the basis of ascribed characteristics (over which they have no control).

chance Something not subject to human will, choice, or effort; it helps determine a person's racial and ethnic classification.

charismatic authority A type of authority that derives from the exceptional and exemplary qualities of the person who issues the commands.

church A group whose members hold the same beliefs about the sacred and the profane, who behave in the same way in the presence of the sacred, and who gather in body or spirit at agreed-on times to reaffirm their commitment to those beliefs and practices.

class A person's overall economic and social status in a system of social stratification.

class system A system of social stratification in which people are ranked on the basis of achieved characteristics, such as merit, talent, ability, or past performance.

censors People whose job is to sift information conveyed through movies, books, letters, TV, the Internet, and other media and to remove or block any material that those in power consider unsuitable or threatening.

censorship A method of preventing information from reaching an audience.

choice The act of choosing from a range of possible behaviors or appearances; a person's choices may evoke associations with a particular race or ethnic group.

civil religion An institutionalized set of beliefs about a nation's past, present, and future and a corresponding set of rituals. Both the beliefs and the rituals take on a sacred quality and elicit feelings of patriotism. Civil religion forges ties between religion and a nation's needs and political interests.

claims makers People who articulate and promote claims and who tend to gain in some way if the targeted audience accepts their claims as true.

claims-making activities Actions taken to draw attention to a claim, such as "demanding services, filling out forms, lodging complaints, filing lawsuits, calling press conferences, writing letters of protest, passing resolutions, publishing exposés, placing ads in newspapers, . . . setting up picket lines or boycotts" (Spector and Kitsuse 1977, p. 79).

coercive organizations Formal organizations that draw in people who have no choice but to participate; such organizations include those dedicated to compulsory socialization or to resocialization or treatment of individuals labeled as deviant.

cohort A group of people born around the same time (such as a specified five-year period) who share common experiences and perspectives by virtue of the time they were born.

collective memory The experiences shared and recalled by significant numbers of people. Such memories are revived, preserved, shared, passed on, and recast in many forms, such as stories, holidays, rituals, and monuments.

colonialism A form of domination in which a foreign power uses superior military force to impose its political, economic, social, and cultural institutions on an indigenous population so it can control their resources, labor, and markets.

colonization A form of domination in which one country imposes its political, economic, social, and cultural institutions on an indigenous population and the land it occupies.

color line A barrier supported by customs and laws separating nonwhites from whites, especially with regard to their place in the division of labor.

commercialization of gender ideals The process of introducing products to the market by using advertising and sales campaigns that draw on socially constructed standards of masculinity and femininity.

concepts Thinking and communication tools used to give and receive complex information efficiently and to frame and focus observations.

conflict In Marxist theory, the major force that drives social change.

conformity Behavior and appearances that follow and maintain the standards of a group. Also, the acceptance of culturally valued goals and the pursuit of those goals through means defined as legitimate.

conglomerate A large corporation that owns "smaller" corporations acquired through merger or acquisition.

constrictive pyramid A population pyramid that is narrower at the base than in the middle. It shows that the population consists disproportionately of middle-aged and older people.

constructionist approach A sociological approach that focuses on the way specific groups, activities, conditions, or artifacts become defined as problems.

context The social setting in which racial and ethnic categories are recognized, constructed, and challenged.

control The guiding or regulating, by planning out in detail, the production or delivery of a service or product.

control variables Variables suspected of causing spurious correlations.

core economies The wealthiest, most highly diversified economies, with strong, stable governments.

corporate crime Crime committed by a corporation as it competes with other companies for market share and profits.

correlation coefficient A mathematical representation that quantifies the extent to which a change in one variable is associated with a change in another variable.

countercultures Subcultures in which the norms, values, beliefs, symbols, and language the members share emphasize conflict or opposition to the larger culture. In fact, rejection of the dominant culture's values, norms, symbols, and beliefs is central to understanding a counterculture.

counterrevolutionary movements Social movements that seek to maintain a social order that reformist and revolutionary movements are seeking to change.

crude birth rate The annual number of births per 1,000 people in a designated geographic area.

crude death rate The annual number of deaths per 1,000 people in a designated geographic area.

cultural genocide An extreme form of ethnocentrism in which the people of one society define the culture of another society not as merely offensive, but as so intolerable that they attempt to destroy it.

cultural relativism The perspective that a foreign culture should not be judged by the standards of a home culture and that a behavior or way of thinking must be examined in its cultural context.

culture The way of life of a people; more specifically, the human created strategies for adjusting to the environment and to those creatures (including humans) that are part of that environment.

culture shock The strain that people from one culture experience when they must reorient themselves to the ways of a new culture.

cults Very small, loosely organized groups, usually founded by a charismatic leader who attracts people by virtue of his or her personal qualities.

cultural base The number of existing innovations, which forms the basis for further inventions.

cultural lag A situation in which adaptive culture fails to adjust in necessary ways to material innovation.

currents of opinion The state of affairs with regard to some way of being expressed through rates (suicide, marriage, savings).

dearth of feedback A situation in which much of the information released or picked up by the popular media is not subjected to honest, constructive criticism, because the critical audience that exists is too small to evaluate the information before it is used.

decolonization A process of undoing colonialism such that the colonized achieves independence from the so-called mother country.

democracy A system of government in which power is vested in the citizen body, and in which members of that citizen body participate directly or indirectly in the decision-making process.

demographic gap The difference between a population's birth rate and death rate.

demography A subspecialty within sociology that focuses on the study of human populations, particularly on their size and rate of growth.

denomination A hierarchical religious organization, led by a professionally trained clergy, in a society in which church and state are usually separate.

dependent variable The variable to be explained or predicted.

deviance Any behavior or physical appearance that is socially challenged or condemned because it departs from the norms and expectations of a group.

deviant subcultures Groups that are part of the larger society but whose members adhere to norms and values that favor violation of the larger society's laws.

differential association A theory of socialization that explains how deviant behavior, especially delinquent behavior, is learned. It states that "when persons become criminal, they do so because of contacts with criminal patterns and also because of isolation from anticriminal patterns" (Sutherland and Cressey 1978, p. 78).

diffusion The process by which an idea, an invention, or some other cultural item is borrowed from a foreign source.

discrimination Intentional or unintentional unequal treatment of individuals or groups because of attributes unrelated to merit, ability, or past performance—treatment that denies equal opportunities to achieve socially valued goals.

disenchantment A great spiritual void accompanied by a crisis of meaning. It occurs when people focus so uncritically on the ways they go about achieving a valued goal that they lose sight of that goal.

dispositional causes Forces over which individuals are supposed to have control—including personal qualities or traits, such as motivation level, mood, and effort.

division of labor Work that is broken down into specialized tasks, each performed by a different set of persons trained to do that task. The persons doing each task often live in different parts of the world. Not only are the tasks specialized, but the parts and materials needed to manufacture products also come from many different regions of the world.

documents Written or printed materials used in research.

domestication The process by which plants and animals were brought under human control.

double consciousness According to DuBois, "this sense of always looking at one's self through the eyes of others, of measuring one's soul by the tape of a world that looks on in amused contempt and pity." The double consciousness includes a sense of two-ness: "an American, a Negro; two souls, two thoughts, two unreconciled strivings; two warring ideals in one dark body, whose dogged strength alone keeps it from being torn asunder."

doubling time The estimated number of years required for a country's population to double in size.

downward mobility A form of vertical mobility in which a person moves down in rank.

dramaturgical model A model in which social interaction is viewed as if it were a theater, people as if they were actors, and roles as if they were performances before an audience in a particular setting.

ecclesia A professionally trained religious organization, governed by a hierarchy of leaders, that claims everyone in a society as a member.

economic system A socially created institution that coordinates human activity in the effort to produce, distribute, and consume goods and services.

education In the broadest sense, the experiences that train, discipline, and shape the mental and physical potentials of the maturing person.

efficiency An organization's claim of offering the "best" products and services, which allow consumers to move quickly from one state of being to another (for example, from hungry to full, from fat to thin, or from uneducated to educated).

egoistic A state in which the ties attaching the individual to others in the society are weak.

egoistic suicide Suicide resulting from weak social ties that fail to attach the person to the group.

emigration The departure of individuals from one country or other geographic area to take up residence elsewhere.

empire A group of countries under the direct or indirect control of a foreign power or government such that the dominant power shapes the subordinate entities' political, economic, and cultural development.

endogamy Norms requiring or encouraging people to choose a partner from the same social category as their own.

established sects Religious organizations, resembling both denominations and sects, that have left denominations or ecclesiae and have existed long enough to acquire a large following and widespread respectability.

esteem The reputation that someone occupying an ascribed or achieved status has earned from people who know and observe them.

ethgender A social category that combines sex, gender, race, and ethnicity.

ethnicity People who share, believe they share, or are believed by others to share a national origin; a common ancestry; a place of birth; distinctive concrete social traits (such as religious practices, style of dress, body adornments, or language); or socially important physical characteristics (such as skin color, hair texture, or body structure).

ethnocentrism A viewpoint that uses one culture, usually the home culture, as the standard for judging the worth of foreign ways.

evolutionary view The idea that human societies progress in stages from primitive to civilized, with each successive stage representing a more complex form of social organization.

exogamy Norms requiring or encouraging people to choose a partner from a social category other than their own.

expansive pyramid A triangular population pyramid that is broadest at the base, with each successive cohort smaller than the one below it. This pyramid shows that the population consists disproportionately of young people.

externality costs Hidden costs of using, making, or disposing of a product that are not figured into the price of the product or paid for by the producer.

extreme poverty The most severe form of poverty, in which people cannot afford the basic human necessities (food, water, clothes, and shelter).

extreme wealth The most excessive form of wealth, in which a very small proportion of people have money, material possessions, and other assets (minus liabilities) in such abundance that a small fraction of it (if spent appropriately) could provide adequate food, safe water, sanitation, and basic health care for the 1 billion poorest people on the planet.

facade of legitimacy An explanation that members of dominant groups give to justify their actions.

falsely accused People who have not broken the rules of a group but are treated as if they have.

family A social institution that binds people together through blood, marriage, law, and/or social norms. Family members are generally expected to care for and support each other.

fatalistic A state in which the ties attaching the individual to the group involve discipline so oppressive it offers no chance of release.

fatalistic suicide Suicide resulting from social ties whose discipline is so oppressive it offers no chance of release.

feeling rules Norms that specify appropriate ways to express internal sensations.

femininity The physical, behavioral, and mental and emotional traits believed to be characteristic of females.

feminism In its most basic sense, a perspective that advocates equality between men and women.

finance aristocracy Bankers and stockholders seemingly detached from the world of "work."

folkways Customary ways of handling the routine matters of everyday life. (ch. 7)

folkways Norms that apply to the mundane aspects or details of daily life. (ch. 3)

foreign-born People living within the political boundaries of a country who were born elsewhere.

formal curriculum The various academic subjects, such as mathematics, science, English, reading, and physical education.

formal dimension The official aspect of an organization, including job descriptions and written rules, guidelines, and procedures established to achieve valued goals.

formal education A systematic, purposeful, planned effort intended to impart specific skills and modes of thought.

formal organization Coordinating mechanisms that bring together people, resources, and technology and then channel human activity toward achieving a specific outcome.

formal sanctions Expressions of approval or disapproval backed by laws, rules, or policies that specify (usually in writing) the conditions under which people should be rewarded or punished and the procedures for allocating rewards and administering punishments.

fortified households Preindustrial arrangements in which a household acts as an armed unit and the head of the household acts as its military commander. The household is characterized by the presence of a nonhouseholder class, consisting of propertyless laborers and servants.

front stage The area of everyday life visible to an audience, where people take care to create and maintain the images and behavior the audience has come to expect.

function The contribution part of a society makes to order and stability within the society.

functionally illiterate Lacking the level of reading, writing, and calculating skills needed to function in the society in which one lives.

fundamentalism A belief in the timelessness of sacred writings and a belief that such writings apply to all kinds of environments.

games Structured, organized activities that usually involve more than one person and a number of constraints, such as established roles, rules, time, place, and outcome.

gender A social distinction based on culturally conceived and learned ideals about appropriate appearance, behavior, and mental and emotional characteristics for males and females.

gender polarization The organizing of social life around male-female ideals, so that people's sex influences every aspect of their life, including how they dress, the time they get up in the morning, what they do before they go to bed at night, the social roles they take on, the things they worry about, and even the ways they express emotion and experience sexual attraction.

gender-schematic A term describing decisions that are influenced by a society's polarized definitions of masculinity and femininity rather than by criteria such as self-fulfillment, interest, ability, and personal comfort.

generalizability The extent to which findings can be applied to the larger population from which a sample is drawn.

generalized other A system of expected behaviors, meanings, and viewpoints that transcend those of the people participating.

global inequality The unequal distribution of income, wealth, or other valued resources across countries and within each country.

global interdependence A situation in which the social, political, financial, and cultural lives of people around the world are so intertwined that one country's problems—such as unemployment, drug abuse, environmental pollution, and the search for national security in the face of terrorism—are part of a larger global situation.

globalization The ever-increasing flow of goods, services, money, people, information, and culture across political borders.

goods Any products manufactured, grown, or extracted from the earth, such as food, clothing, housing, automobiles, coal, computers, and so on.

government The organizational structure that directs and coordinates people's involvement in the political activities of a country or other territory (city, county, state) within that country.

gross domestic product (GDP) The monetary value of the goods and services that a nation's work force produces over the course of a year (or some other time period).

group Two or more people who share a distinct identity, feel a sense of belonging, and interact directly or indirectly with one another.

Hawthorne effect A phenomenon in which research subjects alter their behavior when they learn they are being observed.

hegemony A process by which a power maintains its dominance over other entities.

hidden curriculum All the other activities that go on as students learn subject matter, and the "lessons" that those other activities convey about the value and meaning of what the students are learning.

households All related and unrelated persons who share the same dwelling.

hypothesis A trial explanation put forward as the focus of research; it predicts how independent and dependent variables are related and how a dependent variable will change when an independent variable changes.

ideal A standard against which real cases can be compared.

ideal type A deliberate simplification or caricature that exaggerates defining characteristics, thus establishing a standard against which real cases can be compared.

ideology A set of beliefs taken to be accurate accounts and explanations of why things are as they are. The beliefs are not challenged or subjected to scrutiny by the people who hold them.

illiteracy The inability to understand and use a symbol system, whether it is based on sounds, letters, numbers, pictographs, or some other type of symbol.

immigration The entry of individuals into a country or other geographic area of which they are not natives to take up residence there.

imperialistic power A political entity that exerts control and influence over foreign entities through conquest or force and/or through policies and economic pressures.

impression management The process by which people in social situations manage the setting, their dress, their words, and their gestures to correspond to the impression they are trying to make or the image they are trying to project.

improving innovations Modifications of basic inventions that improve upon the originals—for example, making them smaller, faster, less complicated, more efficient, more attractive, or more profitable.

income The money a person earns, usually on an annual basis through salary or wages.

independent variable The variable that explains or predicts the dependent variable.

individual discrimination Any overt action of an individual that depreciates someone from an outgroup, denies outgroup members opportunities to participate, or does violence to their lives and property.

Industrial Revolution Changes in manufacturing, agriculture, transportation, and mining that transformed virtually every aspect of society.

infant mortality rate The annual number of deaths of infants one year old or younger for every 1,000 such infants born alive.

informal dimension The unofficial aspect of an organization, including behaviors that depart from the formal dimension, such as employee-generated norms that evade, bypass, or ignore official rules, guidelines, and procedures.

informal education Education that occurs in a spontaneous, unplanned way.

informal sanctions Spontaneous, unofficial expressions of approval or disapproval that are not backed by the force of law.

information explosion An unprecedented increase in the amount of stored and transmitted data and messages in all media (including electronic, print, radio, and television).

ingroup A group with which people identify and to which they feel closely attached, particularly when that attachment is founded on hatred or opposition toward an outgroup.

in-migration The movement of people into a designated geographic area, such as a country, region, or city.

innovation (as a response to structural strain) The acceptance of cultural goals but the rejection of the legitimate means to achieve them. (ch. 7)

innovation The invention or discovery of something, such as a new idea, process, practice, device, or tool. (ch. 16)

institution A relatively stable and predictable arrangement among people that has emerged over time to coordinate human interaction and behavior in ways that meet some social need.

institutionally complete subcultures Subcultures whose members do not interact with anyone outside their subculture to shop for food, attend school, receive medical care, or find companionship, because the subculture satisfies these needs.

instrumental-rational action Social action that is pursued by the most efficient means, often without considering the appropriateness or consequences of those means.

insurgents Groups who participate in armed rebellion against some established authority, government, or administration with the hope that those in power will retreat.

intergenerational mobility A form of vertical mobility in which people move upward or downward in rank over two or more generations.

internal migration The movement of people within the boundaries of a single country—from one state, region, or city to another.

internalization The process in which people take as their own and accept as binding the norms, values, beliefs, and language that their socializers are attempting to pass on.

intersexed A broad term used by the medical profession to classify people with some mixture of male and female biological characteristics.

interviews Face-to-face or telephone conversations between an interviewer and a respondent, in which the interviewer asks questions and records the respondent's answers.

intragenerational mobility A form of vertical mobility in which a person moves upward or downward in rank during his or her lifetime.

invention A synthesis of existing innovations.

involuntary minorities Ethnic or racial groups that were forced to become part of a country by slavery, conquest, or colonization.

iron cage of rationality The set of irrationalities that rational systems generate.

Islamic revitalism Responses to the belief that existing political, economic, and social systems have failed—responses that include a disenchantment with, and even a rejection of, the West; soul-searching; a quest for greater authenticity; and a conviction that Islam offers a viable alternative to secular nationalism, socialism, and capitalism.

issue A matter that can be explained only by factors outside an individual's control and immediate environment.

labor-intensive poor economies Economies that have a lower level of industrial production and a lower standard of living than core economies. They differ markedly from core economies on indicators such as doubling time, infant mortality, total fertility, per capita income, and per capita energy consumption.

language A symbol system involving the use of sounds, gestures (signing), and/or characters (such as letters or pictures) to convey meaning.

latent functions Unintended or unanticipated effects that part of a society has on order and stability within the society.

laws of supply and demand Natural laws regulating capitalist economies such that "as demand for an item increases, prices rise." Manufacturers respond by increasing production which in turn "increases competition and drives the price down" (Hirsch, Kett, and Trefil 1993, p. 455).

legal-rational authority A type of authority that rests on a system of impersonal rules that formally specifies the qualifications for occupying a powerful position.

liberation theology A religious movement based on the idea that organized religions have a responsibility to demand social justice for the marginalized peoples of the world, especially landless peasants and the urban poor, and to take an active role at the grassroots level to bring about political and economic justice.

life chances The probability that an individual's life will follow a certain path and will turn out a certain way. (ch. 8)

life chances A critical set of potential social advantages, including the chance to live past the first year of life, to live independently in old age, and everything in between. (ch. 12)

linguistic relativity hypothesis The idea that "no two languages are ever sufficiently similar to be considered as representing the same social reality. The worlds in which different societies live are distinct worlds, not merely the same world with different labels attached."

looking-glass self A process in which a sense of self develops, enabling one to see oneself reflected in others' real or imagined reactions to one's appearance and behaviors.

low-technology tribal societies Hunting-and-gathering societies with technologies that do not permit the creation of surplus wealth.

manifest functions Intended or anticipated effects that part of a society has on order and stability within the society.

masculinity The physical, behavioral, and mental and emotional traits believed to be characteristic of males.

mass media Forms of communication designed to reach large audiences without face-to-face contact between those conveying and those receiving the messages.

master status One status in a status set that is so important to a person's social identity it overshadows all other statuses a person occupies—shaping every aspect of life and dominating social interactions.

master status of deviant An identification marking a rule breaker first and foremost as a deviant.

material culture All the natural and human-created objects to which people have attached meaning.

McDonaldization A process whereby the principles governing the fast-food industry come to dominate other sectors of the American economy, society, and the world.

means of production The land, machinery, buildings, tools, labor, and other resources needed to produce and distribute goods and services.

mechanical solidarity Social order and cohesion based on a common conscience, or uniform thinking and behavior.

mechanization The addition of external sources of power, such as that derived from burning coal and oil, to muscle-powered tools and modes of transportation.

melting pot assimilation Cultural blending in which groups accept many new behaviors and values from one another. The exchange produces a new cultural system, which is a blend of the previously separate systems.

methods of data collection The procedures a researcher follows to gather relevant data.

migration The movement of people from one residence to another.

migration rate A rate based on the difference between the number of people entering and the number of people leaving a designated geographic area in a year. We divide that difference by the size of the relevant population and then multiply the result by 1,000.

militaristic power One that believes military strength, and the willingness to use it, is the source of national—and even global—security.

minority groups Subgroups within a society that can be distinguished from members of the dominant group by visible identifying characteristics, including physical and cultural attributes. These subgroups are systematically excluded, whether consciously or unconsciously, from full participation in society and denied equal access to positions of power, privilege, and wealth.

mixed contacts "The moments when stigmatized normals are in the same 'social situation,' that is, in one another's immediate physical presence, whether in a conversation-like encounter or in the mere co-presence of an unfocused gathering" (Goffman 1963, p. 12).

modern capitalism An economic system that involves careful calculation of costs of production relative to profits, borrowing and lending money, accumulating all forms of capital, and drawing labor from an unrestricted global labor pool.

monopoly A situation in which a single producer dominates a market.

mores Norms that people define as essential to the well-being of their group or nation.

mortality crises Violent fluctuations in the death rate, caused by war, famine, or epidemics.

multinational corporations Enterprises that own, control, or license production or service facilities in countries other than the one where the corporations are headquartered.

mystical religions Religions in which the sacred is sought in states of being that, at their peak, can exclude all awareness of one's existence, sensations, thoughts, and surroundings.

natural increase The number of births minus the number of deaths occurring in a population in a year.

nature Human genetic makeup or biological inheritance.

negative sanction An expression of disapproval for noncompliance.

negatively privileged property class Weber's category for people completely lacking in skills, property, or employment or who depend on seasonal or sporadic employment; they constitute the very bottom of the class system.

neocolonialism A new form of colonialism where more powerful foreign governments and foreign-owned businesses continue to exploit the resources and labor of the post-colonial peoples.

nonmaterial culture Intangible human creations, which we cannot identify directly through the senses.

nonprejudiced discriminators (fair-weather liberals) Persons who believe in equal opportunity but discriminate because doing so gives them an advantage or because they fail to consider the discriminatory consequences of their actions.

nonprejudiced nondiscriminators (all-weather liberals) Persons who accept the creed of equal opportunity and whose conduct conforms to that creed.

norms Written and unwritten rules that specify behaviors appropriate and inappropriate to a particular social situation.

nurture The social environment, or the interaction experiences that make up every individual's life.

objective deprivation The condition of the people who are the worst off or most disadvantaged—those with the lowest incomes, the least education, the lowest social status, the fewest opportunities, and so on.

objectivity A stance in which researchers' personal, or subjective, views do not influence their observations or the outcomes of their research.

observation A research technique in which the researcher watches, listens to, and records behavior and conversations as they happen.

oligarchy Rule by the few, or the concentration of decision-making power in the hands of a few persons, who hold the top positions in a hierarchy.

oligopoly A situation in which a few producers dominate a market.

operational definitions Clear, precise definitions and instructions about how to observe and/or measure the variables under study.

organic solidarity Social order based on interdependence and cooperation among people performing a wide range of diverse and specialized tasks.

outgroup A group toward which members of an ingroup feel a sense of separateness, opposition, or even hatred.

out-migration The movement of people out of a designated geographic area, such as a country, region, or city.

paradigms The dominant and widely accepted theories and concepts in a particular field of study.

participant observation A research technique in which the researcher observes study participants while directly interacting with them.

peripheral economies Economies that rely on a few commodities or even a single commodity (such as coffee, peanuts, or tobacco) or a single mineral resource (such as tin, copper, or zinc). They are exploited by both core and semiperipheral economies.

planned obsolescence A profit-making strategy that involves producing goods that are disposable after a single use, have a shorter life cycle than the industry is capable of producing, or go out of style quickly even though the goods can still serve their purpose.

play A voluntary and often spontaneous activity with few or no formal rules that is not subject to constraints of time or place.

pluralist model A model that views politics as an arena of compromise, alliances, and negotiation among many competing and different special-interest groups, and power as something that is dispersed among those groups.

political action committees (PACs) Committees that raise money to be donated to the political candidates most likely to support their special interests.

political parties According to Weber, "organizations oriented toward the planned acquisition of social power [and] toward influencing social action no matter what its content may be."

political system A socially created institution that regulates the use of and access to power that is essential to articulating and realizing individual, local, regional, national, international, or global interests and agendas.

population pyramid A series of horizontal bar graphs, each representing a different five-year age cohort, that allows us to compare the sizes of the cohorts.

populations The total number of individuals, traces, documents, territories, households, or groups that could be studied.

positive checks Events that increase deaths—including epidemics of infectious and parasitic diseases, war, famine, and natural disasters—and thus keep population size in line with the food supply.

positive sanction An expression of approval and a reward for compliance.

positively privileged property class Weber's category for the people at the very top of the class system.

post-industrial society A society that is dominated by intellectual technologies of telecommunications and computers, not just "large computers but computers on a chip." These intellectual technologies have had a revolutionary effect on virtually every aspect of social life.

power The probability that an individual can achieve his or her will even against another individual's opposition.

power elite Those few people who occupy such lofty positions in the social structure of leading institutions that their decisions have consequences affecting millions of people worldwide.

predestination The belief that God has foreordained all things, including the salvation or damnation of individual souls.

predictability The expectation that a service or product will be the same no matter where or when it is purchased.

prejudice A rigid and usually unfavorable judgment about an outgroup that does not change in the face of contradictory evidence and that applies to anyone who shares the distinguishing characteristics of the outgroup.

prejudiced discriminators (active bigots) Persons who reject the notion of equal opportunity and profess a right, even a duty, to discriminate.

prejudiced nondiscriminators (timid bigots) Persons who reject the creed of equal opportunity but refrain from discrimination primarily because they fear the sanctions they may encounter if they are caught.

primary group A social group that has face-to-face contact and strong emotional ties among its members.

primary sector (of the economy) Economic activities that generate or extract raw materials from the natural environment.

primary sex characteristics The anatomical traits essential to reproduction.

private ownership A situation in which individuals (rather than workers, the government, or communal groups) own the raw materials, machines, tools, labor, trucks, buildings, and other inputs needed to produce and distribute goods and services.

productive work Work that involves "the production of the means of existence, of food, clothing, and shelter and the tools necessary for that production" (Engels 1884, pp. 71–72).

profane A term describing everything that is not sacred, including things opposed to the sacred and things that stand apart from the sacred, albeit not in opposition to it.

professionalization A trend in which organizations hire experts with formal training in a particular subject or activity—training needed to achieve organizational goals.

proletariat A social class composed of workers who own nothing of the production process and who sell their labor to the bourgeoisie. (ch. 2)

proletariat In Marxist theory, individuals who must sell their labor to the bourgeoisie. (ch. 1)

prophetic religions Religions in which the sacred revolves around items that symbolize significant historical events or around the lives, teachings, and writings of great people.

pull factors The conditions that encourage people to move into a geographic area.

pure deviants People who have broken the rules of a group and are caught, punished, and labeled as outsiders.

push factors The conditions that encourage people to move out of a geographic area.

quantification and calculation Numerical indicators that enable customers to evaluate a product or service easily.

race A vast collectivity of people more or less bound together by shared and selected history, ancestors, and physical features; these people are socialized to think of themselves as a distinct group, and they are regarded by others as such.

random sample A type of sample in which every case in the population has an equal chance of being selected.

rate of natural increase The number of births minus the number of deaths occurring in a population in a year, divided by the size of the population at the beginning of the year.

rationalization A process in which thought and action rooted in emotion, superstition, respect for mysterious forces, or tradition is replaced by thought and action grounded in instrumental-rational action.

rebellion The full or partial rejection of both cultural goals and the means of achieving them and the introduction of a new set of goals and means.

reentry shock Culture shock in reverse; it is experienced upon returning home after living in another culture.

regressive or reactionary movements Social movements that seek to turn back the hands of time to an earlier condition or state of being, one sometimes considered a "golden era."

reformist movements Social movements that target a specific feature of society as needing change.

relative deprivation A social condition that is measured not by objective standards, but rather by comparing one group's situation with the situations of groups who are more advantaged.

relative poverty Measured not by some objective standard, but rather by comparing the situation of those at the bottom against an average situation or against the situation of others who are more advantaged.

reliability The extent to which an operational definition gives consistent results.

representative democracy A system of government in which decision making takes place indirectly through elected representatives.

representative sample A type of sample in which those selected for study have the same distribution of characteristics as the population from which it is selected.

reproductive work Work that involves bearing children, caregiving, managing households, and educating children.

research A data-gathering and data-explaining enterprise governed by strict rules.

research design A plan for gathering data that specifies who or what will be studied and the methods of data collection.

research methods Techniques that sociologists and other investigators use to formulate or answer meaningful research questions and to collect, analyze, and interpret data in ways that allow other researchers to verify the results.

resocialization The process of discarding values and behaviors unsuited to new circumstances and replacing them with new, more-appropriate values and norms.

resource mobilization A situation in which a core group of sophisticated strategists works to harness a disaffected group's energies, attract money and supporters, capture the news media's attention, forge alliances with those in power, and develop an organizational structure.

retreatism The rejection of both cultural goals and the means of achieving them.

reverse ethnocentrism A type of ethnocentrism in which the home culture is regarded as inferior to a foreign culture.

revolutionary movements Social movements that seek broad, sweeping, and radical structural changes to a society's basic social institutions or to the world order.

right A behavior that a person assuming a role can demand or expect from another.

ritualism The rejection of cultural goals but a rigid adherence to the legitimate means of achieving them.

rituals Rules that govern how people must behave in the presence of the sacred to achieve an acceptable state of being.

role The behavior, obligations, and rights expected of a social status in relation to another social status.

role conflict A predicament in which the expectations associated with two or more roles in a role set contradict one another.

role obligations The relationship and behavior a person enacting a role must assume toward others occupying a particular social status.

role strain A predicament in which the social role a person is enacting involves contradictory or conflicting expectations.

role taking The process of stepping outside the self and imagining how others view its appearance and behavior from an outsider's perspective.

sacramental religions Religions in which the sacred is sought in places, objects, and actions believed to house a god or spirit.

sacred A domain of experience that includes everything regarded as extraordinary and that inspires in believers deep and absorbing sentiments of awe, respect, mystery, and reverence.

safe debt Debt secured through collateral, such as a house.

samples Portions of the cases from a larger population.

sampling frame A complete list of every case in a population.

sanctions Reactions of approval or disapproval to others' behavior or appearance.

scapegoat A person or group blamed for conditions that (a) cannot be controlled, (b) threaten a community's sense of well-being, or (c) shake the foundations of an important institution.

schooling A program of formal, systematic instruction that takes place primarily in classrooms but also includes extracurricular activities and out-of-classroom assignments.

scientific method An approach to data collection in which knowledge is gained through observation and its truth is confirmed through verification.

secondary groups Impersonal associations among people who interact for a specific purpose.

secondary sector (of the economy) Economic activities that transform raw materials into manufactured goods.

secondary sex characteristics Physical traits not essential to reproduction (such as breast development, quality of voice, distribution of facial and body hair, and skel-

etal form) that result from the action of so-called male (androgen) and female (estrogen) hormones.

secondary sources (archival data) Data that have been collected by other researchers for some other purpose.

secret deviants People who have broken the rules of a group but whose violation goes unnoticed or, if it is noticed, prompts no one to enforce the law.

sect A small community of believers led by a lay ministry, with no formal hierarchy or official governing body to oversee its various religious gatherings and activities. Sects are typically composed of people who broke away from a denomination because they came to view it as corrupt.

secularization A process by which religious influences on thought and behavior are reduced.

secure parental employment A situation in which at least one parent or guardian is employed full-time (35 or more hours per week for at least 50 weeks in the past year).

segregation The physical or social separation of categories of people.

selective perception The process in which prejudiced persons notice only the behaviors or events related to an outgroup that support their stereotypes about the outgroup.

self-administered questionnaire A set of questions given to respondents who read the instructions and fill in the answers themselves.

self-fulfilling prophecy A concept that begins with a false definition of a situation. Despite its falsity, people assume it to be accurate and behave accordingly. The misguided behavior produces responses that confirm the false definition.

semiperipheral economies Economies that are moderately wealthy and diversified but have extreme inequality. They exploit peripheral economies and are in turn exploited by economies.

services Activities performed for others that result in no tangible product, such as entertainment, transportation, financial advice, medical care, spiritual counseling, and education.

sex A biological concept based on primary sex characteristics.

sex ratio The number of females for every thousand males (or another preferred constant, such as 10, 100, or 10,000).

sexism The belief that one sex—and by extension, one gender—is innately superior to another, justifying unequal treatment of the sexes.

significant others People or characters who are important in an individual's life, in that they greatly influence that person's self-evaluation or motivate him or her to behave in a particular manner.

significant symbol A word, gesture, or other learned sign used to convey a meaning from one person to another.

simultaneous independent inventions Situations in which more or less the same invention is produced by two or more persons working independently of one another at about the same time.

situational causes Forces outside an individual's immediate control—such as weather, chance, and others' incompetence.

small groups Groups of 2 to about 20 people who interact with one another in meaningful ways.

social actions Actions people take in response to others.

social change Any significant alteration, modification, or transformation in the organization and operation of social life.

social control Methods used to teach, persuade, or force a group's members, and even nonmembers, to comply with and not deviate from its norms and expectations.

social dynamics The forces that cause societies to change.

social emotions Internal bodily sensations experienced in relationships with other people.

social facts Ideas, feelings, and ways of behaving "that possess the remarkable property of existing outside the consciousness of the individual."

social forces Any human-created ways of doing things that influence, pressure, or force people to behave, interact with others, and think in specified ways.

social interaction An everyday event in which at least two people communicate and respond through language and symbolic gestures to affect one another's behavior and thinking. (ch. 5)

social interaction Everyday events in which the people involved take one another into account by consciously and unconsciously attaching meaning to the situation, interpreting what others are saying, and then responding accordingly. (ch. 1)

social mobility Movement from one social class to another.

social movement A situation in which a substantial number of people organize to make a change, resist a change, or undo a change in some area of society.

social prestige A level of respect or admiration for a status apart from any person who happens to occupy it.

social statics The forces that hold societies together such that they endure over time.

social status A position in a social structure.

social stratification The systematic process of ranking people on a scale of social worth such that the ranking affects life chances in unequal ways.

social structure Two or more people occupying social statuses and interacting in expected ways. Statuses are enacted through roles and are embedded in groups and institutions, also key components of social structures.

socialism An economic system in which the raw materials and the means of producing and distributing goods and services are collectively owned.

socialization The process by which people develop a sense of self and learn the ways of the society in which they live.

society A group of interacting people who share, perpetuate, and create culture.

sociological imagination A quality of mind that allows people to see how larger social forces, especially their place in history and the ways in which society is organized, shape their life stories or biographies.

sociological theory A set of principles and definitions that tell how societies operate and how people in them relate to one another and respond to their surroundings.

sociology The study of human activity as it is affected by social forces emanating from groups, organizations, societies, and even the global community.

solidarity The ties that bind people to one another in a society.

special-interest groups Groups composed of people who share an interest in a particular economic, political, and social issue and who form an organization or join an existing organization with the goal of influencing public opinion and government policy.

spurious correlation A correlation that is coincidental or accidental because the independent and dependent variables are not actually related; rather, some third variable related to both of them makes it seem as though they are.

stationary pyramid A population pyramid in which all cohorts (except the oldest) are roughly the same size.

statistical measures of performance Quantitative (and sometimes qualitative) measures of how well an organization and its members or employees are performing.

status group Weber's term for an amorphous group of people held together both by virtue of a lifestyle that has come to be expected of them and by the level of esteem in which other people hold them.

status value The social value assigned to a status such that people who possess one status (white skin versus brown skin, blonde hair versus dark hair, low income versus high income, single versus married, professional athlete versus high school teacher) are regarded and treated as more valuable or worthy than people who possess another status.

stereotypes Inaccurate generalizations about people who belong to an outgroup.

stigma An attribute defined as deeply discrediting because it overshadows all other attributes that a person might possess.

stupid debt Debt from using credit cards to finance spending sprees and impulse buying.

structural constraints The established and customary rules, policies, and day-to-day practices that affect a person's life chances.

structural strain Any situation in which (1) the goals defined as valuable and legitimate for a society have unclear limits, (2) people are unsure whether the legitimate means that the society provides will allow them to achieve the goals, and (3) legitimate opportunities for reaching the goals remain closed to a significant portion of the population.

structured interview An interview in which the wording and sequence of questions are set in advance and cannot be changed during the interview.

subcultures Groups that share in some parts of the dominant culture but have their own distinctive values, norms, beliefs, symbols, language, or material culture.

subjective secularization A decrease in the number of people who view the world and their place in it from a religious perspective.

suicide The act of severing relationships.

surplus wealth Wealth beyond what is needed to meet basic human needs, such as food and shelter.

surveillance A mechanism of social control that involves monitoring the movements, activities, conversations, and associations of people who are believed likely to engage in wrongdoing; catching those who do engage in it; preventing people from engaging in it; and ensuring that the public is protected from wrongdoers.

survival debt Debt from using credit cards to pay living expenses associated with food, rent, and transportation.

symbols Any kind of physical or conceptual phenomenon—a word, an object, a sound, a feeling, an odor, a gesture or bodily movement, or a concept of time—to which people assign a name and a meaning or value.

symbolic gestures Nonverbal cues, such as tone of voice and body movements, that convey meaning from one person to another.

sympathetic knowledge First-hand knowledge gained by living and working among those being studied.

systems of racial and ethnic classification A systematic process that divides people into racial or ethnic categories that are implicitly or explicitly ranked on a scale of social worth.

technological determinist Someone who believes that human beings have no free will and are controlled entirely by their material innovations.

territories Settings that have borders or that are set aside for particular activities.

terrorism The systematic use of anxiety-inspiring violent acts by clandestine or semi-clandestine individuals, groups, or state-supported actors for idiosyncratic, criminal, or political reasons.

tertiary sector Economic activities related to delivering services such as health care or entertainment and those activities related to creating and distributing information.

theocracy A form of government in which political authority rests in the hands of religious leaders or a theologically trained elite. Under this system, there is no separation of church and state.

theory A framework that can be used to comprehend and explain events.

this-worldly asceticism A belief that people are instruments of divine will and that God determines and directs their activities.

Thomas theorem An assumption focusing on how people construct reality: If people define situations as real, their definitions have real consequences.

tipping points Situations in which a previously rare (or seemingly rare) event, response, or opinion becomes dramatically more common.

total fertility rate The average number of children that women in a specific population bear over their lifetime.

total institutions Institutions in which people surrender control of their lives, voluntarily or involuntarily, to an administrative staff and carry out daily activities with others required to do the same thing.

totalitarianism A system of government characterized by (1) a single ruling party led by a dictator, (2) an unchallenged official ideology that defines a vision of the "perfect" society and the means to achieve that vision, (3) a system of social control that suppresses dissent and opposition, and (4) centralized control over the media and the economy.

traces Materials or other forms of physical evidence that yield information about human activity.

traditional action Social action that pursues a goal because it was pursued in the past.

traditional authority A type of authority that relies on the sanctity of time-honored norms that govern the selection of someone to a powerful position (chief, king, queen) and that specify responsibilities and appropriate conduct for the individual selected.

trained incapacity The inability, because of limited training, to respond to new or unusual circumstances or to recognize when official rules or procedures are outmoded or no longer applicable.

transsexuals People whose primary sex characteristics do not match the sex they perceive themselves to be.

troubles Personal needs, problems, and difficulties that can be explained as individual shortcomings related to motivation, attitude, ability, character, or judgment.

unstructured interview An interview in which the question-and-answer sequence is spontaneous, open-ended, and flexible.

upward mobility A form of vertical mobility in which a person moves up in rank.

urban underclass The group of families and individuals in inner cities who live "outside the mainstream of the American occupational system and [who] consequently represent the very bottom of the economic hierarchy" (Wilson 1983, p. 80).

urbanization An increase in the number of cities in a designated geographic area and growth in the proportion of the area's population living in cities. (ch. 15)

urbanization A transformative process in which people migrate from rural to urban areas and change the way they use land, interact, and make a living. (ch. 16)

utilitarian organizations Formal organizations that draw together people seeking material gain in the form of pay, health benefits, or a new status.

validity The degree to which an operational definition measures what it claims to measure.

value-rational action Social action in which a valued goal is pursued with a deep and abiding awareness of the "symbolic meaning" and purpose of the actions taken to pursue the goal.

values General, shared conceptions of what is good, right, appropriate, worthwhile, and important with regard to conduct, appearance, and states of being.

variable Any trait or characteristic that can change under different conditions or that consists of more than one category.

vertical mobility Change in social class that corresponds to a gain or loss in rank.

voluntary minorities Racial or ethnic groups that come to a country expecting to improve their way of life.

voluntary organizations Formal organizations that draw together people who give time, talent, or treasure to support mutual interests, meet important human needs, or achieve a not-for-profit goal.

wealth The combined value of a person's income *and* other material assets such as stocks, real estate, and savings minus debt.

welfare state A term that applies to an economic system that is a hybrid of capitalism and socialism.

white-collar crime "Crimes committed by persons of respectability and high social status in the course of their occupations" (Sutherland and Cressey 1978, p. 44).

witch hunt A campaign to identify, investigate, and correct behavior that is believed to be undermining a group or country. Usually this behavior is not the real cause of a problem but is used to distract people's attention from the real cause or to make the problem seem manageable.

References

Chapter 1

Abrams, Irwin. 1997. *The Nobel Peace Prize and the Laureates: An Illustrated Biographical History.* The Norwegian Institute.

Addams, Jane. 1910. *Twenty Years at Hull-House: With Autobiographical Notes.* New York: Macmillan.

_____. 1912. *A New Conscience and an Ancient Evil.* 1912. Urbana, IL: University of Illinois Press.

Anderson, Jenny. 2008. "Wall Street Winner Hit a New Jackpot-Bill-Dollar Paydays." *The New York Times* (April 16): A1.

Appellrouth, Scott, and Laura D. Edles. 2007. *Sociological Theory in the Contemporary Era.* Thousand Oaks, CA: Pine Forge Press.

Berger, Peter L. 1963. *Invitation to Sociology: A Humanistic Perspective.* New York: Anchor.

Bernard, Tara Siegel. 2009. "In Grim Job Market, Student Loans Are a Costly Burden." *The New York Times* (April 18), B6.

Bologna, Sergio. 2008. Marx as Correspondent of the New York Daily Tribune, 1856–57. Money and Crisis. http://wildcat-www.de/en/material/cs13bolo.htm.

Bosman, Julie. 2009. "Newly Poor Swell Lines at Food Banks." *The New York Times* (February 19). http://www.nytimes.com/2009/02/20/nyregion/20food.html.

Brecher, Jeremy, John Brown Childs, and Jill Cutler. 1993. *Global Visions: Beyond the New World Order.* Boston: South End Press.

Calhoun, Craig. 2002. Quoted in "A World War among Professors," by Stephen Kotkin. *New York Times* (September 7): A1.

Cassidy, John. 1997. "The Return of Karl Marx." *New Yorker* (October 20): 27.

Deegan, Mary Jo. 1978. "Women and Sociology: 1890–1930." *Journal of the History of Sociology* 1 (Fall 1978): 11–32.

DeGrange, McQuilkin. 1939. "Comte's Sociologies." *American Sociological Review* 4(1): 17–26.

Dinan, Stephen. 2007. "Bush, GOP Senators Trim Border Fence Goal." *Washington Times* (April 3), washingtontimes.com.

DuBois, W. E. B. 1919 [1970]. "Reconstruction and Africa." pp. 372–381 in *W.E.B. DuBois: A Reader.* New York: Harper and Row.

Durkheim, Émile. 1951. *Suicide.* New York: Free Press.

———. 1982. *The Rules of Sociological Method and Selected Texts on Sociology and Its Method,* ed. Steven Lukes, trans. W. D. Halls. New York: Macmillan Press.

Federal Reserve. 2009. "Economic Research and Data," http://www.federalreserve.gov/.

Feeding America. 2009. "Local Impact Survey Results Summary." feedingamerica.org/newsroom/press-release-archive/~/media/Files/research/local-impact-survey/2008-impact-survey.ashx.

Frank, Robert. 2009. "World's Rich Have Lost $10 Trillion in Global Financial Crisis." *The Wall Street Journal* (April 9): http://blogs.wsj.com/wealth/category/investing/.

Gates, Henry Louis. 2003. "Both Sides Now." *New York Times Book Review* (May 4).

Gilbert, M. Gaul. 1989. "The Blood Brokers—America the OPEC of the Global Plasma Industry." *Philadelphia Inquirer* (September 28). http://www.bloodbook.com/part-5.html.

Gordon, John Steele. 1989. "When Our Ancestors Became Us." *American Heritage* (December): 106–221.

Greenhouse, Steven. 2009. "Out of Work, Part Time." *The New York Times* (June 16): B1.

Hamington, Maurice. 2007 "Jane Addams," S*tanford Encyclopedia of Philosophy* (2007) online edition, http://plato.stanford.edu/entries/addams-jane/.

Harvey, Jean, Genevieve Rail, and Lucie Thibault. 1996. "Globalization and Sport: Sketching a Theoretical Model for Empirical Analyses." *Journal of Sport and Social Issues* (August): 258.

Henri 2000.

Kulish, Nicholas. 2009. "Europe Aided by Safety Nets, Resists U.S. Push on Stimulus." *New York Times* (March 27): A1.

Lemert, Charles. 1995. *Sociology after the Crisis.* Boulder, CO: Westview Press.

Lengermann, Patricia M. 1974. *Definitions of Sociology: A Historical Approach.* Columbus, OH: Merrill.

Lewis, David Levering. 1993. *Biography of a Race, 1868–1919.* New York: Holt.

Lewis, Paul. 1998. "Marx's Stock Resurges on 150-Year Tip." *New York Times* (June 27): A17.

Marcus, Steven. 1998. "Marx's Masterpiece at 150." *New York Times Book Review* (April 26): 39.

Marx, Karl. [1881] 1965. "The Class Struggle." pp. 529–535 in *Theories of Society,* ed. T. Parsons, E. Shils, K. D. Naegele, and T. R. Pitts. New York: Free Press.

———. 1987. Quoted in *A Marx Dictionary,* by Terrell Carver. Totowa, NJ: Barnes and Noble.

Mills, C. Wright. 1959. *The Sociological Imagination.* New York: Oxford University Press.

National Geographic. 2002. "Diamonds: The Real Story." www7.nationalgeographic.com/ngm/data/2002/03/01/html/ft_20020301.1.html.

National Plasma Centers. 2009. http://www.nationalplasmacenters.com/.

O'Donoghue, Julia. 2009. "Food Banks See Increase in Need." *Fairfax Connection* (April 16), www.connectionnewspapers.com/article.asp?article=327759&paper=63&cat=104.

Organization for Economic Cooperation and Development. 2007. "Table A. Total Tax Revenue as Percentage of GDP," www.oecd.org/dataoecd/48/27/41498733.pdf.

Proudhon, Pierre-Joseph. 1847. "The Philosophy of Misery." http://www.marxists.org/reference/subject/economics/proudhon/philosophy/intro.htm.

Random House Encyclopedia. 1990. "European Imperialism in the 19th Century." New York: Random House.

Stearns, Scott. 2009. "Cameroon Hit by Global Economic Crisis." *Voice of America* (April 13), http://www.voanews.com/english/2009–04–13-voa31.cfm.

Thrall, Ann. 2007. "A Brief History of Glass Blowing." www.neder.com/glassact/history.html.

Uchitelle, Louis. 2009. "After Recession, Recovery Will Take Years." *New York Times* (April 7): 5B.

U.S. Library of Congress 2008. "Meet Amazing Americans: Jane Addams." http://www.americaslibrary.gov/cgi-bin/page.cgi/aa/addams.

Visser, Margaret. 1986. *Much Depends on Dinner.* New York: Harper Perennial.

Walsh, Mary Williams. 2009. "A.I.G. Lists Firms to Which It Paid Taxpayer Money." *New York Times* (March 16): A1.

Weintraub, Jeff, and Joseph Soares. 2005. HANDOUT #15: Reading Weber—"Types of Social Action for Social & Political Theory." http://jeffweintraub.blogspot.com/2005/06/weber-on-social-action-rationality.html.

Woods, Ngaire. 2009. "Recession Risks Financial 'Tsunami.'" *BBC News* (March 20), news.bbc.co.uk/1/hi/business/7947017.stm.

Zaniello, Tom. 2007. *The Cinema of Globalization: A Guide to Films about the New Economic Order.* New York: Cornell University Press.

Chapter 2

American Society of Microbiology. 1996. "Operation Clean Hands: Press Release." http://www.asmusa.org/pcsrc/och.htm.

Bearden, Tom. 1993. "Focus: Help Wanted." *MacNeil/Lehrer NewsHour.* PBS.

Blumer, Herbert. 1962. "Society as Symbolic Interaction." In *Human Behavior and Social Processes.* Boston: Houghton Mifflin.

Brown, Valerie. 2004. "Reaching Across the Border with the SBRP." *Environmental Health Perspectives* 112(5): 278–280.

Cameron, William B. 1963. *Informal Sociology.* New York: Random House.

Carver, Terrell. 1987. *A Marx Dictionary.* Totowa, NJ: Barnes and Noble.

Conover, Ted. 1987. *Coyotes: A Journey through the Secret World of America's Illegal Aliens.* New York: Vintage.

Cornelius, Wayne. 1981. "Mexican Migration to the United States." *Proceedings of the Academy of Political Science* 34(1):67–77.

Dinan, Stephen. 2007. "Bush, GOP Senators Trim Border Fence Goal." *Washington Times* (April 3) washingtontimes.com.

Dye, Lee. 1995. "Duplication of Research Isn't as Bad as It Sounds." *Los Angeles Times* (April 26):D5.

Gregg, Alan. 1989. Quoted in *Science and the Human Spirit: Contexts for Writing and Learning,* by Fred D. White. Belmont, CA: Wadsworth.

Hacker, Andrew. 1997. "Review of 'The New American Reality: Who We Are, How We Got There' by Reynolds Farley." *Contemporary Sociology* 26(4):478.

Hagan, Frank E. 1989. *Research Methods in Criminal Justice and Criminology.* New York: Macmillan.

Heyman, Josiah M. 1999. "Why Interdiction? Immigration Control at the United States–Mexican Border." *Regional Studies* 33(7):619–30.

Horan, Patrick M. 1995. "Review of 'Working with Archival Data: Studying Lives.'" *Contemporary Sociology* (May):423–424.

Judis, John. 2006. "Border Wars." *New Republic* (January 16).

Kapur, Devesh. 2003. "Remittances: The New Development Mantra?" Paper prepared for the G-24 Technical Group Meeting.

Katzer, Jeffrey, Kenneth H. Cook, and Wayne W. Crouch. 1991. *Evaluating Information: A Guide for Users of Social Science Research,* 3d ed. New York: McGraw-Hill.

Kochhar, Rakesh. 2005. "The Economic Transition to America." *Part Three: Survey of Mexican Migrants.* Pewhispanic.org.

Kruzel, John J. 2007. "President Thanks Guard for Helping Secure U.S.-Mexico Border." *American Forces Press Service* (April 9). www.defenselink.mil/news.

Lowell, Lindsay, Rodolfo de la Garza, and Mike Hogg. 2000. "Remittances, U.S. Latino Communities, and Development in Latin American Countries." *Migration World Magazine* 38(5):13.

Maril, Robert L. 2004. *Patrolling Chaos: The U.S. Border in Deep South Texas.* Lubbock, TX: Texas Tech University Press.

Massey, Douglas. 2006. "The Wall That Keeps Illegal Workers In." *New York Times* (April 4). www.nytimes.com.

Merton, Robert K. 1967. "Manifest and Latent Functions." Pages 73–137 in *On Theoretical Sociology: Five Essays, Old and New.* New York: Free Press.

Mexican Migration Project. 2007. mmp.opr.princeton.edu/home-en.aspx.

Migration Information Source. 2006. "The U.S. Mexico Border." www.migrationinformation.org.

Morrow, David J. 1996. "Trials of Human Guinea Pigs." *New York Times* (September 29):10.

National Academy of Sciences. 1995. "Science Ethics Guide Updated, Expanded for Graduate Students." www2.nas.edu.whatsnew/20fe.html.

Nunez-Neto, Blas. 2006. *Border Security: The Role of the U.S. Border Patrol* (January 25) CRS Report for Congress.

Pew Hispanic Center. 2005. "Survey of Mexican Migrants, Part One." pewhispanic.org

———. 2006a. "Fact Sheet: Modes of Entry for the Unauthorized Migrant Population." pewhispanic.org.

———. 2006b. "The Labor Force Status of Short-Term Unauthorized Workers." pewhispanic.org.

Pomfret, John. 2006. "Fence Meets Wall of Skepticism." washingtonpost.com (October 10).

Robinson, Scott S. 2003. "The Potential Role of IT in International Remittance Transfers." Draft memo (November).

www.ssrc.org/programs/itic/publications/knowledge-report/memos.

Roethlisberger, F. J., and William J. Dickson. 1939. *Management and the Worker.* Cambridge, MA: Harvard University Press.

Rossi, Peter H. 1988. "On Sociological Data." Pages 131–154 in *Handbook of Sociology,* ed. N. Smelser. Newberry Park, CA: Sage.

Secure Fence Act of 2006. 2006. thomas.loc.gov/cgi-bin/query/z?c109:h.r.6061:

Shotola, Robert W. 1992. "Small Groups." Pages 1796–1806 in *Encyclopedia of Sociology,* vol. 4, ed. E. F. Borgatta and M. L. Borgatta. New York: Macmillan.

Singer, Audrey, and Douglas S. Massey. 1998. "The Social Process of Undocumented Border Crossing among Mexican Migrants." *International Migration Review* 3(Fall):561–592.

Singleton, Royce A., Jr., Bruce C. Straits, and Margaret Miller Straits. 1993. *Approaches to Social Research,* 2d ed. New York: Oxford University Press.

Smith, Joel. 1991. "A Methodology for the Twenty-First Century Sociology." *Social Forces* 70(1):1–17.

Strauss, Anselm. 1978. *Negotiations: Varieties, Contexts, Processes, and Social Order.* San Francisco: Jossey-Bass.

Stalker's Guide to International Migration. 2003. "Emigration—Remittances." Pstalker.com/migration.

Stuff.co.nz. 2007. "U.S.-Mexico Volleyball Party Over Border." (April 20). www.stuff.co.nz/print/4028102a4560.html.

Suro, Roberto. 2003a. "Executive Summary." pp. 1–XX in *Billions in Motion: Latino Immigrants, Remittances, and Banking.* Pew Hispanic Center. www.pewhispanic.org.

———. 2003b. "Remittance Senders and Receivers: Tracking the Transnational Channels." Pew Hispanic Center. www.pewhispanic.org.

Thayer, John E., III. 1983. "Sumo." Pages 270–274 in *Kodansha Encyclopedia of Japan,* vol. 7. Tokyo: Kodansha.

Thelen, David. 1992. "Of Audiences, Borderlands, and Comparisons: Toward the Internationalization of American History." *Journal of American History* (September):432–451.

Thompson, Ginger. 2003. "Money Sent Home by Mexicans Is Booming." *New York Times* (October 28):3.

U.S. Department of Homeland Security. 2006a. "Fact Sheet: ICE Accomplishments in Fiscal Year 2006." www.dhs.gov/xnews/releases.

———. 2006b. "Fact Sheet: Securing America's Borders, CBP 2006 Fiscal Year in Review." www.dhs.gov/xnews/releases.

U.S. Department of Justice. 2007. "Background to the Office of the Inspector General Investigation." www.usdoj.gov/oig/special/9807/gkp01.htm.

U.S. Department of Transportation. 2007. "Border Crossing: U.S.-Mexico Border Crossing Data." www.bts.gov/programs/international/border_crossing_entry_data.

Van Doorn, Judith, 2003. "Migration, Remittances, and Small Enterprise Development." International Labor Organization. www.ilo.org/public/english/employment/finance/download/remit2.pdf.

White, Leslie A. 1949. *The Science of Culture: A Study of Man and Civilization.* New York: Farrar, Straus.

Witte, Griff. 2006. "Boeing Wins Deal for Border Security." *Washington Post* (September 20) www.washingtonpost.com.

Chapter 3

Adherents.com. 2002. "The Largest Communities of Jehovah's Witnesses." www.adherents.com

An, Heejung. 1997. "Sports Categories in Korea." http://www.itp.tsoa.nyu.edu/~student/heejung/sports5.htn

Asia News. 2005. "South Korea Sends Food Aid to North Despite Protests by South Korean Activists" (February 28). www.asianews.it

Associated Press. 2006. "South Korea Bulldozes Village to Expand U.S. Military Base." *INQ7.net News* (September 13). newsinfo.inq7.net.

Ballard, Erica. 2006. "Cross-Listed Business Course to Send Students to South Korea." *Indiana Daily Student* (September 26). www.idsnews.com.

BBC News. 2005. "North Korea Slashes Food Rations." (January 24). news.bbc.co.uk.

Benedict, Ruth. 1976. Quoted in *The Person: His and Her Development Throughout the Life Cycle,* by Theodore Lidz. New York: Basic Books.

Berkhofer, Robert F., Jr. 1978. *The White Man's Indian: Images of the American Indian from Columbus to the Present.* New York: Knopf.

Breton, Raymond. 1967. "Institutional Completeness of Ethnic Communities and the Personal Relations of Immigrants." *American Journal of Sociology* 70:193–205.

Brooke, James. 2003a. "Defectors Want to Pry Open North Korea." *New York Times* (January 9):A8.

———. 2003b. "Infiltrators of North Korea: Tiny Radios." *New York Times* (March 5):A32.

———. 2003c. "Trial Runs of a Free Market in North Korea." *New York Times* (March 11):C1.

———. 2003d. James Brooke, "South Opposes Pressuring North Korea, Which Hints It Will Scrap Nuclear Pact," *New York Times,* (January 1): A1.

Brown, Rita Mae. 1988. *Rubyfruit Jungle.* New York: Bantam.

Carey, Peter. 1995. Quoted in interview with Kevin Bacon and Bill Davis. p. 14 in *Glimmer Train Stories,* Glimmer Train Press.

Carroll, John B. (ed.). 1956. *Language, Thought, and Reality: Selected Writings of Benjamin Lee Whorf.* MIT Press.

Caryl, Christian and B.J. Lee. 2006. "Culture Shock: A Flow of Information from the Outside World is Changing the Hermit Kingdom." *Newsweek* (international edition). www.msnbc.msn.com/id/7693649/site/newsweek.

Cherni, Leigh. 1998. Telephone conversation with Andrea Simone Bowers. July 1.

China View. 2003. "Chinese Cultural Festival Opens in South Korea." (September 21). news.xinhuanet.com.

Chung, Ah-young. 2006. "NK Defector Students Face Education Challenges." *Korea Times* (January 18). www.hankooki.com.

Chung, Annie. 2003. "Korean Public Baths." *ThingsAsian.* www.thingsasian.com.

Dale, Steve. 2004. "Pet Owner Survey Reveals Strong Bond between People and Their Pets." www.goodnewsforpets.com/petworld/archive/36013_owner_survey.htm.

Egypt Travel. 1998. "Aswan: The High Dam of Egypt." touregypt.net/highdam.htm.

Fallows, James. 1988. "Trade: Korea Is Not Japan." *Atlantic Monthly* (October): 22–33.

French, Howard. 2003. "Seoul Looks to New Alliances." *New York Times* (January 26). www.nyt.com.

Gordon, Steven L. 1981. "The Sociology of Sentiments and Emotion." Pages 562–592 in *Social Psychology: Sociological Perspectives*, ed. by M. Rosenberg and R. H. Turner. New York: Basic Books.

Halberstam, David. 1986. *The Reckoning.* New York: William Morrow and Company.

Hannerz, Ulf. 1992. *Cultural Complexity: Studies in the Social Organization of Meaning.* New York: Columbia University Press.

Henry, William A. 1988. "No Time for the Poetry: NBC's Cool Coverage Stints on the Drama." *Time* (October 3):80.

Herskovits, Melville J. 1948. *Man and His Works: The Science of Cultural Anthropology.* New York: Knopf.

Hochschild, Arlie R. 1976. "The Sociology of Feeling and Emotion: Selected Possibilities." Pages 280–307 in *Another Voice,* ed. by M. Millman and R. Kanter. New York: Octagon.

———. 1979. "Emotion Work, Feeling Rules, and Social Structure." *American Journal of Sociology* 85:551–575.

Holt International Children's Services. 1998. http://www.holtintl.org/intro.shtml.

Hughes, Everett C. 1984. *The Sociological Eye: Selected Papers.* New Brunswick, NJ: Transaction.

Hunter, Helen-Louise. 1999. Quoted in "A Look at North Korean Society," by J. Winzig. www.winzigconsultingservices.com/files/samples/kg/Helen_ Hunter.html.

Hurst, G. Cameron. 1984. "Getting a Piece of the R.O.K.: American Problems of Doing Business in Korea." *UFSI Reports* 19.

Institute for International Education. 2003a. "American Students Study Abroad in Growing Numbers." (November 17). opendoors .iienetwork.org.

———. 2003b. "International Student Enrollment Growth Slows in 2002/2003, Large Gains from Numerous Countries Offset Numerous Decrease." (November 3). opendoors.iienetwork.org.

International Tourism Association. 2004a. "Market Profile: Korea (Korean Arrivals to the U.S., 2003)." Office of Travel and Tourism Industries. www.ita.doc.gov.

———. 2004b. "2002 Profile of U.S. Resident Travelers Visiting Overseas Destinations." Office of Travel and Tourism Industries. www.ita.doc.gov.

Kang, K. Connie. 1995. *Home Was the Land of Morning Calm: A Saga of a Korean-American Family.* Reading, MA: Addison-Wesley.

Kim, Bo-Kyung, and Kevin Kirby. 1996. Personal correspondence (April 25).

Koehler, Nancy. 1986. "Re-entry Shock." Pages 89–94 in *Cross-Cultural Reentry: A Book of Readings.* Abilene, TX: Abilene Christian University Press.

Korean Herald. 2004. "A Taste of Capitalism in North Korea." (May 24). www.koreaherald.com.

Korea News. 2004. "Climb Every Ice Mountain." www.koreanews .net.

Korean Overseas Information Service. 2006. *Facts about Korea.* Seoul: Korean Overseas Information Agency. www.korea.net.

Koreascope. 1998. "Personality Cult in North Korea." *North Korean Studies.* www.fortunecity.com/meltingpot/champion/65/pers_cult.htm.

Kristof, Nicholas D. 1995. "Where a Culture Clash Lurks Even in the Noodles." *New York Times* (September 4):Y4.

———. 1998. "Big Macs to Go." *New York Times Book Review* (March 22):18.

———. 2005. "The Hermit Nuclear Kingdom." *New York Review of Books* (February 10):25–27.

Ku, J. H., M. E. Kim, N. K. Lee, and Y. H. Park. 2003. "Circumcision Practice Patterns in South Korea: Community Based Survey." *Sexually Transmitted Infections* 79(February):65–68.

Lamb, David. 1987. *The Arabs: Journeys beyond the Mirage.* New York: Random House.

Lee, Jennifer. 1994. "The Invisible Nation of Korean Emigrants." *Korean Culture* (Winter):39–40.

Linton, Ralph. 1936. *The Study of Man: An Introduction.* New York: Appelton-Century-Crofts.

Liu, Hsein-Tung. 1994. "Intercultural Relations in an Emerging World Civilization." *Journal of Global Awareness* 2(1):48–53.

MacKinnon, Rebecca. 2005. "Chinese Cell Phone Breaches North Korean Hermit Kingdom." *YaleGlobal Online* (January 17): yaleglobal.yale.edu

McClane, Daisann. 2000. "Frugal Traveler: Unchartered (Hot) Waters at a Korean Spa." *New York Times* (October 29).

McGeown, Kate. 2003. "On Holiday in North Korea." *BBC News* (September 19). newsvote.bbc.co.uk

Myers, Steven Lee. 2003. "Returning from Iraq War Not so Simple for Soldiers." *New York Times* (September 12):A1.

Oregon Economic and Community Development Department. 2003. "Business and Travel Etiquette in Korea." www.oregon .gov/ECDD.

The Paleontological Research Institution. 2006. "The World of Oil." www.priweb.org/ed/pgws/history/history_home.html.

Park, Myung-Seok. 1979. *Communication Styles in Two Different Cultures: Korean and American.* Seoul: Han Shin.

Quinlan, May Kay. 2004. Review of *Beyond the Shadow of Camptown: Korean Military Brides in America*, by Ji-Yeon Yuh. *Oral History Review* 31(Winter–Spring):97–100.

Redfield, Robert. 1962. "The Universally Human and the Culturally Variable." Pages 439–453 in *Human Nature and the Study of Society: The Papers of Robert Redfield,* vol. 1, ed. by M. P. Redfield. Chicago: University of Chicago Press.

Rohner, Ronald P. 1984. "Toward a Conception of Culture for Cross-Cultural Psychology." *Journal of Cross-Cultural Psychology* 15(2):111–138.

Rokeach, Milton. 1973. *The Nature of Human Values.* New York: Free Press.

Sapir, Edward. 1949. "Selected Writings of Edward Sapir." In *Language, Culture and Personality,* ed. by D. G. Mandelbaum. Berkeley: University of California Press.

Schudson, Michael. 1989. "How Culture Works: Perspectives from Media Studies on the Efficacy of Symbols." *Theory and Society* 18:153.

Sharp, Ari. 2005a. "Ari on the Web: Welcome to the Democratic People's Republic of Korea." www.ariontheweb.blogspot.com

———. 2005b. "Ari on the Web: The Ego of the Kims." www.ariontheweb.blogspot.com.

Smith, Lynn. 1996. "Adoptees Search the World for Their Roots." *New York Times* (June 17):A1.

Smithsonian Magazine. 1998. "On the Frankincense Trail." (October). www.smithsonianmag.si.edu/smithsonian/issues98/oct98/yemen.html.

Sobie, Jane Hipkins. 1986. "The Cultural Shock of Coming Home Again." Pages 95–102 in *The Cultural Transition: Human Experience and Social Transformation in the Third World and Japan,* ed. by M. I. White and S. Pollack. Boston: Routledge & Kegan.

Sumner, William Graham. 1907. *Folkways.* Boston: Ginn.

UNESCO. 2005. "World Heritage: The Criteria for Selection." whc.unesco.org.

UPI. 2006. "South Korea May Send Troops to Lebanon." (September 28). www.upi.org.

U.S. Army. 1998. "Standard Installation Topic Exchange Service." http://www.dmdc.osd.mil/sites/index.html.

U.S. Central Intelligence Agency. 2007. *World Factbook.* www.cia.gov/cia/publications/factbook/.

U.S. Department of Defense. 2003. "Active Duty Military Personnel Strengths by Regional Area and Country." web1.whs.osd.mil/mmid/mo5/hst0309.pdf.

———. 2004. "Military Personnel Statistics." web1.whs.osd.mil/mmid/military/miltop.htm.

U.S. Department of State. 2002. "International Adoption—South Korea." http://travel.state.gov/adoption_korea.html.

U.S. Embassy–Seoul. 2003. "Background on U.S.–Korea Relations." seoul.usembassy.gov.

U.S. Energy Information Administration. 2004. "The Effects of Income on Appliances in U.S. Households." www.eia.gov.

———. 2007. "Quick Stats: Petroleum." www.eia.gov/neic/quickfacts/quickoil.html.

Visser, Margaret. 1986. *Much Depends on Dinner.* New York: HarperPerennial.

Wallace, F. C. 1952. "Notes on Research and Teaching." *American Sociological Review* (December):747–751.

Watchtower Bible and Tract Society of Pennsylvania. 2004. "Jehovah's Witnesses." www.jw-media.org.

Whorf, Benjamin. 1956. *Language, Thought, and Reality: Selected Writings of Benjamin Lee Whorf,* edited by J. B. Carroll. Cambridge: Technology Press of Massachusetts Institute of Technology

Winchester, Simon. 1988. *Korea: A Walk through the Land of Miracles.* New York: Prentice Hall.

Winzig, Jerry. 1999. Review of *Kim Il-song's North Korea* by Helen-Louise Hunter. www.winzigconsultingservices.com/files/samples/kq/Helen_Hunter.html.

World Monitor. 1992. "The Map: Batters Up." (April): 11.

———. 1993. "The Map: Hoop-la." (February): 10–11.

Chapter 4

A. C. Nielsen Co., 2006. Cited in "Television and Health." *The Sourcebook for Teaching Science.* www.csun.edu/science/health/docs/tv&health.html.

Al-Batrawi, Khaled, and Mouin Rabbani. 1991. "Breakup of Families: A Case Study in Creeping Transfer." *Race and Class* 32(4):35–44.

Alexa. 2006a. Top Sites by Country: Palestinian Territory. alexa.com/site/ds/top_500.

———.2006b.TrafficRankings:GlobalTop500.alexa.com/site/ds/top_500.

Amnesty International. July 2002. *Without Distinction: Attacks on Civilians by Palestinian Armed Groups.*

BBC News.1998b. "Israel Celebrates Half-Century." news.bbc.co.uk/hi/english/events/i...at50/israeltoday/newsid85000/85903.stm.

Ben-David, Amith, and Yoav Lavee. 1992. "Families in the Sealed Room: Interaction Patterns of Israeli Families during SCUD Missile Attacks." *Family Process* 31(1):35–44.

Bennet, James. 2003b. "An Israeli's Sorrowful Rule over a Sullen Nablus." *New York Times* (October 3): A1.

Bourne, Jenny. 1990. "The Pending Pain of Reenactment." *Race and Class* 32(2):67–72.

Buñuel, Luis. 1985. Quoted on page 22 in *The Man Who Mistook His Wife for a Hat and Other Clinical Tales,* by Oliver Sacks. New York: Summit Books.

Burns, John F. 2003. "Bomber Left Her Family with a Smile and a Lie." *New York Times* (October 7):A13.

Cole, Charlotte et al. 2003. "The Educational Impact of Rechov Sumsum/Shara'a Simsim: A Sesame Street Television Series to Promote Respect and Understanding Among Children Living in Israel, the West Bank, and Gaza." *International Journal of Behavioral Development* 27(5): 409–422.

Cooley, Charles Horton. 1909. *Social Organization.* New York: Scribner's.

———. 1961. "The Social Self." Pages 822–828 in *Theories of Society: Foundations of Modern Sociological Theory,* ed. by T. Parsons, E. Shils, K. D. Naegele, and J. R. Pitts. New York: Free Press.

Corsaro, William A. 1985. *Friendship and Peer Culture in the Early Years.* Norwood, NJ: Ablex.

Coser, Lewis A. 1992. "The Revival of the Sociology of Culture: The Case of Collective Memory." *Sociological Forum* 7(2):365–373.

Davis, Kingsley. 1940. "Extreme Isolation of a Child." *American Journal of Sociology* 45:554–565.

———. 1947. "Final Note on a Case of Extreme Isolation." *American Journal of Sociology* 3(5):432–437.

Deutsche Welle. 2006. "'Sesame Street' Joins Mideast Peace Process." (October 22). www.dw-world.de.

Durkheim, Émile. 1951 [1888]. *Suicide: A Study in Sociology,* trans. by John A. Spaulding and George Simpson. New York: Free Press.

Dyer, Gwynne. 1985. *War.* New York: Crown.

Elbedour, Salman, David T. Bastien, and Bruce A. Center. 1997. "Identity Formation in a Shadow of Conflict: Projective Drawings by Palestinian and Israeli Arab Children from West Bank and Gaza." *Journal of Peace Research* 34(2):217–232.

Erikson, Erik H. 1950. *Childhood and Society.* New York: Norton.

———. 1982. *The Life Cycle Completed.* New York: Norton.

———. 1988. Quoted in "Erikson In His Own Old Age, Expands His View of Life," by D. Coleman. *The New York Times* (June 14):Y3.

Farnsworth, Elizabeth. 2004. "The Barrier." *Online Newshour* (February 9). www.pbs.org/newshour/bb/middle_east/jan-jun04/barrier_2-9.html.

Figler, Stephen K., and Gail Whitaker. 1991. *Sport and Play in American Life.* Dubuque, IA: Brown.

Freud, Anna, and Sophie Dann. 1958. "An Experiment in Group Upbringing." Pages 127–168 in *The Psychoanalytic Study of the Child, vol. 6,* ed. by R. S. Eissler, A. Freud, H. Hartmann, and E. Kris. New York: Quadrangle.

Goffman, Erving. 1961. *Asylums: Essays on the Social Situation of Mental Patients and Other Inmates.* New York: Anchor.

Grossman, David. 1988. *The Yellow Wind,* trans. by H. Watzman. New York: Farrar, Straus & Giroux.

———. 1998. "Fifty Is a Dangerous Age." *New Yorker* (April 20):55.

Grossman, David. 2006. "The Will To Peace." (November 8) MidEast Web Gateway www.mideastweb.org.

Haaretz. 2006. "Israel Aims to End Decades-Long Reliance on Palestinian Laborers." (June 4, 2005). www.kavlaoved.org.

Halbwachs, Maurice. 1980. *The Collective Memory,* trans. by F. J. Ditter, Jr., and V. Y. Ditter. New York: Harper & Row.

Harel, Amos. 2004a. "Shin Bet: 145 Suicide Bombers since Start of Intifada." *Haaretz* (June 13). www.haaretzdaily.com.

Hellerstein, David. 1988. "Plotting a Theory of the Brain." *New York Times Magazine* (May 22):17+.

Independent Lens. 2006. *The World According to "Sesame Street."* www.pbs.org/independentlens/.

Institute of World Jewish Congress. 2006. *Jewish Communities of the World.* http://www.worldjewishcongress.org/

Kagan, Jerome. 1988. Interview on *The Mind,* PBS (transcript). Boston: WGBH Educational Foundation.

———. 1989. *Unstable Ideas: Temperament, Cognition, and Self.* Cambridge, MA: Harvard University Press.

Kifner, John. 2000. "Out of Place: The Price of Peace Will Be Paid in Dreams." *New York Times* (December 31): Section 4, p. 1.

Lahoud, Lamia. 2001. "76% of Palestinians Support Suicide Attacks." *Jerusalem Post* (June 4). cgis.jpost.com.

Margalit, Avishai. 2003. "The Suicide Bombers." *New York Review of Books* (January 16):36–39.

Mead, George Herbert. 1934. *Mind, Self and Society.* Chicago: University of Chicago Press.

Miniwatts Marketing Group. 2006. "Internet Usage Statistics." www.internetworldstats.com.

Montgomery, Geoffrey. 1989. "Molecules of Memory." *Discover Magazine* (December): 46–55.

Myre, Greg. 2003. "UN Estimates Israeli Barrier Will Disrupt Lives of 600,000." *New York Times* (November 12). www.nyt.com.

Nathan, Susan. 2006. *The Other Side of Israel: My Journey across the Jewish/Arab Divide.* New York: Doubleday.

Natta, Don van, Jr. 2003. "The Terror Industry Fields Its Ultimate Weapon." *New York Times* (August 24):1WK.

Nelsen, Arthur. 2006. *Occupied Minds: A Journey through the Israeli Psyche.* Ann Arbor, MI: Pluto.

Nova. 1986. "Life's First Feelings." WGBH Boston (February 11).

Ornstein, Robert, and Richard F. Thompson. 1984. *The Amazing Brain.* Boston: Houghton Mifflin.

Palestinian Central Bureau of Statistics. 1998. "Projected Mid-Year Population by Age Groups and Sex (1996)." http://www.pcbs.org/english/pop1.htm.

Penfield, Wilder, and P. Perot. 1963. "The Brain's Record of Auditory and Visual Experience: A Final Summary and Discussion." *Brain* 86:595–696.

Peres, Judy. 1998. "A Human Mosaic." *Chicago Tribune* (May 9).

Rodgers, Walter. 1998. "Army Holds Israeli Society Together." http://www.cnn.com/WORLD/meast/9804/29/israel.glue/index.html.

Rose, Peter I., Myron Glazer, and Penina M. Glazer. 1979. "In Controlled Environments: Four Cases of Intensive Resocialization." Pages 320–338 in *Socialization and the Life Cycle,* ed. by P. I. Rose. New York: St. Martin's.

Rowley, Storer. 1998. "After the Army, You Feel More Israeli." *Chicago Tribune* (May 9).

Salamon, Julie. 2002. "Israeli-Palestinian Battles Intrude on 'Sesame Street.'" *New York Times* (July 30). www.nytimes.com.

Satterly, D. J. 1987. "Jean Piaget (1896–1980)." Pages 621–622 in *The Oxford Companion to the Mind,* ed. by R. I. Gregory. Oxford, UK: Oxford University Press.

Shipler, David. 1986. *Arab and Jew: Wounded Spirits in a Promised Land.* New York: Times Books.

Sontag, Deborah. 2001. "Bitter, Stark Souvenirs: Sneakers and Slingshots." *New York Times* (February 21):A4.

Spitz, Rene A. 1951. "The Psychogenic Diseases in Infancy: An Attempt at Their Etiological Classification." Pages 255–278 in *The Psychoanalytic Study of the Child,* vol. 27, ed. by R. S. Eissler and A. Freud. New York: Quadrangle.

Theodorson, George A., and Achilles G. Theodorson. 1979. *A Modern Dictionary of Sociology.* New York: Barnes & Noble.

Townsend, Peter. 1962. Quoted on pages 146–147 in *The Last Frontier: The Social Meaning of Growing Old,* by Andrea Fontana. Beverly Hills, CA: Sage.

U.S. Census Bureau. 2006. "Table 963: Appliances and Office Equipment Used by Region and Household Income: 2001." *Statistical Abstract of the United States.*

U.S. Central Intelligence Agency. 2007. *World Factbook.* www.odci.gov/cia/publications/factbook/.

Usher, Graham. 1991. "Children of Palestine." *Race and Class* 32(4):1–18.

Chapter 5

Abraham, Carolyn. 2006. "The Smartest Virus in History?" *Globe and Mail* (August 12): A9.

Altman, Lawrence K. 1986. "Anxiety on Transfusions." *New York Times* (July 18):A1, B4.

Barr, David. 1990. "What Is AIDS? Think Again." *New York Times* (December 1):Y15.

Centers for Disease Control and Prevention (CDC). 2001. "AIDS Cases by Age Group, Exposure Category, and Sex Reported Through December 2000, United States." www.cdc.gov/hiv/stats.

———. 2004. "HIV and AIDS: Are You at Risk." www.cdc.gov/hiv/pubs.

Conrad, Joseph. 1971. *Heart of Darkness,* rev. and ed. by R. Kimhough. New York: Norton.

Durkheim, Émile. [1933] 1964. *The Division of Labor in Society,* trans. by G. Simpson. New York: Free Press.

Forbath, Peter. 1977. *The River Congo.* New York: Harper and Row.

Frontline. 1993. "AIDS, Blood, and Politics." Boston: WGBH Educational Foundation and Health Quarterly.

GAO. 1997a. *Blood Safety: Enhancing Safeguards Would Strengthen the Nation's Blood Supply*. T-HEHS-97-143 (June 5) www.gao.gov.

Goffman, Erving. 1959. *The Presentation of Self in Everyday Life*. New York: Anchor.

Grover, Jan Zita. 1987. "AIDS: Keywords." *October* 43:17–30.

Halberstam, David. 1986. *The Reckoning*. New York: Morrow.

Henderson, Charles. 1998. "Epidemiology: U.S. Sees AIDS Rise among Older Americans." *AIDS Weekly Plus* (February 9):14.

Hochschild, Adam. 1998. *King Leopold's Ghost*. New York: Houghton Mifflin.

Hurley, Peter, and Glenn Pinder. 1992. "Ethics, Social Forces, and Politics in AIDS-Related Research: Experience in Planning and Implementing a Household HIV Seroprevalence Survey." *Milbank Quarterly* 70(4):605–628.

Johnson, Diane, and John F. Murray. 1988. "AIDS Without End." *New York Review of Books* (August 18):57–63.

Kaptchuk, Ted, and Michael Croucher, with the BBC. 1986. *The Healing Arts: Exploring the Medical Ways of the World*. New York: Summit.

Kornfield, Ruth. 1986. "Dr., Teacher, or Comforter? Medical Consultation in a Zairian Pediatrics Clinic." *Culture, Medicine and Psychiatry* 10:367–387.

Lewin, Tamar. 2000. "Survey Shows Sex Practices of Boys." *New York Times* (December 19):A16.

Lyons, Maryinez. 2002. *The Colonial Disease*. Cambridge: Cambridge University Press.

Mark, Joan. 1995. *The King of the World in the Land of the Pygmies*. Lincoln, NE: University of Nebraska Press.

Markel, Howard. 2003. "HIV Secrecy Is Proving Deadly." *New York Times* (November 23):D6.

Merton, Robert K. 1957. *Social Theory and Social Structure*. Glencoe, IL: Free Press.

Michael, Robert T., John H. Gagnon, Edward O. Lauman, and Gina Kolata. 1994. *Sex in America: A Definitive Survey*. New York: Little, Brown.

Moore, Jim. 2004. "The Puzzling Origins of AIDS." *American Scientist* (92):540–547.

Moss, Sandy. 2006. "Southwest Soldiers Travel to Congo." *New Frontier* 23(17). www.salvationarmy.usawest.org.

National Center for Health Statistics. 2002. Referenced in R. A. Friedman, "Curing and Killing: The Perils of a Growing Medicine Cabinet." *New York Times* (December 12). www.nyt.com.

Noble, Kenneth B. 1989. "More Zaire AIDS Cases Show Less Underreporting." *New York Times* (December 26):J4.

Oliver, Murray. 2006. "The Waiting Is the Hardest Part for Congolese." CTV.ca (November 1). www.ctv.ca.

Online Newshour. 2004. "Importing Drugs" (March 9). www.pbs.org/newshour.

Peterson, Dale and Karl Ammann. 2003. *Eating Apes*. Berkeley: University of California Press.

Reuters Health Information. 1999. "Older HIV-Positive Patients Are Diagnosed at More Advanced Stages." HIV/AIDS Information Center. *Journal of the American Medical Association*. www.ama-assn.org.

Riding, Alan. 2005. "Museum Show Forces Belgium to Ask Hard Questions about Its Colonial Past." *New York Times* (February 9):B3.

Sontag, Susan. 1989. *AIDS and Its Metaphors*. New York: Farrar, Straus & Giroux.

Stein, Michael. 1998. "Sexual Ethics: Disclosure of HIV-Positive Status to Partners." *Archives of Internal Medicine* 158:253.

Stolberg, Sheryl. 1996. "Officials Find Rare HIV Strain in L.A. Woman." *Los Angeles Times* (July 5):A1+.

Thomas, William I., and Dorothy Swain Thomas. [1928] 1970. *The Child in America*. New York: Johnson Reprint.

Turnball, Colin. 1961. *The Forest People*. New York: Simon and Schuster.

UNAIDS. 2006. *Report on the Global AIDS Epidemic*. www.unaids.org.

U.S. Agency for International Development. 2002. "What Happened in Uganda?" (September). www.usaid.gov/

U.S. Central Intelligence Agency. 2005. "Democratic Republic of Congo." *World Factbook*. www.odci.gov/cia/publications/factbook.

U.S. Department of Health and Human Services. 1990. *HIV/AIDS Surveillance Report*. Washington, DC: U.S. Government Printing Office.

U.S. International Trade Administration. 2004. "Arrivals Data for World Regions and Top Markets." tinet.ita.doc.gov.

Villarosa, Linda. 2003. "Raising Awareness about AIDS and the Aging." *New York Times* (July 8). www.nyt.com.

Wilson, Mary E. 1996. "Travel and the Emergence of Infectious Diseases." *Emerging Infectious Disease* (April–June 1995). http://www.cdc.gov/ncidod/EID/_vol1no2/wilson.htm.

Witte, John. 1992. "Deforestation in Zaire: Logging and Landlessness." *Ecologist* 22(2):58.

World Tourism Organization. 2003. *Tourism Highlights*. www.worldtourism.org.

Wrong, Michela. 2000. *In the Footsteps of Mr. Kurtz*. Great Britain: Fourth Estates Limited.

Wrong, Michela. 2002. *In the Footsteps of Mr. Kurtz*. New York: Perennial.

Yasuda, Yikuo. 1994. "Japanese Hemophiliacs Suffering from HIV Infection." www.nmia.com/~mdibble/japan2.html

Zwick, Jim. 2006. "Stereoscopic Visions of War and Empire." Keystone View Company. www.boondocksnet.com/stereo/sv289F.html.

Chapter 6

Aldrich, Howard E., and Peter V. Marsden. 1988. "Environments and Organizations." Pages 361–392 in *Handbook of Sociology*, ed. by N. J. Smelser. Newbury Park, CA: Sage.

Allen, Jason. 2006. "Multi-Million Dollar Kroc Center Confirmed for Green Bay." WBAY (December 1). www.wbay.com.

Angier, Natalie. 2002. "The Most Compassionate Conservative." *New York Times Book Review* (October 27):8–9.

Associated Press. 2004. "McDonald's to Take Credit Cards" (March 26). www.gotriad.com.

Barboza, David. 2003a. "McDonald's Asking Meat Industry to Cut Use of Antibiotics." *New York Times* (June 20):A1.

———. 2003b. "Animal Welfare's Unexpected Allies." *New York Times* (June 25):C1.

Barrionuevo, Alexei. 2007. "Globalization in Every Loaf: Ingredients Come from All Over, but Are They Safe?" *The New York Times* (June 16): B1.

Barnet, Richard J., and Ronald E. Müller. 1974. *Global Reach: The Power of the Multinational Corporations.* New York: Simon & Schuster.

Beech, Alan. 1997. McLibel trial witness statement (October 25). Transcript, http://www.mcspotlight.org/people/witnesses/employment/beech_alan.html.

Bell, Charlie. 2004. Quoted in "McDonald's Reports Record April 2004 Sales." McDonald's press release (May 10). www.mcdonalds.com.

Bharat Book Bureau. 2005. *World Prefabricated Housing to 2004.* www.bharatbook.com.

Blau, Peter M., and Richard A. Schoenherr. 1973. *The Structure of Organizations.* White Plains, NY: Longman.

Brett, Adrian. 1997. McLibel trial witness statement (October 25). Transcript, www.mcspotlight.org/people/witnesses/employment/brett_adrian.html.

Centers for Disease Control and Prevention. 2006. "Public Health Image Library: Food Handling." phil.cdc.gov/Phil/detail.asp.

Chanda, Nayan. 2003. "The New Leviathans: An Atlas of Multinationals Throws Unusual Light on Globalization." Review of *Global Inc: An Atlas of the Multinational Corpora-tion,* by Medard Gabel and Henry Bruner, in *YaleGlobal* (November 12). yaleglobal.yale.edu.

CNN Newsstand and *Fortune* Magazine. 1998. "McMakeover; Exit Strategy." Transcript, www.cnn.com/TRANSCRIPTS/9808/12/nsf.00.html.

Cooley, Charles Horton. 1909. *Social Organization.* New York: Scribner's.

Coton, Ray. 1997. McLibel trial witness statement (October 25). Transcript, www.mcspotlight.org/people/witnesses/employment/cotton_ray.html.

Crecca, Donna. 1997. "Five Models for Success from Foodservice 2005." *Restaurants and Institutions.* www.rimag.com/09/five.htm.

Crispell, Diane. 1995. "Why Working Teens Get into Trouble." *American Demographics.* http://www.demographics.com/publications/ad/95ad/9502ad/9502ab06.htm.

Critser, Greg. 2003. *Fat Land: How Americans Became the Fattest People in the World.* New York: Houghton Mifflin.

Etzioni, Amitai. 1975. *Comparative Analysis of Complex Organi-zations.* New York, NY: Free Press of Glencoe.

Food Institute Report. 2000b. "McDonald's Sets Rules for Raising Hens." (August 28):10.

Foreign Policy. 2001. "McAtlas Shrugged." (Interview with McDonald's CEO Jack Greenberg) (May):26–31.

Fortune Magazine. 2007. "Global 500." Retrieved on August 9 from money.cnn.com/magazines/fortune/global500/2006/full-list.

Freund, Julien. 1968. *The Sociology of Max Weber.* New York: Random House.

Freidman, Thomas. 1996. "Big Mac I." *The New York Times* (December 8):E15.

Gabel, Medard, and Henry Bruner. 2003. *Global Inc: An Atlas of the Multinational Corporation.* New York: New Press.

Gibney, Simon. 1997. McLibel trial witness statement (October 25). Trasnscript, http://www.mcspotlight.org/people/witnesses/employment/gibney_simon.html.

Gilo, Kim, and Tom Welsh. 1997. "McDonald's Marketing of 'Local Burger' Sizzles." *Korea Herald* (November 26). www.koreaherald.co.kr.

Goetzinger, Chester. 2006. "CFIA Carcass Inspection Procedure and Standards: Impact on Producers." *Advances in Pork Production* (Volume 17).

Goldman Environmental Prize. 2004. www.goldmanprize.org.

Greenberg, Jack. 2001. Quoted in "McAtlas Shrugged: An FP Interview with Jack Greenberg." *Foreign Policy* (May/June):26+.

International Potato Center. 1998. "Globalization of French Fries." http://www.cipotato.org/ph&mkt/fries.htm.

Johnson, Brad. 1997. "The World According to French Fries." *Restaurants and Institutions.* http://www.rimag.com/14/bjfry.htm.

Kennedy, Paul. 1993. *Preparing for the Twenty-First Century.* New York: Random House.

Khan, Rahat Nabi. 1986. "Multinational Companies and the World Economy: Economic and Technological Impact." *Impact of Science on Society* 36(141): 15–25.

Koenig, Peter. 1997. "McRevelations." *New Statesmen* 126(4341):20.

Langley, Allison. 2003. "It's a Fat World, After All." *New York Times* (July 20):B1.

Leidner, Robin. 1993. *Fast Food, Fast Talk: Service Work and the Routinization of Everyday Life.* Berkeley, CA: University of California Press.

Lepkowski, Wil. 1985. "Chemical Safety in Developing Countries: The Lessons of Bhopal." *Chemical and Engineering News* 63:9–14.

Lowe, Kimberly. 1997. "The Cost of Employee Theft." *Restaurants and Institutions.* www.rimag.com/07/repthft.htm.

Mannix, Margaret. 1996. "A Big Whopper Stopper?" *U.S. News and World Report* 120(28):60.

MacArthur, Kate. 2007. "Give It Away: Fast Feeders Favor Freebies." *Advertising Age* (June 18) adage.com

McDonald's Corporation. 2001. *McDonald's 2000 Annual Report.* www.mcdonalds.com.

———. 2003. "Summary Annual Report." www.mcdonalds.com.

———. 2004b. "2004 Investor Fact Sheet." www.mcdonalds.com.

———. 2006. "McDonald's USA Laying Hens Guidelines." www.mcdonalds.com/usa/good/products/hen.html.

———. 2007. *McDonald's 2006 Annual Report.* www.mcdonalds.com.

Michels, Robert. 1962. *Political Parties,* trans. by E. Paul and C. Paul. New York: Dover.

National Library of Medicine. 2004. "Your Digestive System and How It Works." www.hlom.nih.gov.

Perrow, Charles. 2000. "An Organizational Analysis of Organiza-tional Theory." *Contemporary Sociology.*

Personick, Martin. 1991. "Profiles in Safety and Health: Eating and Drinking Places." *Monthly Labor Review* (June):19.

Rai, Saritha. 2003. "Tastes of India in U.S. Wrappers." *New York Times* (April 8):BW1.

Ritzer, George. 1993. *The McDonaldization of Society.* Thousand Oaks, CA: Pine Forge Press.

Roach, John. 2003. "Are Plastic Grocery Bags Sacking the Environment?" *National Geographic News.* (September 2). news.nationalgeographic.com.

Rousseau, Rita. 1997. "The Labor-Law Blues." *Restaurants and Institutions.* www.rimag.com/15/buslab.htm.

Sadri, Mahmoud. 1996. "Book Review of Occidentalism: Images of the West." *Contemporary Sociology* (September): 612.

Saul, Stephanie. 2006. "Record Sales of Sleeping Pills Are Causing Worries." *New York Times* (February 7).

Scully, Matthew. 2003. *Dominion: The Power of Man, the Suffering of Animals, and the Call to Mercy.* New York: St. Martin's.

Sekulic, Dusko. 1978. "Approaches to the Study of Informal Organization." *Sociologija* 20(1):27–43.

Snow, Charles P. 1961. *Science and Government.* Cambridge, MA: Harvard University Press.

Spreitzer, Elmer A. 1971. "Organizational Goals and Patterns of Informal Organizations." *Journal of Health and Social Behavior* 12(1):73–75.

Stone, Ann. 1997. "Lean? No Thanks." *Restaurants and Institutions.* www.rimag.com/10/lean.htm.

Tanner, Lindsey. 2006. "Study Slams McDonald's in Hospitals." Associated Press. (December 3). www.aap.org.

Toqueville, Alexis de. 1882. *Democracy in America*, Volume II (7th edition). Boston: John Allyn.

United Egg Producers. 2006. "United Egg Producers Animal Husbandry Guidelines for U.S. Egg Laying Flocks." www.uepcertified.com.

U.S. Central Intelligence Agency. 2007. "Rank Order: GDP (Purchasing Power Parity)." *The World Factbook.* Retrieved on August 9 from www.cia.gov/library/publications/the-world-factbook/rankorder/2001rank.html.

U.S. General Accounting Office. 1978. *U.S. Foreign Relations and Multinational Corporations: What's the Connection?* Washington, DC: U.S. Government Printing Office.

Veblen, Thorstein. 1933. *The Engineers and the Price System.* New York: Viking.

Warner, Melanie. 2006. "Salads or No, Cheap Burgers Revive McDonald's." *New York Times* (April 19) www.nyt.com.

Waters, Jennifer. 1998. "Fractured Franchise." *Restaurants and Institutions.* www.rimag.com/807/busjw.htm.

Watson, James. 1997. *Golden Arches East: McDonald's in East Asia.* Stanford, CA: Stanford University Press.

Weber, Max. 1947. *The Theory of Social and Economic Organization,* ed. and trans. by A. M. Henderson and T. Parsons. New York: Macmillan.

Weiser, Benjamin. 2003. "Big Macs Can Make You Fat." *New York Times* (January 23):A23.

Williams, Geoff. 2006. "Behind the Arches." *Entrepreneur* (January). www.findarticles.com.

Yablen, Marcia. 2000. "Happy Hen, Happy Meal." *U.S. News and World Report* 129(9):46.

Young, T. R. 1975. "Karl Marx and Alienation: The Contributions of Karl Marx to Social Psychology." *Humboldt Journal of Social Relations* 2(2):26–33.

Zuboff, Shoshana. 1988. *In the Age of the Smart Machine: The Future of Work and Power.* New York: Basic Books.

Chapter 7

Author X. 1992. "Mao Fever—Why Now?" Trans. and adapted from the Chinese by R. Terrill. *World Monitor* (December): 22–25.

Baer, Robert. 2003. *Sleeping with the Devil: How Washington Sold Our Soul for Saudi Crude.* New York: Crown.

Barboza, David. 2004. "In Roaring China, Sweaters Are West of Socks City." *New York Times* (December 24):A1.

———. 2005. "China, New Land of Shoppers, Builds Malls on Gigantic Scale." *New York Times* (May 25): A1.

Becker, Howard S. 1963. *Outsiders: Studies in the Sociology of Deviance.* New York: Free Press.

———. 1973. "Labeling Theory Reconsidered." In *Outsiders: Studies in the Sociology of Deviance.* New York: Free Press.

Best, Joel. 1989. *Images of Issues: Typifying Contemporary Social Problems.* New York: Aldine de Gruyter.

Bolido. 1993.

Butterfield, Fox. 1976. "Mao Tse-Tung: Father of Chinese Revolution." *New York Times* (September 10):A13+.

———. 1980. "The Pragmatists Take China's Helm." *New York Times Magazine* (December 28):22–35.

———. 1982. *China: Alive in the Bitter Sea.* New York: Times Books.

Center on Wrongful Convictions. 2003. "The Illinois Exonerated: Anthony Porter." Northwestern University School of Law. www.law.northwestern.edu/depts./clinic/wrongful/exonerations/porter.htm.

Chambliss, William. 1974. "The State, the Law, and the Definition of Behavior as Criminal or Delinquent." Pages 7–44 in *Handbook of Criminology,* ed. by D. Glaser. Chicago: Rand McNally.

Chang Jung. 1991. *Wild Swans: Three Daughters of China.* New York: Simon & Schuster.

Christian Century. 2003. "Capital Offense: Can the Death Penalty System Really Be Sufficiently Reformed?" (February 8):5.

Collins, Randall. 1982. *Sociological Insight: An Introduction to Nonobvious Sociology.* New York: Oxford University Press.

ConsumerAffairs.com. 2005. "Avon Calls On China." (April 12). www.consumeraffairs.com/news04/2005/avon.html.

Doctoroff, Tom. 2006. Interview with Paul Solman for "China's Vast Consumer Class." *Online NewsHour* (December 5). www.pbs.org/newshour/bb/asia/july-dec05/consumers_10-05.html.

Durkheim, Émile. [1901] 1982. *The Rules of Sociological Method and Selected Texts on Sociology and Its Method,* ed. by S. Lukes, trans. by W. D. Halls. New York: Free Press.

Economist. 2002. "The Tale of Two Ryans; Death and Politics in Illinois" (October 19). www.infotrac-college.com.

Erikson, Kai T. 1966. *Wayward Puritans.* New York: Wiley.

Fairbank, John King. 1987. *The Great Chinese Revolution 1800–1985.* New York: Harper & Row.

———. 1989. "Why China's Rulers Fear Democracy." *New York Review of Books* (September 28):32–33.

Federal Bureau of Prisons. 2004. "Quick Facts: Types of Offenses." www.bop.gov/news/quick.jsp.

Field, Scott L. 2002. "On the Emergence of Social Norms." *Contemporary Sociology* 31(6):638–639.

French, Howard W. 2005. "As Girls 'Vanish' Chinese City Battles Tide of Abortions." *New York Times* (February 17):A3.

Gifford, Rob. 2007. *China Road: A Journey into the Future of a Rising Power.* New York: Random House.

Goldman, Merle. 1989. "Vengeance in China." *New York Review of Books* (November 9):5–9.

Gould, Stephen Jay. 1990. "Taxonomy as Politics: The Harm of False Classification." *Dissent* (Winter):73–78.

Governor's Commission on Capital Punishment, State of Illinois. 2002. *Report of the Governor's Commission on Capital Punishment* (April). www.idoc.state.il.us/ccp/ccp/reports/commission_report/summary_recommendations.pdf.

Henriques, Diana B. 1993. "Great Men and Tiny Bubbles: For God, Country and Coca-Cola." *New York Times Book Review* (May 23):13.

Hersh, Seymour. 2004. "Annals of National Security: Torture at Abu Ghraib." *New Yorker* (May 10). www.newyorker.com/fact/content/?040510fa_fact.

Innocence Project. 2004. "About the Innocence Project." http://innocenceproject.org/about/.

Jerome, Richard. 1995. "Suspect Confessions." *New York Times Magazine* (August 13):28–31.

Johnson, Dirk. 2003. "A Leap of Fate: A Governor's Controversial Last Hurrah Clears Out Illinois's Crowded Death Row." *Newsweek.* (January 20):34.

Jones, Jeffrey M. 2001. "Americans Felt Uneasy Toward Arabs Even Before September 1." *Poll Analyses* (September 28). www.gallup.org.

Kahn, Joseph. 2002. "China Has World's Tightest Internet Censorship, Study Finds." *New York Times* (December 4):A15.

———. 2004b. "A Challenge to China's Leaders from a Witness to Brutality." *New York Times* (March 14):p7.

———. 2004c. "The Most Populous Nation Faces a Population Crisis." *New York Times* (May 30):4WK.

———. 2004d. "China Is Filtering Text Messages to Regulate Criticism." *New York Times* (July 3):A8.

Kayal, Michele. 2002. "The Societal Costs of Surveillance." *New York Times* (July 26):A21.

Kaye, Jeffrey. 2001. "Under Suspicion." *Online Newshour* (October 26). www.pbs.org/newshour.

Kelly, Maura. 2003. "Illinois Governor Pardons 4 Inmates on Death Row." *Salt Lake Tribune* (January 11). www.sltrib.com.

Kometani, Foumiko. 1987. "Pictures from Their Nightmare." *New York Times Book Review* (July 19):9–10.

Kristof, Nicholas D. 1989. "China Is Planning 2 Years of Labor for Its Graduates." *New York Times* (August 13):Y1.

———. 2004. "A Little Leap Forward." *New York Review of Books* (June 24):56–58.

Kwong, Julia. 1988. "The 1986 Student Demonstrations in China." *Asian Survey* 28(9):970–985.

Leys, Simon. 1989. "After the Massacres." *New York Review of Books* (October 12):17–19.

MacFarquhar, Roderick, and Michael Schoenhals. 2006. *Mao's Last Revolution.* Belknap/Harvard University.

Mathews, Jay, and Linda Mathews. 1983. *One Billion: A China Chronicle.* New York: Random House.

Merton, Robert K. 1957. *Social Theory and Social Structure.* Glencoe, IL: Free Press.

Milgram, Stanley. 1974. *Obedience to Authority: An Experimental View.* New York: Harper & Row.

———. 1987. "Obedience." Pages 566–568 in *The Oxford Companion to the Mind,* ed. by R. L. Gregory. Oxford, UK: Oxford University Press.

Moss, Michael. 2003. "False Terrorism Tips to F.B.I. Uproot the Lives of Suspects." *New York Times* (June 19):A1.

National Council for Crime Prevention in Sweden. 1985. *Crime and Criminal Policy in Sweden.* Report no. 19. Stockholm: Liber Distribution.

Oxman, Robert. 1993a. "China in Transition." Transcript of interview on *MacNeil/Lehrer Newshour* (December 27). New York: WNET.

Piazza, Alan. 1996. Quoted in "In China's Outlands, Poorest Grow Poorer," by P. E. Tyler. *New York Times* (October 26):A1, A4.

Remez, L. 1991. "China's Fertility Patterns Closely Parallel Recent National Policy Changes." International Family Planning Perspectives, 17(2):75–76

Rosenthal, Elisabeth. 2000. "Rural Flouting of One-Child Policy Undercuts China's Census." *New York Times* (April 14):A10.

Ryan, George H. 2003. Quoted in "Two Days Left in Term, Governor Clears Out Death Row in Illinois" by Jodi Wilgoren. *New York Times* (January 11):A1.

Shapiro, Bruce. 2001. "A Talk with Governor George Ryan." *Nation.* (January 8):17.

Simmons, J. L., with Hazel Chambers. 1965. "Public Stereotypes of Deviants." *Social Problems* 3(2):223–232.

Slavin, Barbara. 2004. "White House Has Final Say on 9/11 Report." Associated Press. www.usatoday.com/news/washington/2004-04-04-terrorreport_x.htm.

Spector, Malcolm, and J. I. Kitsuse. 1977. *Constructing Social Problems.* Menlo Park, CA: Cummings.

Spence, Jonathan. 2006. "China's Great Terror." *New York Review of Books* (September 21):31–34.

Sumner, William Graham. 1907. *Folkways.* Boston: Ginn.

Sutherland, Edwin H., and Donald R. Cressey. 1978. *Principles of Criminology,* 10th ed. Philadelphia: Lippincott.

Tien H. Yuan, Zhang Tianlu, Ping Yu, Li Jingneng, and Liang Zhongtang. 1992. "China's Demographic Dilemmas." *Population Bulletin* 47(1):1–44.

Tobin, Joseph J., David Y. H. Wu, and Dana H. Davidson. 1989. *Preschool in Three Cultures: Japan, China and the United States.* New Haven, CT: Yale University Press.

Tyler, Patrick E. 1996b. "Chinese Maltreatment at Orphanage." *New York Times* (January 9):A4.

U.S. Central Intelligence Agency. 2004. "The People's Republic of China" in *The World Factbook.* www.odci.gov.

U.S. Department of Justice, Bureau of Justice Statistics. 2004. "Prison and Jail Inmates at Midyear, 2003." February 6. www.ojp.usdoj.gov/bjs/glance/tables/drtab.htm.

———. 2005. *Criminal Victimization in the United States.* www.ojp.usdoj.gov/bjs/cvict.htm.

———. *The World Factbook of Criminal Justice Systems.* www.ojp.usdoj.gov/bjs/abstract/wfcj.htm.

———. 2006. *Prison and Jail Inmates at Midyear 2006.* www.ojp.usdoj.gov/bjs/abstract/pjim06.htm

U.S. Department of State. 2003. *Human Rights Reports, 2003.* www.state.gov.

———. 2004. *Supporting Human Rights and Democracy: The U.S. Record 2003–2004.* Released by Bureau of Democracy, Human Rights and Labor. www.state.gov.

———. 2006. *Country Reports on Human Rights Practices, 2005.* Released by the Bureau of Democracy, Human Rights, and Labor. www.state.gov/g/drl/rls/hrrpt/2005/61605.htm.

———. 2007. Immigrant Visas Issued to Orphans Coming to the U.S. travel.state.gov/family/adoption/stats/stats_451.htm.

Wang Shuo. 1997. Quoted in "Bad Boy," by Jamie James. *New Yorker* (April 21):50.

Williams, Terry. 1989. *The Cocaine Kids: The Inside Story of a Teenage Drug Ring.* Reading, MA: Addison-Wesley.

Yardley, Jim. 2005. "Fearing Future, China Starts to Give Girls Their Due." *New York Times* (January 31): A3.

Zogby, James J. 2001. "Lessons Learned." *Jordan Times.* www.jordantimes.org.

Chapter 8

Allen, Walter, and Angie Y. Chung. 2000. "Your Blues Ain't Like My Blues: Race, Ethnicity and Social Inequality in America." *Contemporary Sociology* 29 (6): 796–805.

Asian News International. 2001. "Barefoot College." http://www.aniin.com.

Baker, Peter. 2009. "Obama Presses for Action on Credit Cards." *New York Times* (May 14). http://www.nytimes.com/2009/05/15/us/politics/15obamacnd.html?ref=business.

Barefoot College. 2005. www.barefootcollege.org.

Berkshire, Matt. 2008. "Top 10 NBA Salaries for 2008–2009." *Associated Content* (November 21), http://www.associated-content.com/article/1211086/top_10_nba_salaries_for_20082009.html?cat=14.

Bologna, Sergio. 2008. "Marx as Correspondent of the New York Daily Tribune, 1856–57." *Money and Crisis.* http://wildcat-www.de/en/material/cs13bolo.htm.

Bradsher, Keith. 2006. "Ending Tariffs Is Only the Start." *New York Times* (February 28): C1.

Business Week, 2006, March 27. "How Rising Wages Are Changing the Game in China." http://www.businessweek.com/magazine/content/06_13/b3977049.htm.

BusinessWeek Online. 2006, March 26. "How Rising Wages Are Changing the Game in China." www.businessweek.com/magazine/content/06_13/b3977049.htm.

Bicycle Retailer. 2005, March 16. "Vietnam Bike Makers Challenge Canada's Dumping Lawsuit." http://www.allbusiness.com/retail-trade/miscellaneous-retail-miscellaneous/4143711–1.html.

Centers for Disease Control and Prevention (CDC). 2006. *Health, United States, 2006.* www.cdc.gov/nchs/data/hus/hus06.pdf.

Chaliand, Gerard, and Jean-Pierre Rageau. 1995. *The Penguin Atlas of Diasporas.* New York: Penguin Group.

Chauvin, Lucien. 2005. "Peru Gives Its Poor More Money, but There's a Catch." *Christian Science Monitor* (October 14). www.csmonitor.com/2005/1014/p01s03-woam.html.

Chen, David W. 2004. "What's a Life Worth?" *New York Times* (June 20): 42K.

Cook-Lynn, Elizabeth. 2008. "Deadliest Enemies: Law and the Making of Race Relations on and off Rosebud Reservation, and: Not Without Our Consent: Lakota Resistance to Termination, 1950–59." *Wicazo Sa Review* 23 (1): 155–158. http://muse.jhu.edu/login?uri=/journals/wicazo_sa_review/v023/23.1cook-lynn.html.

Conrad, Robert Edgar. 1996. "Slave Trade." Pages 127–128 in *Encyclopedia of Latin American History and Culture,* edited by B. A. Tenenbaum. New York: Scribner's.

Coser, Lewis A. 1977. *Masters of Sociological Thought,* 2nd ed., ed. by R. K. Merton. New York: Harcourt Brace Jovanovich.

Cowen, Tyler. 2007. "A Contrarian Look at Whether U.S. Chief Executives Are Overpaid." *New York Times* (May 18): C4.

Dash, Eric. 2006. "Off to the Races Again, Leaving Many Behind." *New York Times* (April 9): 3.

Davis, Kingsley, and Wilbert E. Moore. 1945. "Some Principles of Stratification." Pages 413–445 in *Ideological Theory: A Book of Readings,* ed. by L. A. Coser and B. Rosenberg. New York: Macmillan.

Dorsey, David, and Jana Leon. 2000. "Positive Deviant." *Fast Company.* www.fastcompany.com/online/41/sternin.html.

Dugger, Celia W. 2005a. "UN Report Cites U.S. and Japan as the 'Least Generous Donors.'" *New York Times* (September 8): A4.

———. 2005b. "African Food for Africa's Starving Is Roadblocked in Congress." *New York Times* (October 12).

———. 2005c. "Study Finds Small Developing Lands Hit Hardest by 'Brain Drain.'" *New York Times* (October 23): A10.

———. 2006. "Toilets Underused to Fight Disease, UN Study Finds." *New York Times* (November 10): A8.

Barbara Ehrenreich. 2001. *Nickel and Dimed.* New York: Metropolitan.

Eiseley, Loren. 1990. "Man: Prejudice and Personal Choice." Pages 640–943 in *The Random House Encyclopedia,* 3rd ed. New York: Random House.

Environmental Working Group. 2007. *Farm Subsidy Database.* www.ewg.org/16080/farm/.

Frank, Robert. 2008. "The Wealth Report: Tiger Woods to Become First Billionaire Athlete." *Wall Street Journal Blog.* http://blogs.wsj.com/wealth/2008/08/13/tiger-woods-to-become-first-billionaire-athlete/.

Gans, Herbert. 1972. "The Positive Functions of Poverty." *American Journal of Sociology* 78: 275–289.

Garamone, Jim. 2008. *Africa Command Makes Progress with African Allies.* American Forces Press Service. March 31. http://www.defenselink.mil/news/newsarticle.aspx?id=49418.

Gates, William H., Sr., and Chuck Collins. 2003. *Wealth and Commonwealth: Why America Should Tax Accumulated Fortunes.* Boston: Beacon.

Goldstein, Fred. 2009. "GM Restructuring Will Deepen Capitalist Crisis." *Workers World* (May 3). http://www.workers.org/2009/us/gm_restructuring_0507/.

Grameen. 2005. "Grameen Bank at a Glance." www.grameen_info.org.

Gresser, Edward. 2002. "Toughest on the Poor: America's Flawed Tariff System." *Foreign Affairs* (November/December). http://www.foreignaffairs.com/articles/58425/edward-gresser/toughest-on-the-poor-americas-flawed-tariff-system.

Hertz, Tom. 2006. *Understanding Mobility in the United States,* for Center for American Progress. http://www.american-progress.org/issues/2006/04/Hertz_MobilityAnalysis.pdf/.

Holloway, Thomas H. 1996. "Immigration." Pages 239–242 in *Encyclopedia of Latin American History and Culture,* edited by B. A. Tenenbaum. New York: Scribner's.

Hsiang-Ching Kung, Donna L. Hoyert, Jiaquan Xu, and Sherry L. Murphy, *Division of Vital Statistics. 2008.* "Deaths: Final Data for 2005." National Vital Statistics Reports 55 (10). Hyattsville, MD: National Center for Health Statistics. http://www.cdc.gov/nchs/data/nvsr/nvsr56/nvsr56_10.pdf.

Jiang Jingjing. 2004. "Wal-Mart's China Inventory to Hit $18b This Year." *China Business Weekly* (November 29). www.chinadaily.com.cn/english/doc/2004–11/29/content_395728.htm.

Jones, Rachel. 2006. "For First Time, More Poor Live in Suburbs Than Cities." *National Public Radio* (December 6). http://www.npr.org/templates/story/story.php?storyId=6598999.

Kher, Unmesh. 2002. "Sweet Subsidy." *Time* (February 15). www.time.com.

Kirka, Danica. 2004. "Car Bomb, Mortar Fire Kills 5 U.S. Soldiers." *Lexington Herald Leader* (July 9): A3.

Los Angeles Times Blog. 2008. "Miley Cyrus Gets a Crummy Allowance for a Billionaire Teen." (April 24). http://latimes-blogs.latimes.com/alltherage/2008/04/miley-cyrus-is.html.

Marger, Martin. 1991. *Race and Ethnic Relations: American and Global Perspectives.* Belmont, CA: Wadsworth.

Marx, Karl. 1856. "Speech at Anniversary of the People's Paper." http://www.marxists.org/archive/marx/works/1856/04/14.htm.

Mayer, Robert. "Working Paper—One Payday, Many Payday Loans: Short-Term Lending Abuse in Milwaukee County." http://lwvmilwaukee.org/mayer21.pdf.

Mihesuah, Devan A. 2008. *Big Bend Luck.* Booklocker.com.

Morgenson, Gretchen. 2006. "Two Pay Packages, Two Different Galaxies." *New York Times* (April 4): Section 3.

Morrow, David J. 1996. "Trials of Human Guinea Pigs." *New York Times* (May 8): C1.

Murphy, Dean E. 2004. "Imagining Life without Illegal Immigrants." *New York Times* (January 11): 16WK.

Naofusa, Hirai. 1999. "Traditional Cultures and Modernization: Several Problems in the Case of Japan." http://www2.kokugakuin.ac.jp/ijcc/wp/cimac/hirai.html.

Noah, Timothy. 2005. "The Wal-Mart Manifesto." *Slate.* www.slate.com/id/2113954/.

Obadina, Tunde. 2000. "The Myth of Neo-colonialism." *African Economic Analysis.* www.afbis.com.

Obama, Barack. 2009. "The President on Credit Card Tactics: 'Enough is Enough.'" The White House. http://www.whitehouse.gov/blog/The-President-on-Credit-Card-Tactics-Enough-is-Enough/.

O'Brien, Doug. 2004. Quoted in "Hungry in America," in *National Catholic Reporter* (February 15). findarticles.com/p/articles/mi_m1141/is_15_38/ai_83316661.

Office of Management and Budget. 2005. *Budget of the United States Government, FY 2006.* www.whitehouse.gov/omb/budget/fy2006/tables.html.

O'Hare, William P., and Kenneth P. Johnson. 2004. "Facing Child Poverty in Rural America." Population Reference Bureau (January). www.prb.org.

Organization for Economic Cooperation and Development. 2009. "Table 1: Net Official Development Assistance in 2008—Preliminary Evidence for 2008." http://www.adb.org/Documents/Books/Key_Indicators/2008/Part-II.asp.

Oxfam. 2006. "G8 Subsidies Contributing to WTO Crisis." (July 11). www.oxfam.org/en/news/pressreleases2006/pr060711_wto.

Perlez, Jane. 2004. "Asian Maids Often Find Abuse, Not Riches, Abroad." *New York Times* (June 22): A3.

Pew Hispanic Center. 2009. *A Portrait of Unauthorized Immigrants in the United States* (April 14). http://pewhispanic.org/files/reports/107.pdf.

Pew Research Center. 2009. *What Americans Pay For—and How.* (February 7). http://pewresearch.org/pubs/407/what-americans-pay-for---and-how.

Powers, Richard. 2000. "American Dreaming." *New York Times Magazine* (May 7): 66–67.

Proudhon, Pierre-Joseph. 1847. "The Philosophy of Misery." http://www.marxists.org/reference/subject/economics/proudhon/philosophy/intro.htm.

Reuters. 2009. "Wal-Mart paid ex-CEO Scott $30.2 million in FY09" (April 20). http://www.reuters.com/article/GCA-CreditCrisis/idUSTRE53J6OE20090420.

Rostow, W. W. 1960. *The Stages of Economic Growth: A Non-Communist Manifesto.* Cambridge: Cambridge University Press. http://www.mtholyoke.edu/acad/intrel/ipe/rostow.htm.

Sacks, Jeffrey. 2005. "Can We End Global Poverty?" Interview by John Cassidy. *Federal News Service.* www.truthabouttrade.org.

September 11 Victim Compensation Fund. 2001. "Calculating the Losses." *New York Times* (December 21): B6.

Simpson, Richard L. 1956. "A Modification of the Functional Theory of Social Stratification." *Social Forces* 35: 132–137.

Sugrue, Thomas. 2007. "Stamping Out Detroit Postmark Irks 'Michigan Metro' Residents." *Bloomberg News,* October 30, 2007.

SUIE 2008.

Thurow, Roger, and Geoff Winestock. 2005. "Bittersweet: How an Addiction to Sugar Subsidies Hurts Development." aWorld-Connected.org. www.aworldconnected.org/article.php/242.html.

Tumin, Melvin M. 1953. "Some Principles of Stratification: A Critical Analysis." *American Sociological Review* 18: 387–394.

UC Berkeley Labor Center. 2007. "Living Wage Policies at Wal-Mart." http://laborcenter.berkeley.edu/retail/walmart_livingwage_policies07.pdf.

UNICEF. 2009. *The State of the World's Children.* www.unicef.org/sowc09/report/report.php.

United Nations. 1998. "Changing Today's Consumption Patterns—for Tomorrow's Human Development" in *Human Development Report.* www.undp.org/hdro/98.htm.

United Nations. 2008. *The Millennium Development Goals Report.* http://www.un.org/millenniumgoals/pdf/The%20 Millennium%20Development%20Goals%20Report%202008 .pdf.

_____. 2009a. "Statistics in the Human Development Report, 2008." http://hdr.undp.org/en/statistics/.

_____. 2009b. *The Millennium Development Goals Report 2008* http://mdgs.un.org/unsd/mdg/Resources/Static/Products/ Progress2008/MDG_Report_2008_En.pdf.

United Nations General Assembly. 2000. *United Nations Millennium Declaration* (September 8).

United Nations University. 2006. "Pioneering Study Shows Richest Two Percent Own Half World Wealth." http://www .wider.unu.edu/events/past-events/2006-events/en_GB/ 05–12–2006/.

U.S. African Command. 2008. http://www.africom.mil/.

U.S. Bureau of the Census. 2004. "Income Stable, Poverty Up, Numbers of Americans with and without Health Insurance Rise." *U.S. Census Bureau News* (August 26). www.census.gov.

_____. 2005. *American Community Survey.* www.census.gov.

_____. 2008. *American Community Census, 2007.* http://www .census.gov/acs/www/.

U.S. Bureau of Labor Statistics. 2007a. "Table 11: Employed Persons by Detailed Occupation, Sex, Race, and Hispanic or Latino Ethnicity." www.bls.gov/cps/cpsaat11.pdf.

_____. 2007b. "Occupational Employment and Wages: 25–2021 Elementary School Teachers, Except Special Education, May 2005." www.bls.gov/oes/current/oes252021.htm.

_____. 2007c. "Median Weekly Earnings of Full-Time Wage and Salary Workers by Detailed Occupation and Sex, 2005." www .bls.gov/cps/cpsaat39.pdf.

_____. 2008. "39. Median Weekly Earnings of Full-Time Wage And Salary Workers by Detailed Occupation and Sex." http:// www.bls.gov/cps/cpsaat39.pdf.

U.S. Central Intelligence Agency. 2009. *World Factbook* https:// www.cia.gov/library/publications/the-world-factbook/.

U.S. Department of Agriculture. 2009. "National School Lunch Program." http://www.fns.usda.gov/cnd/Lunch/.

U.S. Department of Health and Human Services. 2009. "Poverty Guidelines, Research, and Measurement." http://aspe.hhs.gov/ POVERTY/.

U.S. Bureau of Labor Statistics. 2009. "Table 31. Median Weekly Earnings of Full-Time Wage and Salary Workers by Detailed Occupation and Sex." http://www.bls.gov/cps/cpsaat39.pdf.

U.S. Department of State. 2009. "Budget, Performance and Financial Snapshot Report–Fiscal Year 2008." http://www .state.gov/documents/organization/114174.pdf.

Wacquant, Loic J. D. 1989. "The Ghetto, the State, and the New Capitalist Economy." *Dissent* (Fall): 508–520.

Weber, Max. (1947) 1985. "Social Stratification and Class Structure." Pages 573–576 in *Theories of Society: Foundations of Modern Sociological Theory*, ed. T. Parsons, E. Shils, K. D. Naegele, and J. R. Pitts. New York: Free Press.

_____. 1948. "Class, Status, and Party," in H. Gerth and C. W. Mills. *Essays from Max Weber.* New York: Routledge and Kegan Paul.

_____. 1982. "Status Groups and Classes." Pages 69–73 in *Classical and Contemporary Debates*, ed. A. Giddens and D. Held. Los Angeles: University of California.

Wilson, William Julius. 1983. "The Urban Underclass: Inner-City Dislocations." *Society* 21: 80–86.

_____. 1987. *The Truly Disadvantaged.* Chicago: University of Chicago Press.

_____. 1991. "Studying Inner-City Social Dislocations: The Challenge of Public Agenda Research" (1990 presidential address). *American Sociological Review* (February): 1–14.

_____. 1994. "Another Look at the Truly Disadvantaged." *Political Science Quarterly* 106 (4): 639–656.

World Bank. 2009. "Understanding Poverty." http://web .worldbank.org.

Wright, Erik Olin. 2009. "Erik Olin Wright." http://www.ssc.wisc .edu/~wright/.

Yeutter, Clayton. 1992. "When Fairness Isn't Fair." *New York Times* (March 24): A13.

Chapter 9

About.com. 2004. "Immigration Before and After: Putting It in Perspective." immigration.about.com .

Alba, Richard D. 1992. "Ethnicity." Pages 575–584 in *Encyclopedia of Sociology*, vol. 2, ed. by E. F. Borgatta and M. L. Borgatta. New York: Macmillan.

Angelou, Maya. 1987. "Intra-Racism." Interview on *Oprah Winfrey Show* (Journal Graphics transcript W172).

Arab Institute. 2003. "First Census Report on Arab Ancestry Marks Rising Civic Profile of Arab Americans." www.aaiusa .org.

Carver, Terrell. 1987. *A Marx Dictionary.* Totowa, NJ: Barnes and Noble.

Cornell, Stephen. 1990. "Land, Labour and Group Formation: Blacks and Indians in the United States." *Ethnic and Racial Studies* 13(3):368–388.

Crapanzano, Vincent. 1985. *Waiting: The Whites of South Africa.* New York: Random House.

Darmoni, Stefen. 2005. Olympic Statistics. http://www.darmoni .net/.

Davis, F. James. 1978. *Minority-Dominant Relations: A Sociological Analysis.* Arlington Heights, IL: AHM.

DHS (U.S. Department of Homeland Security). 2003. "Fact Sheet: Changes to National Security Entry/Exit Registration System." www.dhs.gov.

Dunne, John Gregory. 1991. "Law and Disorder in Los Angeles." *New York Review of Books* (October 10):26.

Ferrante, Joan, and Prince Brown Jr. 2001. *The Social Construction of Race and Ethnicity in the United States* (2nd edition) Upper Saddle River, NJ: Prentice-Hall.

Finnegan, William. 1986. *Crossing the Line: A Year in the Land of Apartheid.* New York: Harper and Row.

Franklin, John Hope. 1990. Quoted in "That's History, Not Black History," by Mark McGurl. *New York Times Book Review* (June 3):13.

Gates, Henry Louis. 1995. "The Political Scene: Powell and the Black Elite." *New Yorker* (September 25): 64–80.

Gilmore, Gerry J. 2004. "Service Members Can Apply for Expedited U.S. Citizenship." American Forces Press Service (February 24) http://www.defenselink.mil/news.

Goffman, Erving. 1963. *Stigma: Notes on the Management of Spoiled Identity*. Upper Saddle River, NJ: Prentice Hall.

Goldberg, Carey. 2000. "Accused of Discrimination, Clothing Chain Settles Case." *New York Times* (December 22):C1.

Gordon, Milton M. 1978. *Human Nature, Class, and Ethnicity*. New York: Oxford University Press.

Graham, Lawrence Otis. 2001. "Black Men with a Nose Job." Pages 33–38 in *The Social Construction of Race and Ethnicity in the United States*, ed. by J. Ferrante and P. Brown. Upper Saddle River, NJ: Prentice-Hall.

Greene, Jay P., and Marcus A. Winters. 2005. "The Effect of Residential School Choice on Public High School Graduation Rates." *Education Working Paper No. 9* (April) Manhattan Institute for Policy Research. www.manhattan-institute.org/html/ewp_09.htm.

Hamad, Claudette Shwiry, ed. 2001. "Hate-Based Incidents, September 11–October 10, 2001." Appendix to Arab American Institute Foundation report *Submission to the United States Commission on Civil Rights* by James J. Zogby (October 12).

Haney Lopez, Ian F. 1994. "The Social Construction of Race: Some Observations on Illusion, Fabrication, and Choice." *Harvard Civil Rights—Civil Liberties Law Review* 29: 39–53.

Hertz, Todd. 2003. "Are Most Arab Americans Christian?" *Christianity Today* (March 25). www.christianitytoday.com.

Hoberman, John. 2000. "The Price of Black Dominance." *Society* 37(3): 49–56.

Jencks, Christopher. 2001. "Who Should Get In?" *New York Review of Books* (November 29):57–61.

Lieberman, Leonard. 1968. "The Debate over Race: A Study in the Sociology of Knowledge." *Phylon* 39 (Summer): 127–141.

Lock, Margaret. 1993. "The Concept of Race: An Ideological Construct." *Transcultural Psychiatric Research Review* 30:203–227.

Los Angeles Almanac. 2001. "Profile of Legalized Population under the Immigration Reform and Control Act of 1986." www.losangelesalmanac.com.

Los Angeles Times. 1992. "Probe Finds Pattern of Excessive Force, Brutality by Deputies" (July 21):A18.

Madison, Richard. 2002. "Changes in Immigration Law and Procedures." www.lawcom.com/chngs.html.

Malik, Kenan. 2000. "Why Black Will Beat White at the Olympics," *News Statesman* 129(4504):13–17.

McIntosh, Peggy. 1992. "White Privilege and Male Privilege: A Personal Account of Coming to See Correspondences through Work in Women's Studies." Pages 70–81 in *Race, Class, and Gender: An Anthology*, ed. by M. L. Andersen and P. H. Collins. Belmont, CA: Wadsworth.

Merton, Robert K. 1957. *Social Theory and Social Structure*. New York: Free Press.

———. 1976. "Discrimination and the American Creed." Pages 189–216 in *Sociological Ambivalence and Other Essays*. New York: Free Press.

Microsoft. 2001. "African American History." *Encarta Online Encyclopedia 2001*. http://encarta.msn.

Murray, Christopher et al. (2005) "Eight Americas: Investigating Mortality Disparities across Races, Counties, and Race-Counties in the United States." *PLoS Med* 3(9): e260 doi:10.1371/journal.pmed.0030260.

National Center for Health Statistics. 1993. "Advanced Report of Final Natality Statistics." *Monthly Vital Statistics Report* 43 (9).

Nduru, Moyiga. 2007. "South Africa: Political Sins Continue to Scar the Land." *Inter Press Service* (April 23).

Nile Valley Solutions. 2001. "Jim Crow Laws." www.nilevalley.net/jim_crow_laws.html.

Novas, Himilice. 1994. "What's in a Name?" Pages 2–4 in *Everything You Need to Know About Latino History*. New York: Dutton Signnet.

Obama, Barack. 1995. *Dreams from My Father: A Story of Race and Inheritance*. New York: Times Books.

Ogbu, John U. 1990. "Minority Status and Literacy in Comparative Perspective." *Daedalus* 119(2):141–168.

Page, Clarence. 1996. *Showing My Colors: Impolite Essays on Race and Identity*. New York: Harper Collins.

Parent, Anthony S., Jr. and Susan Brown Wallace. 2002. pp. 451–458 in *The Social Construction of Race and Ethnicity in the United States*, edited by J. Ferrante and P. Brown Jr. Upper Saddle River, NJ: Prentice-Hall.

Pedraza, Silvia. 1999. "Immigration in America at the Turn of the Century: Assimilation or Diasporic Citizenship?" *Contemporary Sociology* 28(4):377–381.

Pollitzer, William S. 1972. "The Physical Anthropology and Genetics of Marginal People in the Southeastern United States." *American Anthropologist* 74(1–2):719–734.

Rawley, James A. 1981. *The Transatlantic Slave Trade: A History*. New York: Norton.

Reeve, R. Penn. 1977. "Race and Social Mobility in a Brazilian Industrial Town." *Luso-Brazilian Review* 14 (2): 236–253.

Reynolds, Larry T. 1992. "A Retrospective on 'Race': The Career of a Concept." *Sociological Focus* 25(1):1–14.

Samhan, Helen Hatab. 1997. "Not Quite White: Race Classification and the Arab American Experience." Paper presented by the Center for Contemporary Arab Studies (April 4). www.aaiusa.org/not_quite_white.htm.

Schmitt, Eric. 2001. "To Fill in Gaps, Shrinking Cities Look Abroad for New Residents." *New York Times* (May 30):A1.

Sowell, Thomas, 1981. *Ethnic America: A History*. New York: Basic Books.

Steele, Claude. 1995. "Black Students Live Down to Expectations." *New York Times* (August 31):A25.

Terkel, Studs. 1992. *Race: How Blacks and Whites Think and Feel about the American Obsession*. New York: New Press.

Toro, Luis Angel. 1995. "'A People Distinct from Others': Race and Identity in Federal Indian Law and the Hispanic Classification in OMB Directive No. 15." *Texas Tech Law Review* 26:1219–1274.

United Nations. 2004. *Cultural Liberty in Today's Diverse World*. hdr.undp.org/reports/global/2004/.

U.S. Bureau of the Census. 1910. "Color or Race, Nativity, and Parentage." www2.census.gov/prod2/decennial/documents/36894832v1ch03.pdf

———. 1860. fisher.lib.virginia.edu/cgi-local/censusbin/census/cen.pl?year=860

———. 1994. *Current Population Survey Interviewing Manual*. Washington, DC: U.S. Government Printing Office.

———. 2000. *The Geography of U.S. Diversity*. www.census.gov/population/www/cen2000/atlas.htm.

———. 2001. "Overview of Race and Hispanic Origin." *Census 2000 Brief*. http://www.census.gov .

———. 2003. *The Arab Population: 2000*. www.census.gov/prod/2003pubs/c2kbr-23.pdf .

———. 2004. *United States Foreign-born Population*. www.census.gov/population/www/socdemo/foreign.html.

———. 2005. "Poverty." www.census.gov/hhes/www/poverty/poverty05/tables05.html

———. 2005a. Pages 68–71 in *American Community Survey: 2005 Subject Definitions*. www.census.gov/acs/www/Downloads/2005/usedata/Subject_Definitions.pdf.

———. 2005b. *We the People of Arab Ancestry in the United States*. www.census.gov/prod/2005pubs/censr-21.pdf.

———. 2006. "Comparability of the 2002 and 1997 SBO Data" in *2002 Economic Census: Survey of Business Owners Methodology*. www.census.gov/econ/census02/guide/.

U.S. Commission on Civil Rights. 1981. *Affirmative Action in the 1980s: Dismantling the Process of Discrimination (A Proposed Statement)*. Clearinghouse Publication 65. Washington, DC: U.S. Government Printing Office.

———. 2003. "The Arab Population: 2000." *Census 200 Brief*. www.census.gov.

———. 2005. *Statistical Abstract of the United States*. www.census.gov/prod/www/statistical- abstract-04.html.

U.S. Department of Justice. 1996. Press Release (October 3). www.usdoj.gov/gopherdata/pressreleases/previous/1999/532cr.htm.

———. 2004. "Justice Department Settles Race Discrimination Lawsuit against Cracker Barrel Restaurant Chain." http://www.usdoj.gov/opa/pr/2004/May/04_crt_288.htm.

U.S. Department of Justice. 2007. Criminal Offenders Statistics. www.ojp.usdoj.gov/bjs/crimoff.htm#prevalence

U.S. Department of Labor. 2006. "Foreign Labor Statistics." www.bls.gov/fls/

U.S. Office of Management and Budget. 1977. "Federal Statistical Directive No. 15: Race and Ethnic Standards for Federal Statistics and Administrative Reporting" (May 12).

———1995. "Standards for the Classification of Federal Data on Race and Notice." *Federal Register* (August 28):44673–44693.

———1997. "Revisions to the Standards for the Classification of Federal Data on Race and Ethnicity." *Federal Register* Notice (October 30). www.whitehouse.gov/omb/fedreg/1997standards.html.

Washington, Booker T. 1901. *Up From Slavery*. www.4literature.net/Booker_T_Washington/Up_From_Slavery/

Waters, Mary C. 1994. "Ethnic and Racial Identities of Second Generation Black Immigrants in New York City." *International Migration Review* 28(4):795–820.

———. 1990. *Ethnic Options: Choosing Identities in America*. Berkeley: University of California.

Weathers, William. 1999. "The Issue Is Clear: 'Redskin' Offensive, Disrespectful." *Kentucky Post* (May 25). www.kypost.com/opinion/weaths052599.html.

Wirth, Louis. 1945. "The Problem of Minority Groups." Pages 347–372 in *The Science of Man*, ed. by R. Linton. New York: Columbia University Press.

Woods, Tiger. 2006. Quoted in excerpts from *Who's Afraid of a Large Black Man?* by C. Barkley. www.wnyc.org/books/45779

Chapter 10

Anspach, Renee R. 1987. "Prognostic Conflict in Life-and-Death Decisions: The Organization as an Ecology of Knowledge." *Journal of Health and Social Behavior* 28(3):215–231.

Anthias, Floya, and Nira Yuval-Davis. 1989. Introduction (pp. 1–15) in *Woman-Nation-State*, ed. by N. Yuval-Davis and F. Anthias. New York: St. Martin's.

Barton, Gina. 2005. "Prisoner Sues for the Right to Sex Change." *Milwaukee Journal Sentinel Online* (January 22). www.jsonline.com/news/state/jan05.

Baumgartner-Papageorgiou, Alice. 1982. *My Daddy Might Have Loved Me: Student Perceptions of Differences between Being Male and Being Female*. Denver: Institute for Equality in Education.

Bem, Sandra Lipsitz. 1993. *The Lenses of Gender: Transforming the Debate on Sexual Inequality*. Binghamton, NY: Vail-Ballou.

Bloom, Amy. 1994. "The Body Lies." *New Yorker* (July 18):38–49.

Boroughs, Don L. 1990. "Valley of the Doll?" *U.S. News & World Report* (December 3):56–59.

Boxer, Barbara (U.S. Senator). 2007. "Historical Timeline for Women's History." http://boxer.senate.gov/whm/time_1.cfm.

Busch, Bill. 2003. Quoted in B. Syken's "Football in Paradise." *Sports Illustrated* (November 3).

Cahill, Betsy, and Eve Adams. 1997. "An Exploratory Study of Early Childhood Teachers' Attitudes Toward Gender Roles." *Sex Roles: A Journal of Research* 36(7–8): 517–530.

Centers for Disease Control. 2006. "Youth Risk Behavior Surveillance—United States" Morbidity & Mortality Weekly Report 55(SS-5):1–108.

CBS News Polls. 2005. "Women's Movement Worthwhile." http://www.cbsnews.com/stories/2005/10/22/opinion/polls/main965224.shtml.

Channell, Carrie. 2002. "The Tatau: A Bridge to Manhood." *Faces: People, Places and Culture* (May):18–22.

Chronicle of Higher Education. 2004. Almanac. chronicle.com/free/almanac/2004.

Cordes, Helen. 1992. "What a Doll! Barbie: Materialistic Bimbo or Feminist Trailblazer?" *Utne Reader* (March/April):46, 50.

Cote, James. 1997. "A Social History of Youth in Samoa: Religion, Capitalism, and Cultural Disenfranchisement." *International Journal of Comparative Sociology* 38(3–4):217.

Daly, Mary. 2006. *Amazon Grace: Re-Calling the Courage to Sin Big*. New York and Hampshire, England: Palgrave Macmillan.

Dewhurst, Christopher J., and Ronald R. Gordon. 1993. Quoted in "How Many Sexes Are There?" *New York Times* (March 12): A15.

Dreifus, Claudia. 2005. "Declaring with Clarity, When Gender Is Ambiguous." *New York Times* (May 31): D2.

Fagot, Beverly, Richard Hagan, Mary Driver Leinbach, and Sandra Kronsberg. 1985. "Differential Reactions to Assertive and Communicative Acts of Toddler Boys and Girls." *Child Development* 56(6):1499–1505.

Ferguson, David. 2005. "James Tackles Samoa's Talent Drain." *Scotsman* (November 18). thescotsman.scotsman.com.

Frank, Nathaniel. 2004. Press release: "Mission-Critical Specialists Discharged for Homosexuality." Center for the Study of Sexual Minorities in the Military (June 21). www.gaymilitary.uscb.edu.

Fraser, Laura. 2002. "The Islands Where Boys Grows Up to Be Girls." *Marie Claire* (December):72–79.

Freeman, Derek. 1996. *Margaret Mead and the Heretic: The Making and Unmaking of an Anthropological Myth.* Australia: Penguin Books.

Garb, Frances. 1991. "Secondary Sex Characteristics." Pages 326–327 in *Women's Studies Encyclopedia. Vol. 1: Views from the Sciences,* ed. by H. Tierney. New York: Bedrick.

Gauguin, Paul. (1919) 1985. *Noa Noa: The Tahitian Journal,* trans. by O. F. Theis. New York: Dover.

Geschwender, James A. 1992. "Ethgender, Women's Waged Labor, and Economic Mobility." *Social Problems* 39(1):1–16.

Grady, Denise. 1992. "Sex Test of Champions." *Discover* (June):78–82.

Grameen. 2005. "Grameen Bank at a Glance." www.grameen_info.org.

Greenhouse, Steven. 2003. "Wal-Mart Faces Lawsuit over Sex Discrimination." *New York Times* (February 16):Y18.

———. 2004. "In-House Audit Says Wal-Mart Violated Labor Laws." *New York Times* (January 13). www.nyt.com.

Hall, Edward T. 1959. *The Silent Language.* New York: Doubleday.

Harris, Gardiner. 2003. "If the Shoe Won't Fit, Fix Foot? Popular Surgery Raises Concerns." *New York Times* (December 7): 24.

Hasbro Toys. 1998. "G.I. Joe: Chronology." http://www.hasbrotoys.com/gijoe/crono.html.

Holmes, L. D., and Holmes, E. R. 1992. *Samoan Village Then and Now.* New York: Harcourt Brace.

Infonautics Corporation. 1998. "American Samoa: Chapter 1. General Information." http://www.elibrary.com.

Kolata, Gina. 1992. "Track Federation Urges End to Gene Test for Femaleness." *New York Times* (February 12):A1, B11.

Lee, Carol E. 2004. "The Evolution of Women's Roles, Chronicled in the Life of a Doll." *New York Times* (March 28):B1.

Lehrman, Sally. 1997. "WO: Forget Men Are from Mars, Women Are from Venus. Gender." *Stanford Today* (May/June):47.

Lemonick, Michael D. 1992. "Genetic Tests Under Fire." *Time* (February 24):65.

Love, David A. 2007. "Walgreens Suit Shows Employment Discrimination Still a Problem." *Progressive Media Project* (March 13). www.progressive.org.

Lyall, Sarah. 2007. "Gay Britons Serve in Military with Little Fuss, as Predicted Discord Does Not Occur." *New York Times* (April 27): A14.

Madrick, Jeff. 2004. "Economic June: The Earning Power of Women Has Really Increased, Right? Take a Closer Look." *New York Times* (June 16):C2.

Mageo, Jeannette. 1992. "Male Transvestism and Cultural Change in Samoa." *American Ethnologist* 19(3):443.

———. 1996. "Hairdos and Don'ts: Hair Symbolism and Sexual History in Samoa." *Frontiers* 17(2):138.

———. 1998. *Theorizing Self in Samoa: Emotions, Genders, and Sexualities.* Ann Arbor: University of Michigan Press.

Malauulu, George. 2003. Quoted in B. Syken's "Football in Paradise." *Sports Illustrated* (November 3).

Matà afa, Tina. 2006. "School Teacher Wins Miss Island Queen." *Samoa News* (May 10). www.samoanews.com.

Mattel. 2005. "Barbie: More Than a Doll." www.mattel.com/our_toys/ot_barb.asp.

Mead, Margaret. 1928. *Coming of Age in Samoa: A Psychological Study of Primitive Youth for Western Civilisation.* New York: William Morrow.

———. 1972. *Blackberry Winter: My Earlier Years.* New York: William Morrow.

Mills, Janet Lee. 1985. "Body Language Speaks Louder Than Words." *Horizons* (February):8–12.

Morawski, Jill G. 1991. "Femininity." Pages 136–139 in *Women's Studies Encyclopedia.* Vol. 1: *Views from the Sciences,* ed. by H. Tierney. New York: Bedrick.

Morgenson, Gretchen. 1991. "Barbie Does Budapest." *Forbes* (January 7):66–69.

Newman, Louise M. 1996. "Coming of Age, but Not in Samoa: Reflections on Margaret Mead's Legacy for Western Liberal Feminism." *American Quarterly* 48(2):223–272.

O'Connor, Marian. 2007. "Corporate Suit Could Change Workplace Politics." City on a Hill Press (February 14). www.cityonahillpress.com.

Online Newshour. 2007. "Pace Remarks Renew 'Don't Ask, Don't Tell' Debate" (March 16). www.pbs.org/newshour/bb/military.

Pion, Alison. 1993. "Accessorizing Ken." *Origins* (November):8.

Population Reference Bureau. 2002. "Children in America Samoa: Results of the 2000 Census." *Kids Count* Report for the Anne E. Casey Foundation.

Raag, Tarja, and Christine Rackliff. 1998. "Preschoolers' Awareness of Social Expectations of Gender: Relationships to Toy Choices." *Sex Roles: A Journal of Research* 38(9–10):685.

Rank, Mark R. 1989. "Fertility among Women on Welfare: Incidence and Determinants." *American Sociological Review* 54(4):296–304.

Rose, Stephen J., and Heidi Hartman. 2004. Cited in "Economic Scene: The Earning Power of Women Has Really Increased, Right? Take a Closer Look," by J. Madrick. *New York Times*: C2.

Sahadi, Jeanne. 2006. "Where Women's Pay Trumps Men's." CNNmoney.com. money.cnn.com/2006/02/28/commentary/everyday/sahadi/index.htm.

Samoa Daily News. 1998c. "Five Women Seek Elected Office in a Male Dominated Arena" (November 3). http://www.ipacific.com/archive/1998/1103.txt.

Schmalz, Jeffrey. 1993. "From Midshipman to Gay-Rights Advocate." *New York Times* (February 4):B1+.

Shalikashvili, John M. 2007. "Second Thought on Gays in the Military." *New York Times* (January 2). www.nyt.com.

Shweder, Richard A. 1994. "What Do Men Want? A Reading List for the Male Identity Crisis." *New York Times Book Review* (January 9):3, 24.

Son, Eugene. 1998. "G.I. Joe—A Real American FAQ." http:www.yojoe.com/faq/gifaq.txt.

Stanner, W. E. H. 1953. South Seas in Transition: A Study of Post-War Rehabilitation and Reconstruction in Three British Pacific Dependencies. AMS Press.

Sturdevant, Saundra Pollock, and Brenda Stoltzfus. 1992. *Let the Good Times Roll: Prostitution and the U.S. Military in Asia.* New York: New Press.

Syken, Bill. 2003. "Football in Paradise." *Sports Illustrated* (November 3).

Tierney, Helen. 1991. "Gender/Sex." Page 153 in *Women's Studies Encyclopedia. Vol. 1: Views from the Sciences*, ed. by H. Tierney. New York: Bedrick.

Time.com. 2007. "Feminism Poll." http://www.time.com/time/polls/feminism.html.

Tomey, Dick. 2003. Quoted in B. Syken's "Football in Paradise." *Sports Illustrated* (November 3).

Tuller, David. 2004. "Gentleman, Start Your Engines?" *New York Times* (June 21):E1.

Turner, George. (1861) 1986. *Samoa: Nineteen Years in Polynesia.* Apia: Western Samoa and Cultural Trust.

U.S. Census Bureau. 2005a. *American Samoa, 2002.* www.census .gov.

———. 2005b. "Indicators of Marriage and Fertility in the United States from the American Community Survey: 2000 to 2003." www.census.gov. http://www.census.gov/population/www/socdemo/fertility/mar-fert-slides.html.

———. 2006. "Median Earnings in the Past 12 Months by Educational Attainment." factfinder.census.gov/.

U.S. Central Intelligence Agency. 2004. "American Samoa." *The World Fact Book.* www.cia.gov/cia/publications/factbook.

———. 2007. "American Samoa." *The World Factbook.* https://www.cia.gov/cia/publications/factbook/fields/2102.html.

U.S. Department of Defense. 1990. "DOD Directive 1332.14." Page 19 in *Gays in Uniform: The Pentagon's Secret Reports*, ed. by K. Dyer. Boston: Alyson.

U.S. Department of Health and Human Services, Centers for Disease Control and Prevention. 2006. *Youth Risk Behavior Surveillance*, http://www.cdc.gov/HealthyYouth/yrbs/.

U.S. Department of the Interior. 1992. "United States Insular Areas and Freely Associated States." www.doe.gov/oia/oiafacts.html#page2.

U.S. Department of Labor, Bureau of Labor Statistics. 2006. "Women's Earnings as a Percentage of Men's. 1979–2005." www.bls.gov.

———. 2007. "Employed Persons by Industry, Sex, Race, and Occupation." www.bls.gov/cps/cpsaat17.pdf.

U.S. House of Representatives. 2004. "Three Samoan NFL Players Make Washington Redskins Squad." Press Release (September 11). www.house.gov/list/press.

U.S. Senate. 2007. "Women in the Senate." http://www.senate.gov.

Vann, Elizabeth. 1995. "Implications of Sex and Gender Differences for Self: Perceived Advantages and Disadvantages of Being the Other Gender." *Sex Roles: A Journal of Research* 33(7–8):531.

Wilgoren, Jodi. 2003. "A New War Brings New Role for Women." *New York Times* (March 30):B1.

Wishart, David. 1995. *The Roles and Status of Men and Women in Nineteenth Century Omaha and Pawnee Societies: Postmodernist Uncertainties and Empirical Evidence.* Lincoln, NE: University of Nebraska Press.

World Health Organization. 2007. "American Samoa: Demographics, Gender, and Poverty." www.wpro.who.int/countries/ams/

Zogby International. 2006. " 'Don't Ask, Don't Tell' Not Working: Survey Indicates Shift in Military Attitudes" (December 18). www.zogby.com.

Chapter 11

Birdsall, Nancy, and Arvind Subramanian. 2004. "Saving Iraq from Its Oil." *Foreign Affairs* 83 (4): 77–89.

Boudon, Raymond, and François Bourricaud. 1989. *A Critical Dictionary of Sociology*, selected and trans. by P. Hamilton. Chicago: University of Chicago Press.

Bradsher, Keith. 2008. "Could Consumers Fill the Export Void?" Chart in "The Virtues of Spending." *New York Times* (October 16): B11.

British Geological Survey. 2008. *Annual Report, 2007–2008.* http://www.bgs.ac.uk/annualreport/0708.html.

Buckley, Alan D. 1998. "Comparing Political Systems." www.smc .edu/_homepage/abuckley/ps2/ps2types.htm.

Bulliet, Richard. 2004. "Whose Theocracy?" *Agence Global* (May 3). www.agenceglobal.com.

Bullock, Alan. 1977. "Democracy." Pages 211–212 in *The Harper Dictionary of Modern Thought*, ed. by A. Bullock, S. Trombley, and B. Eadie. New York: Harper & Row.

Burke, James. 1978. *Connections.* Boston: Little, Brown.

Bush, George W. 2003. "Remarks by the President at the 20th Anniversary of the National Endowment for Democracy" (November 6). www.whitehouse.gov/news/releases/2003/11/20031106–1.html.

———. 2005. "President's Address to the Nation." (December 18). http://www.whitehouse.gov/news/releases/2005/12/20051218–2.html.

———. 2007. "President's Address to the Nation." (January 10). www.whitehouse.gov/news/releases/2007/01/20070110–7 .html.

———. 2007. "May 5 Radio Address." http://www.whitehouse .gov/news/radio/.

Center for Responsive Politics. 2009. "527 Committee Activity: Top 50 Organizations." www.opensecrets.org/527s.

———. 2009. "Top PACs for 2005–2006." www.opensecrets.org/pacs/index.asp.

———. 2009. "Top PACs, 2008 Election Cycle." http://www .opensecrets.org/pacs/toppacs.php.

Chehabi, H. E., and Juan J. Linz (eds.). 1998. *Sultanistic Regimes.* Baltimore and London: Johns Hopkins University Press.

Creighton, Andrew L. 1992. "Democracy." Pages 430–434 in *Encyclopedia of Sociology*, Vol. 1, ed. by E. F. Borgatta and M. L. Borgatta. New York: MacMillan.

Congressional Research Service. 2006. *Conventional Arms Transfers to Developing Nations, 1998–2005.* CRS Report for

Congress prepared by R. R. Grimmett (October 23). www.fas.org/sgp/crs/weapons/RL33696.pdf.

Dash, Eric. 2009. "Credit Card Companies Willing to Deal Over Debt." *The New York Times* (January 2).

Der Spiegel. 2003. Quoted in "Europe Seems to Hear Echoes of Empires Past" by Richard Bernstein. *New York Times* (April 14). www.nyt.com.

Diba, Bahman Aghai. 2004. "Oil Prices: The Fundamental and Temporary Reason." *Payvand's Iran News* (September 4). www.payvand.com/news/04/Sep/1265.htm.

Eckholm, Erik. 2004. "Iraqis Bank on Free Food." *New York Times* (September 13). www.nyt.com.

Eisenhower, Dwight D. 1961. "Farewell Radio and Television Address to the American People." Eisenhower Library. www.eisenhower.archives.gov/speeches/farewell_address.html.

Elamine, Bilal. 2004. "Oil and Empire in the Middle East." *Mathaba News* (October 14). mathaba.net.

Faler, Brian. 2005. "Election Turnout in 2004 Was Highest Since 1968." *Washington Post* (January 14). www.washingtonpost.com.

Felluga, Dino. 2002. "Hegemony." *Introductory Guide to Critical Theory.* Purdue University. http://www.cla.purdue.edu/academic/engl/theory/marxism/terms/.

Galbraith, John K. 1958. *The Affluent Society.* Boston: Houghton Mifflin.

Global Security.org. 2009. "Worldwide Military Expenditures, 2008." http://www.globalsecurity.org/military/world/spending.htm.

Guardian Unlimited. 2007, July 17. "Migration Body Urges More Aid for Displaced Iraqis." www.guardian.co.uk/iraq/story.

Hacker, Andrew. 1971. "Power to Do What?" Pages 134–146 in *The New Sociology: Essays in Social Science and Social Theory in Honor of C. Wright Mills,* ed. by I. L. Horowitz. New York: Oxford University Press.

Hirsch, E. D., Jr., Joseph F. Kett, and James Trefil. 1993. *The Dictionary of Cultural Literacy.* New York: Houghton Mifflin.

Hunt, Mary. 2004. "'Survival Debt' Is the Scariest." *The Cincinnati Post* (October 26): 7B.

Intertanko. 2007. "Tanker Facts: Carrying Oil for the World." www.intertanko.com/tankerfacts/sizes/carrying.htm.

Iraq Body Count Project. 2007. "Reported Civilian Deaths Resulting from the U.S.-Led Military Intervention in Iraq." Retrieved on July 15, 2007, from www.iraqbodycount.org/database/.

Japan Times Online. 2008, December 29. "Editorial: American Capitalism, Battered." http://search.japantimes.co.jp/cgi-bin/ed20081230a1.html.

Joseph, Michael. 1982. *The Timetable of Technology.* London: Marshal Editions.

Judt, Tony. 2004. "Dreams of Empire." *New York Review of Books* (November 4). www.nyrb.com/articles.

Juhasz, Antonia. 2007. "Whose Oil Is It, Anyway?" *New York Times* (March 13). www.nytimes.com/2007/03/13/opinion/13juhasz.html.

Landler, Mark. 2009. "I.M.F Puts Banks Losses from Global Financial Crisis at $4.1 Trillion." *New York Times* (April 22): 2.

Laurier, Joanne. 2004. "U.S. Consumer Debt Reaches Record Levels." *World Socialist Website* (January 15). www.wsws.org/articles2004/jan2004/debt-j15.shtml.

Library of Congress. 2009. "Bill Number H.R.800 for the 110th Congress." thomas.loc.gov/cgi-bin/query/z?c110:H.R.800.

Lockheed Martin Corporation. 2007. "Corporate Overview." Retrieved July 14, 2007, from www.lockheedmartin.com/data/assets/969.pdf.

McNamara, Robert S. 1989. *Out of the Cold: New Thinking for American Foreign and Defense Policy in the 21st Century.* New York: Simon & Schuster.

Metz, Helen Chapin (ed.). 1988. *Iraq: A Country Study.* Library of Congress: Federal Research Division.

Mills, C. Wright. 1956. *The Power Elite.* New York: Oxford University Press.

_____. 1963. "The Structure of Power in American Society." Pages 23–38 in *Power, Politics and People: The Collected Essays of C. Wright Mills,* ed. by I. L. Horowitz. New York: Oxford University Press.

MSNBC. 2007. "Just Who Owns the U.S. National Debt?" msnbc.msn.com/id/17424874.

Obama, Barack. 2009. Speech at Camp Lejeune, N.C. (February 27). http://www.nytimes.com/2009/02/27/us/politics/27obama-text.html.

Refugees International. 2009. Iraq. http://www.refugeesinternational.org/where-we-work/middle-east/iraq

Resolution 661. 1990. Adopted by the UN Security Council at its 2933rd meeting (August 6). www.casi.org.uk/info/undocs/gopher/s90/15.

Roux, George. 1965. *Ancient Iraq.* New York: Penguin.

Sachs, Susan. 2004. "In Iraq's Next Act, Tribes May Play the Lead Role." *New York Times* (June 6): Wk14.

Sanford, Jonathan E. 2003. "Iraq's Economy: Past, Present, Future." *Report for Congress.* Congressional Research Service (June 3).

Stevenson, Richard W. 1991. "Northrop Settles Workers' Suit on False Missile Tests for $8 Million." *New York Times* (June 25): A7.

Story, Louise. 2009. "Lending a Hand, Quietly." *The New York Times* (April 15).

Streitfeld, David. 2009. "Credit Bailout: Issuers Slashing Card Balances." *The New York Times* (June 16): B1.

Tierney, John. 2003. "Iraq's Family Bonds Complicate U.S. Efforts." *New York Times* (September 28): A1.

Trejos, Nancy. 2009. "Toxic Plastic Relief Too Late for Many as Rates Rise, Credit Limits Fall." *Washington Post* (May 17). http://www.washingtonpost.com/wp-dyn/content/article/2009/05/16/AR2009051600055.html.

Tuskowski, Veronika R. 2004. "Iraqi Translator's Service Comes at High Cost." American Forces Press Service, U.S. Department of Defense. http://www.defenselink.mil/news/newsarticle.aspx?id=25244.

Uchitelle, Louis. 2009. "After Recession, Recovery Will Take Years." *New York Times* (April 7): 5B.

United Nations. 2004. "Oil-for-Food." Office of Iraq Programme (October 15). www.un.org/debts/oip/background.

United Nations Development Programme. 2005. *Arab Human Development Report 2004: Toward Freedom in the Arab World.* hdr.undp.org/.

USA.gov. 2009. "Popular New Year's Resolutions." http://www.usa.gov/Citizen/Topics/New_Years_Resolutions.shtml.

U.S. Bureau of the Census. 2009. Voting and Registration in the Election of November 2008. http://www.census.gov/population/www/socdemo/voting/cps2008.html

———. 2009. "U.S. Trade in Goods (Imports, Exports and Balance) by Country." http://www.census.gov/foreign-trade/balance/index.html.

U.S. Bureau of Economic Analysis. 2009. "Personal Income and Outlays." http://www.bea.gov/newsreleases/national/pi/pinewsrelease.htm.

U.S. Bureau of Labor Statistics. 2007. "Union Members Summary." (January 25). www.bls.gov/news.release/union2.nr0.htm.

———. 2009. http://www.bls.gov/.

U.S. Central Intelligence Agency. 2009. "United States." *The World Factbook.* https://www.cia.gov/library/publications/the-world-factbook/geos/us.html#Intro.

U.S. Department of Defense. 2009. "Operation Iraqi Freedom (OIF) U.S. Casualty Status." Retrieved from http://www.defenselink.mil/news/casualty.pdf.

U.S. Department of Defense (2009). Variety of Defense Link News Articles. http://www.defenselink.mil/news.

U.S. Department of Energy. 2003. "Energy Information Sheets: Petroleum Products." www.eia.doe.gov.

U.S. Department of Treasury. 2009. "Major Holders of Treasury Securities." http://www.treas.gov/tic/mfh.txt.

U.S. Energy Information Agency (2008) "Crude Oil Statistics." http://www.eia.doe.gov/pub/oil_gas/natural_gas/data_publications/crude_oil_natural_gas_reserves/current/pdf/ch3.pdf.

U.S. Federal Reserve. 2007. "Consumer Credit." www.federalreserve.gov/releases/g19/current/default.htm.

West, Andrew. 2003. "800 Missiles to Hit Iraq in First 48 Hours." *Sydney Morning Herald* (January 26). www.smh.com.au.

Wordiq.com. 2004. "Famous Viewpoints on Democracy." www.wordiq.com/definition/democracy.

Wyman, Oliver. 2009. "Behind Changing Tactics, Mounting Losses." *New York Times* (January 2).

Chapter 12

Aguilar, Thais. 1999. "Families of the New Millennium." Measure Communication. www.measurecommunication.org.

Al-Badr, Dominic. 2006. "Japan Hit by Huge Rise in Child Abuse." (June 27) *Guardian.* www.guardian.co.uk.

Amerasian Citizenship Initiative. 2007. www.amerasianusa.org/index.php

American Association of Retired Persons. 2001. "Families Discuss Parents' Abilities to Live Independently." www.aarp.org.

AmeriStat. 2000. "Marriage and Family: Women Who Earn More Money Than Their Husbands." Population Reference Bureau and Social Science Data Analysis Network. www.ameristat.org.

Annie E. Casey Foundation. 2007. *Kids Count.* http://www.aecf.org/

Bando, Mariko. 2003. Quoted in "Japan: A Developing Country in Terms of Gender Equality." *Japan Times.* (June 14). www.japantimes.co.jp.

Bianchi, Suzane M., Melissa A. Milkie, Llana C. Sayer, and John P. Robinson. 2000. "Is Anyone Doing the Housework? Division of Household Labor." *Social Forces* 79(1):191+.

Boling, Patricia. 1998. "Family Policy in Japan." *International Journal of Social Policy* 27(2):173–190.

Brasor, Philip. 1999. "Can the Education Escalator Be Derailed?" *Japan Times* (April 1). www.japantimes.co.jp.

Brooke, James. 2004. "Japan Seeks Robotic Help in Caring for the Aged." *New York Times* (March 5):A1.

Burgess, Chris. 2007. "'Multicultural Japan' Remains a Pipe Dream." *Japan Times* (March 27). www.japantimes.co.jp.

Childstats.gov. 2006. "Indicators of Children's Well-Being." *America's Children in Brief: Key National Indicators of Well-Being.* www.childstats.gov/amchildren05/pdf/ac2005/econ.pdf.

College Board. 2005. "Average College Costs 2004–2005." www.collegeboard.com.

Collins, Randall. 1971. "A Conflict Theory of Sexual Stratification." *Social Problems* 19(1):3–21.

Cornell, L. L. 1990. "Constructing a Theory of the Family: From Malinowski through the Modern Nuclear Family to Production and Reproduction." *International Journal of Comparative Sociology* XXXI(1–2):67–78.

Cote, James. 1997. "A Social History of Youth in Samoa: Religion, Capitalism, and Cultural Disenfranchisement." *International Journal of Comparative Sociology* 38(3–4):217.

Davis, Kingsley. 1984. "Wives and Work: The Sex Role Revolution and Its Consequences." *Population and Development Review* 10(3):397–417.

Defense of Marriage Act. 1996. thomas.loc.gov/cgi-bin/bdquery/z?d104:SN01999.

Deutsch, Francine M. 1999. *Having It All: How Equally Shared Parenting Works.* Cambridge, MA: Harvard University Press.

Downing, Karen. 2007. "Statistics on Interracial Issues." www.personal.umich.edu/~kdown/stats.html.

Dychtwald, Ken, and Joe Flower. 1989. *Age Wage: The Challenges and Opportunities of an Aging America.* Los Angeles: Tarcher.

Engels, Friedrich. 1884. *The Origin of the Family, Private Property, and the State,* ed. by E. B. Leacock. New York: International Publisher.

Federal Interagency Forum on Aging-Related Statistics. 2000. "Older Americans 2000: Key Indicators on Well-Being." www.agingstats.gov.

French, Howard W. 2002. "Japan's Neglected Resource: Female Workers." *New York Times* (December 22):A27.

Freud, Sigmund. [1928] 1989. *The Future of an Illusion.* Standard edition, trans. and ed. James Strachey in collaboration with Anna Freud. New York: Norton, p. 53.

Gerson, Kathleen. 1999. "Review of *Gender Vertigo: American Families in Transition.*" *Contemporary Sociology* 28(4):419–420.

Hacker, Andrew. 2002. "Marriage and Divorce: The Fact Gap." *New York Review of Books* (December 5).

Hashizume, Yumi. 2000. "Gender Issues and Japanese Family-Centered Caregiving for Frail Elderly Parents or Parents-in-Law in Modern Japan: From the Sociocultural and Historical Perspectives." *Public Health Nursing* 17(1):25–31.

Hiroko, Kimura. 2003. "Out of the Shadows: Opportunity Knocks for Women in Japan's Climate of Change." *Japan Times* (July 13). www.japantimes.co.jp.

Hoffman, Michael. 2003. "A Simple Solution for Infertile Couples." *Japan Times* (June 22). www.japantimes.co.jp.

Isa, Masako. 2000. "Phenomenological Analysis of the Re-entry Experiences of the Wives of Japanese Corporate Sojourners." *Women and Language* 23(2):26+.

Iwasawa, Miho, James M. Raymo, and Larry Bumpass. 2005. Unmarried Cohabitation and Family Formation in Japan. Paper presented at the annual meeting of the Population Association of America, Los Angeles, March 30, 2006. paa2006.princeton.edu/download.aspx?submissionId=61321.

Jackson, Maggie. 2003. "More Sons Are Juggling Jobs and Care for Patients." *New York Times* (June 15). www.nyt.com.

Japan Information Network. 2000a. "Where Have All the Babies Gone?" (January 31). web-japan.org/trends00/honbun/tj000128.html.

———. 2000b. "'Parasite Singles' Multiply and Parental 'Hosts' Don't Seem to Mind" (May 15). web-japan.org/trends00/honbun/tj000508.html

———. 2007. *Japan in Figures.* www.stat.go.jp/english/data.

Japan Times. 2000. "Blasted Kids Need Scolding: Tokyo Government Says." (August 12) www.japantimes.co.jp.

———. 2002. "20 Local Governments Help Pay for Fertility Treatments" (December 8). www.japantimes.co.jp.

———. 2003a. "Editorial: Japan Needs a Road Map" (November 18). www.japantimes.co.jp.

———. 2003b. "Japan: A Developing Country in Terms of Gender Equality" (June 14). www.japantimes.co.jp.

———. 2004. "Japan and the Immigration Issue." (September 14). www.japantimes.co.jp.

———. 2006. "More Couples Report Domestic Abuse." (April 16). www.japantimes.co.jp.

———. 2007. "Fertility Treatment Coverage Upped." (August 19). www.japantimes.co.jp.

Johansson, S. Ryan. 1987. "Status Anxiety and Demographic Contraction of Privileged Populations." *Population and Development Review* 13(3):439–470.

Kaplan, Matthew, and Leng Leng Thang. 1997. "Intergenerational Programs in Japan: Symbolic Extensions of Family Unity." *Journal of Aging and Identity* 2(4):295–315.

Kingston, Jeff. 2001. "Two Perspectives on a Gray Tomorrow." *Japan Times* (March 6). www.japantimes.co.jp.

Kipnis, Laura. 2004. "The State of the Unions: Should This Marriage Be Saved?" *New York Times* (January 25):25.

Lanier, Shannon, and Jane Feldman. 2000. *Jefferson's Children: The Story of One American Family.* New York: Random House.

Laslett, Barbara, and Johanna Brenner. 1989. "Gender and Social Reproduction: Historical Perspectives." *Annual Review of Sociology.* Volume 15:381–404.

Lewin, Tamar. 1990. "For More People in 20s and 30s, Home Is Where the Parents Are." *New York Times* (December 21). www.nyt.com.

———. 2001. "In Genetic Testing for Paternity, Law Often Lags Behind Science." *New York Times* (March 11):A1.

———. 2003. "For More People in Their 20s and 30s, Going Home Is Easier Because They Never Left." *New York Times* (December 22):A27.

McCurry, Justin. 2006. "Japan's Domestic Abuse Cases Rise." (March 10). *Guardian.* www.guardian.co.uk.

National Alliance for Caregiving and AARP. 2004. *Caregiving in the U.S.* www.caregiving.org/04finalreport.pdf.

National Center for Health Statistics. 2004. *Health, United States, 2004.* www.cdc.gov/nchs/fastats/nursingh.htm.

Newport, Sally F. 2000. "Early Childhood Care, Work, and Family in Japan: Trends in a Society of Smaller Families." *Childhood Education* 77(2):68+.

Ogawa, Naoshiro, and Robert D. Retherford. 1997. "Shifting Costs of Caring for Elderly Back to Families in Japan: Will It Work?" *Population and Development Review* 23(1):59–94.

Online Newshour. 1996. "Education Report Card." (November 21). www.pbs.org/newshour

Ornstein, Peggy. 2001. "Parasites in Prêt-à-Porter." *New York Times Magazine* (July 1):31–35.

PBS. 2000. "Women's Rights and Reform in the 19th Century." Resource guide to *Not for Ourselves Alone: The Story of Elizabeth Cady Stanton and Susan B. Anthony.* http://www.pbs.org/stantonanthony/resources/index.html?body=03activity.html.

Pearl S. Buck International. 2007. www.pearlsbuck.or.kr.

Population Reference Bureau. 2003. "World Population Data Sheet." www.prb.org.

———. 2006. "Living Arrangements of Older Japanese." www.prb.org/presentations/presentations/gb-aging_all.ppt.

Retherford, Robert D., Naoh-iro Ogawa, and Rikiya Matsukura. 2001. "Late Marriage and Less Marriage." *Population and Development Review* 27(1):65+.

Rogers, J.A. 1972. "Remarks on the First Two Volumes of Sex and Race." In *Sex and Race, Volume 3.* St. Petersburg, FL: Helma M. Rogers.

Sato, Minako. 2005. "Cram Schools Cash in on Failure of Public Schools." *Japan Times* (July 28). www.japantimes.co.jp.

Scommegna, Paola. 2002. "Increased Cohabitation Changing Children's Family Settings." *Research on Today's Issues* (Issue No. 13, September). National Institute of Child Health and Human Development. www.nichd.nih.gov/publications/pubs/upload/ti_13.pdf.

Seabrook, John. 2001. "The Tree of Me." *New Yorker* (March 26): 58–71.

Sims, Calvin. 2000. "Japan's Employers Are Giving Bonuses for Having Babies." *New York Times* (May 30). www.nytimes.com.

Soldo, Beth J., and Emily M. Agree. 1988. "America's Elderly." *Population Bulletin* 43(3):5+.

St. George, Donna. 2007. "Survey: Children no longer factor in good marriage." *Washington Post* (July 1) www.freenewmexican.com/news/64101.html

Statistics Bureau, Ministry of Internal Affairs and Communications (Japan). 2005. *Japan in Figures*. www.stat.go.jp/english/data.

————. 2007. *Japan in Figures*. www.stat.go.jp/english/data

Stub, Holger. 1982. *The Social Consequences of Long Life*. Springfield, IL: Thomas.

Takahara, Kanako. 2000. "'New Breed' of Woman Emerges in Japan." *Japan Times* (August 7). www.japantimes.co.jp.

Takahashi, Junko. 1999. "Century of Change: Marriage Sheds Its Traditional Shackles." *Japan Times* (December 13). www.japantimes.co.jp.

Thomas Jefferson Foundation. 2000. "Report of the Research Committee on Thomas Jefferson and Sally Hemings." www.monticello.org.

United Nations Statistics Division. 2004. Demographic Yearbook 2004. New York: United Nations.

U.S. Bureau of Labor Statistics. 2006a. "20 Leading Occupations of Employed Women." www.dol.gov/wb/factsheets/20lead2006.htm.

————. 2006b. "American Time Use Survey—2005 Results Announced." www.bls.gov/tus.

U.S. Census Bureau. 2004a. "Population Pyramid for Japan." www.census.gov.

————. 2004b. "Population Pyramid for the United States." www.census.gov.

————. 2005a. "65+ in the United States." *Current Population Reports*. www.census.gov/prod/2006pubs/p23-209.pdf.

————. 2005b. Family Households. www.census.gov.

————. 2007. "Table F1. Family Households/1, by Type, Age of Own Children, Age of Family Members, and Age, Race and Hispanic Origin/2 of Householder, 2006." www.census.gov/population/socdemo/hh-fam/cps2006/tabF1-all.xls.

U.S. Central Intelligence Agency. 2007. "Japan." *The World Factbook*. https://www.cia.gov/cia/publications/factbook/.

U.S. Department of Agriculture. 2006. *Expenditures on Children by Families*. www.cnpp.usda.gov/Publications/CRC/crc2006.pdf.

U.S. Department of Health and Human Services. 2004. "Child Maltreatment." www.acf.hhs.gov/programs/cb/pubs/cm04/table3_10.htm.

————. 2006. *Health, United States, 2006*. www.cdc.gov/nchs/data/hus/hus06.pdf.

U.S. Department of Labor. 2007. "Highlights of Women's Earnings." http://www.bls.gov/cps.

Vermont Civil Unions Law. 2000. www.leg.state.vt.us/docs/2000.

Waters, Mary C. 1990. *Ethnic Options: Choosing Identities in America*. Berkeley: University of California Press.

Wijers-Hasegawa, Yumi. 2003. "Group to Push for Greater Gender Equality in Japan." *Japan Times* (July 1). www.japantimes.co.jp.

World Congress of Families. 2001. "The Family and Society." www.worldcongress.org.

World Economic Forum 2006. *The Global Gender Gap Report*. www.weforum.org/pdf/gendergap/report2006.pdf.

Yamada, Masahiro. 2000. "The Growing Crop of Spoiled Singles." *Japan Echo* 27(3). www.japanecho.com.

Yamamoto, Noriko, and Margaret I. Wallhagen. 1997. "The Continuation of Family Caregiving in Japan." *Journal of Health and Social Behavior* 38(June):164–176.

Zarit, Steven H., Pamela A. Todd, and Judy M. Zarit. 1986. "Subjective Burden of Husbands and Wives as Caregivers: A Longitudinal Study." *Gerontologist* 26:260–266.

Zelditch, Morris. 1964. "Family, Marriage, and Kinship." Pages 680–733 in *Handbook of Modern Sociology*, ed. by Robert E. L. Faris. Chicago: Rand McNally.

Zielenziger, Michael. 2002. "Young Japanese Prefer 'Parasite Single' Life to Wedding Poverty." Knight-Ridder/Tribune News Service (December 19).

Chapter 13

Ansalone, George. 2004. "Achieving Equity and Excellence in Education: Implications for Educational Policy." *Review of Business* 25 (2):37–43.

Arenson, Karen W. 2006. "Can't Complete High School? Just Go Right Along to College." *New York Times* (May 30): A1.

Baum, S., and O'Malley, M. 2003. "College on Credit: How Borrowers Perceive Their Education Debt." Results of the 2002 National Student Loan Survey, Final Report. Braintree, MA: Nellie Mae Corporation.

Binder, Ramona. 2006. "One Country, One Language." *World Press* (January 5). www.worldpress.org.

Bloom, Benjamin S. 1981. *All Our Children Learning: A Primer for Parents, Teachers and Other Educators*. New York: McGraw-Hill.

Bologna Declaration. 1999. "Joint Declaration of the European Ministries of Education" (June 19). www.bolognaberlin/2003.de/bologna_declaration.pdf.

Celis, William, III. 1992. "A Texas-Size Battle to Teach Rich and Poor Alike." *New York Times* (February 12):B6.

————. 1993. "Study Finds Rising Concentration of Black and Hispanic Students." *New York Times* (December 14):A1+.

Chace, William M. 2006. "A Little Learning Is an Expensive Thing." *New York Times* (September 5):A23.

Civil Rights Project (Harvard University). 2001. "Resegregation in American Schools." *Civil Rights Alert*. www.law.harvard.edu/civilrights.

Coleman, James S. 1960. "The Adolescent Subculture and Academic Achievement." *American Journal of Sociology* 65:337–347.

————. 1966. *Equality of Educational Opportunity*. Washington, DC: U.S. Government Printing Office.

————. 1977. "Choice in American Education." Pages 1–12 in *Parents, Teachers, and Children: Prospects for Choice in American Education*. San Francisco: Institute for Contemporary Studies.

Coleman, James S., John W. C. Johnstone, and Kurt Jonassohn. 1961. *The Adolescent Society*. New York: Free Press.

Combe, George. [1839] 1974. *Notes on the United States of North America during a Phrenological Visit in 1838–40*. New York: Arno Press.

Corporation for National and Community Service. 2007. "2006 President's Higher Education Community Service Honor

Roll." http://www.learnandserve.gov/about/programs/higher_ed_honorroll_2006.asp.

De Lange, Jan. 2006. "Mathematical Literacy for Living, from OECD-PISA Perspective." Freudenthal Institute, Utrecht University—the Netherlands. www.criced.tsukuba.ac.jp/math/sympo_2006/lange.pdf.

Dillon, Sam. 2004. "U.S. Slips in Status as Hub of Higher Education." *New York Times* (December 21):A1.

Dobelle, Evan. 2001. Quoted in "Agents of Change." *Matrix: The Magazine for Leaders in Education* (April). www.findarticles.com.

Durkheim, Émile. 1961. "On the Learning of Discipline." Pages 860–865 in *Theories of Society: Foundations of Modern Sociological Theory*, vol. 2, ed. T. Parsons, E. Shils, K. D. Naegele, and J. R. Pitts. New York: Free Press.

———. 1968. *Education and Sociology*, trans. S. D. Fox. New York: Free Press.

Education International and European Trade Union Committee for Education. 2007. "Study on Stress." www.ei-ie.org.

Eurobarometer. 2005. *Europeans and Languages*. European Commission.

Eurydice. 2001. The Information Network on Education in Europe. http://www.eurydice.org/portal/page/portal/eurydice.

Foster, Jack D. 1991. "The Role of Accountability in Kentucky's Education Reform Act of 1990." *Education Leadership* 34–36.

Friedman, Thomas L. 2006. "Worried about India's and China's Booms? So Are They." *New York Times* (March 24):A21.

Gates, Bill. 2006. Newshour Interview: "Bill and Melinda Gates Reflect on Need for Global Philanthropy." www.pbs.org/newshour/bb/social_issues/july-dec06/gates_12-20.html.

Hallinan, M. T. 1988. "Equality of Educational Opportunity." Pages 249–268 in *Annual Review of Sociology*, vol. 14, ed. W. R. Scott and J. Blake. Palo Alto, CA: Annual Reviews.

———. 1996. "Track Mobility in Secondary School." *Social Forces* 74(3):983.

Hamilton, Thomas. 1883. *Men and Manners in America*. London: Blackwood and Cadell.

Henry, Jules. 1965. *Culture Against Man*. New York: Random House.

Institute of International Education. 2006. *Open Doors 2006: Report on International Educational Exchange*. opendoors.iienetwork.org.

International Association for the Evaluation of Educational Assessment. 2003.

"Trends in Mathematics and Science Study." www.iea.nl/iea.

Kentucky Governor's Education Summit. 2002. *The First Twelve Years—Curriculum, Governance, Finance* (July 30). www.osbd.state.ky.us.

Lee, Jennifer Joan. 2004. "European Universities Unite to Welcome Super Scholars." *International Herald Tribune* (February 17). www.iht.com.

Lewin, Tamar. 2005. "Many Going to College Aren't Ready, Report Finds." *New York Times* (August 17): A13.

Lightfoot, Sara Lawrence. 1988. *Bill Moyers' World of Ideas* (transcript). New York: Public Affairs Television.

Lyall, Sarah. 2003. "Europe Weighs the Unthinkable: High College Fees." *New York Times* (December 25). www.nyt.com.

Mariani, Matthew. 1999. "High-Earning Workers Who Don't Have a Bachelor's Degree." *Occupational Outlook Quarterly* (Fall):9–15.

Merton, Robert K. 1957. *Social Theory and Social Structure*. Glencoe, IL: Free Press.

National Center for Education Statistics. 2004a. "Trends in English and Foreign Language Coursetaking." *The Condition of Education*. nces.ed.gov.

———. 2004b. "Changes in Public Schools Revenue Sources." *The Condition of Education*. nces.ed.gov.

———. 2004c. "Societal Support for Learning: International Comparisons of Expenditures for Education." *The Condition of Education*. nces.ed.gov.

———. 2004d "Contexts of Postsecondary Education: Remedial Course Taking." nces.ed.gov.

National Commission on Excellence in Education. 1983. *A Nation at Risk: The Imperative for Educational Reform*. Washington, DC: U.S. Government Printing Office. http://www.ed.gov/pubs/NatAtRisk/title.html

National Education Association. 2004. "Fast Facts." www.nea.org/edstats.

New York Times. 2003. "Private Colleges: The High End" (November 9). www.nyt.com.

Oakes, Jeannie. 1985. *Keeping Track: How Schools Structure Inequality*. Binghamton, NY: Vail-Ballou.

———. 1986a. "Keeping Track. Part 1: The Policy and Practice of Curriculum Inequality." *Phi Delta Kappan* 67 (September):12–17.

———. 1986b. "Keeping Track. Part 2: Curriculum Inequality and School Reform." *Phi Delta Kappan* 67 (October): 148–154.

Organisation for Economic Co-operation and Development (OECD). 2003. *Education at a Glance*. www.oecd.com

———. 2006. "OECD Education Systems Leave Many Immigrant Children Floundering, Report Shows."

Owen, Richard. 2000. "Voices: The Principled Principal." *Outlook*. www.ucop.edu/outreach.

Phelan, Patricia, and Ann Locke Davidson. 1994. "Looking Across Borders: Students' Investigations of Family, Peer, and School Worlds as Cultural Therapy." Pages 35–59 in *Pathways to Cultural Awareness: Cultural Therapy with Teachers and Students*, ed. George and Louise Spindler. Thousand Oaks, CA: Corwin.

Phelan, Patricia, Ann Locke Davidson, and Hanh Cao Yu. 1991. "Students' Multiple Worlds: Negotiating the Boundaries of Family, Peer, and School Cultures." *Anthropology and Education Quarterly* 22(3):224–250.

———. 1993. "Students' Multiple Worlds: Navigating the Borders of Family, Peer, and School Cultures." Pages 89–107 in *Renegotiating Cultural Diversity in American Schools*, ed. Patricia Phelan and Ann Locke Davidson. New York: Teachers College Press.

Potter, J. Hasloch, and A. E. W. Sheard. 1918. *Catechizings for the Church and Sunday Schools*, 2nd series. London: Skeffington.

Purves, Alan C. 1974. "Divergent Views on the Schools: Some Optimism Justified." *New York Times* (January 16):C74.

Ramirez, Francisco, and John W. Meyer. 1980. "Comparative Education: The Social Construction of the Modern World System." Pages 369–399 in *Annual Review of Sociology*, vol. 6, ed. A. Inkeles, N. J. Smelser, and R. H. Turner. Palo Alto, CA: Annual Reviews.

Resnick, Daniel P. 1990. "Historical Perspectives on Literacy and Schooling." *Daedalus* 119(2):15–32.

Reuters. 2006. "U.S. Teenagers Setting Overly Ambitious Goals." (August 30). www.reuters.com.

Rosenthal, Robert, and Lenore Jacobson. 1968. *Pygmalion in the Classroom*. New York: Holt, Rinehart & Winston.

Schemo, Diana Jean. 2001. "U.S. Schools Turn More Segregated, a Study Finds." *New York Times* (July 20):A12.

———. 2006. "It Takes More Than Schools to Close the Achievement Gap." *New York Times* (August 9):A15.

Schmidt. 2004.

Sciolino, Elaine. 2006. "Higher Learning in France Clings to Its Old Ways." *New York Times* (May 12):A1.

Sowell, Thomas. 1981. *Ethnic America: A History*. New York: Basic Books.

Thomas, William I., and Dorothy Swain Thomas. [1928] 1970. *The Child in America*. New York: Johnson Reprint.

Tyack, David. 1996. "Forming the National Character." *Harvard Educational Review* 36:29–41.

U.S. Bureau of Labor Statistics. 2005. "College Enrollment and Work Activity of 2004 High School Graduates." www.bls.gov/news.release/hsgec.nro.htm.

U.S. Central Intelligence Agency. 2005. "European Union." *The World Factbook*. https://www.cia.gov/cia/publications/factbook/.

U.S. Department of Education. 1993. *Adult Literacy in America: A First Look at the Results of the National Literacy Survey*. Washington, DC: U.S. Government Printing Office.

———. 2006. "Expanding the Advanced Placement Incentive Program." www.ed.gov/about/inits/ed/competitiveness/expanding-apip.html.

———. 2007. *Digest of Education Statistics*. nces.ed.gov/programs/digest/.

Wells, Amy Stuart, and Jeannie Oakes. 1996. "Potential Pitfalls of Systematic Reform: Early Lessons from Research on Detracking." *Sociology of Education* (extra issue).

Chapter 14

Abercrombie, Nicholas, and Bryan S. Turner. 1978. "The Dominant Ideology Thesis." *British Journal of Sociology* 29(2):149–170.

Alston, William P. 1972. "Religion." *The Encyclopedia of Philosophy*, vol. 7, ed. P. Edwards. New York: Macmillan.

Amnesty International. 1996. "Afghanistan: Grave Abuses in the Name of Religion" (November 18). http://web.amnesty.org/library/Index/ENGASA110121996?open&of=ENG-370.

Aron, R. 1969. Quoted in *The Sociology of Max Weber* by Julien Freund. New York: Random House.

Bellah, Robert N. 1992. *The Broken Covenant: American Civil Religion in Time of Trial*. Chicago: University of Chicago Press.

Bin Laden, Osama. 2001a. Quoted in "Britain's Case Against bin Laden: Responsibilities for Terrorist Atrocities in the United States, 11 September 2001." *Online Newshour* (October 4). www.pbs.org/newshour.

———. 2001b. "Bin Laden's Statement: 'The Sword Fell.'" *New York Times* (October 7):B1

Brzezinski, Zbigniew. 2002. "Confronting Anti-American Grievances." *New York Times* (September 1). www.nyt.com.

Bush, George H. W. 1991. Address before a Joint Session of the Congress on the State of the Union (January 29).

Bush, George W. 2001a. "Presidential Address to the Nation: The Treaty Room" (October 7). www.whitehouse.gov.

———. 2001b. "Address to a Joint Session of Congress and the American People." (September 20). www.whitehouse.gov.

———. 2003a. "President Bush Announces Major Combat Operations in Iraq Have Ended." Remarks by the president from the USS *Abraham Lincoln* at sea off the coast of San Diego, California (May 1). www.whitehouse.gov.

———. 2003b. "President Bush at Whitehall Palace in London as Recorded by Federal News Service, Inc." *New York Times* (November 20):A12.

Caplan, Lionel. 1987. "Introduction: Popular Conceptions of Fundamentalism." Pages 1–24 in *Studies in Religious Fundamentalism*, ed. L. Caplan. Albany: State University of New York Press.

CBS News Poll. 2006. (April 6–9). www.pollingreport.com/religion.htm.

Chance, James. 2002. "Tomorrow the World." *New York Review of Books* (November 21):33–35.

Cole, Roberta L. 2002. "Manifest Destiny Adapted for 1990s War Discourse: Mission and Destiny Intertwined." *Sociology of Religion* 63(4).

Davis, Derek. 2002. Panelist Comments for "God Bless America: Reflections on Civil Religion after September 11." The Pew Forum on Religion and Public Life. http://pewforum.org/events/index.php?EventID=R22.

Durkheim, Émile. [1915] 1964. *The Elementary Forms of the Religious Life*, 5th ed., trans. J. W. Swain. New York: Macmillan.

———. 1951. *Suicide: A Study in Sociology*, trans. J. A. Spaulding and G. Simpson. New York: Free Press.

Ebersole, Luke. 1967. "Sacred." *A Dictionary of the Social Sciences*, ed. J. Gould and W. L. Kolb. New York: UNESCO.

Echo-Hawk, Walter. 1979. "Statement of Walter Echo-Hawk before the United States Commission on Civil Rights." Pages 280–287 in *Religious Discrimination: A Neglected Issue by U.S. Commission on Civil Rights*. Washington, DC: U.S. Government Printing Office.

Esposito, John L. 1986. "Islam in the Politics of the Middle East." *Current History* (February):53–57, 81.

———. 1992. *The Islamic Threat: Myth or Reality?* New York: Oxford University Press.

Feldman, Noah. 2003. "A New Democracy, Enshrined in Faith" (November 13). www.nyt.com.

Gall, Carlotta. 2003. "Afghan Motherhood in a Fight for Survival." *New York Times* (May 23):2.

Gallup Poll. 2006. "Religion." www.galluppoll.com/content/?ci=1690&pg=1.

Gibson, David. 2004. "What Did Jesus Really Look Like?" *New York Times* (February 21):A17.

Grimes, William. 1994. "The Man Who Rendered Jesus for the Age of Duplication." *New York Times* (October 12):B1.

Hammond, Phillip E. 1976. "The Sociology of American Civil Religion: A Bibliographic Essay." *Sociological Analysis* 37(2):169–182.

Hedges, Chris. 2002. *War Is the Force That Gives Us Meaning.* New York: Anchor Books/Doubleday.

Halloran, Richard. 2004. "Afghanistan Offers a Glimmer of Hope for U.S. Foreign Policy." *Taipei Times* (August 3):9. www.taipeitimes.com.

Hourani, Albert. 1991. *A History of the Arab Peoples.* Cambridge, MA: Belknap.

Huntington, Samuel. 2001. Quoted in "How a Holy War Against the Soviets Turned on the U.S." *Pittsburgh Post-Gazette* (September 23):A12.

Ibrahim, Youssef M. 1991. "In Kuwait, Ramadan Has a Bitter Taste." *New York Times* (March 19):A1, A7.

Judah, Tim. 2001. "With the Northern Alliance." *New York Review of Books* (November 15):29–33.

Kristof, Nicholas D. 2002. "Saudis in Bikinis." *New York Times* (October 25):A35.

Kurian, George Thomas. 1992. *Encyclopedia of the Third World.* New York: Facts on File.

Lechner, Frank J. 1989. "Fundamentalism Revisited." *Society* (January/February): 51–59.

Lichtblau, Eric. 2003. "Wanted: A Short List of 100,000 Terrorists." *New York Times* (September 21):5WK.

Lincoln, C. Eric, and Lawrence H. Mamiya. 1990. *The Black Church in the African American Experience.* Durham, NC: Duke University Press.

Lyons, Linda. 2005. "Religiosity Measure Shows Stalled Recovery." Gallup Organization. www.gallup.com.

Mahjubah: The Magazine for Moslem Women. 1984 (July).

Mamdani, Mahmood. 2004. *Good Muslim, Bad Muslim: America, the Cold War, and the Roots of Terror.* New York: Pantheon.

McNamara, Robert S. 1989. *Out of the Cold: New Thinking for American Foreign and Defense Policy in the 21st Century.* New York: Simon & Schuster.

Mead, George Herbert. 1940. *Mind, Self and Society,* 3rd ed. Chicago: University of Chicago Press.

National Public Radio. 1984a. "Black Islam." *The World of Islam* (tape). Washington, DC: NPR.

———. 1984b. "Decay or Rebirth: The Plight of Islamic Art." *The World of Islam* (tape). Washington, DC: NPR.

New York Public Library. 1993. The New York Public Library Desk Reference (2nd edition). The New York Public Library and Stonesong Press.

New York Times. 2004. "An Open Letter to President Bush Concerning America's Clean Air." (April 28):A6.

Nottingham, Elizabeth K. 1971. *Religion: A Sociological View.* New York: Random House.

Online NewsHour. 2001. "U.S., Allies Attack Afghanistan" (October 7). www.pbs.org/newshour.

Pelikan, Jaroslav, and Clifton Fadiman, eds. 1990. *The World Treasury of Modern Religious Thought.* Boston: Little, Brown.

Pickering, W. S. F. 1984. *Durkheim's Sociology of Religion.* London: Routledge & Kegan Paul.

Pipes, Daniel. 2003. *Militant Islam Reaches America.* New York: Norton.

Rahman, Saif. 2001. Quoted in "Fighting Fear." *Online Newshour* (October 30). www.pbs.org/newshour.

Rashid, Ahmed. 2001. "How a Holy War Against the Soviets Turned on the U.S." *Pittsburgh Post-Gazette* (September 23): A12.

Robertson, Roland. 1987. "Economics and Religion." *The Encyclopedia of Religion.* New York: Macmillan.

Rubin, Barnett R. 1996. Quoted in *A Nation in Arms* by Karl E. Meyer. *New York Times Book Review* (August 11):21.

Scheiber, Noam. 2003. "5,000 Al Qaeda Operatives in the U.S.: Is There Really an Army of Terrorists in Our Midst?" *New York Times Magazine* (February 16):12.

Smart, Ninian. 1976. *The Religious Experience of Mankind.* New York: Scribner's.

Stark, Rodney, and William S. Bainbridge. 1985. *The Future of Religion: Secularization, Revival and Cult Formation.* Berkeley: University of California Press.

Tocqueville, Alexis de. 1835. *Democracy in America.* xroads. virginia.edu/~HYPER/DETOC/toc_indx.html.

Turner, Bryan S. 1974. *Weber and Islam: A Critical Study.* Boston: Routledge & Kegan Paul.

Turner, Jonathan H. 1978. *Sociology: Studying the Human System.* Santa Monica, CA: Goodyear.

United Nations High Commissioner for Refugees. 2005. Statistical Yearbook Country Data Sheet: Afghanistan. www.unhcr.org/statistics/STATISTICS/464183592.pdf.

U.S. Central Intelligence Agency. 2001. "Afghanistan." *The World Factbook.* www.odci.gov/cia/publications.

———. 2007. "Afghanistan." *The World Factbook.* www.odci.gov/cia/publications.

U.S. Commission on Civil Rights. 1979. *Religious Discrimination: A Neglected Issue.* Washington, DC: U.S. Government Printing Office.

U.S. Committee for Refugees and Immigrants. 2004. *World Refugee Survey Country Report: Afghanistan.* www.refugees.org/countryreports.

Weber, Max. 1922. *The Sociology of Religion,* trans. E. Fischoff. Boston: Beacon.

———. 1958. *The Protestant Ethic and the Spirit of Capitalism,* 5th ed., trans. T. Parsons. New York: Scribner's.

White House press release. 1995. "Remarks by the President on Religious Liberty in America" (July 12). www.whitehouse.gov/WH/EOP/OP/html/book3-plain.html.

Wilmore, Gayraud S. 1972. *Black Religion and Black Radicalism.* Garden City, NY: Doubleday.

Winseman, Albert L. 2005. "Who Has Been Born Again?" Gallup Organization. www.gallup.com.

Yinger, J. Milton. 1971. *The Scientific Study of Religion.* New York: Macmillan.

Zangwill, O. L. 1987. "Isolation Experiments." *The Oxford Companion to the Mind*, ed. R. L. Gregory. New York: Oxford University Press.

Chapter 15

Association of Americans Resident Overseas. 2007. www.aaro.org.

Berelson, Bernard. 1978. "Prospects and Programs for Fertility Reduction: What? Where?" *Population and Development Review* 4:579–616.

Bhopal.net. 2004. "The $195 Million Discrepancy—Where's the Money Gone?" (March 12).

Bowen, Huw V. 2000. "400 Years of the East India Company." *History Today* (July). www.findarticles.com.

Brinkhoff, Thomas. 2001. *The Principal Agglomerations of the World*. www.citypopulation.de.

Brockerhoff, Martin, and Ellen Brennan. 1998. "The Poverty of Cities in Developing Regions." *Population and Development Review* 24(1):75–115.

Brown, Lester R. 1987. "Analyzing the Demographic Trap." *State of the World 1987: A Worldwatch Institute Report on Progress Toward a Sustainable Society*. New York: Norton.

Census of India. 2001. Office of the Registrar General. www.censusindia.net.

Chailiand, Gerard, and Jean-Pierre Rageau. 1995. *The Penguin Atlas of Diasporas*. New York: Viking.

Christian Aid. 2005. "Biotechnology and Genetically Modified Organisms." www.christian- aid.org.uk/indepth/000/ibiot/biotech.htm.

Demangeon, Albert. 1925. *The British Empire: A Study of Colonial Geography*. Trans. E. F. Row. New York: Harcourt, Brace.

Derdak, Thomas. 1988. *International Directory of Company Histories*, vol. 1. Chicago: St. James.

Dugger, Celia W. 2007. "UN Predicts Urban Population Explosion." *New York Times* (June 28): A6.

Economist. 2003. "Missing Sisters." 367 (April 19):36.

Everest, Larry. 1986. *Behind the Poison Cloud: Union Carbide's Bhopal Massacre*. Chicago: Banner.

Indo-Asian News Service. 2005. "Africa Looks to Indian Farmers for Green Revolution." www.newkerala.com/news-daily/news/.

Kapoor, Indira. 2000. "The Billionth Baby: A Time to Reflect." International Planned Parenthood Federation (May 12). www.ippf.org.

Kumar, Sanjay. 1999. "India's Burden of Hunger Predicted to Worsen." *Lancet* 354(October 30):1535.

Lancet. 2002. Editorial: "Abstinence, Monogamy, and Sex." 360(9327):97.

Library of Congress. 2001a. "India: The Village Community." lcweb2.loc.gov.

———. 2001b. "India: Village Unity and Divisiveness." lcweb2.loc.gov.

Light, Ivan. 1983. *Cities in World Perspective*. New York: Macmillan.

Malthus, Thomas R. [1798] 1965. *First Essay on Population*. New York: Kelley.

Marsh, Bill. 2005. "The Vulnerable Become More Vulnerable." *New York Times* (January 2): WK5.

Massing, Michael. 2003. "Does Democracy Avert Famine?" *New York Times* (March 1):A19.

Misra, Neelesh. 2000. "Billionth Baby." *Guardian* (May 12). www.guardianunlimited.co.uk.

Mongabay.com. 2004. "All Cities with Populations Exceeding 100,000."

———. 2007. "Largest Cities in the World." www.mongabay.com/igapo/cities.htm.

New York Times. 2003. "For Sterilization, Target Is Women." (November 7):A1.

Olshansky, S. Jay, and A. Brian Ault. 1986. "The Fourth Stage of the Epidemiologic Transition: The Age of Delayed Degenerative Diseases." *Milbank Quarterly* 64(3):355–391.

Omran, Abdel R. 1971. "The Epidemiologic Transition: A Theory of the Epidemiology of Population Change." *Milbank Quarterly* 49(4):509–538.

Population Reference Bureau. 2001. "Patterns of World Urbanization." www.prb.org.

Queenan, John T. 1999. "Japan Yields to Needs of Women." *Contemporary OB/GYN* 44(7):8.

Rai, Saritha. 2003. "Indian Nurses Sought to Staff U.S. Hospitals." *New York Times* (February 10):A3.

Rhode, David. 2003. "India Steps Up Effort to Halt Abortions of Female Fetuses." *New York Times* (October 26):3.

Samuel, John. 1997. "World Population and Development: Retrospect and Prospects." *Development Express*. wwacdi-cida.gc.ca/express.

Schmitt, Eric. 2001. "Cities and Their Suburbs Are Seen Growing as Units." *New York Times* (July 9): www.nyt.com.

Schumacher, E. F. 1985. Quoted in "Technology Out of Control," by Robert Engler. *Nation* 240(April 27):490.

Sengupta, Somini. 2007. "India Prosperity Creates Paradox; Many Children Are Fat, Even More Are Famished." *New York Times* (December 2006):8.

Sharma, O. P. 2000. "Bhopal Gas Disaster: 16 Long Years of Suffering." *Tribune*, Chandigarh, India (December 10).

Stephenson, Rob, Zoe Matthews, J. W. McDonald. 2003. "The Impact of Rural–Urban Migration on Under-Two Mortality in India." *Journal of Biosocial Science*, 35: 15–31.

Stub, Holger R. 1982. *The Social Consequences of Long Life*. Springfield, IL: Thomas.

Times of India. 2005. "New GM Rice Will Prevent Blindness." (March 29): timesofindia.com/articleshow/1064601.cms.

UN World Food Programme. 2001. www.wfp.org.

U.S. Census Bureau. 2000a. "Percentage of Persons Born in State of Residence." factfinder.census.gov.

———. 2000b. *Current Population Survey* (March).

———. 2005. American Housing Survey. www.census.gov/hhes/www/housing/ahs/ahs05/tab1d1.html.

———. 2007. International Data Base: Population Pyramids. http://www.census.gov/ipc/www/idb/.

U.S. Central Intelligence Agency. 2007. "India." *The World Factbook*. www.cia.gov/library/publications/the-world-factbook/index.html.

U.S. Department of State. 1999. "American Citizens Living Abroad by Country." www.aca.ch/amabroad.pdf.

———. 2002.

———. 2007.

Waldman, Amy. 2002a. "Poor in India Starve as Surplus Wheat Rots." *New York Times* (December 2):A8.

———. 2002b. "Bhopal Seethes, Pained and Poor 18 Years Later." *New York Times* (September 21):A15.

———. 2003a. "India Harvests Fruits of Diaspora." *New York Times* (January 12):A7.

———. 2003b. "Gulf Bounty Is Drying Up in Southern India." *New York Times* (February 24):A11.

Chapter 16

Airports Council International. 2007. "Airports Welcome Record 4.4 Billion Passengers in 2006." (July 18) www.airports.org.

Airport Transport Association. 2001. "The Airline Handbook—Online Version." www.airlines.org.

Archer, Clive, and David Scrivener. 1983. "Frozen Frontiers and Resource Wrangles: Conflict and Cooperation in Northern Waters." *International Affairs* 59(1):59–76.

Arctic Council. 2007. http://arctic-council.org/.

Ash, Timothy Garton. 1989. *The Uses of Adversity: Essays on the Fate of Central Europe.* New York: Random House.

Associated Press. 2005. "Inuit Link Dramatic Lifestyle Changes to Global Warming." (December 4). www.ap.org

———. 2007. "Russians to Claim the Arctic Ocean for Moscow." (July 29):11A.

BBC News. 2007. "Poll Trekker Survives Ice Plunge." (May 30). news.bbc.co.uk/hi/uk_news/England/6706043.htm.

Benford, Robert D. 1992. "Social Movements." *Encyclopedia of Sociology.* Ed. E. F. Borgatta and M. L. Borgatta. New York: McMillan.

Berners-Lee, Wright Tim. 1996. Quoted in "Seek and You Shall Find (Maybe)" by Steve G. Steinberg. *Wired* (May):111.

Brownwell, Ginanne. 2004. " 'We Won't Sink With Our Ice.' " *Newsweek*
(Feb 3) www.msnbc.msn.com/id/6908719/site/newsweek.

Business Wire. 2000. "Iceberg Corporation of America Forms Joint Venture to Export Greenland Water." (August 24). www.businesswire.com.

CNW Telbec. 2005. "Government of Greenland and Alcoa to Study Feasibility of Constructing Hydro-Powered Aluminum Smelter." (May 23) www.newswire.ca/en/releases/archive/May2007/23/c3976.html

Coser, Lewis A. 1973. "Social Conflict and the Theory of Social Change." Pages 114–122 in *Social Change: Sources, Patterns, and Consequences,* ed. E. Etzioni-Halevy and A. Etzioni. New York: Basic Books.

Council on Interchurch Relations. 2007. "The Evangelical Lutheran Church in Denmark." www.interchurch.dk/LutheranChurch/10.htm.

Dahrendorf, Ralf. 1973. "Toward a Theory of Social Conflict." Pages 100–113 in *Social Change: Sources, Patterns, and Consequences,* 2nd ed., ed. by E. Etzioni-Halevy and A. Etzioni. New York: Basic Books.

CNW Telbec. 2005. "Government of Greenland and Alcoa to Study Feasibility of Constructing Hydro-Powered Aluminum Smelter." (May 23). www.cnw.ca/fr/releases.

Economist.com. 2007. "Global Warming's Boom Town." (May 24). www.economist.com/business.

Evening Standard. 2007. "Eskimo Leader Slams Stansted Expansion Plans." (May 31). www.thisislondon.co.uk/news.

Fisk, Umbra. 2007. "One Word: On Oil and Plastic." *Ask Umbra.* www.grist.org/advice/ask/2007/03/14/plastics.

Freund, Julien. 1968. The Sociology of Max Weber. New York: Random House.

Global Water. 2006. www.globalwater.org/index.htm.

Goethe, Johann Wolfgang von. 2004. Quoted in "Biographies: Nicolas Copernicus (1473–1543)" by P. Landry. www.blupete.com/Literature/Biographies/Science/Copernicus.htm.

Goldman, Russell. 2007. "Environmentalists Classified as Terrorists Get Stiff Sentences." ABC News (May 25). abcnews.go.com.

Goode, Erich. 1992. *Collective Behavior.* New York: Harcourt Brace Jovanovich.

Gregory, Paul M. 1947. "A Theory of Purposeful Obsolescence." *Southern Economic Journal* 14(1):24–45.

Greenland Home Rule Government. 2007. www.greenlandexpo.com.

Greenland Tourism and Business Council. 2007. Greenland.com.

Grodzins, Morton. 1958. *The Metropolitan Area as a Racial Problem.* Pittsburgh: University of Pittsburgh Press.

Grossman, Lawrence K. 1999. "An Outlook of Internet-Phobia." *Columbia Journalism Review* 38(3):15+.

Haines, Lester. 2007. "Greenland Zinc Mine Warms to Retreating Ice." *Register* (May 23). www.theregister.co.uk.

Halberstam, David. 1986. *The Reckoning.* New York: Morrow.

Held, D., A. McGrew, D. Goldblatt, and I. Perraton. 1999. *Global Transformations: Politics, Economics, and Culture.* Stanford, CA: Stanford University Press.

Huntington, Henry, and Shari Fox. 2004. *The Arctic Climate Impact Assessment.* www.acia.uaf.edu/.

International Forum on Globalization. 2007. www.ifg.org.

International Organization of Motor Vehicle Manufacturers, 2007. *The World's Automotive Industry.* www.oica.net/htdocs/statistics/statistics.htm.

International Trade Administration. 2007. "Latest Statistics/Outreach." Office of Travel and Tourism Industries. tinet.ita.doc.gov/.

International Whaling Commission. 2007. IWC Information. www.iwcoffice.org/commission/iwcmain.htm.

Jensen Arctic Museum. 2007. www.oregonlink.com.

Kazinform. 2007. "Deadlock at Greenland Whale Plan." (May 30) www.inform.kz.

Kirschbaum, Erik. 2007. "Pelosi Urges U.S. to Come in from the Cold." *Business Day* (May 30) www.businessday.co.za.

Klapp, Orrin E. 1986. *Overload and Boredom: Essays on the Quality of Life in an Information Society.* New York: Greenwood.

Kuhn, Thomas S. 1975. *The Structure of Scientific Revolutions.* Chicago: University of Chicago Press.

Lee, Jennifer. 2003. "Critical Mass: How Protesters Mobilized so Many and so Nimbly." *New York Times* (February 23). www.nyt.com.

Lovgren, Stefen. 2004. "Greenland Melt May Swamp LA, Other Cities, Study Says." *National Geographic News* (April 8). news.nationalgeographic.com/news.

Mahr, Krista. 2004. "Greenland: Colin Powell's Glacier." *Frontline/World*. www.pbs.org/frontlineworld/elections/greenland.

Mandelbaum, Maurice H. 1977. *The Anatomy of Historical Knowledge*. Baltimore: Johns Hopkins University Press.

Martel, Leon. 1986. *Mastering Change: The Key to Business Success*. New York: Simon & Schuster.

Marx, Karl. [1881] 1965. "The Class Struggle." Pages 529–535 in Theories of Society, ed. by T. Parsons, E. Shils, K. D., Naegele, and J. R. Pitts. New York: Free Press.

Mikaelsen, Vittu Maaru. 2007. "Inuit Religion." www.eastgreenland.com.

Mills, C. Wright. 1956. *The Power Elite*. New York: Oxford University Press.

Mouland, Bill. 2007. "Global Warming Sees Polar Bear Stranded on Melting Ice." *Daily Mail* (May 28). www.dailymail.co.uk.

National Association of Homebuilders. 2007. www.nahb.org.

Nicklen, Paul. 2007. "Life at the Edge." *National Geographic* (June): 32–55.

Norman, Donald A. 1988. Quoted in "Management's High-Tech Challenge." *Editorial Research Report* (September 30):482–491.

Ogburn, William F. 1968. "Cultural Lag as Theory." Pages 86–95 in *Culture and Social Change*, 2nd ed., ed. by O. D. Duncan. Chicago: University of Chicago Press.

Online Newshour. 2007. "Pet Food Scare Raises Questions About Food Safety." (June 1). www.pbs.org/newshour/bb/science/jan-june07/foodfears_06-01.html.

Petition Project. 2007. www.oism.org/pproject/s33p357.htm.

PEW Center on Global Climate Change. 2007. http://www.pewclimate.org/about/.

Polar Bears International. 2007. The Polar Bear Population Project. polarbearsinternational.org/pbi-supported-research/population-project.

Reich, Jens. 1989. Quoted in "People of the Year." *Newsweek* (December 25):18–25.

Repair2000.com. 2007. "Average Life Span of Major Appliances." Repair2000.com/lifespan.html.

Ritzer, George. 1993. *The McDonaldization of Society*. Thousand Oaks, CA: Pine Forge Press.

Rosenthal, Elisabeth and Andrew C. Revkin. 2007. "Science Panel Says Global Warming Is 'Unequivocal.'" *New York Times* (February 3):A1.

Sample, Ian. 2005. "Warming Hits 'Tipping Point'". *Guardian* (Aug 11) www.guardian.co.uk/climatechange/story/.

Sierra Club. 2007. "Sprawl Overview." www.sierraclub.org/sprawl/overview/.

Theodorson, George A., and Achilles G. Theordorson. 1969. *A Modern Dictionary of Sociology*. New York: Harper.

UNESCO. 2003. *World Water*. www.unesco.org/water/wwap.

United Nations. 2005. *World Urbanization Prospects*. www.un.org/esa/population/publications/WUP2005/2005wup.htm.

———. 2007. *Climate Change 2007: The Physical Science Basis*. Intergovernmental Panel on Climate Change. www.ipcc.ch/.

UPS (United Parcel Service). 2007. "Fact Sheets." pressroom.ups.com.

U.S. Census Bureau. 2005. "Commuting Characteristics by Sex." factfinder.census.gov/.

———. 2007. IDB Summary Demographic Data. www.census.gov/ipc/www/idbsum.html.

U.S. Central Intelligence Agency. 2007. "Greenland." *The World Factbook*. www.cia.gov/library/publications/the-world-factbook/index.html.

U.S. Department of Energy. 2007. *Energy Use in the United States: 1635–2000*. www.eere.energy.gov.

Wallerstein, Immanuel. 1984. *The Politics of the World-Economy: The States, the Movements and the Civilizations*. New York: Cambridge University Press.

Watt-Cloutier, Sheila. 2005. Interview with Ginanne Brownell for "We Won't Sink with Our Ice." *Newsweek* (February 3). www.nsnbc.msn.com/id/6908719/site/newsweek.

Weber, Bob. 2007. "Canadian Inuit Ally with Tropical Islanders to Fight Climate Change." Canadian Press (May 29).

White, Leslie A. 1949. *The Science of Culture: A Study of Man and Civilization*. New York: Grove.

Whitty, Julia. 2006. "The 13th Tipping Point: 12 Global Disasters and 1 Powerful Antidote." *Mother Jones* (November–December).

Winter, Mark. 2007. "Chemistry: Periodic Table." www.webelements.com/webelements.

World Coal Institute. 2007. "History of Coal Use." www.worldcoal.org.

worldinternetstats.com. 2007. "World Internet Usage Statistics." www.internetworldstats.com/stats.htm.

World Resource Institute. 1996. "What Is An Urban Area?" pubs.wri.org/pubs_content_text.cfm?ContentID=929.

World Tourism Organization. 2007. "The Tourism Market Trends." www.world-tourism.org/newsroom/Releases/2005.

Yergin, Daniel. 1991. *The Prize: The Epic Quest for Oil, Money, and Power*. New York: Simon and Schuster.

Credits

Chapter 1. 2: Lisa Southwick; 6 top right: Lisa Southwick; 6 top left: Missy Gish; 6 bottom left: Lisa Southwick; 6 bottom right: © L. Lartigue/USAID; 7 top right: Lance Cpl. Deanne Travis/U.S. Marine Corps; 7 bottom right: Department of Defense photo by SSGT Lanie McNeal, USAF; 7 left: U.S. photo by Mass Communication Specialist 2nd Class Kimberly Williams/Released; 10: Library of Congress Prints and Photographs Division Washington, D.C. [LC-USZ62-63834]; 11: Library of Congress Prints and Photographs Division Washington, D.C. [LC-USZ6-396]; 13: The Library of Congress Prints and Photographs Division Washington, D.C.; 14: Library of Congress Prints and Photographs Division Washington, D.C. [LC-USZC4-591]; 16: Lance Cpl. Manuel F. Guerrero/United States Marine Corps; 17: The Library of Congress; 19 right: U.S. Navy photo; 20: Vitaliy Voznyak; 21 top: U.S. Customs and Border Protection, Department of Homeland Security; 21 bottom: U.S. Air Force photo by John Hughel Jr.; 23 top: Pfc. Jared Eastman/United States Army; 23 middle: Lisa Southwick; 23 bottom: Lance Cpl. Manuel F. Guerrero/United States Marine Corps; 24 top: Library of Congress Prints and Photographs Division Washington, D.C. [LC-USZ6-396]; 24 middle: Library of Congress Prints and Photographs Division Washington, D.C. [LC-USZ62-7208]; 24 bottom: Earth Sciences and Image Analysis Laboratory, NASA Johnson Space Center

Chapter 2. 26: U.S. Army photo by Sgt. Jim Greenhill; 29: DoD photo by MSG Robert W. Valenca, USAF; 30: DoD photo by Sgt. Jim Greenhill, USA; 31: DoD photo by MCS Seaman Orlando Ramos, USN; 32 top: Tomas Castelazo; 32 bottom: DoD photo by SFC Gordon Hyde, USA; 33: Steve Jurvetson, www.DFJ.com/J; 34 top: © FogStock LLC/Index Open; 34 center: Library of Congress Prints and Photographs Division; 34 bottom: National Archives and Records Administration; 36 top: James F. Hopgood; 36 bottom: DoD photo by TSGT Scott Ree, USAF; 37 top: DoD photo by Sgt. Jim Greenhill, USA; 37 bottom: Jonathan McIntosh; 40 top: U.S. Fish and Wildlife Service photo by Pedro Ramirez, Jr.; 40 bottom: DoD photo by PH3 Bernardo Fuller, USN; 42 top: DoD photo by MCSS Josue Leopoldo Escobosa, USN; 42 bottom: CBP photo by James R. Tourtellotte; 43: CBP photo by James R.

Tourtellotte; 45 top: DoD photo by PH3 Kristopher Wilson, USN; 45 bottom: DoD photo by PH1 (AW) M. Clayton Farrington; 47: CBP photo by Gerald L. Nino; 49: CBP photo by Gerald L. Nino; 52: DoD photo by TSGT Dave Ahlschwede, USAF; 53 top: DoD photo by PH2 M. C. Farrington, USA; 53 center: DoD photo by MSG Robert W. Valenca, USAF; 53 bottom: National Archives and Records Administration; 54 top: DoD photo by TSGT Scott Ree, USAF; 54 top center: U.S. Fish and Wildlife Service photo by Pedro Ramirez, Jr.; 54 bottom center: DoD photo by MCSS Josue Leopoldo Escobosa, USN; 54 bottom: CBP photo by James R. Tourtellotte; 55 top: DoD photo by PH3 Kristopher Wilson, USN; 55 center: CBP photo by Gerald L. Nino; 55 bottom: DoD photo by SSG Dan Heaton, USAF.

Chapter 3. 58: DoD photo by LCPL Adaecus G. Brooks, USMC; 60: Prince Brown, Jr.; 61: DoD photo by CPL Robert J. Thayer, USMC; 62: DoD photo by TSGT James Pearson; 64: © Michael Ponton/Morguefile; 65 left: Jeremy van Bedijk www.beeldenzeggenmeer.nl; 65 right: © Kim Bo-Kyung Kirby and Kevin Kirby; 66: © Lee Jae-Won/Reuters/Landov; 67 top left: National Archives and Records Administration; 67 bottom left: Ray Elfers; 67 bottom right: © Kim Bo-Kyung Kirby and Kevin Kirby; 68: Gregory A. Caldeira; 69: DPRK (North Korea) Archives; 73: © Kim Bo-Kyung Kirby and Kevin Kirby; 74: U. S. Naval Historical Center; 76: © Rod Aydelotte/Bloomberg News/Landov; 77: DoD photo by MSGT Rose Reynolds; 78: U.S. Navy photo by JOC Al Fontenot, CNFK public affairs; 79: DoD photo by Senior Airman Jeffrey Allen, USAF; 80 top: DoD photo by Cpl. Michael S. Cifuentes, USMC; 80 top center: DoD photo by JO3 Monica Miles, USN; 80 bottom center: National Archives and Records Administration; 80 bottom: Gregory A. Caldeira; 81 top: DoD photo by TSGT Curt Eddings; 81 top center: Dr. Terry Pence; 81 bottom center: © Index Open; 81 bottom: © Claro Cortes IV/Reuters/Landov; 82: U.S. Navy photo by JOC Al Fontenot, CNFK public affairs.

Chapter 4. 84: © Justin C. McIntosh; 86: Father Rob Waller; 87 top: Amy Burke; 87 bottom: DoD photo by MCSS Shannon K. Cassidy, USN; 88: © Keith Levit Photography/Index Open; 90: DoD photo by Sgt. Robert Holliway, USA; 92 top left: DoD photo by JO1 Mark D. Faram, USN; 92 top right: National Archives and Records Administration; 92 bottom left: © AP Photo; 92 bottom center: DoD photo by SGT Paul L. Anstine II, USMC; 92 bottom right: © IOM/USAID; 93 top: Nancy Hemminger; 93 bottom: National Park Service, U.S. Department of the Interior; 94: Christopher Brown; 96: Christopher Brown; 97: © Bill Romerhaus/Index Open; 98: © Chris Lowe/Index Open; 99: © Justin C. McIntosh; 100: Michelle Frankfurther for

Ayotte/Released; 328 top: DoD photo by Spc. Jeffrey Sandstrum, U.S. Army; 328 top center: DoD photo by Mass Communication Specialist 1st Class Chad J. McNeeley, U.S. Navy/Released; 328 bottom center: Courtesy Ben F. Schumin/The Schumin Web; 328 bottom: DoD photo by Cpl. Michael S. Cifuentes, USMC.

Chapter 12. 330: © Eriko Koga/Taxi Japan/Getty Images; 332: Sources: U.S. Central Intelligence Agency (2007); Statistics Bureau, Ministry of Internal Affairs and Communications (Japan) (2007); U.S. Department of Health and Human Services (2004, 2006); McCurry (2006); Al-Badr (2006); Scom-megna (2002); Iwasawa, Raymo, and Bumpass (2005), United Nations Statistics Division (2004); 334 top left: Elizabeth McMillan-McCartney; 334 top right: DoD photo by SPC 5 Greg Leary; 334 bottom: Ian Schulze; 335: DoD photo by LCPL Keith Underwood, USMC; 336: Source: Amerasian Citizenship Initiative (2007), Karen Downing (2007), Pearl S. Buck International (2007); 337 top: DoD photo by Pfc. Leslie Angulo, USA; 337 bottom: DoD photo by TSGT Curt Eddings; 337, bottom (figure): Source: Childstats.gov (2006); 337, top (figure): Source: Childstats.gov (2006); 338: Source: U.S. Central Intelligence Agency (2007) 339: DoD photo by Phan Joshua W. Legrand, USN; (table): Source: U.S. Department of Labor (2006); 340, bottom: Source: U.S. Bureau of Labor Statistics (2006); 340, top: Source: Statistics Bureau, Ministry of Internal Affairs and Communications [Japan] (2007); 341: © AP Photo/Richmond Times-Dispatch, Bob Brown; 342 bottom: 343: Sources: U.S. Central Intelligence Agency (2007), U.S. Bureau of the Census (2005, 2007); Rob Williams; 344: Theodor Horydczak Collection, Library of Congress Prints and Photographs Division; 345 left: Farm Security Administration—Office of War Information Photograph Collection, Library of Congress Prints and Photographs Division; 345 right: DoD photo by SSGT Jeffrey A. Wolfe, USA; 347: DoD photo by LCPL Antonio J. Vega, USMC; 348 top & bottom: DoD photo; 348 (table) Source: Japan Ministry of Internal Affairs and Communication (2007); 349: Noriko Ikarashi; 352: Noriko Ikarashi; 355: DoD photo by LCPL John P. McGarity, USMC; 356 left: Farm Security Administration—Office of War Information Photograph Collection, Library of Congress Prints and Photographs Division; 356 right: DoD photo by PO2 Susan Cornell, USN; 358 left: Lori Arviso Alvord, M.D./National Library of Medicine; 358 right: Suharu Ogawa; 359: Source: U.S. Bureau of the Census (2004a, 2004b); 360 left: DoD photo by JO1 Robert Benson; 360 right: DoD photo by CMS Don Sutherland, USAF; 361: Ray Elfers; bottom (figure): Source: U.S. Bureau of the Census (2005); top (figure): Source: U.S. Bureau of the Census (2005); 362: DoD photo by CPL Cory P. Griffith, USMC; (figure): Source: Population Reference Bureau (2006); 363 top: DoD photo by SSGT Michelle Leonard, USAF; 363 center: DoD photo by SPC 5 Greg Leary; 363 bottom: DoD photo by TSGT Curt Eddings; 364 top: Theodor Horydczak Collection, Library of Congress Prints and Photographs Division; 364 center: Lori Arviso Alvord, M.D./National Library of Medicine; 364 bottom: DoD photo by O.J. Sanchez.

Chapter 13. 366: © Age Fotostock; 368 top: Gaijin Bikers; 368 bottom: DoD photo by LCPL Robert D. Fleagle, Jr., USMC; 369: Andy Long; 372: Leif Knutsen; (figure) Source: OECD (2006);

373: Reginald Rogers/U.S. Army; 374 top: DoD photo by GySgt Keith A. Milks, USMC; 374 bottom: Christopher Brown; 375, bottom: Source: Institute of International Education (2006); 375, top: Source: Institute of International Education (2006); 376: Sources: U.S. Central Intelligence Agency (2005); Eurydica (2001); 377: Source: OECD (2003); 378 left: DoD photo by PhoM 2C Daniel J. McLain, USN; 378 right: Stefan Wagner, http://trumpkin .de; 378 (figure) Source: Baum and O'Malley 2003; 380: NASA Johnson Space Center; 381: DoD photo by SSGT Michael R. Holzworth, USAF; 384: DoD photo by A1C Joshua E. Coleman, USAF; 385 left: DoD photo by JO3 S. C. Irwin, USN; 385 right: DoD photo by Lisa Derek M. Poole, 387, left: Source: OECD (2003); 387, right: Source: OECD (2003); USN; 389: Kier Klepzig, Southern Research Station, USDA Forest Service; 390: Source: OECD (2006); 391 left: U.S. Marines in Japan photo by Lance Cpl. Cathryn D. Lindsay; 391 right: Tim Lindenbaum; 393: Tina Collins, Fort Lee School Liaison Officer/U.S. Army; 394: Jason Clarke; (table) Source: OECD (2003); 395 top: Gaijin Bikers; 395 bottom: Joe Mabel; 396 top: DoD photo; 396 center: DoD photo by Jim Gordon, CIV; 396 bottom: White House photo by Paul Morse; 397 top: Christopher Lim Mu Yao; 397 center: DoD photo by Lisa Derek M. Poole, USN; 397 bottom: DoD photo by Sgt. Jimmie Perkins, USMC; 398 top: U.S. News & World Report Magazine Photograph Collection, Library of Congress Prints and Photographs Division; 398 center: Tim Lindenbaum; 398 bottom: César Astudillo.

Chapter 14. 400: DoD photo by TSgt Cecilio M. Ricardo Jr., USAF; 401: DoD photo by SRA Sean Worrell; 402: DoD photo by PH3 William J. Davis, USN; 403 left: DoD photo by SPC Jerry T. Combes, USA; 403 right: DoD photo by Spc. Leslie Angulo, USA; 404: Source: U.S. Central Intelligence Agency (2007); 405 left: Steve Evans; 405 right: DoD photo by SPC Preston E. Cheeks, USA; 406: DoD photo by SSGT Marvin D. Lynchard; 407 top left: Bob Nabor; 407 top right: DoD photo by SSGT Quinton T. Burris, USAF; 407 bottom: DoD photo by SSGT Jennifer C. Wallis, USAF; 408 left & right: Elizabeth Lorenz; 409 left: Tom Kaelin; 409 right: DoD photo by LCPL Kevin C. Quihuis, Jr., USMC; 410: DoD photo by T3C Irving Katz; 411: DoD photo by MCSN Christopher A. Lussier, USN; 412: Library of Congress Prints and Photographs Division; 414 left: DoD photo by SSGT Nolan; 414 right: Farm Security Administration and Office of War Information Collection, Library of Congress Prints and Photographs Division; 415 top left: DoD photo; 415 top right: DoD photo by TSGT Cedric H. Rudisill, USAF; 415 bottom: DoD photo by LCPL Jason L. Andrade, USMC; 416: Source: U.S. Central Intelligence Agency (2007); 417: DoD photo by SPC Christopher Barnhardt, USA; (table) Sources: U.S. Central Intelligence Agency (2007); United Nations High Commissioner for Refugees (2005); 419 top: DoD photo by LCPL Jonathan P. Sotelo, USMC; 419 bottom: © AP Photo; 420: DoD photo by SSGT Jeremy T. Lock, USAF; 421: Sources: Excerpted from "President Bush's Faith-Based and Community Initiative," www.whitehouse.gov/fbci (August 1, 2004); prepared remarks of Attorney General John Ashcroft, White House Faith-Based and Community Initiatives Conference (June 13, 2004). 422 left: DoD photo by TS Lee Harshman, USAF; 422 right: NYWT&S Photograph Collection, Library of Congress Prints and Photographs Division; 423: Library of Congress Prints and Photographs Division; 424: DoD photo by SPC Leslie Angulo, USA; 426: DoD photo by SSGT Cecilio Ricardo, USAF; 428 top: Steve Evans; 428 center: DoD photo by SPC Preston E. Cheeks, USA; 428 bottom: DoD photo by MCSN Christopher A. Lussier, USN; 429 top: DoD photo by TS Lee Harshman, USAF; 429 center: DoD photo by LCPL Jonathan P. Sotelo, USMC; 429 bottom: DoD photo by SSGT Cecilio Ricardo, USAF; 430 top: DoD photo by SSGT Jeremy T. Lock, USAF; 430 center: DoD photo by TS Lee Harshman, USAF; 430 bottom: Library of Congress Prints and Photographs Division; 431: DoD photo by SPC Leslie Angulo, USA.

Chapter 15. 432: Center for Disease Control/Chris Zahniser, B.S.N., R.N., M.P.H.; 433: Center for Disease Control/Pierre Claquin, M.D., B.A.C.; 434: Source: U.S. Central Intelligence Agency 2007 435: DoD photo by PH2 Jeff Viano, USN; (box): Source: Adapted from U.S. Census Bureau (2007); 436: Joe Zachariah; 437 left: DoD photo by SPC Eric D. Moore, WYARNG; 437 right: DoD photo by CMSGT Gonda Moncada, USAF; 438: DoD photo by LCPL Dorian Gardner, USMC; (box): Source: "West African Immigrant Heeds Father's Words, Joins U.S. Marines" by Lance Cpl. Dorian Gardner, USMC Special to American Forces Press Service; 439: Sources: U.S. Department of State (2002), Association of Americans Resident Overseas (2007); 440: Source: U.S. Bureau of the Census (2007); U.S. Central Intelligence Agency (2007); 442: Center for Disease Control/ Greg Knobloch; 443: Source: U.S. Bureau of the Census (2007); 444: Center for Disease Control; 445 left: DoD photo by Cpl. Justin Park, USMC; 445 right: Library of Congress Prints and Photographs Division; 446 left & right: Robert K. Wallace; 448 left & right: Library of Congress Prints and Photographs Division; 449: Center for Disease Control/Chris Zahniser, B.S.N., R.N., M.P.H.; 450, top: Source: U.S. Central Intelligence Agency (2007); 452: Yann Forget; 453: DoD photo by PH3 Jacob J. Kirk, USN; 454: Robert K. Wallace; 455 top: Center for Disease Control/Chris Zahniser, B.S.N., R.N., M.P.H.; 455 bottom: DoD photo by CMSGT Don Sutherland; 456 top: Library of Congress Prints and Photographs Division; 456 center: Center for Disease Control/Chris Zahniser, B.S.N., R.N., M.P.H.; 456 bottom: Yann Forget.

Chapter 16. 458: © Design Pics Inc./Alamy; 460: Library of Congress Prints and Photographs Division; 461: Library of Congress Prints and Photographs Division; (table) Sources: World Coal Institute (2007); U.S. Central Intelligence Agency (2007); 462 left: DoD photo by A1C Isaac G. L. Freeman, USAF; 462 right: © Vstock LLC/Index Open; 464: Missy Gish; 465: Source: U.S. Bureau of the Census (2005); 466 left: © AbleStock/ Index Open; 466 right: DoD photo by SGT Jim Greenhill, USA; 467: DoD photo by SPC Preston E. Cheeks, USA; 468: Source: International Trade Administration (2007); 470: Niko Lang; 471: DoD photo by Wever; 473: Sources: Global Water (2007), Arctic Council (2007) and PEW Center for Global Climate Change (2007); 476 left & right: DoD photo by TSGT Dan Rea, USAF; 478 left: DoD photo by MC2 Andrew Meyers, USN; 478 right: Library of Congress Prints and Photographs Division; 479 left: DoD photo by TSGT Jose Hernandez; 479 right: Library of Congress Prints and Photographs Division; 480: DoD photo by Capt. Mark Harper, USAF; 481: DoD photo by TSGT Jose Hernandez; 482: DoD photo, Thule Air Base, USAF; (table): Source: U.S. Bureau of the Census (2007); 483 top left: DoD photo by JO1 Brent Johnston; 483 top right: © Peter Tyson/Index Stock Imagery; 483 bottom: DoD photo by PH2 Steven Vanderwerff; 484: Source: Greenland Home Rule Government (2007) 486 top: © Design Pics Inc./Alamy; 486 center: © AbleStock/ Index Open; 486 bottom: Library of Congress Prints and Photographs Division; 487 top: Niko Lang; 487 top center: DoD photo by Wever; 487 bottom center: Library of Congress Prints and Photographs Division; 487 bottom: DoD photo by SSGT Staci L. Rosenberger, USAF; 488: Library of Congress Prints and Photographs Division.

Index

DEAR STUDENT:

Thank you for buying *Sociology: A Global Perspective,* Seventh Edition. My goal in writing this book is to enhance your learning experience. Please feel free to contact me at ferrantej@nku.edu.

School and address:_____

Department:_____

Instructor's name:_____

1. What I like most about this book is:_____

_____

_____

2. What I like least about this book is:

_____

_____

3. I would like to say to the author of this book . . .

_____

_____

4. In the space below, or in an email to ferrantej@nku.edu, please write specific suggestions for improving this book and anything else you'd care to share about your experience using it.

_____

_____

_____

_____

_____

_____

_____

_____

_____

_____

_____

FOLD HERE

WADSWORTH
CENGAGE Learning

BUSINESS REPLY MAIL

FIRST-CLASS MAIL   PERMIT NO. 34   BELMONT CA

POSTAGE WILL BE PAID BY ADDRESSEE

Attn: Sociology Editor

Wadsworth

10 Davis Dr

Belmont CA 94002-9801

FOLD HERE

OPTIONAL:

Your name: _____   Date: _____

May we quote you, either in promotion for *Sociology: A Global Perspective,* Tenth
Edition, or in future publishing ventures?

Yes: _____   No:_____

Sincerely yours,

Joan Ferrante